PHYSICS OF
NONNEUTRAL PLASMAS

PHYSICS OF
NONNEUTRAL PLASMAS

Ronald C Davidson

Plasma Physics Laboratory and
Department of Astrophysical Sciences
Princeton University

ICP Imperial College Press

World Scientific

Published by

Imperial College Press
57 Shelton Street
Covent Garden
London WC2H 9HE

and

World Scientific Publishing Co. Pte. Ltd.
5 Toh Tuck Link, Singapore 596224
USA office: 27 Warren Street, Suite 401-402, Hackensack, NJ 07601
UK office: 57 Shelton Street, Covent Garden, London WC2H 9HE

British Library Cataloguing-in-Publication Data
A catalogue record for this book is available from the British Library.

PHYSICS OF NONNEUTRAL PLASMAS

ISBN-13 978-1-86094-302-7
ISBN-10 1-86094-302-0
ISBN-13 978-1-86094-303-4 (pbk)
ISBN-10 1-86094-303-9 (pbk)

To Jean

For as the sun is daily new and old,
So is my love still telling what is told.

William Shakespeare

Author's Foreword

Physics of Nonneutral Plasmas was first published in 1990 by Addison-Wesley (Reading, Massachusetts) when the author was Professor of Physics at the Massachusetts Institute of Technology. During the intervening period, research on nonneutral plasmas has evolved into a well-developed subfield of plasma physics, and I am delighted that World Scientific Publishing Co. has offered to reissue this graduate-level text on nonneutral plasmas, which has become a standard reference in this rapidly evolving area of physics research. *Physics of Nonneutral Plasmas* has been 'field-tested' in numerous graduate-level course offerings, including those given by the author at the Massachusetts Institute of Technology, Princeton University, and the United States Particle Accelerator School.

A nonneutral plasma is a many-body collection of charged particles in which there is not overall charge neutrality. The simplest examples are *one-component* pure ion plasmas, or pure electron plasmas. Since *Physics of Nonneutral Plasmas* was first published, this area of physics research has developed into a sophisticated subfield of pure and applied plasma physics. The diverse areas of application include: the nonlinear dynamics and collective processes in charge bunches in high-intensity accelerators for high energy and nuclear physics applications; investigations of nonlinear vortex dynamics and turbulence in nearly-inviscid two-dimensional fluid flow; the development of precision atomic clocks; coherent electromagnetic wave generation by energetic electrons interacting with magnetic field structures, as occurs in magnetrons, free electron lasers, and cyclotron masers; periodic focusing induction linac accelerators for the acceleration and transport of space-charge-dominated heavy ion beams;

research on properties of strongly correlated (including crystalline) one-component nonneutral plasmas; basic studies of the collective properties and nonlinear dynamics of laboratory-confined nonneutral plasmas in Malmberg-Penning traps and Paul trap configurations; studies of the basic thermal equilibrium and thermodynamic properties of one-component nonneutral plasmas; research on the formation and confinement properties of positron plasmas; investigations of the production of antihydrogen for basic physics studies by the mixing of positron and antiproton plasmas; and the equilibrium and stability of intense nonneutral electron and ion flow in high-voltage diodes, to mention a few examples. *Physics of Nonneutral Plasmas* provides an thorough physics foundation for advanced scientific research in these and related areas.

Physics of Nonneutral Plasmas has been prepared as a graduate-level text which covers a broad range of topics related to the fundamental properties and applications of nonneutral plasmas. The subject matter is treated systematically from first principles using a unified theoretical approach, and the emphasis is on the development of basic concepts that illustrate the underlying physical processes, which are often similar in different application areas. *Physics of Nonneutral Plasmas* includes 138 problems, 143 figures and illustrations, and the results from several classic experiments illustrating fundamental processes in nonneutral plasmas. In view of the book's emphasis on basic physics principles, and the thorough presentation format, it is intended to have a broad and lasting appeal to graduate students and researchers in the field.

Finally, because of the advanced theoretical techniques developed for describing the properties of one-component charged particle systems, *Physics of Nonneutral Plasmas* provides a useful companion volume to *Physics of Intense Charged Particle Beams in High Energy Accelerators* (World Scientific, Singapore, 2001) by Ronald C. Davidson and Hong Qin.

Ronald C. Davidson
Department of Astrophysical Sciences
Princeton University
October, 2001

Preface

A nonneutral plasma is a many-body collection of charged particles in which there is not overall charge neutrality. Such systems are characterized by intense self-electric fields, and in high-current configurations by intense self-magnetic fields. Nonneutral plasmas, like electrically neutral plasmas, exhibit a broad range of collective properties, such as plasma waves, instabilities, and Debye shielding. Moreover, the intense self fields in nonneutral plasmas can have a large influence on detailed plasma behavior, including stability and transport properties.

Since the monograph *Theory of Nonneutral Plasmas* (Benjamin, Reading, Massachusetts) first appeared in 1974, this important area of physics research has developed into a diverse and sophisticated subfield of pure and applied plasma physics. For example, interest in the physics of nonneutral plasmas has increased substantially in such diverse areas as: investigations of basic equilibrium, stability and transport properties; high-intensity accelerators for high energy and nuclear physics applications; phase transitions in strongly coupled, two- and three-dimensional nonneutral plasmas; coherent electromagnetic wave generation by free electrons interacting with applied magnetic field structures; astrophysical studies of large-scale isolated nonneutral plasma regions in the magnetospheres of rotating, magnetized neutron stars; and the development of positron and antiproton sources for antihydrogen production.

In addition to developing a basic physics understanding of many-body charged-particle systems in which there is not overall charge neutrality, there are many practical applications of nonneutral plasmas. These include: the development of precision atomic clocks;

investigations of nonlinear vortex dyamics and turbulence in nearly-inviscid two-dimensional fluid flow; coherent electromagnetic wave generation by energetic electrons, as occurs in magnetrons, free electron lasers, and cyclotron masers; the development of high-intensity accelerators, such as the periodic focusing induction linac accelerators for space-charge-dominated heavy ion beams; the stability of intense nonneutral electron and ion flow in high-voltage diodes; and the stability and transport of intense charged particle beams propagating through background plasma or through the atmosphere, to mention a few examples.

Physics of Nonneutral Plasmas has been prepared as a graduate-level text which covers a broad range of topics related to the fundamental properties and applications of nonneutral plasmas. The subject matter is treated systematically from first principles using a unified theoretical approach, and the emphasis is on the development of basic concepts that illustrate the underlying physical processes, which are often similar in different application areas. The statistical models used to describe the properties of nonneutral plasma are based on the fluid-Maxwell equations, the Vlasov-Maxwell equations, or the Klimontovich-Maxwell equations, as appropriate to the application under consideration.

Physics of Nonneutral Plasmas includes 138 problems, 143 figures and illustrations, and the results from several classic experiments illustrating fundamental processes in nonneutral plasmas. In view of the book's emphasis on basic physics principles, and the thorough presentation format, it is intended to have a broad and lasting appeal to graduate students and researchers in this rapidly developing field of physics research.

Ronald C. Davidson
Princeton, New Jersey
October, 2001

Acknowledgments

This book was first published in 1990 when the author was Professor of Physics at the Massachusetts of Technology. The material for this book evolved over two decades of teaching and research on the physics of nonneutral plasmas and intense charged particle beams. Many colleagues, students and collaborators, too numerous to mention, have contributed to virtually every topic covered in the book.

With regard to the selection of material incorporated in this graduate-level text, I am grateful to George Bekefi, Adam Drobot, David Hammer, Dennis Keefe, Martin Lampe, Tom O'Neil, Martin Reiser, Norman Rostoker, Richard Temkin, Han Uhm, David Wineland and Jonathan Wurtele for many valuable discussions and suggestions.

Preparing the manuscript was a significant undertaking. I am particularly indebted to Vera Sayzew for her skillful typing of the manuscript. I also wish to thank Albe Dawson for assistance in proofreading, Steve Lund for preparing several of the graphical results, and Cathy Lydon for preparation of the figures.

In addition, I am very grateful to Brian Yang for his thorough technical review of the completed manuscript, including the model assumptions and results, equations, and problem sets. I also wish to thank David Chernin, Edward Lee, Martin Reiser and Lloyd Smith for their valuable comments regarding the material on alternating-gradient focusing systems included in Chapter 10.

The preparation of this book was possible only as a result of the encouragement of many individuals, institutions and government agencies. In this regard, I particularly wish to thank Robert Birgeneau, John Deutch and Ronald Parker of the Massachusetts Institute of Technology; Charles Roberson of the United States Office of Naval Research; Timothy Coffey and Sidney Ossakow of the United States Naval Research Laboratory; and David Sutter of the United States Department of Energy, High Energy Physics Division.

Finally, I would like to thank my wife Jean, daughter Cyndy, and son Ronald, Jr., for their strong support and encouragement during the preparation of the manuscript.

Ronald C. Davidson

Princeton, New Jersey

October, 2001

TABLE OF CONTENTS

CHAPTER 1

INTRODUCTION

1.1 Exordium

A nonneutral plasma[1,2] is a many-body collection of charged particles in which there is not overall charge neutrality. Such systems are characterized by intense self-electric fields, and in high-current configurations, by intense self-magnetic fields.[3] Nonneutral plasmas, like electrically neutral plasmas, exhibit a broad range of collective properties, such as plasma waves, instabilities, and Debye shielding. Moreover, the intense self fields in a nonneutral plasma can have a large influence on detailed plasma behavior and stability properties.

Since *Theory of Nonneutral Plasmas*[1] was first published in 1974, interest in the physics of nonneutral plasmas has grown substantially in such diverse areas as: investigations of basic equilibrium, stability and transport properties;[2] high-current electron induction accelerators[4,5] and alternating-gradient accelerators;[6,7] phase transitions[8,9] in strongly coupled, two- and three-dimensional nonneutral plasmas; coherent electromagnetic wave generation by free electrons interacting with applied magnetic field structures;[10-13] astrophysical studies of large-scale isolated nonneutral plasma regions in the magnetospheres of rotating, magnetized neutron stars;[14] and the development of positron[15] and antiproton[16] ion sources. In addition to developing a basic physics understanding of many-body charged-particle systems in which there is not overall charge neutrality, there are many practical applications of nonneutral plasmas. These include: coherent elecromagnetic wave generation by intense electron beams, as in free electron lasers,[10-12] magnetrons and cyclotron masers;[13] the development of advanced accelerator concepts,[5,6] including high-current accelerators such as the modified betatron,[17-19] and periodic focusing accelerators for heavy ions;[20,21] the equilibrium and stability of intense nonneutral electron and ion flow in high-voltage diodes,[22-24] with

1

applications that include particle beam fusion;[25,26] and the stability and transport of intense charged particle beams[27–29] propagating through a background plasma or through the atmosphere, to mention a few examples.

The present volume on nonneutral plasmas has been prepared as a graduate-level text which covers a broad range of topics related to the fundamental properties and applications of nonneutral plasmas. The subject matter is treated systematically from first principles using a unified theoretical approach, and the emphasis is on the development of basic concepts that illustrate the underlying physical processes, which are often similar in different application areas. The statistical models used to describe the properties of nonneutral plasma are based on the fluid-Maxwell equations, the Vlasov-Maxwell equations, or the Klimontovich-Maxwell equations, as appropriate. The book also summarizes the results from several classic experiments illustrating fundamental processes in nonneutral plasmas.

1.2   Historical Background

The very early research on nonneutral plasma predated, by many decades, common usage of the terms "plasma" or "nonneutral plasma" in the lexicon of modern-day physics. [The term "plasma" was introduced by Tonks and Langmuir[30] in 1929 to describe collective electron plasma oscillations in an ionized gas, although widespread use of this descriptor did not occur until the 1950s and 1960s.] Indeed, the classic papers by Child (1911),[31] Langmuir (1923),[32] Lewellyn (1941),[33] Brillouin (1945),[34] McFarlane and Hay (1950),[35] Pierce (1956),[36] Kyhl and Webster (1956),[37] and Buneman (1957),[38] represent some of the earliest efforts to investigate theoretically and experimentally the equilibrium and stability properties of nonneutral electron flow in planar diodes and in geometries with crossed electric and magnetic fields. This research[31–38] and other early research[39–44] on nonneutral plasmas predated the major international development of the theoretical foundations of modern plasma physics, which occurred to a large extent during the 1960s. Moreover, advances in the understanding of nonneutral plasmas during this early period appear to have proceeded largely uninfluenced by the seminal works of Vlasov (1945),[45] Landau (1946)[46] and Bogoliubov (1946)[47] on collective interactions in many-body charged particle systems. This is due, in part, to the fact that the emphasis during this early period was mainly on the practical use and control

of space-charge waves on nonneutral electron beams in microwave generation devices (such as klystrons, traveling wave tubes and magnetrons) and
vacuum tube diodes. Excellent accounts of the early work on microwave
devices and vacuum tube diodes are given by Slater[48] and Okress,[49] and
Birdsall and Bridges.[50]

With the advent of modern plasma theory and improved instrumentation techniques, understanding the fundamental properties of nonneutral
plasmas received new impetus in the late 1960s and early 1970s. Basic theoretical[51−54] and experimental[55−57] studies of one-component pure
electron plasmas showed that many of the equilibrium, stability and collective oscillation properties[51−53] of nonneutral plasmas, including Debye
shielding,[54] are directly analogous to the collective properties of electrically neutral plasmas, appropriately modified by equilibrium self-field effects due to the space charge. In addition, rapid advances in pulsed power
technology during this period, and the improved ability to produce and
accelerate high-current electron beams[58,59] led to increased research on
nonneutral plasmas in such diverse areas as: the development of concepts
for collective-effect acceleration,[60−62] such as the electron ring accelerator,[63−65] which utilizes the intense self fields of an electron cluster to trap
and accelerate ions; novel approaches for the acceleration and stripping of
heavy ions[66−68] in nonneutral electron clouds in toroidal magnetic field
geometry; and the use of intense electron beams to generate high-power
microwaves,[69,70] or to heat plasmas by collective two-stream instabilities.[71] A more complete bibliography of research in these areas prior to
1974 is given by Davidson.[1]

1.3   Technical Advances

In the short space of this introductory chapter, it is not intended to summarize the many technical advances that have occurred in the physics of
nonneutral plasmas since *Theory of Nonneutral Plasmas*[1] was published
in 1974. Nonetheless, it is useful to outline briefly the substantial progress
in selected areas of research and to identify some of the key references.
Many of these topics will be treated in more detail in subsequent chapters.

Experimental studies of the basic equilibrium and stability properties
of nonneutral plasmas have ranged from investigations of plasma waves
in a pure electron plasma,[72] to measurements of the laminar rotation velocity of a pure electron plasma column,[73] to studies of plasma waves in
a pure ion plasma column,[74] to the identification of collective modes in a

two-dimensional, nonneutral ion layer confined below a liquid-helium surface,[75,76] to studies of the linear and nonlinear evolution of the diocotron
instability in an annular electron column,[77] to measurements of the collisional relaxation of anisotropic temperature in a pure electron plasma,[78]
to observations of the transport of magnetically confined pure electron
plasmas to global thermal equilibrium,[79] to mention only a few examples.

Theoretical studies of the basic equilibrium, stability and transport
properties of nonneutral plasmas have ranged from analytical investigations of the influence of intense self fields on the filamentation instability,[80] to development of a confinement theorem for a low-density nonneutral plasma column,[81] to analytical and numerical investigations of the
magnetron instability,[82–84] which is of considerable importance for intense nonneutral electron flow in crossed-field microwave devices[85,86] and
magnetically-insulated high-voltage diodes,[22] to analytical and numerical-
simulation studies of space-charge-induced transverse instabilities in nonneutral heavy ion beams,[87,88] to quasilinear studies of the nonlinear evolution of the diocotron instability for multimode excitation in a nonneutral
electron layer,[89] to theoretical studies of collisional transport processes in
pure electron plasmas,[90,91] to determination of the influence of intense self
fields on the cyclotron maser instability in a relativistic, nonneutral electron beam,[92] to basic theoretical studies of the equilibrium and collective
oscillation properties of a two-dimensional nonneutral ion layer confined
below a liquid-helium surface,[93] again to mention only a few examples.

A particularly fascinating area of research on nonneutral plasmas relates
to phase transitions[8] to the liquid and crystal states when the coupling
parameter

$$\Gamma = \frac{e^2}{k_B T a} \tag{1.1}$$

is sufficiently large, which requires extremely low-temperature conditions.
Here, Γ is the ratio of nearest-neighbor Coulomb energy (e^2/a) to the thermal energy ($k_B T$) of a particle, $a = (3/4\pi\hat{n})^{1/3}$ is the Wigner-Seitz radius,
and $\hat{n}$ is the average particle density. Similar to a neutral plasma, whenever $\Gamma \ll 1$ there are many particles in a Debye interaction sphere and
the correlations are weak. For sufficiently large Γ, however, the plasma
is strongly correlated. Research in this area ranges from experimental
studies of the liquid-to-crystal phase transition in a two-dimensional nonneutral electron layer on a liquid-helium surface,[9] to the production and
laser cooling of a strongly-coupled nonneutral ion plasma confined in a
Penning trap with coupling parameter $\Gamma \sim 20 - 200$,[94–96] to computer

simulation of the phase transition of bounded nonneutral ion plasmas to the liquid and crystal states both in Penning-trap confinement geometries,[97] and in heavy-ion storage rings.[98] Research on phase transitions in nonneutral plasmas has been further stimulated by continued advances in the theoretical understanding of strongly-coupled one-component plasmas.[99-102]

1.4   Outline

As indicated in Sec. 1.1, an important motivation for the present treatise is to provide a unified theoretical treatment of a broad range of topics related to the fundamental properties and applications of nonneutral plasmas. In addition, the results of several classic experiments are summarized which illustrate fundamental processes in nonneutral plasmas. The subject matter is treated systematically from first principles, beginning in Chapter 2 with a review of the statistical frameworks for describing collective and discrete-particle interactions in nonneutral plasma based on the macroscopic fluid-Maxwell equations, the Vlasov-Maxwell equations, and the Klimontovich-Maxwell equations. Chapter 3 describes several fundamental properties of the nonneutral plasma state, ranging from the equilibrium rotation induced by space-charge effects, to thermal equilibrium properties of a nonneutral plasma column, to Debye shielding of the electrostatic potential surrounding a test electron, to phase transitions in strongly-coupled nonneutral plasma. In Chapter 4, the Vlasov-Maxwell equations are used to investigate the basic equilibrium and stability properties of nonneutral plasma in circumstances where both the configuration-space dependence and the momentum-space dependence of the one-particle distribution function $f_j(\mathbf{x}, \mathbf{p}, t)$ play an important role in determining the properties of the system. Such a kinetic description is shown to be remarkably tractable in geometries as diverse as: a nonneutral plasma column confined radially by an axial magnetic field $B_0\hat{\mathbf{e}}_z$; a spheroidal nonneutral plasma confined in Penning-trap geometry, where axial confinement is produced by applied electrostatic potentials on neighboring conductors; and a relativistic nonneutral electron ring confined by the combined toroidal and mirror magnetic fields in a modified betatron. In Chapter 5, the basic equilibrium and stability properties of nonneutral plasma are investigated in circumstances where the plasma is sufficiently cold that the evolution of the system can be described by a macroscopic model based on the cold-fluid-Maxwell equations. Such a model is particularly amenable to determining the detailed influence of boundary effects on collective oscillations and instabilities. Specific examples treated in this chapter include: electrostatic waves and instabilities in a nonneutral plasma column; two-stream instabilities in relativistic beam-plasma systems; the influence of self fields on the electromagnetic filamentation instability; and the equilibrium and collective oscillation properties of a two-dimensional nonneutral ion plasma confined below the surface of liquid helium. Chapter 6 investigates basic properties of the ubiquitous diocotron instability, which is driven by a shear in the flow velocity in low-density nonneutral plasma. In addition to investigating linear stability properties, the nonlinear evolution of the diocotron instability is analyzed for multimode excitation, and the stabilizing influence of electromagnetic and relativistic effects is examined.

The theoretical techniques and basic principles developed in Chapters 2 through 6 are used in Chapters 7 through 10 to investigate several important applications of nonneutral plasmas. Chapter 7 deals with coherent electromagnetic wave generation by the cyclotron maser instability and the free electron laser instability, which can occur when a relativistic electron beam interacts with a uniform magnetic field $B_0\hat{e}_z$ or with a transverse wiggler magnetic field $\mathbf{B}_w(\mathbf{x})$, respectively. These instabilities exhibit a sensitive dependence on the detailed distribution of particles in momentum space, and therefore, the (kinetic) treatment in Chapter 7 is based on the Vlasov-Maxwell equations. In Chapter 8, a macroscopic cold-fluid model is used to investigate the equilibrium and stability properties of intense nonneutral flow in high power diodes. Particular emphasis is placed on: magnetically insulated electron flow; coherent electromagnetic wave generation by the magnetron instability; magnetically insulated ion diodes (also known as applied-B ion diodes); and the ion resonance and transit time instabilities. Chapter 9 deals with the propagation and stability of intense charged particle beams in a solenoidal focusing field $B_0\hat{e}_z$. Here, the term "solenoidal" (or "solenoid") is used to indicate that the applied magnetic field $B_0\hat{e}_z$ is in the axial direction, which is also the direction of beam propagation. The topics covered in this chapter include: the limiting electron current in a cylindrical drift cavity; laminar flow equilibria for an intense nonneutral electron beam; stability of nonneutral electron flow in a one-dimensional drift space; the transverse stability properties of an intense nonneutral ion beam; and the resistive hose, sausage and hollowing instabilities for an intense electron beam propagating through a dense background plasma. Finally, in Chapter 10, kinetic equilibrium and stability properties are investigated for an intense nonneutral ion beam propagating in an alternating-gradient focusing field (e.g., a periodic quadrupole field). The topics covered in this chapter include: analysis of the orbit and envelope equations including self-field effects; the average focusing force and phase advance for a periodic quadrupole field; the stability of the Kapchinskij-Vladimirskij equilibrium[103] for a uniform-density beam; and investigations of the emittance growth due to collective instabilities and space-charge homogenization. As would be expected, for sufficiently high beam current, self-field effects can have a large influence on detailed equilibrium and stability properties.

Chapter 1 References

1. R.C. Davidson, *Theory of Nonneutral Plasmas* (Benjamin, Reading, Massachusetts, 1974), reissued in the Advanced Book Classics Series (Addison-Wesley, Reading, Massachusetts, 1989).

2. *Nonneutral Plasma Physics*, eds., C. W. Roberson and C.F. Driscoll, AIP Conference Proceedings No. 175 (American Institute of Physics, New York, 1988), and papers therein.

3. R.C. Davidson, "Relativistic Electron Beam-Plasma Interaction with Intense Self Fields," in *Handbook of Plasma Physics—Basic Plasma Physics*, eds., M.N. Rosenbluth and R.Z. Sagdeev (North Holland, Amsterdam, 1984), Volume 2, pp. 729-819.

4. S. Humphries, Jr., *Principles of Charged Particle Acceleration* (Wiley, New York, 1986).

5. C.A. Kapetanakos and P. Sprangle, "Ultra-High-Current Electron Induction Accelerators," Physics Today **38** (2), 58 (1985).

6. J.D. Lawson, *The Physics of Charged Particle Beams* (Clarendon Press, Oxford, 1988).

7. M. Reiser, "Periodic Focusing of Intense Beams," Particle Accelerators **8**, 167 (1978).

8. J.H. Malmberg and T.M. O'Neil, "Pure Electron Plasma, Liquid, and Crystal," Phys. Rev. Lett. **39**, 1333 (1977).

9. G.C. Grimes and G. Adams, "Evidence for a Liquid-to-Crystal Phase Transition in a Classical, Two-Dimensional Sheet of Electrons," Phys. Rev. Lett. **42**, 795 (1979).

10. T.C. Marshall, *Free Electron Lasers* (Macmillan, New York, 1985).

11. C.W. Roberson and P. Sprangle, "A Review of Free Electron Lasers," Phys. Fluids **B1**, 3 (1989).

12. *Free Electron Laser Handbook*, eds., W. Colson, C. Pellegrini and A. Renieri (North Holland, Amsterdam, 1989).

13. *High-Power Microwave Sources*, eds., V. Granatstein and I. Alexeff (Artech House, Boston, Massachusetts, 1987), and papers therein.

14. C. Michael, "Nonneutral Plasmas in the Laboratory and Astrophysics," Proceedings of the Astronomical Society of Australia **6**, 127 (1985).

15. C.M. Surko, M. Leventhal, W.S. Crane, A. Passner, F. Wysocki, T.J. Murphy, J. Strachan and W.L. Rowan, "Use of Positrons to Study Transport in Tokamak Plasmas," Rev. Sci. Instrum. **57**, 1862 (1986).

16. G. Gabrielse, X. Fei, K. Helmerson, S.L. Rolston, R. Tjoelker, T.A. Trainor, H. Kalinowsky, J. Haas and W. Kells, "First Capture of Antiprotons in a Penning Trap: A. Kiloelectronvolt Source," Phys. Rev. Lett. **57**, 2504 (1986).

17. P. Sprangle and C.A. Kapetanakos, "Constant Radius Magnetic Acceleration of a Strong Nonneutral Proton Ring," J. Appl. Phys. **49**, 1 (1978).

18. N. Rostoker, "High-Current Betatron," Comments in Plasma Physics and Controlled Fusion **6**, 91 (1980).

19. J.J. Petillo and R.C. Davidson, "Kinetic Equilibrium and Stability Properties of High-Current Betatrons," Phys. Fluids **30**, 2477 (1987).

20. D. Keefe, "Research on High Beam-Current Accelerators," Particle Accelerators **11**, 187 (1987).

21. E.P. Lee and J. Hovingh, "Heavy Ion Induction Linac Drivers for Inertial Confinement Fusion," Fusion Technology **15**, 369 (1989).

22. R.V. Lovelace and E. Ott, "Theory of Magnetic Insulation," Phys. Fluids **17**, 1263 (1974).

23. T.M. Antonsen, Jr. and E. Ott, "Theory of Intense Ion Beam Acceleration," Phys. Fluids **19**, 52 (1976).

24. M.P. Desjarlais, "Impedance Characteristics of Applied-B Ion Diodes," Phys. Rev. Lett. **59**, 2295 (1987).

25. R.B. Miller, *Intense Charged Particle Beams* (Plenum, New York, 1982).

26. J.P. VanDevender and D.L. Cook, "Inertial Confinement Fusion with Light Ion Beams," Science **232**, 831 (1986).

27. E.J. Lauer, R.J. Briggs, T.J. Fessenden, R.E. Hester and E.P. Lee, "Measurements of Hose Instability of a Relativistic Electron Beam," Phys. Fluids **21**, 1344 (1978).

28. H.S. Uhm and M. Lampe, "Return-Current-Driven Instabilities of Propagating Electron Beams," Phys. Fluids **25**, 1444 (1982).

29. R.F. Fernsler, R.F. Hubbard, B. Hui, G. Joyce, M. Lampe and Y.Y. Lau, "Current Enhancement for Hose-Unstable Electron Beams," Phys. Fluids **29**, 3056 (1986).

30. L. Tonks and I. Langmuir, "Oscillations in Ionized Gases," Phys. Rev. **33**, 195 (1929).

31. C.D. Child, "Discharge from Hot CaO," Phys. Rev. **32**, 492 (1911).

32. I. Langmuir, "The Effect of Space Charge and Initial Velocities on the Potential Distribution and Thermionic Current Between Parallel Plane Electrodes," Phys. Rev. **21**, 419 (1923).

33. F.B. Lewellyn, *Electron Inertia Effects* (Cambridge University Press, London, 1941).

34. L. Brillouin, "A Theorem of Larmor and Its Importance for Electrons in Magnetic Fields," Phys. Rev. **67**, 260 (1945).

35. C.C. MacFarlane and H.G. Hay, "Wave Propagation in a Slipping Stream of Electrons; Small-Amplitude Theory," Proc. Phys. Soc. (London) **63B**, 409 (1950).

36. J.R. Pierce, "Instability of Hollow Beams," IRE Trans. Electron Devices **ED-3**, 183 (1956).

37. R.L. Kyhl and H.F. Webster, "Breakup of Hollow Cylindrical Electron Beams," IRE Trans. Electron Devices **ED-3**, 172 (1956).

38. O. Buneman, "Ribbon Beams," J. Electron. Control **3**, 507 (1957).

39. W.W. Rigrod and J.A. Lewis, "Wave Propagation Along a Magnetically-Focussed Electron Beam," Bell System Tech. J. **33**, 399 (1954).

40. G.R. Brewer, "Some Effects of Magnetic Field Strength on Space-Charge Wave Propagation," Proc. IRE **44**, 896 (1956).

41. J. Labus, "Space-Charge Waves Along a Magnetically Focussed Electron Beam," Proc. IRE **45**, 854 (1957).

42. W.W. Rigrod, "Space-Charge Wave Harmonics and Noise Propagating in Rotating Electron Beams," Bell System Tech. J. **38**, 119 (1959).

43. H.F. Webster, "Breakup of Hollow Electron Beams," J. Appl. Phys. **26**, 1386 (1955).

44. G.R. Brewer, "Some Characteristics of a Cylindrical Electron Stream in Immersed Flow," IRE Trans. Electron Devices **ED-4**, 134 (1957).

45. A.A. Vlasov, "On the Kinetic Theory of an Assembly of Particles with Collective Interaction," J. Phys. (USSR) **9**, 25 (1945).

46. L.D. Landau, "On the Vibrations of the Electronic Plasma," J. Phys. (USSR) **10**, 25 (1946).

47. N.N. Bogoliubov, *Problems of a Dynamical Theory in Statistical Physics* (State Technical Press, Moscow, 1946).

48. J. Slater, *Microwave Electronics* (Van Nostrand, New York, 1950).

49. *Cross-Field Microwave Devices*, ed., E. Okress (Academic Press, New York, 1961), Volume 1.

50. C.K. Birdsall and W.B. Bridges, *Electron Dynamics of Diode Regions* (Academic Press, New York, 1966).

51. R.C. Davidson and N.A. Krall, "Vlasov Description of an Electron Gas in a Magnetic Field," Phys. Rev. Lett. **22**, 833 (1969).

52. R.C. Davidson and N.A. Krall, "Vlasov Equilibria and Stability of an Electron Gas," Phys. Fluids **13**, 1543 (1970).

53. B.L. Bogema and R.C. Davidson, "Rotor Equilibria of Nonneutral Plasmas," Phys. Fluids **13**, 2772 (1970).

54. R.C. Davidson, "Electrostatic Shielding of a Test Charge in a Nonneutral Plasma," J. Plasma Phys. **6**, 229 (1971).

55. R.E. Pechacek, C.A. Kapetanakos and A.W. Trivelpiece, "Trapping of a 0.5 MeV Electron Ring in a 15 kG Pulsed Magnetic Mirror Field," Phys. Rev. Lett. **21**, 1436 (1968).

56. C.A. Kapetanakos, R.E. Pechacek, D.M. Spero and A.W. Trivelpiece, "Trapping and Confinement of Nonneutral Hot Electron Clouds in a Magnetic Mirror," Phys. Fluids **14**, 1555 (1971).

57. A.W. Trivelpiece, "Nonneutral Plasmas," Comments in Plasma Physics and Controlled Fusion **1**, 57 (1972).

58. D.A. Hammer and N. Rostoker, "Propagation of High Current Relativistic Electron Beams," Phys. Fluids **13**, 1831 (1970).

59. G. Benford and D.L. Book, "Relativistic Beam Equilibria," in *Advances in Plasma Physics*, eds., A. Simon and W.B. Thompson (Wiley, New York, 1971), Volume 4, p. 125.

60. V.I. Veksler, V.P. Sarantsev, A.G. Bonch-Osmolovskii, G.V. Dolbilov, G.A. Ivanov, I.N. Ivanov, M.L. Iovonich, I.V. Kozhuhov, A.B. Kuznetsov,

V.G. Mahankov, E.A. Perelstein, V.P. Rashevskii, K.A. Reshetnikova, N.B. Rubin, S.B. Rubin, P.I. Ryltsev and O.I. Yarkovov, "Collective Linear Acceleration of Ions," Atomnaya Energiya (USSR) **24**, 317 (1968).

61. J.D. Lawson, "Collective and Coherent Methods of Particle Acceleration," Particle Accelerators **3**, 21 (1972).

62. D. Keefe, "Collective-Effect Accelerators," Scientific American **226**, 22 (1972).

63. D. Keefe, G.R. Lambertson, L.J. Laslett, W.A. Perkins, J.M. Peterson, A.M. Sessler, R.W. Allison, Jr., W.W. Chupp, A.U. Luccio and J.B. Rechen, "Experiments on Forming Intense Rings of Electrons Suitable for Acceleration of Ions," Phys. Rev. Lett. **22**, 558 (1969).

64. D. Keefe, "Research on the Electron Ring Accelerator," Particle Accelerators **1**, 1 (1970).

65. R.C. Davidson and J.D. Lawson, "Self-Consistent Vlasov Description of Relativistic Electron Rings," Particle Accelerators **4**, 1 (1972).

66. G.S. Janes, R.H. Levy, H.A. Bethe and B.T. Feld, "New Type of Accelerator for Heavy Ions," Phys. Rev. **145**, 925 (1966).

67. J.D. Daugherty, L. Grodzins, G.S. Janes and R.H. Levy, "New Source of Highly Stripped Heavy Ions," Phys. Rev. Lett. **20**, 369 (1968).

68. J.D. Daugherty, J.E. Eninger and G.S. Janes, "Experiments on the Injection and Containment of Electron Clouds in a Toroidal Apparatus," Phys. Fluids **12**, 2677 (1969).

69. M. Friedman and M. Herndon, "Microwave Emission Produced by the Interaction of an Intense Relativistic Electron Beam with a Spatially Modulated Magnetic Field," Phys. Rev. Lett. **28**, 210 (1972).

70. J. Nation, "On the Coupling of a High Current Relativistic Electron Beam to a Slow Wave Structure," Appl. Phys. Lett. **17**, 491 (1970).

71. C.A. Kapetanakos and D.A. Hammer, "Plasma Heating by an Intense Relativistic Electron Beam," Appl. Phys. Lett. **23**, 17 (1973).

72. J.H. Malmberg and J.S. deGrassie, "Properties of Nonneutral Plasmas," Phys. Rev. Lett. **35**, 577 (1975).

73. A.J. Theiss, R.A. Mahaffey and A.W. Trivelpiece, "Rigid-Rotor Equilibria of Nonneutral Plasmas," Phys. Rev. Lett. **35**, 1436 (1975).

74. G. Dimonte, "Ion Langmuir Waves in a Nonneutral Plasma," Phys. Rev. Lett. **46**, 26 (1981).

75. M.L. Ott-Rowland, V. Kotsubo, J. Theobald and G.A. Williams, "Two-Dimensional Plasma Resonances in Positive Ions Under the Surface of Liquid Helium," Phys. Rev. Lett. **49**, 1708 (1982).

76. S. Hannahs and G.A. Williams, "Plasma Wave Resonances in Positive Ions Under the Surface of Liquid Helium," Japanese Journal of Applied Physics **26-3**, 741 (1987).

77. G. Rosenthal, G. Dimonte and A.Y. Wong, "Stabilization of the Diocotron Instability in an Annular Plasma," Phys. Fluids **30**, 3257 (1987).

78. A.W. Hyatt, C.F. Driscoll and J.H. Malmberg, "Measurement of the Anisotropic Temperature Relaxation Rate in a Pure Electron Plasma," Phys. Rev. Lett. **59**, 2975 (1987).

79. C.F. Driscoll, J.H. Malmberg and K.S. Fine, "Observation of Transport to Thermal Equilibrium in Pure Electron Plasmas," Phys. Rev. Lett. **60**, 1290 (1988).

80. R.C. Davidson and B.H. Hui, "Influence of Self Fields on the Equilibrium and Stability Properties of Relativistic Electron Beam-Plasma Systems," Annals of Physics **94**, 209 (1975).

81. T.M. O'Neil, "A Confinement Theorem for Nonneutral Plasmas," Phys. Fluids **23**, 2217 (1980).

82. J. Swegle and E. Ott, "Instability of the Brillouin-Flow Equilibrium in Magnetically Insulated Diodes," Phys. Rev. Lett. **46**, 929 (1981).

83. T.M. Antonsen, Jr., E. Ott, C.L. Chang and A.T. Drobot, "Parametric Scaling of the Stability of Relativistic Laminar Flow Magnetic Insulation," Phys. Fluids **28**, 2878 (1985).

84. R.C. Davidson and K.T. Tsang, "Influence of Profile Shape on the Extraordinary-Mode Stability Properties of Relativistic Nonneutral Electron Flow in a Planar Diode with Applied Magnetic Field," Phys. Fluids **28**, 1169 (1985).

85. A. Palevsky and G. Bekefi, "Microwave Emission from Pulsed, Relativistic Electron Beam Diodes II: The Multiresonator Magnetron," Phys. Fluids **22**, 986 (1979).

86. D. Chernin and Y.Y. Lau, "Stability of Laminar Electron Layers," Phys. Fluids **27**, 2319 (1984).

87. I. Hofmann, L.J. Laslett, L. Smith and I. Haber, "Stability of the Kapchinskij-Vladimirskij Distribution in Long Periodic Transport Systems," Particle Accelerators **13**, 145 (1983).

88. J. Struckmeier, J. Klabunde and M. Reiser, "On the Stability and Emittance Growth of Different Particle Phase-Space Distributions in a Long Magnetic Quadrupole Channel," Particle Accelerators **15**, 47 (1984).

89. R.C. Davidson, "Quasilinear Theory of the Diocotron Instability for Nonrelativistic Nonneutral Electron Flow in Planar Geometry," Phys. Fluids **28**, 1937 (1985).

90. T.M. O'Neil, "A New Theory of Transport Due to Like-Particle Collisions," Phys. Rev. Lett. **55**, 943 (1985).

91. D.H.E. Dubin and T.M. O'Neil, "Two-Dimensional Guiding-Center Transport of a Pure Electron Plasma," Phys. Rev. Lett. **60**, 1286 (1988).

92. H.S. Uhm and R.C. Davidson, "Influence of Intense Equilibrium Self Fields on the Cyclotron Maser Instability in High-Current Gyrotrons," Phys. Fluids **29**, 2713 (1986).

93. S.A. Prasad and G.J. Morales, "Equilibrium and Wave Properties of Two-Dimensional Ion Plasmas," Phys. Fluids **30**, 3475 (1987).

94. J.J. Bollinger and D.J. Wineland, "Strongly Coupled Nonneutral Ion Plasma," Phys. Rev. Lett. **53**, 348 (1984).

95. L.R. Brewer, J.D. Prestage, J.J. Bollinger and D.J. Wineland, "A High-Γ, Strongly Coupled, Nonneutral Ion Plasma," in *Strongly Coupled Plasma Physics*, eds., F.J. Rogers and H.E. DeWitt (Plenum, New York, 1987), p. 53.

96. S.L. Gilbert, J.J. Bollinger and D.J. Wineland, "Shell Structure Phase of Magnetically Confined Strongly Coupled Plasmas," Phys. Rev. Lett. **60**, 2022 (1988).

97. D.H.E. Dubin and T.M. O'Neil, "Computer Simulation of Ion Clouds in a Penning Trap," Phys. Rev. Lett. **60**, 511 (1988).

98. A. Rahman and J.P. Schiffer, "Structure of a One-Component Plasma in an External Field: A Molecular-Dynamics Study of Particle Arrangement in a Heavy-Ion Storage Ring," Phys. Rev. Lett. **57**, 1133 (1986).

99. W.L. Slattery, G.D. Doolen and H.E. DeWitt, "Improved Equation of State for the Classical One-Component Plasma," Phys. Rev. **A21**, 2087 (1980).

100. S. Ichimaru, "Strongly Coupled Plasmas: High-Density Classical Plasmas and Degenerate Electron Liquids," Rev. Modern Physics **54**, 1017 (1982).

101. H. Totsuji, "Static and Dynamic Properties of Strongly Coupled Classical One-Component Plasmas: Numerical Experiments on Supercooled Liquid State and Simulation of Ion Plasma in the Penning Trap," in *Strongly Coupled Plasma Physics*, eds., F.J. Rogers and H.E. DeWitt (Plenum, New York, 1987), p. 19.

102. S. Ichimaru, H. Iyetomi and S. Tanaka, "Statistical Physics of Dense Plasmas: Thermodynamics, Transport Coefficients and Dynamic Correlations," Physics Reports **149**, 91 (1987).

103. I.M. Kapchinskij and V.V. Vladimirskij, "Limitations of Proton Beam Current in a Strong-Focusing Linear Accelerator Associated with the Beam

Space Charge," in *Proceedings of the International Conference on High Energy Accelerators and Instrumentation* (CERN Scientific Information Service, Geneva, 1959), p. 274.

CHAPTER 2

THEORETICAL MODELS OF NONNEUTRAL PLASMA

2.1 Introduction

The primary emphasis in this treatise is on collisionless nonneutral plasmas. That is, the equilibrium and stability properties are studied for time scales short in comparison with a binary collision time.[1] Two levels of theoretical description of a collisionless plasma are useful in a practical sense.[2-5] These are: a *macroscopic fluid* description based on the moment-Maxwell equations, and a *kinetic* description based on the Vlasov-Maxwell equations. Both levels of description are used in subsequent chapters to study the equilibrium and stability properties of nonneutral plasmas.

In a *macroscopic fluid* description (Sec. 2.3), we examine the evolution of macroscopic properties of the plasma such as

$$n_j(\mathbf{x}, t) = \text{number density of the } j\text{'th plasma component,}$$

$$\mathbf{V}_j(\mathbf{x}, t) = \text{mean velocity of the } j\text{'th plasma component,}$$

$$\boldsymbol{P}_j(\mathbf{x}, t) = \text{pressure tensor of the } j\text{'th plasma component.} \quad (2.1)$$

These quantities evolve self-consistently in terms of the electric and magnetic fields, $\mathbf{E}(\mathbf{x}, t)$ and $\mathbf{B}(\mathbf{x}, t)$, determined from Maxwell's equations. The advantage of such a description is its relative simplicity. If the nonneutral plasma is cold, variations in the pressure can be neglected, and the approximation, $(\partial/\partial\mathbf{x}) \cdot \boldsymbol{P}_j \simeq 0$, can be made. This approximation results in a closed description of the evolution of $n_j(\mathbf{x}, t)$, $\mathbf{V}_j(\mathbf{x}, t)$, $\mathbf{E}(\mathbf{x}, t)$ and $\mathbf{B}(\mathbf{x}, t)$, based on the continuity equation, the equation of motion for the fluid, and Maxwell's equations. Both equilibrium and stability properties can be investigated using such a model. Since the description is macroscopic, the stability properties depend on gross features of the equilibrium, such as the equilibrium density and velocity profiles, $n_j^0(\mathbf{x})$ and $\mathbf{V}_j^0(\mathbf{x})$. This description of nonneutral plasmas is useful because of its simplicity.[2] Finite-geometry effects can be treated in a relatively straightforward manner using a macroscopic cold-fluid model. There are two main limitations of such an approach. First, it is not straightforward to extend a cold-fluid model to include finite-temperature effects. Second, certain phenomena, such as Landau damping, and collective waves and instabilities associated with the detailed momentum-space distribution of the particles cannot be investigated using a macroscopic fluid description.

To include thermal effects correctly, it is necessary to investigate the equilibrium and stability properties of nonneutral plasmas within a *kinetic* framework (Sec. 2.2). In this case, the one-particle distribution function, $f_j(\mathbf{x}, \mathbf{p}, t)$, and the average electric and magnetic fields, $\mathbf{E}(\mathbf{x}, t)$ and $\mathbf{B}(\mathbf{x}, t)$, evolve self-consistently according to the Vlasov-Maxwell equations.[2-6] Here,

$$f_j(\mathbf{x}, \mathbf{p}, t) d^3x d^3p = \text{probable number of particles of component } j$$
$$\text{located at the phase-space point}(\mathbf{x}, \mathbf{p})$$
$$\text{in the volume element } d^3x d^3p \text{ at time } t.$$

$$(2.2)$$

Self-consistent equilibria are readily constructed in such a kinetic model. Also, there is a broad class of collective waves and instabilities that depend on the detailed momentum-space structure of the equilibrium distribution $f_j^0(\mathbf{x}, \mathbf{p})$. Such waves and instabilities cannot be analyzed using a macroscopic cold-fluid description. Although a broad class of nonuniform equilibria can be constructed using the Vlasov-Maxwell equations, it should be pointed out that the concomitant stability analysis is generally more complicated than a stability analysis based on a macroscopic fluid description.

In circumstances where both collective interactions and *discrete particle effects* (e.g., binary collisions) play an important role in the evolution of a nonneutral plasma, a theoretical model based on the Klimontovich-Maxwell equations[7] provides a complete classical description of the system (Sec. 2.5). In this case, the model follows the evolution of the microscopic phase-space density $N_j(\mathbf{x}, \mathbf{p}, t)$ defined by

$$N_j(\mathbf{x}, \mathbf{p}, t) = \sum_{k=1}^{\bar{N}_j} \delta[\mathbf{x} - \mathbf{x}_k(t)]\delta[\mathbf{p} - \mathbf{p}_k(t)] \, .$$

Here, $\bar{N}_j$ is the total number of j'th component particles, and $\mathbf{x}_k(t)$ and $\mathbf{p}_k(t)$ denote the exact classical orbits of the k'th particle in the total field configuration, including the externally applied fields and the microscopic electromagnetic fields produced by the charge and current densities of all other particles in the system. Analyses based on such a model, which includes both collective and discrete particle interactions, generally require simplifying assumptions that neglect the effects of higher-order discrete particle correlations. Moreover, the coarse-grain statistical average of the microscopic phase-space density corresponds to the (smooth) one-particle distribution function, i.e., $\langle N_j(\mathbf{x}, \mathbf{p}, t) \rangle = f_j(\mathbf{x}, \mathbf{p}, t)$.

In this chapter, we summarize the essential features of theoretical models of collisionless nonneutral plasma based on the Vlasov-Maxwell equations (Sec. 2.2) and the macroscopic fluid-Maxwell equations (Sec. 2.3). The fundamental conservation relations for the momentum and energy of the particles and fields are also derived (Sec. 2.4). Finally, the Klimontovich-Maxwell model that includes discrete particle effects is summarized in the electrostatic approximation (Sec. 2.5), and the gravitational analogue of the Vlasov-Poisson equations is described (Sec. 2.6).

2.2   Kinetic Description

We consider a nonneutral plasma where a particle of species j has charge e_j and rest mass m_j. On time scales short in comparison with a binary collision time, the j'th-component one-particle distribution function $f_j(\mathbf{x}, \mathbf{p}, t)$ evolves according to the Vlasov equation[2-6]

$$\left\{ \frac{\partial}{\partial t} + \mathbf{v} \cdot \frac{\partial}{\partial \mathbf{x}} + e_j \left(\mathbf{E} + \frac{\mathbf{v} \times \mathbf{B}}{c} \right) \cdot \frac{\partial}{\partial \mathbf{p}} \right\} f_j(\mathbf{x}, \mathbf{p}, t) = 0 , \qquad (2.3)$$

where c is the speed of light *in vacuo*, and the velocity $\mathbf{v}$ and momentum $\mathbf{p}$ are related by

$$\mathbf{v} = \frac{\mathbf{p}}{\gamma m_j} = \frac{\mathbf{p}/m_j}{(1 + \mathbf{p}^2/m_j^2 c^2)^{1/2}} \qquad (2.4)$$

when the particle dynamics are relativistic. The electric and magnetic fields, $\mathbf{E}(\mathbf{x}, t)$ and $\mathbf{B}(\mathbf{x}, t)$, in Eq.(2.3) are determined self-consistently from Maxwell's equations

$$\nabla \times \mathbf{E} = -\frac{1}{c} \frac{\partial}{\partial t} \mathbf{B} , \qquad (2.5)$$

$$\nabla \times \mathbf{B} = \frac{4\pi}{c} \sum_j e_j \int d^3p \, \mathbf{v} f_j(\mathbf{x}, \mathbf{p}, t) + \frac{4\pi}{c} \mathbf{J}_{\text{ext}} + \frac{1}{c} \frac{\partial}{\partial t} \mathbf{E} \,, \quad (2.6)$$

$$\nabla \cdot \mathbf{E} = 4\pi \sum_j e_j \int d^3p \, f_j(\mathbf{x}, \mathbf{p}, t) + 4\pi \rho_{\text{ext}} \,, \quad (2.7)$$

$$\nabla \cdot \mathbf{B} = 0 \,. \quad (2.8)$$

Equations (2.6) and (2.7) allow the possibility of external charge and current sources, $\rho_{\text{ext}}(\mathbf{x}, t)$ and $\mathbf{J}_{\text{ext}}(\mathbf{x}, t)$. Equation (2.3) is Liouville's theorem for the incompressible evolution of $f_j(\mathbf{x}, \mathbf{p}, t)$ in the six-dimensional phase space $(\mathbf{x}, \mathbf{p})$ due to the collective interactions of the particles with both the applied fields and the average self-generated fields. It should be noted that the Vlasov-Maxwell equations are highly nonlinear because $f_j(\mathbf{x}, \mathbf{p}, t)$ is modified by the self-generated fields, which in turn evolve as the distribution function changes. Moreover, the intrinsic time scales associated with the Vlasov equation (2.3) are ω_{pj}^{-1} and ω_{cj}^{-1}, where $\omega_{cj} = |e_j \mathbf{B}/m_j c|$ is the cyclotron frequency, and $\omega_{pj} = (4\pi n_j e_j^2/m_j)^{1/2}$ is the plasma frequency. (Here, $n_j = \int d^3p \, f_j$ is the particle density.) Equations (2.3)–(2.8) can be applied to single-component nonneutral plasmas consisting only of electrons ($e_j = -e$) or ions ($e_j = +Z_i e$), or to nonneutral plasmas (or electrically neutral plasmas) in which there is a mixture of electron and ion components, or in which a relativistic electron beam is present.[2]

Under quasi-steady-state conditions, an *equilibrium* analysis of Eq.(2.3) and Eqs.(2.5)–(2.8) proceeds by setting $\partial/\partial t = 0$ and looking for stationary solutions $f_j^0(\mathbf{x}, \mathbf{p})$, $\mathbf{E}^0(\mathbf{x})$, and $\mathbf{B}^0(\mathbf{x})$ that satisfy the equations

$$\left\{ \mathbf{v} \cdot \frac{\partial}{\partial \mathbf{x}} + e_j \left(\mathbf{E}^0 + \frac{\mathbf{v} \times \mathbf{B}^0}{c} \right) \cdot \frac{\partial}{\partial \mathbf{p}} \right\} f_j^0(\mathbf{x}, \mathbf{p}) = 0 \,, \quad (2.9)$$

and

$$\nabla \times \mathbf{E}^0 = 0 \,, \quad (2.10)$$

$$\nabla \times \mathbf{B}^0 = \frac{4\pi}{c} \sum_j e_j \int d^3p \, \mathbf{v} f_j^0(\mathbf{x}, \mathbf{p}) + \frac{4\pi}{c} \mathbf{J}_{\text{ext}}(\mathbf{x}) \,, \quad (2.11)$$

$$\nabla \cdot \mathbf{E}^0 = 4\pi \sum_j e_j \int d^3p \, f_j^0(\mathbf{x}, \mathbf{p}) + 4\pi \rho_{\text{ext}}(\mathbf{x}) \,, \quad (2.12)$$

$$\nabla \cdot \mathbf{B}^0 = 0 \,. \quad (2.13)$$

An analysis of Eq.(2.9) reduces to a determination of the single-particle constants of the motion in the equilibrium fields $\mathbf{E}^0(\mathbf{x})$ and $\mathbf{B}^0(\mathbf{x})$. For the applications of interest here, $\mathbf{E}^0(\mathbf{x})$ is produced by deviations from charge neutrality in equilibrium [i.e., $\sum_j e_j \int d^3p\, f_j^0(\mathbf{x},\mathbf{p}) \neq 0$], and $\mathbf{B}^0(\mathbf{x})$ is produced by external current sources as well as equilibrium currents carried by the plasma. It should be noted that equilibrium self-field effects are incorporated in Eqs.(2.9)–(2.13) in a fully self-consistent manner. Moreover, even the equilibrium equations (2.9)–(2.13) are nonlinear for most applications of practical interest. The term *equilibrium* as used here should not be confused with *thermal equilibrium*. For a given external field configuration there can, in general, be many self-consistent Vlasov equilibria. These equilibria are stationary states that can exist on a time scale less than a binary collision time. A specific equilibrium is unstable if perturbations about that equilibrium grow in time or space.

With regard to a *stability* analysis based on the Vlasov-Maxwell equations (2.3)–(2.8), we consider small-amplitude perturbations about the quasi-steady state $f_j^0(\mathbf{x},\mathbf{p})$, $\mathbf{E}^0(\mathbf{x})$ and $\mathbf{B}^0(\mathbf{x})$, and express

$$f_j(\mathbf{x},\mathbf{p},t) = f_j^0(\mathbf{x},\mathbf{p}) + \delta f_j(\mathbf{x},\mathbf{p},t) \,,$$

$$\mathbf{E}(\mathbf{x},t) = \mathbf{E}^0(\mathbf{x}) + \delta\mathbf{E}(\mathbf{x},t) \,,$$

$$\mathbf{B}(\mathbf{x},t) = \mathbf{B}^0(\mathbf{x}) + \delta\mathbf{B}(\mathbf{x},t) \,. \tag{2.14}$$

Then, the linearized Vlasov equation is given by

$$\left\{ \frac{\partial}{\partial t} + \mathbf{v}\cdot\frac{\partial}{\partial \mathbf{x}} + e_j \left(\mathbf{E}^0 + \frac{\mathbf{v}\times\mathbf{B}^0}{c} \right)\cdot\frac{\partial}{\partial\mathbf{p}} \right\} \delta f_j(\mathbf{x},\mathbf{p},t)$$

$$= -e_j \left(\delta\mathbf{E}(\mathbf{x},t) + \frac{\mathbf{v}\times\delta\mathbf{B}(\mathbf{x},t)}{c} \right)\cdot\frac{\partial}{\partial\mathbf{p}} f_j^0(\mathbf{x},\mathbf{p}) \,. \tag{2.15}$$

Here, the equilibrium quantities $f_j^0(\mathbf{x},\mathbf{p})$, $\mathbf{E}^0(\mathbf{x})$ and $\mathbf{B}^0(\mathbf{x})$ satisfy Eqs.(2.9)–(2.13), and $\delta\mathbf{E}(\mathbf{x},t)$ and $\delta\mathbf{B}(\mathbf{x},t)$ are determined self-consistently from the linearized Maxwell equations

$$\nabla \times \delta\mathbf{E} = -\frac{1}{c}\frac{\partial}{\partial t}\delta\mathbf{B} \,, \tag{2.16}$$

$$\nabla \times \delta\mathbf{B} = \frac{4\pi}{c}\sum_j e_j \int d^3p\,\mathbf{v}\delta f_j(\mathbf{x},\mathbf{p},t) + \frac{1}{c}\frac{\partial}{\partial t}\delta\mathbf{E} \,, \tag{2.17}$$

$$\nabla \cdot \delta\mathbf{E} = 4\pi \sum_j e_j \int d^3p\, \delta f_j(\mathbf{x}, \mathbf{p}, t)\,, \tag{2.18}$$

$$\nabla \cdot \delta\mathbf{B} = 0\,. \tag{2.19}$$

If the perturbations $\delta f_j(\mathbf{x}, \mathbf{p}, t)$, $\delta\mathbf{E}(\mathbf{x}, t)$ and $\delta\mathbf{B}(\mathbf{x}, t)$ grow, then the equilibrium distribution $f_j^0(\mathbf{x}, \mathbf{p})$ is unstable. If the perturbations damp, then the system returns to equilibrium and is stable. For spatially nonuniform equilibria with space charge, a stability analysis based on Eqs.(2.15)–(2.19) is generally difficult. A useful method for solving the linearized Vlasov equation (2.15) for $\delta f_j(\mathbf{x}, \mathbf{p}, t)$ is based on the method of characteristics.[4] We denote by $\mathbf{x}'(t')$ and $\mathbf{p}'(t')$ the particle trajectories in the equilibrium field configuration. That is, $\mathbf{x}'(t')$ and $\mathbf{p}'(t')$ satisfy

$$\frac{d}{dt'}\mathbf{x}'(t') = \mathbf{v}'(t')\,,$$

$$\frac{d}{dt'}\mathbf{p}'(t') = e_j \left\{ \mathbf{E}^0\left[\mathbf{x}'(t')\right] + \frac{\mathbf{v}'(t') \times \mathbf{B}^0\left[\mathbf{x}'(t')\right]}{c} \right\}\,, \tag{2.20}$$

where $\mathbf{v}'(t') = \mathbf{p}'(t')/\gamma'(t')m_j$ and $\gamma'(t') = \left[1 + \mathbf{p}'^2(t')/m_j^2c^2\right]^{1/2}$. We further assume that the trajectories $\mathbf{x}'(t')$ and $\mathbf{p}'(t')$ pass through the phase-space point $(\mathbf{x}, \mathbf{p})$ at time $t' = t$, i.e.,

$$\mathbf{x}'(t' = t) = \mathbf{x}\,,$$

$$\mathbf{p}'(t' = t) = \mathbf{p}\,, \tag{2.21}$$

where $\mathbf{p} = \gamma m_j \mathbf{v} = m_j(1 + \mathbf{p}^2/m_j^2c^2)^{1/2}\mathbf{v}$. Using the chain rule for differentiation, it is readily shown that the linearized Vlasov equation (2.15) is equivalent to

$$\frac{d}{dt'}\delta f_j\left[\mathbf{x}'(t'), \mathbf{p}'(t'), t'\right] = -e_j\left(\delta\mathbf{E}(\mathbf{x}', t') + \frac{\mathbf{v}' \times \delta\mathbf{B}(\mathbf{x}', t')}{c}\right) \cdot \frac{\partial}{\partial\mathbf{p}'}f_j^0(\mathbf{x}', \mathbf{p}') \tag{2.22}$$

evaluated at time $t' = t$. For amplifying perturbations that grow temporally, we integrate Eq.(2.22) from $t' = -\infty$ to $t' = t$, and neglect the 'initial' conditions (at $t' = -\infty$). Making use of Eq.(2.21), this gives

$$\delta f_j(\mathbf{x}, \mathbf{p}, t) = -e_j \int_{-\infty}^{t} dt' \left(\delta\mathbf{E}(\mathbf{x}', t') + \frac{\mathbf{v}' \times \delta\mathbf{B}(\mathbf{x}', t')}{c}\right) \cdot \frac{\partial}{\partial\mathbf{p}'}f_j^0(\mathbf{x}', \mathbf{p}') \tag{2.23}$$

for the perturbed distribution function $\delta f_j(\mathbf{x}, \mathbf{p}, t)$. Equation (2.23) is then substituted into Maxwell's equations (2.17) and (2.18) to determine the self-consistent evolution of the field perturbations, $\delta \mathbf{E}(\mathbf{x}, t)$ and $\delta \mathbf{B}(\mathbf{x}, t)$, in the small-amplitude regime. Note that the integral over t' in Eq.(2.23) requires a determination from Eq.(2.20) of the orbits $\mathbf{x}'(t')$ and $\mathbf{p}'(t') = \gamma'(t') m_j \mathbf{v}'(t')$ in the equilibrium field configuration.

Problem 2.1   Make use of Eqs.(2.20) and (2.21) and the chain rule for differentiation of $\delta f_j(\mathbf{x},' \mathbf{p}', t')$ to show that the linearized Vlasov equation (2.15) is exactly equivalent to Eq.(2.22) evaluated at time $t' = t$. Integrate Eq.(2.22) from $t' = -\infty$ to $t' = t$ to obtain Eq.(2.23) for the case of negligibly small 'initial' perturbation with $[\delta f_j(\mathbf{x}', \mathbf{p}', t')]_{t'=-\infty} = 0$.

Problem 2.2   The method of characteristics can also be applied to obtain a formal solution to the fully nonlinear Vlasov equation (2.3). In this case, $\mathbf{x}'(t')$ and $\mathbf{p}'(t')$ solve the orbit equations

$$\frac{d}{dt'}\mathbf{x}'(t') = \mathbf{v}'(t') \,,$$

$$\frac{d}{dt'}\mathbf{p}'(t') = e_j \left\{ \mathbf{E}\left[\mathbf{x}'(t'), t'\right] + \frac{\mathbf{v}'(t') \times \mathbf{B}\left[\mathbf{x}'(t'), t'\right]}{c} \right\} \,, \qquad (2.2.1)$$

where $\mathbf{p}' = \gamma' m_j \mathbf{v}'$, and $\mathbf{E}(\mathbf{x}, t)$ and $\mathbf{B}(\mathbf{x}, t)$ are the exact fields calculated self-consistently from Maxwell's equations (2.5)–(2.8).

a.   Make use of Eq.(2.2.1) and the boundary conditions $\mathbf{x}'(t' = t) = \mathbf{x}$ and $\mathbf{p}'(t' = t) = \mathbf{p} = \gamma m \mathbf{v}$ to show that the nonlinear Vlasov equation (2.3) is exactly equivalent to

$$\frac{d}{dt'} f_j \left[\mathbf{x}'(t'), \mathbf{p}'(t'), t'\right] = 0 \qquad (2.2.2)$$

evaluated at $t' = t$.

b.   Make use of Eq.(2.2.2) to obtain the formal solution

$$f_j(\mathbf{x}, \mathbf{p}, t) = f_j \left[\mathbf{x}_0'(\mathbf{x}, \mathbf{p}, t), \mathbf{v}_0'(\mathbf{x}, \mathbf{p}, t), 0\right] \,, \qquad (2.2.3)$$

where $\mathbf{x}_0'$ and $\mathbf{p}_0'$ are defined by

$$\mathbf{x}_0'(\mathbf{x}, \mathbf{p}, t) \equiv \mathbf{x}'(t' = 0) \,,$$

$$\mathbf{p}_0'(\mathbf{x}, \mathbf{p}, t) \equiv \mathbf{p}'(t' = 0) \,. \qquad (2.2.4)$$

2.3   Macroscopic Fluid Description

Under some conditions, an adequate description of the equilibrium and stability properties of nonneutral plasma can be provided by a macroscopic fluid description.[2] In this case, we follow the evolution of the j'th component particle density, $n_j(\mathbf{x}, t)$, mean velocity, $\mathbf{V}_j(\mathbf{x}, t)$, mean momentum, $\mathbf{P}_j(\mathbf{x}, t)$, and pressure tensor, $\boldsymbol{P}_j(\mathbf{x}, t)$, defined by

$$n_j(\mathbf{x}, t) \equiv \int d^3 p \, f_j(\mathbf{x}, \mathbf{p}, t) \,, \tag{2.24}$$

$$n_j(\mathbf{x}, t) \mathbf{V}_j(\mathbf{x}, t) \equiv \int d^3 p \, \mathbf{v} f_j(\mathbf{x}, \mathbf{p}, t) \,, \tag{2.25}$$

$$n_j(\mathbf{x}, t) \mathbf{P}_j(\mathbf{x}, t) \equiv \int d^3 p \, \mathbf{p} f_j(\mathbf{x}, \mathbf{p}, t) \,, \tag{2.26}$$

$$\boldsymbol{P}_j(\mathbf{x}, t) \equiv \int d^3 p \, [\mathbf{p} - \mathbf{P}_j(\mathbf{x}, t)] \, [\mathbf{v} - \mathbf{V}_j(\mathbf{x}, t)] \, f_j(\mathbf{x}, \mathbf{p}, t) \,, \tag{2.27}$$

where $\mathbf{v} = (\mathbf{p}/m_j) \left[1 + \mathbf{p}^2/m_j^2 c^2\right]^{-1/2}$. Operating on the Vlasov equation (2.3) with $\int d^3 p \ldots$ and $\int d^3 p \, \mathbf{p} \ldots$ gives the continuity equation

$$\frac{\partial}{\partial t} n_j(\mathbf{x}, t) + \frac{\partial}{\partial \mathbf{x}} \cdot [n_j(\mathbf{x}, t) \mathbf{V}_j(\mathbf{x}, t)] = 0 \,, \tag{2.28}$$

and the force balance equation

$$n_j(\mathbf{x}, t) \left(\frac{\partial}{\partial t} + \mathbf{V}_j(\mathbf{x}, t) \cdot \frac{\partial}{\partial \mathbf{x}} \right) \mathbf{P}_j(\mathbf{x}, t) + \frac{\partial}{\partial \mathbf{x}} \cdot \boldsymbol{P}_j(\mathbf{x}, t)$$

$$= n_j(\mathbf{x}, t) e_j \left(\mathbf{E}(\mathbf{x}, t) + \frac{\mathbf{V}_j(\mathbf{x}, t) \times \mathbf{B}(\mathbf{x}, t)}{c} \right) \,. \tag{2.29}$$

For a cold plasma, the pressure-gradient term, $(\partial/\partial \mathbf{x}) \cdot \boldsymbol{P}_j$, is neglected in Eq.(2.29). However, if finite temperature effects are included in the macroscopic fluid model, then this term is retained and the evolution of $\boldsymbol{P}_j(\mathbf{x}, t)$ is determined by taking the appropriate momentum moments of the Vlasov equation (2.3) (Problem 2.3). The chain of moment equations is then closed by making assumptions regarding the form of the heat flow tensor $\boldsymbol{Q}_j(\mathbf{x}, t)$. Of course, Eqs.(2.28) and (2.29) are to be supplemented by Maxwell's equations for the self-consistent evolution of $\mathbf{E}(\mathbf{x}, t)$ and $\mathbf{B}(\mathbf{x}, t)$, i.e.,

$$\nabla \times \mathbf{E} = -\frac{1}{c}\frac{\partial}{\partial t}\mathbf{B} , \tag{2.30}$$

$$\nabla \times \mathbf{B} = \frac{4\pi}{c}\sum_j e_j n_j(\mathbf{x}, t)\mathbf{V}_j(\mathbf{x}, t) + \frac{4\pi}{c}\mathbf{J}_{\text{ext}} + \frac{1}{c}\frac{\partial}{\partial t}\mathbf{E} , \tag{2.31}$$

$$\nabla \cdot \mathbf{E} = \sum_j 4\pi e_j n_j(\mathbf{x}, t) + 4\pi\rho_{\text{ext}} , \tag{2.32}$$

$$\nabla \cdot \mathbf{B} = 0 . \tag{2.33}$$

As in the Vlasov description, an *equilibrium* analysis of Eqs.(2.28)–(2.33) is carried out by setting $\partial/\partial t = 0$. Dropping the $(\partial/\partial \mathbf{x}) \cdot \mathbf{P}_j$ term in Eq.(2.29), the steady-state continuity and force balance equations become

$$\frac{\partial}{\partial \mathbf{x}} \cdot \left(n_j^0 \mathbf{V}_j^0\right) = 0 , \tag{2.34}$$

$$\mathbf{V}_j^0 \cdot \frac{\partial}{\partial \mathbf{x}}\mathbf{P}_j^0 = e_j \left(\mathbf{E}^0 + \frac{\mathbf{V}_j^0 \times \mathbf{B}^0}{c}\right) , \tag{2.35}$$

and Maxwell's equations can be expressed as

$$\nabla \times \mathbf{E}^0 = 0 , \tag{2.36}$$

$$\nabla \times \mathbf{B}^0 = \frac{4\pi}{c}\sum_j e_j n_j^0 \mathbf{V}_j^0 + \frac{4\pi}{c}\mathbf{J}_{\text{ext}}^0(\mathbf{x}) , \tag{2.37}$$

$$\nabla \cdot \mathbf{E}^0 = \sum_j 4\pi n_j^0 e_j + 4\pi\rho_{\text{ext}}(\mathbf{x}) , \tag{2.38}$$

$$\nabla \cdot \mathbf{B}^0 = 0 . \tag{2.39}$$

Here, $n_j^0(\mathbf{x})$, $\mathbf{V}_j^0(\mathbf{x})$, $\mathbf{P}_j^0(\mathbf{x})$, $\mathbf{E}^0(\mathbf{x})$, and $\mathbf{B}^0(\mathbf{x})$ are the macroscopic equilibrium quantities.

A *stability* analysis based on Eqs.(2.28)–(2.33) proceeds in the following manner. The macroscopic fluid and field quantities are expressed as the sum of their equilibrium values plus a perturbation,

$$\psi(\mathbf{x}, t) = \psi^0(\mathbf{x}) + \delta\psi(\mathbf{x}, t) . \tag{2.40}$$

For small-amplitude perturbations, linearization of Eqs.(2.28)–(2.33) gives

$$\frac{\partial}{\partial t}\delta n_j + \frac{\partial}{\partial \mathbf{x}} \cdot \left(n_j^0 \delta \mathbf{V}_j + \delta n_j \mathbf{V}_j^0\right) = 0 \, , \tag{2.41}$$

$$\frac{\partial}{\partial t}\delta \mathbf{P}_j + \mathbf{V}_j^0 \cdot \frac{\partial}{\partial \mathbf{x}}\delta \mathbf{P}_j + \delta \mathbf{V}_j \cdot \frac{\partial}{\partial \mathbf{x}}\mathbf{P}_j^0 = e_j \left(\delta \mathbf{E} + \frac{\mathbf{V}_j^0 \times \delta \mathbf{B}}{c} + \frac{\delta \mathbf{V}_j \times \mathbf{B}^0}{c}\right) , \tag{2.42}$$

and

$$\nabla \times \delta \mathbf{E} = -\frac{1}{c}\frac{\partial}{\partial t}\delta \mathbf{B} \, , \tag{2.43}$$

$$\nabla \times \delta \mathbf{B} = \frac{4\pi}{c}\sum_j e_j \left(\delta n_j \mathbf{V}_j^0 + n_j^0 \delta \mathbf{V}_j\right) + \frac{1}{c}\frac{\partial}{\partial t}\delta \mathbf{E} \, , \tag{2.44}$$

$$\nabla \cdot \delta \mathbf{E} = 4\pi \sum_j \delta n_j e_j \, , \tag{2.45}$$

$$\nabla \cdot \delta \mathbf{B} = 0 \, . \tag{2.46}$$

Equations (2.41)–(2.46) describe the evolution of the perturbations $\delta n_j(\mathbf{x}, t)$, $\delta \mathbf{V}_j(\mathbf{x}, t)$, $\delta \mathbf{P}_j(\mathbf{x}, t)$, $\delta \mathbf{E}(\mathbf{x}, t)$, and $\delta \mathbf{B}(\mathbf{x}, t)$, assuming negligible pressure variation with $(\partial/\partial \mathbf{x}) \cdot \mathbf{P}_j(\mathbf{x}, t) \simeq 0$. If the perturbations grow in time or space, then the equilibrium is *unstable*.

In concluding Sec. 2.3, it is important to note that a cold-plasma description based on Eqs.(2.28)–(2.33) is equivalent to a kinetic description based on the Vlasov-Maxwell equations provided the distribution function $f_j(\mathbf{x}, \mathbf{p}, t)$ is of the form

$$f_j(\mathbf{x}, \mathbf{p}, t) = n_j(\mathbf{x}, t)\delta \left[\mathbf{p} - \mathbf{P}_j(\mathbf{x}, t)\right] \, . \tag{2.47}$$

Integration of Eq.(2.47) readily gives

$$\int d^3 p \, f_j(\mathbf{x}, \mathbf{p}, t) = n_j(\mathbf{x}, t) \, ,$$

$$\int d^3 p \, \mathbf{p} f_j(\mathbf{x}, \mathbf{p}, t) = n_j(\mathbf{x}, t)\mathbf{P}_j(\mathbf{x}, t) \, , \tag{2.48}$$

$$\int d^3 p \, \mathbf{v} f_j(\mathbf{x}, \mathbf{p}, t) = n_j(\mathbf{x}, t)\mathbf{V}_j(\mathbf{x}, t) \equiv n_j(\mathbf{x}, t)\mathbf{P}_j(\mathbf{x}, t)/\gamma_j(\mathbf{x}, t)m_j \, ,$$

and $P_j(\mathbf{x}, t) = 0$. Here, for a cold fluid element, the flow velocity $\mathbf{V}_j(\mathbf{x}, t)$, momentum $\mathbf{P}_j(\mathbf{x}, t)$, and relativistic mass factor $\gamma_j(\mathbf{x}, t)$ are related by $\mathbf{P}_j = \gamma_j m_j \mathbf{V}_j$, where

$$\gamma_j(\mathbf{x}, t) = \left[1 + \mathbf{P}_j^2(\mathbf{x}, t)/m_j^2 c^2\right]^{1/2} . \tag{2.49}$$

Problem 2.3   Consider a nonrelativistic nonneutral plasma where the particle momentum and velocity are related by $\mathbf{p} = m_j \mathbf{v}$, and the corresponding relation for a fluid element is $\mathbf{P}_j(\mathbf{x}, t) = m_j \mathbf{V}_j(\mathbf{x}, t)$.

a. Take the moment of the nonlinear Vlasov equation (2.3) corresponding to the particle stress tensor $m_j \int d^3p\, \mathbf{vv} f_j(\mathbf{x}, \mathbf{p}, t)$, and make use of Eqs.(2.28) and (2.29) to show that the pressure tensor $\mathbf{P}_j(\mathbf{x}, t)$ evolves according to

$$\frac{\partial}{\partial t}\mathbf{P}_j + \frac{\partial}{\partial \mathbf{x}} \cdot (\mathbf{V}_j \mathbf{P}_j) + \mathbf{P}_j \cdot \left(\frac{\partial}{\partial \mathbf{x}} \mathbf{V}_j\right) + \left(\frac{\partial}{\partial \mathbf{x}} \mathbf{V}_j\right)^T \cdot \mathbf{P}_j + \frac{\partial}{\partial \mathbf{x}} \cdot \mathbf{Q}_j$$

$$= \frac{e_j}{m_j c} (\mathbf{P}_j \times \mathbf{B} - \mathbf{B} \times \mathbf{P}_j) . \tag{2.3.1}$$

Here, $(\ldots)^T$ denotes dyadic transpose, and $\mathbf{P}_j(\mathbf{x}, t)$ and the heat flow tensor $\mathbf{Q}_j(\mathbf{x}, t)$ are defined by

$$\mathbf{P}_j(\mathbf{x}, t) = m_j \int d^3p\, [\mathbf{v} - \mathbf{V}_j(\mathbf{x}, t)][\mathbf{v} - \mathbf{V}_j(\mathbf{x}, t)] f_j(\mathbf{x}, \mathbf{p}, t) , \tag{2.3.2}$$

$$\mathbf{Q}_j(\mathbf{x}, t) = m_j \int d^3p\, [\mathbf{v} - \mathbf{V}_j(\mathbf{x}, t)][\mathbf{v} - \mathbf{V}_j(\mathbf{x}, t)][\mathbf{v} - \mathbf{V}_j(\mathbf{x}, t)] f_j(\mathbf{x}, \mathbf{p}, t) . \tag{2.3.3}$$

b. For $(\partial/\partial \mathbf{x}) \cdot \mathbf{Q}_j \simeq 0$, and sufficiently strong magnetic field that $\omega_{cj} \equiv |e_j \mathbf{B}/m_j c|$ is large in comparison with the $\partial/\partial t$ and $\partial/\partial \mathbf{x}$ variations in Eq.(2.3.1), show that the pressure tensor in Eq.(2.3.1) is necessarily of the (diagonal) form

$$\mathbf{P}_j(\mathbf{x}, t) = P_{j\perp}(\mathbf{x}, t)(\mathbf{I} - \mathbf{nn}) + P_{j\parallel}(\mathbf{x}, t)\mathbf{nn} . \tag{2.3.4}$$

Here, $\mathbf{I}$ is the unit dyadic, and $\mathbf{n}(\mathbf{x}, t) = \mathbf{B}(\mathbf{x}, t)/|\mathbf{B}(\mathbf{x}, t)|$ is a unit vector along the magnetic field.

Problem 2.4   Consider a multicomponent, cold plasma where the relativistic flow velocity $\mathbf{V}_j(\mathbf{x}, t)$ and momentum $\mathbf{P}_j(\mathbf{x}, t)$ are related by $\mathbf{P}_j = \gamma_j m_j \mathbf{V}_j$, and the relativistic mass factor $\gamma_j(\mathbf{x}, t)$ of a fluid element is defined in Eq.(2.49).

a. Show that $\gamma_j(\mathbf{x}, t)$ can be expressed in the equivalent form

$$\gamma_j(\mathbf{x}, t) = \left[1 - \mathbf{V}_j^2(\mathbf{x}, t)/c^2\right]^{-1/2} . \qquad (2.4.1)$$

b. Consider small-amplitude perturbations about the equilibrium flow velocity $\mathbf{V}_j^0(\mathbf{x})$. Show that the perturbed energy $\delta\gamma_j(\mathbf{x}, t)$ is given approximately by

$$\delta\gamma_j(\mathbf{x}, t) = \frac{1}{c^2}\gamma_j^{03}(\mathbf{x})\mathbf{V}_j^0(\mathbf{x}) \cdot \delta\mathbf{V}_j(\mathbf{x}, t) , \qquad (2.4.2)$$

where $\gamma_j^0(\mathbf{x}) = \left[1 - \mathbf{V}_j^{02}(\mathbf{x})/c^2\right]^{-1/2}$.

c. Make use of Eq.(2.4.2) and $\delta\mathbf{P}_j = \gamma_j^0 m_j \delta\mathbf{V}_j + \delta\gamma_j m_j \mathbf{V}_j^0$ to express the perturbed momentum of a fluid element directly in terms of $\mathbf{V}_j^0(\mathbf{x})$ and $\delta\mathbf{V}_j(\mathbf{x}, t)$.

2.4   Conservation Relations

The nonlinear Vlasov-Maxwell equations (2.3) and (2.5)-(2.8) can be used to derive conservation relations that take the form of continuity equations related to the flow of particle and field momentum and energy. For present purposes, we assume that the plasma is isolated from contact with external charge and current sources. Operating on Eq.(2.3) with $\sum_j \int d^3p\,\mathbf{p}\dots$, where $\sum_j$ denotes summation over all plasma components, we obtain

$$\frac{\partial}{\partial t}\left(\sum_j \int d^3p\,\mathbf{p}f_j\right) + \frac{\partial}{\partial \mathbf{x}} \cdot \left(\sum_j \int d^3p\,\mathbf{p}\,\mathbf{v}f_j\right) \qquad (2.50)$$

$$= \left(\sum_j e_j \int d^3p\,f_j\right)\mathbf{E} + \frac{1}{c}\left(\sum_j e_j \int d^3p\,\mathbf{v}f_j\right) \times \mathbf{B} .$$

Here, $f_j(\mathbf{x}, \mathbf{p}, t)$ is the distribution function for component j, and $\mathbf{E}(\mathbf{x}, t)$ and $\mathbf{B}(\mathbf{x}, t)$ are the electric and magnetic fields. In obtaining Eq.(2.50), we have made use of $\mathbf{p} = \gamma m_j \mathbf{v} = m_j(1+\mathbf{p}^2/m_j^2 c^2)^{1/2}\mathbf{v}$ and $(\partial/\partial\mathbf{p})\cdot(\mathbf{v}\times\mathbf{B}) = 0$, and integrated by parts with respect to momentum $\mathbf{p}$. The current and charge densities occurring in Eq.(2.50) can be eliminated in favor of the fields by means of Eqs.(2.6) and (2.7). Some straightforward algebra then gives the momentum balance equation

$$\frac{\partial}{\partial t}\left(\sum_j \int d^3p\, \mathbf{p} f_j + \frac{\mathbf{E} \times \mathbf{B}}{4\pi c}\right) \tag{2.51}$$

$$+ \frac{\partial}{\partial \mathbf{x}} \cdot \left(\sum_j \int d^3p\, \mathbf{p} \mathbf{v} f_j + \frac{|\mathbf{E}|^2 + |\mathbf{B}|^2}{8\pi} \mathbf{I} - \frac{\mathbf{EE} + \mathbf{BB}}{4\pi}\right) = 0 \, ,$$

where $\mathbf{I}$ is the unit dyadic. Equation (2.51) relates the local rate of change of the total (particle plus field) momentum density to the divergence of the particle stress tensor plus the electromagnetic stress tensor.

In a similar manner, we operate on Eq.(2.3) with $\sum_j \int d^3p\, (\gamma-1)m_j c^2 \ldots$ and make use of Eqs.(2.5)–(2.8) to obtain the energy balance equation

$$\frac{\partial}{\partial t}\left(\sum_j \int d^3p\, (\gamma - 1)m_j c^2 f_j + \frac{|\mathbf{E}|^2 + |\mathbf{B}|^2}{8\pi}\right)$$

$$+ \frac{\partial}{\partial \mathbf{x}} \cdot \left(\sum_j \int d^3p\, (\gamma - 1)m_j c^2 \mathbf{v} f_j + \frac{c}{4\pi}\mathbf{E} \times \mathbf{B}\right) = 0 \, . \tag{2.52}$$

Equation (2.52) relates the local rate of change of the total (particle kinetic plus field) energy to the divergence of the flux of particle energy plus field energy.

Equations (2.51) and (2.52) are applicable within the plasma, and in the vacuum region surrounding the plasma where $\rho_{\text{ext}} = 0$ and $\mathbf{J}_{\text{ext}} = 0$. We now integrate Eqs.(2.51) and (2.52) over a volume V containing the entire plasma, which is assumed to be isolated from contact with external current and charge sources. Integrals over the divergence terms proportional to $(\partial/\partial \mathbf{x}) \cdot (\cdots)$ in Eqs.(2.51) and (2.52) can be converted to integrals over the surface S enclosing the volume V. We obtain

$$\frac{\partial}{\partial t}\left\{\int_V d^3x \left(\sum_j \int d^3p\, \mathbf{p} f_j + \frac{\mathbf{E} \times \mathbf{B}}{4\pi c}\right)\right\} = -\oint_S dS \mathbf{n} \cdot \boldsymbol{T},$$

$$\tag{2.53}$$

and

$$\frac{\partial}{\partial t}\left\{\int_V d^3x \left(\sum_j \int d^3p\,(\gamma-1)m_jc^2 f_j + \frac{|\mathbf{E}|^2+|\mathbf{B}|^2}{8\pi}\right)\right\} = -\oint_S dS\mathbf{n}\cdot\mathbf{S}\,.$$

$$(2.54)$$

Here, **n** is a unit vector normal (outward) to the surface S. The electromagnetic stress tensor $-\boldsymbol{T}$ in Eq.(2.53) is defined by

$$\boldsymbol{T} = \frac{|\mathbf{E}|^2+|\mathbf{B}|^2}{8\pi}\boldsymbol{I} - \frac{\mathbf{E}\mathbf{E}+\mathbf{B}\mathbf{B}}{4\pi}\,, \qquad (2.55)$$

and the Poynting vector **S** in Eq.(2.54) is defined by

$$\mathbf{S} = \frac{c}{4\pi}\mathbf{E}\times\mathbf{B}\,. \qquad (2.56)$$

Evidently, $\mathbf{n}\cdot\boldsymbol{T}$ in Eq.(2.53) represents the normal flux of electromagnetic momentum flowing out of the volume V through the surface S. That is, $\mathbf{n}\cdot\boldsymbol{T}$ is the electromagnetic force per unit area transmitted across the surface S.[8] Similarly, $\mathbf{n}\cdot\mathbf{S}$ is the normal flux of electromagnetic energy flowing out of the volume V through the surface S.

In the special case where the plasma is surrounded by a perfectly conducting enclosure, we identify V and S with the (internal) volume and surface of the enclosure. Assuming that the electric field tangential to the perfectly conducting wall satisfies $\mathbf{n}\times\mathbf{E}=0$, the right-hand side of Eq.(2.54) vanishes identically, and Eq.(2.54) reduces to

$$\int_V d^3x \left(\sum_j \int d^3p\,(\gamma-1)m_jc^2 f_j + \frac{|\mathbf{E}|^2+|\mathbf{B}|^2}{8\pi}\right) = \text{const.} \qquad (2.57)$$

Equation (2.57) corresponds to the (global) conservation of particle kinetic energy plus field energy averaged over the volume V.

Finally, it can be shown directly from the Vlasov equation (2.3) that

$$\frac{\partial}{\partial t}\left(\int d^3p\,G(f_j)\right) + \frac{\partial}{\partial\mathbf{x}}\cdot\left(\int d^3p\,\mathbf{v}G(f_j)\right) = 0\,, \qquad (2.58)$$

where $G(f_j)$ is a smooth, differentiable function satisfying $G(f_j\to 0)=0$. The continuity equation (2.28) is a special case of Eq.(2.58), with $G(f_j)=$

f_j. Integrating Eq.(2.58) over a volume V containing the entire plasma gives the global conservation relation

$$\int_V d^3x \int d^3p \, G(f_j) = \text{const.} \tag{2.59}$$

Equation (2.59) corresponds to conservation of the generalized entropy $S_G \equiv \int_V d^3x \int d^3p \, G(f_j)$ by the (collisionless) Vlasov-Maxwell equations.

Problem 2.5   Make use of the nonlinear Vlasov-Maxwell equations (2.3) and (2.5)-(2.8) to provide a detailed derivation of the momentum and energy balance equations (2.51) and (2.52). You will find the vector identities,

$$\nabla \cdot (\mathbf{A} \times \mathbf{B}) = \mathbf{B} \cdot (\nabla \times \mathbf{A}) - \mathbf{A} \cdot (\nabla \times \mathbf{B}), \tag{2.5.1}$$

and

$$\nabla \cdot (\mathbf{A}\mathbf{A}) = (\nabla \cdot \mathbf{A})\mathbf{A} + \mathbf{A} \cdot (\nabla \mathbf{A}) = (\nabla \cdot \mathbf{A})\mathbf{A} + \frac{1}{2}\nabla |\mathbf{A}|^2 - \mathbf{A} \times (\nabla \times \mathbf{A}), \tag{2.5.2}$$

useful in the derivations.

Problem 2.6   A conservation relation for angular momentum analogous to Eqs.(2.51) and (2.52) can also be derived from the nonlinear Vlasov-Maxwell equations (2.3) and (2.5)-(2.8). Operate on Eq.(2.3) with $\sum_j \int d^3p \, \mathbf{x} \times \mathbf{p} \ldots$ and derive the conservation relation

$$\frac{\partial}{\partial t}\left(\sum_j \int d^3p \, \mathbf{x} \times \mathbf{p}f_j + \frac{\mathbf{x} \times (\mathbf{E} \times \mathbf{B})}{4\pi c}\right) \tag{2.6.1}$$

$$+ \frac{\partial}{\partial \mathbf{x}} \cdot \left[\left(-\sum_j \int d^3p \, \mathbf{p} \, \mathbf{v}f_j + \frac{\mathbf{E}\mathbf{E} + \mathbf{B}\mathbf{B}}{4\pi} - \frac{|\mathbf{E}|^2 + |\mathbf{B}|^2}{8\pi}\mathbf{I}\right) \times \mathbf{x}\right] = 0 \, .$$

2.5   Discrete Particle Effects

As indicated in Sec. 2.1, in circumstances where both collective interactions and discrete particle effects play an important role in the evolution of a nonneutral plasma, a theoretical model based on the Klimontovich-Maxwell equations[7] provides a complete classical description of the system. In this case, we introduce the microscopic phase-space density $N_j(\mathbf{x}, \mathbf{p}, t)$ defined by

$$N_j(\mathbf{x}, \mathbf{p}, t) = \sum_{k=1}^{\bar{N}_j} \delta\left[\mathbf{x} - \mathbf{x}_k(t)\right] \delta[\mathbf{p} - \mathbf{p}_k(t)] \,. \tag{2.60}$$

Here, $\bar{N}_j$ is the total number of j'th component particles, and $\mathbf{x}_k(t)$ and $\mathbf{p}_k(t)$ denote the exact classical orbits of the k'th particle in the total field configuration, including the externally applied fields and the microscopic electromagnetic fields produced by the charge and current densities of all other particles in the system.

For purposes of a brief description of the Klimontovich model,[7] we assume that the microscopic magnetic field $\mathbf{B}^M(\mathbf{x}, t)$ can be neglected in comparison with the applied magnetic field $\mathbf{B}^0(\mathbf{x})$ (assumed steady) determined from $\nabla \times \mathbf{B}^0(\mathbf{x}) = (4\pi/c)\mathbf{J}_{\text{ext}}(\mathbf{x})$, and that the microscopic electric field $\mathbf{E}^M(\mathbf{x}, t)$ satisfies $\nabla \times \mathbf{E}^M(\mathbf{x}, t) = 0$. The particle orbits in Eq.(2.60) are determined from

$$\frac{d}{dt}\mathbf{x}_k(t) = \mathbf{v}_k(t) \,, \tag{2.61}$$

$$\frac{d}{dt}\mathbf{p}_k(t) = e_j\left[\mathbf{E}^M\left[\mathbf{x}_k(t), t\right] + \frac{\mathbf{v}_k(t) \times \mathbf{B}^0\left[\mathbf{x}_k(t)\right]}{c}\right] \,, \tag{2.62}$$

where e_j and m_j are the charge and rest mass of a j'th component particle, and the momentum and velocity are related by $\mathbf{p}_k(t) = \gamma_k(t)m_j\mathbf{v}_k(t)$, where $\gamma_k(t) = [1 + \mathbf{p}_k^2(t)/m_j^2 c^2]^{1/2}$ is the relativistic mass factor of the k'th particle. Some straightforward algebra that makes use of Eqs.(2.60)–(2.62) shows that the microscopic phase-space density evolves in phase space according to (Problem 2.7)

$$\left\{\frac{\partial}{\partial t} + \mathbf{v} \cdot \frac{\partial}{\partial \mathbf{x}} + e_j\left(\mathbf{E}^M + \frac{\mathbf{v} \times \mathbf{B}^0}{c}\right) \cdot \frac{\partial}{\partial \mathbf{p}}\right\} N_j(\mathbf{x}, \mathbf{p}, t) = 0 \,. \tag{2.63}$$

Here, the microscopic electric field $\mathbf{E}^M(\mathbf{x}, t) = -\nabla\phi^M(\mathbf{x}, t)$ is determined self-consistently from Poisson's equation

$$\nabla \cdot \mathbf{E}^M = 4\pi \sum_j e_j \int d^3p \, N_j(\mathbf{x}, \mathbf{p}, t) + 4\pi\rho_{\text{ext}} \,, \tag{2.64}$$

where ρ_{ext} denotes the external charge density, and $\mathbf{p} = \gamma m_j \mathbf{v}$ is the momentum in phase space, where $\gamma = (1 + \mathbf{p}^2/m_j^2 c^2)^{1/2}$. In Eqs.(2.63) and (2.64), and throughout the subsequent analysis, when the microscopic electric field $\mathbf{E}^M(\mathbf{x}, t)$ is evaluated at $\mathbf{x} = \mathbf{x}_s$ (the location of the s'th particle), the self-field contribution to $\mathbf{E}^M(\mathbf{x}, t)$ from the s'th particle should be neglected. Moreover, the generalization of Eqs.(2.61)–(2.64) to include the microscopic magnetic field $\mathbf{B}^M(\mathbf{x}, t)$ is carried out in a straightforward manner by making the replacement $\mathbf{B}^0(\mathbf{x}) \to \mathbf{B}^0(\mathbf{x}) + \mathbf{B}^M(\mathbf{x}, t)$, where $\mathbf{B}^M(\mathbf{x}, t)$ is determined self-consistently from $\nabla \times \mathbf{E}^M = -c^{-1}\partial\mathbf{B}^M/\partial t$ and $\nabla \times \mathbf{B}^M = (4\pi/c)\mathbf{J}^M + c^{-1}\partial\mathbf{E}^M/\partial t$, and $\mathbf{J}^M(\mathbf{x}, t) = \sum_j e_j \int d^3p\, \mathbf{v} N_j(\mathbf{x}, \mathbf{p}, t)$ is the microscopic current density.

In the electrostatic approximation, the Klimontovich-Poisson equations (2.63) and (2.64) for $N_j(\mathbf{x}, \mathbf{p}, t)$ and $\mathbf{E}^M(\mathbf{x}, t)$ are identical in form to the Vlasov-Poisson equations (2.3) and (2.7) for $f_j(\mathbf{x}, \mathbf{p}, t)$ and $\mathbf{E}(\mathbf{x}, t)$. The primary difference, of course, is that the microscopic phase-space density $N_j(\mathbf{x}, \mathbf{p}, t)$ defined in Eq.(2.60) is a singular function, unlike the probability density in phase-space described by the (smooth) distribution function $f_j(\mathbf{x}, \mathbf{p}, t)$. Indeed, a precise specification of the microscopic phase-space density $N_j(\mathbf{x}, \mathbf{p}, t)$ requires a complete determination of all of the particle orbits $\{\mathbf{x}_k(t), \mathbf{p}_k(t)\}$ in the applied magnetic field $\mathbf{B}^0(\mathbf{x})$ and the microscopic electric field $\mathbf{E}^M(\mathbf{x}, t)$ generated self-consistently by all of the plasma particles. In addition, the electric field $\mathbf{E}(\mathbf{x}, t)$ occurring in the Vlasov-Poisson equations (2.3) and (2.7) corresponds to the average electric field due to collective interactions. The two descriptions are related by taking a coarse-grain statistical average (denoted by $\langle \ldots \rangle$) of the Klimontovich-Poisson equations and making the identification[7]

$$\langle N_j(\mathbf{x}, \mathbf{p}, t) \rangle = f_j(\mathbf{x}, \mathbf{p}, t) \,,$$
$$\langle \mathbf{E}^M(\mathbf{x}, t) \rangle = \mathbf{E}(\mathbf{x}, t) \,. \qquad (2.65)$$

That is, the one-particle distribution function $f_j(\mathbf{x}, \mathbf{p}, t)$ and the electric field $\mathbf{E}(\mathbf{x}, t)$ occurring in the Vlasov-Poisson equations correspond to the ensemble averages of $N_j(\mathbf{x}, \mathbf{p}, t)$ and $\mathbf{E}^M(\mathbf{x}, t)$, respectively.

Because the Klimontovich-Poisson equations (2.63) and (2.64) are identical in form to the Vlasov-Poisson equations (2.3) and (2.7), the derivation of the conservation relations analogous to Eqs.(2.51), (2.52), and (2.6.1) proceeds in a similar manner. As in Sec. 2.4, we assume that the nonneutral plasma is isolated from physical contact with external charge and current sources and the surrounding enclosure. Moreover, particle losses from the system are neglected, as well as recombination and ionization processes. Without presenting algebraic details, the following conservation relations can be derived from Eqs.(2.63) and (2.64) in the electrostatic approximation:

A.   Number Conservation

$$\frac{\partial}{\partial t}\left(\int d^3p\, N_j\right) + \frac{\partial}{\partial \mathbf{x}}\cdot\left(\int d^3p\, \mathbf{v} N_j\right) = 0 , \qquad (2.66)$$

B.   Momentum Conservation

$$\frac{\partial}{\partial t}\left(\sum_j \int d^3p\, \mathbf{p} N_j + \frac{\mathbf{E}^M \times \mathbf{B}^0}{4\pi c}\right) \qquad (2.67)$$

$$+ \frac{\partial}{\partial \mathbf{x}}\cdot\left(\sum_j \int d^3p\, \mathbf{p}\, \mathbf{v} N_j + \frac{\left|\mathbf{E}^M\right|^2 + \left|\mathbf{B}^0\right|^2}{8\pi}\mathbf{I} - \frac{\mathbf{E}^M \mathbf{E}^M + \mathbf{B}^0 \mathbf{B}^0}{4\pi}\right) = 0 ,$$

C.   Energy Conservation

$$\frac{\partial}{\partial t}\left(\sum_j \int d^3p\,(\gamma-1)m_j c^2 N_j + \frac{\left|\mathbf{E}^M\right|^2}{8\pi}\right) \qquad (2.68)$$

$$+ \frac{\partial}{\partial \mathbf{x}}\cdot\left(\sum_j \int d^3p\,(\gamma-1)m_j c^2 \mathbf{v} N_j + \frac{c}{4\pi}\mathbf{E}^M \times \mathbf{B}^0\right) = 0 ,$$

D.   Angular Momentum Conservation

$$\frac{\partial}{\partial t}\left(\sum_j \int d^3p\, \mathbf{x} \times \mathbf{p} N_j + \frac{\mathbf{x} \times (\mathbf{E}^M \times \mathbf{B}^0)}{4\pi c}\right) \qquad (2.69)$$

$$+ \frac{\partial}{\partial \mathbf{x}}\cdot\left[\left(-\sum_j \int d^3p\, \mathbf{p}\, \mathbf{v} N_j + \frac{\mathbf{E}^M \mathbf{E}^M + \mathbf{B}^0 \mathbf{B}^0}{4\pi} - \frac{\left|\mathbf{E}^M\right|^2 + \left|\mathbf{B}^0\right|^2}{8\pi}\mathbf{I}\right) \times \mathbf{x}\right]$$

$$= 0 .$$

Here, $\mathbf{p} = \gamma m_j \mathbf{v}$ where $\gamma = (1 + \mathbf{p}^2/m_j^2 c^2)^{1/2}$, and the arguments of $N_j(\mathbf{x}, \mathbf{p}, t)$, $\mathbf{E}^M(\mathbf{x}, t)$ and $\mathbf{B}^0(\mathbf{x})$ have been suppressed in Eqs.(2.66)–(2.69). Because of the form of $N_j(\mathbf{x}, \mathbf{p}, t)$ in Eq.(2.60), the conservation relations in Eqs.(2.66)–(2.69) involve summations over all (discrete) particles in the system. As in Sec. 2.4, global (spatially-averaged) conservation relations can be derived from Eqs.(2.66)–(2.69) by integrating over the volume of the enclosure containing the nonneutral plasma and making the appropriate assumptions regarding wall conditions (e.g., perfectly conducting walls). For example, for a cylindrical plasma column contained inside a perfectly conducting cylinder, use will be made in Sec. 4.6 of the global conservation of energy and canonical angular momentum calculated from Eqs.(2.68) and (2.69) to develop a general confinement theorem[9] for one-component nonneutral plasmas.

To conclude Sec. 2.5, the general procedure for calculating the modification of the one-particle distribution function $f_j(\mathbf{x}, \mathbf{p}, t)$ due to discrete particle collisions can be summarized as follows. The microscopic phase-space density $N_j(\mathbf{x}, \mathbf{p}, t)$ and electric field $\mathbf{E}^M(\mathbf{x}, t)$ are expressed as

$$N_j(\mathbf{x}, \mathbf{p}, t) = \langle N_j(\mathbf{x}, \mathbf{p}, t) \rangle + \delta N_j(\mathbf{x}, \mathbf{p}, t),$$
$$\mathbf{E}^M(\mathbf{x}, t) = \langle \mathbf{E}^M(\mathbf{x}, t) \rangle + \delta \mathbf{E}^M(\mathbf{x}, t), \tag{2.70}$$

where $\langle N_j(\mathbf{x}, \mathbf{p}, t) \rangle = f_j(\mathbf{x}, \mathbf{p}, t)$ and $\langle \mathbf{E}^M(\mathbf{x}, t) \rangle = \mathbf{E}(\mathbf{x}, t)$ denote ensemble average values, and $\delta N_j(\mathbf{x}, \mathbf{p}, t)$ and $\delta \mathbf{E}^M(\mathbf{x}, t)$ denote fluctuations about the average values induced by discrete particle effects. Taking the ensemble average of the Klimontovich-Poisson equations (2.63) and (2.64), and making use of $\langle \delta N_j \rangle = 0 = \langle \delta \mathbf{E}^M \rangle$, we obtain

$$\left\{ \frac{\partial}{\partial t} + \mathbf{v} \cdot \frac{\partial}{\partial \mathbf{x}} + e_j \left(\mathbf{E} + \frac{\mathbf{v} \times \mathbf{B}^0}{c} \right) \cdot \frac{\partial}{\partial \mathbf{p}} \right\} f_j(\mathbf{x}, \mathbf{p}, t) = \left(\frac{\partial f_j}{\partial t} \right)_{coll},$$
$$\tag{2.71}$$

and

$$\nabla \cdot \mathbf{E} = 4\pi \sum_j e_j \int d^3 p \, f_j(\mathbf{x}, \mathbf{p}, t) + 4\pi \rho_{\text{ext}}, \tag{2.72}$$

where $(\partial f_j / \partial t)_{coll}$ is defined by

$$\left(\frac{\partial f_j}{\partial t} \right)_{coll} = -e_j \frac{\partial}{\partial \mathbf{p}} \cdot \langle \delta \mathbf{E}^M(\mathbf{x}, t) \delta N_j(\mathbf{x}, \mathbf{p}, t) \rangle. \tag{2.73}$$

Here, $(\partial f_j / \partial t)_{coll}$ describes the change in $f_j(\mathbf{x}, \mathbf{p}, t)$ induced by discrete particle collisions. The evolution of the fluctuations $\delta N_j(\mathbf{x}, \mathbf{p}, t)$ and $\delta \mathbf{E}^M(\mathbf{x}, t)$ occurring in Eq.(2.73) is obtained by subtracting Eqs.(2.71) and (2.72) from Eqs.(2.63) and (2.64), respectively. This gives

$$\left\{ \frac{\partial}{\partial t} + \mathbf{v} \cdot \frac{\partial}{\partial \mathbf{x}} + e_j \left(\mathbf{E} + \frac{\mathbf{v} \times \mathbf{B}^0}{c} \right) \cdot \frac{\partial}{\partial \mathbf{p}} \right\} \delta N_j(\mathbf{x}, \mathbf{p}, t) + e_j \delta \mathbf{E}^M \cdot \frac{\partial}{\partial \mathbf{p}} f_j$$

$$= e_j \left\langle \delta \mathbf{E}^M \cdot \frac{\partial}{\partial \mathbf{p}} \delta N_j \right\rangle - e_j \delta \mathbf{E}^M \cdot \frac{\partial}{\partial \mathbf{p}} \delta N_j , \qquad (2.74)$$

and

$$\nabla \cdot \delta \mathbf{E}^M = 4\pi \sum_j e_j \int d^3 p \, \delta N_j(\mathbf{x}, \mathbf{p}, t). \qquad (2.75)$$

Equations (2.71)–(2.75) are exactly equivalent to the original Klimontovich-Poisson equations (2.63) and (2.64). If the collision term on the right-hand side of Eq.(2.71) is neglected, then Eqs.(2.71) and (2.72) reduce, as expected, to the nonlinear Vlasov-Poisson equations, which describe the evolution of the average distribution function $f_j(\mathbf{x}, \mathbf{p}, t)$ and electric field $\mathbf{E}(\mathbf{x}, t)$ due to collective processes. On the other hand, if $(\partial f_j / \partial t)_{coll}$ is retained in Eq.(2.71), then the resulting equation for the evolution of $f_j(\mathbf{x}, \mathbf{p}, t)$ includes collisional effects as well as collective interactions. To evaluate $(\partial f_j / \partial t)_{coll}$ to leading order in Eq.(2.73), the usual procedure is to calculate $\delta N_j(\mathbf{x}, \mathbf{p}, t)$ by neglecting the right-hand side of Eq.(2.74), which includes ternary (and higher-order) collisional effects. Because the (nonlinear) Vlasov propagator operates on $\delta N_j(\mathbf{x}, \mathbf{p}, t)$ in Eq.(2.74) and because the average electric field $\mathbf{E}(\mathbf{x}, t)$ is non-zero in a nonneutral plasma (even close to thermal equilibrium), a detailed calculation of $(\partial f_j / \partial t)_{coll}$ is generally formidable. In the limit of a low-density, strongly magnetized nonneutral plasma, however, a formal expression for $(\partial f_j / \partial t)_{coll}$ can be obtained[10-12] using techniques similar to those employed in calculating $(\partial f_j / \partial t)_{coll}$ for a neutral plasma.[1] Collisional transport coefficients can then be determined in a similar manner.

Problem 2.7

a. Make use of the orbit equations (2.61) and (2.62) and the definition of the microscopic phase-space density in Eq.(2.60) to show that $N_j(\mathbf{x}, \mathbf{p}, t)$ evolves according to the Klimontovich equation (2.63).

b.   Take the ensemble average of the Klimontovich-Poisson equations (2.63) and (2.64) and show that the average distribution function $f_j(\mathbf{x}, \mathbf{p}, t)$ and electric field $\mathbf{E}(\mathbf{x}, t)$ evolve according to Eqs.(2.71) and (2.72), where $(\partial f_j / \partial t)_{coll}$ is defined in Eq.(2.73).

c.   Subtract Eqs.(2.71) and (2.72) from Eqs.(2.63) and (2.64), respectively, and show that the fluctuations $\delta N_j(\mathbf{x}, \mathbf{p}, t)$ and $\delta \mathbf{E}^M(\mathbf{x}, t)$ evolve (exactly) according to Eqs.(2.74) and (2.75).

2.6   Gravitational Analogue

In concluding Chapter 2, it should be pointed out that the theoretical models for describing the evolution of a self-gravitating system[13,14] of point masses are similar in form to the theoretical models for describing the evolution of a one-component nonneutral plasma consisting only of electrons (say) which interact electrostatically. For the case of electrons, the (repulsive) Coulomb potential is $\phi_{ij}\left(|\mathbf{x}_i - \mathbf{x}_j|\right) = e^2 / |\mathbf{x}_i - \mathbf{x}_j|$, where $-e$ is the electron charge. On the other hand, for the case of interacting masses of mass m, the (attractive) gravitational potential is $\phi_{ij}\left(|\mathbf{x}_i - \mathbf{x}_j|\right) = -Gm^2 / |\mathbf{x}_i - \mathbf{x}_j|$, where G is the universal gravitational constant. Then, for example, the gravitational analogue of the Vlasov equation (2.3) describing collective interactions in a nonrelativistic self-gravitating system is given by

$$\left\{ \frac{\partial}{\partial t} + \mathbf{v} \cdot \frac{\partial}{\partial \mathbf{x}} - m \frac{\partial \phi}{\partial \mathbf{x}} \cdot \frac{\partial}{\partial \mathbf{p}} \right\} f(\mathbf{x}, \mathbf{p}, t) = 0 \,. \tag{2.76}$$

Here, $f(\mathbf{x}, \mathbf{p}, t)$ is the (smooth) one-particle distribution function, and $\phi(\mathbf{x}, t)$ is the average gravitational potential determined self-consistently from

$$\nabla^2 \phi(\mathbf{x}, t) = 4\pi G m \int d^3 p \, f(\mathbf{x}, \mathbf{p}, t) \,, \tag{2.77}$$

which plays the role of Poisson's equation (2.7).

Chapter 2 References

1.   R. Balescu, *Transport Processes in Plasmas* (North Holland, Amsterdam, 1988), Volumes 1 and 2.

2.   R.C. Davidson, *Theory of Nonneutral Plasmas* (Addison-Wesley, Reading, Massachusetts, 1989), Chapter 1.

3. D.R. Nicholson, *Introduction to Plasma Theory* (Wiley, New York, 1983).

4. N.A. Krall and A.W. Trivelpiece, *Principles of Plasma Physics* (San Francisco Press, San Francisco, California, 1986).

5. S. Ichimaru, *Basic Principles of Plasma Physics—A Statistical Approach* (Benjamin, Reading, Massachusetts, 1973).

6. A.A. Vlasov, "On the Kinetic Theory of an Assembly of Particles with Collective Interaction," J. Phys. (USSR) **9**, 25 (1945).

7. Y.L. Klimontovich, *Statistical Theory of Nonequilibrium Processes in Plasmas* (MIT Press, Cambridge, Massachusetts, 1967).

8. J.D. Jackson, *Classical Electrodynamics* (Wiley, New York, 1975).

9. T.M. O'Neil, "A Confinement Theorem for Nonneutral Plasmas," Phys. Fluids **23**, 2217 (1980).

10. T.M. O'Neil and C.F. Driscoll, "Transport to Thermal Equilibrium of a Pure Electron Plasma," Phys. Fluids **22**, 266 (1979).

11. T.M. O'Neil, "A New Theory of Transport Due to Like-Particle Collisions," Phys. Rev. Lett. **55**, 943 (1985).

12. D.H.E. Dubin and T.M. O'Neil, "Two-Dimensional Guiding Center Transport of a Pure Electron Plasma," Phys. Rev. Lett. **60**, 1286 (1988).

13. A.M. Fridman and V.L. Polyachenko, *Physics of Gravitating Systems I: Equilibrium and Stability* (Springer-Verlag, New York, 1984).

14. A.M. Fridman and V.L. Polyachenko, *Physics of Gravitating Systems II: Nonlinear Collective Processes* (Springer Verlag, New York, 1984).

Nonneutral plasmas exhibit many collective properties and interaction processes which are similar to those in electrically neutral plasmas. The following references, while not cited directly in the text, are relevant to the general subject matter of this chapter.

R. Balescu, *Statistical Mechanics of Charged Particles* (Wiley Interscience, New York, 1963).

G. Bekefi, *Radiation Processes in Plasmas* (Wiley, New York, 1966).

F.F. Chen, *Introduction to Plasma Physics* (Plenum, New York, 1984).

R.C. Davidson, *Methods in Nonlinear Plasma Theory* (Academic Press, New York, 1972).

J.P. Freidberg, *Ideal Magnetohydrodynamics* (Plenum, New York, 1987).

S. Humphries, Jr., *Principles of Charged Particle Acceleration* (Wiley, New York, 1986).

B.B. Kadomtsev, *Plasma Turbulence* (Academic Press, New York, 1968).

J.D. Lawson, *The Physics of Charged Particle Beams* (Clarendon Press, Oxford, 1988).

A.B. Mikhailovskii, *Theory of Plasma Instabilities* (Consultants Bureau, New York, 1974).

D.C. Montgomery, *Theory of Unmagnetized Plasma* (Gordon and Breach, New York, 1971).

D.C. Montgomery and D.A. Tidman, *Plasma Kinetic Theory* (McGraw Hill, New York, 1964).

R.Z. Sagdeev and A.A. Galeev, *Nonlinear Plasma Theory*, revised and edited by T.M. O'Neil and D.L. Book (Benjamin, Reading, Massachusetts, 1969).

A.G. Sitenko, *Electromagnetic Fluctuations in Plasma* (Academic Press, New York, 1967).

T.H. Stix, *Theory of Plasma Waves* (McGraw Hill, New York, 1962).

V.N. Tsytovich, *Nonlinear Effects in Plasmas* (Plenum, New York, 1970).

Handbook of Plasma Physics—Basic Plasma Physics, eds., M.N. Rosenbluth and R.Z. Sagdeev (North Holland, Amsterdam, 1983 and 1984), Volumes 1 and 2.

CHAPTER 3

FUNDAMENTAL PROPERTIES OF NONNEUTRAL PLASMA

In Chapters 4–10, the kinetic and macroscopic formalisms described in Chapter 2 will be used to investigate detailed equilibrium and stability properties of a wide range of nonneutral plasma systems. To orient the reader, in this chapter we summarize several fundamental properties of nonneutral plasmas that clearly delineate the strong influence of intense self-electric fields.[1,2] The topics covered here include: the equilibrium rotation of a nonneutral plasma column (Sec. 3.1); analysis of the single-particle trajectories (Sec. 3.2); investigation of thermal equilibrium properties (Sec. 3.3); calculation of the Debye shielding of the electrostatic potential surrounding a test electron (Sec. 3.4); determination of the spontaneous emission of electromagnetic radiation from a test electron (Sec. 3.5); and discussion of the properties of strongly coupled nonneutral plasma (Sec. 3.6).

For simplicity, with the exception of Sec. 3.6, we specialize in Chapter 3 to the case of a nonrelativistic, one-component, nonneutral plasma consisting only of electrons. The analysis and results are readily extended to a pure ion plasma or to a multicomponent nonneutral plasma.

3.1 Cold-Fluid Equilibrium Rotation

As illustrated in Fig. 3.1, we consider an infinitely-long nonneutral plasma column confined radially by a uniform magnetic field $B_0\hat{e}_z$. Under steady-state conditions ($\partial/\partial t = 0$), the equilibrium density profile $n_e^0(r)$ is assumed to have the rectangular form given by

$$n_e^0(r) = \begin{cases} \hat{n}_e = \text{const.} & 0 \leq r < r_b \,, \\ 0 \,, & r > r_b \,, \end{cases} \tag{3.1}$$

where r_b is the radius of the plasma column. In the absence of a neutralizing ion background, the electron space charge produces an equilibrium electric field $\mathbf{E}^0(\mathbf{x}) = E_r(r)\hat{\mathbf{e}}_r$ in the radial direction. Integrating Poisson's equation, $r^{-1}(\partial/\partial r)\left[rE_r(r)\right] = -4\pi e n_e^0(r)$, we find that the radial self-electric field can be expressed as

$$E_r(r) = \begin{cases} -\dfrac{m_e}{2e}\omega_{pe}^2 r , & 0 \le r < r_b , \\ -\dfrac{m_e}{2e}\omega_{pe}^2 \dfrac{r_b^2}{r} , & r > r_b , \end{cases} \qquad (3.2)$$

where $\omega_{pe}^2 = 4\pi \hat{n}_e e^2/m_e = \text{const.}$ is the electron plasma frequency-squared, and $r = (x^2 + y^2)^{1/2}$ is the radial distance from the axis of symmetry. Assuming that the electrons can be treated macroscopically as a cold fluid (Sec. 2.3), radial force balance on an electron fluid element can be expressed as

$$-\frac{m_e V_{\theta e}^{02}(r)}{r} = -eE_r(r) - \frac{e}{c}V_{\theta e}^0(r)B_0 . \qquad (3.3)$$

Here, $-e$ is the electron charge, and $V_{\theta e}^0(r)$ is the equilibrium azimuthal velocity of an electron fluid element. Note that Eq.(3.3) is a statement of the balance between the outward centrifugal and electric forces and the inward magnetic force (Fig. 3.2). Introducing the angular rotation velocity $\omega_{re}(r)$ defined by

$$V_{\theta e}^0(r) = \omega_{re}(r)r , \qquad (3.4)$$

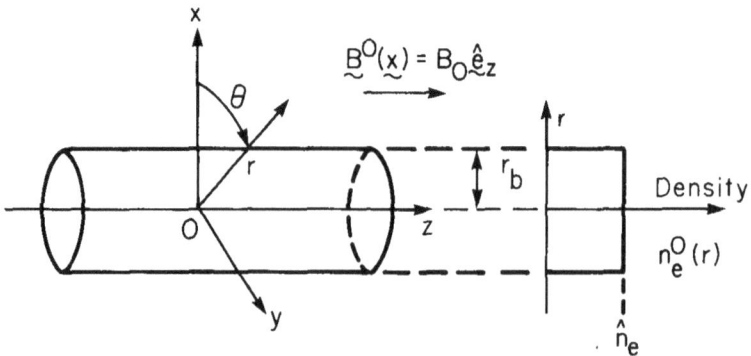

Figure 3.1.   Nonneutral plasma column confined radially by a uniform applied magnetic field $B_0\hat{\mathbf{e}}_z$.

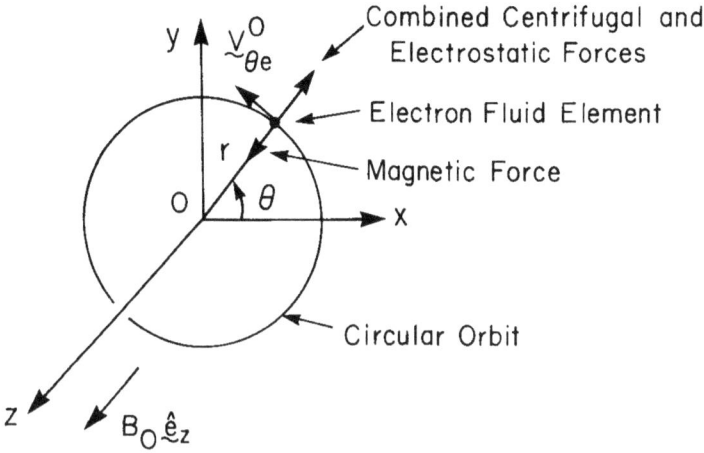

Figure 3.2.   Balance between the outward centrifugal and electric forces and the inward magnetic force on an electron fluid element [Eq.(3.3)].

and making use of Eq.(3.2), it is straightforward to show that Eq.(3.3) can be expressed as

$$- \omega_{re}^2 = \frac{1}{2}\omega_{pe}^2 - \omega_{re}\omega_{ce} \ . \tag{3.5}$$

Here, $\omega_{ce} = eB_0/m_e c$ is the electron cyclotron frequency, and Eq.(3.5) is applicable within the plasma column ($0 \leq r < r_b$). Solving Eq.(3.5) for ω_{re} gives the two cold-fluid rotation velocities

$$\omega_{re} = \omega_{re}^{\pm} \equiv \frac{1}{2}\omega_{ce}\left\{1 \pm \left(1 - \frac{2\omega_{pe}^2}{\omega_{ce}^2}\right)^{1/2}\right\} , \tag{3.6}$$

where $\omega_{re}^-(\omega_{re}^+)$ corresponds to a slow (fast) rotation of the plasma column. Because ω_{re}^+ and ω_{re}^- are constant (independent of r), the average azimuthal motion of the plasma column corresponds to a rigid rotation about the axis of symmetry.

Figure 3.3 shows plots of ω_{re}^+ and ω_{re}^- versus the self-field parameter $s_e = 2\omega_{pe}^2/\omega_{ce}^2$, which measures the relative strengths of the (defocusing) space-charge force and the (focusing) magnetic force on a fluid element.[3,4] For a low-density plasma with $2\omega_{pe}^2/\omega_{ce}^2 \ll 1$, we find from Eq.(3.6) and Fig. 3.3 that $\omega_{re}^+ \simeq \omega_{ce}$, which corresponds to a fast rotation of the plasma column at the electron cyclotron frequency. On the other hand, at low density, ω_{re}^- can be approximated by

$$\omega_{re}^- = \frac{\omega_{pe}^2}{2\omega_{ce}} \equiv \omega_D \text{ , for } 2\omega_{pe}^2/\omega_{ce}^2 \ll 1 \text{ ,} \qquad (3.7)$$

which corresponds to a slow $E_r \hat{e}_r \times B_0 \hat{e}_z$ rotation. The quantity ω_D defined in Eq.(3.7) is referred to as the *diocotron frequency*. By contrast, for sufficiently large density that $2\omega_{pe}^2/\omega_{ce}^2 = 1$, we find from Eq.(3.6) and Fig. 3.3 that

$$\omega_{re}^\pm = \frac{1}{2}\omega_{ce}, \text{ for } 2\omega_{pe}^2/\omega_{ce}^2 = 1 \text{ .} \qquad (3.8)$$

The condition $2\omega_{pe}^2/\omega_{ce}^2 = 1$ is referred to as the *Brillouin density limit*.[5] It is clear from Eq.(3.6) and Fig. 3.3 that radially confined equilibria do not exist for $2\omega_{pe}^2/\omega_{ce}^2 > 1$, because the (defocusing) space-charge force is too

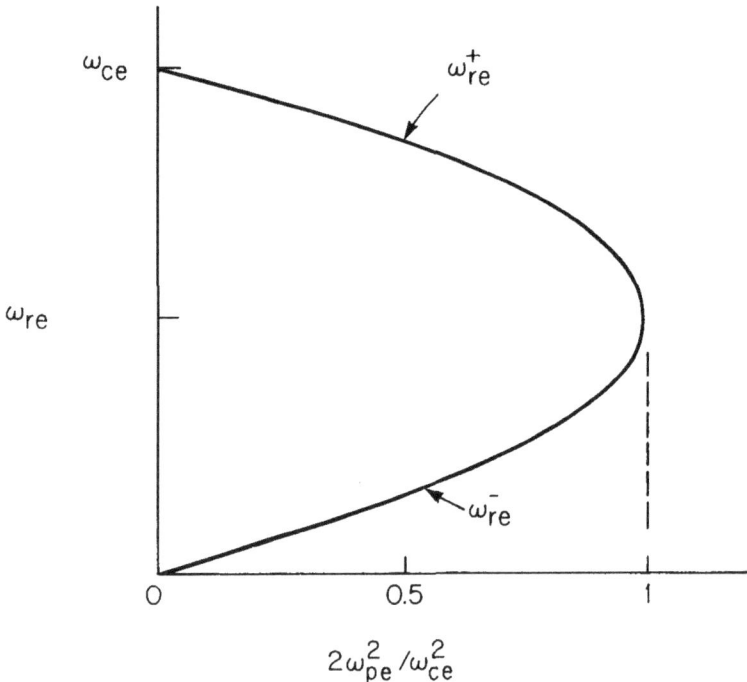

Figure 3.3. Plots of the cold-fluid angular rotation velocities ω_{re}^+ and ω_{re}^- versus the self-field parameter $s_e = 2\omega_{pe}^2/\omega_{ce}^2$ [Eq.(3.6)].

large. Moreover, the allowed values of angular rotation velocity in Eq.(3.6) exhibit a strong dependence on the self-field parameter $s_e = 2\omega_{pe}^2/\omega_{ce}^2$. Generally speaking, whether the plasma column is rotating with either slow (ω_{re}^-) or fast (ω_{re}^+) rotation velocity, depends on the way in which the nonneutral plasma is formed (e.g., injection conditions).

The two rotational equilibria in Eq.(3.6) have been measured experimentally by Theiss, Mahaffey and Trivelpiece[6] for a cold, constant-density, nonneutral plasma column confined by a uniform axial magnetic field. The experimental set-up is illustrated in Fig. 3.4. A $30\,\mu s$-pulse electron beam with diameter $2r_b = 1$ cm, axial kinetic energy $E_b = 100 - 450\,\text{eV}$, and current $I_b = 1 - 10\,\text{mA}$, is injected through a magnetic-field step into a steady magnetic field $B_0 = 20 - 50\,\text{G}$. The magnetic-field step in Fig. 3.4 induces a rotation of the electron beam. For specified values of the beam current and energy (and therefore the electron density $\hat{n}_e$), the size of the magnetic-field step is adjusted so that the beam is launched in either a slow (ω_{re}^-) or fast (ω_{re}^+) rotational equilibrium consistent with Eq.(3.6). To measure the rotation velocity, the beam is intercepted by a thin tungsten needle which produces a (rotated) shadow on a moveable phosphor screen located further downstream (Fig. 3.4). By varying the separation (Δz) of the needle and the screen, and measuring the angle of rotation ($\Delta\theta$), the average angular velocity is determined from $\langle\dot{\theta}\rangle = (\Delta\theta/\Delta z)V_{ze}^0$, where V_{ze}^0 is the axial velocity of the electron beam. Typical experimental results[6] are illustrated in Fig. 3.5 for a slow rotational equilibrium (ω_{re}^-) with density $\hat{n}_e \simeq 1.4 \times 10^6$ cm^{-3} and self-field parameter $s_e = 2\omega_{pe}^2/\omega_{ce}^2 \simeq 0.094$. Repeating the measurements for several choices of magnetic-field step, electron density ($\hat{n}_e$), and magnetic field (B_0), good agreement is obtained with the predicted values of ω_{re}^- and ω_{re}^+ in Eq.(3.6) for both slow and fast rotational equilibria (Fig. 3.6).

The simple equilibrium analysis presented earlier in Sec. 3.2 neglects the influence of the diamagnetic depression in axial magnetic field $B_z(r) = B_0 + B_z^s(r)$ produced by the rotation of the nonneutral plasma column. In this regard, the $\nabla \times \mathbf{B}^0$ Maxwell equation gives $\partial B_z(r)/\partial r = (4\pi e/c)\hat{n}_e \times \omega_{re}(r)r$ within the plasma column where the electron density is equal to $\hat{n}_e = \text{const}$. If we treat the angular rotation velocity $\omega_{re}(r)$ as independent of r to leading order, then solving for $\Delta B_z \equiv B_z(r = r_b) - B_z(r = 0)$ gives

$$\frac{\Delta B_z}{B_0} = \frac{1}{2}\frac{\omega_{pe}^2 r_b^2}{c^2}\frac{\omega_{re}}{\omega_{ce}}. \tag{3.9}$$

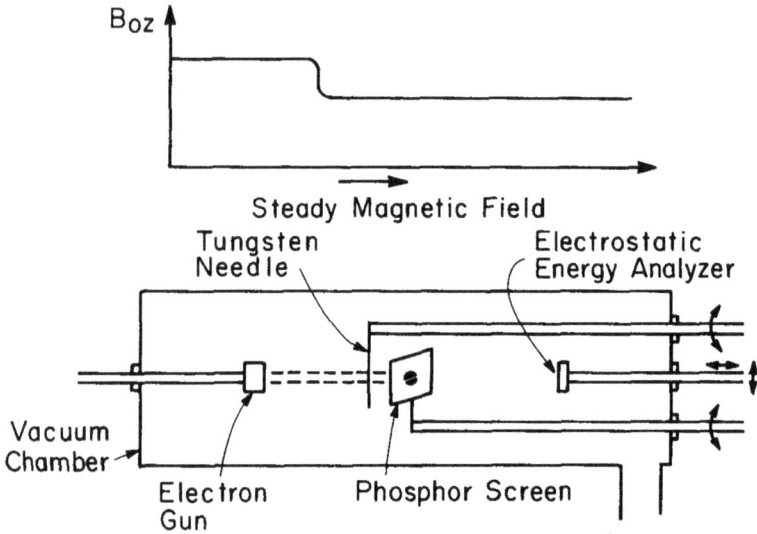

Figure 3.4. Schematic of the experimental set-up used to investigate the rotating equilibria of a nonneutral plasma column [A.J. Theiss, R.A. Mahaffey and A.W. Trivelpiece, Phys. Rev. Lett. **35**, 1436 (1975)].

To neglect the diamagnetic variation of $B_z(r)$ and approximate $B_z(r) \simeq B_0$ within the plasma column, we require $(\Delta B_z)/B_0 = (\omega_{pe}^2 r_b^2/2c^2) \times (\omega_{re}/\omega_{ce}) \ll 1$. At the maximum density limit where $\omega_{re} = \omega_{re}^{\pm} = \omega_{ce}/2$ (Fig. 3.3), this inequality is equivalent to

$$\nu \equiv N_e \frac{e^2}{m_e c^2} = \frac{1}{4} \frac{\omega_{pe}^2 r_b^2}{c^2} \ll 1 . \tag{3.10}$$

Here, $e^2/m_e c^2$ is the classical electron radius, $N_e = 2\pi \int_0^{r_b} dr r n_e^0(r) = \hat{n}_e \pi r_b^2$ is the number of electrons per unit axial length of the plasma column, and $\nu = N_e e^2/m_e c^2$ is Budker's parameter.[7] Therefore, $\nu \ll 1$ assures that $(\Delta B_z)/B_0 \ll 1$, and the axial magnetic field $B_z(r)$ can be approximated by the applied field B_0. Note that Eq.(3.10) requires that the collisionless skin depth c/ω_{pe} be large in comparison with the column radius r_b.

To summarize, $\nu = \omega_{pe}^2 r_b^2/4c^2 \ll 1$ assures that the axial self-magnetic field $B_z^s(r) = B_z(r) - B_0$ is weak. However, the self-electric field $E_r(r)$ can have a large influence on the equilibrium rotation of the plasma column, depending on the value of $s_e = 2\omega_{pe}^2/\omega_{ce}^2$.

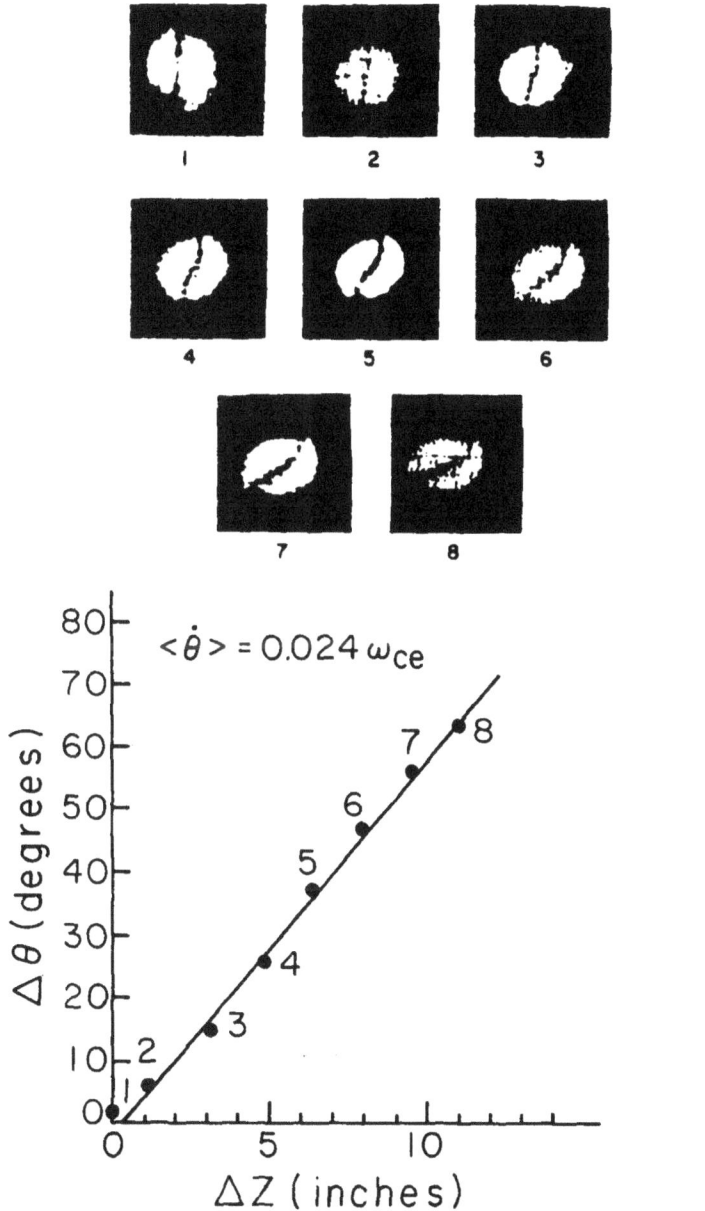

Figure 3.5. Photographs of the phosphor screen show the (rotating) shadow of the tungsten needle. The angular velocity inferred from the rotation of the shadow is $\langle \dot{\theta} \rangle = 0.024\,\omega_{ce}$ [A.J. Theiss, R.A. Mahaffey and A.W. Trivelpiece, Phys. Rev. Lett. **35**, 1436 (1975)].

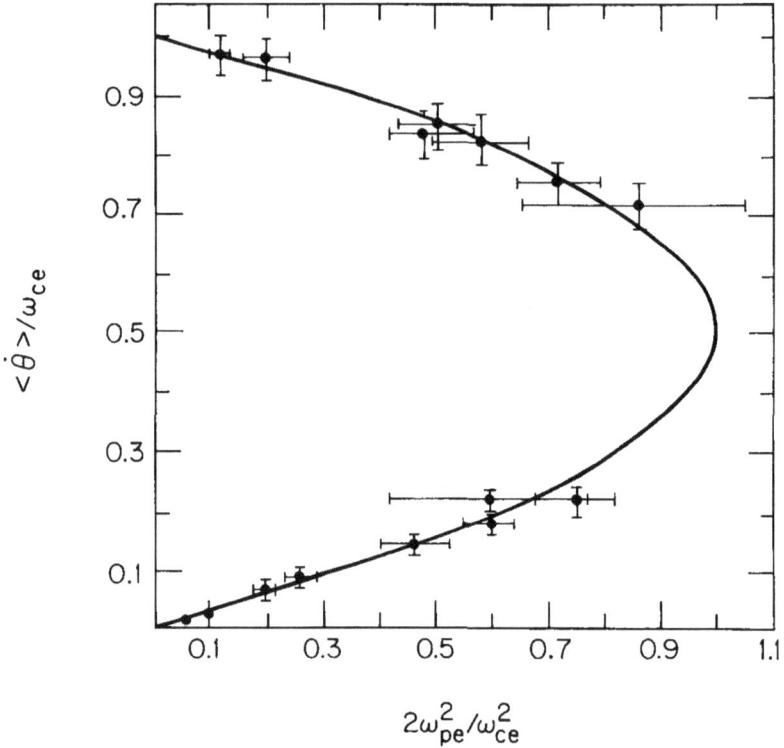

Figure 3.6. Measured values of the normalized angular velocity $\langle \dot{\theta} \rangle / \omega_{ce}$ are plotted versus the self-field parameter $s_e = 2\omega_{pe}^2 / \omega_{ce}^2$. The solid curve corresponds to the predicted values of ω_{re}^+ and ω_{re}^- in Eq.(3.6) [A.J. Theiss, R.A. Mahaffey and A.W. Trivelpiece, Phys. Rev. Lett. **35**, 1436 (1975)].

Problem 3.1   Consider a nonneutral plasma column with uniform density and cold-fluid equilibrium properties described by Eqs.(3.1)–(3.6). The rotational kinetic energy (U_K) and electric field energy (U_E) per unit length of the plasma column are defined by

$$U_K = 2\pi \int_0^{r_b} dr\, r \frac{1}{2} n_e^0(r) m_e \omega_{re}^2 r^2 \ , \tag{3.1.1}$$

and

$$U_E = 2\pi \int_0^{r_b} dr\, r \frac{|E_r(r)|^2}{8\pi} \ , \tag{3.1.2}$$

where r_b is the column radius. Show that U_K and U_E can be expressed as

$$U_K = \frac{1}{2} \frac{\omega_{pe}^2 r_b^2}{c^2} \frac{\omega_{re}^2}{\omega_{ce}^2} U_B ,$$
(3.1.3)

and

$$U_E = \frac{1}{8} \frac{\omega_{pe}^2 r_b^2}{c^2} \frac{\omega_{pe}^2}{\omega_{ce}^2} U_B ,$$
(3.1.4)

where $U_B = (B_0^2/8\pi)\pi r_b^2$ is the (applied) magnetic field energy per unit length of the plasma column.

For $\nu = \omega_{pe}^2 r_b^2/4c^2 \ll 1$, it follows from Eqs.(3.1.3) and (3.1.4) that $U_K, U_E \ll U_B$ over the entire range of $\omega_{re}^2/\omega_{ce}^2$ and $2\omega_{pe}^2/\omega_{ce}^2$ allowed by Eq.(3.6) and Fig. 3.3. Moreover, at the Brillouin density limit, where $s_e = 2\omega_{pe}^2/\omega_{ce}^2 = 1$ and $\omega_{re} = \omega_{re}^{\pm} = \omega_{ce}/2$, Eqs.(3.1.3) and (3.1.4) give $U_E = (1/2)U_K$. That is, the self-electric-field energy contained within the plasma column is equal to one-half of the rotational kinetic energy.

Problem 3.2   Consider the nonneutral plasma column with uniform density profile and radius r_b illustrated in Fig. 3.1. Assume that the plasma column is enclosed by a conducting cylinder with radius $r = b > r_b$.

a. Make use of the expression for the equilibrium radial electric field $E_r(r) = -\partial\phi_0(r)/\partial r$ in Eq.(3.2) to determine the electrostatic potential $\phi_0(r)$ in the two regions, $0 \leq r < r_b$ and $r_b < r \leq b$. In obtaining the solutions, take the zero of potential to be $\phi_0(r = 0) = 0$, and enforce the continuity of $\phi_0(r)$ at the surface of the plasma column ($r = r_b$).

b. Show that the electrostatic potential $V = \phi_0(r = b)$ at the surface of the conducting cylinder can be expressed as

$$\frac{eV}{m_e c^2} = \frac{1}{4} \frac{\omega_{pe}^2 r_b^2}{c^2} \left[1 + 2\ln \left(\frac{b}{r_b} \right) \right] .$$
(3.2.1)

For $\nu = \omega_{pe}^2 r_b^2/4c^2 \ll 1$, it follows from Eq.(3.2.1) that $eV/m_e c^2 \ll 1$.

3.2   Single-Particle Trajectories

It is informative to examine the motion of an individual test electron confined within the nonneutral plasma column illustrated in Fig. 3.1. Assuming uniform electron density, an individual electron experiences the forces produced by the radial self-electric field $\mathbf{E}^0(\mathbf{x}) = E_r(r)\hat{\mathbf{e}}_r$ in Eq.(3.2), and the axial magnetic field $B_0\hat{\mathbf{e}}_z$. The motion of an electron is determined from

$$m_e \frac{d\mathbf{v}}{dt} = -e \left(\mathbf{E}^0(\mathbf{x}) + \frac{\mathbf{v} \times B_0\hat{\mathbf{e}}_z}{c} \right) ,$$
(3.11)

where $\mathbf{v}(t) = d\mathbf{x}(t)/dt$ is the particle velocity, and

$$\mathbf{E}^0(\mathbf{x}) = -\frac{m_e}{2e}\omega_{pe}^2(x\hat{\mathbf{e}}_x + y\hat{\mathbf{e}}_y) \tag{3.12}$$

is the equilibrium electric field within the plasma column $(0 \le r < r_b)$. From Eq.(3.11), the motion along the magnetic field $B_0\hat{\mathbf{e}}_z$ is free streaming with axial velocity $v_z = dz/dt = \text{const}$. On the other hand, the motion perpendicular to the magnetic field is in crossed electric and magnetic fields. In Cartesian coordinates, the components of Eq.(3.11) perpendicular to $B_0\hat{\mathbf{e}}_z$ can be expressed as

$$\frac{d^2x(t)}{dt^2} = \frac{1}{2}\omega_{pe}^2 x(t) - \omega_{ce}\frac{dy(t)}{dt}, \tag{3.13a}$$

$$\frac{d^2y(t)}{dt^2} = \frac{1}{2}\omega_{pe}^2 y(t) + \omega_{ce}\frac{dx(t)}{dt}, \tag{3.13b}$$

where $\omega_{ce} = eB_0/m_e c$ and $\omega_{pe}^2 = 4\pi\hat{n}_e e^2/m_e$.

To analyze the electron motion, it is useful to transform Eq.(3.13) to a frame of reference rotating with angular velocity

$$\omega_{re} = \omega_{re}^+ \text{ or } \omega_{re} = \omega_{re}^- , \tag{3.14}$$

which are the two possible average rotation velocities of the plasma column [Eq.(3.6)]. We introduce the orbits, $x_r'(t)$ and $y_r'(t)$, in the rotating frame defined by[1,2]

$$x_r'(t) = x(t)\cos(\omega_{re}t) + y(t)\sin(\omega_{re}t) , \tag{3.15a}$$

$$y_r'(t) = y(t)\cos(\omega_{re}t) - x(t)\sin(\omega_{re}t) . \tag{3.15b}$$

After some straightforward algebra, the equations of motion in the rotating frame can be expressed as

$$\frac{d^2x_r'(t)}{dt^2} = -(\omega_{ce} - 2\omega_{re})\frac{dy_r'(t)}{dt} + \left(\omega_{re}^2 - \omega_{ce}\omega_{re} + \frac{1}{2}\omega_{pe}^2\right)x_r'(t) , \tag{3.16a}$$

$$\frac{d^2y_r'(t)}{dt^2} = (\omega_{ce} - 2\omega_{re})\frac{dx_r'(t)}{dt} + \left(\omega_{re}^2 - \omega_{ce}\omega_{re} + \frac{1}{2}\omega_{pe}^2\right)y_r'(t) , \tag{3.16b}$$

in the region where the electron density is nonzero. Because $\omega_{re} = \omega_{re}^+$ or $\omega_{re} = \omega_{re}^-$ solves the equilibrium force balance equation (3.5), the final terms in Eq.(3.16) are identically zero. Moreover, for $\omega_{re} = \omega_{re}^{\pm}$, we express

$$\omega_{ce} - 2\omega_{re} = \mp\omega_{ce}(1 - 2\omega_{pe}^2/\omega_{ce}^2)^{1/2} = \mp(\omega_{re}^+ - \omega_{re}^-) , \qquad (3.17)$$

where use has been made of Eq.(3.6). Equation (3.16) then becomes

$$\frac{d^2x_r'(t)}{dt^2} = \pm (\omega_{re}^+ - \omega_{re}^-) \frac{dy_r'(t)}{dt} , \qquad (3.18a)$$

$$\frac{d^2y_r'(t)}{dt^2} = \mp (\omega_{re}^+ - \omega_{re}^-) \frac{dx_r'(t)}{dt} , \qquad (3.18b)$$

where the upper (lower) sign in Eq.(3.18) corresponds to a fast (slow) rotational equilibrium with $\omega_{re} = \omega_{re}^+(\omega_{re} = \omega_{re}^-)$. Therefore, the electron motion as viewed in the *rotating* frame consists of circular gyrations with period

$$T = \frac{2\pi}{(\omega_{re}^+ - \omega_{re}^-)} = \frac{2\pi/\omega_{ce}}{(1 - 2\omega_{pe}^2/\omega_{ce}^2)^{1/2}} . \qquad (3.19)$$

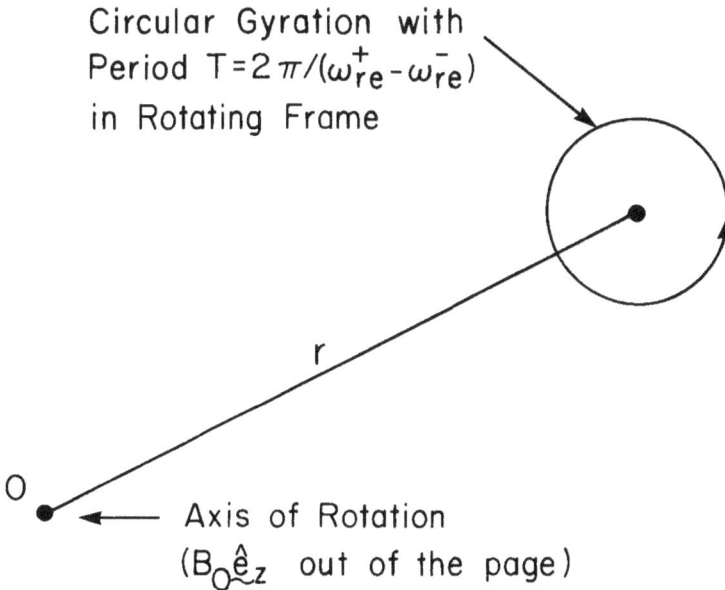

Figure 3.7.   Right-circular electron motion in the rotating frame for a nonneutral plasma column with slow rotational equilibrium ($\omega_{re} = \omega_{re}^-$).

Electron

Trajectory in ⎯⎯⎯⎯⎯⎯⎯⎯⎯⎯

Laboratory Frame

Mean Angular
Velocity $= \omega_{re}^{-}$

r

O ⎯ Axis of Rotation

$(B_0 \hat{e}_z$ out of the page)

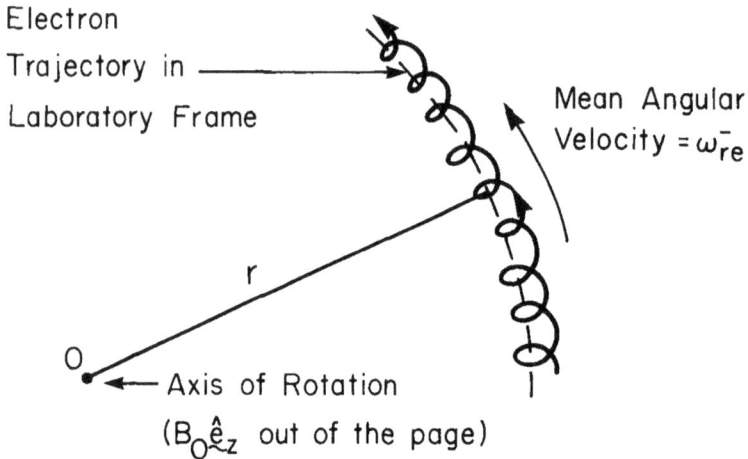

Figure 3.8. Trochoidal electron motion in the laboratory frame for a nonneutral plasma column with slow rotational equilibrium ($\omega_{re} = \omega_{re}^{-}$).

As illustrated in Fig. 3.7, for a slow rotational equilibrium with $\omega_{re} = \omega_{re}^{-}$, the electron motion in the rotating frame is *right-hand* circular. The corresponding motion in the laboratory frame, however, is trochoidal (Fig. 3.8). In contrast, for a fast rotational equilibrium with $\omega_{re} = \omega_{re}^{+}$, the electron motion in the rotating frame is *left-hand* circular.

To summarize, the equilibrium self-electric field in a nonneutral plasma column can have a large influence on the trajectories of individual particles, depending on the value of the self-field parameter $s_e = 2\omega_{pe}^2/\omega_{ce}^2$. Indeed, the gyrating motion as seen in the rotating frame is *slowed* by the self-electric field [Eq.(3.19)], and the perpendicular orbits become free-streaming at the Brillouin density limit, where $s_e = 1$ and $\omega_{re}^{+} = \omega_{re}^{-} = \omega_{ce}/2$ [Eq.(3.18)]. As would be expected, equilibrium self-electric fields have a corresponding large influence on detailed collective properties such as plasma waves and instabilities.[1,2] As a final point, the effect of self-magnetic fields must also be included when the motion is relativistic and the electron current is sufficiently high.

Problem 3.3   Make use of the transformation to rotating coordinates in Eq.(3.15) and the equilibrium force balance equation (3.5) to show that the laboratory-frame orbit equations (3.13) can be transformed to the orbit equations (3.18) in the rotating frame.

Problem 3.4   Consider the nonrelativistic motion of a test electron in the (general) cylindrically symmetric equilibrium configuration with crossed electric and magnetic fields, $E_r(r)\hat{e}_r$ and $B_z(r)\hat{e}_z$. Here, $r = (x^2 + y^2)^{1/2}$ is the radial distance from the axis of symmetry, and

$$E_r(r) = -\frac{\partial}{\partial r}\phi_0(r)\,,$$

$$B_z(r) = \frac{1}{r}\frac{\partial}{\partial r}\left[rA_\theta^0(r)\right]\,, \tag{3.4.1}$$

where $\phi_0(r)$ is the electrostatic potential, and $A_\theta^0(r)$ is the vector potential for the total axial magnetic field $B_z(r)$. In cylindrical coordinates, the total energy (H) and canonical angular momentum (P_θ) of the test electron in the laboratory frame can be expressed as

$$H = \frac{1}{2}m_e\left(v_r^2 + v_\theta^2 + v_z^2\right) - e\phi_0(r),$$

$$P_\theta = m_e r v_\theta - (e/c)rA_\theta^0(r)\,, \tag{3.4.2}$$

where $v_r = dr/dt$ and $v_\theta = rd\theta/dt$ are the perpendicular velocity components, and $v_z = dz/dt = $ const. is the axial velocity. In addition, the perpendicular orbit equations in cylindrical coordinates are given by

$$m_e\left[\frac{d^2r}{dt^2} - r\left(\frac{d\theta}{dt}\right)^2\right] = -e\left[E_r(r) + \frac{r}{c}\frac{d\theta}{dt}B_z(r)\right]\,,$$

$$m_e\left[r\frac{d^2\theta}{dt^2} + 2\frac{dr}{dt}\frac{d\theta}{dt}\right] = \frac{e}{c}\frac{dr}{dt}B_z(r)\,. \tag{3.4.3}$$

a.   Show directly from Eqs.(3.4.1)–(3.4.3) that

$$\frac{dH}{dt} = 0 \text{ and } \frac{dP_\theta}{dt} = 0\,. \tag{3.4.4}$$

That is, the energy H and canonical angular momentum P_θ are exact single-particle constants of the motion in the cylindrically symmetric equilibrium field configuration $E_r(r)\hat{e}_r$ and $B_z(r)\hat{e}_z$.

b.   For uniform magnetic field $B_0\hat{e}_z$ and uniform-density nonneutral plasma column (Fig. 3.1), show that the single-particle constants of the motion defined in Eq.(3.4.2) reduce to

$$H = \frac{1}{2}m_e\left(v_r^2 + v_\theta^2 + v_z^2\right) - \frac{1}{4}m_e\omega_{pe}^2 r^2\,,$$

$$P_\theta = r\left(m_e v_\theta - m_e r\omega_{ce}/2\right)\,, \tag{3.4.5}$$

where $\omega_{ce} = eB_0/m_e c$ and $\omega_{pe}^2 = 4\pi\hat{n}_e e^2/m_e$.

c. Extend the analysis in Part (a) to the case of relativistic electron dynamics. Show that the single-particle constants of the motion are given by

$$H = \left(m_e^2 c^4 + c^2 p_r^2 + c^2 p_\theta^2 + c^2 p_z^2\right)^{1/2} - m_e c^2 - e\phi_0(r) = \text{const.},$$

$$P_\theta = r\left[p_\theta - (e/c)A_\theta^0(r)\right] = \text{const.},$$

$$p_z = \text{const.}, \tag{3.4.6}$$

where $\mathbf{p} = \gamma m_e \mathbf{v}$ is the mechanical momentum, and $\gamma = (1 + \mathbf{p}^2/m_e^2 c^2)^{1/2}$ is the relativistic mass factor.

3.3   Thermal Equilibrium

We now remove the assumption that the plasma is cold, and make use of the steady-state Vlasov-Poisson equations (Sec. 2.2) to examine the equilibrium properties of an axisymmetric nonneutral plasma column confined radially by a uniform magnetic field $B_0\hat{\mathbf{e}}_z$. The electron space charge produces an equilibrium radial self-electric field, $\mathbf{E}^0(\mathbf{x}) = E_r(r)\hat{\mathbf{e}}_r$, where $E_r(r) = -\partial\phi_0(r)/\partial r$. The single-particle constants of the motion in the equilibrium field configuration correspond to the total energy,

$$H = \frac{1}{2m_e}\left(p_r^2 + p_\theta^2 + p_z^2\right) - e\phi_0(r), \tag{3.20}$$

the canonical angular momentum,

$$P_\theta = r(p_\theta - m_e r\omega_{ce}/2), \tag{3.21}$$

and the axial momentum p_z (see Problem 3.4). Here, $r = (x^2 + y^2)^{1/2}$ is the radial distance from the axis of symmetry, $\omega_{ce} = eB_0/m_e c$ is the electron cyclotron frequency, and nonrelativistic electron motion with $\mathbf{p} = m_e\mathbf{v}$ is assumed. It can be verified by direct substitution that the distribution function

$$f_e^0(r, \mathbf{p}) = f_e^0(H, P_\theta, p_z) \tag{3.22}$$

exactly solves the steady-state Vlasov equation (2.9).[1,2] In Eq.(3.20), the electrostatic potential $\phi_0(r)$ is determined self-consistently in terms of

$f_e^0(r, \mathbf{p})$ from the equilibrium Poisson equation (2.12), which can be expressed as

$$\frac{1}{r}\frac{\partial}{\partial r}r\frac{\partial}{\partial r}\phi_0(r) = 4\pi e n_e^0(r) \equiv 4\pi e \int d^3p\, f_e^0(H, P_\theta, p_z) . \tag{3.23}$$

Depending on how the nonneutral plasma is formed, there is considerable latitude in specifying the functional form of the equilibrium distribution function $f_e^0(H, P_\theta, p_z)$. One interesting class of equilibria corresponds to distribution functions that depend on H and P_θ only through the linear combination $H - \omega_{re}P_\theta$, i.e.,

$$f_e^0(r, \mathbf{p}) = f_e^0(H - \omega_{re}P_\theta, p_z) , \tag{3.24}$$

where ω_{re} is a constant (independent of r). It is readily shown from Eqs.(3.20) and (3.21) that the combination $H - \omega_{re}P_\theta$ can be expressed as

$$H - \omega_{re}P_\theta = \frac{1}{2m_e}\left[p_r^2 + (p_\theta - m_e\omega_{re}r)^2 + p_z^2\right] + \psi(r) , \tag{3.25}$$

where $\psi(r)$ is an effective potential defined by

$$\psi(r) = \frac{m_e}{2}r^2\left(\omega_{re}\omega_{ce} - \omega_{re}^2\right) - e\phi_0(r) . \tag{3.26}$$

Making use of Eqs.(3.24)–(3.26), we find that the mean azimuthal velocity of an electron fluid element, $V_{\theta e}^0(r) = (\int d^3p\, v_\theta f_e^0)/(\int d^3p\, f_e^0)$, is given by

$$V_{\theta e}^0(r) = \omega_{re}r . \tag{3.27}$$

That is, the mean azimuthal motion for the class of self-consistent Vlasov equilibria $f_e^0(H - \omega_{re}P_\theta, p_z)$ corresponds to a rigid rotation about the axis of symmetry with angular velocity $\omega_{re} = $ const.

Thermal equilibrium corresponds to the special choice of rigid-rotor distribution function[3,4]

$$f_e^0(r, \mathbf{p}) = \frac{\hat{n}_e}{(2\pi m_e k_B T_e)^{3/2}}\exp\left\{-\frac{(H - \omega_{re}P_\theta)}{k_B T_e}\right\} , \tag{3.28}$$

where $T_e = $ const. is the electron temperature, and k_B is Boltzmann's constant. Equation (3.28) is the one-particle equilibrium distribution function to which an isolated nonneutral plasma column would relax through binary collision processes.[8,9] Integrating over momentum, the equilibrium density profile $n_e^0(r) = \int d^3p \, f_e^0$ associated with Eq.(3.28) is given by

$$n_e^0(r) = \hat{n}_e \exp\left\{ -\frac{m_e}{2k_B T_e} \left[r^2 \left(\omega_{re}\omega_{ce} - \omega_{re}^2 \right) - \frac{2e}{m_e}\phi_0(r) \right] \right\}$$

$$\equiv \hat{n}_e \exp\left\{ -\frac{\psi(r)}{k_B T_e} \right\} . \tag{3.29}$$

We take $\phi_0(r = 0) = 0$ without loss of generality, so that $\hat{n}_e$ can be identified with the on-axis $(r = 0)$ value of electron density. Substituting Eq.(3.29) into Eq.(3.23), Poisson's equation can be expressed as

$$\frac{1}{r}\frac{\partial}{\partial r} r \frac{\partial}{\partial r}\phi_0(r) = 4\pi e \hat{n}_e \exp\left\{ -\frac{m_e}{2k_B T_e} \left[r^2 \left(\omega_{re}\omega_{ce} - \omega_{re}^2 \right) - \frac{2e}{m_e}\phi_0(r) \right] \right\} . \tag{3.30}$$

It is evident from Eq.(3.30) that Poisson's equation for $\phi_0(r)$ is highly nonlinear. However, for the choice of distribution function f_e^0 in Eq.(3.28), and (more generally) for the entire class of rigid-rotor equilibria $f_e^0(H - \omega_{re}P_\theta)$, it can be shown that the plasma column is radially confined $[n_e^0(r \to \infty) = 0]$ provided[3,4]

$$\omega_{ce}\omega_{re} - \omega_{re}^2 - \omega_{pe}^2/2 > 0 , \tag{3.31}$$

where $\omega_{pe}^2 = 4\pi \hat{n}_e e^2/m_e$. Equation (3.31) is the necessary and sufficient condition for the function $\psi(r)$ defined in Eq.(3.26) to be a monotonically increasing function of r, and therefore for the density profile $n_e^0(r) = \hat{n}_e \exp\{-\psi(r)/k_B T_e\}$ to be a decreasing function of r (Problem 3.5).

Equation (3.31) is equivalent to the condition

$$\omega_{re}^- < \omega_{re} < \omega_{re}^+ , \tag{3.32}$$

where ω_{re}^+ and ω_{re}^- are the cold-fluid rotation velocities defined in Eq.(3.6). The range of allowed values of angular rotation velocity ω_{re}, consistent with Eqs.(3.31) and (3.32) and required for radial confinement of the nonneutral plasma column, is illustrated by the shaded region in Fig. 3.9.

If ω_{re} is well removed from the cold-fluid rotation velocities ω_{re}^+ or ω_{re}^-, then the (bell-shaped) density profile in Eq.(3.29) is typically a few times the thermal Debye length $\lambda_{De} = (k_B T_e/4\pi \hat{n}_e e^2)^{1/2}$ in radial dimension. On the other hand, if ω_{re} is closely tuned to either ω_{re}^+ or ω_{re}^-, then the density profile $n_e^0(r)$ is approximately uniform (and equal to $\hat{n}_e$) over

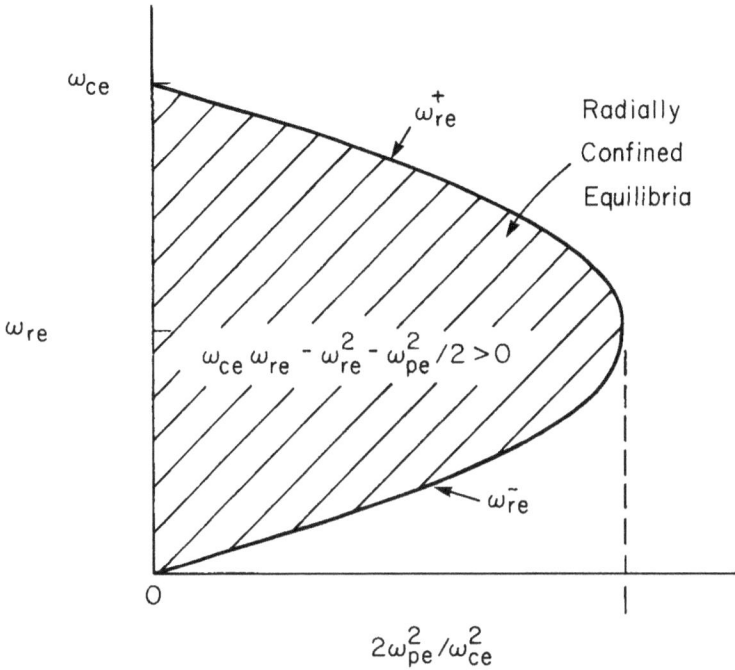

Figure 3.9. The shaded area corresponds to the region of allowed angular rotation velocity ω_{re} consistent with Eq.(3.32) [or Eq.(3.31)]. Here, ω_{re}^{+} and ω_{re}^{-} are the cold-fluid rotation velocities defined in Eq.(3.6).

many Debye lengths out to some radius $r_b(\gg \lambda_{De})$, and then $n_e^0(r)$ falls off abruptly over a distance of a few times λ_{De}. That is, if

$$\omega_{re} = \omega_{re}^{+}(1 - \varepsilon) \text{ or } \omega_{re} = \omega_{re}^{-}(1 + \varepsilon) \text{ for positive } \varepsilon \ll 1 , \qquad (3.33)$$

then the density profile is given (approximately) by

$$n_e^0(r) \simeq \begin{cases} \hat{n}_e , & 0 \leq r \lesssim r_b(\gg \lambda_{De}) , \\ 0 , & r > r_b . \end{cases} \qquad (3.34)$$

Making use of Eqs.(3.30) and (3.34), it readily follows that the equilibrium electrostatic potential is given approximately by $\phi_0(r) \simeq (m_e/4e)\omega_{pe}^2 r^2$ within the plasma column ($0 \leq r \lesssim r_b$). The electron density profile $n_e^0(r)$, calculated numerically from Eqs.(3.29) and (3.30) for $2\omega_{pe}^2/\omega_{ce}^2 = 0.5361$ and $\omega_{re}/\omega_{re}^{-} - 1 = 0.938 \times 10^{-4}$, is illustrated in Fig. 3.10. For the system parameters chosen in Fig. 3.10, the characteristic diameter of the nonneutral plasma column is $2r_b \approx 16\lambda_{De}$.

One-component nonneutral plasmas exhibit the remarkable property that they can relax to global thermal equilibrium (through binary collision processes) and still remain confined.[10] Experimental measurements of the relaxation of a nonneutral electron plasma column to global thermal equilibrium have been carried out by Driscoll, Malmberg and Fine.[11] The plasmas are contained in the cylindrical charged-particle trap illustrated in Fig. 3.11. The experiment is operated in an inject, hold, and dump-and-measure cycle. For injection, cylinder A is briefly grounded, resulting in a column of electrons from the negatively biased filament to cylinder C. When a negative voltage $-V$ is applied to cylinder A, the

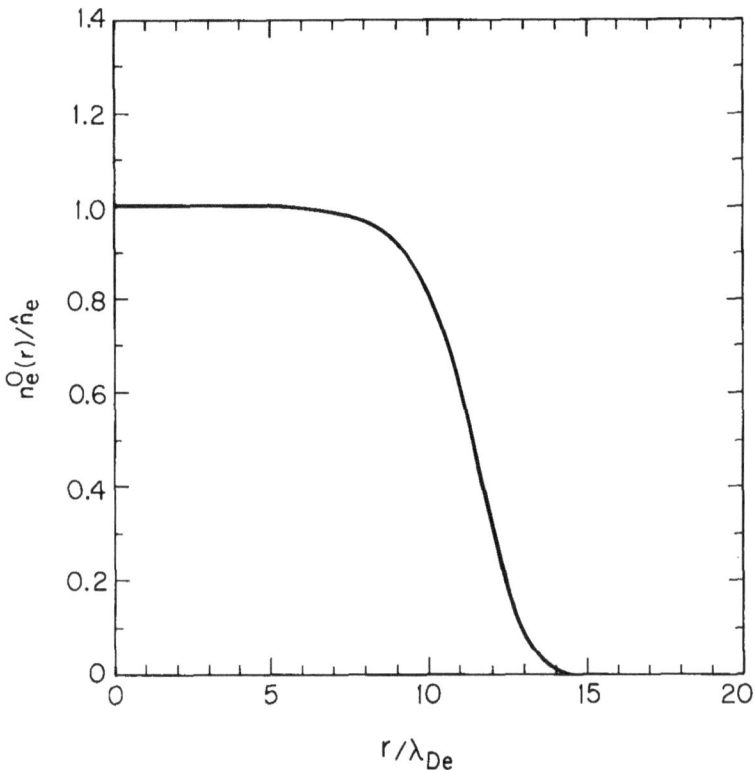

Figure 3.10. Plot of the thermal equilibrium density profile $n_e^0(r)$ versus r/λ_{De} calculated numerically from Eqs.(3.29) and (3.30) for $2\omega_{pe}^2/\omega_{ce}^2 = 0.5361$ and $\omega_{re}/\omega_{re}^- - 1 = 0.938 \times 10^{-4}$.

plasma is trapped axially within cylinder B, which has length and diameter equal to 7.6 cm. The applied axial magnetic field $B_0\hat{e}_z$ provides radial confinement of the electrons, and the overall shape of the plasma is that of an elongated spheroid with relatively weak variation of profiles in the z-direction. Typical plasma parameters in the experiment[11] are: electron density $\hat{n}_e \simeq 10^7$ cm^{-3}; electron temperature $T_e \simeq 0.8$ eV; column diameter $2r_b \simeq 4$ cm; and column length $L_b \simeq 5$ cm. After some time t_0 has elapsed, cylinder C is pulsed to ground potential, and the electrons stream along the magnetic field to the collimator, velocity analyzer, and collector, permitting detailed measurements of the electron density and temperature profiles. Repetition of the cycle with varying containment times t_0 in cylinder B and with varying radius r of the collimator hole, permits a detailed reconstruction of the (r, t) evolution of the plasma. Shot-to-shot variations in the charge collected at any radius are less than one percent, indicating a high degree of azimuthal symmetry $(\partial/\partial\theta \simeq 0)$.

Typical experimental results[11] are illustrated in Fig. 3.12 for bias voltage equal to -80 V, applied magnetic field $B_0 = 188$ G, and self-field parameter $s_e = 2\omega_{pe}^2/\omega_{ce}^2 \simeq 0.0052$. Here, the electron density $n_e(r, t)$ and mean angular rotation velocity $\omega_{re}(r, t)$ are plotted versus r at times $t = 0$ and $t = 10$ s. Initially, both the electron density profile and the angular velocity profile are peaked near the outer edge of the plasma column. As the system evolves, however, some of the electrons are transported inward to fill the central region, whereas other electrons are transported outward in radius. By $t = 10$ s, both the angular velocity ω_{re} and the

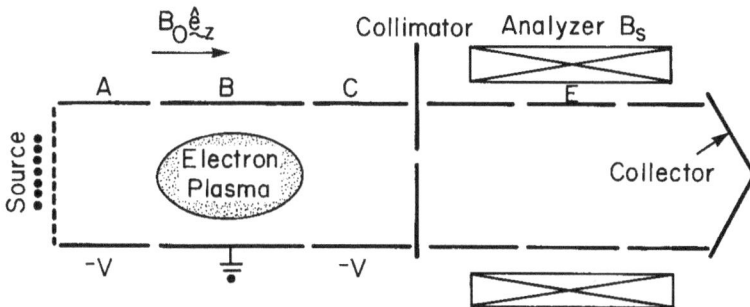

Figure 3.11. Schematic of the experimental set-up used to investigate the approach to global thermal equilibrium of a nonneutral electron plasma shaped like an elongated spheroid [C.F. Driscoll, J.H. Malmberg and K.S. Fine, Phys. Rev. Lett. **60**, 1290 (1988)].

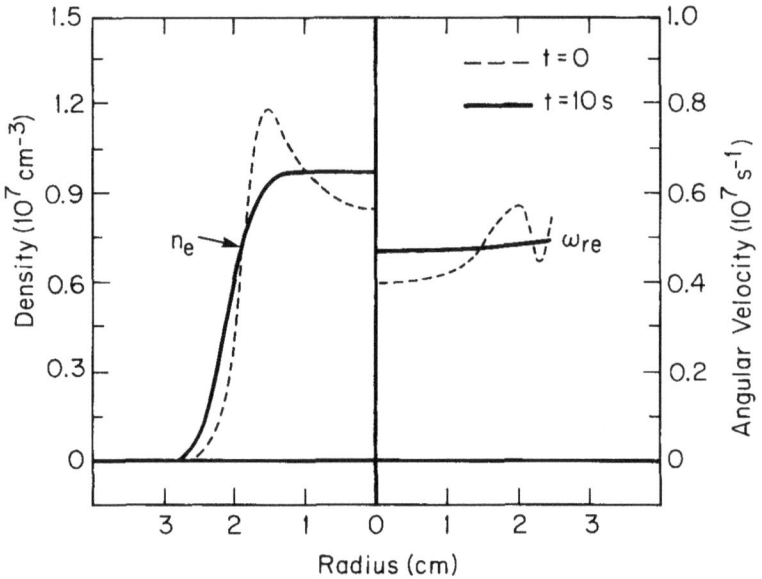

Figure 3.12. Plots versus radius r of the electron density profile $n_e(r,t)$ and angular rotation velocity $\omega_{re}(r,t)$ at times $t = 0$ and $t = 10$ s [C.F. Driscoll, J.H. Malmberg and K.S. Fine, Phys. Rev. Lett. **60**, 1290 (1988)].

electron density n_e are relatively uniform within the plasma column, consistent with the thermal equilibrium distribution function $f_e^0(H - \omega_{re}P_\theta)$ in Eq.(3.28). During the evolution sequence in Fig. 3.12, the electron temperature profile $T_e(r,t)$ is initially uniform at 0.8 eV, and remains uniform throughout the evolution, but increases to 1.1 eV by $t = 10$ s. For the system parameters in Fig. 3.12, note that the thermal Debye length is $\lambda_{De} = (k_B T_e/4\pi\hat{n}_e e^2) \simeq 0.25$ cm and the column diameter is $2r_b \simeq 4$ cm $\gg \lambda_{De}$.

Finally, although these experiments demonstrate relaxation to the rigid-rotor thermal equilibrium distribution function in Eq.(3.28), the measurements[11] also indicate that the density profile relaxes towards thermal equilibrium up to 5000 times faster than predicted by collisional transport theory.[8,9] Moreover, the time required for equilibration of the density profile scales linearly with B_0 over a decade range in magnetic field, rather than as B_0^4.[11] It is possible that the collisional relaxation process is being enhanced by (mild) macroinstabilities or microinstabilities associated with the detailed form of the distribution of particles in phase space, $f_e(\mathbf{x}, \mathbf{p}, t)$, at intermediate times. Most notably, the initial density profile in Fig. 3.12 may drive a weak form of the diocotron instability, which will also reduce the velocity shear through nonlinear collective interactions (Chapter 6).

Problem 3.5   Consider the thermal equilibrium distribution function f_e^0 specified by Eq.(3.28).

a. Make use of Poisson's equation (3.30) to show that the effective potential $\psi(r) = (m_e/2)r^2(\omega_{re}\omega_{ce} - \omega_{re}^2) - e\phi_0(r)$ satisfies

$$\frac{1}{r}\frac{\partial}{\partial r}r\frac{\partial}{\partial r}\psi(r) = 2m_e\left[(\omega_{re}\omega_{ce} - \omega_{re}^2) - \frac{1}{2}\omega_{pe}^2\exp\left\{-\frac{\psi(r)}{k_BT_e}\right\}\right] , \quad (3.5.1)$$

where $\omega_{pe}^2 = 4\pi\hat{n}_e e^2/m_e$, and the boundary conditions on $\psi(r)$ are $\psi(r = 0)$ $= 0$ and $\partial\psi(r)/\partial r|_{r=0} = 0$.

b. Show from Eq.(3.5.1) that $\psi(r)$ is a monotonically increasing function of r for $r > 0$ provided

$$\omega_{re}\omega_{ce} - \omega_{re}^2 - \omega_{pe}^2/2 > 0 . \quad (3.5.2)$$

The inequality in Eq.(3.5.2) ensures radial confinement of the electrons with $n_e^0(r \to \infty) = \hat{n}_e\exp\{-\psi(r \to \infty)/k_BT_e\} = 0$.

c. Make use of Eqs.(2.27), (3.25), and (3.28) to show that the electron pressure tensor is isotropic in thermal equilibrium with

$$\boldsymbol{P}_e^0(\mathbf{x}) = n_e^0(r)k_BT_e\boldsymbol{I} . \quad (3.5.3)$$

Here, $\boldsymbol{I}$ is the unit dyadic, $n_e^0(r)$ is defined in Eq.(3.29), and

$$\boldsymbol{P}_e^0(\mathbf{x}) = m_e\int d^3p\,(\mathbf{v} - \omega_{re}r\hat{\mathbf{e}}_\theta)(\mathbf{v} - \omega_{re}r\hat{\mathbf{e}}_\theta)f_e^0(H - \omega_{re}P_\theta) \quad (3.5.4)$$

is the pressure tensor.

3.4   Debye Shielding

To illustrate the *plasma* nature of a one-component collection of electrons, we describe here the Debye shielding of the electrostatic potential surrounding a test electron in a nonneutral plasma column.[12] In particular, we consider the thermal equilibrium distribution of electrons in Eq.(3.28), and assume that the angular rotation velocity ω_{re} is closely tuned to either ω_{re}^+ or ω_{re}^-, so that $n_e^0(r)$ is approximately constant in the column interior [Eq.(3.34)]. A test electron (charge $= -e$) is introduced at radius r_0' well inside the column ($r_0' \ll r_b$) in the region where $n_e^0(r) \simeq \hat{n}_e =$

const. For simplicity, it is assumed that the test electron is rotating about the axis of symmetry with the same angular velocity ω_{re} as the average rotation of the column, i.e., the test electron is *at rest* relative to the plasma column. (In the general case, there will be a dynamic screening of the potential surrounding the test electron by plasma dielectric effects.) Denoting the location of the test electron by $\mathbf{x}_0'$, Poisson's equation for the total electrostatic potential $\phi = \phi_0(r) + \delta\phi(\mathbf{x}, t)$ can be expressed as

$$\frac{1}{r}\frac{\partial}{\partial r}r\frac{\partial}{\partial r}\phi_0 + \nabla^2\delta\phi = 4\pi e\hat{n}_e \exp\left(\frac{e\delta\phi}{k_BT_e}\right) + 4\pi e\delta(\mathbf{x} - \mathbf{x}_0') , \qquad (3.35)$$

where $\delta\phi$ is the perturbed potential associated with introducing the test electron. In obtaining Eq.(3.35), it has been assumed that the thermal equilibrium distribution of plasma electrons adjusts adiabatically to the presence of the test electron and has the form

$$f_e^0(r, \mathbf{p}) = \frac{\hat{n}_e}{(2\pi m_e k_B T_e)^{3/2}} \exp\left\{-\frac{(H_0 - e\delta\phi - \omega_{re}P_\theta)}{k_BT_e}\right\} , \qquad (3.36)$$

where $H_0 - e\delta\phi = (2m_e)^{-1}(p_r^2 + p_\theta^2 + p_z^2) - e\phi_0 - e\delta\phi$. It is also assumed that $\omega_{re} \simeq \omega_{re}^+$ or $\omega_{re} \simeq \omega_{re}^-$, so that the electron density profile in the absence of the test electron is approximately uniform with

$$n_e^0(r) = \hat{n}_e \exp\left\{-\frac{m_e}{2k_BT_e}\left[r^2(\omega_{re}\omega_{ce} - \omega_{re}^2) - \frac{2e}{m_e}\phi_0(r)\right]\right\} \simeq \hat{n}_e \quad (3.37)$$

in the column interior ($0 \le r < r_b$).

For $|e\delta\phi/k_BT_e| \ll 1$, we expand the exponential factor in Eq.(3.35), and make use of the fact that $\phi_0(r) = (m_e/4e)\omega_{pe}^2 r^2$ solves the equilibrium Poisson equation $r^{-1}(\partial/\partial r)\,[r\partial\phi_0/\partial r] = 4\pi e\hat{n}_e$ in the column interior. Retaining terms linear in $\delta\phi$, we obtain[12]

$$\nabla^2\delta\phi = \frac{1}{\lambda_{De}^2}\delta\phi + 4\pi e\delta(\mathbf{x} - \mathbf{x}_0') , \qquad (3.38)$$

where $\lambda_{De} = (k_B T_e/4\pi\hat{n}_e e^2)^{1/2}$ is the thermal Debye length. The solution to Eq.(3.38) gives the familiar shielded potential

$$\delta\phi = -\frac{e}{|\mathbf{x} - \mathbf{x}_0'|} \exp\left\{-\frac{|\mathbf{x} - \mathbf{x}_0'|}{\lambda_{De}}\right\} \qquad (3.39)$$

surrounding the test electron at $\mathbf{x} = \mathbf{x}_0'$.

As in a neutral plasma, the electrons surrounding the test electron redistribute so as to screen the bare Coulomb potential for distances $|\mathbf{x} - \mathbf{x}_0'| > \lambda_{De}$. The main difference in the nonneutral case is that $\delta\phi$ is superimposed on the dc potential $\phi_0(r) = (m_e/4e)\omega_{pe}^2 r^2$ produced by equilibrium space-charge effects.

3.5 Spontaneous Emission from a Test Electron

It is well known that an isolated test electron gyrating in a uniform magnetic field $B_0\hat{\mathbf{e}}_z$ emits electromagnetic radiation. For example, in the forward direction (along $\hat{\mathbf{e}}_z$), radiation is emitted at the electron cyclotron frequency ω_{ce}.[13] How is this spontaneous emission affected when the test electron is located inside a nonneutral plasma column? To address this question, we again consider a nonneutral plasma column with uniform electron density (Fig. 3.1) and equilibrium radial electric field $E_r(r) = -(m_e/2e)\omega_{pe}^2 r$. The spontaneous emission $\eta(\omega)$ in the forward direction from an individual test electron is given in the far-field region by[13]

$$\eta(\omega) = \frac{1}{T}\frac{d^2 I}{d\omega d\Omega}$$

$$= \frac{e^2\omega^2}{4\pi^2 c^3 T}\left| \int_0^T d\tau\, \hat{\mathbf{e}}_z \times [\hat{\mathbf{e}}_z \times \mathbf{v}'(\tau)] \exp\left[ik_z z'(\tau) - i\omega\tau\right]\right|^2 , \quad (3.40)$$

where ω is the frequency and k_z is the wavenumber. Here, $d^2 I/d\omega d\Omega$ is the energy radiated per unit frequency interval per unit solid angle; $z'(\tau) = z + v_z\tau$ is the free-streaming orbit along $B_0\hat{\mathbf{e}}_z$; $T = L/v_z$ is the length of time that the electron is in the interaction region (of length L); and $\mathbf{v}'(\tau)$ is the electron velocity in the laboratory frame.

A detailed analysis[14] of the electron motion in the crossed electric and magnetic fields, $E_r(r)\hat{\mathbf{e}}_r$ and $B_0\hat{\mathbf{e}}_z$, shows that the perpendicular orbits $v_x'(\tau)$ and $v_y'(\tau)$ in the laboratory frame are *biharmonic*, with oscillatory components at the two frequencies ω_{re}^+ and ω_{re}^- defined in Eq.(3.6). In particular, $v_x'(\tau) + iv_y'(\tau)$ can be expressed as (Problem 3.6)

$$v'_x(\tau) + iv'_y(\tau) \tag{3.41}$$

$$= \frac{1}{(\omega_{re}^+ - \omega_{re}^-)} \left\{ -\omega_{re}^+ \left[i\,\omega_{re}^- r \exp(i\theta) - v_\perp \exp(i\phi) \right] \exp(i\omega_{re}^+\tau) \right.$$

$$\left. + \omega_{re}^- \left[i\,\omega_{re}^+ r \exp(i\theta) - v_\perp \exp(i\phi) \right] \exp(i\omega_{re}^-\tau) \right\} \; .$$

Here, $\tau = t' - t$, and "initial" conditions are chosen as in the method of characteristics (Sec. 2.2). That is,

$$x'(\tau = 0) + iy'(\tau = 0) = x + iy \equiv r \exp(i\theta) \; ,$$

$$v'_x(\tau = 0) + iv'_y(\tau = 0) = v_x + iv_y \equiv v_\perp \exp(i\phi) \; , \tag{3.42}$$

where $v'_x = dx'/d\tau$ and $v'_y = dy'/d\tau$, and $(r, \theta, v_\perp, \phi)$ denote perpendicular phase-space variables. We introduce the shifted frequencies $\Omega^\pm$ defined by

$$\Omega^\pm \equiv \omega - k_z v_z - \omega_{re}^\pm \; . \tag{3.43}$$

Making use of Eq.(3.41) to carry out the τ-integration in Eq.(3.40), and integrating over the perpendicular velocity phase with $(2\pi)^{-1} \int_0^{2\pi} d\phi \dots$ gives[14]

$$\eta(\omega) = \frac{e^2\omega^2 T}{8\pi^2 c^3} \left\{ \frac{\sin^2(\Omega^+ T/2)}{(\Omega^+ T/2)^2} \frac{\omega_{re}^{+2}}{(\omega_{re}^+ - \omega_{re}^-)^2} \left(v_\perp^2 + \omega_{re}^{-2} r^2 \right) \right.$$

$$+ \frac{\sin^2(\Omega^- T/2)}{(\Omega^- T/2)^2} \frac{\omega_{re}^{-2}}{(\omega_{re}^+ - \omega_{re}^-)^2} \left(v_\perp^2 + \omega_{re}^{+2} r^2 \right)$$

$$- \frac{2\omega_{re}^+ \omega_{re}^-}{(\omega_{re}^+ - \omega_{re}^-)^2} \frac{\sin(\Omega^+ T/2)\sin(\Omega^- T/2)}{(\Omega^+ T/2)(\Omega^- T/2)}$$

$$\left. \times \left[(v_\perp^2 + \omega_{re}^+ \omega_{re}^- r^2) \cos \left(\frac{(\Omega^+ - \Omega^-)T}{2} \right) \right] \right\} \; , \tag{3.44}$$

where $\omega_{re}^\pm$ and $\Omega^\pm$ are defined in Eqs.(3.6) and (3.43), respectively.

Not surprisingly, since the perpendicular orbits are biharmonic with oscillatory components at the two frequencies ω_{re}^+ and ω_{re}^-, Eq.(3.44) pre-

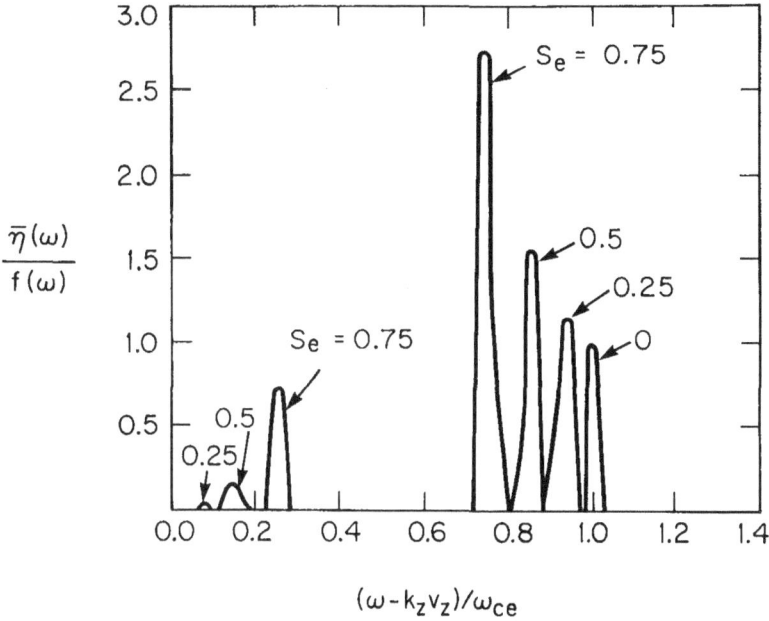

Figure 3.13. Plot of the normalized spontaneous emission from a test electron versus $(\omega - k_z v_z)/\omega_{ce}$ for several values of the self-field parameter $s_e = 2\omega_{pe}^2/\omega_{ce}^2$. Here, $\omega_{ce}T = 200$ and $r_L^2/r_b^2 = v_\perp^2/\omega_{ce}^2 r_b^2 = 1/10$ are assumed [R.C. Davidson and W.A. McMullin, Phys. Fluids **27**, 1268 (1984)].

dicts that the spontaneous emission spectrum $\eta(\omega)$ has two maxima located at $\Omega^+ = 0$ and $\Omega^- = 0$, or equivalently,

$$\omega - k_z v_z \simeq \omega_{re}^\pm = \frac{1}{2}\omega_{ce}\left\{1 \pm \left(1 - \frac{2\omega_{pe}^2}{\omega_{ce}^2}\right)^{1/2}\right\}. \qquad (3.45)$$

Moreover, the relative strengths of the two emission peaks in Eq.(3.44) are proportional to ω_{re}^{-2} (for $\omega - k_z v_z \simeq \omega_{re}^-$) and ω_{re}^{+2} (for $\omega - k_z v_z \simeq \omega_{re}^+$). For $s_e = 2\omega_{pe}^2/\omega_{ce}^2 \to 0$, note from Eq.(3.45) that $\omega_{re}^- \to 0$ and $\omega_{re}^+ \to \omega_{ce}$. Therefore, as expected, for $s_e = 0$ only the high-frequency emission peak is present with $\omega - k_z v_z \simeq \omega_{ce}$, and Eq.(3.44) reduces to the familiar single-particle result

$$[\eta(\omega)]_{s_e=0} = \frac{e^2 \omega^2 T v_\perp^2}{8\pi^2 c^3} \frac{\sin^2[(\omega - k_z v_z - \omega_{ce})T/2]}{[(\omega - k_z v_z - \omega_{ce})T/2]^2} . \tag{3.46}$$

However, as the self-field parameter s_e is increased, the high-frequency emission peak at $\omega - k_z v_z \simeq \omega_{re}^+$ shifts downward (from ω_{ce}), and the low-frequency emission peak emerges at $\omega - k_z v_z \simeq \omega_{re}^-$ and shifts upward (from zero frequency). This behavior is illustrated quantitatively[14] in Fig. 3.13, where $\bar\eta(\omega)/f(\omega)$ is plotted versus $(\omega - k_z v_z)/\omega_{ce}$ for several values of the self-field parameter $s_e = 2\omega_{pe}^2/\omega_{ce}^2$. Here, $f(\omega)$ is the form factor defined by $f(\omega) = e^2 \omega^2 T v_\perp^2/8\pi^2 c^3$, and $\bar\eta(\omega)$ denotes the spatial average, $\bar\eta(\omega) = r_b^{-1} \int_0^{r_b} dr\eta(\omega)$, over the radial extent of the plasma column. It is clear from Fig. 3.13 that the strength of the self-electric field (as measured by $s_e = 2\omega_{pe}^2/\omega_{ce}^2$) can have a large influence on the spontaneous emission from an individual test electron in a nonneutral plasma column. Of course a more precise calculation of radiation emission would include plasma dielectric effects, and allow for stimulated emission.

Problem 3.6   The method of characteristics was introduced in Sec. 2.2 to solve the linearized Vlasov equation. Consider the nonrelativistic motion of an electron in a uniform-density nonneutral plasma column (Fig. 3.1) with crossed magnetic and electric fields, $B_0 \hat e_z$ and $\mathbf{E}^0(\mathbf{x}) = -(m_e/2e)\omega_{pe}^2(x\hat e_x + y\hat e_y)$.

a. In terms of the "primed" variables introduced in Eq.(2.20), show that $x'(t') + iy'(t')$ satisfies

$$\left(\frac{d^2}{dt'^2} - i\omega_{ce}\frac{d}{dt'} - \frac{1}{2}\omega_{pe}^2\right)\left[x'(t') + iy'(t')\right] = 0 . \tag{3.6.1}$$

Here, $x'(t') + iy'(t')$ is a convenient complex representation of the particle orbit perpendicular to $B_0\hat e_z$.

b. Solve Eq.(3.6.1) subject to the "initial" conditions

$$x'(t' = t) + iy'(t' = t) = x + iy \equiv r\exp(i\theta) ,$$
$$v_x'(t' = t) + iv_y'(t' = t) = v_x + iv_y \equiv v_\perp \exp(i\phi) , \tag{3.6.2}$$

where $v_x' = dx'/dt'$ and $v_y' = dy'/dt'$. From Eq.(3.6.2), the orbit (x', y', v_x', v_y') passes through the phase-space point (x, y, v_x, v_y) at time $t' = t$.

Show that the solutions for $x' + iy'$ and $v'_x + iv'_y$ can be expressed as

$$x'(t') + iy'(t')$$

$$= \left(\omega^+_{re} - \omega^-_{re}\right)^{-1} \left\{ - \left[\omega^-_{re} r \exp(i\theta) + iv_\perp \exp(i\phi)\right] \exp(i\omega^+_{re} \tau) \right.$$

$$\left. + \left[\omega^+_{re} r \exp(i\theta) + iv_\perp \exp(i\phi)\right] \exp(i\omega^-_{re} \tau) \right\} , \qquad (3.6.3)$$

and

$$v'_x(t') + iv'_y(t')$$

$$= \left(\omega^+_{re} - \omega^-_{re}\right)^{-1} \left\{ - \left[i\omega^+_{re}\omega^-_{re} r \exp(i\theta) - \omega^+_{re} v_\perp \exp(i\phi)\right] \exp(i\omega^+_{re} \tau) \right.$$

$$\left. + \left[i\omega^-_{re}\omega^+_{re} r \exp(i\theta) - \omega^-_{re} v_\perp \exp(i\phi)\right] \exp(i\omega^-_{re} \tau) \right\} , \qquad (3.6.4)$$

where $\tau = t' - t$, and $\omega^\pm_{re}$ are the rotation frequencies solving Eq.(3.5) and defined in Eq.(3.6). Evidently, the particle motion perpendicular to $B_0 \hat{e}_z$ in the laboratory frame is biharmonic at the frequencies ω^+_{re} and ω^-_{re}.

Problem 3.7   Substitute the perpendicular velocity orbit in Eq.(3.41) into the expression for $\eta(\omega)$ in Eq.(3.40) and average over the perpendicular velocity phase with $(2\pi)^{-1} \int_0^{2\pi} d\phi$ .... Show that the resulting expression for the spontaneous emission spectrum can be expressed in the form given in Eq.(3.44).

3.6   Strongly Coupled Nonneutral Plasma

The emphasis in this treatise is on nonneutral plasmas in which correlation effects are weak. For example, in Sec. 3.3, thermal equilibrium properties of a nonneutral electron plasma were summarized for the case in which the plasma is treated as a classical electron gas in which the distance of closest Coulomb approach $(e^2/k_B T_e)$ is much smaller than the characteristic interparticle spacing $(\hat{n}_e^{-1/3})$, i.e.,

$$\frac{e^2}{k_B T_e} \ll \hat{n}_e^{-1/3}. \qquad (3.47)$$

Equation. (3.47) also implies that the plasma expansion parameter satisfies $g = 1/\hat{n}_e \lambda^3_{De} \ll 1$, where $\lambda_{De} = (k_B T_e/4\pi \hat{n}_e e^2)^{1/2}$ is the thermal Debye length. For sufficiently cold electrons (small λ_{De}) it was shown [see Fig. 3.10 and Eq.(3.34)] that the electron density is approximately uniform in the column interior, and that $\hat{n}_e$ is related to the rotation frequency ω_{re} and the confining field B_0 by the force-balance condition $\omega_{pe}^2/2 \simeq \omega_{re}(\omega_{ce} - \omega_{re})$, where $\omega_{pe}^2 = 4\pi\hat{n}_e e^2/m_e$ [see also Eq.(3.5)]. Indeed, for sufficiently small λ_{De}, it follows from Eq.(3.26) and Problem 3.5 that the effective electrostatic potential defined by $-\psi(r)/e \equiv \phi_0(r) - (m_e/2e)r^2(\omega_{re}\omega_{ce} - \omega_{re}^2)$ is approximately equal to zero in the column interior, where $\phi_0(r) \simeq (m_e/4e)\omega_{pe}^2 r^2$. Therefore, the effect of the column rotation on properties of the electron equilibrium is similar[15] to that produced by a hypothetical cylinder of singly-charged positive ions with uniform density equal to $(m_e/2\pi e^2)\omega_{re}(\omega_{ce} - \omega_{re})$.

A particularly fascinating area of research on nonneutral plasmas relates to phase transitions[15] to the liquid and crystal states when the coupling parameter

$$\Gamma_e = \frac{e^2}{k_B T_e a_e} = \frac{e^2}{k_B T_e (3/4\pi\hat{n}_e)^{1/3}} \tag{3.48}$$

is sufficiently large, which requires very-low-temperature conditions. Here, Γ_e is the ratio of nearest-neighbor Coulomb energy (e^2/a_e) to the characteristic thermal energy of a particle $(k_B T_e)$, and a_e is the Wigner-Seitz radius defined by $4\pi\hat{n}_e a_e^3/3 = 1$. [For a pure ion plasma, the definition of Γ_i is similar to Eq.(3.48) with the replacements $T_e \rightarrow T_i, \hat{n}_e \rightarrow \hat{n}_i$, and $e^2 \rightarrow Z_i^2 e^2$, where Z_i is the degree of ionization.] Note that the inequality in Eq.(3.47) corresponds to a weakly coupled nonneutral plasma with $\Gamma_e \ll 1$. On the other hand, as Γ_e is increased to the regime where $\Gamma_e > 1$ (by cooling the electrons,[15] say), then correlation effects become increasingly important, and the plasma is said to be strongly coupled.[16-18] As Γ_e is increased to large enough values, it would be expected that the long-range order increases to the point where a nonneutral plasma exhibits liquid-like behavior, and at still higher values of the coupling parameter the system undergoes a phase transition to a crystal-like structure.[16-18] Research on strongly coupled nonneutral plasma includes experimental studies of the liquid-to-crystal phase transition in a two-dimensional nonneutral electron layer on a liquid-helium surface,[19,20] the production and laser cooling of a strongly-coupled nonneutral ion plasma confined in a Penning trap with coupling parameter $\Gamma_i \sim 20 - 200$,[21-23] and computer simulations of the phase transition of bounded nonneutral ion plasmas to the liquid and crystal states both in Penning-trap confinement geometries,[24,25] and in heavy-ion storage rings.[26] Research on phase transitions in nonneutral plasmas has been further stimulated by continued advances in the theoretical understanding of strongly coupled one-component neutral plasmas.[16-18] Here, a "one-component" neutral plasma refers to a neutral plasma where the positive ions (say) form a fixed, charge-neutralizing background, and only the electrons are treated as an active dynamic component that interacts electrostatically with the electrons and fixed ion background. This model has been studied extensively because of its application to correlation effects in such diverse systems as neutron stars, metals, dielectric solutions, and colloidal suspensions.

As further background, we consider a nonneutral plasma column consisting of $\bar{N}_e$ electrons treated as classical particles. The plasma column is confined radially by a uniform axial magnetic field $B_0\hat{e}_z$, and a perfectly conducting wall is located at radius $r = b$, which is large in comparison with the radius r_b of the plasma column. As discussed in Secs. 2.5 and 4.6, neglecting effects associated with electron-neutral collisions and finite wall resistivity, the total energy (particle kinetic energy plus field energy) and the total canonical angular momentum, summed over all $\bar{N}_e$ particles in the system, are conserved quantities. Therefore, for a canonical ensemble in thermal equilibrium at temperature T_e, the $\bar{N}_e$-particle phase-space density $\rho(\mathbf{x}_1, \mathbf{p}_1, \ldots, \mathbf{x}_{\bar{N}_e}, \mathbf{p}_{\bar{N}_e})$ is given nonrelativistically by[27]

$$\rho = \frac{1}{Z} \exp\left\{ -\frac{1}{k_B T_e} \left[\sum_{k=1}^{\bar{N}_e} \frac{\mathbf{p}_k^2}{2m_e} + U\left(\mathbf{x}_1, \mathbf{x}_2, \ldots, \mathbf{x}_{\bar{N}_e}\right) \right.\right.$$

$$\left.\left. - \omega_{re} \sum_{k=1}^{\bar{N}_e} r_k \left(p_{\theta k} - eB_0 r_k/c\right) \right] \right\}. \tag{3.49}$$

Here, $U(\mathbf{x}_1, \ldots, \mathbf{x}_{\bar{N}_e})$ is the electrostatic potential energy required to assemble the electrons (transverse electromagnetic effects are neglected), and $Z(\bar{N}_e, \omega_{re}, T_e) = \int \cdots \int d^3 x_1 d^3 p_1 \cdots d^3 x_{\bar{N}_e} d^3 p_{\bar{N}_e} \rho$ is the partition function. Moreover, the constants T_e and ω_{re} are related to the total kinetic energy and the total canonical angular momentum of the particles. As was done in Eq.(3.25) for the one-particle distribution function, the terms in Eq.(3.49) can be combined to give the equivalent expression[15]

$$\rho = \frac{1}{Z} \exp\left\{ -\frac{1}{k_B T_e} \left[\sum_{k=1}^{\bar{N}_e} \frac{1}{2m_e} \left[p_{rk}^2 + (p_{\theta k} - m_e \omega_{re} r_k)^2 + p_{zk}^2 \right] \right.\right.$$

$$\left.\left. + U\left(\mathbf{x}_1, \ldots, \mathbf{x}_{\bar{N}_e}\right) + \frac{1}{2} m_e \omega_{re} (\omega_{ce} - \omega_{re}) \sum_{k=1}^{\bar{N}_e} r_k^2 \right] \right\}, \tag{3.50}$$

where $\omega_{ce} = eB_0/m_e c$. The momentum dependence for each electron in Eq.(3.50) is simply a shifted Maxwellian rotating with angular velocity $\omega_{re} =$ const. about the axis of symmetry. As expected, for sufficiently large B_0 the final term in the exponent in Eq.(3.50) ensures that the average electron density is exponentially small at large r.

In models where the system is treated as infinite in spatial extent, the potential energy in Eq.(3.50) takes the simple Coulomb interaction form, $U = \sum_{k<\ell}^{\bar{N}_e} e^2/|\mathbf{x}_k - \mathbf{x}_\ell|$. In practical circumstances, however, the form of $U(\mathbf{x}_1, \ldots, \mathbf{x}_{\bar{N}_e})$ will generally be more complicated. For example, in the segmented cylinder and Penning trap confinement geometries widely used in experimental studies of one-component nonneutral plasmas (see Figs. 3.11 and 4.3, and Sec. 4.3.1), axial confinement of the electrons along $B_0 \hat{\mathbf{e}}_z$ is provided by external potentials applied to neighboring conductors, which produce an axial focusing of the electron orbits. As discussed by Malmberg and O'Neil,[15] whatever the form of $U(\mathbf{x}_1, \ldots, \mathbf{x}_{\bar{N}_e})$, the (factorable) contribution proportional to $(m_e \omega_{re}/2)(\omega_{ce} - \omega_{re}) \sum_{k=1}^{\bar{N}_e} r_k^2$ in Eq.(3.50) plays the role of a (hypothetical) cylinder of singly-charged positive ions with uniform density equal to $(m_e/2\pi e^2)\omega_{re}(\omega_{ce} - \omega_{re})$. Indeed, replacement of the uniform magnetic field $B_0 \hat{\mathbf{e}}_z$ by a cylinder of uniform positive charge would leave the expression for $\rho(\mathbf{x}_1, \mathbf{p}_1, \ldots, \mathbf{x}_{\bar{N}_e}, \mathbf{p}_{\bar{N}_e})$ in Eq.(3.50) unchanged, except for the rigid-body rotation manifest in the momentum dependence of the phase-space density ρ. Because this rotation does not enter into the partition function Z, the thermodynamic properties of a magnetically confined nonneutral electron plasma are the same as the thermodynamic properties of electrons immersed in a cylinder of uniform positive charge with density $(m_e/2\pi e^2)\omega_{re}(\omega_{ce} - \omega_{re})$. The spatial correlation functions for the electrons are also the same in the two cases.

To summarize, for small values of the coupling parameter ($\Gamma_e \ll 1$), the one-particle distribution function analyzed in Sec. 3.3 (infinitely long plasma cylinder) and Sec. 4.3.1 (axially-confined plasmoid) provides an adequate theoretical model for determining thermal equilibrium properties, such as the density profile and the particle pressure tensor. On the other hand, as Γ_e is increased, correlation effects entering through the interaction potential $U(\mathbf{x}_1, \ldots, \mathbf{x}_{\bar{N}_e})$ in Eq.(3.50) become increasingly strong. If boundary effects are ignored, then Monte Carlo calculations show that the pair correlation function begins to exhibit the oscillations characteristic of a liquid[18] for $\Gamma_e \gtrsim 2$, and a phase transition to a crystal-like structure[18] occurs for Γ_e in the range of 150–180.

The analysis leading to Eq.(3.50) assumes classical electron dynamics. However, as the electron temperature T_e is reduced (Γ_e is increased), the system can enter a regime where $k_B T_e < \hbar\omega_{ce}$, where $h = 2\pi\hbar$ is Planck's constant. In this regime, the (quantized) Landau energy levels become important, and the electron spins become aligned with the magnetic field $B_0 \hat{e}_z$. At somewhat lower temperatures, where $k_B T_e < \hbar\omega_{pe}$, quantum effects become important in describing the collective interaction of the electrons. When both $k_B T_e < \hbar\omega_{pe}$ and $\hbar\omega_{pe} \ll e^2/a_e$ are satisfied (which requires $\Gamma_e \gg 1$), Malmberg and O'Neil[15] were the first to propose that the electrons in a nonneutral plasma may arrange themselves into an ordered lattice, using arguments similar to those presented by Wigner[28] for an unmagnetized system. The inequality $e^2/a_e \gg \hbar\omega_{pe}$ is more familiar in the form

$$\frac{a_e}{a_B} \gg 1 , \qquad (3.51)$$

where $a_B = \hbar^2/m_e e^2$ is the Bohr radius.

The criteria noted in the previous paragraph are collected together in Fig. 3.14.[15] Here, the vertical axis is $\log_{10}(T_e)$, where T_e is in degrees kelvin; the lower abscissa is $\log_{10}(\hat{n}_e)$, where $\hat{n}_e$ is in cm^{-3}; and the upper abscissa is $\log_{10}(B_0)$, where B_0 is in gauss. The two horizontal scales are related by the Brillouin density limit, $2\omega_{pe}^2/\omega_{ce}^2 = 1$. That is, the value of $\hat{n}_e$ directly below a particular value of B_0 is the highest density that can be confined by that value of magnetic field. Various degrees of increasing correlation are illustrated in Fig. 3.14 by the straight lines corresponding to $\hat{n}_e \lambda_{De}^3 = 1, \Gamma_e = 2$ and $\Gamma_e = 155$. Moreover, quantum behavior occurs for values of T_e below the curves $k_B T_e = \hbar\omega_{ce}$ and $k_B T_e = \hbar\omega_{pe}$. The largest value of $a_e/a_B = (m_e e^2/\hbar^2)(3/4\pi\hat{n}_e)^{1/3}$ in Fig. 3.14 occurs for $\hat{n}_e = 10^{16}$ cm^{-3}, where $a_e/a_B \simeq 10^3$. The dashed lines in the figure correspond to $B_0 = 100$ kG and $T_e = 10^{-2}$K, which are rough measures of technological limits.

The preceding discussion in Sec. 3.6 is readily extended to the case of a one-component nonneutral ion plasma by making the replacements $m_e \rightarrow m_i, -e \rightarrow +Z_i e, T_e \rightarrow T_i, \Gamma_e \rightarrow \Gamma_i = (Z_i^2 e^2/k_B T_i)(4\pi\hat{n}_i/3)^{1/3}, \omega_{re} \rightarrow \omega_{ri}$, and $-\omega_{ce} \rightarrow \omega_{ci} = Z_i e B_0/m_i c$. Here, we have chosen the sign convention such that the force-balance condition for $\lambda_{Di} \rightarrow 0$ is given by $\omega_{pi}^2/2 = -\omega_{ri}(\omega_{ci} + \omega_{ri})$, which has the solutions $\omega_{ri} = \omega_{ri}^{\pm} = -(\omega_{ci}/2) \times \{1 \pm (1 - 2\omega_{pi}^2/\omega_{ci}^2)^{1/2}\}$, corresponding to $\omega_{ri}^{\pm} < 0$. Of course the analogue of Fig. 3.14 for a pure ion plasma has a shifted lower abscissa scale (see Fig. 3.15) because the Brillouin density limit, $2\omega_{pi}^2/\omega_{ci}^2 = 1$, gives a lower value of $\hat{n}_i$ by a factor of m_e/m_i. Moreover, for protons

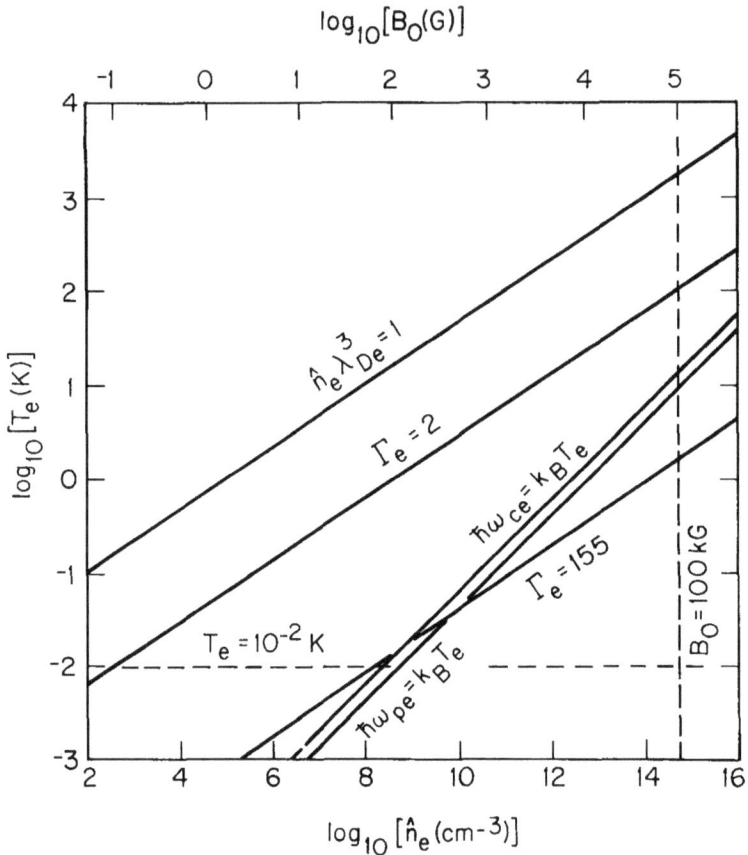

Figure 3.14. Plots in $(\hat{n}_e, T_e, B_0)$ parameter space showing the various degrees of increasing correlation in a nonneutral electron plasma in thermal equilibrium [J.H. Malmberg and T.M. O'Neil, Phys. Rev. Lett. **39**, 1333 (1977)].

with $Z_i = 1$ and $m_i/m_e = 1836$, the largest value of $a_i/a_{Bi} = (m_i e^2/\hbar^2)(3/4\pi\hat{n}_i)^{1/3}$ in Fig. 3.15 occurs for $\hat{n}_i = 5.4 \times 10^{12}$ cm^{-3}, where $a_i/a_{Bi} \simeq 2.2 \times 10^7$.

The ability to cool collections of electrons or ions to temperatures where $\Gamma_e \gg 1$ or $\Gamma_i \gg 1$ and strong correlations may produce a spatially ordered state has stimulated interest in several fields of physics. Detailed experiments have been carried out in two-dimensional electron and ion disks,[20] and in systems where the charged particles interact through a shielded Coulomb potential.[29-31] Crystal-like structures of small numbers

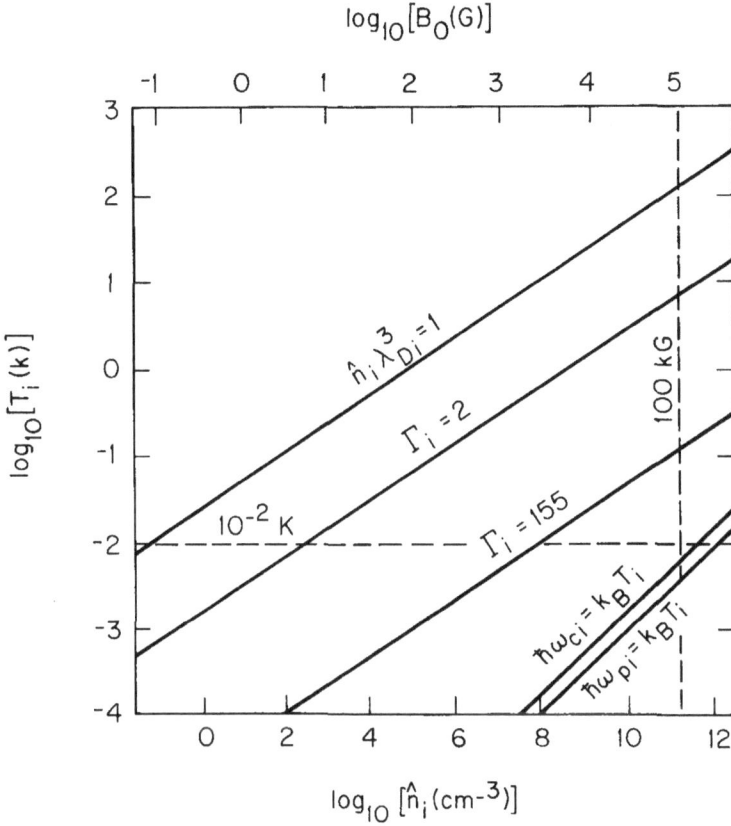

Figure 3.15. Plots in $(\hat{n}_i, T_i, B_0)$ parameter space showing the various degrees of increasing correlation in a nonneutral ion plasma in thermal equilibrium. Here, $Z_i = 1$ and $m_i/m_e = 1836$ are assumed, corresponding to a nonneutral proton plasma.

$(\bar{N}_i < 100)$ of atomic ions in RF traps have been observed.[32] The possibility also exists for the observation of ordered ion structures in storage rings.[26] Experimentally, strong coupling has been reported in nonneutral ion[21-23] and electron[33] plasmas confined in the types of Penning trap geometries illustrated in Fig. 4.3.

For present purposes, we conclude this section on strongly coupled nonneutral plasma by a brief description of computer simulation studies[25] and experimental results[23] on the observation of shell structure in bounded nonneutral ion plasmas cooled to sufficiently low temperatures that $\Gamma_i \gg 1$.

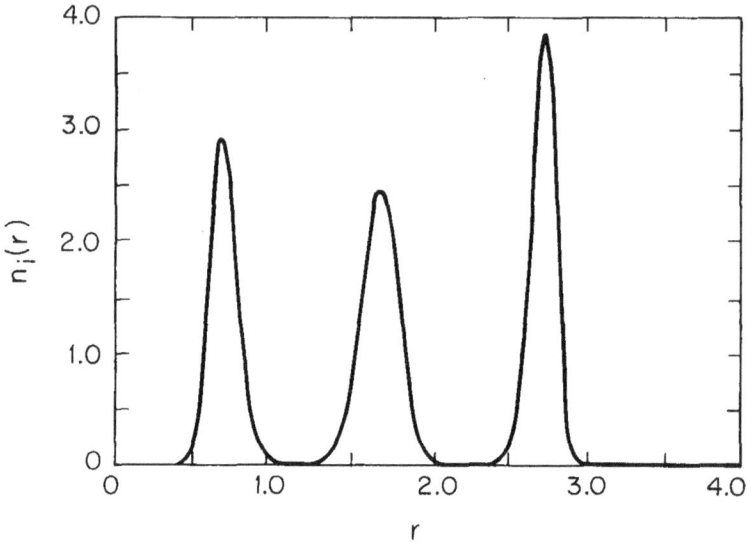

Figure 3.16. Plot of the average ion density $n_i(r)$ versus radius r obtained in molecular-dynamics simulations of a spherical ion cloud with $\bar{N}_i = 100$ and $\Gamma_i = 140$ [D.H.E. Dubin and T.M. O'Neil, Phys. Rev. Lett. **60**, 511 (1988)].

Dubin and O'Neil[25] have carried out a series of molecular-dynamics simulations of ions confined in a Penning trap. The ion orbits are calculated in the guiding-center approximation, and the electrostatic potential energy is approximated by

$$U\left(\mathbf{x}_1, \ldots, \mathbf{x}_{\bar{N}_i}\right) = \sum_{k<\ell}^{\bar{N}_i} \frac{e^2}{|\mathbf{x}_k - \mathbf{x}_\ell|} + \frac{1}{2} m_i \omega_{zi}^2 \sum_{k=1}^{\bar{N}_i} (z_k^2 - r_k^2/2) \qquad (3.52)$$

for $Z_i = 1$. In Eq.(3.52), ω_{zi} is the axial betatron frequency for a single ion in the trap (Sec. 4.3.1), and the ion cloud is assumed to be sufficiently far removed from external boundaries that the quadrupole approximation is valid for the applied electrostatic potential. The simulations[25] have been performed for $\bar{N}_i = 100$ and $\bar{N}_i = 256$ in the large-Γ_i regime (Γ_i up to 400, and T_i in the mK range) for the case of a spherical plasmoid. The observed behavior of the (bounded) ion cloud[25] is unlike that observed in simulations of unbounded homogeneous systems,[16] where a liquid phase occurs for $\Gamma_i \gtrsim 2$ and the transition to a body-centered cubic lattice occurs for $\Gamma_i \simeq 170$. In contrast, Dubin and O'Neil find at sufficiently large values of the coupling parameter ($\Gamma_i = 140$ in Fig. 3.16) that the ion cloud develops a radial structure corresponding to concentric spheroidal shells. In the surface of the shells, however, the ions move freely. This behavior is similar to that observed in smectic-liquid crystals.[17] That is, for the system parameters in Fig. 3.16, the nonneutral ion cloud may be characterized as a crystal in the direction perpendicular to the shells, and as a liquid on the shells. As Γ_i is further increased in the simulations,[25] diffusion on the shells decreases, and a two-dimensional hexagonal lattice forms on the outermost shells. However, even for $\Gamma_i \sim 300 - 400$, the lattice is imperfect and a low level of diffusion persists.

Gilbert, Bollinger and Wineland[23] have observed this shell structure experimentally in a nonneutral ion plasma consisting of up to 15,000 $^9Be^+$ ions confined in a Penning trap. Ion clouds with densities in the range of 10^8 cm^{-3} are confined radially by a magnetic field $B_0 = 19.2$ kG, and axially by an applied potential ranging up to 100 V (Fig. 3.17). The ion clouds are laser cooled to temperatures of about 10 mK, and the value of the coupling parameter Γ_i ranges from 20 to 200 in the experiments.[23] The cooling radiation (30 μW at 313 nm) can be directed perpendicular to the magnetic field $B_0 \hat{e}_z$ and/or along a diagonal of the plasmoid (Fig. 3.17). In addition to cooling the ions, the laser radiation also exerts

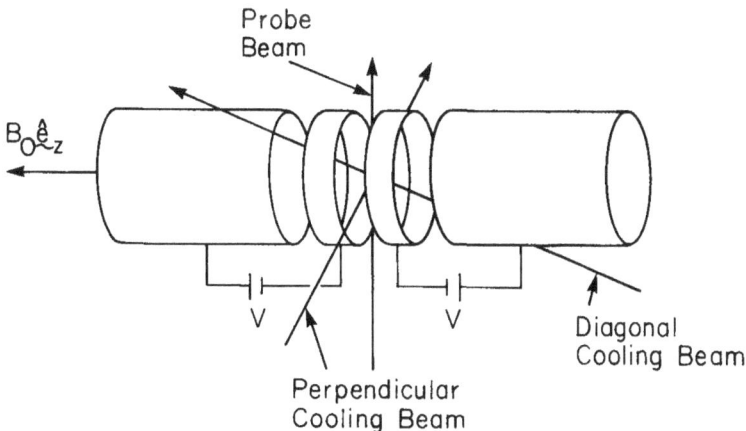

Figure 3.17. Schematic of the segmented-cylinder Penning trap geometry used to investigate the shell structure of a strongly coupled nonneutral $^9Be^+$ plasma. The orientation of the diagonal cooling laser beam with respect to $B_0 \hat{e}_z$ is 51°. The overall length of the trap is 10.2 cm, and the inner diameter of the two end cylinders and the two electrically connected central cylinders is 2.5 cm. The ion clouds are typically less than 1 mm in diameter and axial length [S.L. Gilbert, J.J. Bollinger and D.J. Wineland, Phys. Rev. Lett. **60**, 2022 (1988)].

(a)

y

→ x

200 μm

(b)

Diagonal
Cooling

Probe

Perpendicular
Cooling

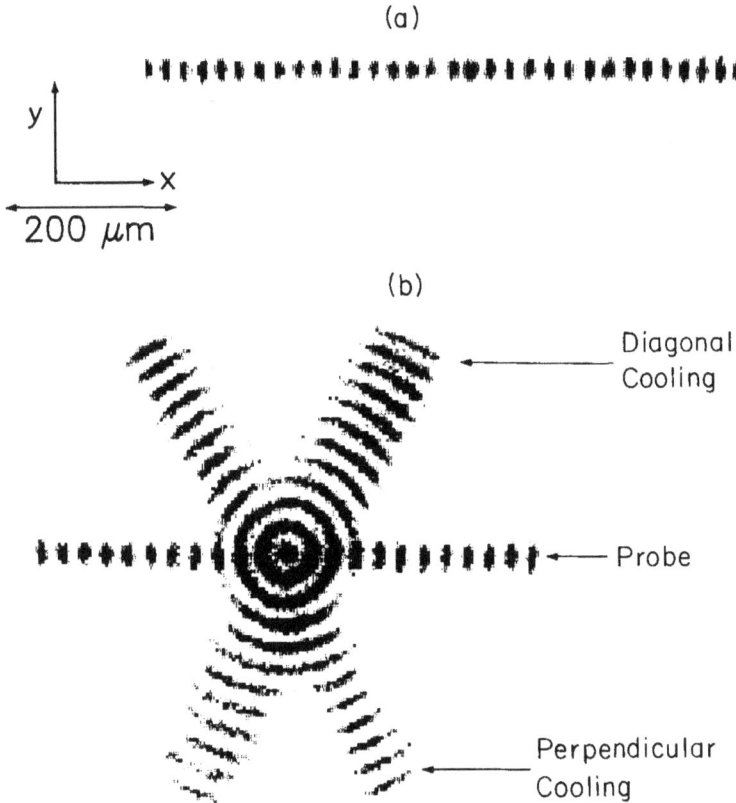

Figure 3.18. Images of the shell structure in a strongly coupled non-neutral $^9Be^+$ plasma. (a) Sixteen shells are measured in an ion cloud with aspect ratio (axial length/diameter) of 0.8 containing 15,000 ions, using only the probe beam ion fluorescence. (b) Eleven shells plus a central column are measured in an ion cloud with aspect ratio of 2.4 containing 15,000 ions, using ion fluorescence of all three laser beams [S.L. Gilbert, J.J. Bollinger and D.J. Wineland, Phys. Rev. Lett. **60**, 2022 (1988)].

a net torque, which permits a controlled expansion or compression of the ion cloud. A second, lower-power probe laser, directed through the ion cloud perpendicular to $B_0\hat{e}_z$, is used to map the spatial structure of the cloud, as well as measure the cloud rotation frequency and ion temperature, and thereby infer the average ion density and the total number of ions. Typical experimental measurements[23] of the shell structure at large values of Γ_i are illustrated in Fig. 3.18, which shows the images of the laser-induced ion fluorescence. In Fig. 3.18(a) the trap voltage is 100 V; the cloud aspect ratio (axial length/diameter) is approximately 0.8; and the total number of ions is about 15,000. Sixteen shells using only the probe beam ion fluorescence are evident in Fig. 3.18(a). In Fig. 3.18(b), the trap voltage is 28 V; the cloud aspect ratio is approximately 2.4; the total number of ions is about 15,000; and the images are obtained from ion fluorescence of all three laser beams. Figure 3.18(b) clearly shows the signature of eleven shells plus a central ion column at the origin. For a spherical ion cloud, approximately $(\bar{N}_i/4)^{1/3}$ shells are predicted.[26] For the nearly spherical cloud in Fig. 3.18(a) ($\bar{N}_i \sim 15{,}000$), this formula predicts 15.5 shells, and the measurements[23] show 16 shells. The experimental data[23] agree qualitatively with the numerical simulations, except in some cases the experiments show an open-cylinder shell structure, in contrast with the predicted closed spheroids.

To summarize, both numerical simulations[25] and experimental studies[23] of bounded nonneutral ion plasmas at sufficiently large values of the coupling parameter Γ_i are characterized by smectic-like crystal properties. Undoubtedly, future work in this area will investigate lattice properties in ion clouds with larger volumes and larger values of Γ_i.

Chapter 3 References

1. R.C. Davidson, *Theory of Nonneutral Plasmas* (Addison-Wesley, Reading, Massachusetts, 1989).

2. R.C. Davidson, "Waves and Instabilities in Nonneutral Plasmas," in *Nonneutral Plasma Physics*, eds., C.W. Roberson and C.F. Driscoll, AIP Conference Proceedings No. 175 (American Institute of Physics, New York, 1988), pp. 139-208.

3. R.C. Davidson and N.A. Krall, "Vlasov Description of an Electron Gas in a Magnetic Field," Phys. Rev. Lett. **22**, 833 (1969).

4. R.C. Davidson and N.A. Krall, "Vlasov Equilibria and Stability of an Electron Gas," Phys. Fluids **13**, 1543 (1970).

5. L. Brillouin, "A Theorem of Larmor and Its Importance for Electrons in Magnetic Fields," Phys. Rev. **67**, 260 (1945).

6. A.J. Theiss, R.A. Mahaffey and A.W. Trivelpiece, "Rigid-Rotor Equilibria of Nonneutral Plasmas," Phys. Rev. Lett. **35**, 1436 (1975).

7. G.J. Budker, "Relativistic Stabilized Electron Beam," in *Proceedings of the Symposium on High-Energy Accelerators and Pion Physics* (CERN Scientific Information Service, Geneva, 1956), Volume 1, p. 68.

8. T.M. O'Neil, "New Theory of Transport Due to Like-Particle Collisions," Phys. Rev. Lett. **55**, 943 (1985).

9. D.H.E. Dubin and T.M. O'Neil, "Two-Dimensional Guiding-Center Transport of a Pure Electron Plasma," Phys. Rev. Lett. **60**, 1286 (1988).

10. T.M. O'Neil, "A Confinement Theorem for Nonneutral Plasmas," Phys. Fluids **23**, 2217 (1980).

11. C.F. Driscoll, J.H. Malmberg and K.S. Fine, "Observation of Transport to Thermal Equilibrium in Pure Electron Plasmas," Phys. Rev. Lett. **60**, 1290 (1988).

12. R.C. Davidson, "Electrostatic Shielding of a Test Charge in a Nonneutral Plasma," J. Plasma Phys. **6**, 229 (1971).

13. J.D. Jackson, *Classical Electrodynamics* (Wiley, New York, 1975).

14. R.C. Davidson and W.A. McMullin, "Influence of Intense Equilibrium Self Fields on the Spontaneous Emission from a Test Electron in a Relativistic Nonneutral Electron Beam," Phys. Fluids **27**, 1268 (1984).

15. J.H. Malmberg and T.M. O'Neil, "Pure Electron Plasma, Liquid, and Crystal," Phys. Rev. Lett. **39**, 1333 (1977).

16. W.L. Slattery, G.D. Doolen and H.E. DeWitt, "Improved Equation of State for the Classical One-Component Plasma," Phys. Rev. **A21**, 2087 (1980).

17. S. Ichimaru, "Strongly Coupled Plasmas: High-Density Classical Plasmas and Degenerate Electron Liquids," Rev. Modern Physics **54**, 1017 (1982).

18. S. Ichimaru, H. Iyetomi and S. Tanaka, "Statistical Physics of Dense Plasmas: Thermodynamics, Transport Coefficients and Dynamic Correlations," Physics Reports **149**, 91 (1987).

19. G.C. Grimes and G. Adams, "Evidence for a Liquid-to-Crystal Phase Transition in a Classical, Two-Dimensional Sheet of Electrons," Phys. Rev. Lett. **42**, 795 (1979).

20. A.J. Dahm and W.F. Viven, "Electrons and Ions at the Helium Surface," Physics Today **40** (2), 43 (1987).

21. J.J. Bollinger and D.J. Wineland, "Strongly Coupled Nonneutral Ion Plasma," Phys. Rev. Lett. **53**, 348 (1984).

22. L.R. Brewer, J.D. Prestage, J.J. Bollinger and D.J. Wineland, "A High- Γ, Strongly Coupled, Nonneutral Ion Plasma," in *Strongly Coupled Plasma Physics*, eds., F.J. Rogers and H.E. DeWitt (Plenum, New York, 1987), p. 53.

23. S.L. Gilberg, J.J. Bollinger and D.J. Wineland, "Shell Structure Phase of Magnetically Confined Strongly Coupled Plasmas," Phys. Rev. Lett. **60**, 2022 (1988).

24. H. Totsuji, "Static and Dynamic Properties of Strongly Coupled Classical One-Component Plasmas: Numerical Experiments on Supercooled Liquid

State and Simulation of Ion Plasma in the Penning Trap," in *Strongly Coupled Plasma Physics*, eds., F.J. Rogers and H.E. DeWitt (Plenum, New York, 1987), p. 19.

25. D.H.E. Dubin and T.M. O'Neil, "Computer Simulation of Ion Clouds in a Penning Trap," Phys. Rev. Lett. **60**, 511 (1988).

26. A. Rahman and J.P. Schiffer, "Structure of a One-Component Plasma in an External Field: A Molecular-Dynamics Study of Particle Arrangement in a Heavy-Ion Storage Ring," Phys. Rev. Lett. **57**, 1133 (1986).

27. L. Landau and E. Lifshitz, *Statistical Physics* (Pergamon, Oxford, 1980).

28. E. Wigner, "Effects of the Electron Interaction on the Energy Levels of Electrons in Metals," Trans. Faraday Soc. **34**, 678 (1938).

29. D.J. Aastuen, N.A. Clark and L.K. Cotter, "Nucleation and Growth of Colloidal Crystals," Phys. Rev. Lett. **57**, 1733 (1986).

30. J.M. di Meglio, D.A. Weitz and P.M. Chaikin, "Competition Between Shear-Melting and Taylor Instabilities in Colloidal Crystals," Phys. Rev. Lett. **58**, 136 (1987).

31. C.A. Murray and D.H. Van Winkle, "Experimental Observation of Two-Stage Melting in a Classical Two-Dimensional Screened Coulomb System," Phys. Rev. Lett. **58**, 1200 (1987).

32. D.J. Wineland, J.C. Bergquist, W.M. Itano, J.J. Bollinger and C.H. Manney, "Atomic Ion Coulomb Clusters in an Ion Trap," Phys. Rev. Lett. **59**, 2935 (1987).

33. J.H. Malmberg, T.M. O'Neil, A.W. Hyatt and C.F. Driscoll, in *Proceedings of the 1984 Symposium on Nonlinear Plasma Phenomena* (Tohoku University, Sendai, Japan, 1984), p. 31.

The following references, while not cited directly in the text, are relevant to the general subject matter of this chapter.

D.H.E. Dubin, "Correlation Energies of Simple Bounded Coulomb Lattices," Phys. Rev. Lett. **A40**, 1140 (1989).

R.A. Smith, "Phase-Transition Behavior in a Negative-Temperature Guiding-Center Plasma," Phys. Rev. Lett. **63**, 1479 (1989).

T.M. O'Neil, "Plasmas with a Single Sign of Charge," in *Nonneutral Plasma Physics*, eds., C.W. Roberson and C.F. Driscoll, AIP Conference Proceedings No. 175 (American Institute of Physics, New York, 1988), p. 1.

D.L. Eagleston and J.H. Malmberg, "Observation of an Induced Scattering Instability Driven by Static Field Asymmetries in a Pure Electron Plasma," Phys. Rev. Lett. **59**, 1675 (1987).

D.H.E. Dubin and T.M. O'Neil, "Thermal Equilibrium of a Cryogenic Magnetized Pure Electron Plasma," Phys. Fluids **29**, 11 (1986).

D.H.E. Dubin and T.M. O'Neil, "Adiabatic Expansion of a Strongly Correlated Pure Electron Plasma," Phys. Rev. Lett. **56**, 728 (1986).

C.F. Driscoll, K.S. Fine and J.H. Malmberg, "Reduction of Radial Losses in a Pure Electron Plasma," Phys. Fluids **29**, 2015 (1986).

T.M. O'Neil, "Collision Operator for a Strongly Magnetized Pure Electron Plasma," Phys. Fluids **26**, 2128 (1983).

G. Tsakiris, D. Boyd, D. Hammer, A.W. Trivelpiece and R.C. Davidson, "Electron Energy Distribution of a Mirror-Confined Relativistic Plasma Inferred from Synchrotron Radiation Measurements," Phys. Fluids **21**, 2050 (1978).

G.D. Tsakiris and R.C. Davidson, "Influence of Canonical Angular Momentum Spread on the Synchrotron Radiation Spectrum for a Relativistic Plasma," Phys. Fluids **20**, 436 (1977).

R.C. Davidson, "Synchrotron Radiation Spectrum for a Relativistic Plasma Column," Phys. Fluids **18**, 1143 (1975).

S.F. Nee, A.W. Trivelpiece and R.E. Pechacek, "Synchrotron Radiation from a Magnetically Confined Nonneutral Hot Electron Plasma," Phys. Fluids **16**, 502 (1973).

CHAPTER 4

KINETIC EQUILIBRIUM AND STABILITY PROPERTIES

This chapter makes use of the Vlasov-Maxwell equations discussed in Sec. 2.2 to investigate the equilibrium and stability properties of non-neutral plasmas[1-6] in a wide variety of confinement geometries. Such a kinetic model provides a complete classical description of collective plasma processes, including stability properties that depend on the detailed momentum-space and configuration-space features of the equilibrium distribution function $f_j^0(\mathbf{x}, \mathbf{p})$. Following a general discussion of the equilibrium Vlasov-Maxwell equations for axisymmetric confinement geometries (Sec. 4.1), nonrelativistic Vlasov equilibria are developed (Sec. 4.2) for infinitely long nonneutral plasma columns and electron layers confined radially by a uniform applied magnetic field $B_0\hat{\mathbf{e}}_z$. Examples of kinetic equilibria confined both radially and axially by externally applied focusing fields are then discussed (Sec. 4.3), including the Penning trap and segmented cylinder geometries,[7,8] as well as mirror-confined nonneutral electron layers.[9] Next, the influence of intense self-magnetic fields (and self-electric fields) on kinetic equilibrium properties are investigated for the case of a relativistic electron beam[10-14] propagating parallel to a uniform guide field $B_0\hat{\mathbf{e}}_z$ through an ion background that provides partial charge neutralization (Sec. 4.4).

Before analyzing the kinetic stability properties of specific equilibrium configurations, use is made of global conservation constraints satisfied by the nonlinear Vlasov-Maxwell equations and the Klimontovich-Maxwell equations to derive a kinetic stability theorem (Sec. 4.5)[1-5] and a particle confinement theorem (Sec. 4.6)[15] for a one-component nonneutral plasma confined radially by a uniform magnetic field $B_0\hat{\mathbf{e}}_z$. These theorems represent very powerful results, including a sufficient condition for stability of spatially nonuniform equilibria $f_j^0(r, \mathbf{p})$ to small-amplitude perturbations, as well as a nonlinear bound on the number of particles lost radially from the system through collective and/or discrete particle interactions. Following a derivation of the formal eigenvalue equation (Sec. 4.7) for electrostatic perturbations about a nonneutral plasma column with rigid-rotor Vlasov equilibrium $f_j^0(H_\perp - \omega_{rj} P_\theta, p_z)$, the dispersion relation[1-5] for body-wave perturbations (perturbations localized to the column interior) about a uniform-density nonneutral plasma is derived and analyzed in various regimes where kinetic effects play an important role in determining detailed stability behavior (Sec. 4.8). Kinetic stability properties are then investigated for electrostatic flute perturbations $(\partial/\partial z = 0)$ about a uniform-density plasma column including the full influence of finite radial geometry (Sec. 4.9).[16,17] Finally, the Vlasov-Maxwell equations are used to analyze the equilibrium and stability properties of a high-current relativistic ring[18-22] confined in a modified betatron (Sec. 4.10),[23-41] where an externally applied toroidal magnetic field together with a mirror (betatron) field provide confinement of the electron ring at the midplane. Particular emphasis in the stability studies is placed on the negative-mass and surface-kink instabilities.[41] Given the overall complexity of the geometric configuration relative to the conventional betatron (which has no toroidal confining field),[42,43] and the importance of both electric and magnetic self fields at high electron current, the modified betatron represents a particularly instructive application of state-of-the-art techniques using the Vlasov-Maxwell equations to investigate the equilibrium and stability properties of magnetically confined nonneutral plasma. As a final point (not treated in Sec. 4.10), it is found that the introduction of helical (stellarator) windings[44,45] to the modified betatron configuration provides improved equilibrium and stability properties[44-48] relative to the modified betatron.[23-41]

4.1   Equilibrium Vlasov-Maxwell Equations for Axisymmetric Nonneutral Plasma

4.1.1   Axisymmetric Confinement Geometries

In this section, kinetic equilibrium properties are investigated for the axisymmetric, nonneutral plasma configurations illustrated in Figs. 4.1–4.3. The analysis is based on the steady-state Vlasov-Maxwell equations (2.9)–(2.13), and therefore naturally incorporates thermal effects through the equilibrium distribution function $f_j^0(\mathbf{x}, \mathbf{p})$. In this regard, a kinetic description of the equilibrium provides an important basis for investigating stability properties that depend on the detailed distribution of particles in both momentum $(\mathbf{p})$ and configuration $(\mathbf{x})$ space.[1-6]

The equilibrium configuration illustrated in Fig. 4.1 corresponds to a nonneutral plasma column of infinite axial extent, confined radially by a uniform applied magnetic field $\mathbf{B}_0^{ext}(\mathbf{x}) = B_0 \hat{\mathbf{e}}_z$. Equilibrium properties are assumed to be independent of $z(\partial/\partial z = 0)$ and azimuthally symmetric $(\partial/\partial\theta = 0)$ about an axis of symmetry parallel to $B_0 \hat{\mathbf{e}}_z$. For the equilibrium configuration illustrated in Fig. 4.1, the steady-state $(\partial/\partial t = 0)$ Vlasov-Maxwell equations are discussed in Sec. 4.1.2, and specific examples of nonrelativistic and relativistic nonneutral plasma equilibria are analyzed in Secs. 4.2 and 4.4.

The equilibrium configurations illustrated in Figs. 4.2 and 4.3 correspond to axisymmetric nonneutral plasmas which are confined axially as well as radially. In Fig. 4.2, axial confinement of a ring or layer of electrons is provided by an externally applied mirror (or betatron) magnetic field $\mathbf{B}_0^{ext}(\mathbf{x})$ which produces sufficient axial focusing of the particle orbits to assure confinement in the z-direction. Such confinement geometries are

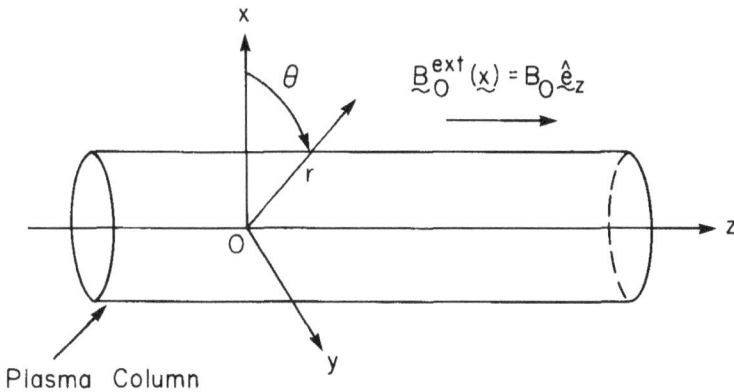

Figure 4.1. Axisymmetric equilibrium configuration for a nonneutral plasma column of infinite axial extent, aligned parallel to a uniform applied magnetic field $\mathbf{B}_0^{ext}(\mathbf{x}) = B_0 \hat{\mathbf{e}}_z$. Cylindrical polar coordinates (r, θ, z) are introduced with the z-axis coinciding with the axis of symmetry; θ is the polar angle in the x-y plane, and $r = (x^2 + y^2)^{1/2}$ is the radial distance from the z-axis.

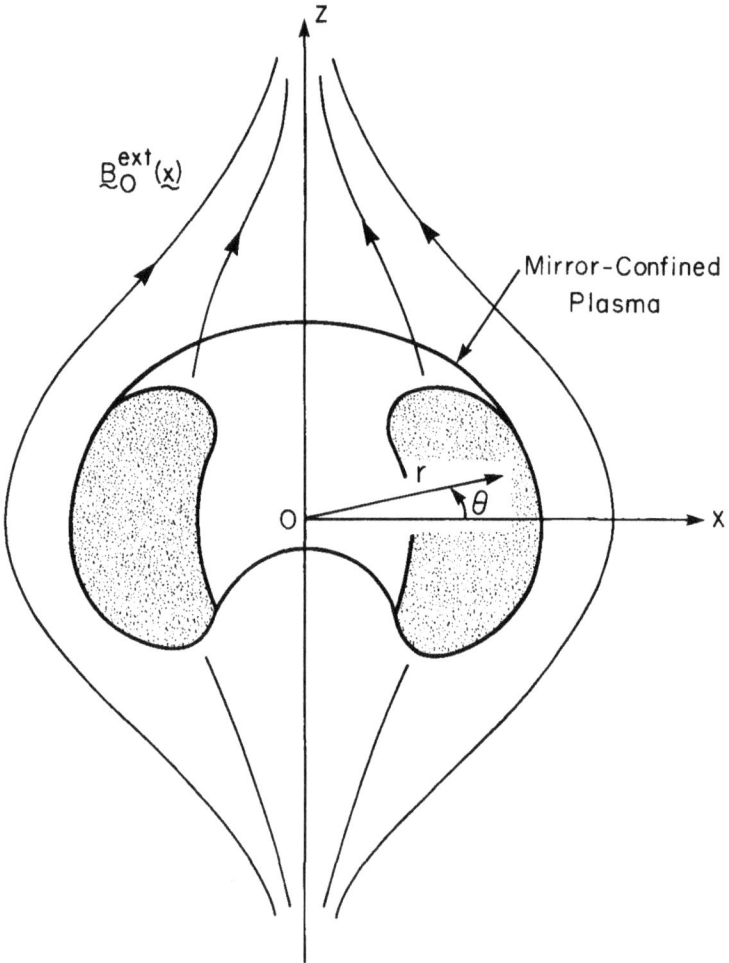

Figure 4.2. Axisymmetric equilibrium configuration for a toroidal non-neutral plasma radially and axially confined by an externally applied mirror or betatron field $\mathbf{B}_0^{ext}(\mathbf{x})$. Cylindrical polar coordinates (r, θ, z) are introduced with the z-axis coinciding with the axis of symmetry, and $z = 0$ at the mirror midplane; r is the radial distance from the axis of symmetry, and θ is the polar angle.

(a)

(b)

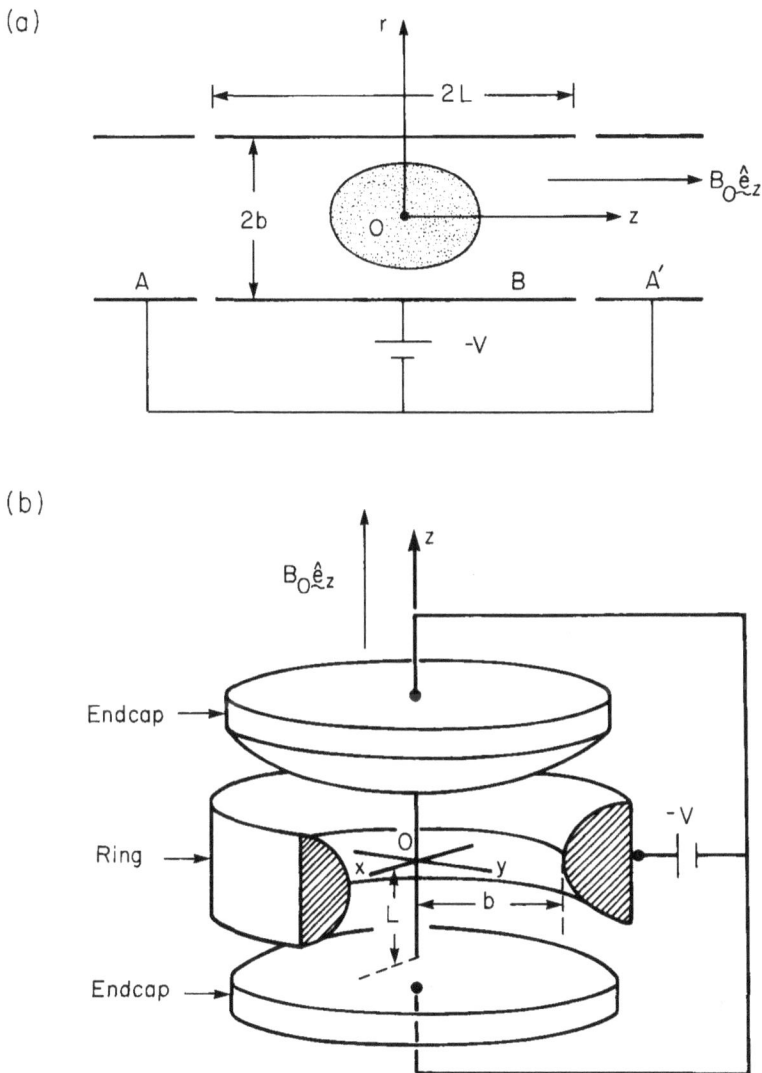

Figure 4.3. Schematics of (a) the segmented cylinder and (b) Penning trap confinement geometries. Radial confinement of a nonneutral electron plasma (say) is provided by a uniform applied magnetic field $B_0\hat{e}_z$, and axial confinement is provided by applied electrostatic potentials on neighboring conductors. In the segmented cylinder geometry [Fig. 4.3(a)], the end cylinders A and A' are biased at a potential $-V$ relative to the central cylinder B. In the Penning trap [Fig. 4.3(b)], the end caps are biased at a potential $-V$ relative to the ring electrode. In both cases, the applied potential near the origin has the form of a quadrupole electrostatic potential.

used in relativistic betatron accelerators, and in experiments that investigate the basic equilibrium and stability properties of mirror-confined nonneutral plasmas.

The two confinement geometries illustrated in Fig. 4.3 provide radial confinement of the nonneutral plasma by a uniform applied magnetic field $B_0 \hat{e}_z$, and axial confinement by applied electrostatic potentials on neighboring conductors.[7,8] In the segmented cylinder geometry in Fig. 4.3(a), axial confinement of a nonneutral electron plasma (say) is provided by biasing the end cylinders A and A' at a potential $-V$ relative to the central cylinder B of length $2L$. Figure 4.3(b) is a schematic of the Penning trap, which consists of two endcaps and one ring electrode, biased at a relative potential $-V$. The hyperbolic surfaces of the electrodes produce a quadrupole electrostatic potential which confines the electrons in the z-direction. The confinement geometries illustrated in Fig. 4.3 are widely used for experimental investigations of the basic equilibrium and stability properties of nonneutral electron and ion plasmas (see Chapter 3).

For the confinement geometries illustrated in Figs. 4.2 and 4.3, the steady-state Vlasov-Maxwell equations are discussed in Sec. 4.1.3, and specific examples of nonrelativistic and relativistic nonneutral plasma equilibria are analyzed in Secs. 4.3 and 4.10.

4.1.2   Nonneutral Plasma Column

For the axisymmetric equilibrium configuration illustrated in Fig. 4.1, the uniform applied magnetic field $B_0 \hat{e}_z$ provides radial confinement of a multicomponent nonneutral plasma where particles of species j have charge e_j and rest mass m_j. The following simplifying assumptions are made in analyzing the steady-state $(\partial/\partial t = 0)$ Vlasov-Maxwell equations (2.9)–(2.13).[1-6]

(a) The plasma is uniform in the z-direction with $\partial f_j^0(\mathbf{x}, \mathbf{p})/\partial z = 0$, and the equilibrium electric field parallel to $B_0 \hat{e}_z$ is equal to zero, i.e., $\mathbf{E}^0(\mathbf{x}) \cdot \hat{e}_z = 0$.

(b) All equilibrium quantities are assumed to be azimuthally symmetric about the magnetic axis, e.g., $\partial f_j^0(\mathbf{x}, \mathbf{p})/\partial \theta = 0$. Here, cylindrical polar coordinates (r, θ, z) are introduced as in Fig. 4.1.

(c) In general, the azimuthal plasma current $J_\theta^0(r) = \sum_j e_j \int d^3 p \times v_\theta f_j^0(\mathbf{x}, \mathbf{p})$ induces an axial self-magnetic field $B_z^s(r)$, and the axial plasma current $J_z^0(r) = \sum_j e_j \int d^3 p\, v_z f_j^0(\mathbf{x}, \mathbf{p})$ induces an azimuthal self-magnetic field $B_\theta^s(r)$. Deviations from equilibrium charge neutrality, $\sum_j e_j \int d^3p\, f_j^0(\mathbf{x}, \mathbf{p}) \neq 0$, produce a radial self-electric field $E_r(r)$.

Within the context of the above assumptions, the equilibrium field components are

$$\mathbf{E}^0(\mathbf{x}) = E_r(r)\hat{\mathbf{e}}_r \ ,$$

$$\mathbf{B}^0(\mathbf{x}) = [B_0 + B_z^s(r)]\,\hat{\mathbf{e}}_z + B_\theta^s(r)\hat{\mathbf{e}}_\theta. \tag{4.1}$$

For future reference, it is convenient to introduce the potentials $\phi_0(r)$, $A_z^s(r)$ and $A_\theta^s(r)$, related to the equilibrium field components $E_r(r)$, $B_\theta^s(r)$ and $B_z^s(r)$ by

$$E_r(r) = -\frac{\partial}{\partial r}\phi_0(r) \ ,$$

$$B_\theta^s(r) = -\frac{\partial}{\partial r}A_z^s(r) \ ,$$

$$B_z^s(r) = \frac{1}{r}\frac{\partial}{\partial r}\left[rA_\theta^s(r)\right] \ . \tag{4.2}$$

Here, $\mathbf{E}^0(\mathbf{x}) = -\nabla\phi_0(\mathbf{x})$ and $\mathbf{B}^s(\mathbf{x}) = \nabla \times \mathbf{A}^s(\mathbf{x})$, and the *total* azimuthal component of the equilibrium vector potential is $A_\theta^0(r) = A_\theta^{\text{ext}}(r) + A_\theta^s(r)$, where $A_\theta^{\text{ext}}(r) = rB_0/2$ is the vector potential for the uniform applied magnetic field $B_0\hat{\mathbf{e}}_z$.

For the equilibrium field configuration in Eq.(4.1), the steady-state Vlasov equation (2.9) can be expressed as

$$\left\{\mathbf{v} \cdot \frac{\partial}{\partial \mathbf{x}} + e_j \left[E_r\hat{\mathbf{e}}_r + \frac{1}{c}\mathbf{v} \times ([B_0 + B_z^s]\,\hat{\mathbf{e}}_z + B_\theta^s\hat{\mathbf{e}}_\theta)\right] \cdot \frac{\partial}{\partial \mathbf{p}}\right\} f_j^0(r, \mathbf{p}) = 0 \ . \tag{4.3}$$

Allowing for relativistic particle motion, the velocity $\mathbf{v}$ and momentum $\mathbf{p}$ are related by $\mathbf{p} = \gamma m_j \mathbf{v}$, where the relativistic mass factor γ is defined by

$$\gamma = \left(1 + \mathbf{p}^2/m_j^2 c^2\right)^{1/2} \ . \tag{4.4}$$

It is straightforward to show that any function $f_j^0(r, \mathbf{p})$ that is a function only of the single-particle constants of the motion in the equilibrium field configuration (4.1) exactly solves the steady-state Vlasov equation (4.3).

That is, the equilibrium distribution function $f_j^0(r, \mathbf{p})$ is of the form (Problem 4.1)

$$f_j^0(r, \mathbf{p}) = f_j^0(H, P_\theta, P_z) \, , \tag{4.5}$$

where H is the energy,

$$H = \left(m_j^2 c^4 + c^2 \mathbf{p}^2\right)^{1/2} + e_j \phi_0(r) - m_j c^2 \, , \tag{4.6}$$

P_θ is the canonical angular momentum,

$$P_\theta = r \left[p_\theta + \frac{1}{2} \frac{e_j}{c} B_0 r + \frac{e_j}{c} A_\theta^s(r)\right] \, , \tag{4.7}$$

and P_z is the axial canonical momentum,

$$P_z = p_z + \frac{e_j}{c} A_z^s(r) \, . \tag{4.8}$$

Although general $f_j^0(H, P_\theta, P_z)$ solves Eq.(4.3), in specific applications the choice of the functional form for $f_j^0(H, P_\theta, P_z)$ is guided by several practical considerations, such as the method of plasma formation, the degree of energy spread, etc. Once the functional form of $f_j^0(H, P_\theta, P_z)$ is specified, the procedure is to calculate $\phi_0(r)$, $A_z^s(r)$ and $A_\theta^s(r)$ self-consistently from the steady-state Maxwell equations

$$\frac{1}{r} \frac{\partial}{\partial r} r \frac{\partial}{\partial r} \phi_0(r) = -\sum_j 4\pi e_j \int d^3 p \, f_j^0(H, P_\theta, P_z) \, , \tag{4.9}$$

$$\frac{1}{r} \frac{\partial}{\partial r} r \frac{\partial}{\partial r} A_z^s(r) = -\sum_j \frac{4\pi e_j}{c} \int d^3 p \, v_z f_j^0(H, P_\theta, P_z) \, , \tag{4.10}$$

and

$$\frac{\partial}{\partial r} \frac{1}{r} \frac{\partial}{\partial r} r A_\theta^s(r) = -\sum_j \frac{4\pi e_j}{c} \int d^3 p \, v_\theta f_j^0(H, P_\theta, P_z) \, . \tag{4.11}$$

Because H, P_θ and P_z are functions of $\phi_0(r)$, $A_\theta^s(r)$ and $A_z^s(r)$, the Maxwell equations (4.9)–(4.11) are generally *nonlinear* equations for the self-field potentials.

For specified equilibrium distribution function $f_j^0(H, P_\theta, P_z)$, macroscopic equilibrium properties can also be calculated. These include, for example, the density profile,

$$n_j^0(r) = \int d^3p\, f_j^0(H, P_\theta, P_z) \,, \tag{4.12}$$

the mean azimuthal velocity,

$$V_{\theta j}^0(r) = \frac{\int d^3p\, v_\theta f_j^0(H, P_\theta, P_z)}{\int d^3p f_j^0(H, P_\theta, P_z)} \,, \tag{4.13}$$

the mean axial velocity,

$$V_{zj}^0(r) = \frac{\int d^3p\, v_z f_j^0(H, P_\theta, P_z)}{\int d^3p\, f_j^0(H, P_\theta, P_z)} \,, \tag{4.14}$$

and the pressure tensor,

$$\boldsymbol{P}_j^0(\mathbf{x}) = \int d^3p \left[\mathbf{v} - \mathbf{V}_j^0(\mathbf{x})\right] \left[\mathbf{p} - \mathbf{P}_j^0(\mathbf{x})\right] f_j^0(H, P_\theta, P_z) \,. \tag{4.15}$$

Here, $\mathbf{p} = \gamma m_j \mathbf{v}$ and $\mathbf{P}_j^0(\mathbf{x}) = (\int d^3p \mathbf{p} f_j^0)/(\int d^3p\, f_j^0)$ is the mean momentum of a fluid element. Because H is an even function of p_r, note that $V_{rj}^0 = (\int d^3p\, v_r f_j^0)/(\int d^3p f_j^0) = 0$ and $P_{rj}^0 = (\int d^3p\, p_r f_j^0)/(\int d^3p\, f_j^0) = 0$ follow exactly for general $f_j^0(H, P_\theta, P_z)$.

To summarize, Eqs.(4.5)–(4.15) can be used to investigate the equilibrium properties of a multicomponent nonneutral plasma for a broad range of distribution functions $f_j^0(H, P_\theta, P_z)$. Specific equilibrium examples are considered in Secs. 4.2 and 4.4. The analysis simplifies considerably in circumstances where the particle motion is nonrelativistic $(\mathbf{p}^2/m_j^2 c^2 \ll 1$ and $\gamma \simeq 1)$ and the self-magnetic fields are sufficiently weak that

$$|B_z^s(r)| \ll B_0 \,,$$

$$|(v_z/c)B_\theta^s(r)| \ll |E_r(r)| \,. \tag{4.16}$$

In this case, the constants of the motion defined in Eqs.(4.6)–(4.8) can be approximated by

$$H = \frac{1}{2m_j}\mathbf{p}^2 + e_j\phi_0(r),$$

$$P_\theta = r\left(p_\theta + \frac{1}{2}\varepsilon_j m_j \omega_{cj} r\right),$$

$$P_z = p_z, \tag{4.17}$$

where $\varepsilon_j = \mathrm{sgn}\, e_j$, $\omega_{cj} = |e_j| B_0/m_j c$ is the cyclotron frequency, and $\mathbf{p} = m_j\mathbf{v}$ is the mechanical momentum. For specified equilibrium distribution function $f_j^0(H, P_\theta, p_z)$, the electrostatic potential $\phi_0(r)$ is determined self-consistently from Poisson's equation (4.9).

Problem 4.1   Consider the axisymmetric confinement geometry illustrated in Fig. 4.1 and described by Eq.(4.1). Show by direct calculation that

$$\left\{\mathbf{v} \cdot \frac{\partial}{\partial \mathbf{x}} + e_j\left[E_r\hat{\mathbf{e}}_r + \frac{1}{c}\mathbf{v} \times ([B_0 + B_z^s]\hat{\mathbf{e}}_z + B_\theta^s\hat{\mathbf{e}}_\theta)\right] \cdot \frac{\partial}{\partial \mathbf{p}}\right\} C_\ell = 0, \tag{4.1.1}$$

where C_ℓ denotes the constants of the motion, H, P_θ and P_z, defined in Eqs.(4.6)–(4.8). It follows directly from Eq.(4.1.1) and application of the chain rule for differentiation that the distribution function $f_j^0(H, P_\theta, P_z)$ solves the steady-state Vlasov equation (4.3).

4.1.3   Axially Confined Equilibria

The kinetic equilibrium formalism discussed in Sec. 4.1.2 is readily extended to the case of axially confined equilibria for the confinement geometries illustrated in Figs. 4.2 and 4.3.[7-9] In particular, the equilibrium potentials $\phi_0(r, z)$ and $A_\theta^0(r, z)$ now depend on axial coordinate z, although $\partial/\partial\theta = 0$ is still assumed. For equilibria with $\partial/\partial z \neq 0$, there is no constant of the motion analogous to the axial canonical momentum P_z. However, the single-particle constants of the motion, H and P_θ defined in Eqs.(4.6) and (4.7), are modified to become

$$H = \left(m_j^2 c^4 + c^2 \mathbf{p}^2\right)^{1/2} + e_j \phi_0(r, z) - m_j c^2 , \qquad (4.18)$$

and

$$P_\theta = r \left[p_\theta + \frac{e_j}{c} A_\theta^{\mathrm{ext}}(r, z) + \frac{e_j}{c} A_\theta^s(r, z)\right] . \qquad (4.19)$$

In Eq.(4.19), $A_\theta^{\mathrm{ext}}(r, z)$ is the vector potential for the externally applied mirror magnetic field in Fig. 4.2. For axially confined equilibria, the electric and magnetic fields generally have both radial and axial components, i.e.,

$$\mathbf{E}^0(\mathbf{x}) = E_r(r, z)\hat{\mathbf{e}}_r + E_z(r, z)\hat{\mathbf{e}}_z ,$$

$$\mathbf{B}^0(\mathbf{x}) = \mathbf{B}_0^{\mathrm{ext}}(\mathbf{x}) + B_r^s(r, z)\hat{\mathbf{e}}_r + B_z^s(r, z)\hat{\mathbf{e}}_z , \qquad (4.20)$$

where $\mathbf{B}_0^{\mathrm{ext}}(\mathbf{x}) = B_r^{\mathrm{ext}}(r, z)\hat{\mathbf{e}}_r + B_z^{\mathrm{ext}}(r, z)\hat{\mathbf{e}}_z$ is the externally applied magnetic field (Sec. 4.3.2).

Equations (4.18) and (4.19) are relativistically correct expressions for the single-particle constants of the motion. Within the framework of an equilibrium theory based on the Vlasov-Maxwell equations, any distribution function $f_j^0(r, z, \mathbf{p})$ that depends only on H and P_θ is a solution to the steady-state ($\partial/\partial t = 0$) Vlasov equation. That is, distribution functions of the form

$$f_j^0(r, z, \mathbf{p}) = f_j^0(H, P_\theta) \qquad (4.21)$$

are exact solutions to Eq.(2.9), which can be verified by direct substitution. Making use of $\mathbf{E}^0 = -\nabla\phi_0$ and $\mathbf{B}^s = \nabla \times \mathbf{A}^s$, the equilibrium potentials, $\phi_0(r, z)$ and $A_\theta^s(r, z)$, are determined self-consistently from

$$\left\{\frac{1}{r}\frac{\partial}{\partial r}r\frac{\partial}{\partial r} + \frac{\partial^2}{\partial z^2}\right\} \phi_0(r, z) = -\sum_j 4\pi e_j \int d^3p\, f_j^0(H, P_\theta) , \qquad (4.22)$$

and

$$\left\{\frac{\partial}{\partial r}\frac{1}{r}\frac{\partial}{\partial r}r + \frac{\partial^2}{\partial z^2}\right\} A_\theta^s(r, z) = -\sum_j \frac{4\pi e_j}{c} \int d^3p\, v_\theta f_j^0(H, P_\theta) . \qquad (4.23)$$

where $v_\theta = p_\theta/m_j(1 + \mathbf{p}^2/m_j^2 c^2)^{1/2}$. Because H and P_θ depend on $\phi_0(r, z)$ and $A_\theta^s(r, z)$, Eqs.(4.22) and (4.23) are generally *nonlinear* equations for the potentials. Once $\phi_0(r,z)$ and $A_\theta^s(r,z)$ are determined from Eqs.(4.22) and (4.23) consistent with appropriate boundary conditions, the corresponding field components are given by

$$E_r(r,z) = -\frac{\partial}{\partial r}\phi_0(r,z) , \quad E_z(r,z) = -\frac{\partial}{\partial z}\phi_0(r,z) ,$$

$$B_r^s(r,z) = -\frac{\partial}{\partial z}A_\theta^s(r,z) , \quad B_z^s(r,z) = \frac{1}{r}\frac{\partial}{\partial r}\left[rA_\theta^s(r,z)\right] . \quad (4.24)$$

For specified equilibrium distribution function $f_j^0(H, P_\theta)$, the procedure is to calculate $\phi_0(r,z)$ and $A_\theta^s(r,z)$ self-consistently from Eqs.(4.22) and (4.23). Other equilibrium properties can then be calculated in a manner similar to that discussed in Sec. 4.1.2.

If the particle motion is nonrelativistic and $|\mathbf{B}_0^s(\mathbf{x})| \ll |\mathbf{B}_0^{\text{ext}}(\mathbf{x})|$, then the single-particle constants of the motion defined in Eqs.(4.18) and (4.19) can be approximated by

$$H = \frac{1}{2m_j}\mathbf{p}^2 + e_j\phi_0(r,z) ,$$

$$P_\theta = r\left[p_\theta + \frac{e_j}{c}A_\theta^{\text{ext}}(r,z)\right] . \quad (4.25)$$

For an electrically nonneutral equilibrium with negligibly small self-magnetic field, the only equilibrium equation to solve is Poisson's equation (4.22), where H and P_θ are approximated by Eq.(4.25). Here, the components of $\mathbf{B}_0^{\text{ext}}(\mathbf{x})$ are $B_r^{\text{ext}}(r,z) = -(\partial/\partial z)A_\theta^{\text{ext}}(r,z)$ and $B_z^{\text{ext}}(r,z) = r^{-1}(\partial/\partial r)\left[rA_\theta^{\text{ext}}(r,z)\right]$.

For the case where $A_\theta^{\text{ext}}(r,z) = rB_0/2$ and axial confinement is provided by externally applied electric potentials (Fig. 4.3), Poisson's equation (4.22) must of course be solved subject to the appropriate boundary conditions at the conducting walls. Specific examples of axisymmetric nonneutral plasma equilibria that are confined axially as well as radially are presented in Secs. 4.3 and 4.10.

4.2   Nonrelativistic Kinetic Equilibria

In this section, use is made of the kinetic formalism developed in Sec. 4.1 to investigate several examples of kinetic equilibria for the nonneutral plasma column illustrated in Fig. 4.1. For present purposes, the particle motion is assumed to be nonrelativistic, and the self-magnetic fields are assumed to be sufficiently weak that $|B_z^s(r)| \ll B_0$ and $|(v_z/c)B_\theta^s(r)| \ll |E_r(r)|$. The examples considered in Sec. 4.2 range from rigid-rotor Vlasov equilibria (Secs. 4.2.1–4.2.3) to an annular electron layer (Sec. 4.2.4).

4.2.1   Rigid-Rotor Vlasov Equilibria

For nonrelativistic particle dynamics, the single-particle constants of the motion are defined in Eq.(4.17). We consider here the class of self-consistent Vlasov equilibria of the form[1–6]

$$f_j^0(r, \mathbf{p}) = \left(\frac{\hat{n}_j}{2\pi m_j} \right) F_j(H_\perp - \omega_{rj}P_\theta)G_j(p_z) . \qquad (4.26)$$

Here, $\hat{n}_j$ and ω_{rj} are constants (independent of r), $p_z = m_j v_z$ is the axial momentum, and $P_\theta = r(p_\theta + \varepsilon_j m_j \omega_{cj} r/2)$ is the canonical angular momentum, where ε_j and ω_{cj} are defined by $\varepsilon_j = \mathrm{sgn}\, e_j$ and $\omega_{cj} = |e_j|\, B_0/m_j c$. Moreover, $H_\perp$ is the perpendicular energy defined by

$$H_\perp = \frac{1}{2m_j}\left(p_r^2 + p_\theta^2\right) + e_j \phi_0(r) . \qquad (4.27)$$

The (yet unspecified) functions F_j and G_j in Eq.(4.26) are normalized according to

$$\int_{-\infty}^{\infty} dp_z\, G(p_z) = 1 ,$$

$$\frac{1}{2\pi m_j}\left[\int_{-\infty}^{\infty} dp_r \int_{-\infty}^{\infty} dp_\theta F_j(H_\perp - \omega_{rj}P_\theta)\right]_{r=0} = 1 . \qquad (4.28)$$

Assuming $\phi_0(r = 0) = 0$ without loss of generality, and making use of $n_j^0(r) = \int d^3 p f_j^0(r, \mathbf{p})$, it follows from Eqs.(4.26) and (4.28) that the constant $\hat{n}_j \equiv n_j^0(r = 0)$ can be identified with the on-axis ($r = 0$) value of the density of the j'th plasma component.

It is convenient to express the combination $H_\perp - \omega_{rj}P_\theta$ occurring in Eq.(4.26) in the form

$$H_\perp - \omega_{rj}P_\theta = \frac{1}{2m_j}\left[p_r^2 + (p_\theta - m_j \omega_{rj} r)^2\right] + \psi_j(r) , \qquad (4.29)$$

where the effective potential $\psi_j(r)$ is defined by

$$\psi_j(r) = \frac{m_j}{2}\left(-\varepsilon_j \omega_{cj}\omega_{rj} - \omega_{rj}^2\right)r^2 + e_j\phi_0(r) . \tag{4.30}$$

From Eqs.(4.26) and (4.29), it is evident that the average azimuthal flow velocity $V_{\theta j}^0(r) = \left(\int d^3p\, v_\theta f_j^0\right) / \left(\int d^3p\, f_j^0\right)$ reduces to

$$V_{\theta j}^0(r) = \omega_{rj}r. \tag{4.31}$$

Therefore, for the class of equilibria described by Eq.(4.26), the average azimuthal motion of component j corresponds to a rigid rotation with angular velocity $\omega_{rj} = $ const. (hence, the terminology "rigid-rotor" equilibria). To calculate the various momentum moments of Eq.(4.26), it is convenient to change variables first to $p_r' = p_r$ and $p_\theta' = p_\theta - m_j r\omega_{rj}$, and then to $U = (2m_j)^{-1}(p_r'^2 + p_\theta'^2)$. Then, for example, the density profile $n_j^0(r) = \int d^3p\, f_j^0(r, \mathbf{p})$ can be expressed as

$$n_j^0(r) = \hat{n}_j \int_0^\infty dU\, F_j\left[U + \psi_j(r)\right] , \tag{4.32}$$

where $\psi_j(r)$ is defined in Eq.(4.30), and use has been made of Eqs.(4.26) and (4.28). Because $\psi_j(r = 0) = 0$, note that the normalization of $F_j(U)$ is $\int_0^\infty dU\, F_j(U) = 1$.

For specified $F_j(H_\perp - \omega_{rj}P_\theta)$, the density profile $n_j^0(r)$ is calculated in terms of $\psi_j(r)$, and therefore $\phi_0(r)$, from Eq.(4.32). The electrostatic potential $\phi_0(r)$ is then determined self-consistently from the equilibrium Poisson equation (4.9), which can be expressed as

$$\frac{1}{r}\frac{\partial}{\partial r}r\frac{\partial}{\partial r}\phi_0(r) = -\sum_j 4\pi e_j\hat{n}_j \int_0^\infty dU\, F_j\left[U + \psi_j(r)\right] , \tag{4.33}$$

where $\psi_j(r) = (m_j/2)(-\varepsilon_j\omega_{cj}\omega_{rj} - \omega_{rj}^2)r^2 + e_j\phi_0(r)$. In general, Eq.(4.33) is a nonlinear differential equation for $\phi_0(r)$ which depends on the specific form assumed for $F_j(H_\perp - \omega_{rj}P_\theta)$. In the absence of positive ions ($\hat{n}_i = 0$), note that the thermal equilibrium distribution function for the electrons defined in Eq.(3.28) corresponds to the particular choices $F_e(H_\perp - \omega_{re}P_\theta) = (k_B T_e)^{-1}\exp\left\{-(H_\perp - \omega_{re}P_\theta)/k_B T_e\right\}$ and $G_e(p_z) = (2\pi m_e k_B T_e)^{-1/2}\exp(-p_z^2/2m_e k_B T_e)$.

Near the axis of the plasma column ($r = 0$), Eq.(4.33) can be integrated to give $\phi_0(r) = -\left(\sum_k \pi \hat{n}_k e_k\right) r^2$, where use has been made of $\psi_j(r = 0) = 0$ and $\int_0^\infty dU\, F_j(U) = 1$. Substituting into Eq.(4.30) gives

$$\psi_j(r) = \frac{m_j}{2}\left(-\varepsilon_j \omega_{cj}\omega_{rj} - \omega_{rj}^2 - \frac{2e_j}{m_j}\sum_k \pi \hat{n}_k e_k\right) r^2$$

$$= \frac{m_j}{2}\left(\omega_{rj}^+ - \omega_{rj}\right)\left(\omega_{rj} - \omega_{rj}^-\right) r^2 \qquad (4.34)$$

for small values of r. Here, ω_{rj}^+ and ω_{rj}^- are the laminar rotation frequencies defined by

$$\omega_{rj}^\pm \equiv -\frac{\varepsilon_j \omega_{cj}}{2}\left\{1 \pm \left(1 - \sum_k \frac{8\pi e_j \hat{n}_k e_k}{m_j \omega_{cj}^2}\right)^{1/2}\right\}. \qquad (4.35)$$

In Eqs.(4.34) and (4.35), note that the strength of the equilibrium self-electric field acting on component j is measured by the dimensionless self-field parameter $s_j \equiv (8\pi e_j/m_j\omega_{cj}^2)\sum_k \hat{n}_k e_k$. Note also that Eq.(4.35) is the natural generalization of the electron rotation frequencies $\omega_{re}^\pm$ defined in Eq.(3.6) to the case of a multicomponent nonneutral plasma column (Problem 4.2).

For the sake of definiteness, we consider a two-component nonneutral plasma column consisting of electrons ($j = e$) and positive ions ($j = i$). It is further assumed that the plasma is electron rich with $\sum_k \hat{n}_k e_k < 0$. Then the polarities in Eq.(4.35) correspond to s_e, ω_{re}^+, ω_{re}^-, $\omega_{ri}^- > 0$ and s_i, $\omega_{ri}^+ < 0$. Examination of the expression for $n_j^0(r)$ in Eq.(4.32) shows that the density profile decreases monotonically with r and that $n_j^0(r \to \infty) = 0$ only if $\psi_j(r)$ is positive and $\psi_j(r \to \infty) \to \infty$. This necessarily requires $\psi_j''(r = 0) > 0$, where use has been made of $\psi_j(r = 0) = 0$ and $\psi_j'(r = 0) = 0$. Therefore, examination of Eq.(4.34) for an electron-rich plasma with $\sum_k \hat{n}_k e_k < 0$ shows that the necessary and sufficient conditions for radially confined electron and ion equilibria can be expressed as

$$\omega_{re}^- < \omega_{re} < \omega_{re}^+ ,$$

$$\omega_{ri}^+ < \omega_{ri} < \omega_{ri}^- . \qquad (4.36)$$

That is, $\partial n_j^0/\partial r \leq 0$ and $n_j^0(r \to \infty) = 0$ for $j = e, i$, whenever the angular rotation velocities, ω_{re} and ω_{ri}, lie within the intervals defined in Eq.(4.36). Equation (4.36) generalizes the inequality in Eq.(3.32) to the case of a two-component nonneutral plasma and general rigid-rotor equilibrium $F_j(H_\perp - \omega_{rj}P_\theta)$.

If ω_{rj} is very closely tuned to either ω_{rj}^+ or ω_{rj}^-, it also follows from Eqs.(4.32) and (4.34) that the density profile $n_j^0(r)$ is approximately uniform (and equal to $\hat{n}_j$) over a broad interior region of the plasma column, and decreases abruptly to zero at some radius r_b. That is,

$$n_j^0(r) \simeq \begin{cases} \hat{n}_j , & 0 \leq r \lesssim r_b , \\ 0 , & r > r_b , \end{cases} \tag{4.37}$$

whenever

$$\omega_{re} = \omega_{re}^+(1 - \varepsilon) \text{ or } \omega_{re} = \omega_{re}^-(1 + \varepsilon) ,$$

$$\omega_{ri} = \omega_{ri}^-(1 - \varepsilon) \text{ or } \omega_{ri} = \omega_{ri}^+(1 + \varepsilon) ,$$

$$\text{for positive } \varepsilon \ll 1 . \tag{4.38}$$

Equation (4.37) follows from Eqs.(4.32) and (4.34) because $\psi_j(r) \simeq 0(\varepsilon)$ within the plasma column whenever Eq.(4.38) is satisfied. The class of rigid-rotor Vlasov equilibria described by Eq.(4.26) and characterized by uniform density in the column interior [Eqs.(4.37) and (4.38)] is particularly tractable for detailed investigations of kinetic stability behavior (Secs. 4.6–4.9).

Other equilibrium properties are readily calculated for the class of rigid-rotor Vlasov equilibria described by Eq.(4.26). For example, because $H_\perp$ is an even function of p_r, it is readily shown that the average radial velocity is $V_{jr}^0(r) = \left(\int d^3p\, v_r f_j^0\right) / \left(\int d^3p\, f_j^0\right) = 0$. Similarly, making use of the definition of the pressure tensor $\boldsymbol{P}_j^0(\mathbf{x})$ in Eq.(4.15), it can be shown from Eq.(4.26) that all off-diagonal elements of the pressure tensor are identically zero. Some straightforward algebra shows that the pressure tensor for the class of rigid-rotor Vlasov equilibria in Eq.(4.26) can be expressed as

$$\boldsymbol{P}_j^0(\mathbf{x}) = n_j^0(r)T_{j\perp}(r)\left[\hat{\mathbf{e}}_r\hat{\mathbf{e}}_r + \hat{\mathbf{e}}_\theta\hat{\mathbf{e}}_\theta\right] + n_j^0(r)T_{jz}\hat{\mathbf{e}}_z\hat{\mathbf{e}}_z .$$

Here, the effective perpendicular and parallel temperatures, $T_{j\perp}(r)$ and T_{jz}, are defined by

$$T_{j\perp}(r) = \frac{\int_0^\infty dU\, U F_j\, [U + \psi_j(r)]}{\int_0^\infty dU\, F_j\, [U + \psi_j(r)]} \,, \qquad (4.39)$$

and

$$T_{jz} = m_j^{-1} \int_{-\infty}^\infty dp_z \left(p_z - \langle p_z \rangle_j\right)^2 G_j(p_z) \,, \qquad (4.40)$$

where $\langle p_z \rangle_j = \int_{-\infty}^\infty dp_z p_z G_j(p_z)$ is the average axial momentum of component j, and a variable change to $U = (2m_j)^{-1}\left[p_r^2 + (p_\theta - m_j\omega_{rj}r)^2\right]$ has been made in obtaining Eq.(4.39). In Eqs.(4.39) and (4.40), note that $T_{j\perp}(r)$ and T_{jz} have units of kinetic energy per particle. (In effect, Boltzmann's constant has been absorbed into the definitions of $T_{j\perp}$ and T_{jz}.)

In conclusion, for specified distribution function $F_j(H_\perp - \omega_{rj}P_\theta)$, we reiterate that Poisson's equation (4.33) is generally a nonlinear differential equation for the electrostatic potential $\phi_0(r)$ which must be solved numerically. Once $\phi_0(r)$ is determined, however, the function $\psi_j(r)$ is known [Eq.(4.30)], and other equilibrium properties can be calculated self-consistently, such as the perpendicular temperature profile $T_{j\perp}(r)$ defined in Eq.(4.39). As a final point, some straightforward algebra shows that radial force balance on a fluid element for the class of rigid-rotor Vlasov equilibria in Eq.(4.26) can be expressed as (Problem 4.4)

$$-n_j^0(r)m_j\left[\left(\omega_{rj}^2 + \varepsilon_j\omega_{cj}\omega_{rj}\right)r + \frac{e_j}{m_j}\sum_k \frac{4\pi e_k}{r}\int_0^r dr'\, r' n_k^0(r')\right]$$

$$+ \frac{\partial}{\partial r}\left[n_j^0(r)T_{j\perp}(r)\right] = 0 \,. \qquad (4.41)$$

Here, $\sum_k(4\pi e_k/r)\int_0^r dr'\, r' n_k^0(r') = E_r(r) = -\partial\phi_0/\partial r$ is the radial electric field, and $n_j^0(r)$ and $T_{j\perp}(r)$ are defined in Eqs.(4.32) and (4.39), respectively. Equation (4.41) is simply a statement of radial force balance on a fluid element among the centrifugal, magnetic, electric, and pressure-gradient forces. In the region where the pressure-gradient force is weak (e.g., small $T_{j\perp}$) and the density profile is approximately uniform with $n_j^0(r) \simeq \hat{n}_j$, it is evident that Eq.(4.41) can be approximated by $\omega_{rj}^2 + \varepsilon_j\omega_{cj}\omega_{rj} + (e_j/m_j)\sum_k 2\pi\hat{n}_k e_k = (\omega_{rj} - \omega_{rj}^+)(\omega_{rj} - \omega_{rj}^-) \simeq 0$, where

ω_{rj}^+ and ω_{rj}^- are the laminar (cold-fluid) rotation frequencies defined in Eq.(4.35).

Problem 4.2   Consider a two-component nonneutral plasma consisting of electrons ($j = e$) and singly-ionized ions ($j = i$) that provide partial charge neutralization with $\hat{n}_i = f\hat{n}_e$ on axis ($r = 0$). Here, $f = $ const. is the fractional charge neutralization, and $f \leq 1$ is assumed.

a.   Show that the laminar rotation frequencies $\omega_{rj}^\pm$ defined in Eq.(4.35) can be expressed as

$$\omega_{re}^\pm = \frac{\omega_{ce}}{2}\left\{1 \pm \left(1 - \frac{2\omega_{pe}^2}{\omega_{ce}^2}(1 - f)\right)^{1/2}\right\} , \qquad (4.2.1)$$

and

$$\omega_{ri}^\pm = -\frac{\omega_{ci}}{2}\left\{1 \pm \left(1 + \frac{2\omega_{pe}^2}{\omega_{ce}^2}(1 - f)\frac{m_i}{m_e}\right)^{1/2}\right\} , \qquad (4.2.2)$$

where $\omega_{pj}^2 = 4\pi\hat{n}_j e_j^2/m_j$, $\omega_{cj} = |e_j| B_0/m_j c$, $e_j = +e$ for the ions, and $e_j = -e$ for the electrons.

b.   Comparing Eqs.(3.6) and (4.2.1), it is evident that the plots of ω_{re}^+ and ω_{re}^- versus normalized density are identical in form to those in Figs. 3.3 and 3.9 with the abscissa replaced by the self-field parameter $s_e = (2\omega_{pe}^2/\omega_{ce}^2) \times (1 - f)$. The maximum allowed density in Eq.(4.2.1) corresponds to $s_e = 1$. For $f \leq 1$, it follows from Eq.(4.2.2) that $\omega_{ri}^+ < 0$ whereas $\omega_{ri}^- > 0$. For $s_e \ll m_e/m_i$ show that

$$\omega_{ri}^- \simeq \frac{m_i}{4m_e}s_e\omega_{ci} ,$$

$$\omega_{ri}^+ \simeq -\omega_{ci} . \qquad (4.2.3)$$

On the other hand, at the maximum allowed density ($s_e = 1$), show that $\omega_{ri}^\pm$ can be approximated by

$$\omega_{ri}^\pm \simeq \mp\frac{1}{2}\left(\omega_{ce}\omega_{ci}\right)^{1/2} . \qquad (4.2.4)$$

Problem 4.3

a.   Make use of Poisson's equation (4.33) and $\psi_j(r) = (m_j/2) \times (-\varepsilon_j\omega_{cj}\omega_{rj} - \omega_{rj}^2)r^2 + e_j\phi_0(r)$ to show that $\psi_j(r)$ satisfies

$$\frac{1}{r}\frac{\partial}{\partial r}r\frac{\partial}{\partial r}\psi_j(r) = 2m_j(\omega_{rj}^+ - \omega_{rj})(\omega_{rj} - \omega_{rj}^-)$$

$$+ \sum_k 4\pi\hat{n}_k e_k e_j\left[1 - \int_0^\infty dU\, F\left[U + \psi_k(r)\right]\right] . \qquad (4.3.1)$$

Here, ω_{rj}^+ and ω_{rj}^- are defined in Eq.(4.35).

b. Make use of $\psi_j(r=0)=0$, $\partial\psi_j/\partial r|_{r=0}=0$, and $\int_0^\infty dU\, F_j(U)=1$ to show from Eq.(4.3.1) that

$$\left[\frac{\partial^2}{\partial r^2}\psi_j(r)\right]_{r=0} = 2m_j\left(\omega_{rj}^+ - \omega_{rj}\right)\left(\omega_{rj} - \omega_{rj}^-\right). \qquad (4.3.2)$$

Therefore, for an electron-rich plasma (say), it follows from Eq.(4.3.2) that the effective potential $\psi_j(r)$ increases away from $r=0$ for both electrons $(j=e)$ and ions $(j=i)$ provided ω_{re} and ω_{ri} lie in the intervals defined in Eq.(4.36).

Problem 4.4   Equations (4.32) and (4.39) can be combined to give

$$n_j^0(r)T_{j\perp}(r) = \hat{n}_j \int_0^\infty dU\, U F_j\left[U + \psi_j(r)\right]. \qquad (4.4.1)$$

Equation (4.4.1) constitutes a formal expression for the perpendicular pressure $n_j^0(r)T_{j\perp}(r)$ in terms of the rigid-rotor equilibrium $F_j(H_\perp - \omega_{rj}P_\theta)$.

a. Take the derivative of Eq.(4.4.1) with respect to r, and integrate by parts with respect to U enforcing $F_j(U\to\infty)=0$. Show that

$$\frac{\partial}{\partial r}\left[n_j^0(r)T_{j\perp}(r)\right] = -\frac{\partial\psi_j(r)}{\partial r}\hat{n}_j \int_0^\infty dU\, F_j\left[U + \psi_j(r)\right]. \qquad (4.4.2)$$

b. Make use of the definitions of $\psi_j(r)$, $n_j^0(r)$, etc., to show that the radial force balance equation (4.41) follows directly from Eq.(4.4.2).

4.2.2   Density Inversion Theorem

For the class of rigid-rotor Vlasov equilibria in Eq.(4.26), the dependence of $f_j^0(r,\mathbf{p})$ on radius r and momentum $\mathbf{p}$ are strongly interconnected. Indeed, a measurement or specification of the radial dependence of a macroscopic profile, such as the density profile $n_j^0(r)$, is sufficient information to reconstruct the detailed functional form of the perpendicular distribution function $F_j(H_\perp - \omega_{rj}P_\theta)$.[6]

To illustrate this property, we specialize to the case of a one-component nonneutral plasma consisting only of electrons with charge $e_j = -e$. Assume that the values of $\omega_{ce} = eB_0/m_e c$ and $\omega_{re} = $ const. are known. Also assume that $n_e^0(r)$ is a known (measured) function of r. Then from

Poisson's equation, $r^{-1}(\partial/\partial r)\,[r\partial\phi_0/\partial r] = 4\pi e n_e^0(r)$, the electrostatic potential profile is given by

$$\phi_0(r) = 4\pi e \int_0^r \frac{dr'}{r'} \int_0^{r'} dr''\, r'' n_e^0(r'') \, . \tag{4.42}$$

Therefore, for specified $n_e^0(r)$, the effective potential $\psi_e(r)$ defined in Eq.(4.30) has the known radial dependence given by

$$\psi_e(r) = \frac{m_e}{2}\left(\omega_{ce}\omega_{re} - \omega_{re}^2\right)r^2 - 4\pi e^2 \int_0^r \frac{dr'}{r'} \int_0^{r'} dr''\, r'' n_e^0(r'') \, . \tag{4.43}$$

Furthermore, for $\psi_e(r)$ a monotonically increasing function of r, Eq.(4.43) can be inverted to give $r = r(\psi_e)$, where $r(\psi_e)$ is a single-valued function of ψ_e. This implies that $n_e^0(r)$, or equivalently $n_e^0(\psi_e)$, is a single-valued function of ψ_e.

Now consider Eq.(4.32), which relates the perpendicular distribution function F_e to the density profile $n_e^0(r)$. Taking the derivative of Eq.(4.32) with respect to ψ_e gives

$$\frac{\partial n_e^0}{\partial \psi_e} = \hat{n}_e \int_0^\infty dU\, \frac{\partial}{\partial U} F_e(U + \psi_e)$$

$$= -\hat{n}_e F_e(\psi_e) \, . \tag{4.44}$$

In obtaining Eq.(4.44), we have integrated by parts with respect to U and made use of $F_e(U \to \infty) = 0$. Because $n_e^0(r)$ is a known function of r and therefore of ψ_e, Eq.(4.44) gives a closed expression for the perpendicular distribution function $F_e(H_\perp - \omega_{re}P_\theta)$. In particular,

$$\hat{n}_e F_e(H_\perp - \omega_{re}P_\theta) = \left[-\frac{\partial n_e^0}{\partial \psi_e}\right]_{\psi_e = H_\perp - \omega_{re}P_\theta} \tag{4.45}$$

follows directly from Eq.(4.44), which completes the proof.

Specific applications of the inversion theorem in Eq.(4.45) are described in Problems 4.5 and 4.6. For example, if $n_e^0(r)$ corresponds to a uniform density profile out to radius r_b, then it can be shown that the equilibrium distribution function is necessarily of the form

$$F_e(H_\perp - \omega_{re}P_\theta) = \delta(H_\perp - \omega_{re}P_\theta - C_\perp)$$

where $C_\perp$ is a positive constant (see Problem 4.6).

Problem 4.5   Consider a one-component nonneutral plasma consisting only of electrons. Assume that the density profile is prescribed by

$$
n_e^0(r) =
\begin{cases}
\hat{n}_e \left(1 - \dfrac{r^2}{r_b^2}\right) , & 0 \leq r < r_b , \\[2mm]
0 , & r > r_b ,
\end{cases}
\tag{4.5.1}
$$

a. Calculate the electrostatic potential $\phi_0(r)$ within the plasma column ($0 \leq r < r_b$) from Poisson's equation and Eq.(4.5.1). Show that the effective potential $\psi_e(r) = (m_e/2)(\omega_{ce}\omega_{re} - \omega_{re}^2)r^2 - e\phi_0(r)$ can be expressed as

$$
\psi_e(r) = \frac{1}{2} m_e r_b^2 \left\{ \left(\omega_{re}\omega_{ce} - \omega_{re}^2\right) \frac{r^2}{r_b^2} - \frac{1}{2}\omega_{pe}^2 \frac{r^2}{r_b^2}\left(1 - \frac{1}{4}\frac{r^2}{r_b^2}\right) \right\}
\tag{4.5.2}
$$

for $0 \leq r < r_b$. Here, $\omega_{pe}^2 = 4\pi\hat{n}_e e^2/m_e$.

b. Invert Eq.(4.5.2) and show that the solution for $r^2(\psi_e)$ is given by

$$
\frac{r^2(\psi_e)}{r_b^2} = \left\{ \left[\frac{(\omega_{re}^+ - \omega_{re})(\omega_{re} - \omega_{re}^-)}{\omega_{pe}^2/4} \right]^2 + \frac{16\psi_e}{m_e\omega_{pe}^2 r_b^2} \right\}^{1/2}
$$

$$
- \left[\frac{(\omega_{re}^+ - \omega_{re})(\omega_{re} - \omega_{re}^-)}{\omega_{pe}^2/4} \right] .
\tag{4.5.3}
$$

Here, $\omega_{re}^- < \omega_{re} < \omega_{re}^+$ and $\omega_{re}^\pm \equiv (\omega_{ce}/2)\left\{1 \pm (1 - 2\omega_{pe}^2/\omega_{ce}^2)^{1/2}\right\}$.

c. Make use of Eqs.(4.45), (4.5.1), and (4.5.3) to obtain a closed expression for the equilibrium distribution function $F_e(H - \omega_{re}P_\theta)$ consistent with the density profile $n_e^0(r)$ in Eq.(4.5.1).

4.2.3   Uniform-Density Plasma Column

As an example of a rigid-rotor Vlasov equilibrium, consider a nonneutral plasma column consisting only of electrons ($j = e$) in which the perpendicular distribution function occurring in Eq.(4.26) is specified by

$$
F_e\left(H_\perp - \omega_{re}P_\theta\right) = \delta\left(H_\perp - \omega_{re}P_\theta - \hat{T}_{e\perp}\right) ,
\tag{4.46}
$$

where $\hat{T}_{e\perp}$ is a positive constant. In Eq.(4.46), $H_\perp - \omega_{re}P_\theta = U + \psi_e(r)$, where $\psi_e(r)$ is the effective potential defined in Eq.(4.30), and $U = (2m_e)^{-1}\left[p_r^2 + (p_\theta - m_e\omega_{re}r)^2\right]$ is the perpendicular kinetic energy in the rotating frame. The form of F_e in Eq.(4.46) corresponds to a loss-cone equilibrium in which the distribution in U is highly peaked at $\hat{T}_{e\perp} - \psi_e(r)$. From Eqs.(4.32) and (4.46), the electron density profile is given by

$$n_e^0(r) = \hat{n}_e \int_0^\infty dU \, \delta \left[U + \psi_e(r) - \hat{T}_{e\perp} \right] , \qquad (4.47)$$

where $\psi_e(r) = (m_e/2)(\omega_{ce}\omega_{re} - \omega_{re}^2)r^2 - e\phi_0(r)$. The integral over U in Eq.(4.47) is equal to unity when $\psi_e(r) < \hat{T}_{e\perp}$ and equal to zero when $\psi_e(r) > \hat{T}_{e\perp}$. Therefore Eq.(4.47) gives the rectangular density profile (Fig. 4.4)

$$n_e^0(r) = \begin{cases} \hat{n}_e = \text{const.} , & 0 \leq r < r_b , \\ 0 , & r > r_b , \end{cases} \qquad (4.48)$$

where the radius r_b of the plasma column is determined self-consistently from $\psi_e(r_b) = \hat{T}_{e\perp}$, or equivalently

$$\frac{m_e}{2} \left(\omega_{re}\omega_{ce} - \omega_{re}^2 \right) r_b^2 - e\phi_0(r_b) = \hat{T}_{e\perp} . \qquad (4.49)$$

For the uniform density profile in Eq.(4.48), Poisson's equation (4.33) gives $\phi_0(r) = (m_e/4e)\omega_{pe}^2 r^2$ within the plasma column ($0 \leq r < r_b$). Here, $\omega_{pe}^2 = 4\pi\hat{n}_e e^2/m_e$ is the electron plasma frequency-squared. Substituting into Eq.(4.49) and solving for r_b^2 gives the closed expression

$$\begin{aligned} r_b^2 &= \frac{(2\hat{T}_{e\perp}/m_e)}{(\omega_{re}\omega_{ce} - \omega_{re}^2 - \omega_{pe}^2/2)} \\ &= \frac{(2\hat{T}_{e\perp}/m_e)}{(\omega_{re}^+ - \omega_{re})(\omega_{re} - \omega_{re}^-)} . \end{aligned} \qquad (4.50)$$

Here, $\omega_{re}^\pm \equiv (\omega_{ce}/2) \left\{ 1 \pm (1 - 2\omega_{pe}^2/\omega_{ce}^2)^{1/2} \right\}$ are the laminar rotation frequencies defined in Eq.(4.35) [or Eq.(3.6)] for a uniform-density non-neutral electron plasma. As expected, it is required that ω_{re} lie in the interval $\omega_{re}^- < \omega_{re} < \omega_{re}^+$ for existence of a radially confined equilibrium. Moreover, whenever ω_{re} is closely tuned to ω_{re}^+ or ω_{re}^- according to Eq.(4.38) [or Eq.(3.33)], it follows from Eq.(4.50) that $r_b \gg \lambda_{De}$, where $\lambda_{De} = (2\hat{T}_{e\perp}/m_e\omega_{pe}^2)^{1/2}$ is the effective Debye length.

Other equilibrium properties are readily calculated for the choice of perpendicular distribution function in Eq.(4.46). Making use of $\phi_0(r) = (m_e/4e)\omega_{pe}^2 r^2$ within the plasma column, it follows from Eqs.(4.30) and (4.50) that $\psi_e(r) = \hat{T}_{e\perp} r^2/r_b^2$ for $0 \leq r < r_b$. Substituting this

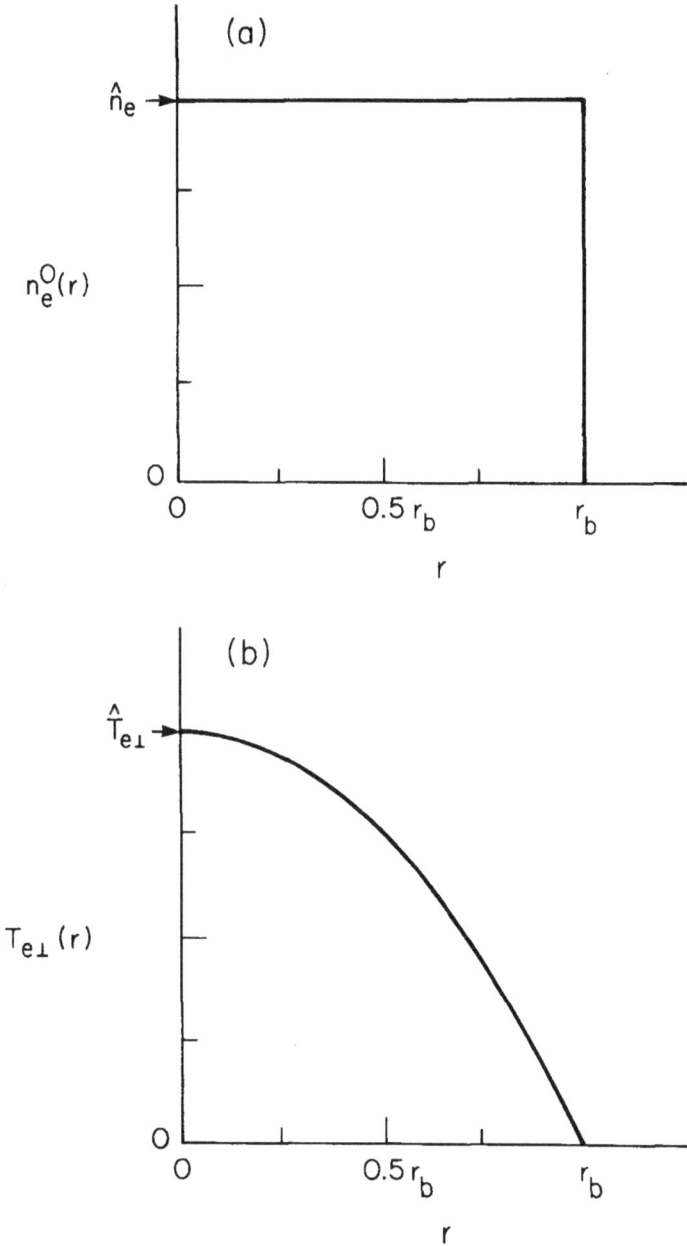

Figure 4.4. Plots versus radius r of (a) the electron density profile [Eq.(4.48)] and (b) the perpendicular temperature profile [Eq.(4.51)] obtained for the choice of rigid-rotor Vlasov equilibrium in Eq.(4.46).

expression for $\psi_e(r)$ into Eq.(4.46) and evaluating the perpendicular temperature profile $T_{e\perp}(r)$ in Eq.(4.39) readily gives (Fig. 4.4)

$$T_{e\perp}(r) = \hat{T}_{e\perp} \left(1 - \frac{r^2}{r_b^2} \right) \qquad (4.51)$$

for $0 \leq r < r_b$. Therefore, for the choice of distribution function in Eq.(4.46), the (parabolic) temperature profile decreases monotonically from a maximum value of $\hat{T}_{e\perp}$ at $r = 0$ to zero at the boundary of the plasma column $(r = r_b)$. This is expected because the delta-function in Eq.(4.46) selects $U = \hat{T}_{e\perp} - \psi_e(r) = T_{e\perp}(1 - r^2/r_b^2)$, so that $r = r_b$ is the envelope of turning points (where $U = 0$) for the perpendicular orbits.

Problem 4.6   Consider a one-component nonneutral plasma consisting only of electrons. Assume that the equilibrium density profile is prescribed by the rectangular form in Eq.(4.48).

a.   Show that $\psi_e(r) = (m_e/2)(\omega_{re}\omega_{ce} - \omega_{re}^2) - e\phi_0(r)$ can be expressed as

$$\psi_e(r) = \frac{1}{2}m_e \left(\omega_{re}^+ - \omega_{re} \right) \left(\omega_{re} - \omega_{re}^- \right) r^2 \qquad (4.6.1)$$

within the plasma column $(0 \leq r < r_b)$. Here, $\omega_{re}^- < \omega_{re} < \omega_{re}^+$ and $\omega_{re}^\pm \equiv (\omega_{ce}/2) \left\{ 1 \pm (1 - 2\omega_{pe}^2/\omega_{ce}^2)^{1/2} \right\}$.

b.   Make use of Eqs.(4.45), (4.48), and (4.6.1) to show that the equilibrium distribution function is necessarily of the form

$$F_e \left(H_\perp - \omega_{re}P_\theta \right) = \delta \left(H_\perp - \omega_{re}P_\theta - C_\perp \right), \qquad (4.6.2)$$

where $C_\perp = (m_e/2)(\omega_{re}^+ - \omega_{re})(\omega_{re} - \omega_{re}^-)r_b^2 = $ const. [compare with Eqs.(4.46) and (4.50)].

4.2.4   Annular Electron Layer

The steady-state Vlasov-Maxwell formalism in Sec. 4.1.2 can also be used to describe the equilibrium properties of nonneutral plasmas with annular (hollow) density profiles. Equilibrium profiles have been constructed ranging from sharp-boundary, relativistic electron layers[49-51] with a strong diamagnetic field component $B_z^s(r)$, to nonrelativistic equilibria[1-6] with sharp-boundary or diffuse density profiles peaked off-axis. In this section, we present a simple example corresponding to an annular electron layer with sharp radial boundaries. No positive ions are present $(f_i^0 = 0)$, and

the equilibrium distribution function for the electrons is assumed to be of
the form

$$f_e^0(r, \mathbf{p}) = \frac{\hat{n}_e r_b^-}{2\pi m_e}\delta(H - V_e p_z - H_0 + m_e V_e^2/2)\delta(P_\theta - P_0) , \qquad (4.52)$$

where $\hat{n}_e, r_b^-$ and H_0 are positive constants, V_e is a constant, and $\omega_{ce} P_0 <$
H_0 is assumed. In Eq.(4.52), $p_z = m_e v_z$ is the axial momentum, and the
total energy H and canonical angular momentum P_θ are defined by

$$H = \frac{1}{2m_e}\left(p_r^2 + p_\theta^2 + p_z^2\right) - e\phi_0(r) ,$$

$$P_\theta = r\left(p_\theta - m_e r\omega_{ce}/2\right) , \qquad (4.53)$$

where $\omega_{ce} = eB_0/m_e c$. A distribution function of the form in Eq.(4.52)
can be formed by injecting a low-energy electron beam through a magnetic
cusp field[52] into a uniform guide-field region with $\mathbf{B}^0(\mathbf{x}) = B_0 \hat{\mathbf{e}}_z$. The
cusp field induces a rotation of the electron beam, and Eq.(4.52) assumes a
negligible spread in canonical angular momentum P_θ and energy $H - V_e p_z$
(in a frame of reference moving with axial velocity V_e).

For $P_\theta = P_0$, we eliminate $p_\theta = P_0/r + m_e r\omega_{ce}/2$ and express

$$H - V_e p_z - H_0 + m_e V_e^2/2$$

$$= \frac{1}{2m_e}\left[p_r^2 + (p_z - m_e V_e)^2\right] - \frac{1}{2m_e}\psi_e(r) . \qquad (4.54)$$

Here, the envelope function $\psi_e(r)$ [not to be confused with Eq.(4.30)] is
defined by

$$(2m_e)^{-1}\psi_e(r) = H_0 + e\phi_0(r) - \frac{1}{2}m_e\left(\frac{P_0}{m_e r} + \frac{r\omega_{ce}}{2}\right)^2 . \qquad (4.55)$$

To calculate momentum moments, the distribution function in Eq.(4.52)
can be expressed as

$$f_e^0(r, \mathbf{p}) = \frac{\hat{n}_e}{\pi}\frac{r_b^-}{r}\delta\left[p_r^2 + (p_z - m_e V_e)^2 - \psi_e(r)\right]$$

$$\times \delta\left[p_\theta - \left(\frac{P_0}{r} + \frac{m_e r\omega_{ce}}{2}\right)\right] . \qquad (4.56)$$

The moments of Eq.(4.56) corresponding to the average flow velocities, $V_{\theta e}^0(r) = (\int d^3p \, v_\theta f_e^0)/(\int d^3p \, f_e^0)$ and $V_{ze}^0(r) = (\int d^3p \, v_z f_e^0)/(\int d^3p \, f_e^0)$, readily give

$$V_{\theta e}^0(r) = \frac{P_0}{m_e r} + \frac{1}{2}r\omega_{ce} \, , \tag{4.57}$$

and

$$V_{ze}^0(r) = V_e = \text{const.} \, , \tag{4.58}$$

over the radial extent of the electron layer. For $P_0 \neq 0$, note from Eq.(4.57) that the angular velocity shear is nonzero for the choice of distribution function in Eq.(4.56), i.e., $\partial\omega_{re}(r)/\partial r \neq 0$, where $\omega_{re}(r) \equiv V_{\theta e}^0(r)/r$.

To calculate the electron density profile $n_e^0(r) = \int d^3p \, f_e^0$, we first integrate Eq.(4.56) over p_θ and make the variable change to $p_\perp^2 = p_r^2 + (p_z - m_e V_e)^2$ and $\int dp_r \int dp_z \cdots = 2\pi \int_0^\infty dp_\perp p_\perp \cdots$. (Here, the subscript $\perp$ means perpendicular to the θ-direction.) This readily gives

$$n_e^0(r) = \hat{n}_e \frac{r_b^-}{r} \int_0^\infty dp_\perp^2 \, \delta\left[p_\perp^2 - \psi_e(r)\right] \, , \tag{4.59}$$

where the envelope function $\psi_e(r)$ is defined in Eq.(4.55). The integral over $p_\perp^2$ in Eq.(4.59) is equal to unity whenever $\psi_e(r) > 0$ and zero whenever $\psi_e(r) < 0$. That is, electron motion is allowed only for $\psi_e(r) \geq 0$, with the condition

$$\psi_e(r_b) = 0 \tag{4.60}$$

determining the envelope of radial turning points in the r-z motion. For $0 < \omega_{ce}P_0 < H_0$, there are two values of r_b which solve Eq.(4.60). The smallest value of r_b (denote by r_b^-) that solves Eq.(4.60) is the inner radius of the electron layer, and the largest value of r_b (denote by r_b^+) is the outer radius (Fig. 4.5). Carrying out the integration in Eq.(4.59) gives the annular density profile

$$n_e^0(r) = \begin{cases} 0 \, , & 0 \leq r < r_b^- \, , \\[2mm] \hat{n}_e \dfrac{r_b^-}{r} \, , & r_b^- < r < r_b^+ \, , \\[2mm] 0 \, , & r > r_b^+ \, , \end{cases} \tag{4.61}$$

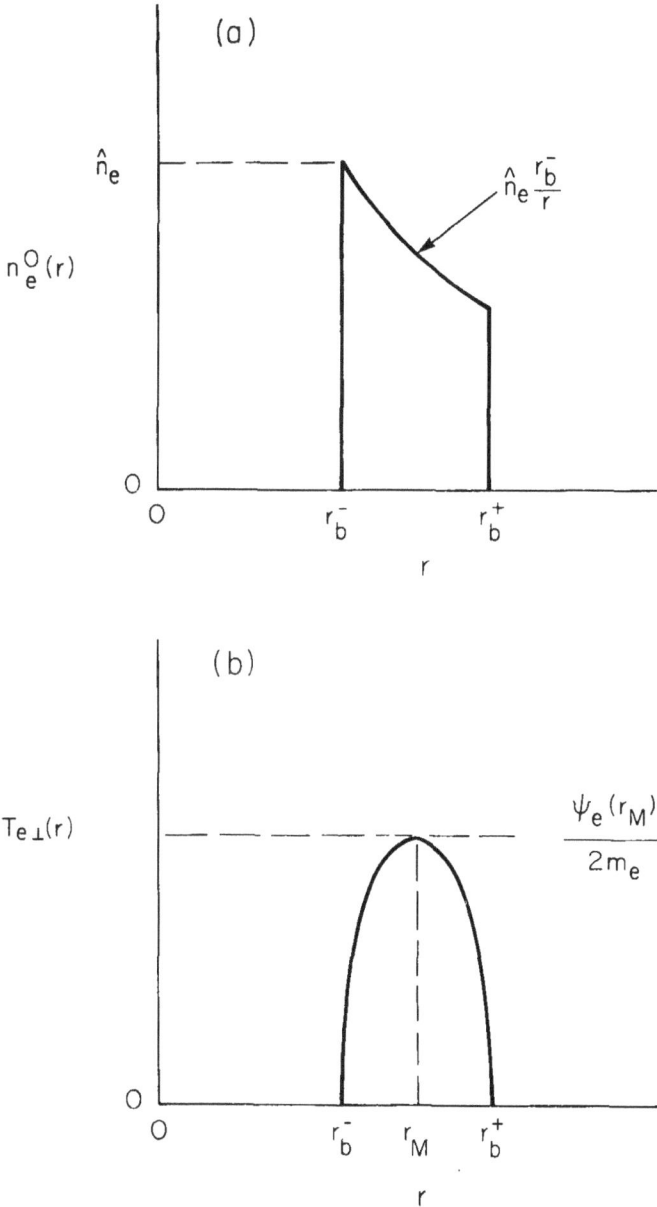

(a)

$\hat{n}_e$

$n_e^0(r)$

$\hat{n}_e \dfrac{r_b^-}{r}$

0

r_b^- r_b^+

r

(b)

$T_{e\perp}(r)$

$\dfrac{\psi_e(r_M)}{2m_e}$

0

r_b^- r_M r_b^+

r

Figure 4.5.   Schematic plots versus radius r of (a) the annular electron density profile [Eq.(4.61)] and (b) the perpendicular temperature profile [Eq.(4.65)] obtained for the choice of Vlasov equilibrium in Eq.(4.52).

where r_b^- and r_b^+ solve Eq.(4.60), and $n_e^0(r = r_b^-) = \hat{n}_e$. Note that the density profile in Eq.(4.61) has sharp radial boundaries at r_b^- and r_b^+, and falls off as $1/r$ within the electron layer. Substituting Eq.(4.61) into the equilibrium Poisson equation, $r^{-1}(\partial/\partial r)\left[r\partial\phi_0/\partial r\right] = 4\pi e n_e^0(r)$, and solving for the electrostatic potential $\phi_0(r)$, we obtain

$$
\phi_0(r) = \begin{cases}
0\,, & 0 \le r < r_b^-\,, \\[2mm]
4\pi e \hat{n}_e r_b^- \left[r - r_b^- - r_b^- \ln(r/r_b^-)\right]\,, & r_b^- < r < r_b^+\,, \\[2mm]
4\pi e \hat{n}_e r_b^- \left[r_b^+ - r_b^- - r_b^- \ln(r_b^+/r_b^-)\right. \\[1mm]
\left. + (r_b^+ - r_b^-)\ln(r/r_b^+)\right]\,, & r > r_b^+\,.
\end{cases}
\tag{4.62}
$$

In obtaining Eq.(4.62), continuity of $\phi_0(r)$ and $\partial\phi_0(r)/\partial r$ at r_b^- and r_b^+ have been enforced, and $\phi_0(r) = 0$ has been assumed for $0 \le r < r_b^-$.
To complete the specification of the equilibrium, Eq.(4.62) is substituted into the definition of $\psi_e(r)$ in Eq.(4.55). The solutions for the boundary locations ($r_b = r_b^-$ and $r_b = r_b^+$) are then determined from the envelope equation $\psi_e(r_b) = 0$, which can be expressed as

$$
0 = (2m_e)^{-1}\psi_e(r_b) = H_0 + m_e\omega_{pe}^2 r_b^- \left[r_b - r_b^- - r_b^- \ln(r_b/r_b^-)\right]
$$
$$
- \frac{1}{2}m_e\left(\frac{P_0}{m_e r_b} + \frac{r_b\omega_{ce}}{2}\right)^2\,.
\tag{4.63}
$$

The term proportional to $\omega_{pe}^2 = 4\pi\hat{n}_e e^2/m_e$ in Eq.(4.63) describes the influence of self-electric fields on the solutions for r_b. Note that this term vanishes at the inner boundary ($r_b = r_b^-$) because $\phi_0(r_b^-) = 0$. Generally speaking, the envelope equation (4.63) is highly nonlinear and must be solved numerically for the location of the outer boundary ($r_b = r_b^+$). For a tenuous electron layer with $m_e\omega_{pe}^2 r_b^-(r_b^+ - r_b^-) \ll H_0$, however, Eq.(4.63) can be solved iteratively, treating the contribution proportional to ω_{pe}^2 as a small effect.

The equilibrium pressure tensor $\boldsymbol{P}_e^0(\mathbf{x})$ defined in Eq.(4.15) can also be calculated for the choice of distribution function in Eq.(4.52) [or equivalently, Eq.(4.56)]. All off-diagonal elements and the θ-θ element of $\boldsymbol{P}_e^0(\mathbf{x})$ are identically zero. Some straightforward algebra shows that

$$
\boldsymbol{P}_e^0(\mathbf{x}) = n_e^0(r)T_{e\perp}(r)(\hat{\mathbf{e}}_r\hat{\mathbf{e}}_r + \hat{\mathbf{e}}_z\hat{\mathbf{e}}_z)\,.
\tag{4.64}
$$

Here, the effective temperature

$$T_{e\perp}(r) \equiv \frac{\int d^3p \left[p_r^2 + (p_z - m_e V_e)^2\right] f_e^0/2m_e}{\int d^3p \, f_e^0}$$

can be expressed as

$$T_{e\perp}(r) = \frac{1}{2m_e}\psi_e(r) \tag{4.65}$$

where $\psi_e(r)$ is defined in Eq.(4.55). The envelope function $\psi_e(r)$, and therefore $T_{e\perp}(r)$, is zero at the layer boundaries r_b^- and r_b^+, and achieves a maximum value at some radius r_M intermediate between r_b^- and r_b^+ (Fig. 4.5).

Thus far, the polarity of $P_0 < H_0/\omega_{ce}$ has not been specified. Generally speaking, however, it follows from Eq.(4.57) that equilibria with $P_0 > 0$ have a "fast" rotation with $\omega_{re}(r) > \omega_{ce}/2$, whereas equilibria with $P_0 < 0$ have a "slow" rotation with $\omega_{re}(r) < \omega_{ce}/2$. For present purposes, we assume $P_0 > 0$, and introduce the positive quantities r_0 and Δ defined in terms of H_0 and P_0 by

$$r_0 \equiv \left(\frac{2H_0}{m_e\omega_{ce}^2}\right)^{1/2}, \qquad \Delta \equiv \left(1 - \frac{P_0\omega_{ce}}{H_0}\right)^{1/2}, \tag{4.66}$$

where $0 < P_0\omega_{ce} < H_0$ and $\Delta < 1$. From Eq.(4.63) the location of the inner boundary r_b^- is determined from $\psi(r_b^-) = 0$, which can be expressed as

$$\left(\frac{2H_0}{m_e\omega_{ce}^2}\right)^{1/2} = \frac{P_0}{m_e\omega_{ce}r_b^-} + \frac{1}{2}r_b^- . \tag{4.67}$$

Equation (4.67) is a quadratic equation for r_b^-. The smallest solution to Eq.(4.67) is given by

$$r_b^- = r_0\left(1 - \Delta\right) \tag{4.68}$$

where r_0 and Δ are defined in Eq.(4.66). Equation (4.68) represents a closed expression for r_b^- in terms of the equilibrium quantities H_0, P_0 and ω_{ce}.

As indicated earlier, a determination of the location of the outer boundary r_b^+ is generally more difficult because of the self-electric-field contributions in Eq.(4.63). In the limiting case of a tenuous electron layer with $\omega_{pe}^2 r_b^-(r_b^+ - r_b^-) \ll H_0/m_e = \omega_{ce}^2 r_0^2/2$, however, Eq.(4.63) can be solved iteratively, first for the case where $\omega_{pe}^2 \to 0$. Denoting the $\omega_{pe}^2 \to 0$ solution for r_b^+ by r_{b0}^+, we find from Eq.(4.63) that r_{b0}^+ solves an equation identical in form to Eq.(4.67). This gives

$$r_{b0}^+ = r_0 \left(1 + \Delta\right) \tag{4.69}$$

for the leading-order estimate of the outer radius of the electron layer. Comparing Eqs.(4.68) and (4.69), the average radius of the electron layer is $(1/2)(r_{b0}^+ + r_b^-) = r_0$, and the layer thickness is $r_{b0}^+ - r_b^- = 2r_0\Delta$.

The simplest iterative procedure to calculate the self-field correction to Eq.(4.69) is to approximate $H_0 + e\phi_0(r_b^+)$ by $H_0 + e\phi_0(r_{b0}^+)$ when solving

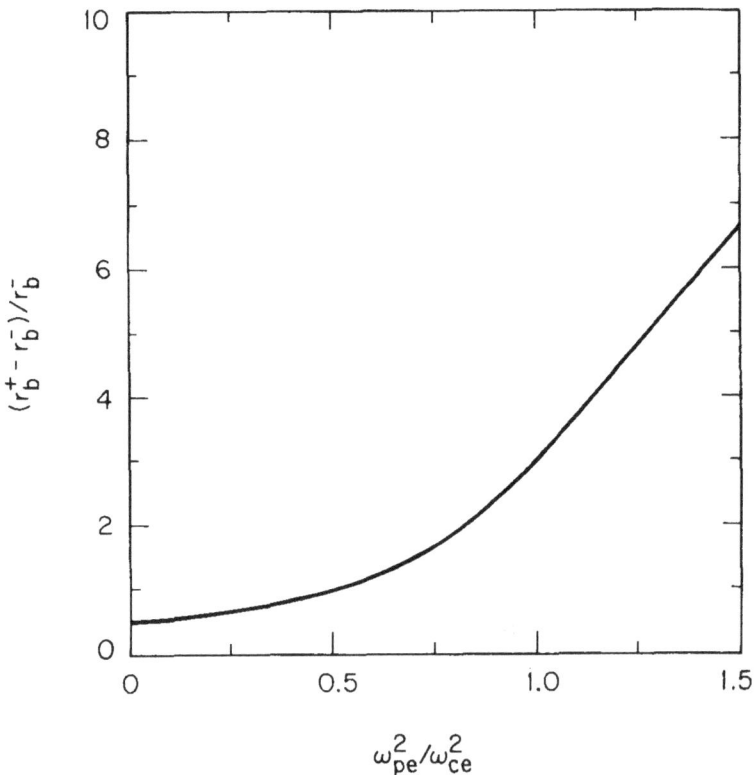

Figure 4.6. Plot of the normalized layer thickness $(r_b^+ - r_b^-)/r_b^-$ versus $\omega_{pe}^2/\omega_{ce}^2$ obtained numerically from the exact envelope equation (4.63) for fixed $\Delta = (1 - P_0\omega_{ce}/H_0)^{1/2} = 0.2$.

the envelope equation $\psi(r_b^+) = 0$. From Eqs.(4.63), (4.68), and (4.69), we obtain

$$\left(\frac{2H_0}{m_e\omega_{ce}^2}\right)^{1/2}\left[1 + \frac{e\phi_0(r_{b0}^+)}{H_0}\right]^{1/2} = \frac{P_0}{m_e\omega_{ce}r_b^+} + \frac{1}{2}r_b^+ , \qquad (4.70)$$

where

$$\frac{e\phi_0(r_{b0}^+)}{H_0} = \frac{2\omega_{pe}^2}{\omega_{ce}^2}(1-\Delta)\left[2\Delta - (1-\Delta)\ln\left(\frac{1+\Delta}{1-\Delta}\right)\right] \qquad (4.71)$$

is treated as a small parameter. If $e\phi_0(r_{b0}^+)/H_0$ is neglected, then Eq.(4.70) gives the solution $r_b^+ = r_{b0}^+$ in Eq.(4.69), as expected. Retaining $e\phi_0(r_{b0}^+)/H_0$ as a small correction, however, Eq.(4.70) can be solved to give an improved analytical estimate of r_b^+ (Problem 4.7).

The exact envelope equation (4.63) has been solved numerically for the location of the outer boundary r_b^+ assuming fixed values of P_0, H_0 and ω_{ce} (and therefore fixed r_0 and r_b^-) but variable electron density. Typical numerical results are illustrated in Fig. 4.6, where the normalized layer thickness $(r_b^+ - r_b^-)/r_b^-$ is plotted versus $\omega_{pe}^2/\omega_{ce}^2$ for $\Delta = 0.2$. Evidently, the influence of self-electric field effects on the layer thickness can be substantial for moderate values of $\omega_{pe}^2/\omega_{ce}^2$.

The kinetic equilibrium in Eq.(4.52) is generalized in Sec. 4.3.2 to the case of an electron layer confined axially by a magnetic mirror field $\mathbf{B}_0^{ext}(x)$, which provides axial focusing of the electron orbits as well as radial confinement.

Problem 4.7   Consider the quadratic equation (4.70) for r_b^+, treating $\varepsilon \equiv e\phi_0(r_{b0}^+)/2H_0 \ll 1$ as a small parameter. Assume $\varepsilon \ll \Delta$ and show that the solution for r_b^+ can be approximated by

$$r_b^+ = r_0\left[1 + \Delta + \varepsilon\left(1 + \frac{1}{\Delta}\right)\right] , \qquad (4.7.1)$$

where leading-order self-field corrections (proportional to ε) are retained. In Eq.(4.7.1), r_0 and Δ are defined in Eq.(4.66).

4.3   Axially Confined Kinetic Equilibria

In this section, examples of axisymmetric nonneutral plasmas are considered which are confined axially as well as radially. In Sec. 4.3.1, we investigate the confinement geometries illustrated in Fig. 4.3, which provide radial confinement by a uniform magnetic field $B_0 \hat{e}_z$ and axial confinement by applied electrostatic potentials on neighboring conductors.[7,8] The confinement geometry illustrated in Fig. 4.2 is examined in Sec. 4.3.2. Here, radial and axial confinement of an electron layer is provided by an externally applied mirror magnetic field $\mathbf{B}_0^{\mathrm{ext}}(\mathbf{x})$ which produces sufficient axial focusing of the particle orbits to assure confinement in the z-direction.

4.3.1 Penning Trap and Segmented Cylinder Geometries

The segmented cylinder and Penning trap confinement geometries illustrated in Fig. 4.3 are widely used for basic experimental investigations of the equilibrium and stability properties of nonneutral electron or ion plasmas, including studies of phase transitions[53-55] in strongly coupled nonneutral ion plasmas as $T_j \to 0$ (Sec. 3.6). Axial confinement is provided by biasing the external conductors in such a way as to produce an applied electrostatic potential $\phi_0^a(r, z)$ that confines the electrons or ions axially.

For present purposes, we consider a nonrelativistic, one-component nonneutral plasma consisting of electrons ($e_j = -e$) or ions ($e_j = +Z_i e$). It is further assumed that the steady-state distribution function $f_j^0(r, z, \mathbf{p})$ corresponds to thermal equilibrium (Sec. 3.3). That is, $f_j^0(r, z, \mathbf{p})$ is assumed to be of the form[1-8]

$$f_j^0(r, z, \mathbf{p}) = \frac{\hat{n}_j}{(2\pi m_j k_B T_j)^{3/2}} \exp\left\{ -\frac{H - \omega_{rj} P_\theta}{k_B T_j} \right\} . \qquad (4.72)$$

Here, $\omega_{rj} = $ const. is the angular rotation velocity, $H = (2m_j)^{-1} \times (p_r^2 + p_\theta^2 + p_z^2) + e_j \phi_0(r, z)$ is the energy, $P_\theta = r(p_\theta + \varepsilon_j m_j \omega_{cj} r/2)$ is the canonical angular momentum, and the notation is the same as in Eq.(4.17). Paralleling the analysis in Secs. 3.3 and 4.2, it is readily shown that the equilibrium density profile $n_j^0(r, z) = \int d^3 p\, f_j^0(r, z, \mathbf{p})$ can be expressed as

$$n_j^0(r, z) = \hat{n}_j \exp\left\{ -\frac{m_j}{2 k_B T_j} \left[-r^2 \left(\varepsilon_j \omega_{cj} \omega_{rj} + \omega_{rj}^2 \right) + \frac{2 e_j}{m_j} \phi_0(r, z) \right] \right\} . \qquad (4.73)$$

Moreover, the equilibrium Poisson equation (4.22) for a one-component nonneutral plasma becomes[7,8]

$$\left\{ \frac{1}{r}\frac{\partial}{\partial r}r\frac{\partial}{\partial r} + \frac{\partial^2}{\partial z^2} \right\} \phi_0(r,z)$$

$$= -4\pi\hat{n}_j e_j \exp\left\{ -\frac{m_j}{2k_B T_j}\left[-r^2\left(\varepsilon_j\omega_{cj}\omega_{rj} + \omega_{rj}^2\right) + \frac{2e_j}{m_j}\phi_0(r,z) \right] \right\} .$$

$$\text{(4.74)}$$

Referring to the confinement geometries in Fig. 4.3, the coordinate system is chosen so that the z-axis coincides with the axis of symmetry parallel to $B_0\hat{e}_z$, and the origin ($z = 0$) is chosen at the midplane of the configuration so that all equilibrium properties are symmetric under the reflection $z \to -z$, e.g., $n_j^0(r,z) = n_j^0(r,-z)$. For future reference, the characteristic radial dimension of the plasmoid is denoted by $2r_b$ and the characteristic axial dimension is denoted by $2z_b$. That is, the plasmoid extends throughout the region

$$0 \le r < r_b \,, \qquad \text{for } z = 0 \,,$$

$$-z_b < z < z_b \,, \qquad \text{for } r = 0 \,. \qquad \text{(4.75)}$$

Of course, the density profile in Eq.(4.73) generally has diffuse boundaries, except in the limit of small Debye length with $\lambda_{Dj} = (k_B T_j/4\pi\hat{n}_j e_j^2)^{1/2} \to 0$. Without loss of generality, the zero of potential in Eq.(4.74) is chosen such that $\phi_0(r = 0, z = 0) = 0$. Therefore, $\hat{n}_j = n_j^0(r = 0, z = 0)$ can be identified with the plasma density at the origin.

As in the case where $\partial/\partial z = 0$ (Sec. 3.3), the equilibrium Poisson equation (4.74) is a highly nonlinear differential equation for the electrostatic potential $\phi_0(r,z)$. Equation (4.74) has been solved numerically by Prasad and O'Neil[7] for the segmented cylinder geometry in Fig. 4.3(a). Here, for electrons, axial confinement is provided by biasing the end cylinders A and A' at a potential $-V$ relative to the central cylinder B of length $2L$. The applied (vacuum) potential $\phi_0^a(r,z)$ can be calculated analytically[7] in closed form (Problem 4.8). Moreover, the total electrostatic potential $\phi_0(r,z)$ can be divided into two contributions

$$\phi_0(r,z) = \phi_0^a(r,z) + \phi_0^s(r,z) \,, \qquad \text{(4.76)}$$

where $\nabla^2 \phi_0^a = 0$, and $\phi_0^s(r, z)$ is the self-field contribution determined from $\nabla^2 \phi_0^s = -4\pi e_j n_j^0(r, z)$. Numerical solutions to Eq.(4.74) are then obtained iteratively[7] over a wide range of dimensionless system parameters.

Only in the limits where the plasmoid has small physical dimensions ($r_b \ll b$ and $z_b \ll L$ in Fig. 4.3) and the temperature T_j (Debye length λ_{Dj}) approaches zero does the solution to Eq.(4.74) take a simple analytical form, which corresponds to a plasmoid shaped like a spheroid (Fig. 4.7). We illustrate this feature by considering the Penning trap geometry in Fig. 4.3(b). For the sake of definiteness, it is assumed that a nonneutral ion plasma ($e_j = +Z_i e$) is confined in the midplane region with small physical dimensions ($r_b \ll b$ and $z_b \ll L$). The two encaps and one ring electrode are biased at a relative potential $+V$, and the hyperbolic surfaces of the electrodes produce a quadrupole electrostatic potential which can be approximated by[8]

$$\phi_0^a(r, z) = V \left(\frac{2z^2 - r^2}{2L^2 + b^2} \right) \tag{4.77}$$

over the extent of the ion plasmoid. The potential in Eq.(4.77) provides axial confinement of the ions. Indeed, for an individual ion, it is readily shown from Eq.(4.77) that the frequency of axial betatron oscillations along the axis of symmetry ($r = 0$) is given by

$$\omega_{zi}^2 = \frac{4Z_i eV}{m_i(2L^2 + b^2)} . \tag{4.78}$$

Therefore, Eq.(4.77) can be expressed in the equivalent form

$$\phi_0^a(r, z) = \frac{1}{4} \frac{m_i}{Z_i e} \omega_{zi}^2 \left(2z^2 - r^2 \right) . \tag{4.79}$$

Note from Eq.(4.77) [or Eq.(4.79)] that $\nabla^2 \phi_0^a(r, z) = 0$.

We now return to Eqs.(4.73) and (4.74) to examine properties of the density profile $n_i^0(r, z)$ and electrostatic potential $\phi_0(r, z)$ in the limit where $T_i \to 0$. This regime is particularly relevant in experimental studies[53,54] of strongly coupled ion plasmas (e.g., $^9Be^+$) which are cooled (by laser cooling) to temperatures in the mK range in order to investigate phase transitions to the liquid and crystalline states. As in Sec. 3.3 (e.g., see Fig. 3.10), in the limit where $T_i \to 0$ and $\lambda_{Di} = (k_B T_i / 4\pi \hat{n}_i Z_i^2 e^2)^{1/2} \to 0$, the ion density profile $n_i^0(r, z)$ in Eq.(4.73) is approximately uniform

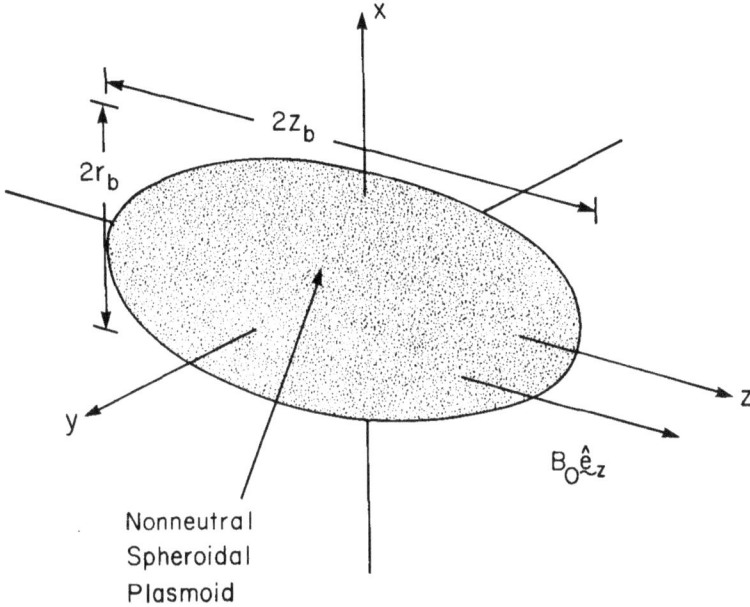

Figure 4.7. The thermal equilibrium distribution function in Eq.(4.72) is confined radially and axially in the segmented cylinder or Penning trap geometries shown in Fig. 4.3. Assuming $\lambda_{Dj} \to 0$ and small plasma volume ($r_b \ll b$, $z_b \ll L$), the nonneutral plasmoid has the shape of a spheroid with axial length $2z_b$ and radius r_b revolved about the z-axis.

(and equal to $\hat{n}_i$) within the plasmoid, and decreases abruptly to zero in a thin surface region which is a few λ_{Di} in thickness. Examination of the argument of the exponential in Eq.(4.73) shows for $T_i \to 0$ that

$$Z_i e \phi_0(r, z) - \frac{m_i}{2} \left(\omega_{ci} \omega_{ri} + \omega_{ri}^2 \right) r^2 \to 0 \qquad (4.80)$$

within the plasmoid. In Eq.(4.80), note that the *total* electrostatic potential $\phi_0(r, z)$ within the plasmoid is independent of z. Substituting $\phi_0(r, z) = \phi_0^a(r, z) + \phi_0^s(r, z)$, where the applied potential $\phi_0^a(r, z)$ is defined in Eq.(4.79), we solve Eq.(4.80) for the self-field potential $\phi_0^s(r, z)$. This gives

$$Z_i e \phi_0^s(r, z) = \frac{m_i}{2} \left\{ \frac{1}{2} \omega_{zi}^2 + \omega_{ri} \omega_{ci} + \omega_{ri}^2 \right\} r^2 - \frac{m_i}{2} \omega_{zi}^2 z^2 \qquad (4.81)$$

within the plasmoid. Strictly speaking, if the potential at the electrodes induced by the plasma ions is neglected (because $b \gg r_b$ and $L \gg z_b$), then $\phi_0^s(r, z)$ in Eq.(4.81) can be viewed as the self-field potential in the absence of trap walls.

Within the plasmoid, where $n_i^0(r, z) \simeq \hat{n}_i = $ const., Poisson's equation is $\nabla^2(\phi_0^a + \phi_0^s) = -4\pi Z_i e \hat{n}_i$, where $\nabla^2 \phi_0^a = 0$. Substituting $\phi_0^s(r, z)$ from Eq.(4.81) then gives the familiar force-balance equation

$$\omega_{ri}^2 + \omega_{ri}\omega_{ci} + \frac{1}{2}\omega_{pi}^2 = 0 , \qquad (4.82)$$

which should be compared with Eq.(3.5) for a nonneutral electron plasma. Here, $\omega_{pi}^2 = 4\pi \hat{n}_i Z_i^2 e^2 / m_i$ is the ion plasma frequency-squared, and the solutions to Eq.(4.82) are $\omega_{ri} = \omega_{ri}^{\pm} = -(\omega_{ci}/2)\left\{1 \pm (1 - 2\omega_{pi}^2/\omega_{ci}^2)^{1/2}\right\}$. (Note that the allowed rotation frequencies, ω_{ri}^+ and ω_{ri}^-, are both negative.) Substituting Eq.(4.82) into Eq.(4.81) then gives the simple expression[8]

$$Z_i e \phi_0^s(r, z) = -\frac{m_i}{4}\left(\omega_{pi}^2 - \omega_{zi}^2\right) r^2 - \frac{m_i}{2}\omega_{zi}^2 z^2 \qquad (4.83)$$

where $\omega_{zi}^2 < \omega_{pi}^2$ is required for radially confined ion orbits.

Equation (4.83) should be compared with the potential inside an isolated, uniformly charged spheroid (Fig. 4.7) with uniform charge density $Z_i e \hat{n}_i$ in the region $r^2/r_b^2 + z^2/z_b^2 < 1$.[56,57] The corresponding potential $\Phi(r, z)$ for $r^2/r_b^2 + z^2/z_b^2 < 1$ is given by (Problem 4.9)

$$Z_i e \Phi(r, z) = -\frac{1}{4}m_i\omega_{pi}^2\left(\alpha r^2 + \beta z^2\right) , \qquad (4.84)$$

where the constants α and β are defined in Eqs.(4.9.6) and (4.9.7) of Problem 4.9. For example, when $z_b < r_b$, the constants α and β are defined by

$$\alpha = -\frac{1}{(r_b^2/z_b^2 - 1)} + \frac{r_b^2/z_b^2}{(r_b^2/z_b^2 - 1)^{3/2}}\tan^{-1}\left(\frac{r_b^2}{z_b^2} - 1\right)^{1/2} ,$$

$$\frac{1}{2}\beta = \frac{r_b^2/z_b^2}{(r_b^2/z_b^2 - 1)} - \frac{r_b^2/z_b^2}{(r_b^2/z_b^2 - 1)^{3/2}}\tan^{-1}\left(\frac{r_b^2}{z_b^2} - 1\right)^{1/2} . \qquad (4.85)$$

On the other hand, when $z_b > r_b$, the constants α and β are defined by

Eq.(4.9.7). Note that $\alpha + \beta/2 = 1$ in both cases. Furthermore, comparing Eqs.(4.83) and (4.84) gives

$$\alpha = 1 - \frac{\omega_{zi}^2}{\omega_{pi}^2}, \qquad \beta = \frac{2\omega_{zi}^2}{\omega_{pi}^2}. \qquad (4.86)$$

Evidently, the special case where $\omega_{zi}^2 = (1/3)\omega_{pi}^2$ corresponds to a *spherical* plasmoid with $z_b = r_b$ and $\alpha = \beta = 2/3$. The case where $\omega_{pi}^2 > \omega_{zi}^2 > (1/3)\omega_{pi}^2$, however, corresponds to an *oblate spheroid* compressed in the axial direction with $z_b < r_b$, $\alpha < 2/3$ and $\beta > 2/3$. On the other hand, for $\omega_{zi}^2 < (1/3)\omega_{pi}^2$, it follows that $\alpha > 2/3$, $\beta < 2/3$ and $z_b > r_b$, which corresponds to an *elongated spheroid*, with elongation along the z-axis. As a general remark, if the total number of ions N_i in the trap is known, then the ion density $\hat{n}_i = N_i/V$ where $V = (4\pi/3)r_b^2 z_b$ is the volume of the spheroid. For specified ω_{zi}^2 and N_i, Eqs.(4.85) and (4.86) [or Eqs.(4.9.7) and (4.86)] can then be used to calculate r_b and z_b.

Problem 4.8   Consider the segmented cylinder geometry in Fig. 4.3(a), where cylinders A and A' are biased at a potential $-V$ relative to cylinder B (for axial confinement of electrons). The cylinders have radius b, and the length of cylinder B is $2L$.

a. Solve the vacuum Poisson equation

$$\left\{ \frac{1}{r}\frac{\partial}{\partial r}r\frac{\partial}{\partial r} + \frac{\partial^2}{\partial z^2} \right\} \phi_0^a(r, z) = 0 \qquad (4.8.1)$$

in the region $0 \leq r \leq b$, subject to the boundary conditions

$$\phi_0^a(r = b, z) = 0, \qquad -L < z < L,$$
$$\phi_0^a(r = b, z) = -V, \qquad |z| > L. \qquad (4.8.2)$$

Show within cylinder B $(0 \leq r \leq b, |z| < L)$ that $\phi_0^a(r, z)$ can be expressed as

$$\phi_0^a(r, z) = -2V \sum_{n=1}^{\infty} \frac{J_0(k_{0n}r)}{k_{0n}b J_1(k_{0n}b)} \frac{\cosh(k_{0n}z)}{[\cosh(k_{0n}L) + \sinh(k_{0n}L)]}, \qquad (4.8.3)$$

where k_{0n} is the n'th zero of $J_0(k_{0n}b) = 0$, and $J_0(x)$ is the Bessel function of the first kind of order zero.

b. Obtain a leading-order (quadrupole approximation) expression for $\phi_0^a(r, z)$ valid near the origin $(r \ll b, |z| \ll L)$ in Fig. 4.3(a).

With appropriate redefinition of $\omega_{z_j}^2$, the $T_j \rightarrow 0$ analysis in Sec. 4.3.1 for a small-volume spheroid confined in a Penning trap is readily extended to the case of the segmented cylinder in Fig. 4.3(a).

Problem 4.9   Consider an isolated, uniformly charged spheroid with uniform charge density $Z_i e \hat{n}_i$ in the region $r^2/r_b^2 + z^2/z_b^2 < 1$ (Fig. 4.7). Poisson's equation for $\Phi(r, z)$ can be expressed as

$$\left\{ \frac{1}{r} \frac{\partial}{\partial r} r \frac{\partial}{\partial r} + \frac{\partial^2}{\partial z^2} \right\} \Phi(r, z) = \begin{cases} -4\pi Z_i e \hat{n}_i, & r^2/r_b^2 + z^2/z_b^2 < 1, \\ 0, & r^2/r_b^2 + z^2/z_b^2 > 1. \end{cases} \tag{4.9.1}$$

a. The complete solution to Eq.(4.9.1) has been given by Landau and Lifshitz [*The Classical Theory of Fields* (Pergamon, New York, 1971) p. 281] and by Chrien, et al. [Phys. Fluids **29**, 1675 (1986)] including the region outside the spheroid. Show that the complete solution to Eq.(4.9.1) can be expressed as

$$\Phi(r, z) = -\pi Z_i e \hat{n}_i r_b^2 z_b \left[\int_0^\xi \frac{ds}{R_s} + \int_\xi^\infty \frac{ds}{R_s} \left(\frac{r^2}{r_b^2 + s} + \frac{z^2}{z_b^2 + s} \right) \right]. \tag{4.9.2}$$

Here, R_s is defined by

$$R_s = \left(r_b^2 + s \right) \left(z_b^2 + s \right)^{1/2}, \tag{4.9.3}$$

and the quantity ξ is defined by

$$\frac{r^2}{r_b^2 + \xi} + \frac{z^2}{z_b^2 + \xi} = 1 \tag{4.9.4}$$

outside the spheroid, and $\xi = 0$ inside the spheroid $(r^2/r_b^2 + z^2/z_b^2 < 1)$.

b. Within the spheroid, make use of Eq.(4.9.2) to show that the potential reduces to

$$Z_i e \Phi(r, z) = -\frac{1}{4} m_i \omega_{pi}^2 (\alpha r^2 + \beta z^2) \tag{4.9.5}$$

for $r^2/r_b^2 + z^2/z_b^2 < 1$. Here, $\omega_{pi}^2 = 4\pi \hat{n}_i Z_i^2 e^2/m_i$ is the ion plasma frequency-squared, and the constants α and β are defined by

$$\alpha = -\frac{1}{(r_b^2/z_b^2 - 1)} + \frac{r_b^2/z_b^2}{(r_b^2/z_b^2 - 1)^{3/2}} \tan^{-1} \left(\frac{r_b^2}{z_b^2} - 1 \right)^{1/2},$$

$$\frac{1}{2}\beta = \frac{r_b^2/z_b^2}{(r_b^2/z_b^2 - 1)} - \frac{r_b^2/z_b^2}{(r_b^2/z_b^2 - 1)^{3/2}} \tan^{-1} \left(\frac{r_b^2}{z_b^2} - 1 \right)^{1/2}, \tag{4.9.6}$$

for an *oblate spheroid* with $z_b < r_b$, and by

$$\alpha = \frac{1}{(1 - r_b^2/z_b^2)} - \frac{r_b^2/z_b^2}{2(1 - r_b^2/z_b^2)^{3/2}} \ln \left| \frac{1 + (1 - r_b^2/z_b^2)^{1/2}}{1 - (1 - r_b^2/z_b^2)^{1/2}} \right| ,$$

$$\frac{1}{2}\beta = -\frac{r_b^2/z_b^2}{(1 - r_b^2/z_b^2)} + \frac{r_b^2/z_b^2}{2(1 - r_b^2/z_b^2)^{3/2}} \ln \left| \frac{1 + (1 - r_b^2/z_b^2)^{1/2}}{1 - (1 - r_b^2/z_b^2)^{1/2}} \right| ,$$

$$(4.9.7)$$

for an *elongated spheroid* with $z_b > r_b$. Note from Eqs.(4.9.6) and (4.9.7) that

$$\alpha + \frac{1}{2}\beta = 1 . \tag{4.9.8}$$

c.   For a spherical plasmoid with $z_b = r_b$, show that

$$\alpha = \beta = \frac{2}{3} . \tag{4.9.9}$$

4.3.2   Mirror-Confined Electron Layer

In this section, we extend the equilibrium analysis in Sec. 4.2.4 to the case of a nonrelativistic electron layer confined axially (and radially) by a magnetic mirror field $\mathbf{B}_0^{\text{ext}}(\mathbf{x})$ (Fig. 4.2).[9] For simplicity, the equilibrium electron distribution function $f_e^0(r, z, \mathbf{p})$ is taken to be

$$f_e^0(r, z, \mathbf{p}) = \frac{\hat{n}_e \hat{r}_b^-}{2\pi m_e} \delta(H - H_0)\delta(P_\theta - P_0) , \tag{4.87}$$

where $\hat{n}_e$, $\hat{r}_b^-$, H_0 and P_0 are positive constants. Here, H and P_θ are defined by

$$H = \frac{1}{2m_e}\left(p_r^2 + p_\theta^2 + p_z^2\right) - e\phi_0(r, z) ,$$

$$P_\theta = r\left[p_\theta - \frac{e}{c}A_\theta^{\text{ext}}(r, z)\right] , \tag{4.88}$$

where $A_\theta^{\text{ext}}(r, z)$ is the vector potential for the externally applied mirror field. (Self-magnetic fields are neglected in the present analysis.) Referring to Fig. 4.8, in the region of configuration space satisfying $r \ll L$, $A_\theta^{\text{ext}}(r, z)$ can be approximated by[9]

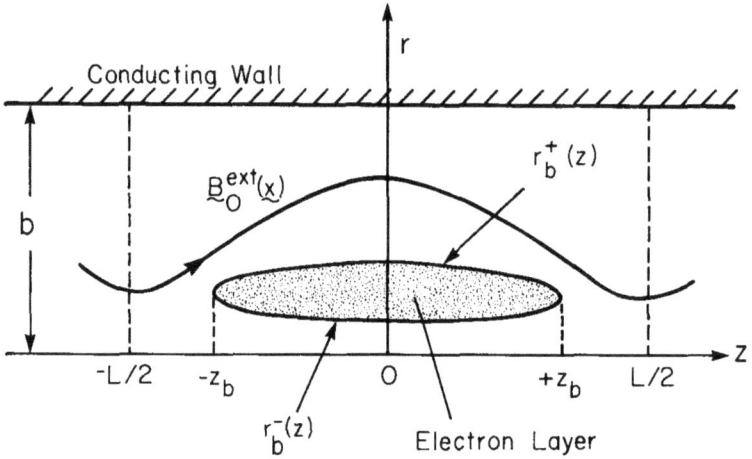

Figure 4.8. Sketch of the r-z cross section of the mirror-confined, non-neutral electron layer described by the Vlasov equilibrium in Eq.(4.87). The distance between mirrors is L and the length of the layer is $2z_b$. For sufficiently low electron density, the outer and inner boundaries of the layer, $r_b^+(z)$ and $r_b^-(z)$, are defined in Eq.(4.102).

$$A_\theta^{\text{ext}}(r, z) = \frac{1}{2}B_0 \left[r - \frac{L}{\pi}\frac{R-1}{R+1}I_1\left(\frac{2\pi r}{L}\right)\cos\left(\frac{2\pi z}{L}\right) \right] . \tag{4.89}$$

In Eq.(4.89), $I_1(x)$ is the modified Bessel function of the first kind of order one, L is the distance between magnetic mirrors (located at $z = \pm L/2$), B_0 is the value of the axial magnetic field on axis ($r = 0$) at $z = \pm L/4$, and R is the mirror ratio defined by

$$R \equiv \frac{B_{\text{max}}}{B_{\text{min}}} = \frac{B_z^{\text{ext}}(r = 0,\, z = \pm L/2)}{B_z^{\text{ext}}(r = 0,\, z = 0)} . \tag{4.90}$$

From Eq.(4.89), the radial and axial components of the magnetic field are

$$B_r^{\text{ext}}(r, z) = -\frac{\partial}{\partial z}A_\theta^{\text{ext}}(r, z) = -B_0\left(\frac{R-1}{R+1}\right)I_1\left(\frac{2\pi r}{L}\right)\sin\left(\frac{2\pi z}{L}\right) , \tag{4.91}$$

and

$$B_z^{\text{ext}}(r, z) = \frac{1}{r}\frac{\partial}{\partial r}\left[rA_\theta^{\text{ext}}(r, z)\right] = B_0\left[1 - \frac{R-1}{R+1}I_0\left(\frac{2\pi r}{L}\right)\cos\left(\frac{2\pi z}{L}\right)\right] . \tag{4.92}$$

A detailed analysis of the equilibrium properties for the choice of distribution function in Eq.(4.87) proceeds in a manner completely analogous to Sec. 4.2.4.[9] For $P_\theta = P_0$, we eliminate $p_\theta = P_0/r + (e/c)A_\theta^{\text{ext}}(r, z)$ and express

$$H - H_0 = \frac{1}{2m_e}\left(p_r^2 + p_z^2\right) - \frac{1}{2m_e}\psi_e(r, z) , \tag{4.93}$$

where the envelope function $\psi_e(r, z)$ is defined by

$$(2m_e)^{-1}\psi_e(r, z) = H_0 + e\phi_0(r, z) - \frac{1}{2m_e}\left[\frac{P_0}{r} + \frac{e}{c}A_\theta^{\text{ext}}(r, z)\right]^2 . \tag{4.94}$$

Equation (4.94) is identical in form to Eq.(4.55), but now includes the z-dependence of $\phi_0(r, z)$ and $A_\theta^{\text{ext}}(r, z)$. Therefore, to calculate momentum moments, the distribution function in Eq.(4.87) can be expressed as

$$f_e^0(r, z, \mathbf{p}) = \frac{\hat{n}_e}{\pi}\frac{\hat{r}_b^-}{r}\delta\left[p_r^2 + p_z^2 - \psi_e(r, z)\right]\delta\left[p_\theta - \left(\frac{P_0}{r} + \frac{e}{c}A_\theta^{\text{ext}}(r, z)\right)\right] , \tag{4.95}$$

where $A_\theta^{\text{ext}}(r, z)$ and $\psi_e(r, z)$ are defined in Eqs.(4.89) and (4.94), respectively.

The moments of Eq.(4.95) corresponding to the average flow velocities, $V_{ze}^0(r, z) = \left(\int d^3p\, v_z f_e^0\right)/\left(\int d^3p\, f_e^0\right)$ and $V_{\theta e}^0(r, z) = \left(\int d^3p\, v_\theta f_e^0\right)/\left(\int d^3p\, f_e^0\right)$, readily give $V_{ze}^0(r, z) = 0$ and

$$V_{\theta e}^0(r, z) = \frac{P_0}{m_e r} + \frac{e}{m_e c}A_\theta^{\text{ext}}(r, z) \tag{4.96}$$

over the radial and axial extents of the electron layer. Similarly, the electron density profile $n_e^0(r, z) = \int d^3p\, f_e^0$ can be expressed as

$$n_e^0(r, z) = \hat{n}_e\frac{\hat{r}_b^-}{r}\int_0^\infty dp_\perp^2\,\delta\left[p_\perp^2 - \psi_e(r, z)\right]$$

$$= \hat{n}_e\frac{\hat{r}_b^-}{r}U\left[\psi_e(r, z)\right] . \tag{4.97}$$

Here, $U(x)$ is the Heaviside step function defined by $U(x) = 0$ for $x < 0$, and $U(x) = +1$ for $x > 0$. As in Sec. 4.2.4, electron motion is allowed in the region where $\psi_e(r, z) \geq 0$ and forbidden in the region where $\psi_e(r, z) < 0$. Moreover, the boundary of the electron layer, which separates the vacuum region in Fig. 4.8 from the region where $n_e^0(r,z) = \hat{n}_e \hat{r}_b^- / r$, is determined from the envelope equation

$$\psi_e(r,z) = 0 \qquad (4.98)$$

where $\psi_e(r,z)$ is defined in Eq.(4.94). Equation (4.98) is to be solved in conjunction with Poisson's equation

$$\left\{ \frac{1}{r} \frac{\partial}{\partial r} r \frac{\partial}{\partial r} + \frac{\partial^2}{\partial z^2} \right\} \phi_0(r,z) = 4\pi e \hat{n}_e \frac{\hat{r}_b^-}{r} U\left[\psi_e(r,z)\right] . \qquad (4.99)$$

Because $\psi_e(r,z)$ depends on the electrostatic potential $\phi_0(r,z)$ [Eq.(4.94)], Poisson's equation (4.99) must be solved as a nonlinear boundary-value problem. That is, the location of the boundary depends on the solution for $\phi_0(r,z)$, and vice versa. Approximate solutions to Eqs.(4.98) and (4.99) have been obtained by Davidson, Kapetanakos and Drobot,[9] treating $e\phi_0(r,z)/H_0$ as a small parameter (tenuous electron layer) and expressing $\phi_0(r,z)$ in terms of the Green's function for an infinitely long conducting cylinder. The basic procedure is to express Eq.(4.98) as $\psi_e^{(0)}(r,z) + \psi_e^{(1)}(r,z) = 0$, where $(2m_e)^{-1}\psi_e^{(1)}(r,z) = e\phi_0(r,z)$ and $\psi_e^{(0)}(r,z)$ is defined by

$$(2m_e)^{-1}\psi_e^{(0)}(r,z) = H_0 - \frac{1}{2m_e}\left[\frac{P_0}{r} + \frac{e}{c}A_\theta^{\text{ext}}(r,z)\right]^2 . \qquad (4.100)$$

The zero-order location of the layer boundary is determined from $\psi_e^{(0)}(r,z) = 0$, and then $\phi_0^{(0)}(r,z)$ is calculated from Poisson's equation (4.99). An improved estimate of the location of the layer boundary is then obtained[9] from $(2m_e)^{-1}\psi_e^{(0)}(r,z) + e\phi_0^{(0)}(r,z) = 0$, and so on.

For present purposes, the zero-order location of the layer boundary is calculated from $\psi_e^{(0)}(r,z) = 0$, where $\psi_e^{(0)}(r,z)$ is defined in Eq.(4.100). For $r \ll L$, we approximate $I_1(2\pi r/L) \simeq \pi r/L$ in Eq.(4.89). From Eqs.(4.89) and (4.100), the envelope equation $\psi_e^{(0)}(r,z) = 0$ gives

$$\left(\frac{2H_0}{m_e\omega_{ce}^2}\right)^{1/2} = \frac{P_0}{m_e\omega_{ce}r_b} + \frac{1}{2}r_b\left[1 - \frac{R-1}{R+1}\cos\left(\frac{2\pi z}{L}\right)\right] \qquad (4.101)$$

where $\omega_{ce} \equiv eB_0/m_ec$. The (two) solutions to Eq.(4.101) for $r_b(z)$ determine the inner and outer boundaries of the electron layer (Fig. 4.8). Defining $r_0 \equiv (2H_0/m_e\omega_{ce}^2)^{1/2}$, the solutions to Eq.(4.101) are

$$r_b(z) = r_b^{\pm}(z) \equiv r_0 \left[1 \pm \left\{ 1 - \frac{\omega_{ce}P_0}{H_0} \left[1 - \frac{R-1}{R+1}\cos\left(\frac{2\pi z}{L}\right) \right] \right\}^{1/2} \right]$$

$$\times \left[1 - \frac{R-1}{R+1}\cos\left(\frac{2\pi z}{L}\right) \right]^{-1} . \qquad (4.102)$$

As expected, for a uniform guide field with $R = 1$, the electron layer described by Eq.(4.102) is infinitely long, and the expressions for $r_b^{\pm}$ reduce exactly to Eqs.(4.68) and (4.69). For $R > 1$, however, the mirror field provides axial focusing of the electrons, and the electron layer has a finite length extending from $z = -z_b$ to $z = +z_b$. The length $2z_b$ of the layer is determined from the intersection condition $r_b^{+}(z_b) = r_b^{-}(z_b)$, which gives

$$\frac{R-1}{R+1}\cos\left(\frac{2\pi z_b}{L}\right) = 1 - \frac{H_0}{\omega_{ce}P_0} . \qquad (4.103)$$

In order for the equilibrium to be physically meaningful, it is required that the radical in Eq.(4.102) be real for all values of z in the interval $-z_b < z < z_b$, and that the length of the layer be shorter than the distance between mirrors, i.e., $2z_b < L$. This is the case whenever $H_0/\omega_{ce}P_0$ lies in the interval

$$\frac{2}{R+1} < \frac{H_0}{\omega_{ce}P_0} < \frac{2R}{R+1} . \qquad (4.104)$$

For example, if the mirror ratio is $R = 1.5$, then the inequality in Eq.(4.104) reduces to $0.8 < H_0/\omega_{ce}P_0 < 1.2$.

If the mirror ratio R is sufficiently removed from $R = 1$, and if the injection conditions in forming the layer are such that $H_0/\omega_{ce}P_0$ is closely tuned to the value $2/(R + 1)$ [the lower limit in Eq.(4.104)], then the solution to Eq.(4.103) satisfies $z_b^2 \ll L^2$ and the electron layer corresponds to a ring with elliptical cross section. Solving Eq.(4.103) for the case $z_b^2 \ll L^2$ gives

$$z_b^2 = \frac{L^2}{2\pi^2}\left(\frac{H_0}{\omega_{ce}P_0} - \frac{2}{R+1}\right)\left(\frac{R+1}{R-1}\right) . \qquad (4.105)$$

From Eq.(4.102), we calculate $(r - r_0)^2$ for $z^2 \ll L^2$, where $r = r_b$ is on the surface of the layer. This gives the elliptical surface

$$\frac{(r - r_0)^2}{\Delta_b^2} + \frac{z^2}{z_b^2} = 1 \,, \tag{4.106}$$

where z_b is defined in Eq.(4.105), and Δ_b is defined by

$$\Delta_b^2 = \frac{1}{4} r_0^2 \left(\frac{H_0}{\omega_{ce} P_0} - \frac{2}{R + 1} \right) \frac{\omega_{ce} P_0}{H_0} (R + 1)^2 \,. \tag{4.107}$$

As indicated earlier, $H_0/\omega_{ce} P_0 \simeq 2/(R + 1)$ has been assumed in deriving Eqs.(4.105)–(4.107). Therefore, it follows that the electron layer corresponds to a thin ring ($\Delta_b \ll r_0$, $z_b \ll L$) with elliptical cross section centered at $(r, z) = (r_0, 0)$.

Relativistic thin-ring equilibria are considered in Sec. 4.10 for a high-current modified betatron. The analysis includes the important influence of intense self-electric and self-magnetic fields as well as an applied toroidal magnetic field $B_\theta^{\text{ext}}(r, z)\hat{e}_\theta$ which provides additional confinement of the ring electrons.

Problem 4.10   Consider the kinetic equilibrium described by Eq.(4.87) [or Eq.(4.95)] for a mirror-confined, nonneutral electron layer. Show that the pressure tensor $\boldsymbol{P}_e^0(\mathbf{x})$ defined in Eq.(4.15) reduces to

$$\boldsymbol{P}_e^0(\mathbf{x}) = \frac{1}{2m_e} n_e^0(r, z)\psi_e(r, z)(\hat{e}_r\hat{e}_r + \hat{e}_z\hat{e}_z) \,. \tag{4.10.1}$$

Here, the density profile $n_e^0(r, z)$ and envelope function $\psi_e(r, z)$ are defined in Eqs.(4.94) and (4.97).

4.4   Intense Relativistic Electron Beam Equilibrium

In this section, we make use of Eqs.(4.5)–(4.11) to investigate the kinetic equilibrium properties[10-14] of an intense relativistic electron beam (denote by $j = b$) propagating in the z-direction through a stationary ion background ($m_i \to \infty$ and $V_{zi}^0 = 0$) that provides partial charge neutralization with

$$Z_i n_i^0(r) = f n_b^0(r) \,. \tag{4.108}$$

Here, $+Z_i e$ is the ion charge, and the fractional charge neutralization $f = $ const. is assumed to be uniform over the beam cross section. As an example that is analytically tractable, consider the Hammer-Rostoker equilibrium[10] specified by the beam distribution function

$$f_b^0(r, \mathbf{p}) = \frac{\hat{n}_b}{2\pi \gamma_b m_e} \delta \left[H - (\gamma_b - 1) m_e c^2 \right] \delta \left(P_z - \gamma_b m_e \beta_b c \right) . \quad (4.109)$$

In Eq.(4.109), the energy H and axial canonical momentum P_z are defined in Eqs.(4.6) and (4.8), and $\hat{n}_b$, γ_b and β_b are positive constants. Note from Eq.(4.109) that all of the beam electrons are assumed to have the same total energy ($H = $ const.) and the same axial canonical momentum ($P_z = $ const.). Because $f_b^0(r, \mathbf{p})$ does not depend explicitly on the canonical angular momentum P_θ, and because H is an even function of p_θ, it follows that the average beam rotation is $V_{\theta b}^0(r) = \left(\int d^3 p \, v_\theta f_b^0 \right) / \left(\int d^3 p \, f_b^0 \right) = 0$. Therefore, no axial diamagnetic field $B_z^s(r)$ is produced by the beam equilibrium in Eq.(4.109).

Making use of Eqs.(4.108) and (4.109), the steady-state Poisson equation (4.9) can be expressed as

$$\frac{1}{r} \frac{\partial}{\partial r} r \frac{\partial}{\partial r} \phi_0(r) = 4\pi e \left(1 - f \right) n_b^0(r) , \quad (4.110)$$

where the beam density $n_b^0(r) = \int d^3 p \, f_b^0$ is given by

$$n_b^0(r) = \frac{\hat{n}_b}{\gamma_b m_e} \int_0^\infty dp_\perp \, p_\perp \delta \left\{ \left[c^2 p_\perp^2 + m_e^2 c^4 + c^2 \left(\gamma_b m_e \beta_b c + \frac{e}{c} A_z^s(r) \right)^2 \right]^{1/2} \right.$$

$$\left. - \gamma_b m_e c^2 \left(1 + \frac{e\phi_0(r)}{\gamma_b m_e c^2} \right) \right\} \quad (4.111)$$

with $p_\perp^2 = p_r^2 + p_\theta^2$. Carrying out the integration over $p_\perp$ in Eq.(4.111) gives

$$n_b^0(r) = \hat{n}_b \left[1 + \frac{e\phi_0(r)}{\gamma_b m_e c^2} \right] U \left[\psi_b(r) \right] , \quad (4.112)$$

where the envelope function $\psi_b(r)$ is defined by

$$\psi_b(r) = \gamma_b m_e c^2 + e\phi_0(r) - \left[m_e^2 c^4 + c^2 \left(\gamma_b m_e \beta_b c + \frac{e}{c} A_z^s(r) \right)^2 \right]^{1/2} .$$

$$(4.113)$$

Here, $U(x)$ is the Heaviside step function defined by $U(x) = +1$ for $x > 0$ and $U(x) = 0$ for $x < 1$. As in Secs. 4.2.4 and 4.3.2, electron motion is allowed in the region where $\psi_b(r) \geq 0$ and forbidden in the region where $\psi_b(r) < 0$. The density profile in Eq.(4.112) extends from $r = 0$ to $r = r_b$, where the outer radius r_b of the electron beam is determined from the envelope equation

$$\psi(r_b) = 0 . \tag{4.114}$$

To complete the description, we calculate $n_b^0(r)V_{zb}^0(r) = \int d^3p\, v_z f_b^0$ for the choice of distribution function in Eq.(4.109) and substitute the resulting expression for the axial beam current into the $\nabla \times B_\theta^s \hat{e}_\theta$ Maxwell equation (4.10). Some straightforward algebra that makes use of $v_z = p_z/\gamma m_e$ and $\gamma = (1 + \mathbf{p}^2/m_e^2 c^2)^{1/2}$ yields

$$\frac{1}{r}\frac{\partial}{\partial r}r\frac{\partial}{\partial r}A_z^s(r) = \frac{4\pi\hat{n}_b e}{\gamma_b m_e c}\left[\gamma_b m_e \beta_b c + \frac{e}{c}A_z^s(r)\right]U\left[\psi_b(r)\right] . \tag{4.115}$$

Here, the azimuthal self-magnetic field is $B_\theta^s(r) = -(\partial/\partial r)A_z^s(r)$. Moreover, the average axial velocity of the electron beam is given by[10]

$$V_{ze}^0(r) = \frac{\left[\beta_b c + (e/\gamma_b m_e c)A_z^s(r)\right]}{\left[1 + e\phi_0(r)/\gamma_b m_e c^2\right]} \tag{4.116}$$

for $0 \leq r < r_b$. In the special case of a charge-neutralized electron beam with $f = 1$, note that Eq.(4.116) reduces to

$$V_{ze}^0(r) = \beta_b c + \frac{e}{\gamma_b m_e c}A_z^s(r) .$$

Evidently, the axial velocity profile is uniform with $V_{ze}^0(r) \simeq \beta_b c$ only when $|eA_z^s(r)/\gamma_b m_e c| \ll 1$.

Note from Eqs.(4.113) and (4.114) that the beam radius r_b cannot be evaluated explicitly until $\phi_0(r)$ and $A_z^s(r)$ have been determined from Eqs.(4.110) and (4.115). Furthermore, the solutions for $\phi_0(r)$ and $A_z^s(r)$ depend on r_b. Therefore, the condition $\psi(r_b) = 0$ that determines r_b is, in effect, nonlinear. The solutions to Eqs.(4.110) and (4.115) that are continuous with continuous first derivatives at $r = r_b$, and satisfy

$\phi_0(r = 0) = 0 = A_z^s(r = 0)$, are given by

$$\frac{e\phi_0(r)}{\gamma_b m_e c^2} = \begin{cases} -\left[1 - I_0\left(\frac{\alpha_f r}{\delta}\right)\right] , & 0 \leq r < r_b , \\ -\left[1 - I_0\left(\frac{\alpha_f r_b}{\delta}\right) - \left(\frac{\alpha_f r_b}{\delta}\right) I_1\left(\frac{\alpha_f r_b}{\delta}\right) \ln\left(\frac{r}{r_b}\right)\right] , & r > r_b , \end{cases}$$

(4.117)

and

$$\frac{e A_z^s(r)}{\gamma_b m_e c^2 \beta_b} = \begin{cases} -\left[1 - I_0\left(\frac{r}{\delta}\right)\right] , & 0 \leq r < r_b , \\ -\left[1 - I_0\left(\frac{r_b}{\delta}\right) - \left(\frac{r_b}{\delta}\right) I_1\left(\frac{r_b}{\delta}\right) \ln\left(\frac{r}{r_b}\right)\right] , & r > r_b , \end{cases}$$

(4.118)

In Eqs.(4.117) and (4.118), $I_n(x)$ is the modified Bessel function of the first kind of order n, and α_f and δ are defined by

$$\alpha_f = (1 - f)^{1/2} ,$$

$$\delta = \gamma_b^{1/2} c / \omega_{pb} ,$$

(4.119)

where $\omega_{pb}^2 = 4\pi \hat{n}_b e^2 / m_e$ is the nonrelativistic plasma frequency-squared. In Eq.(4.119), δ is the collisionless skin depth, and $\alpha_f = 0$ corresponds to a charge-neutralized electron beam with $f = 1$.

The beam radius r_b is determined from the envelope equation $\psi_b(r_b) = 0$, where $\psi_b(r)$ is defined in Eq.(4.113). Substituting Eqs.(4.117) and (4.118) into Eq.(4.113) gives the envelope equation

$$1 = \gamma_b^2 \left[I_0^2\left(\frac{\alpha_f r_b}{\delta}\right) - \beta_b^2 I_0^2\left(\frac{r_b}{\delta}\right)\right] .$$

(4.120)

Note that Eq.(4.120) determines r_b in terms of $\alpha_f = (1-f)^{1/2}$ and parameters that characterize the equilibrium distribution function in Eq.(4.109). Once the values of f, γ_b and β_b are specified, the ratio r_b/δ can be calculated numerically from Eq.(4.120). Because r_b/δ is required to be real, there are restrictions on the allowed values of f, γ_b and β_b in Eq.(4.120).

Substituting Eqs.(4.117) and (4.118) into Eqs.(4.112) and (4.116), the electron density profile and axial velocity profile can be expressed as[10]

$$
n_b^0(r) = \begin{cases} \hat{n}_b I_0 \left(\dfrac{\alpha_f r}{\delta} \right), & 0 \le r < r_b, \\ 0, & r > r_b, \end{cases}
\tag{4.121}
$$

and

$$
V_{zb}^0(r) = \beta_b c \frac{I_0(r/\delta)}{I_0(\alpha_f r/\delta)}, \quad 0 \le r < r_b.
\tag{4.122}
$$

Because $I_0(x = 0) = 1$, it follows from Eqs.(4.121) and (4.122) that the constants $\hat{n}_b$ and $\beta_b c$ can be identified with the on-axis ($r = 0$) values of beam density and axial velocity, respectively. The electron distribution in Eq.(4.109) selects $H = (\gamma_b - 1)m_e c^2 = \text{const}$. Therefore, the kinetic energy of an electron fluid element at radius r is $\left[\gamma_b^0(r) - 1 \right] m_e c^2 = e\phi_0(r) + (\gamma_b - 1)m_e c^2$. Making use of Eq.(4.117), it follows that $\gamma_b^0(r)$ can be expressed as

$$
\gamma_b^0(r) = \gamma_b I_0 \left(\frac{\alpha_f r}{\delta} \right)
\tag{4.123}
$$

for $0 \le r < r_b$. Note that $\gamma_b^0(r)$ has the same functional form as the density profile $n_b^0(r)$, and that $(\gamma_b - 1)m_e c^2 = \text{const}$. can be identified with the electron kinetic energy as it passes through $r = 0$. It is also straightforward to calculate the total current carried by the electron beam for the equilibrium distribution function in Eq.(4.109). Making use of Eqs.(4.121) and (4.122), the magnitude of the total beam current can be expressed as

$$
I = 2\pi e \int_0^{r_b} dr\, r n_b^0(r) V_{zb}^0(r) = \beta_b \gamma_b \left(\frac{mc^3}{e} \right) \left(\frac{r_b}{2\delta} \right) I_1 \left(\frac{r_b}{\delta} \right).
\tag{4.124}
$$

Evidently, the current I is very sensitive to the value of r_b/δ.

For a completely neutralized electron beam ($f = 1$) it is instructive to express the right-hand side of Eq.(4.124) in units of the Alfvén limiting current I_A.[10] Using Lawson's definition[58,59] of I_A for a nonuniform beam, I_A is equal to the beam current at which the Larmor radius of an electron in the maximum self-magnetic field (at $r = r_b$) is equal to one-half of the beam radius, i.e.,

$$
\gamma_b^0(r_b) m_e c V_{zb}^0(r_b)/e B_\theta^s(r_b) = \frac{1}{2} r_b.
\tag{4.125}
$$

Making use of Eqs.(4.118) and (4.122)-(4.125), I_A can be expressed as

$$I_A \equiv \left(\frac{mc^3}{e}\right) \beta_b \gamma_b I_0 \left(\frac{r_b}{\delta}\right) \simeq 17000 \, \beta_b \gamma_b I_0 \left(\frac{r_b}{\delta}\right) A \, . \tag{4.126}$$

Equation (4.124) becomes

$$I = I_A \frac{r_b}{2\delta} \frac{I_1(r_b/\delta)}{I_0(r_b/\delta)} \, . \tag{4.127}$$

Making use of asymptotic expansions of $I_n(x)$, it follows from Eq.(4.127) that

$$I \simeq \frac{r_b}{2\delta} I_A \gg I_A, \text{ for } r_b \gg \delta \, ,$$

$$I \simeq \frac{r_b^2}{4\delta^2} I_A \ll I_A, \text{ for } r_b \ll \delta \, . \tag{4.128}$$

Equation (4.128) states that the beam current I exceeds the Alfvén current I_A by a large amount whenever the beam radius is much larger than the collisionless skin depth, $r_b \gg \delta$. However, if $r_b \ll \delta$, then the beam current is much less than the Alfvén current.

The ratio of beam radius to collisionless skin depth, r_b/δ, can be related to ν/γ_b, where $\nu = N_b e^2/m_e c^2$ is Budker's parameter,[60] and $N_b = 2\pi \int_0^{r_b} dr \, r n_b^0(r)$ is the number of electrons per unit length of the beam. Making use of Eq.(4.121), it is straightforward to show that

$$\frac{\nu}{\gamma_b} = \frac{r_b}{2\alpha_f \delta} I_1 \left(\frac{\alpha_f r_b}{\delta}\right) \, . \tag{4.129}$$

If the beam is completely charged neutralized ($f = 1$ and $\alpha_f = 0$), then Eq.(4.129) reduces to

$$\frac{\nu}{\gamma_b} = \frac{r_b^2}{4\delta^2} = \frac{1}{4} \frac{\omega_{pb}^2 r_b^2}{\gamma_b c^2} \, . \tag{4.130}$$

Evidently, $r_b \gtrless 2\delta$ accordingly as $\nu/\gamma_b \gtrless 1$. It follows from Eqs.(4.128) and (4.130) that $\nu/\gamma_b \gg 1$ is required for the beam current I to exceed the Alfvén current I_A by a large amount.

We now examine the envelope equation (4.120). Keeping in mind that $(\gamma_b - 1)m_e c^2$ and $\beta_b c$ are the kinetic energy and the axial velocity, respectively, of an electron at $r = 0$, it follows that the transverse speed $\beta_\perp c$ of an electron as it passes through the axis $(r = 0)$ of the beam is related to γ_b and β_b by

$$\beta_\perp^2 = 1 - \beta_b^2 - 1/\gamma_b^2 \, , \tag{4.131}$$

where $1 - \beta_b^2 > 1/\gamma_b^2$ is required to assure $\beta_\perp^2 > 0$. Equation (4.120) can be analyzed in several limiting cases. For example, for $r_b/\delta \ll 1$, it is readily shown that $\beta_b^2 > 1 - f$ is required for existence of the equilibrium. This is the familiar condition that the magnetic focusing force exceeds the electrostatic repulsive force (see also Sec. 5.2.3). As a limiting case in which the analysis simplifies, consider Eq.(4.120) in circumstances where the beam charge is fully neutralized $(f = 1)$. For $f = 1$, Eq.(4.120) reduces exactly to

$$\frac{1 - 1/\gamma_b^2}{\beta_b^2} = \frac{\beta_\perp^2 + \beta_b^2}{\beta_b^2} = I_0^2 \left(\frac{r_b}{\delta}\right) \, , \tag{4.132}$$

which determines the normalized beam radius r_b/δ for specified γ_b and β_b. Evidently, $r_b \gg \delta(r_b \ll \delta)$ requires $\beta_\perp^2 \gg \beta_b^2(\beta_\perp^2 \ll \beta_b^2)$. Furthermore, for $f = 1$, it follows from Eqs.(4.121) and (4.122) that the beam density is constant,

$$n_b^0(r) = \begin{cases} \hat{n}_b \, , & 0 \leq r < r_b \, , \\ 0 \, , & r > r_b \, , \end{cases} \tag{4.133}$$

and the axial velocity profile can be expressed as

$$V_{zb}^0(r) = \beta_b c I_0 \left(\frac{r}{\delta}\right) \, , \qquad 0 \leq r < r_b \, . \tag{4.134}$$

Note that $V_{zb}^0(r)$ increases monotonically from the value $\beta_b c$ (at $r = 0$) to the value $V_{zb}^0(r_b) = \beta_b c I_0(r_b/\delta)$ at $r = r_b$. If the beam radius r_b is much larger than the collisionless skin depth δ, then the axial current is strongly peaked near the surface of the electron beam.

As a numerical example, for $\beta_\perp^2/\beta_b^2 = 126.73$, Eq.(4.132) gives $r_b \simeq 4.00\,\delta$. The corresponding axial velocity profile $V_{zb}^0(r)$ obtained from Eq.(4.134) is plotted versus r/δ in Fig. 4.9. Note that the axial velocity profile is strongly peaked at $r = r_b$ with $V_{zb}^0(r_b) = 11.30\,\beta_b c$. For this example, $\nu/\gamma_b = 4$, and the total current carried by the electron beam is $I = 1.73\,I_A$.

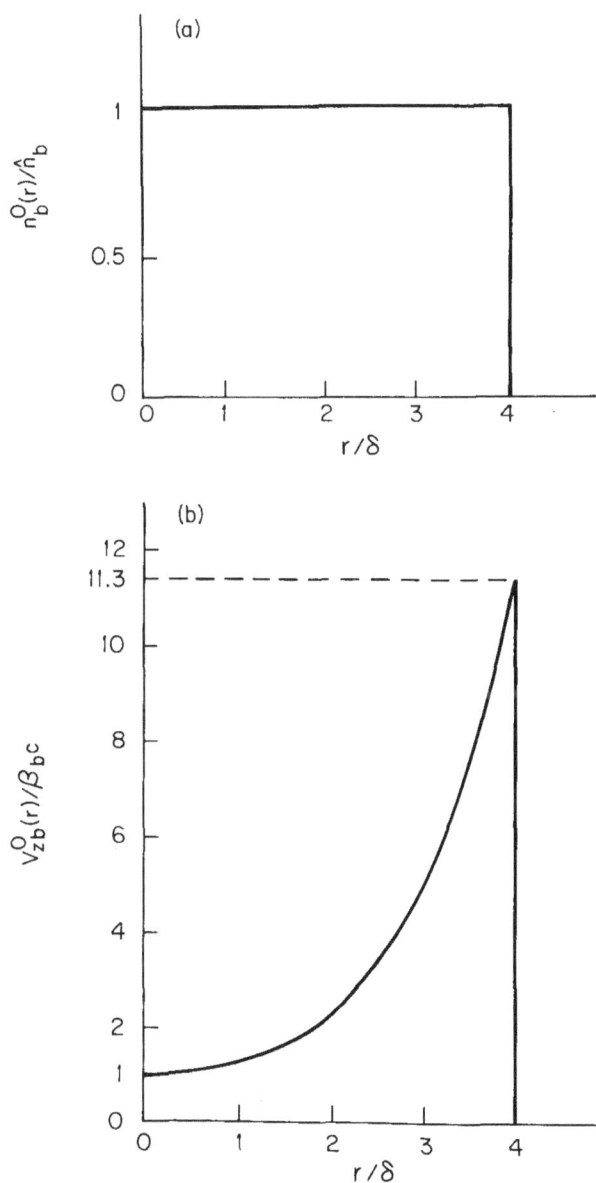

Figure 4.9. Plots versus normalized radius r/δ of (a) the electron density profile [Eq.(4.133)], and (b) the axial velocity profile [Eq.(4.134)], obtained for $f = 1$, $(1 - 1/\gamma_b^2)/\beta_b^2 = 127.73$ and $r_b/\delta = 4$ and the choice of Vlasov equilibrium in Eq.(4.109).

Problem 4.11   Consider Eqs.(4.132) and (4.134), which are valid for a charge-neutralized electron beam $(f = 1)$. Assume $\beta_\perp^2 \ll \beta_b^2$, and show that the beam radius is approximately

$$r_b^2 = \frac{2\beta_\perp^2}{\beta_b^2}\delta^2 , \tag{4.11.1}$$

where $r_b \ll \delta$. Also show that the axial velocity profile is approximately uniform with

$$V_{zb}^0(r) = \beta_b c \tag{4.11.2}$$

for $0 \leq r < r_b$.

Problem 4.12   The equilibrium pressure tensor $\boldsymbol{P}_j^0(\mathbf{x})$ is defined in Eq.(4.15). For the choice of beam distribution function in Eq.(4.109), make use of the solutions for $\phi_0(r)$ and $A_z^s(r)$ in Eqs.(4.117) and (4.118) to show that

$$\boldsymbol{P}_b^0(\mathbf{x}) = n_b^0(r)T_{b\perp}(r)\left(\hat{\mathbf{e}}_r\hat{\mathbf{e}}_r + \hat{\mathbf{e}}_\theta\hat{\mathbf{e}}_\theta\right) , \tag{4.12.1}$$

where the effective $r - \theta$ temperature of the beam electrons is given by

$$T_{b\perp}(r) = \frac{1}{2}\gamma_b m_e c^2 \left[I_2^2\left(\frac{\alpha_f r}{\delta}\right) - \frac{1}{\gamma_b^2} - \beta_b^2 I_0^2\left(\frac{r}{\delta}\right)\right]\left[I_0\left(\frac{\alpha_f r}{\delta}\right)\right]^{-1} . \tag{4.12.2}$$

Here, α_f, δ and $n_b^0(r)$ are defined in Eqs.(4.119) and (4.121). Comparing Eqs.(4.120) and (4.12.2), note that $T_{b\perp}(r = r_b) = 0$ at the boundary of the electron beam, as expected.

4.5   Kinetic Stability Theorem for Nonneutral Plasma

One-component nonneutral plasmas have remarkably robust stability and confinement properties (Secs. 4.5 and 4.6), as evident from experimental results such as those summarized in Sec. 3.3. The detailed stability properties of spatially nonuniform kinetic equilibria with equilibrium self fields are usually difficult to ascertain analytically. Therefore, even a *sufficient condition for stability* is a welcome result since it provides valuable information regarding the class of equilibrium distribution functions that may be unstable. In this section, a sufficient condition for stability is derived for the nonrelativistic Vlasov equilibria discussed in Sec. 4.2.[1-5] The equilibrium configuration consists of a nonneutral plasma column aligned parallel to a uniform external magnetic field $\mathbf{B}_0^{ext}(\mathbf{x}) = B_0\hat{\mathbf{e}}_z$. As in Sec. 4.2, the equilibrium properties are assumed to be independent of $z(\partial/\partial z = 0)$ and azimuthally symmetric $(\partial/\partial\theta = 0)$ about an axis of symmetry parallel to $B_0\hat{e}_z$. Furthermore, it is assumed that the particle motions are non-relativistic, and that the axial and azimuthal self-magnetic fields, $B_z^s(r)$ and $B_\theta^s(r)$, are negligibly small. For present purposes, it is also assumed that the system is fully nonneutral and composed only of electrons. In general, the equilibrium distribution function for the electrons is of the form

$$f_e^0(r, \mathbf{p}) = f_e^0(H, P_\theta, p_z) \,, \tag{4.135}$$

where $p_z = m_e v_z$ is the axial momentum, and

$$H = \frac{1}{2m_e}\left(p_r^2 + p_\theta^2 + p_z^2\right) - e\phi_0(r) \,,$$

$$P_\theta = r\left[p_\theta - m_e r\omega_{ce}/2\right] \,, \tag{4.136}$$

are the energy and canonical angular momentum, respectively. Here, $\omega_{ce} = eB_0/m_e c$, and the notation used in Eqs.(4.135) and (4.136) is the same as in Sec. 4.2. The equilibrium electrostatic potential $\phi_0(r)$ is determined self-consistently in terms of $f_e^0(H, P_\theta, p_z)$ from the equilibrium Poisson equation.

To determine the stability properties of the equilibrium distribution function $f_e^0(H, P_\theta, p_z)$, the evolution of perturbations about equilibrium is examined within the framework of the Vlasov-Maxwell equations. For the equilibrium configuration considered here, the electron distribution function $f_e(\mathbf{x}, \mathbf{p}, t)$, the electric field $\mathbf{E}(\mathbf{x}, t)$, and the magnetic field $\mathbf{B}(\mathbf{x}, t)$ can be expressed as

$$f_e(\mathbf{x}, \mathbf{p}, t) = f_e^0(H, P_\theta, p_z) + \delta f_e(\mathbf{x}, \mathbf{p}, t) \,,$$

$$\mathbf{E}(\mathbf{x}, t) = E_r(r)\hat{e}_r + \delta\mathbf{E}(\mathbf{x}, t) \,,$$

$$\mathbf{B}(\mathbf{x}, t) = B_0\hat{e}_z + \delta\mathbf{B}(\mathbf{x}, t) \,, \tag{4.137}$$

where $\hat{e}_r$ and $\hat{e}_z$ are unit vectors in the r- and z-directions, and $E_r(r) = -\partial\phi_0(r)/\partial r$ is the equilibrium radial electric field. In the present analysis, a sufficient condition for stability of the equilibrium distribution function $f_e^0(H, P_\theta, p_z)$ is derived in the *electrostatic approximation*. That is, it is assumed that the perturbed magnetic field, $\delta\mathbf{B}(\mathbf{x}, t)$, remains negligibly small as the system evolves, and the $\nabla \times \mathbf{E}$ Maxwell equation is approximated by

$$\nabla \times \mathbf{E}(\mathbf{x}, t) = 0 \,. \tag{4.138}$$

The analysis can be extended in a relatively straightforward manner to include the perturbed magnetic field $\delta\mathbf{B}(\mathbf{x}, t)$. Approximating $\mathbf{B}(\mathbf{x}, t) \simeq B_0\hat{\mathbf{e}}_z$, the Vlasov equation for $f_e(\mathbf{x}, \mathbf{p}, t)$ can be expressed as

$$\left\{ \frac{\partial}{\partial t} + \mathbf{v} \cdot \frac{\partial}{\partial \mathbf{x}} - e\left(\mathbf{E}(\mathbf{x}, t) + \frac{\mathbf{v} \times B_0\hat{\mathbf{e}}_z}{c} \right) \cdot \frac{\partial}{\partial \mathbf{p}} \right\} f_e(\mathbf{x}, \mathbf{p}, t) = 0 , \quad (4.139)$$

where $\mathbf{v} = \mathbf{p}/m_e$, since the particle motions are nonrelativistic. The electric field $\mathbf{E}(\mathbf{x}, t) = -\nabla\phi(\mathbf{x}, t)$ in Eq.(4.139) is determined self-consistently in terms of $f_e(\mathbf{x}, \mathbf{p}, t)$ from Poisson's equation. Since no ions are present, Poisson's equation is given by

$$\nabla \cdot \mathbf{E}(\mathbf{x}, t) = -4\pi e \int d^3p\, f_e(\mathbf{x}, \mathbf{p}, t) , \quad (4.140)$$

where $-e$ is the electron charge. Note that Eq.(4.139) is fully nonlinear. That is, no small-amplitude approximation has been made.

In the analysis of Eqs.(4.139) and (4.140), it is assumed that the non-neutral plasma column is located inside a perfectly conducting cylinder with radius $r = b$. It is also assumed that the plasma column is not in physical contact with the conducting wall. For a perfect conductor, the boundary conditions

$$\delta E_\theta(r = b, z, \theta, t) = 0 = \delta E_z(r = b, z, \theta, t) \quad (4.141)$$

are imposed, which correspond to zero tangential electric field at $r = b$. For present purposes, perturbed quantities are assumed to satisfy

$$\delta\psi(r, \theta, z + L, t) = \delta\psi(r, \theta, z, t) \quad (4.142)$$

as well as $\delta\psi(r, \theta + 2\pi, z, t) = \delta\psi(r, \theta, z, t)$. Here, L is the fundamental periodicity length of the perturbations in the z-direction. (A proof of the stability theorem for isolated pulse-like disturbances in the z-direction can be obtained in a completely analogous manner.) Within the context of the above assumptions, it is readily shown that Eqs.(4.139) and (4.140) possess the following global (spatially-averaged) nonlinear conservation constraints

$$\Delta U = \int d^3x \left\{ \frac{|\mathbf{E}|^2 - |\mathbf{E}^0|^2}{8\pi} + \int d^3p\, \frac{\mathbf{p}^2}{2m_e}\left(f_e - f_e^0 \right) \right\} = \text{const.}, \quad (4.143a)$$

$$\Delta C_\theta = \int d^3x \int d^3p\, P_\theta \left(f_e - f_e^0\right) = \text{const.,} \tag{4.143b}$$

$$\Delta C_z = \int d^3x \int d^3p\, p_z \left(f_e - f_e^0\right) = \text{const.,} \tag{4.143c}$$

$$\Delta C_G = \int d^3x \int d^3p \left[G(f_e) - G(f_e^0)\right] = \text{const.,} \tag{4.143d}$$

where $(d/dt)\Delta U = 0$ and $(d/dt)\Delta C_j = 0$. Here, $f_e^0(H, P_\theta, p_z)$ is the (unspecified) equilibrium distribution function, and $f_e(\mathbf{x}, \mathbf{p}, t)$ solves the nonlinear Vlasov-Poisson equations (4.139) and (4.140). In Eq.(4.143), the volume integral covers the region $\int d^3x \cdots = \int_0^b r\,dr \int_0^{2\pi} d\theta \int_0^L dz \cdots$, and $G(f_e)$ is a smooth, differentiable, but otherwise unspecified function satisfying $G(f_e \to 0) = 0$. The constraint conditions, $(d/dt)\Delta U = 0$ and $(d/dt)\Delta C_G = 0$, follow directly from the energy and entropy conservation relations in Eqs.(2.57) and (2.59), respectively. The constraint conditions, $(d/dt)\Delta C_\theta = 0$ and $(d/dt)\Delta C_z = 0$, follow upon using Eq.(4.139) to eliminate $\partial f_e/\partial t$, and integrating by parts with respect to $\mathbf{x}$ and $\mathbf{p}$. For example, some straightforward algebra gives

$$\frac{d}{dt}\Delta C_\theta = -e \int d^3x \int d^3p\, r E_\theta f_e = \frac{1}{4\pi} \int d^3x\, r E_\theta \nabla \cdot \mathbf{E}\,, \tag{4.144}$$

where use has been made of Poisson's equation (4.140). Expressing $\mathbf{E}(\mathbf{x}, t) = -\nabla\phi(r, \theta, z, t)$, Eq.(4.144) gives

$$\frac{d}{dt}\Delta C_\theta = \frac{1}{4\pi} \int d^3x\, \frac{\partial\phi}{\partial\theta} \left[\frac{1}{r}\frac{\partial}{\partial r} r\frac{\partial\phi}{\partial r} + \frac{1}{r^2}\frac{\partial^2\phi}{\partial\theta^2} + \frac{\partial^2\phi}{\partial z^2}\right]\,, \tag{4.145}$$

where $\int d^3x \cdots = \int_0^b dr\, r \int_0^{2\pi} d\theta \int_0^L dz \cdots$. Integrating by parts with respect to r, θ and z in Eq.(4.145), and making use of $[\partial\phi/\partial\theta]_{r=b} = 0$ and periodicity of ϕ in the θ- and z-directions, it follows that $(d/dt)\Delta C_\theta = 0$. Therefore, canonical angular momentum is conserved in a global (spatially averaged) sense. In a similar manner, it can be shown that $(d/dt)\Delta C_z = 0$ (Problem 4.13).

To derive a sufficient condition for stability, we construct the effective Helmholtz free energy function ΔF defined by $\Delta F = \Delta U - \omega_{re}\Delta C_\theta - V_e\Delta C_z - \Delta C_G$, where ω_{re} and V_e are constants. Making use of Eq.(4.143), it follows that

$$\Delta F = \int d^3x \left\{ \frac{|\mathbf{E}|^2 - |\mathbf{E}^0|^2}{8\pi} + \int d^3p \left[\left(\frac{\mathbf{p}^2}{2m_e} - \omega_{re}P_\theta - V_e p_z \right) (f_e - f_e^0) \right. \right.$$

$$\left. \left. + G(f_e) - G(f_e^0) \right] \right\} = \text{const.}, \tag{4.146}$$

where $f_e^0(H, P_\theta, p_z)$ is the equilibrium distribution function, and $f_e(\mathbf{x}, \mathbf{p}, t)$ solves the fully nonlinear Vlasov-Poisson equations (4.139) and (4.140). Now consider small-amplitude perturbations. Taylor expanding $G(f_e) = G(f_e^0 + \delta f_e)$ for small δf_e gives

$$G(f_e) = G(f_e^0) + G'(f_e^0)\delta f_e + G''(f_e^0)\frac{(\delta f_e)^2}{2} + \cdots . \tag{4.147}$$

Correct to quadratic order in the perturbation amplitude, $\Delta F^{(2)}$ can be expressed as

$$\Delta F^{(2)} = \int d^3x \left\{ \frac{|\delta\mathbf{E}|^2}{8\pi} + \int d^3p \left[\left(\frac{\mathbf{p}^2}{2m_e} - e\phi_0(r) - \omega_{re}P_\theta \right. \right. \right.$$

$$\left. \left. \left. -V_e p_z + G'(f_e^0) \right) \delta f_e + G''(f_e^0)\frac{(\delta f_e)^2}{2} \right] \right\}, \tag{4.148}$$

where $\delta\mathbf{E} = \mathbf{E} - \mathbf{E}^0$ and $\delta f_e = f_e - f_e^0$. In obtaining Eq.(4.148) from Eq.(4.146), use has been made of Eq.(4.147) and the identity

$$\int \frac{d^3x}{4\pi}\mathbf{E}^0 \cdot \delta\mathbf{E} = \int \frac{d^3x}{4\pi}\phi_0 \nabla \cdot \delta\mathbf{E} = -e\int d^3x\,\phi_0\int d^3p\,\delta f_e , \tag{4.149}$$

which follows from $\mathbf{E}^0 = -\nabla\phi_0$ and $\nabla\cdot\delta\mathbf{E} = -4\pi e\int d^3p\,\delta f_e$. The function $G(f_e^0)$, which has been arbitrary up to this point, is now chosen to satisfy

$$G'(f_e^0) = -(H - \omega_{re}P_\theta - V_e p_z) , \tag{4.150}$$

where $H = \mathbf{p}^2/2m_e - e\phi_0(r)$. The choice of $G'(f_e^0)$ in Eq.(4.150) implies that f_e^0 depends on H, P_θ and p_z through the linear combination $H - \omega_{re}P_\theta - V_e p_z$. That is, the present analysis is restricted to equilibria with

$$f_e^0 (H, P_\theta, p_z) = f_e^0 (H - \omega_{re} P_\theta - V_e p_z) \ . \tag{4.151}$$

Note that the class of equilibria in Eq.(4.151) has average macroscopic motion corresponding to a rigid rotation of the plasma column $\left[V_{\theta e}^0 (r) = \omega_{re} r \right]$ and uniform velocity in the axial direction $[V_{ze}^0 (r) = V_e = \text{const}]$. Making use of Eq.(4.150), the term linear in δf_e vanishes in Eq.(4.148) and $\Delta F^{(2)}$ reduces to

$$\Delta F^{(2)} = \int d^3 x \left[\frac{|\delta \mathbf{E}|^2}{8\pi} + \int d^3 p \, G''(f_e^0) \frac{(\delta f_e)^2}{2} \right] \ . \tag{4.152}$$

Differentiating Eq.(4.150) with respect to f_e^0 gives

$$G''(f_e^0) = -\frac{1}{\partial f_e^0 / \partial (H - \omega_{re} P_\theta - V_e p_z)} \ . \tag{4.153}$$

Substituting Eq.(4.153) into Eq.(4.152), $\Delta F^{(2)}$ can be expressed in the equivalent form[1-5]

$$\Delta F^{(2)} = \int d^3 x \left\{ \frac{|\delta \mathbf{E}|^2}{8\pi} + \int d^3 p \frac{(\delta f_e)^2}{2} \left[\frac{-1}{\partial f_e^0 / \partial (H - \omega_{re} P_\theta - V_e p_z)} \right] \right\} \ . \tag{4.154}$$

If

$$\frac{\partial}{\partial (H - \omega_{re} P_\theta - V_e p_z)} f_e^0 (H - \omega_{re} P_\theta - V_e p_z) \leq 0 \ , \tag{4.155}$$

then it follows from Eq.(4.154) that $\Delta F^{(2)}$ is a sum of non-negative terms. Because ΔF is a constant, the perturbations $\delta \mathbf{E}(\mathbf{x}, t)$ and $\delta f_e(\mathbf{x}, \mathbf{p}, t)$ cannot grow without bound when Eq.(4.155) is satisfied. Therefore, a sufficient condition for stability can be stated as follows:

If $f_e^0 (H - \omega_{re} P_\theta - V_e p_z)$ is a monotonic decreasing function of $H - \omega_{re} P_\theta - V_e p_z$, then the equilibrium is stable to small-amplitude electrostatic perturbations. (4.156)

Generalization of Eq.(4.156) to a one-component nonneutral ion plasma is trivial.

Equation (4.156) is the generalization of Newcomb's theorem to a fully nonneutral plasma column. The sufficient condition for stability stated in Eq.(4.156) is especially significant since it is applicable to *spatially nonuniform* equilibria characterized by an equilibrium self-electric field, $E^0(x) = -\hat{e}_r \partial \phi_0(r)/\partial r$. The stability theorem is applicable to surface perturbations as well as perturbations interior to the plasma column. As an example, note that the thermal equilibrium distribution function (Sec. 3.3)

$$f_e^0(H - \omega_{re} P_\theta - V_e p_z) = \frac{\hat{n}_e}{(2\pi m_e k_B T_e)^{3/2}}$$

$$\times \exp\left\{ -\frac{(H - \omega_{re} P_\theta - V_e p_z + m_e V_e^2/2)}{k_B T_e} \right\} \qquad (4.157)$$

is electrostatically stable within the context of the present analysis. However, an equilibrium with an inverted population in $H - \omega_{re} P_\theta$, or an energy anisotropy, such as

$$f_e^0(H, P_\theta, p_z) = \frac{\hat{n}_e}{(2\pi m_e)}\delta\left(H - \omega_{re} P_\theta - \hat{T}_{e\perp}\right) G(p_z), \qquad (4.158)$$

may be unstable since it is not a monotonic decreasing function of $H - \omega_{re} P_\theta - V_e p_z$.

Several important generalizations of the preceding analysis can be made. First, if the analysis is extended to include a perturbed magnetic field $\delta B(x,t)$, then the same stability condition is obtained. That is, the condition stated in Eq.(4.155) [or Eq.(4.156)] is a sufficient condition for stability of $f_e^0(H - \omega_{re} P_\theta - V_e p_z)$ to small-amplitude electromagnetic perturbations with arbitrary polarization. Second, the stability theorem is readily extended to a multicomponent nonneutral plasma provided each component is rotating with the same angular velocity, $\omega_{rj} = \omega_0 =$ const., and translating axially with the same average velocity $V_j = V_0 =$ const.[3] It is found that $\partial f_j^0(H - \omega_0 P_\theta - V_0 p_z)/\partial(H - \omega_0 P_\theta - V_0 p_z) \leq 0$, for each species j, is a sufficient condition for stability of $f_j^0(H - \omega_0 P_\theta - V_0 p_z)$ to small-amplitude perturbations. Moreover, following Gardner,[61] the stability theorem can be extended to show that $\partial f_j^0(H - \omega_0 P_\theta - V_0 p_z)/\partial(H - \omega_0 P_\theta - V_0 p_z) \leq 0$, for each plasma component j, is a sufficient condition for *nonlinear* stability of $f_j^0(H - \omega_0 P_\theta - V_0 p_z)$ to arbitrary-amplitude perturbations.

Finally, the stability theorem in Eq.(4.156) can be extended to the case of a relativistic, nonneutral electron beam, allowing for electromagnetic perturbations with arbitrary polarization. In this case, Eq.(4.156) is generalized to become:

If $f_e^0(H - \omega_{re}P_\theta - V_e P_z)$ is a monotonic decreasing function of $H - \omega_{re}P_\theta - V_e P_z$, then the equilibrium is stable to small-amplitude electromagnetic perturbations with arbitrary polarization.   (4.159)

Here, $H = (m_e^2 c^4 + c^2 \mathbf{p}^2)^{1/2} - m_e c^2 - e\phi_0(r)$ is the energy, $P_\theta = rp_\theta - (eB_0/2c)r^2$ is the canonical angular momentum, and $P_z = p_z - (e/c)A_z^s(r)$ is the axial canonical momentum. The $B_z^s(r)$ self-magnetic field produced by beam rotation has been neglected in comparison with the applied axial field B_0 in deriving Eq.(4.159). The effects of the azimuthal self-magnetic field $B_\theta^s(r) = -\partial A_z^s(r)/\partial r$ and the radial self-electric field $E_r(r) = -\partial \phi_0(r)/\partial r$, however, are included.

Problem 4.13

a. Make use of the nonlinear Vlasov-Poisson equations (4.139) and (4.140) to show that

$$\frac{d}{dt}\Delta C_z = -e \int d^3x \int d^3p \, E_z f_e$$

$$= \frac{1}{4\pi}\int d^3x \frac{\partial \phi}{\partial z}\left[\frac{1}{r}\frac{\partial}{\partial r}r\frac{\partial \phi}{\partial r} + \frac{1}{r^2}\frac{\partial^2 \phi}{\partial \theta^2} + \frac{\partial^2 \phi}{\partial z^2}\right] ,   (4.13.1)$$

where $\Delta C_z = \int d^3x \int d^3p \, p_z(f_e - f_e^0)$ and the notation and boundary conditions are the same as in Sec. 4.5.

b. Make use of $[\partial\phi/\partial z]_{r=b} = 0$ and the periodicity of $\phi(r, \theta, z, t)$ in the θ- and z-directions to integrate by parts in Eq.(4.13.1). Show that

$$\frac{d}{dt}\Delta C_z = 0 ,   (4.13.2)$$

which is the required result.

4.6   Kinetic Confinement Theorem for Nonneutral Plasma

As intimated in Secs. 3.3 and 4.5, one-component nonneutral plasmas also have remarkable confinement properties.[15] To illustrate this point, we again consider the cylindrical plasma configuration investigated in Sec. 4.5. It is assumed that the nonneutral plasma column consists only of electrons confined inside a perfectly conducting cylinder with radius $r = b$. In the present analysis, the electrons are described as discrete particles with microscopic phase-space density (Sec. 2.5)

$$N_e(\mathbf{x}, \mathbf{p}, t) = \sum_{k=1}^{\bar{N}_e} \delta\left[\mathbf{x} - \mathbf{x}_k(t)\right] \delta\left[\mathbf{p} - \mathbf{p}_k(t)\right]. \qquad (4.160)$$

Here, the summation is over all electrons in the system, and $N_e(\mathbf{x}, \mathbf{p}, t)$ evolves according to the Klimontovich equation (2.63), which is identical in form to the nonlinear Vlasov equation (4.139) with the microscopic electric field $\mathbf{E}^M(\mathbf{x}, t)$ determined self-consistently in the electrostatic approximation from

$$\nabla \cdot \mathbf{E}^M(\mathbf{x}, t) = -4\pi e \int d^3 p \, N_e(\mathbf{x}, \mathbf{p}, t).$$

The nonlinear electron trajectories $\mathbf{x}_k(t)$ and $\mathbf{p}_k(t), k = 1, 2, \cdots, \bar{N}_e$, in Eq.(4.160) are calculated in terms of $\mathbf{E}^M(\mathbf{x}, t)$, and therefore $N_e(\mathbf{x}, \mathbf{p}, t)$, from the orbit equations (2.61) and (2.62). Conservation relations analogous to those in Eq.(4.143) are readily derived from the Klimontovich-Poisson equations. The two used in the present analysis correspond to energy conservation,

$$U(t) = \int d^3 x \left\{ \frac{\left|\mathbf{E}^M(\mathbf{x}, t)\right|^2}{8\pi} + \int d^3 p \frac{\mathbf{p}^2}{2m_e} N_e(\mathbf{x}, \mathbf{p}, t) \right\} = \text{const.}, \qquad (4.161)$$

and the conservation of canonical angular momentum,

$$C_\theta(t) = \int d^3 x \int d^3 p \, P_\theta N_e(\mathbf{x}, \mathbf{p}, t) = \text{const.}, \qquad (4.162)$$

where $P_\theta = r(p_\theta - m_e r \omega_{ce}/2)$. Note that Eqs.(4.161) and (4.162) incorporate the effects of discrete-particle collisions as well as the interaction of the particles with the average (collective) fields.

Following O'Neil,[15] the nonlinear constraints conditions in Eqs.(4.161) and (4.162) can be used to place a limit on the maximum radial excursion of the particles for given initial conditions. This permits an estimate of the upper bound on the number of particles lost radially from the system, i.e., particles that achieve radial excursions with $r_k(t) \geq b$, where $r = b$ is the radial location of the conducting wall. We carry out the integrations over $\mathbf{x}$ and $\mathbf{p}$ in Eqs.(4.161) and (4.162), and introduce the definitions

$$K(t) = \sum_{k=1}^{\bar{N}_e} \frac{1}{2} m_e v_k^2(t) \,,$$

$$\mathcal{E}_F(t) = \int d^3x \frac{\left| \mathbf{E}^M(\mathbf{x}, t) \right|^2}{8\pi} \,. \tag{4.163}$$

Here, $v_k^2(t) \equiv v_{rk}^2(t) + v_{\theta k}^2(t) + v_{zk}^2(t)$, where $\mathbf{p}_k(t) = m_e \mathbf{v}_k(t)$. Equations (4.161) and (4.162) then become

$$K(t) = U(0) - \mathcal{E}_F(t) \,, \tag{4.164}$$

and

$$\frac{1}{2} m_e \omega_{ce} \sum_k r_k^2(t) = \frac{1}{2} m_e \omega_{ce} \sum_k r_k^2(0) + m_e \sum_k [r_k(t) v_{\theta k}(t) - r_k(0) v_{\theta k}(0)] \,, \tag{4.165}$$

where $U(0) = \mathcal{E}_F(0) + K(0)$, and $\sum_k \cdots$ denotes $\sum_{k=1}^{\bar{N}_e} \cdots$. Equations (4.164) and (4.165) are exactly equivalent to Eqs.(4.161) and (4.162), respectively. From Eq.(4.164), it follows trivially that the kinetic energy $K(t)$ is bounded by

$$K(t) \leq U(0) \,. \tag{4.166}$$

Moreover, because $|v_{\theta k}(t)| \leq [v_{rk}^2(t) + v_{\theta k}^2(t) + v_{zk}^2(t)]^{1/2} \equiv v_k(t)$, it is readily shown that $\sum_k r_k^2(t)$ is bounded by

$$\frac{1}{2} m_e \omega_{ce} \sum_k r_k^2(t) \leq \frac{1}{2} m_e \omega_{ce} \sum_k r_k^2(0) + m_e \sum_k [r_k(t) v_k(t) + r_k(0) v_k(0)] \,, \tag{4.167}$$

Consider the term $m_e \sum_k r_k(t) v_k(t)$ on the right-hand side of Eq.(4.167). We now choose the values of v_k so that $m_e \sum_k r_k v_k$ is a maximum for given values of r_k and given values of $K = \sum_k m_e v_k^2/2$. That is, the values of v_k are chosen so that[15]

$$\delta \sum_k \left(m_e r_k v_k - \alpha m_e v_k^2/2 \right) = 0 \,. \tag{4.168}$$

Here, α is a Lagrange multiplier, and δ denotes variation with respect to v_k. Equation (4.168) becomes

$$\sum_k m_e(r_k - \alpha v_k)\delta v_k = 0 . \tag{4.169}$$

Because the δv_k's are treated as independent, Eq.(4.169) gives $v_k = \Omega r_k$, where the notation $\Omega = 1/\alpha$ has been introduced. It is readily verified that the extremum corresponds to a maximum when $v_k = \Omega r_k$. Therefore,

$$m_e \sum_k r_k(t)v_k(t) \leq m_e\Omega \sum_k r_k^2(t) . \tag{4.170}$$

But $K = (m_e/2)\sum_k v_k^2 = \Omega^2(m_e/2)\sum_k r_k^2$ gives $\Omega = (2K/m_e\sum_k r_k^2)^{1/2}$. Hence, Eq.(4.170) reduces to

$$m_e \sum_k r_k(t)v_k(t) \leq \left[2K(t)m_e \sum_k r_k^2(t)\right]^{1/2} . \tag{4.171}$$

Making use of $K(t) \leq U(0)$ [Eq.(4.166)], it then follows that

$$m_e \sum_k r_k(t)v_k(t) \leq \left[2U(0)m_e \sum_k r_k^2(t)\right]^{1/2} . \tag{4.172}$$

A similar inequality applies at $t = 0$.

Making use of Eqs.(4.172) and (4.167), we obtain

$$\sum_k r_k^2(t) \leq \sum_k r_k^2(0) + \left[\frac{8U(0)}{m_e\omega_{ce}^2}\sum_k r_k^2(t)\right]^{1/2} + \left[\frac{8U(0)}{m_e\omega_{ce}^2}\sum_k r_k^2(0)\right]^{1/2} . \tag{4.173}$$

Equation (4.173) can be solved to give the nonlinear bound[15]

$$\left[\sum_k r_k^2(t)\right]^{1/2} \leq \left[\sum_k r_k^2(0)\right]^{1/2} (1 + \eta) , \tag{4.174}$$

where η is defined by

$$\eta = \left[\frac{8U(0)}{m_e\omega_{ce}^2\sum_k r_k^2(0)}\right]^{1/2} . \tag{4.175}$$

Note that η^2 is equal to the ratio of the initial internal energy, $U(0) = \mathcal{E}_F(0) + K(0)$, to the kinetic energy $(m_e/2) \sum_k r_k^2(0)(\omega_{ce}/2)^2$ that would exist if all of the electrons were initially rotating with angular velocity $\omega_{ce}/2$. In the special case of a low-density plasma $(\omega_{pe}^2 \ll \omega_{ce}^2)$ with low rotational kinetic energy, it follows that $\eta \ll 1$, and Eq.(4.174) reduces to the expected result, $\sum_k r_k^2(t) \leq \sum_k r_k^2(0)$, for strongly magnetized electrons.

The maximum number of electrons $(\Delta \bar{N}_e)$ that could reach the conducting wall at $r = b$ and therefore be lost from the system is determined from $(\Delta \bar{N}_e)b^2 = \sum_k r_k^2$. Defining the mean-square initial radius (r_0^2) by $\bar{N}_e r_0^2 = \sum_k r_k^2(0)$, the inequality in Eq.(4.174) readily gives

$$\frac{\Delta \bar{N}_e}{\bar{N}_e} \leq \frac{r_0^2}{b^2} \left(1 + \eta\right)^2 , \qquad (4.176)$$

where η is defined in Eq.(4.175). As an example, for $\eta \ll 1$ and $r_0/b = 1/4$, Eq.(4.176) predicts that less than 6.25% of the electrons are lost radially from the system. In the present model, this is true no matter how complex the nonlinear evolution of the system due to collective processes and discrete-particle collisions. For example, the initial radial distribution of particles could correspond to a hollow annulus of cold electrons, which exhibits a strong collective instability known as the diocotron instability (Chapter 6).

Even for a stable plasma in which collective processes are weak, the behavior predicted by Eqs.(4.174) and (4.176) for a one-component nonneutral plasma is very different from that calculated for a neutral plasma consisting of electrons and ions. After sufficient time, radial transport due to collisions between electrons and ions will result in all of the particles being lost from the system. For a one-component nonneutral plasma, however, the conservation of canonical angular momentum in Eq.(4.165), which includes the effects of (like-particle) collisions, places a very stringent constraint on the evolution of $\sum_k r_k^2(t)$ and the fraction of particles lost from the system. Of course, the conclusions in Eqs.(4.174) and (4.176) will be modified by effects not incorporated in the present analysis, such as collisions with low-density background ions or neutral gas, or small (but finite) wall resistivity. In these circumstances, the quantity C_θ defined in Eq.(4.162) will not be conserved exactly.

To summarize, the conclusions in Secs. 4.5 and 4.6 are consistent with the robust stability and confinement properties of one-component nonneutral plasmas observed experimentally.[63] As a final point, the analysis in Sec. 4.6 can be extended[15] to include relativistic particle dynamics and the effects of fully electromagnetic field components, $\mathbf{B}^M(\mathbf{x}, t)$ and $\mathbf{E}^M(\mathbf{x}, t)$, associated with the microscopic current $\mathbf{J}^M(\mathbf{x}, t) = -e \sum_{k=1}^{\bar{N}_e} \times \mathbf{v}_k(t)\delta[\mathbf{x} - \mathbf{x}_k(t)]$.

Problem 4.14   The analysis in Sec. 4.6 can also be carried out for relativistic particle dynamics [T.M. O'Neil, Phys. Fluids **23**, 2216 (1980)]. Consider the combination

$$L = m_e \sum_k \gamma_k r_k v_k - \alpha K \,, \tag{4.14.1}$$

where α is a Lagrange multiplier, $\gamma_k = (1 - v_k^2/c^2)^{-1/2}$ is the relativistic mass factor, and $K = m_e c^2 \sum_k (\gamma_k - 1)$ is the kinetic energy.

a. Take the variation of Eq.(4.14.1) with respect to the v_k's and show that $\delta L = 0$ can be expressed as

$$m_e \sum_k (r_k - \alpha v_k)\gamma_k^3 \delta v_k = 0 \,. \tag{4.14.2}$$

Because the δv_k's are to be treated as independent, there is only one extremum, which occurs for $v_k = \Omega r_k$, where $\Omega = 1/\alpha$.

b. Show that the extremum is a maximum, and obtain the bound

$$m_e \sum_k \gamma_k r_k v_k \leq m_e \Omega \sum_k r_k^2 \left(1 - \Omega^2 r_k^2/c^2\right)^{-1/2} \,. \tag{4.14.3}$$

Here, the Lagrange multiplier ($\Omega = 1/\alpha$) is determined in terms of K by

$$K = m_e c^2 \sum_k \left[\left(1 - \Omega^2 r_k^2/c^2\right)^{-1/2} - 1\right]. \tag{4.14.4}$$

c. Make use of Taylor expansions to show that

$$x(1 - x)^{-1/2} \leq 2\left[(1 - x)^{-1/2} - 1\right] \tag{4.14.5}$$

for $0 < x < 1$. Use this result to obtain the bound

$$m_e \sum_k \gamma_k r_k v_k \leq \frac{2K}{\Omega} \,. \tag{4.14.6}$$

For nonrelativistic electron motion with $\Omega^2 r_k^2/c^2 \ll 1$, make use of Eq.(4.14.4) to show that Eq.(4.14.6) reduces to Eq.(4.170).

4.7   Electrostatic Eigenvalue Equation

In this section, use is made of the linearized Vlasov-Poisson equations to derive the formal eigenvalue equation[1-5] for electrostatic perturbations about the general class of nonrelativistic rigid-rotor Vlasov equilibria (Sec. 4.2) described by

$$f_e^0(r, \mathbf{p}) = f_e^0(H_\perp - \omega_{re} P_\theta, p_z) \, . \tag{4.177}$$

Here, the equilibrium configuration corresponds to an infinitely long nonneutral plasma column (Fig. 4.1) aligned parallel to a uniform applied magnetic field $B_0 \hat{e}_z$. For simplicity, the analysis is presented for a one-component nonneutral plasma consisting only of electrons. The final eigenvalue equation (4.195) is readily generalized to a multicomponent nonneutral plasma column. In Eq.(4.177), $p_z = m_e v_z$ is the axial momentum, and the perpendicular energy $H_\perp$ and canonical angular momentum P_θ are defined by

$$H_\perp = \frac{1}{2m_e} \left(p_r^2 + p_\theta^2 \right) - e\phi_0(r) \, ,$$

$$P_\theta = r \left(p_\theta - m_e r \omega_{ce}/2 \right) \, , \tag{4.178}$$

where $-e$ is the electron charge, and $\omega_{ce} = eB_0/m_e c$ is the electron cyclotron frequency. Evidently, the combination $H_\perp - \omega_{re} P_\theta$ can be expressed as

$$H_\perp - \omega_{re} P_\theta = \frac{1}{2m_e} \left[p_r^2 + (p_\theta - m_e \omega_{re} r)^2 \right] + \psi_e(r) \, , \tag{4.179}$$

where $\omega_{re} = $ const. is the average angular rotation velocity of the plasma column, and $\psi_e(r)$ is the effective potential defined by

$$\psi_e(r) = \frac{1}{2} m_e \left(\omega_{ce} \omega_{re} - \omega_{re}^2 \right) r^2 - e\phi_0(r) \, . \tag{4.180}$$

For specified equilibrium distribution function f_e^0, the electrostatic potential $\phi_0(r)$ in Eq.(4.180) is determined self-consistently from the steady-state Poisson equation

$$\frac{1}{r} \frac{\partial}{\partial r} r \frac{\partial}{\partial r} \phi_0(r) = 4\pi e \int d^3 p \, f_e^0(H_\perp - \omega_{re} P_\theta, p_z) \, . \tag{4.181}$$

A linear stability analysis of the Vlasov-Poisson equations (4.139) and (4.140) in the electrostatic approximation proceeds by expressing $f_e(\mathbf{x}, \mathbf{p}, t) = f_e^0(H_\perp - \omega_{re}P_\theta, p_z) + \delta f_e(\mathbf{x}, \mathbf{p}, t)$ and $\delta \mathbf{E}(\mathbf{x}, t) = -\nabla \delta\phi(\mathbf{x}, t)$. For small-amplitude perturbations, the linearized equations become (Sec. 2.2)

$$\left\{ \frac{\partial}{\partial t} + \mathbf{v} \cdot \frac{\partial}{\partial \mathbf{x}} - e\left(\mathbf{E}^0(\mathbf{x}) + \frac{\mathbf{v} \times B_0 \hat{\mathbf{e}}_z}{c} \right) \cdot \frac{\partial}{\partial \mathbf{p}} \right\} \delta f_e(\mathbf{x}, \mathbf{p}, t)$$

$$= -e \nabla \delta\phi(\mathbf{x}, t) \cdot \frac{\partial}{\partial \mathbf{p}} f_e^0(H_\perp - \omega_{re}P_\theta, p_z) , \qquad (4.182)$$

$$\nabla^2 \delta\phi(\mathbf{x}, t) = 4\pi e \int d^3 p \, \delta f_e(\mathbf{x}, \mathbf{p}, t) , \qquad (4.183)$$

where $\mathbf{E}^0(\mathbf{x}) = E_r(r)\hat{\mathbf{e}}_r$ is the equilibrium radial electric field, and $E_r(r) = -\partial\phi_0(r)/\partial r$. Equation (4.182) can be integrated formally in the laboratory frame using the method of characteristics (Problem 4.15). As an alternate approach, which simplifies the derivation of the dispersion relation for body-wave perturbations about a uniform-density plasma column (Sec. 4.8), we transform Eq.(4.182) from the independent variables $(\mathbf{x}, \mathbf{p}, t)$ in the laboratory frame to the independent variables $(\mathbf{x}', \mathbf{p}', t')$ in a frame of reference rotating with angular velocity $\omega_{re} = $ const. about the axis of symmetry. In cylindrical polar coordinates, the transformation is given by (Fig. 4.10)[1-5]

$$r'=r , \qquad \theta'=\theta - \omega_{re}t , \qquad z'=z ,$$
$$\qquad\qquad\qquad\qquad\qquad\qquad\qquad\qquad\qquad\qquad (4.184)$$
$$p_r'=p_r , \qquad p_\theta'=p_\theta - m_e r \omega_{re} , \qquad p_z'=p_z , \qquad t'=t .$$

Some straightforward algebra shows that Eqs.(4.182) and (4.183) can be expressed in the rotating frame as

$$\left\{ \frac{\partial}{\partial t'} + \mathbf{v}' \cdot \frac{\partial}{\partial \mathbf{x}'} - e\left(\mathbf{E}_{\text{eff}}^0(\mathbf{x}') + \frac{m_e}{e}(\omega_{ce} - 2\omega_{re}) \mathbf{v}' \times \hat{\mathbf{e}}_z \right) \cdot \frac{\partial}{\partial \mathbf{p}'} \right\} \delta f_e(\mathbf{x}', \mathbf{p}', t')$$

$$= -e \nabla' \delta\phi(\mathbf{x}', t') \cdot \frac{\partial}{\partial \mathbf{p}'} f_e^0(H_\perp' - \omega_{re}P_\theta', p_z') , \qquad (4.185)$$

$$\nabla'^2 \delta\phi(\mathbf{x}', t') = 4\pi e \int d^3 p' \delta f_e(\mathbf{x}', \mathbf{p}', t') . \qquad (4.186)$$

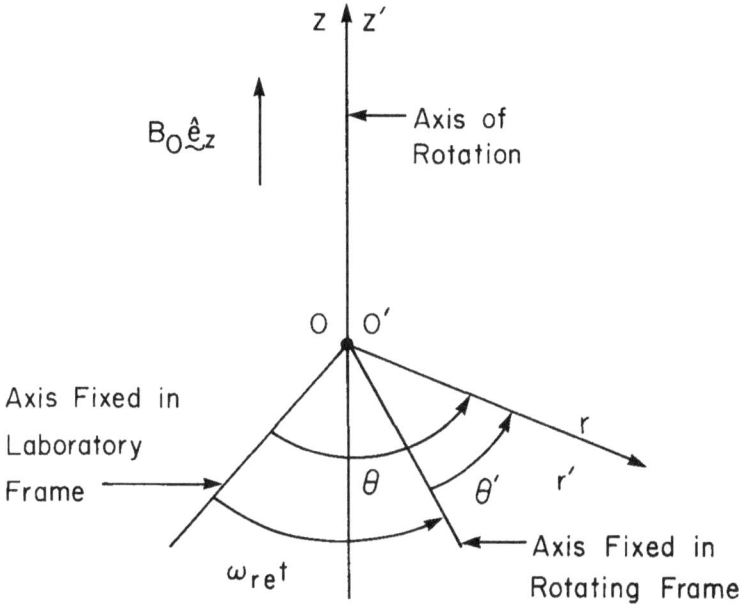

Figure 4.10. Equations (4.182) and (4.183) are transformed to a frame of reference rotating with angular velocity ω_{re} = const. about the axis of symmetry. The variables (r', θ', z') in the rotating frame are related to the variables (r, θ, z) in the laboratory frame by $r' = r$, $\theta' = \theta - \omega_{re}t$, and $z' = z$.

In Eqs.(4.185) and (4.186), $\mathbf{v}' = \mathbf{p}'/m_e$, $\nabla' = \partial/\partial\mathbf{x}'$, and $\mathbf{E}^0_{\text{eff}}(\mathbf{x}')$ is the effective electric field in the rotating frame defined by

$$e\mathbf{E}^0_{\text{eff}}(\mathbf{x}') = e\mathbf{E}^0(\mathbf{x}') + m_e \left(\omega_{ce}\omega_{re} - \omega_{re}^2\right)\mathbf{x}'_\perp$$

$$= \frac{\partial}{\partial r'}\psi_e(r')\mathbf{e}'_r . \tag{4.187}$$

Here, $\mathbf{x}'_\perp = r'\hat{\mathbf{e}}'_r$ where $\hat{\mathbf{e}}'_r$ is a unit vector in the r'-direction (in the rotating frame), and $\psi_e(r')$ is the effective potential defined in Eq.(4.180). The term $2m_e\omega_{re}\mathbf{v}' \times \hat{\mathbf{e}}'_z \cdot (\partial/\partial\mathbf{p}')\delta f_e$ in Eq.(4.185) arises from the Coriolis acceleration of the electrons in the rotating frame. Moreover, from Eqs.(4.179) and (4.184), the combination $H'_\perp - \omega_{re}P'_\theta$ occurring in the argument of f_e^0 in Eq.(4.185) can be expressed as

$$H'_\perp - \omega_{re}P'_\theta = \frac{1}{2m_e}\left(p_r'^2 + p_\theta'^2\right) + \psi_e(r') . \tag{4.188}$$

It follows from Eqs.(4.187) and (4.188) that $-\psi_e(r)/e$ plays the role of the effective electrostatic potential in the rotating frame.

Before proceeding with a formal integration of Eq.(4.185) for general equilibrium distribution function f_e^0, it is useful to point out the major simplification that occurs in circumstances where the equilibrium density is uniform with $n_e^0(r') = \hat{n}_e = $ const. and $\phi_0(r') = (m_e/4e)\omega_{pe}^2 r'^2$. In this case, it follows from Eq.(4.180) that $\psi_e(r) = (m_e/2)(\omega_{re}^+ - \omega_{re})(\omega_{re} - \omega_{re}^-)r'^2$, where $\omega_{re}^\pm = (\omega_{ce}/2)\left\{1 \pm (1 - 2\omega_{pe}^2/\omega_{ce}^2)^{1/2}\right\}$ are the laminar rotation frequencies defined in Eq.(3.6). Therefore, whenever $\omega_{re} \simeq \omega_{re}^-$ or $\omega_{re} \simeq \omega_{re}^+$ it follows that $\psi_e(r') \simeq 0$, which gives $\mathbf{E}_{\text{eff}}^0(\mathbf{x}') \simeq 0$ and $H_\perp' - \omega_{re}P_\theta' \simeq (2m_e)^{-1}(p_r'^2 + p_z'^2)$ in the rotating frame. The analysis of Eq.(4.185) then simplifies considerably (Sec. 4.8) because the only equilibrium force acting on the electrons is an effective magnetic force with $\omega_{ce} \to \omega_{ce} - 2\omega_{re}$.

We now return to Eqs.(4.185) and (4.186) for perturbations about general equilibrium distribution function $f_e^0(H_\perp' - \omega_{re}P_\theta', p_z')$. Using the method of characteristics, the formal solution to Eq.(4.185) in the rotating frame can be expressed as (Sec. 2.2)

$$\delta f_e(\mathbf{x}',\mathbf{p}',t') = -e\int_{-\infty}^{t'} dt'' \frac{\partial}{\partial \mathbf{x}''}\delta\phi(\mathbf{x}'',t'') \cdot \frac{\partial}{\partial \mathbf{p}''}f_e^0(H_\perp'' - \omega_{re}P_\theta'',p_z'') \ .$$

(4.189)

Here, amplifying perturbations have been assumed and the time integration is from $t'' = -\infty$ (where the perturbations have negligible amplitude) to $t'' = t'$. The orbits $\mathbf{x}''(t'')$ and $\mathbf{p}''(t'')$ occurring in the integrand in Eq.(4.189) correspond to the particle trajectories in the equilibrium field configuration in the rotating frame. That is, $\mathbf{x}''(t'')$ and $\mathbf{p}''(t'')$ are the solutions to

$$\frac{d}{dt''}\mathbf{x}''(t'') = \mathbf{v}''(t'') \ ,$$

$$\frac{d}{dt''}\mathbf{p}''(t'') = -e\left\{\mathbf{E}_{\text{eff}}^0\left[\mathbf{x}''(t'')\right] + \frac{m_e}{e}(\omega_{ce} - 2\omega_{re})\,\mathbf{v}''(t'') \times \hat{\mathbf{e}}_z'\right\} \ . \quad (4.190)$$

Here, $\mathbf{p}''(t'') = m_e\mathbf{v}''(t'')$ is the momentum, $\mathbf{E}_{\text{eff}}^0(\mathbf{x}'')$ is defined in Eq.(4.187), and the "initial" conditions are chosen such that

$$\mathbf{x}''(t'' = t') = \mathbf{x}' \ ,$$

$$\mathbf{p}''(t'' = t') = \mathbf{p}' \ , \quad (4.191)$$

where $(\mathbf{x}', \mathbf{p}')$ are the phase space variables in the rotating frame. For amplifying perturbations, $\delta\phi(\mathbf{x}', t')$ is expressed as

$$\delta\phi(\mathbf{x}', t') = \delta\phi(\mathbf{x}') \exp(-i\omega' t') \,, \tag{4.192}$$

where $\mathrm{Im}\,\omega' > 0$ corresponds to temporal growth, and $\delta\phi(\mathbf{x}')$ is the (complex) amplitude. Some straightforward algebra that makes use of Eq.(4.188) gives

$$-e\frac{\partial}{\partial\mathbf{x}''}\delta\phi(\mathbf{x}'') \cdot \frac{\partial}{\partial\mathbf{p}''}f_e^0(H_\perp'' - \omega_{re}P_\theta'', p_z'')$$

$$= -e\left[\mathbf{v}_\perp'' \cdot \frac{\partial}{\partial\mathbf{x}''}\delta\phi(\mathbf{x}'')\right]\frac{\partial}{\partial H_\perp''}f_e^0(H_\perp'' - \omega_{re}P_\theta'', p_z'')$$

$$- e\left[\frac{\partial}{\partial z''}\delta\phi(\mathbf{x}'')\right]\frac{\partial}{\partial p_z''}f_e^0(H_\perp'' - \omega_{re}P_\theta'', p_z'') \,, \tag{4.193}$$

where $\mathbf{v}_\perp''(t'') = \mathbf{p}_\perp''(t'')/m_e$ is the velocity perpendicular to $B_0\hat{\mathbf{e}}_z'$. Because $p_z'' = p_z'$ and $H_\perp'' - \omega_{re}P_\theta'' = H_\perp' - \omega_{re}P_\theta'$ are exactly conserved quantities by the orbit equations (4.190), the derivatives of $f_e(H_\perp'' - \omega_{re}P_\theta'', p_z'')$ in Eq.(4.193) can be taken outside the integral over t'' in Eq.(4.189). Expressing $\delta f_e(\mathbf{x}', \mathbf{p}', t') = \delta f_e(\mathbf{x}', \mathbf{p}')\exp(-i\omega' t')$, we obtain from Eq.(4.189)

$$\delta f_e(\mathbf{x}', \mathbf{p}') = -e\frac{\partial f_e^0}{\partial H_\perp'}\int_{-\infty}^{t'}dt'' \exp\left[-i\omega'(t'' - t')\right]\mathbf{v}_\perp'' \cdot \frac{\partial}{\partial\mathbf{x}''}\delta\phi(\mathbf{x}'')$$

$$- e\frac{\partial f_e^0}{\partial p_z'}\int_{-\infty}^{t'}dt'' \exp\left[-i\omega'(t'' - t')\right]\frac{\partial}{\partial z''}\delta\phi(\mathbf{x}'') \,. \tag{4.194}$$

Here, $f_e^0 = f_e^0(H_\perp' - \omega_{re}P_\theta', p_z')$, and the orbits $\mathbf{x}''(t'')$ and $\mathbf{v}''(t'')$ solve Eq.(4.190) subject to the initial conditions in Eq.(4.191). Substituting Eqs.(4.192) and (4.194) into Poisson's equation (4.186) then gives the desired eigenvalue equation

$$\nabla'^2\delta\phi(\mathbf{x}') = -4\pi e^2\int d^3p'\left\{\frac{\partial f_e^0}{\partial H_\perp'}\int_{-\infty}^0 d\tau \exp(-i\omega'\tau)\mathbf{v}_\perp'' \cdot \frac{\partial}{\partial\mathbf{x}''}\delta\phi(\mathbf{x}'')\right.$$

$$\left. + \frac{\partial f_e^0}{\partial p_z''}\int_{-\infty}^0 d\tau \exp(-i\omega'\tau)\frac{\partial}{\partial z''}\delta\phi(\mathbf{x}'')\right\} \,, \tag{4.195}$$

where $\tau \equiv t'' - t'$.

The eigenvalue equation (4.195), derived in a frame of reference rotating about the axis of symmetry with angular velocity $\omega_{re} =$ const., can be used to investigate detailed stability properties for electrostatic perturbations about the class of self-consistent Vlasov equilibria described by Eq.(4.177). For specified equilibrium distribution function f_e^0, the procedure is to solve Eq.(4.195) for both the eigenfunction $\delta\phi(\mathbf{x}')$ and the (complex) eigenfrequency ω'. Generally speaking, because the orbit $\mathbf{x}''(t'')$ occurs in the argument of $\delta\phi(\mathbf{x}'')$, analytical solutions to Eq.(4.195) are often difficult to obtain. Two examples that are analytically tractable are discussed in Secs. 4.8 and 4.9.

Problem 4.15

a. Make use of the method of characteristics to integrate Eq.(4.182) directly in the laboratory frame. Assume perturbations of the form $\delta\phi(\mathbf{x}, t) = \delta\phi(\mathbf{x}) \exp(-i\omega t)$ and $\delta f_e(\mathbf{x}, \mathbf{p}, t) = \delta f_e(\mathbf{x}, \mathbf{p}) \exp(-i\omega t)$, and show that

$$\delta f_e(\mathbf{x}, \mathbf{p}) = -e \frac{\partial f_e^0}{\partial H_\perp} \int_{-\infty}^0 d\tau \, \exp(-i\omega\tau) \left(\mathbf{v}'_\perp \cdot \frac{\partial}{\partial \mathbf{x}'} - \omega_{re} \frac{\partial}{\partial \theta'} \right) \delta\phi(\mathbf{x}')$$

$$- e \frac{\partial f_e^0}{\partial p_z} \int_{-\infty}^0 d\tau \, \exp(-i\omega\tau) \frac{\partial}{\partial z'} \delta\phi(\mathbf{x}') \,, \tag{4.15.1}$$

where $\tau = t' - t$ and $f_e^0 = f_e^0(H_\perp - \omega_{re} P_\theta, p_z)$. In the integrand in Eq.(4.15.1), $\mathbf{x}'(t')$ and $\mathbf{p}'(t') = m_e \mathbf{v}'(t')$ solve the *laboratory-frame* orbit equations

$$\frac{d}{dt'} \mathbf{x}'(t') = \mathbf{v}'(t') \,,$$

$$\frac{d}{dt'} \mathbf{p}'(t') = -e \left\{ \mathbf{E}^0 \left[\mathbf{x}'(t') \right] + \frac{\mathbf{v}'(t') \times B_0 \hat{\mathbf{e}}_z}{c} \right\} \,, \tag{4.15.2}$$

subject to the "initial" conditions $\mathbf{x}'(t' = t) = \mathbf{x}$ and $\mathbf{p}'(t' = t) = \mathbf{p}$. [The "primed" orbit notation in Eqs.(4.15.1) and (4.15.2) should not be confused with the coordinate transformation to the rotating frame in Eq.(4.184).]

b. Make use of Eqs.(4.183) and (4.15.1) to derive the formal eigenvalue equation for $\delta\phi(\mathbf{x})$ in the laboratory frame.

4.8   Dispersion Relation for Body-Wave Perturbations

4.8.1   Electrostatic Dispersion Relation

We now examine the eigenvalue equation (4.195) in circumstances where the equilibrium density is approximately uniform in the column interior $(0 \leq r \lesssim r_b)$ with $n_e^0(r) = \int d^3p \, f_e^0 = \hat{n}_e = $ const., and the electrostatic potential is parabolic with $\phi_0(r) = (m_e/4e)\omega_{pe}^2 r^2$, where $\omega_{pe}^2 = 4\pi\hat{n}_e e^2/m_e$. In this case, the effective potential $\psi_e(r)$ defined in Eq.(4.180) can be approximated by

$$\psi_e(r) = \frac{1}{2}m_e(\omega_{re}^+ - \omega_{re})(\omega_{re} - \omega_{re}^-)r^2 \qquad (4.196)$$

for $0 \leq r \lesssim r_b$, where ω_{re}^+ and ω_{re}^- are the laminar rotation frequencies defined by

$$\omega_{re}^{\pm} = \frac{1}{2}\omega_{ce}\left\{1 \pm \left(1 - \frac{2\omega_{pe}^2}{\omega_{ce}^2}\right)^{1/2}\right\} . \qquad (4.197)$$

As discussed in Sec. 4.2.1, for the entire class of rigid-rotor Vlasov equilibria described by Eq.(4.177), whenever ω_{re} is closely tuned to either ω_{re}^+ or ω_{re}^- the density profile is approximately uniform over a broad interior region of the plasma column. For $\omega_{re} \simeq \omega_{re}^{\pm}$ it also follows from Eqs.(4.187) and (4.196) that $\psi_e(r') \simeq 0$ and $\mathbf{E}_{\text{eff}}^0(\mathbf{x}') = -(\partial\psi_e/\partial r')\hat{\mathbf{e}}_r' \simeq 0$ in rotating-frame variables. Of course the analysis of the orbit equations (4.190) and the eigenvalue equation (4.195) simplifies accordingly.

We therefore assume $\omega_{re} \simeq \omega_{re}^{\pm}$ and approximate

$$\psi_e(r') = 0 , \quad \mathbf{E}_{\text{eff}}^0(\mathbf{x}') = 0 ,$$

$$H_{\perp}' - \omega_{re}P_{\theta}' = \frac{1}{2m_e}\left(p_r'^2 + p_{\theta}'^2\right) , \qquad (4.198)$$

in the column interior $(0 \leq r \lesssim r_b)$. Moreover, in the region of uniform density, the distribution function $f_e^0(H_{\perp}' - \omega_{re}P_{\theta}', p_z')$ occurring in the eigenvalue equation (4.195) is expressed as

$$f_e^0(H_{\perp}' - \omega_{re}P_{\theta}', p_z') = \hat{n}_e F_e(p_{\perp}'^2, p_z') , \qquad (4.199)$$

where $p_\perp'^2 = p_r'^2 + p_\theta'^2$, and the normalization of F_e is $\int d^3 p' \, F_e(p_\perp'^2, p_z') = 1$. In the present analysis of Eqs.(4.190) and (4.195), it is further assumed that the potential perturbation*

$$\delta\phi(\mathbf{x}', t') = \int d^3 k' \, \delta\phi(\mathbf{k}') \exp(i\mathbf{k}' \cdot \mathbf{x}' - i\omega' t') \qquad (4.200)$$

is localized to the column interior with $k_\perp'^2 r_b^2 \gg 1$. Substituting Eqs.(4.199) and (4.200) into Eq.(4.195), some straightforward algebra gives[1-5]

$$D(k_\perp', k_z', \omega')\delta\phi(\mathbf{k}') = 0 \qquad (4.201)$$

where the plasma dielectric function $D(k_\perp', k_z', \omega')$ is defined by

$$D(k_\perp', k_z', \omega') = 1 - \frac{\omega_{pe}^2}{(k_\perp'^2 + k_z'^2)}$$

$$\times \int d^3 p' \left\{ \frac{1}{v_\perp'} \frac{\partial F_e}{\partial v_\perp'} \int_{-\infty}^0 d\tau \, i k_\perp' \cdot \mathbf{v}_\perp''(t'') \exp\{i\mathbf{k}' \cdot [\mathbf{x}''(t'') - \mathbf{x}'] - i\omega'\tau\} \right.$$

$$\left. + i k_z' \frac{\partial F_e}{\partial v_z'} \int_{-\infty}^0 d\tau \, \exp\{i\mathbf{k}' \cdot [\mathbf{x}''(t'') - \mathbf{x}'] - i\omega'\tau\} \right\}. \qquad (4.202)$$

Here, $\tau = t'' - t'$ is the time integration variable, ω' is the (complex) oscillation frequency with $\operatorname{Im}\omega' > 0$ corresponding to instability, $\mathbf{k}_\perp' = \mathbf{k}' - k_z' \hat{\mathbf{e}}_z'$ is the perpendicular wavenumber in the rotating frame, and v_z' and $v_\perp'$ are defined by $v_z' = p_z'/m_e$ and $v_\perp'^2 = p_\perp'^2/m_e^2$.

---

*As an alternate representation, paralleling the analysis by Davidson and Krall,[5] it should be noted that Eq.(4.200) can be expressed in the equivalent form

$$\delta\phi(\mathbf{x}', t') = \int_{-\infty}^\infty dk_z' \int_0^\infty dk_\perp' \, k_\perp' \sum_{\ell = -\infty}^\infty \delta\phi_\ell(k_\perp', k_z') J_\ell(k_\perp' r') \exp\left[i(\ell\theta' + k_z' z' - \omega' t')\right],$$

where

$$\delta\phi_\ell(k_\perp', k_z') = (i)^\ell \int_{-\pi}^\pi d\alpha' \delta\phi(\mathbf{k}') \exp(-i\ell\alpha').$$

Here, we have expressed $\mathbf{k}_\perp' \cdot \mathbf{x}_\perp' = k_\perp' r'(\cos\alpha'\cos\theta' + \sin\alpha'\sin\theta') = k_\perp' r'\cos(\theta' - \alpha')$, where $k_x' = k_\perp' \cos\alpha'$, $k_y' = k_\perp' \sin\alpha'$, and $r' = (x'^2 + y'^2)^{1/2}$. Moreover, use has been made of the Bessel function identity in Eq.(4.206) to expand

$$\exp\left[ik_\perp' r' \cos(\theta' - \alpha')\right] = \sum_{\ell = -\infty}^\infty (i)^\ell J_\ell(k_\perp' r') \exp\left[i\ell(\theta' - \alpha')\right].$$

The orbits $\mathbf{x}''(t'')$ and $\mathbf{v}''(t'')$ occurring in the integrals in Eq.(4.202) are determined from Eq.(4.190) with $\mathbf{E}_{\text{eff}}^0(\mathbf{x}'') = 0$. In Cartesian coordinates, the axial motion is free-streaming with $v_z'' = v_z'$ and $z'' = z' + v_z'(t'' - t')$, and the perpendicular orbits correspond to circular gyrations determined from (see also Sec. 3.2)

$$\frac{d}{dt''}v_x'' = -(\omega_{ce} - 2\omega_{re})v_y'' \, ,$$

$$\frac{d}{dt''}v_y'' = (\omega_{ce} - 2\omega_{re})v_x'' \, . \tag{4.203}$$

Here, for $\omega_{re} = \omega_{re}^{\pm}$, it follows that

$$\omega_{ce} - 2\omega_{re} = \omega_{ve} = \mp \left(\omega_{re}^+ - \omega_{re}^-\right) \, , \tag{4.204}$$

where $\omega_{ve} = +(\omega_{re}^+ - \omega_{re}^-)$ corresponds to a *slow* rotational equilibrium with $\omega_{re} = \omega_{re}^-$, and $\omega_{ve} = -(\omega_{re}^+ - \omega_{re}^-)$ corresponds to a *fast* rotational equilibrium with $\omega_{re} = \omega_{re}^+$. Solving Eq.(4.203) subject to the boundary conditions $\mathbf{x}''(t'' = t') = \mathbf{x}'$ and $\mathbf{v}''(t'' = t') = \mathbf{v}'$ gives

$$v_x'' = v_\perp' \cos(\phi' + \omega_{ve}\tau) \, ,$$

$$v_y'' = v_\perp' \sin(\phi' + \omega_{ve}\tau) \, ,$$

$$x'' = x' + \left(\frac{v_\perp'}{\omega_{ve}}\right)[\sin(\phi' + \omega_{ve}\tau) - \sin\phi'] \, ,$$

$$y'' = y' - \left(\frac{v_\perp'}{\omega_{ve}}\right)[\cos(\phi' + \omega_{ve}\tau) - \cos\phi'] \, , \tag{4.205}$$

where $v_x' = v_\perp' \cos\phi'$, $v_y' = v_\perp' \sin\phi'$, and ϕ' is the perpendicular velocity phase in the rotating frame. Equation (4.205) is substituted into the orbit integrals in Eq.(4.202), and use is made of the Bessel function representation

$$\exp(ib\sin\theta) = \sum_{n=-\infty}^{\infty} J_n(b)\exp(in\theta) \, , \tag{4.206}$$

where $J_n(b)$ is the Bessel function of the first kind of order n. Paralleling the analysis in the neutral plasma case,[64] we express $\int d^3p' \cdots = \int_{-\infty}^{\infty} dp_z' \int_0^{\infty} dp_\perp' \, p_\perp' \int_0^{2\pi} d\phi' \cdots$, and carry out the integrations over ϕ' in

Eq.(4.202). Use is also made of

$$\int_{-\infty}^{0} d\tau \, \exp[-i(\omega' - k_z' v_z' - n\omega_{ve})\tau] = \frac{i}{(\omega' - k_z' v_z' - n\omega_{ve})} \,, \qquad (4.207)$$

where $\text{Im}\,\omega' > 0$ is assumed. Some straightforward algebra gives the desired electrostatic dispersion relation[1-5]

$$0 = D(k_\perp', k_z', \omega') = 1 + \frac{\omega_{pe}^2}{(k_\perp'^2 + k_z'^2)} \sum_{n=-\infty}^{\infty} \int d^3p' \, J_n^2 \left(\frac{k_\perp' v_\perp'}{\omega_{re}^+ - \omega_{re}^-} \right)$$

$$\times \frac{\left[\frac{n(\omega_{re}^+ - \omega_{re}^-)}{v_\perp'} \frac{\partial}{\partial v_\perp'} + k_z' \frac{\partial}{\partial v_z'} \right] F_e(p_\perp'^2, p_z')}{\omega' - k_z' v_z' - n(\omega_{re}^+ - \omega_{re}^-)} \,, \qquad (4.208)$$

where $\int d^3p' \cdots = 2\pi \int_{-\infty}^{\infty} dp_z' \int_0^{\infty} dp_\perp' \, p_\perp' \cdots$. In Eq.(4.208), $\text{Im}\,\omega' > 0$ is assumed, and use has been made of $\omega_{ce} - 2\omega_{re} = \mp(\omega_{re}^+ - \omega_{re}^-)$ for $\omega_{re} = \omega_{re}^\pm$. [Note that the integrand in Eq.(4.208) remains unchanged for $(\omega_{re}^+ - \omega_{re}^-) \to -(\omega_{re}^+ - \omega_{re}^-)$ and $n \to -n$.] To summarize other definitions: $k_z' = \mathbf{k}' \cdot \hat{\mathbf{e}}_z'$, $\mathbf{k}_\perp' = \mathbf{k}' - k_z' \hat{\mathbf{e}}_z'$, $k_\perp' = |\mathbf{k}_\perp'|$, $v_z' = \mathbf{v}' \cdot \hat{\mathbf{e}}_z'$, $\mathbf{v}_\perp' = \mathbf{v}' - v_z' \hat{\mathbf{e}}_z'$, and $v_\perp' = |\mathbf{v}_\perp'|$.

The dispersion relation (4.208) determines the (complex) oscillation frequency ω' in terms of $k_\perp'$, k_z' and properties of the equilibrium distribution function $F_e(p_\perp'^2, p_z')$. In particular, Eq.(4.208) is the electrostatic dispersion relation in a frame of reference rotating with angular velocity $\omega_{re}(\simeq \omega_{re}^\pm)$ about the axis of symmetry (Fig. 4.10), that is, in a frame of reference corotating with the equilibrium. To obtain the corresponding dispersion relation in the laboratory frame, it is necessary to relate $(k_\perp', k_z', \omega')$ in the rotating frame to $(k_\perp, k_z, \omega)$ in the laboratory frame. For perturbations with azimuthal mode number ℓ in the laboratory frame, the relation is

$$\omega' = \omega - \ell\omega_{re} \,,$$

$$k_\perp' = k_\perp \,,$$

$$k_z' = k_z \,. \qquad (4.209)$$

Since the equilibrium is azimuthally symmetric, the only effect is to Doppler shift the frequency by $\ell\omega_{re}$. Dropping the prime notation on

$\mathbf{p}'$ and $\mathbf{v}'$ in Eq.(4.208), the corresponding dispersion relation in the laboratory frame can be expressed as[1-5]

$$0 = D_\ell(k_\perp, k_z, \omega) = 1 + \frac{\omega_{pe}^2}{(k_\perp^2 + k_z^2)} \sum_{n=-\infty}^{\infty} \int d^3p \, J_n^2 \left(\frac{k_\perp v_\perp}{\omega_{re}^+ - \omega_{re}^-} \right)$$

$$\times \frac{\left[\frac{n(\omega_{re}^+ - \omega_{re}^-)}{v_\perp} \frac{\partial}{\partial v_\perp} + k_z \frac{\partial}{\partial v_z} \right] F_e(p_\perp^2, p_z)}{\omega - \ell\omega_{re} - k_z v_z - n(\omega_{re}^+ - \omega_{re}^-)}. \tag{4.210}$$

Equation (4.210) is the electrostatic dispersion relation for body-wave perturbations in a nonneutral plasma column. Keep in mind the range of applicability of Eq.(4.210). First, it has been assumed that $\omega_{re} \simeq \omega_{re}^\pm$ and that the electron density is uniform in the column interior $(0 \leq r \lesssim r_b)$. Second, boundary effects for $r \gtrsim r_b$ have been neglected, and the analysis is restricted to electrostatic perturbations localized to the column interior with $r < r_b$ and $k_\perp^2 r_b^2 \gg 1$ (so-called "body-wave" perturbations). Finally, Eq.(4.210) has been derived for a pure electron plasma. The analogous dispersion relation for a pure ion plasma is identical to Eq.(4.210) with the replacements: $F_e(p_\perp^2, p_z) \rightarrow F_i(p_\perp^2, p_z)$; $\omega_{pe}^2 \rightarrow \omega_{pi}^2 = 4\pi \hat{n}_i Z_i^2 e^2/m_i$; $\omega_{re} \rightarrow \omega_{ri} = \omega_{ri}^\pm = -(\omega_{ci}/2)[1 \pm (1 - 2\omega_{pi}^2/\omega_{ci}^2)^{1/2}]$; and $\omega_{re}^+ - \omega_{re}^- \rightarrow \omega_{ri}^+ - \omega_{ri}^-$.

It is straightforward to extend the previous analysis in Secs. 4.7 and 4.8 to a multicomponent nonneutral plasma column, assuming that the equilibrium distribution function for each plasma component is of the form $f_j^0(H_\perp - \omega_{rj} P_\theta, p_z)$, and that ω_{rj} is closely tuned to the laminar rotation velocity ω_{rj}^+ or ω_{rj}^- defined by [see Eq.(4.35)]

$$\omega_{rj}^\pm = -\frac{\varepsilon_j \omega_{cj}}{2} \left\{ 1 \pm \left(1 - \frac{\sum_k 8\pi e_j \hat{n}_k e_k}{m_j \omega_{cj}^2} \right)^{1/2} \right\}. \tag{4.211}$$

Here, $\varepsilon_j = \mathrm{sgn}\, e_j$ and $\omega_{cj} = |e_j| B_0/m_j c$. If $\omega_{rj} \simeq \omega_{rj}^+$ or $\omega_{rj} \simeq \omega_{rj}^-$, the density of each plasma component is approximately uniform $[n_j^0(r) \simeq \hat{n}_j = \mathrm{const.}]$ in the column interior $(0 \leq r \lesssim r_b)$, and $H_\perp' - \omega_{rj} P_\theta' \simeq p_\perp'^2/2m_j$ in a frame of reference rotating with angular velocity ω_{rj}. In the multicomponent case, the electrostatic dispersion relation for body-wave perturbations in the column interior can be expressed as

$$0 = D_\ell(k_\perp, k_z, \omega) = 1 + \sum_j \frac{\omega_{pj}^2}{(k_\perp^2 + k_z^2)} \sum_{n=-\infty}^{\infty} \int d^3p \, J_n^2 \left(\frac{k_\perp v_\perp}{\omega_{rj}^+ - \omega_{rj}^-} \right)$$

$$\times \frac{\left[\dfrac{n(\omega_{rj}^+ - \omega_{rj}^-)}{v_\perp} \dfrac{\partial}{\partial v_\perp} + k_z \dfrac{\partial}{\partial v_z} \right] F_j(p_\perp^2, p_z)}{\omega - \ell\omega_{rj} - k_z v_z - n(\omega_{rj}^+ - \omega_{rj}^-)} . \qquad (4.212)$$

Equation (4.212) is the appropriate generalization of Eq.(4.210) to a multicomponent nonneutral plasma. It relates the wavevector $\mathbf{k}$ and complex oscillation frequency ω (in the laboratory frame) for spatial perturbations with azimuthal mode number ℓ. If no ions are present in the system, and the plasma consists of a single component of electrons (rotating with mean angular velocity $\omega_{re} = \omega_{re}^+$ or $\omega_{re} = \omega_{re}^-$), then Eq.(4.212) reduces to Eq.(4.210). On the other hand, if the plasma is electrically neutral $\sum_k \hat{n}_k e_k = 0$, and each component is in the slow rotational mode $(\omega_{rj} = \omega_{rj}^-)$, Eq.(4.212) reduces to the familiar dispersion relation for electrostatic perturbations in a uniform neutral plasma.[64] This follows because $\omega_{rj}^+ - \omega_{rj}^- = -\varepsilon_j \omega_{cj}$ and $\omega_{rj}^- = 0$ for $\sum_k \hat{n}_k e_k = 0$ [see Eq.(4.211)].

Because Eqs.(4.210) and (4.212) are similar in form to the corresponding dispersion relations for a neutral plasma, many of the waves and instabilities that depend on the detailed momentum-space structure of $F_j(p_\perp^2, p_z)$ have their analogues in a uniform-density nonneutral plasma in circumstances where the present analysis is applicable. In fact, the results of a large body of neutral plasma literature[64] can be applied virtually intact with the replacements $\omega \to \omega - \ell\omega_{rj}$ and $-\varepsilon_j \omega_{cj} \to \pm(\omega_{rj}^+ - \omega_{rj}^-)$. Specific examples are discussed in Sec. 4.8.2.

As a final point, the electrostatic dispersion relation (4.212) can be extended[65] to the case of a relativistic, multicomponent, nonneutral plasma, allowing for the propagation of a relativistic electron beam $(j = b)$ with $\nu/\gamma_b \ll 1$ through a uniform background plasma $(j = e, i)$ parallel to the axial guide field $B_0\hat{\mathbf{e}}_z$. Here, ν is Budker's parameter defined in Eq.(4.129). The generalization[65] of Eq.(4.212) includes the influence of the azimuthal self-magnetic field $B_\theta^s(r)\hat{\mathbf{e}}_\theta$ produced by the net axial current $J_z^0(r) = \sum_j n_j^0(r)e_j V_{zj}^0(r)$, assumed uniform over the beam-plasma cross section, as well as the effects of the radial self-electric field $E_r(r)\hat{\mathbf{e}}_r$ associated with the lack of overall charge neutrality.

Problem 4.16   To simplify the orbit integrals in Eq.(4.202), assume $\mathbf{k}' = k'_\perp \hat{\mathbf{e}}_x + k'_z \hat{\mathbf{e}}_z$.

a. Make use of Eqs.(4.205) and (4.206) to express

$$\exp\{i\mathbf{k}' \cdot [\mathbf{x}''(t'') - \mathbf{x}'] - i\omega'(t'' - t')\}$$

$$= \sum_{n=-\infty}^{\infty} \sum_{m=-\infty}^{\infty} J_n\left(\frac{k'_\perp v'_\perp}{\omega_{ve}}\right) J_m\left(\frac{k'_\perp v'_\perp}{\omega_{ve}}\right) \exp[i(n-m)\phi']$$

$$\times \exp[-i(\omega' - k'_z v'_z - n\omega_{ve})\tau] , \qquad (4.16.1)$$

where $\tau = t'' - t'$, and $\omega_{ve} = \omega_{ce} - 2\omega_{re}$.

b. Show from Eqs.(4.205) and (4.16.1) that

$$\int_0^{2\pi} \frac{d\phi'}{2\pi} \exp\{i\mathbf{k}' \cdot [\mathbf{x}''(t'') - \mathbf{x}'] - i\omega'\tau\}$$

$$= \sum_{n=-\infty}^{\infty} J_n^2\left(\frac{k'_\perp v'_\perp}{\omega_{ve}}\right) \exp\{-i[\omega' - k'_z v'_z - n\omega_{ve}]\tau\} , \qquad (4.16.2)$$

and

$$\int_0^{2\pi} \frac{d\phi'}{2\pi} \mathbf{k}'_\perp \cdot \mathbf{v}''_\perp(t'') \exp\{i\mathbf{k}' \cdot [\mathbf{x}''(t'') - \mathbf{x}'] - i\omega'\tau\}$$

$$= \sum_{n=-\infty}^{\infty} n\omega_{ve} J_n^2\left(\frac{k'_\perp v'_\perp}{\omega_{ve}}\right) \exp[-i(\omega' - k'_z v'_z - n\omega_{ve})\tau] . \quad (4.16.3)$$

Here, ϕ' is the perpendicular momentum (velocity) phase occurring in Eq.(4.205) and $\int d^3 p' \cdots = \int_0^{2\pi} d\phi' \int_0^\infty dp'_\perp \, p'_\perp \int_{-\infty}^{\infty} dp'_z \cdots$, and use has been made of $J_{n+1}(x) + J_{n-1}(x) = (2n/x) J_n(x)$.

c. Substitute Eqs.(4.16.2) and (4.16.3) into Poisson's equation (4.202) and carry out the integrations over τ according to Eq.(4.207) with $\operatorname{Im}\omega' > 0$. Show that the resulting electrostatic dispersion relation can be expressed in the form given in Eq.(4.208).

4.8.2   Examples of Electrostatic Waves and Instabilities

In this section, some of the electrostatic waves and instabilities characteristic of body-wave perturbations in a nonneutral electron plasma are discussed.[1-5] Use is made of the dispersion relation (4.210) and the algorithms for obtaining stability information for a pure electron plasma from the corresponding results for a neutral plasma, i.e.,

$$\omega \to \omega - \ell\omega_{re} \,, \quad \omega_{ce} \to \pm \left(\omega_{re}^+ - \omega_{re}^-\right) \,, \quad m_i \to \infty \,. \qquad (4.213)$$

A. Electron Plasma Oscillations at Brillouin Flow

Equation (4.210) is valid for electron density in the range $0 < 2\omega_{pe}^2/\omega_{ce}^2 < 1$. In the limit of Brillouin flow, where $2\omega_{pe}^2/\omega_{ce}^2 = 1$, note from Eq.(4.197) and Fig. 3.3 that $\omega_{re}^\pm = \omega_{ce}/2$ and $\omega_{re}^+ - \omega_{re}^- = \omega_{ce}(1 - 2\omega_{pe}^2/\omega_{ce}^2)^{1/2} = 0$. From Eq.(4.213), the analogous limit in the neutral plasma case is the zero magnetic field limit ($\omega_{ce} \to 0$). Therefore, at Brillouin flow, the dispersion relation in Eq.(4.202) can be expressed as

$$0 = 1 + \frac{\omega_{pe}^2}{k^2} \int d^3p \, \frac{\mathbf{k} \cdot (\partial/\partial\mathbf{v})}{\omega' - \mathbf{k} \cdot \mathbf{v}} F_e(p_\perp^2, p_z) \,, \qquad (4.214)$$

where $\omega' = \omega - \ell\omega_{ce}/2$, $k^2 = k_\perp^2 + k_z^2$, and $\mathbf{v} = \mathbf{p}/m_e$. Except for the Doppler shift in frequency by $\ell\omega_{ce}/2$, Eq.(4.214) is identical to the dispersion relation for electrostatic perturbations in a uniform, *unmagnetized* neutral plasma,[64,66] ignoring the ion dynamics ($m_i \to \infty$). Depending on the detailed form of $F_e(p_\perp^2, p_z)$, Eq.(4.214) can support solutions corresponding to instability (Im $\omega > 0$). These include the gentle bump-in-tail instability, and the classical two-stream instability [e.g., if the p_z-dependence of $F_e(p_\perp^2, p_z)$ corresponds to two cold, counterstreaming electron components.][67]

As an example corresponding to stable oscillations (Im $\omega < 0$), consider the isotropic thermal equilibrium distribution function specified by [Eq.(3.28)]

$$f_e^0(H_\perp - \omega_{re}P_\theta, p_z) = \frac{\hat{n}_e}{(2\pi m_e k_B T_e)^{3/2}} \exp\left[-\frac{(H - \omega_{re}P_\theta)}{k_B T_e}\right] \,, \qquad (4.215)$$

where $H = H_\perp + p_z^2/2m_e$. Comparing Eqs.(4.199) and (4.215) for $\omega_{re} \simeq \omega_{re}^\pm$ and $0 \le r \lesssim r_b$, it follows that

$$F_e(p_\perp^2, p_z) = \frac{1}{(2\pi m_e k_B T_e)^{3/2}} \exp\left[-\frac{(p_\perp^2 + p_z^2)}{2m_e k_B T_e}\right] \,, \qquad (4.216)$$

where the "prime" notation has been dropped in Eq.(4.216). Dividing $\omega = \text{Re}\,\omega + i\,\text{Im}\,\omega$ into real and imaginary parts, and assuming long-wavelength perturbations with $k^2\lambda_{De}^2 = k^2 k_B T_e/4\pi \hat{n}_e e^2 \ll 1$, it is readily shown[64,66] from Eq.(4.214) that

$$(\text{Re}\,\omega - \ell\omega_{ce}/2)^2 = \omega_{pe}^2 \left(1 + 3k^2\lambda_{De}^2 + \cdots\right) , \tag{4.217}$$

and

$$\text{Im}\,\omega = -\left(\frac{\pi}{8}\right)^{1/2} \frac{\omega_{pe}}{(|k|\lambda_{De})^3} \exp\left(-\frac{1}{2k^2\lambda_{De}^2} - \frac{3}{2}\right) . \tag{4.218}$$

In other words, for wavelengths long in comparison with the electron thermal Debye length λ_{De}, the system supports weakly damped electron plasma oscillations.

B.   Stable Oscillations in Thermal Equilibrium Plasma

We now allow for general value of the self-field parameter $s_e = 2\omega_{pe}^2/\omega_{ce}^2$ in the interval $0 < 2\omega_{pe}^2/\omega_{ce}^2 < 1$ with $\omega_{re}^+ - \omega_{re}^- = \omega_{ce}(1 - 2\omega_{pe}^2/\omega_{ce}^2)^{1/2} \neq 0$. It is further assumed that $\omega_{re} = \omega_{re}^+$ or $\omega_{re} = \omega_{re}^-$ and that the electrons are distributed according to the thermal equilibrium distribution in Eq.(4.216) [or Eq.(4.215)]. It is convenient to introduce the plasma dispersion function $Z(\xi_n)$ defined by

$$Z(\xi_n) = \frac{1}{\sqrt{\pi}} \int_{-\infty}^{\infty} \frac{dx\,\exp(-x^2)}{x - \xi_n} , \tag{4.219}$$

where

$$\xi_n = \frac{[\omega - \ell\omega_{re} - n(\omega_{re}^+ - \omega_{re}^-)]}{k_z v_{Te}} , \tag{4.220}$$

and $v_{Te} = (2k_B T_e/m_e)^{1/2}$ is the electron thermal speed. Substituting Eq.(4.216) into Eq.(4.210) and making use of Eq.(4.219), the integrations over $p_\perp$ and p_z can be carried out to give the electrostatic dispersion relation (Problem 4.17)

$$0 = D_\ell(k_\perp, k_z, \omega)$$

$$= 1 + \frac{1}{k^2\lambda_{De}^2}\left[1 + \frac{(\omega - \ell\omega_{re})}{k_z v_{Te}} \sum_{n=-\infty}^{\infty} \exp(-\lambda_e)I_n(\lambda_e)Z(\xi_n)\right] . \tag{4.221}$$

Here, $\lambda_{\mathrm{De}}^2 = k_B T_e / 4\pi \hat{n}_e e^2$, $k^2 = k_\perp^2 + k_z^2$, $\lambda_e = k_\perp^2 k_B T_e / m_e (\omega_{re}^+ - \omega_{re}^-)^2$, $I_n(\lambda_e)$ is the modified Bessel function of the first kind of order n, and use has been made of $\sum_{n=-\infty}^{\infty} J_n^2(x) = 1$ in deriving Eq.(4.221).

The kinetic dispersion relation (4.221) can be used to investigate detailed wave properties for electrostatic perturbations about thermal equilibrium for a broad range of dimensionless system parameters $k^2 \lambda_{\mathrm{De}}^2$, λ_e, $k_z^2/k_\perp^2$ and ξ_n. Because the thermal equilibrium distribution in Eq.(4.216) [or Eq.(4.215)] is monotonic decreasing, it follows from the stability theorem developed in Sec. 4.5 that all solutions to Eq.(4.221) are stable, i.e., the solutions satisfy either $\mathrm{Im}\,\omega = 0$ (pure oscillatory solutions) or $\mathrm{Im}\,\omega < 0$ (damped solutions). Particularly useful for the analysis of Eq.(4.221) in limiting regimes corresponding to small $k_z v_{\mathrm{Te}}$ or large $k_z v_{\mathrm{Te}}$ are the asymptotic expansions[64]

$$Z(\xi_n) = -\frac{1}{\xi_n} - \frac{1}{2\xi_n^3} - \frac{3}{4\xi_n^5} - \cdots, \text{ for } |\xi_n| \gg 1 , \qquad (4.222)$$

and

$$Z(\xi_n) = -2\xi_n + \frac{4}{3}\xi_n^3 - \cdots + i\sqrt{\pi}\frac{k_z}{|k_z|}\exp(-\xi_n^2), \text{ for } |\xi_n| \ll 1 . \quad (4.223)$$

For present purposes, Eq.(4.221) is simplified in the limiting cases corresponding to: (a) parallel propagation ($k_z \neq 0$, $k_\perp = 0$); (b) perpendicular propagation ($k_\perp \neq 0$, $k_z = 0$); and (c) arbitrary propagation angle in a cold plasma ($v_{\mathrm{Te}} \to 0$).

C.   Wave Propagation Parallel to $B_0\hat{e}_z$

In the limit where $k_\perp \to 0$, it follows that $\lambda_e = k_\perp^2 k_B T_e / m_e (\omega_{re}^+ - \omega_{re}^-)^2 \to 0$ and $I_n(\lambda_e \to 0) = 0$ for $n \neq 0$. Making use of $I_0(0) = 1$, Eq.(4.221) reduces to

$$0 = 1 + \frac{1}{k_z^2 \lambda_{\mathrm{De}}^2}\left[1 + \xi_0 Z(\xi_0)\right] , \qquad (4.224)$$

where $\xi_0 = (\omega - \ell\omega_{re})/k_z v_{\mathrm{Te}}$. As expected, for parallel propagation, the electron motion perpendicular to $B_0\hat{e}_z$ does not enter into Eq.(4.224), and the electrostatic dispersion relation is identical to that for an unmagnetized plasma (with ω replaced by $\omega - \ell\omega_{re}$).[64] For long-wavelength perturbations with $k_z^2 \lambda_{\mathrm{De}}^2 \ll 1$, Eq.(4.224) gives weakly damped electron plasma oscillations with $(\mathrm{Re}\,\omega - \ell\omega_{re})^2 = \omega_{pe}^2(1 + 3k_z^2 \lambda_{\mathrm{De}}^2 + \cdots)$ and damping increment specified by Eq.(4.218).

As a cautionary remark, the derivation of the dispersion relation (4.210) [and therefore Eq.(4.221)] assumed body-wave perturbations localized to the column interior with $k_\perp^2 r_b^2 \gg 1$. Therefore, strictly speaking, the dispersion relation (4.224) is valid in the limit where $k_z^2 \gg k_\perp^2 \gg r_b^{-2}$ (not $k_\perp^2 \to 0$).

D.   Wave Propagation Perpendicular to $B_0 \hat{e}_z$

In the limit where $k_z \to 0$, Eqs.(4.221) and (4.222) give

$$0 = 1 + \frac{1}{k_\perp^2 \lambda_{De}^2} \left[1 - \sum_{n=-\infty}^{\infty} \frac{(\omega - \ell \omega_{re}) \exp(-\lambda_e) I_n(\lambda_e)}{\omega - \ell \omega_{re} - n(\omega_{re}^+ - \omega_{re}^-)} \right] , \qquad (4.225)$$

where $\lambda_e = k_\perp^2 k_B T_e / m_e (\omega_{re}^+ - \omega_{re}^-)^2$. Rearranging terms, Eq.(4.225) can be expressed in the equivalent form

$$0 = 1 - \frac{\omega_{pe}^2}{(\omega_{re}^+ - \omega_{re}^-)^2} \sum_{n=-\infty}^{\infty} \frac{n^2 (\omega_{re}^+ - \omega_{re}^-)^2}{(\omega - \ell \omega_{re})^2 - n^2 (\omega_{re}^+ - \omega_{re}^-)^2} \frac{\exp(-\lambda_e) I_n(\lambda_e)}{\lambda_e} .$$

$$(4.226)$$

The dispersion relation (4.226) is the analogue of the Bernstein-mode dispersion relation[64,68] for a nonneutral electron plasma. The solutions to Eq.(4.226) correspond to pure oscillations (Im $\omega = 0$). If $\omega_{pe}^2 / (\omega_{re}^+ - \omega_{re}^-)^2 < 1$, then the solutions to Eq.(4.226) can be approximated by

$$(\omega - \ell \omega_{re})^2 = n^2 (\omega_{re}^+ - \omega_{re}^-)^2 [1 + \alpha_n(\lambda_e)], \quad n = \pm 1, \pm 2, \cdots , \qquad (4.227)$$

where $\alpha_n(\lambda_e)$ is defined by

$$\alpha_n(\lambda_e) = \frac{2\omega_{pe}^2}{(\omega_{re}^+ - \omega_{re}^-)^2} \frac{\exp(-\lambda_e) I_n(\lambda_e)}{\lambda_e} . \qquad (4.228)$$

Therefore, for $\omega_{pe}^2 < (\omega_{re}^+ - \omega_{re}^-)^2$ and wave propagation perpendicular to $B_0 \hat{e}_z$, the electrostatic dispersion relation (4.226) supports oscillations near harmonics of $\omega_{re}^+ - \omega_{re}^-$. For sufficiently high electron density that $\omega_{pe}^2 > (\omega_{re}^+ - \omega_{re}^-)^2$, however, there is harmonic overlap in Eq.(4.226), and several (adjacent) terms in the summation must be retained when solving the dispersion relation.

E.   Wave Propagation in a Cold Plasma

The dispersion relation (4.221) also simplifies in the limit of a cold plasma with

$$v_{\text{Te}}^2 \to 0 \,, \quad \lambda_{\text{De}}^2 \to 0 \,, \quad \lambda_e \to 0 \,, \tag{4.229}$$

and arbitrary $k_z^2/k_\perp^2$. A careful examination of Eq.(4.221) that makes use of Eqs.(4.222) and (4.229) gives the cold-plasma dispersion relation

$$0 = 1 - \frac{k_\perp^2}{k^2} \frac{\omega_{pe}^2}{(\omega - \ell\omega_{re})^2 - (\omega_{re}^+ - \omega_{re}^-)^2} - \frac{k_z^2}{k^2} \frac{\omega_{pe}^2}{(\omega - \ell\omega_{re})^2} \,, \tag{4.230}$$

where $k^2 = k_\perp^2 + k_z^2$. Equation (4.230) also follows directly from Eq.(4.210) for the choice of distribution function $F_e = (2\pi p_\perp)^{-1}\delta(p_\perp)\delta(p_z)$. As expected, Eq.(4.230) is identical in form to the corresponding dispersion relation for a neutral plasma (with $m_i \to \infty$) provided the replacements $\omega \to \omega - \ell\omega_{re}$ and $\omega_{ce} \to \pm(\omega_{re}^+ - \omega_{re}^-)$ are made in the neutral plasma dispersion relation.[64] The exact solutions to Eq.(4.230) can be expressed as

$$(\omega - \ell\omega_{re})^2 = \frac{1}{2} \left[\omega_{pe}^2 + (\omega_{re}^+ - \omega_{re}^-)^2\right]$$

$$\times \left\{1 \pm \left[1 - \frac{k_z^2/k_\perp^2}{1+k_z^2/k_\perp^2} \frac{4\omega_{pe}^2(\omega_{re}^+ - \omega_{re}^-)^2}{[\omega_{pe}^2 + (\omega_{re}^+ - \omega_{re}^-)^2]^2}\right]^{1/2}\right\}. \tag{4.231}$$

For $k_z^2/k_\perp^2 \gg 1$, the two solutions in Eq.(4.231) asymptote at ω_{pe}^2 and $(\omega_{re}^+ - \omega_{re}^-)^2$. For $k_z^2/k_\perp^2 \ll 1$, the high-frequency solution in Eq.(4.231) reduces to $\omega_{pe}^2 + (\omega_{re}^+ - \omega_{re}^-)^2$, which is the nonneutral analogue of the upper hybrid frequency-squared.

F.   Examples of Electrostatic Instabilities

The electrostatic dispersion relation (4.210) also supports a variety of unstable solutions (Im $\omega > 0$) which depend on the detailed form of the equilibrium distribution function $F_e(p_\perp^2, p_z)$. Since the instability analysis closely parallels the neutral plasma treatment[64] by making the replacements in Eq.(4.213), it is adequate for present purposes to list some

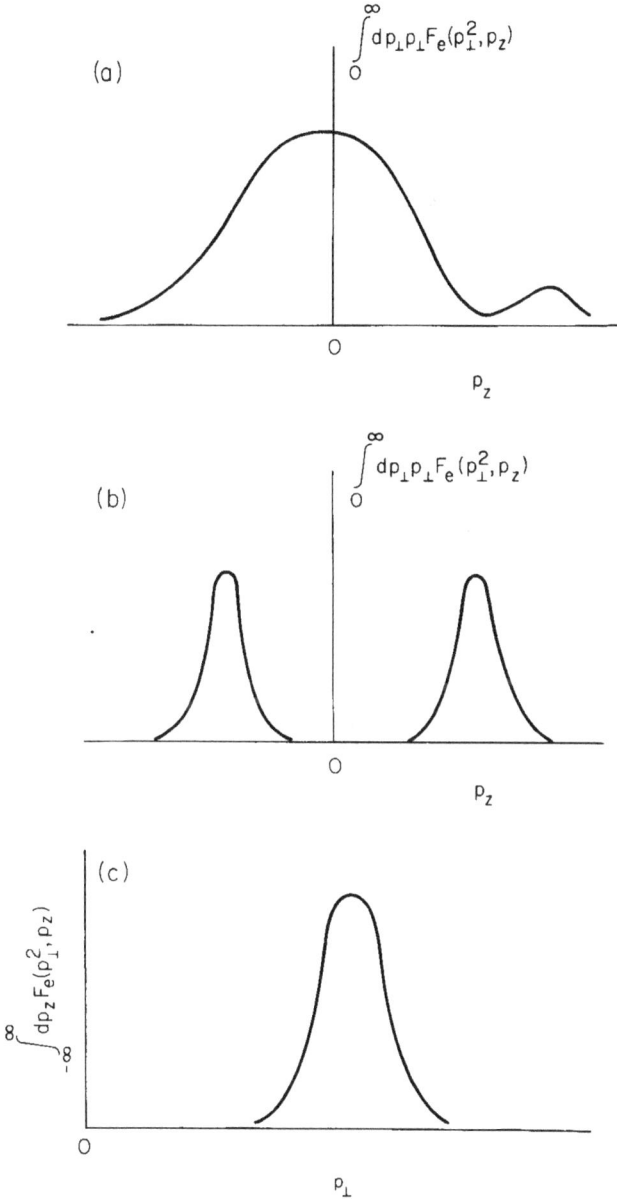

Figure 4.11. Sketches of illustrative electron distribution functions $F_e(p_\perp^2, p_z)$ which can lead to (a) the gentle bump-in-tail instability, (b) the strong axial two-stream instability, and (c) the electrostatic loss-cone instability, when the body-wave dispersion relation (4.210) is used to investigate electrostatic stability behavior.

illustrative examples (Fig. 4.11):

(a) Gentle bump-in-tail instability[64] for a low-density electron beam propagating parallel to $B_0\hat{e}_z$ through a warm electron background ($k_z^2 \gg k_\perp^2$).

(b) Strong two-stream instability[67] for cold electron components counter-streaming along $B_0\hat{e}_z(k_z^2 \gg k_\perp^2)$.

(c) Electrostatic loss-cone instability[69] for an electron distribution function $F_e(p_\perp^2, p_z)$ with an inverted population in perpendicular momentum $p_\perp(k_\perp^2 \gg k_z^2)$.

Problem 4.17   Consider the electrostatic dispersion relation (4.210) for the choice of thermal equilibrium distribution in Eq.(4.216).

a. Derive the form of the dispersion relation given in Eq.(4.221), where $Z(\xi_n)$ is the plasma dispersion function defined in Eq.(4.219). The identities $\sum_{n=-\infty}^{\infty} J_n^2(x) = 1$, and

$$\frac{1}{m_e k_B T_e} \int_0^\infty dp_\perp \, p_\perp J_n^2 \left(\frac{k_\perp p_\perp/m_e}{\omega_{re}^+ - \omega_{re}^-} \right) \exp \left(-\frac{p_\perp^2}{2m_e k_B T_e} \right) = \exp(-\lambda_e) I_n(\lambda_e) \,,$$

(4.17.1)

will be useful in the derivation. Here, $\lambda_e = k_\perp^2 k_B T_e / m_e (\omega_{re}^+ - \omega_{re}^-)^2$, and $I_n(x)$ is the modified Bessel function of the first kind of order n.

b. In the limit of perpendicular propagation ($k_z \to 0$), show that Eq.(4.221) reduces to the form of the dispersion relation given in Eq.(4.226).

c. In the limit of zero electron temperature ($\lambda_e \to 0$), show that the solution to Eq.(4.226) is given by

$$(\omega - \ell\omega_{re})^2 = \omega_{pe}^2 + (\omega_{re}^+ - \omega_{re}^-)^2 \,.$$

(4.17.2)

Equation (4.17.2) corresponds to oscillations at the upper hybrid frequency, modified by self-electric fields.

4.8.3   Dispersion Relation for Transverse Electromagnetic Waves

For a nonneutral plasma, an analysis of the linearized Vlasov-Maxwell equations that includes electromagnetic perturbations with arbitrary polarization proceeds along the lines outlined in Secs. 4.7 and 4.8 for the electrostatic case. For present purposes, it is sufficient to consider the appropriate body-wave dispersion relations for two configurations in which the polarizations are *purely transverse*. These are illustrated in Figs. 4.12 and 4.13. As in Secs. 4.8.1 and 4.8.2, it is assumed that perturbations are about uniform-density rigid-rotor equilibria with $\omega_{re} \simeq \omega_{re}^+$ or $\omega_{re} \simeq \omega_{re}^-$ and equilibrium distribution function of the form $f_e^0(H_\perp - \omega_{re}P_\theta, p_z)$. It is also assumed that the perturbations are localized to the column interior $(0 \leq r \lesssim r_b)$, with characteristic perpendicular wavelength $|k_\perp|^{-1}$ small in comparison with the column radius r_b.

The perturbed field configuration illustrated in Fig. 4.12 corresponds to transverse electromagnetic waves propagating perpendicular to $B_0\hat{e}_z$ with $k_\perp \neq 0$, $k_z = 0$, $\delta E \parallel \hat{e}_z$, and $\delta B \perp \hat{e}_z$. For a nonneutral electron plasma, omitting algebraic details, the dispersion relation for electromagnetic waves with this polarization can be expressed as[1-5]

$$0 = D_\ell^T(k_\perp, \omega) = \omega^2 - c^2 k_\perp^2 - \omega_{pe}^2 + \omega_{pe}^2 \sum_{n=-\infty}^{\infty} \frac{n(\omega_{re}^+ - \omega_{re}^-)}{\omega - \ell\omega_{re} - n(\omega_{re}^+ - \omega_{re}^-)}$$

$$\times \int d^3p\, J_n^2\left(\frac{k_\perp v_\perp}{\omega_{re}^+ - \omega_{re}^-}\right) \frac{v_z^2}{v_\perp} \frac{\partial}{\partial v_\perp} F_e(p_\perp^2, p_z) , \qquad (4.232)$$

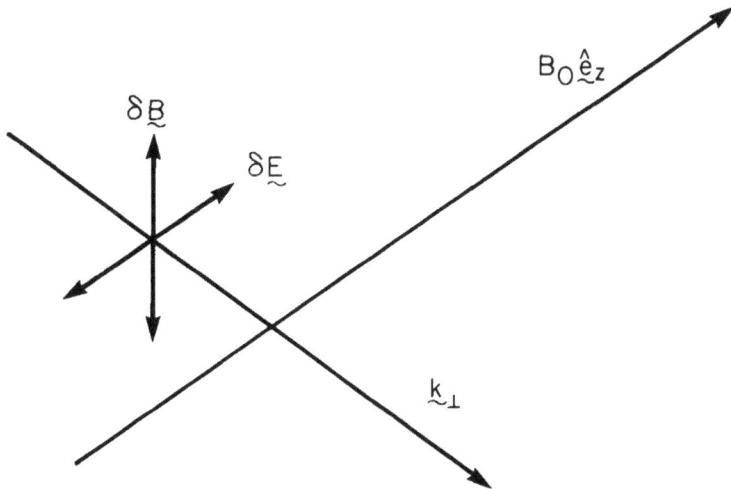

Figure 4.12.   Ordinary-mode polarization for transverse electromagnetic waves propagating perpendicular to $B_0\hat{e}_z$. The perturbed field amplitudes are oriented with $\delta B \cdot \hat{e}_z = 0$, δE parallel to $\hat{e}_z$, and $\delta E \cdot \delta B = 0$.

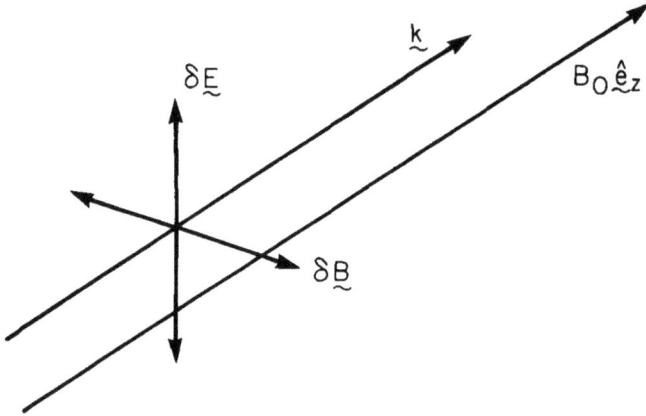

Figure 4.13. Transverse electromagnetic waves propagating parallel to $B_0 \hat{e}_z$. The perturbed field amplitudes are oriented with $\delta \mathbf{B} \cdot \hat{e}_z = 0 = \delta \mathbf{E} \cdot \hat{e}_z$ and $\delta \mathbf{E} \cdot \delta \mathbf{B} = 0$.

where the notation is the same as in Sec. 4.8.1. If the plasma is cold parallel to $B_0 \hat{e}_z$, with $\int d^3 p\, v_z^2 F_e = 0$, then Eq.(4.232) reduces to the familiar result

$$\omega^2 = c^2 k_\perp^2 + \omega_{pe}^2 \ . \tag{4.233}$$

The ordinary-mode dispersion relation (4.233) provides a useful basis for measuring the line-averaged electron density in both nonneutral and neutral plasmas. For a warm nonneutral plasma with thermal equilibrium distribution function specified by Eq.(4.216) [or Eq.(4.215)], the ordinary-mode dispersion relation (4.232) reduces to

$$0 = D_\ell^T(k_\perp, \omega) = \omega^2 - c^2 k_\perp^2 - \omega_{pe}^2 \tag{4.234}$$

$$+ \omega_{pe}^2 \sum_{n=-\infty}^{\infty} \frac{n^2 (\omega_{re}^+ - \omega_{re}^-)^2}{(\omega - \ell \omega_{re})^2 - n^2 (\omega_{re}^+ - \omega_{re}^-)^2} \exp(-\lambda_e) I_n(\lambda_e) \ ,$$

where $\lambda_e = k_\perp^2 k_B T_e / m_e (\omega_{re}^+ - \omega_{re}^-)^2$. The solutions to Eq.(4.234) satisfy $\text{Im}\,\omega = 0$ and exhibit an intricate harmonic structure for $(\omega - \ell \omega_{re})^2$ in the vicinity of $n^2 (\omega_{re}^+ - \omega_{re}^-)^2$, $n = \pm 1, \pm 2 \cdots$.

For a multicomponent nonneutral plasma, such as a relativistic electron beam $(j = b)$ propagating parallel to $B_0 \hat{e}_z$ through a background plasma

$(j = e, i)$, the generalization[65] of the ordinary-mode dispersion relation (4.232) can lead to a Weibel-like instability known as the filamentation instability (see Sec. 5.8).[70-73] The free energy source for this transverse electromagnetic instability is the excess kinetic energy of the beam-plasma components parallel to $B_0\hat{e}_z$ relative to the kinetic energy perpendicular to $B_0\hat{e}_z$.

The perturbed field configuration illustrated in Fig. 4.13 corresponds to transverse electromagnetic waves propagating parallel to $B_0\hat{e}_z$ with $k_z \neq 0$, $k_\perp \simeq 0$, $\delta\mathbf{E} \perp \hat{e}_z$, and $\delta\mathbf{B} \perp \hat{e}_z$. For a nonneutral electron plasma rotating in the *slow* $(\omega_{re} \simeq \omega_{re}^-)$ rotational mode, the dispersion relation for electromagnetic waves with this polarization can be expressed as

$$0 = D_-^T(k_z, \omega) = \omega^2 - c^2 k_z^2 + \omega_{pe}^2 \int d^3p \frac{v_\perp}{2} \qquad (4.235)$$

$$\times \frac{[k_z v_\perp \partial/\partial v_z + (\omega - k_z v_z)\partial/\partial v_\perp]}{\omega - k_z v_z \mp \omega_{re}^+} F_e(p_\perp^2, p_z) \,,$$

where $\omega_{re}^+ = (\omega_{ce}/2)[1 + (1 - 2\omega_{pe}^2/\omega_{ce}^2)^{1/2}]$ and the notation is the same as in Sec. 4.8.1. The upper $(-)$ and lower $(+)$ signs in Eq.(4.235) correspond to electromagnetic waves with right-hand-circular polarization $[\delta E_x(k_z) = -i\delta E_y(k_z)]$ and left-hand-circular polarization $[\delta E_x(k_z) = +i\delta E_y(k_z)]$, respectively. The dispersion relation (4.235) is identical in form to the corresponding result for a neutral plasma[64] with $m_i \to \infty$, provided the replacement $\omega_{ce} \to \omega_{re}^+$ is made in the neutral plasma dispersion relation. For a nonneutral electron plasma rotating in the *fast* $(\omega_{re} \simeq \omega_{re}^+)$ rotational mode, the resulting dispersion relation[5] is identical in form to Eq.(4.235) with the denominator $\omega - k_z v_z \mp \omega_{re}^+$ replaced by $\omega - k_z v_z \mp \omega_{re}^-$.

As in the case of a neutral plasma, if there is an anisotropy in kinetic energy with $\int d^3p\, (m_e v_\perp^2/2) F_e(p_\perp^2, p_z)$ exceeding $\int d^3p\, m_e v_z^2 F_e(p_\perp^2, p_z)$ by a sufficient amount, then Eq.(4.235) supports unstable solutions with $\mathrm{Im}\,\omega > 0$. Examples of unstable equilibria include the loss-cone distribution in Fig. 4.11(c), and a bi-Maxwellian distribution with perpendicular temperature $T_{e\perp}$ exceeding the temperature $T_{e\|}$ parallel to $B_0\hat{e}_z$. The corresponding growth rates and oscillation frequencies for a nonneutral electron plasma can be derived by direct analogy with the neutral plasma results.[64,74,75]

As an example of stable oscillations with $\text{Im}\,\omega = 0$, consider Eq.(4.235) in the limit of cold electrons with $|k_z^2 v_{\text{Te}}^2/(\omega \mp \omega_{re}^+)^2| \to 0$. For $F_e(p_\perp^2, p_z) = (1/2\pi p_\perp)\delta(p_\perp)\delta(p_z)$, the dispersion relation (4.235) reduces to[4,5]

$$0 = \omega^2 - c^2 k_z^2 - \omega_{pe}^2 \frac{\omega}{\omega \mp \omega_{re}^+} . \tag{4.236}$$

For high-frequency perturbations with $|\omega| \gg \omega_{re}^+$, Eq.(4.236) reduces to the expected result, $\omega^2 = c^2 k_z^2 + \omega_{pe}^2$. On the other hand, for slow-wave perturbations with $\omega^2 \ll c^2 k_z^2$, Eq.(4.236) gives

$$\omega = \pm \omega_{re}^+ \frac{c^2 k_z^2}{c^2 k_z^2 + \omega_{pe}^2} , \tag{4.237}$$

which corresponds to the electron whistler branch. For $c^2 k_z^2 \gg \omega_{pe}^2$, the solution in Eq.(4.237) asymptotes at $\omega = \pm \omega_{re}^+ = \pm(\omega_{ce}/2) \times [1 + (1 - 2\omega_{pe}^2/\omega_{ce}^2)^{1/2}]$. On the other hand, for $c^2 k_z^2 \ll \omega_{pe}^2$, Eq.(4.237) gives

$$\omega = \pm \omega_{re}^+ \frac{c^2 k_z^2}{\omega_{pe}^2} . \tag{4.238}$$

In contrast to the neutral plasma case, where $\omega^2 = k_z^2 V_A^2$ at low frequencies (here, V_A is the Alfvén velocity),[64] the electron whistler mode persists in a pure electron plasma down to zero frequency, and the mode is dispersive ($\omega \sim k_z^2$) as $k_z \to 0$.

As an example corresponding to instability, consider the case of extreme energy anisotropy where

$$F_e(p_\perp^2, p_z) = G(p_\perp^2)\delta(p_z) . \tag{4.239}$$

Here, $T_{e\perp} = \int d^3p \, (p_\perp^2/2m_e)F_e(p_\perp^2, p_z) \neq 0$ is assumed, and the normalization of $F_e(p_\perp^2, p_z)$ is $2\pi \int_0^\infty dp_\perp \, p_\perp \int_{-\infty}^\infty dp_z \, F_e(p_\perp^2, p_z) = 1$. Substituting Eq.(4.239) into Eq.(4.235) gives the dispersion relation

$$0 = \omega^2 - c^2 k_z^2 - \omega_{pe}^2 \frac{\omega}{\omega \mp \omega_{re}^+} - \omega_{pe}^2 \frac{k_z^2 T_{e\perp}/m_e}{(\omega \mp \omega_{re}^+)^2} . \tag{4.240}$$

For $T_{e\perp} \to 0$, Eq.(4.240) reduces to the cold plasma result in Eq.(4.236). For $T_{e\perp} \neq 0$, however, the slow-wave branch ($|\omega|^2 \ll c^2 k_z^2$) in Eq.(4.240) exhibits instability with $\operatorname{Im} \omega > 0$. This is easily verified in the short-wavelength limit $(c^2 k_z^2/\omega_{pe}^2 \gg 1)$ where Eq.(4.240) reduces to

$$(\omega \mp \omega_{re}^+)^2 = -\left(\frac{T_{e\perp}}{m_e c^2}\right) \omega_{pe}^2 \, . \tag{4.241}$$

Solving Eq.(4.241) for the unstable branch gives

$$\operatorname{Re} \omega = \pm \omega_{re}^+ \, ,$$

$$\operatorname{Im} \omega = \left(\frac{T_{e\perp}}{m_e c^2}\right)^{1/2} \omega_{pe} \, . \tag{4.242}$$

Equation (4.242) gives the familiar Weibel-like growth rate $\operatorname{Im} \omega = (v_{Te}/c)\omega_{pe}$, where $v_{Te} = (T_{e\perp}/m_e)^{1/2}$. Depending on the value of v_{Te}/c, the normalized growth rate $(\operatorname{Im} \omega)/\omega_{pe}$ in Eq.(4.242) can be sizeable.

In Eq.(4.240), keep in mind that the source of free energy that drives the instability is the electron energy anisotropy $(T_{e\perp} > T_{e\parallel} = 0)$. For an isotropic equilibrium $F_e(p_\perp^2 + p_z^2)$ that is a monotonically decreasing function of $p_\perp^2 + p_z^2$ [e.g., Eq.(4.216)], the system is completely stable (Sec. 4.5).

To summarize, the analysis in Sec. 4.8 illustrates that a one-component nonneutral plasma supports a wide range of collective waves and instabilities analogous to a neutral plasma, appropriately modified by self-electric fields. It should be emphasized that the dispersion relations and stability analyses are readily generalized to the case of a multicomponent nonneutral plasma. Accordingly, there are additional waves and instabilities in the multicomponent case. For example, the relative rotation of electrons and ions can provide the free energy source to drive a two-rotating-stream instability known as the ion resonance instability (Sec. 5.6).[76,77]

Problem 4.18   Make use of the linearized Vlasov-Maxwell equations to derive the ordinary-mode dispersion relation (4.232) for body-wave perturbations about a uniform-density nonneutral plasma column. The assumptions regarding the equilibrium are similar to those made in Sec. 4.8.1, and the wave polarization is illustrated in Fig. 4.12.

Hint: The electric field perturbation can be expressed as

$$\delta E_z(\mathbf{x}_\perp, t) = \int d^2 k \, \delta E_z(\mathbf{k}_\perp) \exp(i \mathbf{k}_\perp \cdot \mathbf{x}_\perp - i\omega t) \, , \tag{4.18.1}$$

or equivalently

$$\delta E_z(\mathbf{x}_\perp, t) = \int_0^\infty dk_\perp k_\perp \sum_{\ell=-\infty}^\infty \delta E_z^\ell(k_\perp) J_\ell(k_\perp r) \exp[i(\ell\theta - \omega t)] . \qquad (4.18.2)$$

Here,

$$\delta E_z^\ell(k_\perp) = (i)^\ell \int_{-\pi}^\pi d\alpha \, \delta E_z(\mathbf{k}_\perp) \exp(-i\ell\alpha) , \qquad (4.18.3)$$

and use is made of $\mathbf{k}_\perp \cdot \mathbf{x}_\perp = k_\perp r \cos(\theta - \alpha)$, where $k_x = k_\perp \cos\alpha$, $k_y = k_\perp \sin\alpha$, and $r = (x^2 + y^2)^{1/2}$.

4.9   Dispersion Relation for Electrostatic Flute Perturbations

In this section, the linearized Vlasov-Poisson equations (Sec. 4.7) are analyzed for the case of electrostatic flute perturbations ($\partial/\partial z = 0$) about a uniform-density plasma column including the full influence of finite radial geometry. For variety, we consider a one-component nonneutral plasma consisting only of ions with charge $+Z_i e$ and rigid-rotor equilibrium distribution function

$$f_i^0(H_\perp - \omega_{ri} P_\theta, p_z) = \frac{\hat{n}_i}{2\pi m_i} \delta(H_\perp - \omega_{ri} P_\theta - \hat{T}_{i\perp}) G_i(p_z) . \qquad (4.243)$$

Here, $H_\perp = (2m_i)^{-1}(p_r^2 + p_\theta^2) + Z_i e\phi_0(r)$ is the perpendicular energy; $P_\theta = r(p_\theta + m_i r\omega_{ci}/2)$ is the canonical angular momentum; the normalization of $G_i(p_z)$ is $\int_{-\infty}^\infty dp_z \, G_i(p_z) = 1$; and $\hat{n}_i$ and $\hat{T}_{i\perp}$ are positive constants. Paralleling the analysis in Sec. 4.2.3, Eq.(4.243) gives the rectangular density profile

$$n_i^0(r) = \begin{cases} \hat{n}_i = \text{const.} , & 0 \le r < r_b , \\ 0 , & r_b < r \le b , \end{cases} \qquad (4.244)$$

where $r = b$ is the radial location of a perfectly conducting wall (Fig. 4.14), and the column radius r_b is determined self-consistently from

$$r_b^2 = \frac{(2\hat{T}_{i\perp}/m_i)}{(\omega_{ri}^- - \omega_r)(\omega_r - \omega_{ri}^+)} . \qquad (4.245)$$

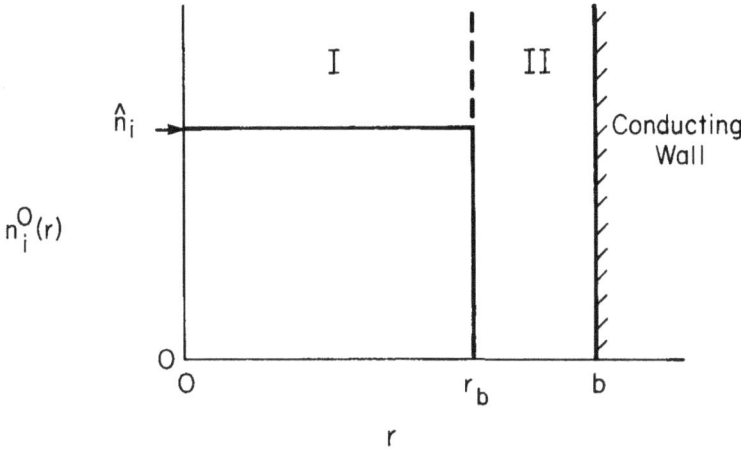

Figure 4.14.   Plot versus radius r of the ion density profile [Eq.(4.244)] obtained for the choice of Vlasov equilibrium in Eq.(4.243). A perfectly conducting wall is located at $r = b$.

Here, ω_{ri}^+ and ω_{ri}^- are the laminar rotation frequencies defined by

$$\omega_{ri}^\pm = -\frac{\omega_{ci}}{2} \left\{ 1 \pm \left(1 - \frac{2\omega_{pi}^2}{\omega_{ci}^2} \right)^{1/2} \right\} , \qquad (4.246)$$

where $\omega_{pi}^2 = 4\pi \hat{n}_i Z_i^2 e^2 / m_i$ and $\omega_{ci} = Z_i e B_0 / m_i c$. Radially confined solutions exist provided ω_{ri} lies in the interval $\omega_{ri}^+ < \omega_{ri} < \omega_{ri}^-$, where $\omega_{ri}^\pm < 0$. For the choice of loss-cone equilibrium in Eq.(4.243), the ion density profile is exactly uniform in the column interior, and the electrostatic potential is $\phi_0(r) = -(m_i/4Z_i e)\omega_{pi}^2 r^2$ for $0 \le r < r_b$. Moreover, the combination $H_\perp - \omega_{ri} P_\theta$ can be expressed as

$$H_\perp - \omega_{ri} P_\theta = \frac{1}{2m_i} \left[p_r^2 + (p_\theta - m_i \omega_{ri} r)^2 \right] + \psi_i(r) , \qquad (4.247)$$

where

$$\psi_i(r) = \frac{m_i}{2} \left(\omega_{ri}^- - \omega_{ri} \right) \left(\omega_{ri} - \omega_{ri}^+ \right) r^2 \equiv \hat{T}_{i\perp} \frac{r^2}{r_b^2} \qquad (4.248)$$

for $0 \le r < r_b$.

We now analyze the linearized Vlasov-Poisson equations (4.182) and (4.183), extended to the case of a nonneutral ion plasma, assuming flute perturbations with $\partial/\partial z = 0$. Again for variety, the analysis is carried out in the *laboratory frame*, and no *a priori* assumption is made that ω_{ri} is closely tuned to either ω_{ri}^+ or ω_{ri}^- because Eq.(4.243) automatically gives a uniform density profile out to radius $r = r_b$ [Eq.(4.244)]. The perturbed electrostatic potential $\delta\phi(r, \theta, t)$ is expressed as

$$\delta\phi(r, \theta, t) = \sum_{\ell=-\infty}^{\infty} \delta\phi^\ell(r) \exp(i\ell\theta - i\omega t), \qquad (4.249)$$

where $\text{Im}\,\omega > 0$ corresponds to instability (temporal growth). Making use of the method of characteristics in the laboratory frame (Sec. 2.2 and Problem 4.15), the linearized Vlasov-Poisson equations give

$$\frac{1}{r}\frac{\partial}{\partial r}r\frac{\partial}{\partial r}\delta\phi^\ell(r) - \frac{\ell^2}{r^2}\delta\phi^\ell(r) = -4\pi Z_i e \int d^3p\,\delta f_i^\ell(r, \mathbf{p}), \qquad (4.250)$$

where $\delta f_i(\mathbf{x}_\perp, \mathbf{p}, t) = \sum_{\ell=-\infty}^{\infty} \delta f_i(r, \mathbf{p})\exp(i\ell\theta - i\omega t)$, and the amplitude of the perturbed ion distribution function can be expressed as (Problem 4.19)

$$\delta f_i^\ell(r, \mathbf{p}) = Z_i e \frac{\partial f_i^0}{\partial H_\perp}\left[\delta\phi^\ell(r) + i(\omega - \ell\omega_{ri})\right.$$

$$\left. \times \int_\infty^t dt'\,\delta\phi^\ell[r'(t')]\exp\{i\ell[\theta'(t') - \theta] - i\omega(t' - t)\}\right]. \qquad (4.251)$$

In Eq.(4.251), use has been made of the constancy of $H_\perp' - \omega_{ri}P_\theta' = H_\perp - \omega_{ri}P_\theta$ and $p_z' = p_z$ to factor $(\partial/\partial H_\perp)f_i^0(H_\perp - \omega_{ri}P_\theta, p_z)$ from the orbit integration over t'. Moreover, the "primed" orbits $r'(t')$ and $\theta'(t')$ occurring in the integrand in Eq.(4.251) correspond to the particle trajectories in the equilibrium field configuration that pass through the point (r, θ) at time $t' = t$.

To determine the particle trajectories in the equilibrium field configuration, it is convenient to use the (complex) Cartesian representation $x'(t') + iy'(t') = r'(t')\exp[i\theta'(t')]$ and $v_x'(t') + iv_y'(t') = (d/dt')[x'(t') + iy'(t')]$. For $\mathbf{E}^0(\mathbf{x}') = (m_i/2Z_ie)\omega_{pi}^2(x'\hat{\mathbf{e}}_x + y'\hat{\mathbf{e}}_y)$ and $\mathbf{B}^0(\mathbf{x}') = B_0\hat{\mathbf{e}}_z$, the orbit equations for the ion motion perpendicular to $B_0\hat{e}_z$ are given by

$$\frac{d^2}{dt'^2}x'(t') = \frac{1}{2}\omega_{pi}^2 x'(t') + \omega_{ci}\frac{d}{dt'}y'(t') ,$$

$$\frac{d^2}{dt'^2}y'(t') = \frac{1}{2}\omega_{pi}^2 y'(t') - \omega_{ci}\frac{d}{dt'}x'(t') , \qquad (4.252)$$

for $0 \leq r' < r_b$. The equations for $x'(t')$ and $y'(t')$ can be combined to give

$$\left[\frac{d^2}{dt'^2} + i\omega_{ci}\frac{d}{dt'} - \frac{1}{2}\omega_{pi}^2\right][x'(t') + iy'(t')] = 0 . \qquad (4.253)$$

Equation (4.253) is solved subject to the "initial" conditions

$$x'(t' = t) + iy'(t' = t) = x + iy \equiv r\exp(i\theta) ,$$

$$v_x'(t' = t) + \omega_{ri}y'(t' = t) = v_x + \omega_{ri}y = V_x \equiv V_\perp\cos\phi ,$$

$$v_y'(t' = t) - \omega_{ri}x'(t' = t) = v_y - \omega_{ri}x = V_y \equiv V_\perp\sin\phi . \qquad (4.254)$$

That is, the perpendicular orbit (x', y', v_x', v_y') passes through the phase-space point (x, y, v_x, v_y) at time $t' = t$. In Eq.(4.254) note that (V_x, V_y) or $(V_\perp, \phi)$ are velocity-space variables appropriate to the *rotating frame*. For example, making use of $\mathbf{p}_\perp = m_i\mathbf{v}_\perp$, it follows from Eqs.(4.247) and (4.254) that $H_\perp - \omega_{ri}P_\theta = (m_i/2)[(v_x + \omega_{ri}y)^2 + (v_y - \omega_{ri}x)^2] + \psi_i(r)$ can be expressed as

$$H_\perp - \omega_{ri}P_\theta = \frac{m_i}{2}V_\perp^2 + \psi_i(r) , \qquad (4.255)$$

where $\psi_i(r)$ is defined in Eq.(4.248). Solving Eq.(4.253) subject to the boundary conditions in Eq.(4.254), some straightforward algebra gives

$$x'(t') + iy'(t') = (\omega_{ri}^- - \omega_{ri}^+)^{-1}\left\{iV_\perp\exp(i\phi)\left[\exp(i\omega_{ri}^+\tau) - \exp(i\omega_{ri}^-\tau)\right]\right.$$
$$\left. + r\exp(i\theta)\left[(\omega_{ri} - \omega_{ri}^+)\exp(i\omega_{ri}^-\tau) + (\omega_{ri}^- - \omega_{ri})\exp(i\omega_{ri}^+\tau)\right]\right\} , \qquad (4.256)$$

where $\tau = t' - t$. The perpendicular velocity $v_x' + iv_y' = (d/dt')(x' + iy')$ can be calculated directly from Eq.(4.256). As expected, it is evident from Eq.(4.256) that the perpendicular ion motion in the laboratory frame is biharmonic at the frequencies ω_{ri}^+ and ω_{ri}^- (see also Secs. 3.2 and 3.5).

The expression for the perturbed ion distribution function $\delta f_i^\ell(r, \mathbf{p})$ in Eq.(4.251) is substituted into Poisson's equation (4.250). In this regard, it is convenient to express $(\partial/\partial H_\perp) f_i^0 (H_\perp - \omega_{ri} P_\theta, p_z) = (m_i V_\perp)^{-1} (\partial/\partial V_\perp) \times f_i^0 (m_i V_\perp^2/2 + \psi_i(r), p_z)$, where $m_i V_\perp^2/2$ is the perpendicular ion kinetic energy in the rotating frame [Eqs.(4.254) and (4.255)]. Defining the phase-averaged (over ϕ) orbit integral I_ℓ by

$$I_\ell = i \int_0^{2\pi} \frac{d\phi}{2\pi} \int_{-\infty}^0 d\tau \, \delta\phi^\ell [r'(\tau)] \exp\{i\ell[\theta'(\tau) - \theta] - i\omega\tau\} \,, \qquad (4.257)$$

it is straightforward to show that Poisson's equation (4.250) can be expressed as

$$\frac{1}{r} \frac{\partial}{\partial r} r \frac{\partial}{\partial r} \delta\phi^\ell(r) - \frac{\ell^2}{r^2} \delta\phi^\ell(r)$$

$$= -\frac{4\pi Z_i^2 e^2}{m_i} \int d^3 p \, \frac{1}{V_\perp} \frac{\partial f_i^0}{\partial V_\perp} \left[\delta\phi^\ell(r) + (\omega - \ell\omega_{ri}) I_\ell \right] \,, \qquad (4.258)$$

where $\int d^3 p \cdots = 2\pi \int_0^\infty d(m_i V_\perp) \, m_i V_\perp \int_{-\infty}^\infty dp_z \cdots$. The choice of ion distribution function $f_i^0 = (\hat{n}_i/2\pi m_i) \delta[m_i V_\perp^2/2 - \hat{T}_{i\perp}(1 - r^2/r_b^2)] G_i(p_z)$ in Eq.(4.243) gives the rectangular density profile $n_i^0(r)$ illustrated in Fig. 4.14 and permits the integration over momentum to be carried out on the right-hand side of Eq.(4.258). Some straightforward integration by parts for this choice of f_i^0 gives the identity (Problem 4.20)

$$\int d^3 p \, \frac{1}{V_\perp} \frac{\partial f_i^0}{\partial V_\perp} \chi(\omega, r, V_\perp)$$

$$= \int d^3 p \left[\frac{1}{V_\perp} \frac{\partial}{\partial V_\perp} \left(f_i^0 \chi \right) - f_i^0 \frac{1}{V_\perp} \frac{\partial}{\partial V_\perp} \chi \right]$$

$$= -\frac{\hat{n}_i r_b}{v_{Ti}^2} \delta(r - r_b) \left[\chi\right]_{V_\perp = 0} - n_i^0(r) \left[\frac{1}{V_\perp} \frac{\partial \chi}{\partial V_\perp} \right]_{V_\perp^2 = v_{Ti}^2 (1 - r^2/r_b^2)} \qquad (4.259)$$

for general function $\chi(\omega, r, V_\perp)$. Here, $v_{Ti}^2 = 2\hat{T}_{i\perp}/m_i$ is the effective ion thermal speed-squared, and $n_i^0(r) = \int d^3 p f_i^0$ is the rectangular density profile in Eq.(4.244) and Fig. 4.14. For $\chi(\omega, r, V_\perp) = \delta\phi^\ell(r) + (\omega - \ell\omega_{ri}) I_\ell$, we make use of Eq.(4.259) to express Poisson's equation (4.258) in the form

$$\frac{1}{r}\frac{\partial}{\partial r}r\frac{\partial}{\partial r}\delta\phi^{\ell}(r) - \frac{\ell^2}{r^2}\delta\phi^{\ell}(r)$$

$$= \frac{\omega_{pi}^2 r_b}{v_{Ti}^2}\left\{\delta\phi^{\ell}(r) + (\omega - \ell\omega_{ri})[I_\ell]_{V_\perp=0}\right\}\delta(r - r_b)$$

$$+\omega_{pi}^2(r)(\omega - \ell\omega_{ri})\left[\frac{1}{V_\perp}\frac{\partial I_\ell}{\partial V_\perp}\right]_{V_\perp^2=v_{Ti}^2(1-r^2/r_b^2)}, \qquad (4.260)$$

where $\omega_{pi}^2 = 4\pi\hat{n}_i Z_i^2 e^2/m_i$, $\omega_{pi}^2(r) = 4\pi n_i^0(r)Z_i^2 e^2/m_i$, and the orbit integral I_ℓ is defined in Eq.(4.257).

The term proportional to $\delta(r - r_b)$ in Eq.(4.260) corresponds to a surface-charge perturbation localized at $r = r_b$, whereas the term proportional to $\omega_{pi}^2(r)$ corresponds to a body-charge perturbation extending over the interval $0 \leq r < r_b$. In circumstances where the orbit integral I_ℓ is independent of $V_\perp$, note that the body-charge contribution in Eq.(4.260) is identically zero.

The general solution to Eq.(4.260) for $0 \leq r < r_b$ is of the form[77,78]

$$\delta\phi^{\ell}(r) = Ar^{\ell}\sum_{j=0}^{n} a_j(\omega)(r/r_b)^{2j},$$

where n is a positive integer, and $a_0(\omega) = 1$ without loss of generality. For $n = 0$, the eigenfunction $\delta\phi^{\ell}(r) = Ar^{\ell}$ is peaked at the outer surface ($r = r_b$) of the plasma column. For $n \geq 1$, however, the eigenfunction has a more complicated radial structure, and $\delta\phi^{\ell}(r)$ can be peaked within the plasma column, corresponding to a strong body-wave perturbation (Sec. 9.3). For present purposes, the analysis in Sec. 4.9 is limited to perturbations with $n = 0$ and $\delta\phi^{\ell}(r) = Ar^{\ell}$ for $0 \leq r < r_b$. In this case, it is found that the orbit integral I_ℓ is independent of $V_\perp$, and therefore the body-charge contribution on the right-hand side of Eq.(4.260) is equal to zero.

Substituting $\delta\phi^{\ell}[r'(\tau)] = A[r'(\tau)]^{\ell}$ into Eq.(4.257) gives

$$I_\ell = iA\int_{-\infty}^{0} d\tau \exp(-i\ell\theta - i\omega\tau)\int_0^{2\pi}\frac{d\phi}{2\pi}[x'(\tau) + iy'(\tau)]^{\ell}, \qquad (4.261)$$

where use has been made of $r' \exp(i\theta') = x' + iy'$. To evaluate I_ℓ in Eq.(4.261), use is made of Eq.(4.256) and the binomial theorem to take the ℓ'th power of $x' + iy'$. All of the terms explicitly involving $V_\perp \exp(i\phi)$ average to zero because of the ϕ integration, and Eq.(4.261) reduces to

$$I_\ell = \frac{i\delta\phi^\ell(r)}{(\omega_{ri}^- - \omega_{ri}^+)^\ell} \int_{-\infty}^0 d\tau \, \exp(-i\omega\tau)$$

$$\times \left[(\omega_{ri} - \omega_{ri}^+) \exp(i\omega_{ri}^-\tau) + (\omega_{ri}^- - \omega_{ri}) \exp(i\omega_{ri}^+\tau) \right]^\ell , \quad (4.262)$$

where $\delta\phi^\ell(r) = Ar^\ell$. Note from Eq.(4.262) that $\partial I_\ell/\partial V_\perp = 0$. Carrying out the τ integration in Eq.(4.262) for $\text{Im}\,\omega > 0$ then gives

$$\delta\phi^\ell(r) + (\omega - \ell\omega_{ri})I_\ell = \Gamma_\ell(\omega)\delta\phi^\ell(r) , \quad (4.263)$$

where $\Gamma_\ell(\omega)$ is defined by[77]

$$\Gamma_\ell(\omega) = 1 - \left(\frac{\omega_{ri} - \omega_{ri}^+}{\omega_{ri}^- - \omega_{ri}^+} \right)^\ell$$

$$\times \sum_{m=0}^\ell \frac{\ell!}{m!(\ell-m)!} \frac{\omega - \ell\omega_{ri}}{\omega - \ell\omega_{ri}^- + m(\omega_{ri}^- - \omega_{ri}^+)} \left(\frac{\omega_{ri}^- - \omega_{ri}}{\omega_{ri} - \omega_{ri}^+} \right)^m . \quad (4.264)$$

In Eq.(4.264), keep in mind that $\omega_{ri}^+ < \omega_{ri} < \omega_{ri}^-$ is required for existence of a radially confined equilibrium. Moreover, $\omega_{ri}^\pm < 0$ follows from Eq.(4.246).

Substituting Eq.(4.263) into Poisson's equation (4.260) gives

$$\frac{1}{r}\frac{\partial}{\partial r}r\frac{\partial}{\partial r}\delta\phi^\ell(r) - \frac{\ell^2}{r^2}\delta\phi^\ell(r) = \frac{\omega_{pi}^2 r_b}{v_{Ti}^2}\Gamma_\ell(\omega)\delta\phi^\ell(r)\delta(r - r_b) . \quad (4.265)$$

The right-hand side of the eigenvalue equation (4.265) is equal to zero except at the surface of the plasma column $(r = r_b)$. Moreover, the eigenfunction $\delta\phi^\ell(r)$ satisfies the vacuum Poisson equation except at $r = r_b$. Therefore, the solution to Eq.(4.265) can be expressed as (see Fig. 4.14)

$$\delta\phi_I^\ell(r) = Ar^\ell , \; 0 \leq r < r_b , \quad (4.266)$$

inside the plasma column, and

$$\delta\phi_{II}^{\ell}(r) = Ar^{\ell}\frac{(1 - b^{2\ell}/r^{2\ell})}{(1 - b^{2\ell}/r_b^{2\ell})}, \; r_b < r \le b, \qquad (4.267)$$

in the vacuum region between the surface of the plasma column and the perfectly conducting wall at $r = b$. Note that $\delta\phi^{\ell}(r)$ is continuous at $r = r_b$, and that $\delta\phi^{\ell}(r = b) = 0$, which corresponds to zero tangential electric field at the conducting wall, i.e., $\delta E_{\theta}(r = b) = -(i\ell/b) \times \delta\phi_{II}^{\ell}(r = b) = 0$. The dispersion relation that determines the eigenfrequency ω is obtained by multiplying Eq.(4.265) by r and integrating across the surface of the plasma column from $r_b(1 - \varepsilon)$ to $r_b(1 + \varepsilon)$ with $\varepsilon \to 0_{+}$. This gives

$$\left[r\frac{\partial}{\partial r}\delta\phi_{II}^{\ell}\right]_{r=r_b} - \left[r\frac{\partial}{\partial r}\delta\phi_{I}^{\ell}\right]_{r=r_b} = \frac{\omega_{pi}^2 r_b^2}{v_{Ti}^2}\Gamma_{\ell}(\omega)\delta\phi^{\ell}(r = r_b). \qquad (4.268)$$

Equation (4.268) relates the discontinuity in radial electric field at $r = r_b$ to the perturbed surface charge density. Substituting Eqs.(4.266) and (4.267) into Eq.(4.268) and rearranging terms, we obtain

$$0 = D_{\ell}(\omega) = 1 + \frac{\omega_{pi}^2 r_b^2}{2\ell v_{Ti}^2}\left[1 - \left(\frac{r_b}{b}\right)^{2\ell}\right]\Gamma_{\ell}(\omega)$$

$$= 1 + \frac{\omega_{pi}^2 r_b^2}{2\ell v_{Ti}^2}\left[1 - \left(\frac{r_b}{b}\right)^{2\ell}\right]\left\{1 - \left(\frac{\omega_{ri} - \omega_{ri}^{+}}{\omega_{ri}^{-} - \omega_{ri}^{+}}\right)^{\ell}\right. \qquad (4.269)$$

$$\left.\times \sum_{m=0}^{\ell}\frac{\ell!}{m!(\ell - m)!}\frac{\omega - \ell\omega_{ri}}{\omega - \ell\omega_{ri}^{-} + m(\omega_{ri}^{-} - \omega_{ri}^{+})}\left(\frac{\omega_{ri}^{-} - \omega_{ri}}{\omega_{ri} - \omega_{ri}^{+}}\right)^{m}\right\}.$$

Here, a nontrivial solution $(A \ne 0)$ has been assumed.

Equation (4.269) is the desired electrostatic dispersion relation for flute perturbations $(\partial/\partial z = 0)$ about the equilibrium ion distribution function in Eq.(4.243). Several points are noteworthy regarding the dispersion relation (4.269). First, the effects of finite radial geometry are included in a fully self-consistent manner. Second, the analysis leading to Eq.(4.269) is readily extended to a multicomponent nonneutral plasma in circumstances where the other plasma components have distribution functions similar in form to Eq.(4.243). Third, Eq.(4.269) is a fully kinetic dispersion relation. That is, no a priori assumption has been made that ion thermal effects are weak or that ω_{ri} is closely tuned to either of the laminar rotation frequencies, ω_{ri}^+ or ω_{ri}^-. Indeed, introducing the effective ion Larmor radius r_{Li} defined by $r_{Li} = v_{Ti}/\omega_{ci} = (2\hat{T}_{i\perp}/m_i\omega_{ci}^2)^{1/2}$, the expression for r_b^2 in Eq.(4.245) can be expressed in the equivalent form

$$(\omega_{ri}^- - \omega_{ri})(\omega_{ri} - \omega_{ri}^+) = \frac{r_{Li}^2}{r_b^2}\omega_{ci}^2 \,, \tag{4.270}$$

where $\omega_{ri}^\pm = -(\omega_{ci}/2)[1 \pm (1 - 2\omega_{pi}^2/\omega_{ci}^2)^{1/2}]$. For fixed values of r_b, ω_{ci} and $2\omega_{pi}^2/\omega_{ci}^2$, it follows from Eq.(4.270) that $\omega_{ri} \simeq \omega_{ri}^+$ or $\omega_{ri} \simeq \omega_{ri}^-$ whenever the ions are sufficiently cold that $r_{Li}^2 \ll r_b^2$. On the other hand, for moderate values of r_{Li}/r_b, the rotation frequency ω_{ri} is not closely tuned to either ω_{ri}^+ or ω_{ri}^-. In any case, the dispersion relation (4.269) is fully kinetic and allows for ion orbits with arbitrary size (as measured by r_{Li}/r_b) consistent with Eq.(4.270).

An important instability that involves the relative rotation of electrons and ions in a nonneutral plasma column is the so-called ion resonance instability.[76,77] This instability is analyzed in Sec. 5.6 using a macroscopic cold-fluid model, assuming electrostatic perturbations about a uniform-density plasma column. The ion response in the corresponding dispersion relation can be inferred directly from Eq.(5.87). In this regard, a careful examination of Eq.(4.269) in the cold-fluid limit with r_b fixed but $r_{Li}^2/r_b^2 \to 0$ and $\omega_{ri} \to \omega_{ri}^+$ or $\omega_{ri} \to \omega_{ri}^-$ shows that

$$\lim_{\substack{v_{Ti}^2 \to 0 \\ \omega_{ri} \to \omega_{ri}^\pm}} \left[\frac{\omega_{pi}^2 r_b^2}{2\ell v_{Ti}^2}\Gamma_\ell(\omega)\right] = -\frac{\omega_{pi}^2}{2(\omega - \ell\omega_{ri})[\omega - \ell\omega_{ri} + (\omega_{ci} + 2\omega_{ri})]} \,.$$

$$\tag{4.271}$$

By comparing Eqs.(4.271) and (5.87), it is evident that the usual cold-fluid ion response is recovered from the kinetic expression for $\Gamma_\ell(\omega)$ in the limit $v_{Ti}^2 \to 0$. In a hybrid-kinetic model[77] of the ion resonance instability, the electrons are treated as a macroscopic cold fluid (as in Sec. 5.6), but the ions are treated kinetically (as in Sec. 4.9). As r_{Li}/r_b is increased, it is found[77] that finite ion Larmor radius effects have a strong stabilizing influence on the ion resonance instability for azimuthal mode number $\ell \geq 2$.

For present purposes, the kinetic dispersion relation $D_\ell(\omega) = 0$ in Eq.(4.269) is analyzed for the case of (stable) oscillations in a pure ion plasma. Typical numerical solutions to Eq.(4.269) are illustrated in Fig. 4.15, where the normalized frequency ω/ω_{ci} is plotted versus the azimuthal mode number for the choice of system parameters $r_b/b = 1/2$

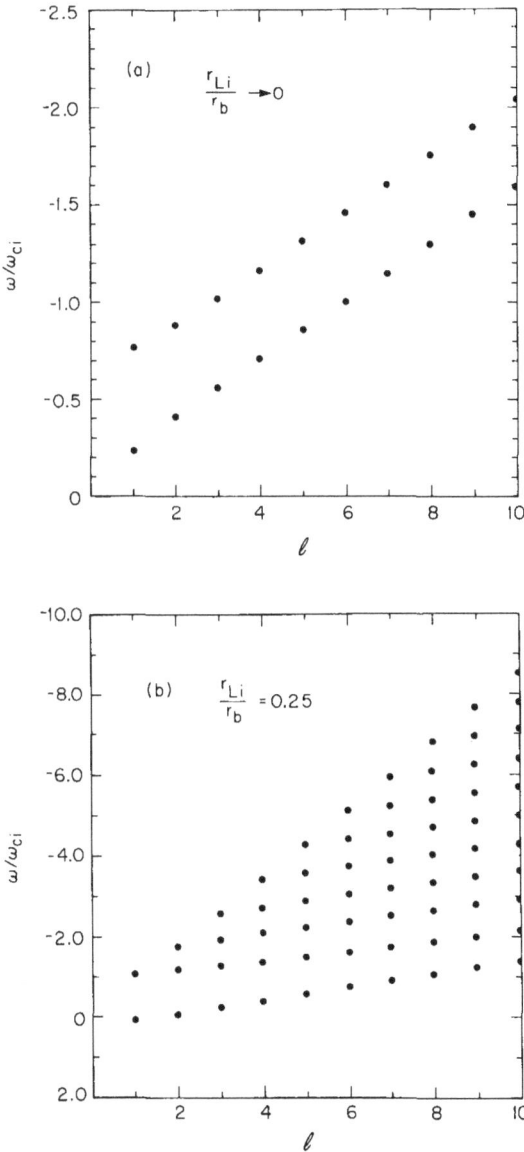

Figure 4.15.   Plot of normalized frequency ω/ω_{ci} versus azimuthal mode number ℓ obtained from Eq.(4.269) for the choice of system parameters $r_b/b = 0.5$ and $2\omega_{pi}^2/\omega_{ci}^2 = 0.5$. The two cases corresponding to (a) $r_{Li}/r_b \to 0$, and (b) $r_{Li}/r_b = 0.25$ are shown in the figure.

and $2\omega_{pi}^2/\omega_{ci}^2 = 1/2$. In Fig. 4.15, the ions are assumed to be in a slow rotational equilibrium with

$$\omega_{ri} = -\frac{\omega_{ci}}{2}\left\{1 - \left(1 - \frac{2\omega_{pi}^2}{\omega_{ci}^2} - \frac{4r_{Li}^2}{r_b^2}\right)^{1/2}\right\}. \qquad (4.272)$$

Here, ω_{ri} has been calculated from Eq.(4.270) including ion pressure gradient effects, i.e., the term proportional to $(r_{Li}^2/r_b^2)\omega_{ci}^2$. The two cases illustrated in Fig. 4.15 correspond to the cold-plasma limit with $r_{Li}/r_b \to 0$ [Fig. 4.15(a)], and a warm-plasma equilibrium with $r_{Li}/r_b = 0.25$ [Fig. 4.15(b)]. In the cold-plasma limit, we note from Fig. 4.15(a) and Eqs.(4.269) and (4.271) that there are two solutions for ω for each value of azimuthal mode number ℓ. On the other hand, for the warm-plasma equilibrium with $r_{Li}/r_b = 0.25$ in Fig. 4.15(b), there are $\ell+1$ solutions for ω for each value of ℓ. Not only is the number of modes increased by the inclusion of thermal effects, the frequency range excited in Fig. 4.15(b) covers a much broader interval than in Fig. 4.15(a).

Problem 4.19   For electrostatic flute perturbations ($\partial/\partial z = 0$) about a nonneutral ion equilibrium $f_i^0(H_\perp - \omega_{ri}P_\theta, p_z)$, the linearized Vlasov equation can be integrated to give (Problem 4.15)

$$\delta f_i(\mathbf{x}_\perp, \mathbf{p}) = Z_i e \frac{\partial f_i^0}{\partial H_\perp} \int_{-\infty}^t dt'\, \exp[-i\omega(t' - t)]$$

$$\times \left(\mathbf{v}_\perp'(t') \cdot \frac{\partial}{\partial \mathbf{x}_\perp'} - \omega_{ri}\frac{\partial}{\partial \theta'}\right)\delta\phi[\mathbf{x}_\perp'(t')], \qquad (4.19.1)$$

where $\text{Im}\,\omega > 0$ is assumed, and the perpendicular orbits $\mathbf{x}_\perp'(t')$ and $\mathbf{v}_\perp'(t')$ in the equilibrium field configuration pass through the phase-space point (x, y, v_x, v_y) at time $t' = t$.

a. Make use of $(d/dt')\delta\phi(\mathbf{x}_\perp') = \mathbf{v}_\perp' \cdot (\partial/\partial \mathbf{x}_\perp')\delta\phi(\mathbf{x}_\perp')$ to integrate by parts with respect to t' in Eq.(4.19.1). Show that

$$\delta f_i(\mathbf{x}_\perp, \mathbf{p}) = Z_i e \frac{\partial f_i^0}{\partial H_\perp}\left[\delta\phi(\mathbf{x}_\perp)\right.$$

$$\left. + \int_{-\infty}^t dt'\, \exp[-i\omega(t' - t)]\left(i\omega - \omega_{ri}\frac{\partial}{\partial \theta'}\right)\delta\phi[\mathbf{x}_\perp'(t')]\right]. \qquad (4.19.2)$$

b. Express $\delta\phi(\mathbf{x}_\perp) = \sum_{\ell=-\infty}^{\infty}\delta\phi^\ell(r)\exp(i\ell\theta)$ and $\delta f_i(\mathbf{x}_\perp, \mathbf{p}) = \sum_{\ell=-\infty}^{\infty}\delta f_i^\ell(r, \mathbf{p})\exp(i\ell\theta)$ in Eq.(4.19.2). Show that the ℓ'th harmonic amplitude $\delta f_i^\ell(r, \mathbf{p})$ can be expressed in the form given in Eq.(4.251).

Problem 4.20   The equilibrium distribution function f_i^0 in Eq.(4.243) can be expressed as

$$f_i^0 = \frac{\hat{n}_i}{\pi m_i^2} \delta \left[V_\perp^2 - v_{Ti}^2 (1 - r^2/r_b^2) \right] G_i(p_z) , \qquad (4.20.1)$$

where $v_{Ti}^2 = 2\hat{T}_{i\perp}/m_i$, and $m_i V_\perp^2/2$ is the kinetic energy in the rotating frame.

a. Show from Eq.(4.20.1) that the ion density profile $n_i^0(r) = \int d^3p \, f_i^0$ has the rectangular form illustrated in Fig. 4.14. Here, $\int_0^\infty d^3p \cdots = 2\pi \int_0^\infty d(m_i V_\perp)$ $\times m_i V_\perp \int_{-\infty}^\infty dp_z \cdots$.

b. Prove the identity in Eq.(4.259) for the choice of ion distribution function f_i^0 in Eq.(4.20.1).

4.10   Kinetic Equilibrium and Stability Properties of the Modified Betatron

4.10.1   Introduction and Assumptions

The topics covered in Secs. 4.1–4.9 have illustrated the considerable versatility in using the Vlasov-Maxwell equations to investigate the detailed equilibrium and stability properties of nonneutral plasmas. Indeed, the equilibrium configurations considered here have ranged from nonneutral plasma columns (Sec. 4.2) and relativistic electron beams (Sec. 4.4), to kinetic equilibria confined both axially and radially by externally applied focusing fields (Sec. 4.3). Moreover, the stability studies have ranged from the derivation of a kinetic stability theorem (Sec. 4.5), to investigations of the detailed stability properties for electrostatic body-wave and flute perturbations about a uniform-density plasma column (Secs. 4.7–4.9). In this section, we make use of several of the techniques developed earlier in Chapter 4 to investigate the kinetic equilibrium and stability properties of an intense relativistic electron ring in a modified betatron (Fig. 4.16).[23-41] Given the overall complexity of the configuration and the importance of both electric and magnetic self fields at high electron current, the modified betatron represents a particularly instructive application of state-of-the-art techniques using the Vlasov-Maxwell equations to investigate the equilibrium and stability properties of magnetically confined nonneutral plasma.

The development of high-energy accelerators capable of producing high-current electron and ion beams has been an active and growing area of research, including work on novel accelerator concepts as well as research to improve and modify the technology of existing acceleration schemes. Cyclic induction accelerators,[79] which include the conventional high-current betatron as well as the modified betatron, represent one

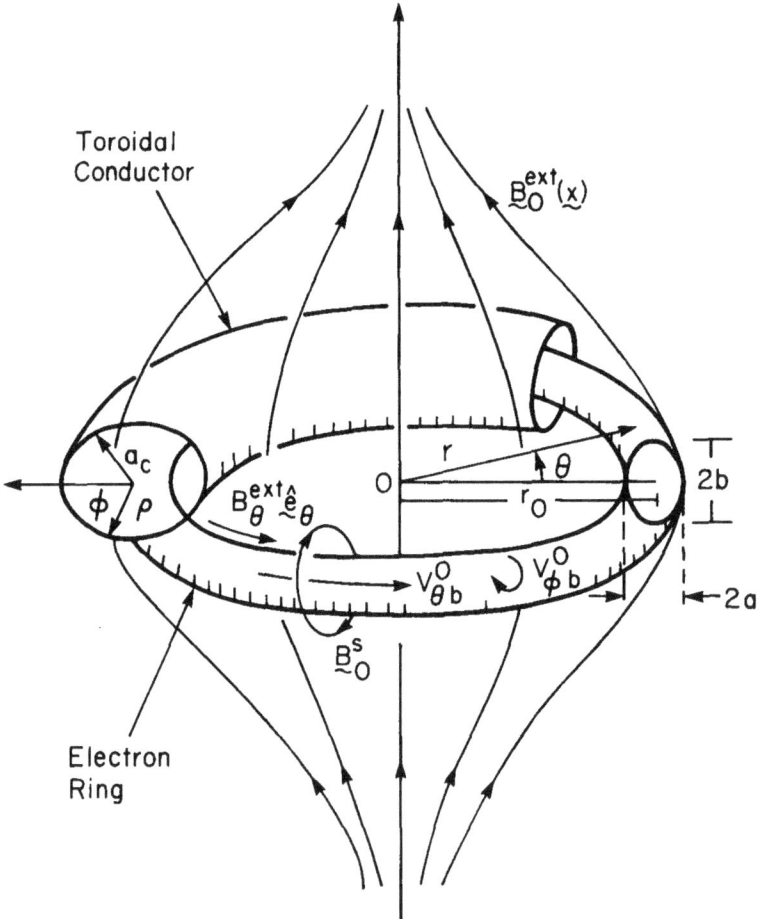

Figure 4.16. Modifed betatron geometry showing an intense relativistic electron ring confined at the midplane of an applied betatron (mirror) field $B_r^{ext}(r,z)\hat{e}_r + B_z^{ext}(r,z)\hat{e}_z$ and an applied toroidal field $B_\theta^{ext}(r,z)\hat{e}_\theta$.

promising approach. In the conventional betatron,[42,43] a relativistic toroidal electron ring is confined by an external mirror (or betatron) magnetic field, and the change of this magnetic field with time is responsible for the acceleration. In the modified betatron,[23-41] a strong applied toroidal magnetic field is added to the mirror field of the conventional betatron. This has the effect of considerably increasing the limiting beam current over that of the conventional betatron. Denoting the applied vertical and toroidal magnetic fields at the beam center $(r, z) = (r_0, 0)$ by $\hat{B}_z$ and $\hat{B}_\theta$, respectively, it is found[23] that the beam current is increased by a factor of $(1/2)(\hat{B}_\theta/\hat{B}_z)^2$, when $\hat{B}_\theta \gg \hat{B}_z$. Also, as will become evident in Sec. 4.10.4, the stability of the accelerated beam is improved substantially by the presence of the toroidal field.

Detailed analyses of the modified betatron configuration are relatively recent. However, theoretical studies have been carried out using kinetic, single-particle, and macroscopic fluid models. In this section, we use a kinetic model[25,36,41] based on the Vlasov-Maxwell equations to investigate longitudinal stability properties over a wide parameter range. Included in the analysis are the important effects of intense self-electric and self-magnetic fields. Although not discussed in the present analysis, it should be noted that the introduction of helical (stellarator) windings[44,45] to the modified betatron configuration in Fig. 4.16 provides improved equilibrium and stability properties[44-48] relative to the modified betatron configuration[23-41] considered here.

The equilibrium configuration used to model the modified betatron is illustrated in Fig. 4.16. It consists of a relativistic electron ring located at the midplane of an externally applied betatron (mirror) magnetic field $B_r^{\text{ext}}(r, z)\hat{e}_r + B_z^{\text{ext}}(r, z)\hat{e}_z$. Moreover, there is an external toroidal magnetic field $B_\theta^{\text{ext}}(r, z)\hat{e}_\theta$, together with the external betatron field, which act to confine the electrons both axially and radially. The equilibrium radius of the electron ring is denoted by r_0 and the minor dimensions of the ring are denoted by $2a$ (radial dimension) and $2b$ (axial dimension). In addition, the electron ring is located concentrically inside a toroidal conductor with minor radius a_c. The electron current in the toroidal direction produces a poloidal self-magnetic field. Furthermore, for $B_\theta^{\text{ext}}(r, z) = 0$ in Fig. 4.16, the conventional betatron configuration is recovered.

The equilibrium fields provide both focusing and defocusing forces on the electrons in the ring. The electrons travel at relativistic velocities in the positive θ-direction. This gives an associated ring current in the negative θ-direction, which produces a poloidal self-magnetic field $B_r^s(r, z)\hat{e}_r +$

$B_z^s(r, z)\hat{e}_z$ with the polarity indicated in Fig. 4.16. This self-magnetic field, by virtue of the Lorentz force on the electrons, produces a focusing force that acts to compress the ring in the minor dimensions. The electron charge is assumed to be partially neutralized by a positive ion background. The excess electrons form a potential well for the ions. For the electrons, however, the electrostatic forces are repulsive. Therefore, the self-electric field produced by the nonneutral ring $(f < 1)$ acts as a defocusing field that tends to increase the minor dimensions of the ring.

The modified betatron equilibrium possesses an average equilibrium poloidal rotation of the ring electrons. This rotation, together with the externally applied toroidal magnetic field, provides an additional Lorentz focusing force acting to confine the electrons.

To make the theoretical analysis tractable, we make the following simplifying assumptions, which are consistent with the operating parameters for present and planned modified betatron experiments.

(a) The relativistic electron ring $(j = b)$ is partially charge neutralized by an immobile $(m_i \to \infty)$ ion background $(j = i)$. The equilibrium ion density is

$$n_i^0(r, z) = fn_b^0(r, z) \,, \qquad (4.273)$$

where $f =$ const. is the fractional charge neutralization, and $n_b^0(r, z)$ is the equilibrium density of the beam electrons. This approximation may be somewhat idealized, but it provides a good qualitative indication of the effects of a partially neutralizing ion background on equilibrium and stability properties of the electron ring. The case in which $f = 0$ is not excluded.

(b) The minor dimensions of the electron ring and the conducting wall are small compared to the major radius, i.e.,

$$a, b \ll r_0 \,, \qquad a_c \ll r_0 \,. \qquad (4.274)$$

(c) Furthermore, it is assumed that

$$\frac{\nu}{\gamma_b} = \frac{\bar{N}_b}{2\pi r_0} \frac{e^2}{m_e c^2} \frac{1}{\gamma_b} \ll 1 \,, \qquad (4.275)$$

where ν is Budker's parameter, $\bar{N}_b$ is the total number of electrons in the ring, $-e$ is the electron charge, m_e is the electron rest mass,

c is the speed of light *in vacuo*, $e^2/m_e c^2$ is the classical electron radius, and $\gamma_b m_e c^2$ is the characteristic energy of the beam electrons. For an electron beam with uniform density $\hat{n}_b = \bar{N}_b/2\pi^2 ab r_0$, it can be shown that

$$\frac{\nu}{\gamma_b} = \frac{1}{4\gamma_b} \frac{\omega_{pb}^2 ab}{c^2} , \qquad (4.276)$$

where $\omega_{pb}^2 = 4\pi \hat{n}_b e^2/m_e$ is the nonrelativistic plasma frequency-squared.

(d) The characteristic transverse (r, z) kinetic energy of the beam electrons is small compared to the characteristic azimuthal energy, i.e.,

$$\frac{\left(p_r^2 + p_z^2\right)}{2\gamma_b m_e} \ll \gamma_b m_e c^2 , \qquad (4.277)$$

where $p_\theta \simeq \gamma_b m_e \beta_b c$ is the characteristic azimuthal momentum.

(e) The spread in canonical angular momentum $\delta P_\theta = P_\theta - P_0$ is assumed to be sufficiently small that $|\delta P_\theta| \ll \gamma_b m_e \beta_b c r_0$. For the modified betatron, it is further assumed that

$$\frac{|\delta P_\theta|}{\gamma_b m_e \beta_b c r_0} \ll \frac{\hat{B}_\theta}{\hat{B}_z} \left| \frac{p_\perp}{p_\theta} \right| , \qquad (4.278)$$

where $p_\perp = (p_r^2 + p_z^2)^{1/2}$ is the characteristic transverse momentum, p_θ is the characteristic azimuthal momentum, and $\hat{B}_z = B_z^{\text{ext}}(r_0, 0)$ and $\hat{B}_\theta = B_\theta^{\text{ext}}(r_0, 0)$ are the vertical and toroidal magnetic fields at the beam center. For the modified betatron, Eq.(4.278) and the assumption of circular cross section, i.e.,

$$a = b , \quad n = \frac{1}{2} , \qquad (4.279)$$

are sufficient to assure that the poloidal canonical angular momentum P_ϕ defined in Eq.(4.285) is an approximate single-particle constant of the motion in the equilibrium field configuration (Problem 4.22). In Eq.(4.279), the external field index n is defined by

$$n = - \left[\frac{r}{B_z^{\text{ext}}(r, z)} \frac{\partial B_z^{\text{ext}}(r, z)}{\partial r} \right]_{(r_0, 0)} . \qquad (4.280)$$

4.10.2   Self-Consistent Vlasov Equilibrium

For the equilibrium configuration illustrated in Fig. 4.16, the relativistic electron ring is located at the midplane of the applied mirror field $B_r^{\text{ext}}(r,z)\hat{e}_r + B_z^{\text{ext}}(r,z)\hat{e}_z$, and the beam is centered at $(r,z) = (r_0, 0)$ where r_0 is the equilibrium radius of the ring. The applied toroidal magnetic field

$$\mathbf{B}_\theta^{\text{ext}}(\mathbf{x}) = \hat{B}_\theta \frac{r_0}{r}\hat{e}_\theta \ , \tag{4.281}$$

and the mirror field act together to confine the ring both axially and radially. Here, $\hat{e}_r$, $\hat{e}_\theta$ and $\hat{e}_z$ are unit vectors in the r-, θ- and z-directions, respectively. In addition to the cylindrical polar coordinates (r, θ, z) we introduce the toroidal polar coordinate system (ρ, ϕ, θ) illustrated in Fig. 4.16 and defined by

$$r - r_0 = \rho \cos \phi \ ,$$
$$z = -\rho \sin \phi \ , \tag{4.282}$$

where ρ is measured from the equilibrium radius r_0. The electrons composing the ring undergo a large orbit gyration in the applied mirror field with characteristic mean azimuthal velocity $V_{\theta b}^0 = \beta_b c$ in the positive θ-direction. For azimuthally symmetric equilibria $(\partial/\partial\theta = 0)$ with both r and z dependence, there are two *exact* single-particle constants of the motion. These are the total energy H,

$$H = \left(m_e^2 c^4 + c^2 \mathbf{p}^2\right)^{1/2} - m_e c^2 - e\phi_0(r, z) \ , \tag{4.283}$$

and the canonical angular momentum P_θ,

$$P_\theta = r \left[p_\theta - \frac{e}{c}A_\theta^0(r, z)\right] \ . \tag{4.284}$$

Here, $\mathbf{p} = \gamma m_e \mathbf{v}$ is the mechical momentum, $\gamma = (1 + \mathbf{p}^2/m_e^2 c^2)^{1/2}$ is the relativistic mass factor, $\phi_0(r, z)$ is the equilibrium electrostatic potential, and $A_\theta^0(r, z) = A_\theta^{\text{ext}}(r, z) + A_\theta^s(r, z)$ is the θ-component of the vector potential for the total (external plus self) equilibrium magnetic field. Without loss of generality, $\phi_0(r_0, 0) = 0 = A_\theta^s(r_0, 0)$ is assumed in Eqs.(4.283) and (4.284). Within the context of Eqs.(4.277)–(4.280), it can be shown (Problem 4.22) that the canonical angular momentum

$$P_\phi = \rho p_\phi - \frac{e}{2c}\hat{B}_\theta \rho^2 \ , \qquad (4.285)$$

in the plane perpendicular to the toroidal magnetic field $B_\theta^{\text{ext}}\hat{e}_\theta$ is an approximate single-particle constant of the motion for a thin circular beam with $a = b$ and $n = 1/2$.[25] Here, p_ϕ is the mechanical momentum in the ϕ-direction (Fig. 4.16), and $\rho^2 = (r-r_0)^2 + z^2$ is defined in Eq.(4.282). For a thin ring, the toroidal magnetic field $B_\theta^{\text{ext}} = \hat{B}_\theta r_0/r$ is approximately uniform over the minor cross section of the ring.

Any distribution function f_b^0 that is a function only of the single-particle constants of the motion in the equilibrium field configuration satisfies the steady-state $(\partial/\partial t = 0)$ Vlasov equation. For present purposes, we consider the electron distribution function specified by[25]

$$f_b^0(H, P_\phi, P_\theta) = \frac{\hat{n}_b r_0 \Delta}{2\pi^2 \gamma_b m_e} \frac{\delta[H - \omega_{rb} P_\phi - (\hat{\gamma}_b - 1)m_e c^2]}{[(P_\theta - P_0)^2 + \Delta^2]} \ , \qquad (4.286)$$

where $\hat{n}_b = n_b^0(r_0, 0)$ is the electron density at the equilibrium radius $(r, z) = (r_0, 0)$, $\omega_{rb} = $ const. is the angular velocity of mean rotation in the ϕ-direction, Δ is the characteristic spread in the canonical angular momentum P_θ, and γ_b and $\hat{\gamma}_b$ are constants. The equilibrium Poisson equation can be expressed as

$$\left(\frac{1}{r}\frac{\partial}{\partial r}r\frac{\partial}{\partial r} + \frac{\partial^2}{\partial z^2}\right)\phi_0(r, z) = 4\pi e(1 - f)\int d^3p\, f_b^0(H, P_\phi, P_\theta) \ . \quad (4.287)$$

Furthermore, the θ-component of the $\nabla \times \mathbf{B}_0^s(\mathbf{x})$ Maxwell equation can be expressed as

$$\left(\frac{1}{r}\frac{\partial}{\partial r}r\frac{\partial}{\partial r} + \frac{\partial^2}{\partial z^2}\right)A_\theta^s(r, z) = \frac{4\pi e}{c}\int d^3p\, v_\theta f_b^0(H, P_\phi, P_\theta) \ , \quad (4.288)$$

where $v_\theta = (p_\theta/m_e)(1 + \mathbf{p}^2/m_e^2 c^2)^{-1/2}$ is the azimuthal electron velocity.

Consistent with the thin-ring approximation [Eq.(4.274)] is the requirement that the $r - z$ kinetic energy be small in comparison with the effective azimuthal energy

$$\gamma_\theta(r, z)m_e c^2 = \left\{m_e^2 c^4 + c^2\left(\frac{P_0}{r} + \frac{e}{c}A_\theta^0(r, z)\right)^2\right\}^{1/2} \qquad (4.289)$$

defined for $p_r^2 + p_z^2 = 0$ and $P_\theta = P_0$. It is also assumed that the spread in canonical angular momentum is small with $|\delta P_\theta| = |P_\theta - P_0| \ll \gamma_b m_e \beta_b c r_0$. Taylor expanding for small values of $p_r^2 + p_z^2$ and $\delta P_\theta = P_\theta - P_0$, the total energy H defined in Eq.(4.283) can be approximated by[25]

$$
H = \frac{p_r^2 + p_z^2}{2\gamma_\theta(r,z)m_e} + [\gamma_\theta(r,z) - 1]\, m_e c^2 - e\phi_0(r,z)
$$
$$
+ \frac{V_{\theta b}^0(r,z)}{r}(\delta P_\theta) + \frac{(\delta P_\theta)^2}{2\gamma_\theta^3(r,z)m_e r^2} \, , \tag{4.290}
$$

where the mean azimuthal velocity of an electron fluid element is defined by

$$
V_{\theta b}^0(r,z) = \frac{\int d^3 p\, v_\theta f_b^0}{\int d^3 p\, f_b^0} \simeq \frac{1}{\gamma_\theta(r,z)m_e}\left[\frac{P_0}{r} + \frac{e}{c}A_\theta^0(r,z)\right] \, . \tag{4.291}
$$

For future reference, it is convenient to introduce the characteristic azimuthal energy of a beam electron defined by

$$
\gamma_b m_e c^2 = \gamma_\theta(r_0, 0)m_e c^2 \, . \tag{4.292}
$$

In addition, we choose

$$
P_0 = \frac{e\hat{B}_z}{2c}r_0^2 \, , \quad A_\theta^0(r_0, 0) = \frac{1}{2}\hat{B}_z r_0 \, , \tag{4.293}
$$

where $\hat{B}_z \equiv B_z^{\text{ext}}(r_0, 0)$.

The mean equilibrium radius r_0 of the electron ring is determined from the condition

$$
\left[m_e c^2 \frac{\partial}{\partial r}\gamma_\theta(r,z) - e\frac{\partial}{\partial r}\phi_0(r,z)\right]_{(r_0,0)} = 0 \, . \tag{4.294}
$$

Making use of Eqs.(4.289) and (4.294), we obtain

$$
-\frac{\gamma_b m_e \beta_b^2 c^2}{r_0} = -e\left[E_r(r_0, 0) + \beta_b B_z(r_0, 0)\right] \, , \tag{4.295}
$$

where $\beta_b c \equiv V_{\theta b}^0(r_0, 0)$ is the azimuthal velocity, $E_r(r,z) = -\partial\phi_0/\partial r$ is the radial self-electric field, and $B_z(r,z) = (1/r)(\partial/\partial r)(rA_\theta^0)$ is the

axial magnetic field. Equation (4.295) is simply a statement of radial
force balance on an electron fluid element at $(r, z) = (r_0, 0)$. Due to the
symmetry of the equilibrium configuration (Fig. 4.16), it is clear that
the axial electric field and the radial magnetic field at the midplane are
identically zero. We therefore conclude that

$$
\left(\frac{\partial}{\partial z} \gamma_\theta (r, z) \right)_{(r_0, 0)} = 0 = \left(\frac{\partial}{\partial z} \phi_0 (r, z) \right)_{(r_0, 0)} , \tag{4.296}
$$

$$
\left(\frac{\partial^2}{\partial r \partial z} \gamma_\theta (r, z) \right)_{(r_0, 0)} = 0 = \left(\frac{\partial^2}{\partial r \partial z} \phi_0 (r, z) \right)_{(r_0, 0)} . \tag{4.297}
$$

Making use of Eqs.(4.293)–(4.297), the expression for the total energy
H in Eq.(4.290) can be further simplified by Taylor expanding the expres-
sions for $\gamma_\theta(r, z)$, $\phi_0(r, z)$, etc., about $(r, z) = (r_0, 0)$. It is convenient to
introduce the radial betatron frequency ω_r defined by

$$
\omega_r^2 = \frac{1}{\gamma_b m_e} \left(\frac{\partial^2}{\partial r^2} \left(\gamma_\theta m_e c^2 - e\phi_0 \right) \right)_{(r_0, 0)} , \tag{4.298}
$$

and the axial betatron frequency ω_z defined by

$$
\omega_z^2 = \frac{1}{\gamma_b m_e} \left(\frac{\partial^2}{\partial z^2} \left(\gamma_\theta m_e c^2 - e\phi_0 \right) \right)_{(r_0, 0)} . \tag{4.299}
$$

Taylor expanding Eq.(4.290) about $(r, z) = (r_0, 0)$ for a thin ring and
retaining terms to quadratic order in $r - r_0$ and z then gives[25]

$$
H = (\gamma_b - 1)m_e c^2 + \frac{p_r^2 + p_z^2}{2\gamma_b m_e} + \frac{\omega_{cz}}{\gamma_b} \left(1 - \frac{(r - r_0)}{r_0} \right) \delta P_\theta
$$

$$
+ \frac{(\delta P_\theta)^2}{2\gamma_b^3 m_e r_0^2} + \frac{1}{2} \gamma_b m_e \left[\omega_r^2 (r - r_0)^2 + \omega_z^2 z^2 \right] . \tag{4.300}
$$

Here, ω_{cz} is the nonrelativistic electron cyclotron frequency in the axial
magnetic field

$$
\omega_{cz} = \frac{e \hat{B}_z}{m_e c} . \tag{4.301}
$$

Making use of Eqs.(4.289), (4.298), and (4.299) and the approximate expressions for the equilibrium potentials, it follows that ω_r^2 and ω_z^2 can be expressed as (Problem 4.21)

$$\omega_r^2 = \frac{\omega_{cz}^2}{\gamma_b^2}(1-n) + \frac{\omega_{pb}^2}{\gamma_b}\frac{b}{a+b}\left[\beta_b^2 - (1-f)\right] , \qquad (4.302)$$

and

$$\omega_z^2 = \frac{\omega_{cz}^2}{\gamma_b^2}n + \frac{\omega_{pb}^2}{\gamma_b}\frac{a}{a+b}\left[\beta_b^2 - (1-f)\right] , \qquad (4.303)$$

where $\gamma_b = (1 - \beta_b^2)^{-1/2}$. In Eqs.(4.302) and (4.303), $\omega_{pb}^2 = 4\pi\hat{n}_b e^2/m_e$ is the nonrelativistic plasma frequency-squared, n is the external field index at $(r,z) = (r_0, 0)$ defined in Eq.(4.280), and $\hat{n}_b$ and β_b are defined by $\hat{n}_b = n_b^0(r_0, 0)$ and $\beta_b c = V_{\theta b}^0(r_0, 0)$.

In the remainder of Sec. 4.10, we specialize to the case of a circular electron beam with $a = b$ and $n = 1/2$. Equations (4.302) and (4.303) then reduce to

$$\omega_r^2 = \omega_z^2 = \omega_\beta^2 , \qquad (4.304)$$

where

$$\omega_\beta^2 = \frac{1}{2}\frac{\omega_{cz}^2}{\gamma_b^2} + \frac{1}{2}\frac{\omega_{pb}^2}{\gamma_b}\left[\beta_b^2 - (1-f)\right] . \qquad (4.305)$$

It is further assumed that the spread Δ in canonical angular momentum P_θ is sufficiently small that $\Delta \ll \gamma_b^2 m_e\omega_\beta a r_0,\ \gamma_b^2 m_e\omega_r a^2\omega_\beta/\omega_{cz}$. For $\delta P_\theta = 0$ and $n = 1/2$, the expression for H in Eq.(4.300) reduces to[25]

$$H = (\gamma_b - 1)m_e c^2 + \frac{p_r^2 + p_z^2}{2\gamma_b m_e} + \frac{1}{2}\gamma_b m_e\omega_\beta^2\rho^2 , \qquad (4.306)$$

where $\rho^2 = (r-r_0)^2 + z^2$. For an electron beam with circular cross section and $P_\theta = P_0$, some straightforward algebra that makes use of Eqs.(4.285) and (4.306) shows that the combination $H - \omega_{rb}P_\phi$ occurring in Eq.(4.286) can be expressed in the approximate form

$$H - \omega_{rb}P_\phi = (\gamma_b - 1)m_e c^2 + \frac{p_\perp^2}{2\gamma_b m_e} + \psi_b(r, z) . \qquad (4.307)$$

Here, $p_\perp^2 = p_\rho^2 + (p_\phi - \gamma_b m_e\omega_{rb}\rho)^2$ is the transverse momentum-squared in a frame of reference rotating with angular velocity ω_{rb} about the toroidal axis. The effective potential $\psi_b(r, z)$ in Eq.(4.307) is defined by

$$\psi_b(r, z) = \frac{1}{2}\gamma_b m_e \Omega_\beta^2 \rho^2 , \qquad (4.308)$$

for a thin ring with circular cross section $(a = b)$. Moreover, the effective focusing frequency Ω_β^2 occurring in Eq.(4.308) is defined by

$$\Omega_\beta^2 = \frac{\omega_{rb}\omega_{c\theta}}{\gamma_b} - \omega_{rb}^2 + \omega_\beta^2 , \qquad (4.309)$$

where ω_β^2 is defined in Eq.(4.305), and $\omega_{c\theta} = e\hat{B}_\theta/m_e c$ is the nonrelativistic cyclotron frequency in the applied toroidal field at the equilibrium orbit $(r, z) = (r_0, 0)$.

Substituting Eq.(4.307) into Eq.(4.286) and evaluating the electron density profile $n_b^0(r, z) = \int d^3p \, f_b^0(H, P_\phi, P_\theta)$, we obtain

$$n_b^0(r, z) = \begin{cases} \hat{n}_b , & 0 \le \rho < a , \\ 0 , & a < \rho \le a_c , \end{cases} \qquad (4.310)$$

where the minor radius a of the electron ring is defined by

$$a = \left[\frac{2c^2(\hat{\gamma}_b - \gamma_b)}{\gamma_b \Omega_\beta^2}\right]^{1/2} . \qquad (4.311)$$

For the particular choice of equilibrium distribution function in Eq.(4.286), note from Eq.(4.310) that the electron density is uniform over the ring cross section. For the equilibrium to exist, it is necessary that $\hat{\gamma}_b > \gamma_b$ and $\Omega_\beta^2 > 0$. In addition, $\hat{\gamma}_b - \gamma_b \ll \gamma_b$ is assumed, which assures a thin ring with $a \ll r_0$ and $\nu/\gamma_b \ll 1$. It is convenient to express $\Omega_\beta^2 = (\omega_{rb}^+ - \omega_{rb})(\omega_{rb} - \omega_{rb}^-)$ where the cold-fluid rotation frequencies $\omega_{rb}^\pm$ are defined by

$$\omega_{rb}^\pm = \frac{\omega_{c\theta}}{2\gamma_b}\left\{1 \pm \left(1 + \frac{4\gamma_b^2\omega_\beta^2}{\omega_{c\theta}^2}\right)^{1/2}\right\} . \qquad (4.312)$$

Therefore, the condition $\Omega_\beta^2 > 0$ can be expressed in the equivalent form

$$\omega_{rb}^- < \omega_{rb} < \omega_{rb}^+ . \qquad (4.313)$$

It is readily shown (Problem 4.23) that the frequencies ω_{rb}^+ and ω_{rb}^- defined in Eq.(4.312) are the natural orbital oscillation frequencies for transverse electron oscillations about the equilibrium orbit $(r, z) = (r_0, 0)$.

It can be shown that the equilibrium pressure tensor in the (ρ, ϕ) plane perpendicular to the θ-direction is isotropic with the perpendicular pressure $P_{b\perp}^0(\rho) = n_b^0(\rho)T_{b\perp}^0(\rho)$ given by

$$n_b^0(\rho)T_{b\perp}^0(\rho) = 2\pi \int_0^\infty dp_\perp \, p_\perp \int_{-\infty}^\infty dp_\theta \frac{[p_\rho^2 + (p_\phi - \gamma_b m_e \omega_{rb}\rho)^2]}{2\gamma_b m_e} f_b^0 ,$$

(4.314)

where $T_{b\perp}^0(\rho)$ is the effective transverse temperature profile. Defining

$$\hat{T}_{b\perp} \equiv \frac{1}{2}\gamma_b m_e \Omega_\beta^2 a^2 = \frac{1}{2}\gamma_b m_e \left(\frac{\omega_{c\theta}}{\gamma_b}\right)^2 r_{Lb}^2 ,$$

(4.315)

and substituting Eq.(4.286) into Eq.(4.314) gives the parabolic temperature profile

$$T_{b\perp}^0(\rho) = \hat{T}_{b\perp} \left(1 - \frac{\rho^2}{a^2}\right)$$

(4.316)

for $0 \leq \rho < a$. For the choice of equilibrium distribution function in Eq.(4.286), note from Eq.(4.316) that $T_{b\perp}^0(\rho)$ decreases monotonically from the maximum value $\hat{T}_{b\perp}$ at $\rho = 0$ to zero at $\rho = a$. In Eq.(4.315), $r_{Lb} = (2\gamma_b \hat{T}_{b\perp}/m_e \omega_{c\theta}^2)^{1/2}$ is the characteristic thermal Larmor radius of the beam electrons in the azimuthal magnetic field $\hat{B}_\theta$. Making use of Eq.(4.315) to eliminate Ω_β in Eq.(4.309) in favor of r_{Lb}, Eq.(4.309) can be solved for the rotation frequency ω_{rb}. This gives[25]

$$\omega_{rb} = \frac{\omega_{c\theta}}{2\gamma_b} \left\{1 \pm \left[1 + \frac{4\gamma_b^2 \omega_\beta^2}{\omega_{c\theta}^2} - \left(\frac{2r_{Lb}}{a}\right)^2\right]^{1/2}\right\} ,$$

(4.317)

which relates ω_{rb} to the thermal Larmor radius r_{Lb}. The two signs ($\pm$) in Eq.(4.317) represent *fast* ($+$) and *slow* ($-$) rotational equilibria. Whenever $r_{Lb}/a \to 0$, the rotation frequencies defined in Eq.(4.317) approach the laminar (cold-fluid) rotation frequencies defined in Eq.(4.312).

Substituting the definition of ω_β^2 into Eq.(4.317), the requirement that the radical be real can be expressed as[25]

$$0 \leq \left(\frac{2r_{Lb}}{a}\right)^2 \leq 1 - \frac{2\gamma_b \omega_{pb}^2}{\omega_{c\theta}^2}\left(1 - f - \beta_b^2\right) + \frac{2\omega_{cz}^2}{\omega_{c\theta}^2} .$$

(4.318)

Equation (4.318) is required for existence of the ring equilibrium. For example, if $f = 0$ and $r_{Lb}^2/a^2 \ll 1$, then $\omega_\beta^2 = \omega_{cz}^2/2\gamma_b^2 - \omega_{pb}^2/2\gamma_b^3$ and Eq.(4.318) gives the restriction on beam density

$$\frac{\omega_{pb}^2}{\gamma_b\omega_{cz}^2} < 1 + \frac{1}{2}\frac{\omega_{c\theta}^2}{\omega_{cz}^2} , \qquad (4.319)$$

where $\gamma_b = (1 - \beta_b^2)^{-1/2}$, $\omega_{cz} = e\hat{B}_z/m_ec$, $\omega_{c\theta} = e\hat{B}_\theta/m_ec$, and $\omega_{pb}^2 = 4\pi\hat{n}_be^2/m_e$. It is evident from Eq.(4.319) that the presence of the toroidal magnetic field in the modified betatron increases the maximum allowed beam density (or current) by a factor of $(1 + \omega_{c\theta}^2/2\omega_{cz}^2)$ relative to the case of a conventional betatron (where $\omega_{c\theta} = 0$). For $\omega_{c\theta}^2/\omega_{cz}^2 \gg 1$, this enhancement factor is large, which accounts for the considerable appeal of the modified betatron as a high-current accelerator.[23-25]

Problem 4.21   Consider the effective potential

$$V(r,z) = [\gamma_\theta(r,z) - 1]m_ec^2 - e\phi_0(r,z) , \qquad (4.21.1)$$

where $\gamma_\theta(r,z)$ is defined in Eq.(4.289). Note that $V(r,z)$ occurs in the expression (4.290) for the total energy H. For a thin ring, Taylor expand Eq.(4.21.1) about the equilibrium orbit $(r,z) = (r_0,0)$, retaining terms to quadratic order in $r - r_0$ and z. In the analysis, make use of the definitions $\gamma_b = \gamma_\theta(r_0,0)$ and $\beta_b = V_{\theta b}^0(r_0,0)/c = (1 - 1/\gamma_b^2)^{1/2}$, Maxwell's equations (4.287) and (4.288), and the choices of P_0 and $A_\theta^0(r_0,0)$ in Eq.(4.293).

a. Show that the condition $[\partial V(r,z)/\partial r]_{(r_0,0)} = 0$ can be expressed as

$$-\frac{\gamma_b m_e\beta_b^2 c^2}{r_0} = -e\left[E_r(r_0,0) + \beta_b B_z(r_0,0)\right] , \qquad (4.21.2)$$

where $E_r = -\partial\phi_0/\partial r$ and $B_z = (1/r)(\partial/\partial r)(rA_\theta^0)$. Equation (4.21.2) is a statement of radial force balance on an electron fluid element at $(r,z) = (r_0,0)$. In effect, Eq.(4.21.2) determines the equilibrium ring radius r_0. If the radial electric field $E_r(r_0,0)$ is neglected in Eq.(4.21.2), we obtain the familiar result $r_0 = \gamma_b\beta_bc/\omega_{cz}$, where $\omega_{cz} = e\hat{B}_z/m_ec$ and $\hat{B}_z = B_z(r_0,0)$.

b. Use symmetry arguments to show that $[\partial V(r,z)/\partial z]_{(r_0,0)} = 0$ and $[\partial^2 V(r,z)/\partial r\partial z]_{(r_0,0)} = 0$ [see also Eqs.(4.296) and (4.297)]. Correct to quadratic order, $V(r,z)$ can then be expressed as

$$V(r,z) = (\gamma_b - 1)m_ec^2 + \frac{1}{2}\gamma_b m_e\left[\omega_r^2(r - r_0)^2 + \omega_z^2 z^2\right] , \qquad (4.21.3)$$

where use has been made of $\phi_0(r_0, 0) = 0$, and ω_r^2 and ω_z^2 are defined by

$$\omega_r^2 = \frac{1}{\gamma_b m_e} \left[\frac{\partial^2}{\partial r^2} V(r, z) \right]_{(r_0, 0)} ,$$

$$\omega_z^2 = \frac{1}{\gamma_b m_e} \left[\frac{\partial^2}{\partial z^2} V(r, z) \right]_{(r_0, 0)} . \qquad (4.21.4)$$

c. Show from Eqs.(4.21.1) and (4.21.4) that

$$\gamma_b m_e \omega_r^2 = \frac{1}{\gamma_b m_e} \left(1 - \beta_b^2 \right) \left[-\frac{P_0}{r^2} + \frac{e}{c} \frac{\partial A_\theta^0}{\partial r} \right]_{(r_0, 0)}^2$$

$$+ \beta_b c \left[\frac{2 P_0}{r^3} + \frac{e}{c} \frac{\partial^2 A_\theta^0}{\partial r^2} \right]_{(r_0, 0)} - e \left[\frac{\partial^2 \phi_0}{\partial r^2} \right]_{(r_0, 0)} , \qquad (4.21.5)$$

$$\gamma_b m_e \omega_z^2 = \frac{1}{\gamma_b m_e} \left(1 - \beta_b^2 \right) \left[\frac{e}{c} \frac{\partial A_\theta^0}{\partial z} \right]_{(r_0, 0)}^2$$

$$+ \beta_b c \left[\frac{e}{c} \frac{\partial^2 A_\theta^0}{\partial z^2} \right]_{(r_0, 0)} - e \left[\frac{\partial^2 \phi_0}{\partial z^2} \right]_{(r_0, 0)} , \qquad (4.21.6)$$

where use is made of Eqs.(4.291) and (4.292).

d. For a thin ring, assume that the electron charge density and current density occurring in the equilibrium Maxwell equations (4.287) and (4.288) are constant on elliptical surfaces with $(r - r_0)^2/a^2 + z^2/b^2 = \text{const.}$ (see Fig. 4.16). To leading order, the electrostatic potential $\phi_0(r, z)$ and vector potential $A_\theta^0(r, z)$ within the electron ring can be approximated by

$$\phi_0(r, z) = 2\pi e (1 - f) \frac{\hat{n}_b}{(a + b)} \left[b(r - r_0)^2 + az^2 \right] , \qquad (4.21.7)$$

$$r A_\theta^0(r, z) = \frac{1}{2} r_0^2 \hat{B}_z + r_0 \hat{B}_z (r - r_0) + \frac{1}{2} \hat{B}_z \left[(1 - n)(r - r_0)^2 + nz^2 \right]$$

$$+ 2\pi e r_0 \beta_b \frac{\hat{n}_b}{(a + b)} \left[b(r - r_0)^2 + az^2 \right] . \qquad (4.21.8)$$

Here, $\hat{n}_b = n_b^0(r_0, 0)$ is the electron density at $(r, z) = (r_0, 0)$, $n = -[r \partial \ln B_z^{\text{ext}}(r, z)/\partial r]_{(r_0, 0)}$ is the external field index, $\hat{B}_z$ is defined by $\hat{B}_z = B_z^{\text{ext}}(r_0, 0)$, and the components of the total equilibrium magnetic field are determined from $B_r = -(\partial/\partial z) A_\theta^0$ and $B_z = r^{-1}(\partial/\partial r)[r A_\theta^0]$. Note that $E_r(r_0, 0) = -[\partial \phi_0/\partial r]_{(r_0, 0)} = 0$ is assumed in Eq.(4.21.7). Make use of Eqs.(4.293), (4.21.7), and (4.21.8) to show that

$$\left[-\frac{P_0}{r^2} + \frac{e}{c} \frac{\partial A_\theta^0}{\partial r} \right]_{(r_0, 0)} = 0 = \left[\frac{e}{c} \frac{\partial A_\theta^0}{\partial z} \right]_{(r_0, 0)} . \qquad (4.21.9)$$

Show that the radial and axial betatron frequencies defined in Eqs.(4.21.5) and (4.21.6) can be expressed as

$$\omega_r^2 = \frac{\beta_b c}{\gamma_b r_0} \frac{e\hat{B}_z}{m_e c}(1-n) + \frac{4\pi\hat{n}_b e^2}{\gamma_b m_e} \frac{b}{a+b}\left[\beta_b^2 - (1-f)\right] \; , \quad (4.21.10)$$

and

$$\omega_z^2 = \frac{\beta_b c}{\gamma_b r_0} \frac{e\hat{B}_z}{m_e c}n + \frac{4\pi\hat{n}_b e^2}{\gamma_b m_e} \frac{a}{a+b}\left[\beta_b^2 - (1-f)\right] \; , \quad (4.21.11)$$

Making use of $\gamma_b m_e \beta_b^2 c^2 / r_0 = e\hat{B}_z \beta_b$ to eliminate β_b / r_0, Eqs.(4.21.10) and (4.21.11) reduce directly to the expressions for ω_r^2 and ω_z^2 given in Eqs.(4.302) and (4.303), where $\omega_{cz} = e\hat{B}_z / m_e c$ and $\omega_{pb}^2 = 4\pi\hat{n}_b e^2 / m_e$.

Problem 4.22 Consider the transverse (r, z) motion of an electron about the equilibrium orbit $(r, z) = (r_0, 0)$ in Fig. 4.16. The radial and axial equations of motion are

$$\frac{d}{dt}p_r - \frac{v_\theta p_\theta}{r} = -e\left(E_r + \frac{1}{c}v_\theta B_z - \frac{1}{c}v_z B_\theta^{\text{ext}}\right) \; , \quad (4.22.1)$$

$$\frac{d}{dt}p_z = -e\left(E_z + \frac{1}{c}v_r B_\theta^{\text{ext}} - \frac{1}{c}v_\theta B_r\right) \; , \quad (4.22.2)$$

where the azimuthal momentum p_θ is related to the (constant) canonical angular momentum $P_\theta = P_0 + \delta P_\theta$ by

$$p_\theta = \frac{P_0}{r} + \frac{e}{c}A_\theta^0(r, z) + \frac{\delta P_\theta}{r} \; . \quad (4.22.3)$$

For a thin ring with $a, b \ll r_0$, the equilibrium field components in Eqs.(4.22.1) and (4.22.2) are determined approximately from Eqs.(4.21.7) and (4.21.8), which give

$$E_r = -\frac{\partial \phi_0}{\partial r} = -4\pi e\hat{n}_b(1-f)\frac{b}{a+b}(r-r_0) \; , \quad (4.22.4)$$

$$E_z = -\frac{\partial \phi_0}{\partial z} = -4\pi \hat{n}_b e(1-f)\frac{a}{a+b}z \; , \quad (4.22.5)$$

$$B_r = -\frac{\partial A_\theta^0}{\partial z} = -n\hat{B}_z \frac{z}{r_0} - 4\pi\hat{n}_b e\beta_b \frac{a}{a+b}z \; , \quad (4.22.6)$$

and

$$B_z = \frac{1}{r}\frac{\partial}{\partial r}(rA_\theta^0) = \hat{B}_z - n\hat{B}_z \frac{(r-r_0)}{r_0} + 4\pi\hat{n}_b e\beta_b \frac{b}{a+b}(r-r_0) \; . \quad (4.22.7)$$

Furthermore, the applied toroidal magnetic field $B_\theta^{\text{ext}} = \hat{B}_\theta r_0 / r$ can be approximated by

$$B_\theta^{\text{ext}} = \hat{B}_\theta - \hat{B}_\theta \frac{(r-r_0)}{r_0} \; . \quad (4.22.8)$$

Equations (4.22.4)–(4.22.8) are good approximations to the equilibrium field components in the ring interior ($|r - r_0| \lesssim a$ and $|z| \lesssim b$) provided the ring is thin with a, $b \ll r_0$.

a. Make use of Eqs.(4.22.1)–(4.22.8) to examine the (r, z) electron motion for the case of small transverse kinetic energy with $(p_r^2 + p_z^2)/2\gamma_b m_e \ll \gamma_b m_e c^2$ [Eq.(4.277)]. Neglecting terms of cubic order and higher [$z(r-r_0)v_r$, $(r - r_0)^2 v_r$, etc.], show that

$$\frac{d}{dt}[zp_r - (r - r_0)p_z] =$$

$$\frac{1}{c}e\hat{B}_\theta\left[zv_z + (r - r_0)v_r\right] + 4\pi\hat{n}_b e^2(1 - f)\frac{b - a}{a + b}z(r - r_0)$$

$$- 4\pi\hat{n}_b e^2\beta_b\frac{v_\theta}{c}\frac{b - a}{a + b}z(r - r_0) + v_\theta\left[\frac{p_\theta}{r} - \frac{1}{c}e\hat{B}_z + 2n\frac{e\hat{B}_z}{r_0 c}(r - r_0)\right]z . \quad (4.22.9)$$

b. Assume a circular electron beam with $a = b$. The self-field contributions in Eq.(4.22.9) then vanish identically. Make use of Eqs.(4.293), (4.21.8), and (4.22.3) to show that

$$\frac{p_\theta}{r} - \frac{1}{c}e\hat{B}_z + 2n\frac{e\hat{B}_z}{r_0 c}(r - r_0) = \frac{1}{c}(2n - 1)e\hat{B}_z\frac{(r - r_0)}{r_0} + \frac{\delta P_\theta}{r_0^2} \quad (4.22.10)$$

to leading order. Strictly speaking, a circular beam with $a = b$ requires external field index $n = 1/2$, so that Eq.(4.22.9) reduces to

$$\frac{d}{dt}[zp_r - (r - r_0)p_z] = \frac{1}{c}e\hat{B}_\theta[zv_z + (r - r_0)v_r] + zv_\theta\frac{\delta P_\theta}{r_0^2} . \quad (4.22.11)$$

c. Compare the two terms on the right-hand side of Eq.(4.22.11). Estimating $|z| \sim |r - r_0| \sim a$, $v_z \sim v_r \sim p_\perp/\gamma_b m_e$, and $\gamma_b m_e v_\theta \sim e\hat{B}_z r_0/c$, show that the δP_θ contribution in Eq.(4.22.11) is negligibly small whenever the inequality

$$\frac{|\delta P_\theta|}{\gamma_b m_e \beta_b c r_0} \ll \frac{\hat{B}_\theta}{\hat{B}_z}\left|\frac{p_\perp}{p_\theta}\right| \quad (4.22.12)$$

is satisfied [see Eq.(4.278)]. Within the context of Eq.(4.22.12), it follows from Eq.(4.22.11) that

$$\frac{d}{dt}\left\{[zp_r - (r - r_0)p_z] - \frac{1}{2c}e\hat{B}_\theta\left[z^2 + (r - r_0)^2\right]\right\} = \frac{d}{dt}P_\phi = 0 . \quad (4.22.13)$$

Here, $P_\phi = \rho p_\phi - (e\hat{B}_\theta/2c)\rho^2$ is the poloidal canonical angular momentum defined in Eq.(4.285), and $\rho^2 = (r - r_0)^2 + z^2$ [Fig. 4.16].

4.10.3   Linearized Vlasov-Maxwell Equations

In this section, use is made of the linearized Vlasov-Maxwell equations
(Sec. 2.2) to investigate electromagnetic stability properties of the equilibrium ring configuration described in Sec. 4.10.2. In the stability analysis,
a normal-mode approach is adopted in which all perturbed quantities are
assumed to vary according to

$$\delta\psi(\mathbf{x}, t) = \delta\psi(\mathbf{x}) \exp(-i\omega t) , \qquad (4.320)$$

where ω is the complex oscillation frequency, and $\mathrm{Im}\,\omega > 0$ is assumed.
Integrating the linearized Vlasov equation from $t' = -\infty$ to $t' = t$ and
neglecting initial perturbations, the perturbed distribution function can
be expressed as $\delta f_b(\mathbf{x}, \mathbf{p}, t) = \delta f_b(\mathbf{x}, \mathbf{p}) \exp(-i\omega t)$, where

$$\delta f_b(\mathbf{x}, \mathbf{p}) = e \int_{-\infty}^{0} d\tau \exp(-i\omega\tau) \left[\delta\mathbf{E}(\mathbf{x}') + \frac{\mathbf{v}' \times \delta\mathbf{B}(\mathbf{x}')}{c} \right] \cdot \frac{\partial}{\partial \mathbf{p}'} f_b^0(\mathbf{x}', \mathbf{p}') .$$

$$(4.321)$$

In Eq.(4.321), $\delta\mathbf{E}(\mathbf{x})$ and $\delta\mathbf{B}(\mathbf{x})$ are the perturbed electromagnetic field
amplitudes, τ is defined by $\tau = t' - t$, and the particle trajectories $\mathbf{x}'(t')$
and $\mathbf{p}'(t')$ satisfy $d\mathbf{x}'/dt' = \mathbf{v}'$ and $d\mathbf{p}'/dt' = -e(\mathbf{E}^0 + \mathbf{v}' \times \mathbf{B}^0/c)$ with
"initial" conditions $\mathbf{x}'(t' = t) = \mathbf{x}$ and $\mathbf{p}'(t' = t) = \mathbf{p}$. The Maxwell
equations for $\delta\mathbf{E}(\mathbf{x})$ and $\delta\mathbf{B}(\mathbf{x})$ are given by

$$\nabla \times \delta\mathbf{E}(\mathbf{x}) = \frac{i\omega}{c}\delta\mathbf{B}(\mathbf{x}) , \qquad (4.322)$$

$$\nabla \times \delta\mathbf{B}(\mathbf{x}) = \frac{4\pi}{c}\delta\mathbf{J}_b(\mathbf{x}) - \frac{i\omega}{c}\delta\mathbf{E}(\mathbf{x}) , \qquad (4.323)$$

where $\delta\mathbf{J}_b(\mathbf{x}) = -e \int d^3p\, \mathbf{v} \delta f_b(\mathbf{x}, \mathbf{p})$ is the perturbed current density. From
Eqs.(4.322) and (4.323), it is readily shown that

$$\left(\nabla^2 + \frac{\omega^2}{c^2} \right) \delta\mathbf{E}(\mathbf{x}) = 4\pi \left[\nabla\delta\rho_b(\mathbf{x}) - i\frac{\omega}{c^2}\delta\mathbf{J}_b(\mathbf{x}) \right] , \qquad (4.324)$$

where $\delta\rho_b(\mathbf{x}) = -e \int d^3p\, \delta f_b(\mathbf{x}, \mathbf{p})$ is the perturbed charge density, and
use has been made of $\nabla \cdot \delta\mathbf{E} = 4\pi\delta\rho_b$. Equation (4.324) is the form of
Maxwell's equations used in the present analysis.

As indicated in Sec. 4.10.1, large aspect ratio is assumed, with a, $a_c \ll r_0$. It is evident that Eqs.(4.321) and (4.324) generally support solutions over a wide frequency range. For present purposes, we examine Eqs.(4.321) and (4.324) for $\omega \simeq \ell\omega_{cz}/\gamma_b$, where ℓ is the azimuthal mode number, and $\omega_{cz} = e\hat{B}_z/m_e c$ is the nonrelativistic cyclotron frequency in the vertical (betatron) field. It is assumed that the wave perturbations are far removed from resonance with the transverse oscillatory (r, z) motion of the electrons. That is, it is assumed that[25,36,41]

$$\left| \frac{\omega_{rb}^{\pm}}{\omega - \ell\omega_{cz}/\gamma_b} - 1 \right| , \quad \left| \frac{\omega_{rb}^{\pm}}{\omega} \right| \gg \frac{a}{r_0} , \qquad (4.325)$$

where ω_{rb}^{+} and ω_{rb}^{-} are the characteristic single-particle orbital oscillation frequencies defined in Eq.(4.312). For perturbations with θ-dependence proportional to $\exp(i\ell\theta)$, it is further assumed that the toroidal mode number ℓ is sufficiently small that

$$\frac{\ell a}{r_0} \ll 1 . \qquad (4.326)$$

To lowest order, consistent with Eq.(4.325), the azimuthal and radial orbits calculated from the Hamiltonian H in Eq.(4.300) can be approximated by (Problem 4.23)

$$\theta'(\tau) = \theta + \left(\frac{\omega_{cz}}{\gamma_b} - \frac{\mu}{\gamma_b m_e r_0^2} \delta P_\theta \right) \tau ,$$

$$r'(\tau) = r_0 + \frac{\omega_{cz}}{\gamma_b^2 m_e \omega_\beta^2 r_0} \delta P_\theta , \qquad (4.327)$$

where $\delta P_\theta = P_\theta - P_0$, and the oscillatory contributions (at frequencies ω_{rb}^{+} and ω_{rb}^{-}) have been neglected in Eq.(4.327). Here, the negative-mass parameter μ is defined by

$$\mu = \frac{\omega_{cz}^2}{\gamma_b^2 \omega_\beta^2} - \frac{1}{\gamma_b^2} , \qquad (4.328)$$

where $\omega_\beta^2 = \omega_{cz}^2/2\gamma_b^2 + (\omega_{pb}^2/2\gamma_b)[\beta_b^2 - (1 - f)]$. For $\mu > 0$, note from Eq.(4.327) that the azimuthal motion behaves as if the electron has a "negative" mass, i.e., if the canonical angular momentum is *increased* ($\delta P_\theta > 0$), then $d\theta'/d\tau$ *decreases* whenever $\mu > 0$.[80] Within the context of Eqs.(4.325) and (4.326) and the assumption of a thin electron ring with $\nu/\gamma_b \ll 1$ and $p_r^2 + p_z^2 \ll p_\theta^2$, we approximate[36,41]

$$\delta \mathbf{E}(\mathbf{x}') = \sum_{\ell=-\infty}^{\infty} \delta E_\theta^\ell(r', z') \exp(i\ell\theta') \hat{\mathbf{e}}_\theta' ,$$

$$\delta \mathbf{B}(\mathbf{x}') = \sum_{\ell=-\infty}^{\infty} \left\{ \frac{ic}{\omega} \frac{\partial}{\partial z'} \delta E_\theta^\ell(r', z') \hat{\mathbf{e}}_r' \right.$$

$$\left. - \frac{ic}{\omega r'} \frac{\partial}{\partial r'} \left[r' \delta E_\theta^\ell(r', z') \right] \hat{\mathbf{e}}_z \right\} \exp(i\ell\theta') , \tag{4.329}$$

on the right-hand side of Eq.(4.321). In this regard, the perturbed electromagnetic force in Eq.(4.321) is expressed as

$$-e \left[\delta \mathbf{E}(\mathbf{x}') + \frac{\mathbf{v}' \times \delta \mathbf{B}(\mathbf{x}')}{c} \right] =$$

$$-e \sum_{\ell=-\infty}^{\infty} \exp(i\ell\theta') \left\{ -\frac{iv_\theta'}{\omega r'} \frac{\partial}{\partial r'} \left(r' \delta E_\theta^\ell \right) \hat{\mathbf{e}}_r' \right. \tag{4.330}$$

$$\left. + \left[\delta E_\theta^\ell + \frac{i}{\omega} \left(v_z' \frac{\partial}{\partial z'} \delta E_\theta^\ell + \frac{v_r'}{r'} \frac{\partial}{\partial r'} \left(r' \delta E_\theta^\ell \right) \right) \right] \hat{\mathbf{e}}_\theta' - \frac{iv_\theta'}{\omega} \frac{\partial}{\partial z'} \delta E_\theta^\ell \hat{\mathbf{e}}_z \right\} ,$$

where $\delta E_\theta^\ell = \delta E_\theta^\ell(r', z')$. Substituting Eq.(4.330) into Eq.(4.321) and making use of $\partial f_b^0/\partial \mathbf{p}' = \mathbf{v}'(\partial f_b^0/\partial H) + r'\hat{\mathbf{e}}_\theta'(\partial f_b^0/\partial P_\theta)$ for small $\omega_{rb}^2 a^2 \ll \beta_b^2 c^2$, it is straightforward to show that the perturbed distribution function for the ℓ'th harmonic component can be expressed as[36,41]

$$\delta f_b^\ell(r, z, \mathbf{p}) = e \frac{\partial f_b^0}{\partial P_\theta} \int_{-\infty}^{0} d\tau \, r' \left\{ \delta E_\theta^\ell + \frac{i}{\omega} \left[v_z' \frac{\partial}{\partial z'} \delta E_\theta^\ell + \frac{v_r'}{r'} \frac{\partial}{\partial r'} \left(r' \delta E_\theta \right) \right] \right\}$$

$$\times \exp[i\ell(\theta' - \theta) - i\omega\tau]$$

$$+ e \frac{\partial f_b^0}{\partial H} \int_{-\infty}^{0} d\tau \, v_\theta' \delta E_\theta^\ell \exp[i\ell(\theta' - \theta) - i\omega\tau] . \tag{4.331}$$

Here, the orbits θ' and r' are defined in Eq.(4.327) and use has been made of the fact that $\partial f_b^0/\partial H$ and $\partial f_b^0/\partial P_\theta$ are independent of t' (i.e., constant along a particle trajectory in the equilibrium field configuration).

To further reduce Eq.(4.331), we make use of the identity

$$\frac{d}{d\tau}\left[r'\delta E_\theta^\ell(r',z')\exp(i\ell\theta'-i\omega\tau)\right] \tag{4.332}$$

$$= \exp(i\ell\theta'-i\omega\tau)r'\left\{-i\omega\delta E_\theta^\ell+\frac{i\ell v_\theta'}{r'}\delta E_\theta^\ell+v_z'\frac{\partial}{\partial z'}\delta E_\theta^\ell+\frac{v_r'}{r'}\frac{\partial}{\partial r'}\left(r'\delta E_\theta^\ell\right)\right\} .$$

The integrand in the term multiplying $\partial f_b^0/\partial P_\theta$ in Eq.(4.331) can then be expressed in the equivalent form

$$r'\left\{\delta E_\theta^\ell+\frac{i}{\omega}\left[v_z'\frac{\partial}{\partial z'}\delta E_\theta^\ell+\frac{v_r'}{r'}\frac{\partial}{\partial r'}\left(r'\delta E_\theta^\ell\right)\right]\right\}\exp(i\ell\theta'-i\omega\tau)$$

$$= \frac{\ell v_\theta'}{\omega}\delta E_\theta^\ell\exp(i\ell\theta'-i\omega\tau)+\frac{i}{\omega}\frac{d}{d\tau}\left[r'\delta E_\theta^\ell\exp(i\ell\theta'-i\omega\tau)\right] . \tag{4.333}$$

It is convenient to introduce the effective perturbed potential $\Phi^\ell(r,z)$ defined by

$$\Phi^\ell(r,z)\equiv\frac{ir}{\ell}\delta E_\theta^\ell(r,z) , \tag{4.334}$$

and the orbit integral I defined by

$$I = -i\ell\int_{-\infty}^0 d\tau\dot\theta'\Phi^\ell(r',z')\exp[i\ell(\theta'-\theta)-i\omega\tau] . \tag{4.335}$$

Here, use has been made of $v_\theta'=r'd\theta'/dt'$. Substituting Eq.(4.333) into Eq.(4.331), and integrating by parts with respect to τ then gives, for the perturbed distribution function,[36,41]

$$\delta f_b^\ell(r,z,\mathbf{p}) = \frac{\ell e}{\omega}\frac{\partial f_b^0}{\partial P_\theta}\left[\Phi^\ell(r,z)+I\right]+e\frac{\partial f_b^0}{\partial H}I , \tag{4.336}$$

where use has been made of $\mathrm{Im}\,\omega>0$. The expression for $\delta f_b^\ell(r,z,\mathbf{p})$ in Eq.(4.336) is fully equivalent to Eq.(4.331).

Within the context of Eq.(4.327), it is valid to approximate

$$\Phi^\ell(r',z') = \Phi^\ell(r_0,0)+\frac{\omega_{cz}\delta P_\theta}{\gamma_b^2 m_e r_0\omega_\beta^2}\left[\frac{\partial}{\partial r}\Phi^\ell\right]_{(r_0,0)} \tag{4.337}$$

in Eqs.(4.335) and (4.336), where the small-amplitude oscillatory modulations in the r' and z' orbits have been neglected (Problem 4.21). Substituting Eqs.(4.327) and (4.337) into Eq.(4.335), we find that the orbit integral I can be expressed as[36,41]

$$I = \ell \left(\frac{\omega_{cz}}{\gamma_b} - \frac{\mu}{\gamma_b m_e r_0^2} \delta P_\theta \right) \frac{[\hat{\Phi}_0^\ell + \hat{\Phi}_0^{\ell\prime}(\omega_{cz}/\gamma_b^2 m_e r_0^2 \omega_\beta^2)\delta P_\theta]}{[\omega - \ell\omega_{cz}/\gamma_b + (\ell\mu/\gamma_b m_e r_0^2)\delta P_\theta]} , \quad (4.338)$$

where the abbreviated notation

$$\hat{\Phi}_0^\ell \equiv \Phi^\ell(r_0, 0) , \quad \hat{\Phi}_0^{\ell\prime} \equiv \left[r\frac{\partial}{\partial r}\Phi^\ell \right]_{(r_0,0)} \quad (4.339)$$

has been introduced. In Eq.(4.336), we define

$$I_\theta = \frac{\ell}{\omega} \left[I + \Phi^\ell(r, z) \right] , \quad (4.340)$$

and approximate $\Phi^\ell(r, z) = \hat{\Phi}_0^\ell + \hat{\Phi}_0^{\ell\prime}(\omega_{cz}/\gamma_b^2 m_e r_0^2 \omega_\beta^2)\delta P_\theta$. Making use of Eqs.(4.338) and (4.340), it is straightforward to show that I_θ can be expressed as

$$I_\theta = \frac{\ell[\hat{\Phi}_0^\ell + \hat{\Phi}_0^{\ell\prime}(\omega_{cz}/\gamma_b^2 m_e r_0^2 \omega_\beta^2)\delta P_\theta]}{[\omega - \ell\omega_{cz}/\gamma_b + (\ell\mu/\gamma_b m_e r_0^2)\delta P_\theta]} . \quad (4.341)$$

Therefore, the perturbed distribution function in Eq.(4.336) is given by

$$\delta f_b^\ell(r, z, \mathbf{p}) = e\frac{\partial f_b^0}{\partial P_\theta}I_\theta + e\frac{\partial f_b^0}{\partial H}I , \quad (4.342)$$

where I and I_θ are defined in Eqs.(4.338) and (4.341), respectively.

We now return to the eigenvalue equation (4.324), where $\delta\rho_b$ and δJ_b are calculated self-consistently for the perturbed distribution function in Eq.(4.342). Taking the θ-component of Eq.(4.324), and making use of the large-aspect-ratio assumption in Eq.(4.274), we obtain the eigenvalue equation[36,41]

$$\left(\frac{1}{\rho}\frac{\partial}{\partial\rho}\rho\frac{\partial}{\partial\rho} + \frac{1}{\rho^2}\frac{\partial^2}{\partial\phi^2} - q^2 \right) \delta E_\theta^\ell(\rho, \phi) = 4\pi \left[ik\delta\rho_b^\ell(\rho, \phi) - \frac{i\omega}{c^2}\delta J_{\theta b}^\ell(\rho, \phi) \right] , \quad (4.343)$$

where k and q^2 are defined by

$$k = \frac{\ell}{r_0} , \quad q^2 = k^2 - \frac{\omega^2}{c^2} . \quad (4.344)$$

Here, ∇^2 has been approximated by $\nabla^2 = \rho^{-1}(\partial/\partial\rho)(\rho\partial/\partial\rho) + \rho^{-2}(\partial^2/\partial\phi^2)$
$- k^2$ in the limit of large aspect ratio, and (ρ, ϕ) is the toroidal polar
coordinate system defined in Eq.(4.282) and Fig. 4.16. To complete
the description, we evaluate $4\pi(ik\delta\rho_b^\ell - i\omega\delta J_{\theta b}^\ell/c^2) = -4\pi eik \int d^3p\, \delta f_b^\ell \times$
$[1 - (\omega/c^2 k)p_\theta/\gamma m_e]$ on the right-hand side of Eq.(4.343). To the
required accuracy, $(1 - \omega p_\theta/\gamma m_e c^2 k)$ is approximated in the integrand
by $(1 - \omega\beta_b/ck) - (\omega/c^2 k)(\delta P_\theta/\gamma_b^3 m_e r_0)$, where $\ell\omega_{cz}/\gamma_b = k\beta_b c$. Making
use of Eqs.(4.342) and (4.343) then gives the eigenvalue equation for
$\Phi^\ell(\rho, \phi) = (i/k)\delta E_\theta^\ell(\rho, \phi)$,

$$\left(\frac{1}{\rho}\frac{\partial}{\partial\rho}\rho\frac{\partial}{\partial\rho} + \frac{1}{\rho^2}\frac{\partial^2}{\partial\phi^2} - q^2\right)\Phi^\ell(\rho, \phi)$$

$$= 4\pi e^2 \int d^3p \left[\left(1 - \beta_b\frac{\omega}{ck}\right) - \frac{\omega\delta P_\theta}{c^2 k\gamma_b^3 m_e r_0}\right]\left(\frac{\partial f_b^0}{\partial P_\theta}I_\theta + \frac{\partial f_b^0}{\partial H}I\right) . \quad (4.345)$$

To carry out the momentum integrations in Eq.(4.345), it is convenient
to integrate by parts with respect to p_θ, making use of the identity

$$\alpha\frac{\partial}{\partial P_\theta}f_b^0(H - \omega_b P_\phi, P_\theta) = \frac{1}{r}\frac{\partial}{\partial p_\theta}\left[\alpha f_b^0(H - \omega_b P_\phi, P_\theta)\right]$$

$$- \left\{f_b^0(H - \omega_b P_\phi, P_\theta)\frac{\partial\alpha}{\partial P_\theta} + \frac{v_\theta}{r}\frac{\partial}{\partial H}\left[\alpha f_b^0(H - \omega_b P_\phi, P_\theta)\right]\right\} , \quad (4.346)$$

where α is an arbitrary function of P_θ, and $v_\theta = \partial H/\partial p_\theta$ is the azimuthal
velocity. Equation (4.345) then becomes[36,41]

$$\left(\frac{1}{\rho}\frac{\partial}{\partial\rho}\rho\frac{\partial}{\partial\rho} + \frac{1}{\rho^2}\frac{\partial^2}{\partial\phi^2} - q^2\right)\Phi^\ell(\rho, \phi)$$

$$= 4\pi e^2 \int d^3p \left\{\left(1 - \beta_b\frac{\omega}{ck} - \frac{\omega\delta P_\theta}{c^2 k\gamma_b^3 m_e r_0}\right)\left(I - \frac{v_\theta}{r}I_\theta\right)\frac{\partial f_b^0}{\partial H}\right.$$

$$\left. - f_b^0\frac{\partial}{\partial P_\theta}\left[\left(1 - \beta_b\frac{\omega}{ck} - \frac{\omega\delta P_\theta}{c^2 k\gamma_b^3 m_e r_0}\right)I_\theta\right]\right\} , \quad (4.347)$$

where use has been made of Eq.(4.346) to integrate by parts with respect
to p_θ.

The angular velocity of an electron at radius r is given by $\dot{\theta} = v_\theta/r = \partial H/\partial P_\theta$. Assuming that the spread in canonical angular momentum is small, the angular velocity is approximated by

$$\left(\frac{v_\theta}{r}\right)_{P_\theta = P_0} \simeq \frac{\omega_{cz}}{\gamma_b}\left(1 - \frac{(r - r_0)}{r_0}\right) . \qquad (4.348)$$

To evaluate the momentum integrals in Eq.(4.347), use is made of Eqs.(4.286), (4.307), and (4.308). After some straightforward algebra, we obtain

$$\int dp_r\, dp_z \delta\left[H - \omega_{rb}P_\phi - (\hat{\gamma}_b - 1)m_e c^2\right] = 2\pi\gamma_b m_e U(a - \rho) , \qquad (4.349)$$

and

$$\int dp_r\, dp_z \frac{\partial}{\partial H}\delta\left[H - \omega_{rb}P_\phi - (\hat{\gamma}_b - 1)m_e c^2\right] = -\frac{2\pi}{\Omega_\beta^2 a}\delta(\rho - a) . \qquad (4.350)$$

Here, $U(x)$ is the Heaviside step function defined by $U(x > 0) = +1$, and $U(x) = 0$ for $x < 0$, and the quantity $\Omega_\beta^2 = (\omega_{rb}^+ - \omega_{rb})(\omega_{rb} - \omega_{rb}^-)$ is defined in Eq.(4.309).

Substituting Eq.(4.348) into Eq.(4.347), and making use of Eqs.(4.349) and (4.350), the eigenvalue equation (4.347) reduces to[36,41]

$$\left(\frac{1}{\rho}\frac{\partial}{\partial\rho}\rho\frac{\partial}{\partial\rho} + \frac{1}{\rho^2}\frac{\partial^2}{\partial\phi^2} - q^2\right)\Phi^\ell(\rho, \phi)$$

$$= -\frac{S(k,\omega)\hat{\Phi}_0^\ell}{a}\cos\phi\, \delta(\rho - a)$$

$$+ \frac{1}{a^2}\left[N_0(k,\omega)\hat{\Phi}_0^\ell - N_1(k,\omega)\hat{\Phi}_0^{\ell\prime}\right]U(a - \rho) , \qquad (4.351)$$

where $\hat{\Phi}_0^\ell \equiv \Phi^\ell(r_0, 0)$, $\hat{\Phi}_0^{\ell\prime} \equiv [r\partial\Phi^\ell/\partial r]_{(r_0,0)}$, and use has been made of Eq.(4.282). The (complex) dielectric coefficients occurring in Eq.(4.351) are defined by

$$S(k,\omega) = \frac{\omega_{pb}^2\omega_{cz}}{\gamma_b^2\Omega_\beta^2}\frac{ka(1 - \beta_b\omega/ck)}{(\omega - \ell\omega_{cz}/\gamma_b + i|\mu k\Delta|/\gamma_b m_e r_0)} , \qquad (4.352)$$

$$N_0(k,\omega) = \frac{\omega_{pb}^2}{\gamma_b}\frac{k^2 a^2[\mu(1 - \beta_b\omega/ck) + \omega(\omega - \ell\omega_{cz}/\gamma_b)/c^2 k^2\gamma_b^2]}{(\omega - \ell\omega_{cz}/\gamma_b + i|\mu k\Delta|/\gamma_b m_e r_0)^2} , \qquad (4.353)$$

$$N_1(k, \omega) = \frac{\omega_{pb}^2 \omega_{cz} a}{\gamma_b^2 \omega_\beta^2 r_0} \frac{ka(1 - \beta_b \omega/ck)}{(\omega - \ell \omega_{cz}/\gamma_b + i|\mu k\Delta|/\gamma_b m_e r_0)} . \qquad (4.354)$$

It should be noted that the term in Eq.(4.351) proportional to $\delta(\rho - a)$ corresponds to a surface-perturbation at the boundary $(\rho = a)$ of the electron ring. This term, which is absent in standard treatments[80] of the negative-mass instability, originates from the perturbed charge density contribution in Eq.(4.347) proportional to $\partial f_b^0/\partial H$. We further note that the final two terms on the right-hand side of Eq.(4.351) are proportional to $U(a - \rho)$ and correspond to a body-wave perturbation which extends throughout the electron ring $(0 \leq \rho < a)$. Finally, the terms proportional to ω/ck in Eqs.(4.352)–(4.354) are related to electromagnetic effects. In this regard, it is customary in conventional treatments of longitudinal stability properties to approximate terms such as $1 - \beta_b \omega/ck$ by $1 - \beta_b \omega/ck \simeq 1 - \beta_b^2$, where use is made of $\omega \simeq \ell \omega_{cz}/\gamma_b = k\beta_b c$. To be more precise, we should express

$$1 - \beta_b \frac{\omega}{ck} = \left(1 - \beta_b^2\right) - \beta_b \frac{(\omega - \ell \omega_{cz}/\gamma_b)}{ck} , \qquad (4.355)$$

which is the procedure followed in analyzing the dispersion relation in Sec. 4.10.4. Indeed, retaining the contributions proportional to $\beta_b(\omega - \ell \omega_{cz})/ck$, it is found that the inclusion of the concomitant "electromagnetic" effects can have an important influence on detailed stability behavior,[36,41] at least in some parameter regimes of betatron operation.

Problem 4.23   Consider the "primed" orbits $\mathbf{x}'(t')$ and $\mathbf{p}'(t)$ in the equilibrium field configuration required to evaluate the expression for $\delta f_b(\mathbf{x}, \mathbf{p})$ in Eq.(4.321). For a thin electron ring with circular cross section $(a = b$ and $n = 1/2)$, the Hamiltonian H in Eq.(4.300) can be expressed as

$$H = (\gamma_b - 1)m_e c^2 + \frac{1}{2\gamma_b m_e} \left\{ \left[P_r' + \frac{1}{2} m_e \omega_{c\theta} z' \right]^2 \right.$$

$$\left. + \left[P_z' - \frac{1}{2} m_e \omega_{c\theta} (r' - r_0) \right]^2 \right\} + \frac{(\delta P_\theta)^2}{2\gamma_b^3 m_e r_0^2}$$

$$+ \frac{\omega_{cz}}{\gamma_b} \left[1 - \frac{(r' - r_0)}{r_0} \right] (\delta P_\theta) + \frac{1}{2} \gamma_b m_e \omega_\beta^2 \left[(r' - r_0)^2 + z'^2 \right] . \qquad (4.23.1)$$

Here, $\delta P_\theta = P_\theta - P_0$, $\omega_{cz} = e\hat{B}_z/m_e c$, $\omega_{c\theta} = e\hat{B}_\theta/m_e c$, $\omega_\beta^2 = \omega_{cz}^2/2\gamma_b^2 + (\omega_{pb}^2/2\gamma_b)[\beta_b^2 - (1 - f)]$, and (P_r', P_z') are the transverse canonical momenta defined by

$$P_r' = p_r' - \frac{1}{2}m_e\omega_{c\theta}z' , \qquad P_z' = p_z' + \frac{1}{2}m_e\omega_{c\theta}(r' - r_0) . \qquad (4.23.2)$$

Moreover, $p_r' = \gamma_b m_e v_r'$ and $p_z' = \gamma_b m_e v_z'$ for small transverse kinetic energy $(p_r'^2 + p_z'^2)/2\gamma_b m_e \ll \gamma_b m_e c^2$.

a. Make use of Hamilton's equations of motion,

$$\frac{d}{dt'}P_r' = -\frac{\partial H}{\partial r'} , \qquad \frac{d}{dt'}P_z' = -\frac{\partial H}{\partial z'} , \qquad \frac{d\theta'}{dt'} = \frac{\partial H}{\partial P_\theta} , \qquad (4.23.3)$$

to show that

$$\frac{d^2}{dt'^2}\left(r' - r_0\right) - \frac{\omega_{c\theta}}{\gamma_b}\frac{dz'}{dt'} + \omega_\beta^2\left[\left(r' - r_0\right) - \frac{\omega_{cz}}{\gamma_b^2 m_e \omega_\beta^2 r_0}\delta P_\theta\right] = 0 , \quad (4.23.4)$$

$$\frac{d^2}{dt'^2}z' + \frac{\omega_{c\theta}}{\gamma_b}\frac{dr'}{dt'} + \omega_\beta^2 z' = 0 , \qquad (4.23.5)$$

and

$$\frac{d\theta'}{dt'} = \frac{\omega_{cz}}{\gamma_b}\left[1 - \frac{(r' - r_0)}{r_0} + \frac{\delta P_\theta}{\omega_{cz}\gamma_b^2 m_e r_0^2}\right] . \qquad (4.23.6)$$

b. Show that the solutions to Eqs.(4.23.4) and (4.23.5) can be expressed as

$$r' = r_0 + \frac{\omega_{cz}}{\gamma_b^2 m_e \omega_\beta^2 r_0}\delta P_\theta + A_+ \cos\left(\omega_{re}^+ \tau + \phi_+\right) + A_- \cos\left(\omega_{re}^- \tau + \phi_-\right) , \quad (4.23.7)$$

and

$$z' = -A_+ \sin\left(\omega_{re}^+ \tau + \phi_+\right) - A_- \sin\left(\omega_{re}^- \tau + \phi_-\right) . \qquad (4.23.8)$$

Here, $\tau = t' - t$, and the constants (A_+, ϕ_+) and (A_-, ϕ_-) can be related to the phase-space coordinates (p_r, p_z, r, z) at time $t' = t$. Moreover, the orbital oscillation frequencies ω_{re}^+ and ω_{re}^- are defined by [see also Eq.(4.312)]

$$\omega_{re}^\pm = \frac{\omega_{c\theta}}{2\gamma_b}\left\{1 \pm \left(1 + \frac{4\gamma_b^2\omega_\beta^2}{\omega_{c\theta}^2}\right)^{1/2}\right\} . \qquad (4.23.9)$$

c. Substitute Eq.(4.23.7) into Eq.(4.23.6) and integrate with respect to t'. Show that

$$\theta'(t') = \theta + \left(\frac{\omega_{cz}}{\gamma_b} - \frac{\mu}{\gamma_b m_e r_0^2}\delta P_\theta\right)\tau + \cdots , \qquad (4.23.10)$$

where $\theta = \theta'(t' = t)$, and the negative-mass factor μ is defined by

$$\mu = \frac{\omega_{cz}^2}{\gamma_b^2 \omega_\beta^2} - \frac{1}{\gamma_b^2} \,. \tag{4.23.11}$$

In Eq.(4.23.10), $+ \cdots$ denotes the small oscillatory contributions (of order a/r_0) at frequencies ω_{re}^+ and ω_{re}^-. Depending on the sign of $\beta_b^2 - (1 - f)$ and the size of γ_b and ω_{pb}/ω_{cz}, note that μ can assume positive or negative values.

4.10.4   Negative-Mass and Surface-Kink Instabilities

The eigenvalue equation (4.351) is solved in Appendix A, which leads to the dispersion relation (A.20) for toroidal mode numbers ℓ satisfying $\ell a_c/r_0 = ka_c \ll 1$. Substituting the definitions of $S(k,\omega)$, $N_0(k,\omega)$ and $N_1(k,\omega)$ given in Eqs.(4.352)–(4.354) into Eq.(A.20), we obtain the dispersion relation[36,41]

$$0 = D_\ell(\omega) = 1 + \frac{\nu}{\gamma_b}\left[1 + 2\ln\left(\frac{a_c}{a}\right)\right]\frac{k^2 c^2}{(\omega - \ell\omega_{cz}/\gamma_b + i|\mu k\Delta|/\gamma_b m_e r_0)^2}$$

$$\times \left[\mu\left(1 - \frac{\omega}{ck}\beta_b\right) + \frac{\omega(\omega - \ell\omega_{cz}/\gamma_b)}{\gamma_b^2 c^2 k^2} - \frac{\omega_{pb}^2 \omega_{cz}^2}{2\gamma_b^3 \omega_\beta^2 \Omega_\beta^2}\left(1 - \frac{\omega}{ck}\beta_b\right)^2\left(1 - \frac{a^2}{a_c^2}\right)\right] \,,$$

$$\tag{4.356}$$

valid for perturbations with frequency $\omega \simeq \ell\omega_{cz}/\gamma_b$ about a thin, circular electron ring with $a \ll r_0$. The various quantities occurring in Eq.(4.356) are defined by $\nu/\gamma_b = \omega_{pb}^2 a^2/4c^2\gamma_b \ll 1$, $k = \ell/r_0$, $\omega_{cz} = e\hat{B}_z/m_e c$, $\gamma_b = (1 - \beta_b^2)^{-1/2}$, and $\omega_{pb}^2 = 4\pi\hat{n}_b e^2/m_e$. Moreover, $\mu = \omega_{cz}^2/\gamma_b^2\omega_\beta^2 - 1/\gamma_b^2$ is the negative-mass factor defined in Eq.(4.328), $\omega_\beta^2 = \omega_{cz}^2/2\gamma_b^2 + (\omega_{pb}^2/2\gamma_b)[\beta_b^2 - (1 - f)]$ is defined in Eq.(4.305), and $\Omega_\beta^2 = (\omega_{rb}^+ - \omega_{rb})(\omega_{rb} - \omega_{rb}^-)$ is the effective focusing frequency defined in Eq.(4.309). It should be pointed out that transverse electromagnetic effects are included in the term proportional to $(1 - \omega\beta_b/ck)$ in Eq.(4.356), and the terms proportional to μ and $\omega(\omega - \ell\omega_{cz}/\gamma_b)$ represent contributions from body-wave perturbations within the electron ring ($\rho < a$). On the other hand, the term proportional to $(1 - a^2/a_c^2)$ in Eq.(4.356) represents the contribution localized at the surface of the electron ring ($\rho = a$). The factor proportional to Δ in Eq.(4.356) arises from the spread in canonical angular momentum.

Moreover, because the effective transverse temperature is proportional to Ω_β^2 [see Eq.(4.315)], transverse thermal effects are incorporated in the last term on the right-hand side of Eq.(4.356).

The dispersion relation (4.356) can be expressed as a quadratic equation for $\omega - \ell\omega_{cz}/\gamma_b$. Making use of $\beta_b = \omega_{cz}r_0/\gamma_b c$ and $k = \ell/r_0$, it is convenient to express $1 - \beta_b\omega/ck = 1 - \beta_b^2 - \beta_b^2[\omega - \ell\omega_{cz}/\gamma_b]/(\ell\omega_{cz}/\gamma_b)$. Introducing the dimensionless frequency

$$\chi = \frac{\omega - \ell\omega_{cz}/\gamma_b}{\ell\omega_{cz}/\gamma_b} , \qquad (4.357)$$

the dispersion relation (4.356) becomes

$$A\chi^2 + \left(B + 2i|\mu|\hat{\Delta}\right)\chi + \left(C - \mu^2\hat{\Delta}^2\right) = 0 . \qquad (4.358)$$

Here, the coefficients A, B, C and $\hat{\Delta}$ are defined by

$$A = 1 + \frac{1}{4}\frac{\beta_b^2}{\gamma_b}\frac{\omega_{pb}^2}{\omega_{cz}^2}\frac{a^2}{r_0^2}\left[2\ln\left(\frac{a_c}{a}\right)+1\right]\left[1 - \frac{\beta_b^2\omega_{pb}^2\omega_{cz}^2}{2\gamma_b\omega_\beta^2\Omega_\beta^2}\left(1 - \frac{a^2}{a_c^2}\right)\right] , (4.359)$$

$$B = \frac{1}{4}\frac{\beta_b^2}{\gamma_b}\frac{\omega_{pb}^2}{\omega_{cz}^2}\frac{a^2}{r_0^2}\left[2\ln\left(\frac{a_c}{a}\right)+1\right]\left\{1 - \gamma_b^2\left[\mu - \frac{\omega_{pb}^2\omega_{cz}^2}{\gamma_b^5\omega_\beta^2\Omega_\beta^2}\left(1 - \frac{a^2}{a_c^2}\right)\right]\right\} ,$$

$$\qquad (4.360)$$

$$C = \frac{1}{4\gamma_b}\frac{\omega_{pb}^2}{\omega_{cz}^2}\frac{a^2}{r_0^2}\left[2\ln\left(\frac{a_c}{a}\right)+1\right]\left[\mu - \frac{\omega_{pb}^2\omega_{cz}^2}{\gamma_b^5\omega_\beta^2\Omega_\beta^2}\left(1 - \frac{a^2}{a_c^2}\right)\right] , \qquad (4.361)$$

$$\hat{\Delta} = \frac{\Delta}{\gamma_b m_e\beta_b cr_0} , \qquad (4.362)$$

where the various quantities in Eqs.(4.359)–(4.362) are defined following Eq.(4.356), and positive ℓ is assumed. Solving Eq.(4.358) for the complex eigenfrequency ω gives[41]

$$\omega = \frac{\ell\omega_{cz}}{\gamma_b}\left\{1 - \frac{(B + 2i|\mu|\hat{\Delta})}{2A} \pm \frac{[(B + 2i|\mu|\hat{\Delta})^2 - 4A(C - \mu^2\hat{\Delta}^2)]^{1/2}}{2A}\right\} . \qquad (4.363)$$

Equation (4.363) can be used to investigate detailed stability behavior of the modified (or conventional) betatron over a wide range of system parameters. For present purposes, we examine Eq.(4.363) in several limiting regimes which illustrate important aspects of stability behavior. As a general remark, in deriving Eqs.(4.358) and (4.363), keep in mind that we have expressed $1 - \beta_b \omega/ck = 1 - \beta_b^2 - \beta_b^2 \chi$, where $-\beta_b^2 \chi$ is the "electromagnetic" correction to $1 - \beta_b^2 = \gamma_b^{-2}$. In this regard, the term in the definition of A [Eq.(4.359)] linearly proportional to $\beta_b^2 \omega_{pb}^2$ arises from this electromagnetic correction, as does the contribution in the definition of B [Eq.(4.360)] proportional to $\beta_b^2 \omega_{pb}^2 (1 - \gamma_b^2 \mu)$.

A. Influence of Canonical Angular Momentum Spread on the Negative-Mass Instability

For zero spread in canonical angular momentum ($\Delta = 0$), the necessary and sufficient condition for instability (Im $\omega > 0$) obtained from Eq.(4.363) is $4AC - B^2 > 0$. If we assume sufficiently small (nonzero) Δ that

$$\left(\frac{\Delta}{\gamma_b m_e \beta_b cr_0}\right)^2 \ll \left|\frac{B^2 - 4AC}{4\mu^2(A-1)}\right| , \qquad (4.364)$$

then it can be shown from Eq.(4.363) that the necessary and sufficient condition for stability is given by

$$\frac{\Delta}{\gamma_b m_e \beta_b cr_0} > \left(\frac{AC}{\mu^2} - \frac{B^2}{4\mu^2}\right)^{1/2} . \qquad (4.365)$$

Equation (4.365) assures that the solutions in Eq.(4.363) satisfy Im $\omega \le 0$.

As a simple limiting case, we further assume a tenuous electron ring and negligibly small surface-wave contributions, i.e.,

$$\frac{\beta_b^2 \omega_{pb}^2 a^2}{\gamma_b \omega_{cz}^2 r_0^2} \ll 1 , \quad \frac{\beta_b^2 \omega_{pb}^2 \omega_{cz}^2}{2\gamma_b \omega_\beta^2 \Omega_\beta^2} \left(1 - \frac{a^2}{a_c^2}\right) \ll 1 . \qquad (4.366)$$

Then the condition for instability when $\Delta = 0 \, (4AC - B^2 > 0)$ reduces to $4C > 0$, or equivalently,

$$\mu = \frac{\omega_{cz}^2}{\gamma_b^2 \omega_\beta^2} - \frac{1}{\gamma_b^2} > 0 , \qquad (4.367)$$

which is the familiar condition for the azimuthal motion of an electron to
have an effective "negative" mass [see Eq.(4.327)]. Within the context of
Eq.(4.366), the condition for stabilization can be expressed as

$$\left(\frac{\Delta}{\gamma_b m_e \beta_b c r_0}\right)^2 > \frac{1}{4\gamma_b \mu} \frac{\omega_{pb}^2}{\omega_{cz}^2} \frac{a^2}{r_0^2} \left[2 \ln \left(\frac{a_c}{a}\right) + 1\right] . \tag{4.368}$$

Equation (4.368) states that a small spread in canonical angular momentum indeed has a strong stabilizing influence on the negative-mass
instability, at least in the regime described by Eq.(4.366) and $\mu > 0$.

B. Influence of Intense Self Fields on the Negative-Mass Instability

To feature the important influence of intense self-electric and self-magnetic
fields on the negative-mass instability, we consider the exact solution for
ω in Eq.(4.363) in circumstances where $\Delta = 0$ and the surface-wave contributions are negligibly small, i.e.,

$$\frac{\beta_b^2 \omega_{pb}^2 \omega_{cz}^2}{2\gamma_b \omega_\beta^2 \Omega_\beta^2} \left(1 - \frac{a^2}{a_c^2}\right) \ll 1 . \tag{4.369}$$

Making use of Eq.(4.369), the necessary and sufficient condition for
Eq.(4.363) to have stable solutions with $\text{Im}\,\omega = 0$ $(B^2 > 4AC)$ can be
expressed as[41]

$$\frac{1}{4}\beta_b^2 \left(1 - \gamma_b^2 \mu\right)^2 \frac{\beta_b^2 \omega_{pb}^2 a^2}{\gamma_b \omega_{cz}^2 r_0^2} \left[2 \ln \left(\frac{a_c}{a}\right) + 1\right] > 4\mu \left\{1 + \frac{1}{4} \frac{\beta_b^2 \omega_{pb}^2 a^2}{\gamma_b \omega_{cz}^2 r_0^2} \left[2 \ln \left(\frac{a_c}{a}\right) + 1\right]\right\}. \tag{4.370}$$

From Eq.(4.370), there are two ranges of μ that are sufficient for stability:
μ negative, and μ greater than some positive value.

Negative values of μ correspond to an effective "positive" mass for
the azimuthal motion of an electron [Eq.(4.327)]. For $f = 0$, shown in
Fig. 4.17 is a plot of $\mu = \omega_{cz}^2/\gamma_b^2\omega_\beta^2 - 1/\gamma_b^2$ versus $\omega_{pb}^2/\gamma_b\omega_{cz}^2$, where use has
been made of $\omega_\beta^2 = \omega_{cz}^2/2\gamma_b^2 - \omega_{pb}^2/2\gamma_b^3$. Note from Fig. 4.17 that $\mu > 0$ for
$0 < \omega_{pb}^2/\gamma_b\omega_{cz}^2 < 1$, which is the region of operation of the conventional
negative-mass instability described in the previous section. On the other
hand, it follows from Fig. 4.17 that $\mu < 0$ for $\omega_{pb}^2/\gamma_b\omega_{cz}^2 > 1$. Therefore,

for the modified betatron, the stability condition $\mu < 0$ combined with the condition for existence of the equilibrium in Eq.(4.318) become

$$1 < \frac{\omega_{pb}^2}{\gamma_b \omega_{cz}^2} \leq 1 + \frac{\omega_{c\theta}^2}{2\omega_{cz}^2}\left[1 - \left(\frac{2r_{Lb}}{a}\right)^2\right], \qquad (4.371)$$

where it has been assumed that $f = 0$. It is evident that the inequality in Eq.(4.371) can be satisfied provided $\omega_{pb}^2/\omega_{cz}^2$ is sufficiently large. Therefore, the negative-mass instability in a modified betatron can be completely stabilized provided the equilibrium density and therefore the equilibrium self fields are sufficiently strong that $\mu < 0$, which corresponds to an effective "positive" mass.

On the other hand, in the conventional betatron with $\hat{B}_\theta = 0$, the condition $\mu < 0$ and the condition for existence of the equilibrium cannot be satisfied simultaneously when $f = 0$. However, for $f \neq 0$, the condition

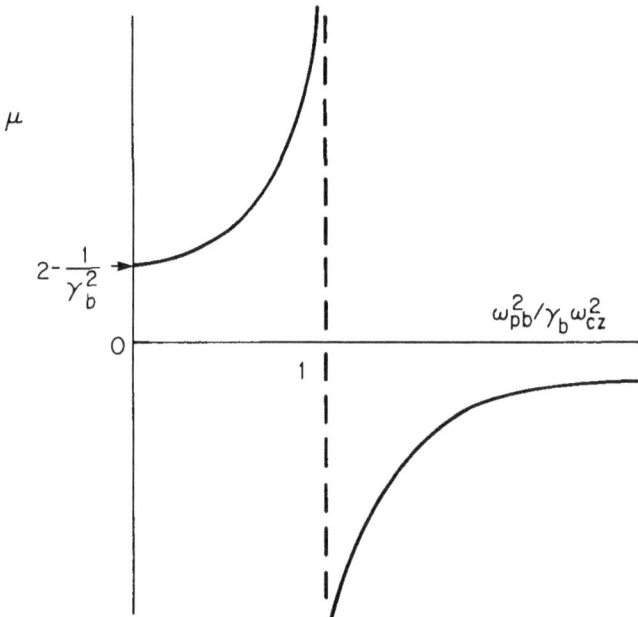

Figure 4.17. Sketch of the negative-mass parameter $\mu = \omega_{cz}^2/\gamma_b^2\omega_\beta^2 - 1/\gamma_b^2$ versus the normalized beam density $\omega_{pb}^2/\gamma_b\omega_{cz}^2$ for $f = 0$ and $\omega_\beta^2 = \omega_{cz}^2/2\gamma_b^2 - \omega_{pb}^2/2\gamma_b^3$.

$\mu < 0$ combined with the condition for existence of the equilibrium can be expressed as

$$\frac{\omega_{pb}^2}{\gamma_b \omega_{cz}^2} \left(\gamma_b^2 f - 1\right) > 2\gamma_b^2 - 1 . \tag{4.372}$$

Equation (4.372) states that $f > 1/\gamma_b^2 = 1 - \beta_b^2$ is required for both equilibrium and stability when $\mu < 0$. Therefore, a partially neutralizing ion background with $f > 1/\gamma_b^2$ will stabilize the negative-mass instability in the conventional betatron. Assuming that there is sufficient charge neutralization, Eq.(4.372) can be satisfied provided $\omega_{pb}^2/\omega_{cz}^2$ is sufficiently large. Thus, the conventional negative-mass instability is completely stabilized in a conventional betatron provided the focusing effect of the equilibrium self-magnetic field exceeds the defocusing effect of the self-electric field $(\beta_b^2 > 1 - f)$, and the net self-focusing force is sufficiently strong.

In the regime where $\mu > 0$, there is a radical departure from the usual negative-mass stability criterion.[80] In particular, due to transverse electromagnetic effects, we find from Eq.(4.370) that a threshold value of μ exists above which instability is absent. For a given positive value of μ, the necessary and sufficient condition for stability in Eq.(4.370) can be expressed as[41]

$$\frac{\nu}{\gamma_b} > \frac{4\mu/(\gamma_b^2 - 1)}{(\mu - 1/\gamma_b^2)^2 [2\ln(a_c/a) + 1]} , \tag{4.373}$$

where $\nu = \omega_{pb}^2 a^2/4c^2$ is Budker's parameter. For an ultrarelativistic electron beam with $\gamma_b \gg 1$, Eq.(4.373) can be satisfied easily for both the modified and conventional betatron configurations provided the beam current (proportional to ν/γ_b) is sufficiently large.

C. Influence of Transverse Beam Temperature on the Surface-Kink Instability

The influence of surface-wave perturbations on stability behavior is contained in the factors proportional to $(\omega_{pb}^2/\Omega_\beta^2)(1 - a^2/a_c^2)$ in the definitions of A, B and C in Eqs.(4.359)–(4.361). When the minor radius of the electron ring $(r = a)$ is sufficiently far removed from the conducting wall $(r = a_c)$, the corresponding surface-wave contributions in Eq.(4.363) can lead to a kink-like instability and distortion of the electron beam. To emphasize these effects, we consider a fully nonneutral modified betatron configuration with $f = 0$ and $\hat{B}_\theta \neq 0$, and no stabilizing canonical angular momentum spread ($\Delta = 0$).

From Eq.(4.315) the transverse temperature at the center of the minor cross section of the ring is related to Ω_β^2 and r_{Lb}^2 by $\hat{T}_{b\perp} = \gamma_b m_e \Omega_\beta^2 a^2/2 = m_e \omega_{c\theta}^2 r_{Lb}^2/2\gamma_b$. A detailed examination of Eq.(4.363) shows that the necessary and sufficient condition for the dispersion relation to have only stable solutions with $\text{Im}\,\omega = 0$ ($B^2 > 4AC$) can be expressed as[41]

$$\left(\frac{\gamma_b 2\hat{T}_{b\perp}}{m_e \omega_{cz}^2 a^2}\right)\left\{4\mu - \frac{1}{4}\frac{\beta_b^2 \omega_{pb}^2 a^2}{\gamma_b \omega_{cz}^2 r_0^2}\left[2\ln\left(\frac{a_c}{a}\right)+1\right]\left[(\mu^2\gamma_b^4 + 1)\,\beta_b^2 - 2\mu(\gamma_b^2+1)\right]\right\}$$

$$< \frac{2}{\gamma_b^3}\frac{\omega_{pb}^2}{\omega_\beta^2}\left(1 - \frac{a^2}{a_c^2}\right)\,, \tag{4.374}$$

where $\Delta = 0$ and $f = 0$ are assumed.

In the regime where $\omega_{pb}^2/\gamma_b\omega_{cz}^2 > 1$, it follows that $\omega_\beta^2 = \omega_{cz}^2/2\gamma_b^2 - \omega_{pb}^2/2\gamma_b^3 < 0$ and $\mu = \omega_{cz}^2/\gamma_b^2\omega_\beta^2 - 1/\gamma_b^2 < 0$ when $f = 0$. Equation (4.374) therefore shows that the surface-kink instability is stabilized whenever the transverse beam temperature *exceeds* a certain critical value.

On the other hand, in the regime where $\omega_{pb}^2/\gamma_b\omega_{cz}^2 < 1$, it follows that $\omega_\beta^2 > 0$ and $\mu > 0$ when $f = 0$. For moderate beam density, a detailed analysis[41] of Eq.(4.374) shows that a sufficiently *low* transverse beam temperature is required for stabilization. However, for beam density above some critical value, Eq.(4.374) shows that the system is stable for *all* values of transverse temperature.

To summarize, the dispersion relation (4.356) can be used to investigate detailed properties of the negative-mass and surface-kink instabilities over a wide range of system parameters. Particularly noteworthy is the strong influence of intense self fields and kinetic effects (transverse beam temperature $\hat{T}_{b\perp}$ and canonical angular momentum spread Δ) on stability behavior. Typical numerical results are illustrated in Figs. 4.18 and 4.19 for a modified betatron with $f = 0$ and $\Delta = 0$. Here, we choose $\gamma_b m_e c^2 = 3.3\,\text{MeV}$, $a = 1.5\,\text{cm}$, $a_c = 15\,\text{cm}$, $r_0 = 100\,\text{cm}$ and $I_b = 10\,\text{kA}$, which correspond to $\gamma_b = 7.5$, $a/a_c = 0.1$, $a/r_0 = 0.15$ and $\omega_{pb}/\omega_{cz} = 13.9$. Figure 4.18 shows plots of the equilibrium boundary (dashed curve) and stability boundary (solid curve) calculated from Eqs.(4.318) and (4.374), respectively, in ($r_{Lb}/a, \omega_{c\theta}/\omega_{cz}$) parameter space. [In Eq.(4.374), use has been made of $2\hat{T}_{b\perp}\gamma_b/m_e\omega_{cz}^2 a^2 = (\omega_{c\theta}/\omega_{cz})^2(r_{Lb}/a)^2$.] Evidently, equilib-

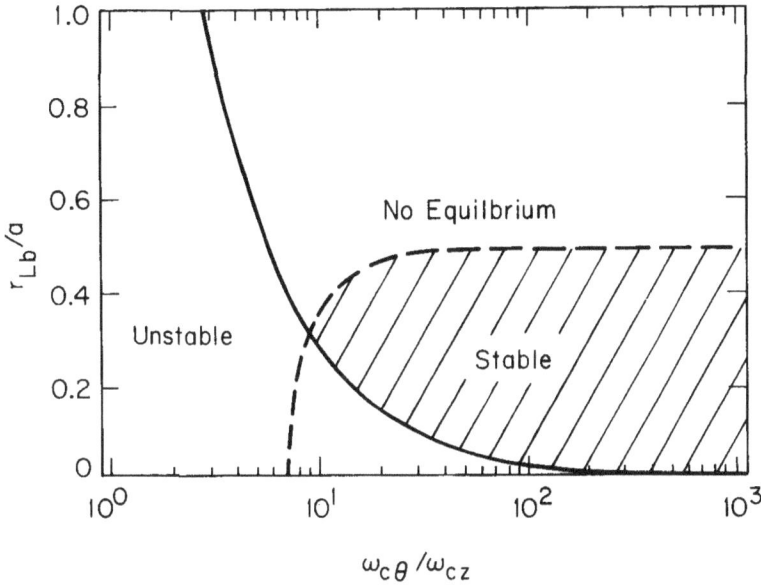

Figure 4.18. Plots of the regions of equilibrium and stability in $(r_{Lb}/a, \omega_{c\theta}/\omega_{cz})$ parameter space for a modified betatron with $f = 0$ and $\Delta = 0$. The equilibrium boundary (dashed curve) is calculated from Eq.(4.318) and the stability boundary (solid curve) is calculated from Eq.(4.374) for the choice of system parameters corresponding to $\gamma_b m_e c^2 = 3.3\,\mathrm{MeV}$, $I_b = 10\,\mathrm{kA}$, $a = 1.5\,\mathrm{cm}$, $a_c = 15\,\mathrm{cm}$, $r_0 = 1\,\mathrm{m}$, $\gamma_b = 7.5$ and $\omega_{pb}/\omega_{cz} = 13.9$ [J.J. Petillo and R.C. Davidson, Phys. Fluids **30**, 2477 (1987)].

rium exists for sufficiently large values of $\omega_{c\theta}/\omega_{cz}$ and sufficiently small values of r_{Lb}/a [Eq.(4.318)]. Furthermore, the system is stable (Im $\omega = 0$) for sufficiently large $\omega_{c\theta}/\omega_{cz}$ [Eq.(4.374)]. That is, stable equilibrium solutions exist in the shaded region of Fig. 4.18. For the fundamental $\ell = 1$ mode, Fig. 4.19 shows plots of the normalized growth rate $\gamma_b \operatorname{Im} \omega/\omega_{cz}$ versus $\omega_{c\theta}/\omega_{cz}$ calculated numerically from Eq.(4.363) for values of r_{Lb}/a ranging from 0.01 to 0.225 and system parameters otherwise identical to Fig. 4.18. For specified value of r_{Lb}/a, the instability is stabilized by increasing $\omega_{c\theta}/\omega_{cz}$ to sufficiently large values. On the other hand, for specified value of $\omega_{c\theta}/\omega_{cz}$, the instability is stabilized by increasing $\hat{T}_{b\perp}$ (and therefore r_{Lb}/a) to sufficiently large values.

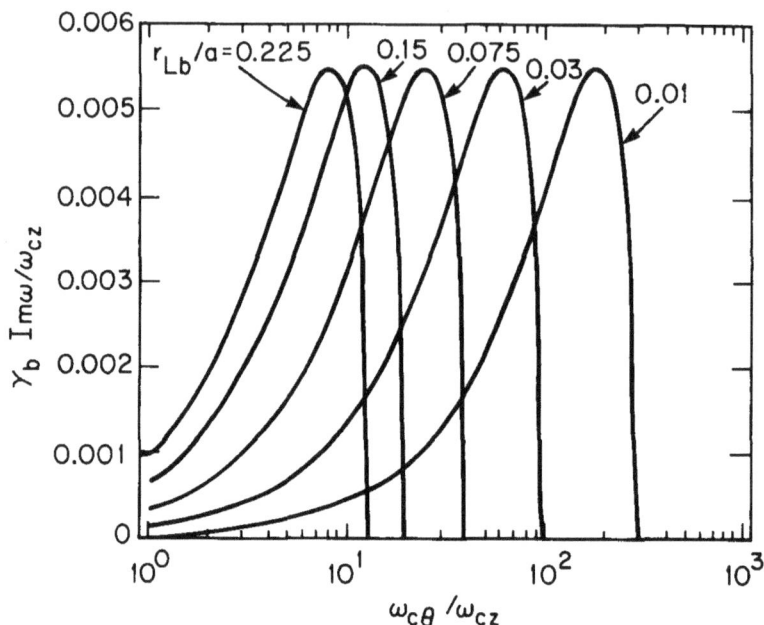

Figure 4.19. Plots of the normalized growth rate $\gamma_b \, \mathrm{Im}\, \omega/\omega_{cz}$ versus $\omega_{c\theta}/\omega_{cz}$ obtained from Eq.(4.363) for values of r_{Lb}/a ranging from 0.01 to 0.225 and system parameters otherwise identical to Fig. 4.18 [J.J. Petillo and R.C. Davidson, Phys. Fluids **30**, 2477 (1987)].

Chapter 4 References

1. R.C. Davidson, *Theory of Nonneutral Plasmas* (Addison-Wesley, Reading, Massachusetts, 1989), Chapter 3.

2. R.C. Davidson, "Waves and Instabilities in Nonneutral Plasmas," in *Nonneutral Plasma Physics*, eds., C.W. Roberson and C.F. Driscoll, AIP Conference Proceedings No. 175 (American Institute of Physics, New York, 1988), pp. 139-208.

3. R.C. Davidson, "Relativistic Electron Beam-Plasma Interaction with Intense Self Fields," in *Handbook of Plasma Physics—Basic Plasma Physics*, eds., M.N. Rosenbluth and R.Z. Sagdeev (North Holland, Amsterdam, 1984), Volume 2, pp. 729-819.

4. R.C. Davidson and N.A. Krall, "Vlasov Description of an Electron Gas in a Magnetic Field," Phys. Rev. Lett. **22**, 833 (1969).

5. R.C. Davidson and N.A. Krall, "Vlasov Equilibria and Stability of an Electron Gas," Phys. Fluids **13**, 1543 (1970).

6. B.L. Bogema and R.C. Davidson, "Rotor Equilibria of Nonneutral Plasmas," Phys. Fluids **13**, 2772 (1970).

7. S.A. Prasad and T.M. O'Neil, "Finite-Length Thermal Equilibria of a Pure Electron Plasma," Phys. Fluids **22**, 278 (1979).

8. L.R. Brewer, J.D. Prestage, J.J. Bollinger and D.J. Wineland, "A High-Γ, Strongly Coupled, Nonneutral Ion Plasma," in *Strongly Coupled Plasma Physics*, eds., F.J. Rogers and H.E. DeWitt (Plenum, New York, 1987), p. 53.

9. R.C. Davidson, A. Drobot and C.A. Kapetanakos, "Equilibrium and Stability of Mirror-Confined Nonneutral Plasmas," Phys. Fluids **16**, 2199 (1973).

10. D.A. Hammer and N. Rostoker, "Propagation of High-Current Relativistic Electron Beams," Phys. Fluids **13**, 1831 (1970).

11. G. Benford, D.L. Book and R.N. Sudan, "Relativistic Beam Equilibria with Back Currents," Phys. Fluids **13**, 2621 (1970).

12. M.E. Rensink, "Self-Consistent Relativistic Beam Equilibria," Phys. Fluids **14**, 2241 (1971).

13. G. Benford and D.L. Book, "Relativistic Beam Equilibria," in *Advances in Plasma Physics*, eds., A. Simon and W.B. Thompson (Wiley, New York, 1971), Volume 4, p. 125.

14. R.C. Davidson and H.S. Uhm, "Thermal Equilibrium Properties of an Intense Relativistic Electron Beam," Phys. Fluids **22**, 1375 (1979).

15. T.M. O'Neil, "A Confinement Theorem for Nonneutral Plasmas," Phys. Fluids **23**, 2216 (1980).

16. R.C. Davidson and H.S. Uhm, "Influence of Finite Ion Larmor Radius Effects on the Ion Resonance Instability in a Nonneutral Plasma Column," Phys. Fluids **21**, 265 (1978).

17. H.S. Uhm and R.C. Davidson, "Low-Frequency Flute Perturbations in Intense Nonneutral Electron and Ion Beams," Phys. Fluids **23**, 1586 (1980).

18. E. Ott, "Toroidal Equilibria of Electrically Unneutralized Intense Relativistic Electron Beams," Plasma Physics **13**, 529 (1971).

19. E. Ott and R.N. Sudan, "Finite Beta Equilibria of Relativistic Electron Beams in Toroidal Geometry," Phys. Fluids **14**, 1226 (1971).

20. G. Schmidt, "Self-Consistent Field Theory of Relativistic Electron Rings," Phys. Rev. Lett. **26**, 952 (1971).

21. R.C. Davidson and J.D. Lawson, "Self-Consistent Vlasov Description of Relativistic Electron Rings," Particle Accelerators **4**, 1 (1972).

22. R.C. Davidson and S. Mahajan, "A Relativistic Electron Ring Equilibrium with Thermal Energy Spread," Particle Accelerators **4**, 53 (1972).

23. P. Sprangle and C.A. Kapetanakos, "Constant Radius Magnetic Acceleration of a Strong Nonneutral Proton Ring," J. Appl. Phys. **49**, 1 (1978).

24. N. Rostoker, "High-Current Betatron," Comments in Plasma Physics and Controlled Fusion **6**, 91 (1980).

25. R.C. Davidson and H.S. Uhm, "Stability Properties of an Intense Relativistic Nonneutral Electron Ring in a Modified Betatron Accelerator," Phys. Fluids **25**, 2089 (1982).

26. H.S. Uhm and R.C. Davidson, "Ion Resonance Instability in a Modified Betatron Accelerator," Phys. Fluids **25**, 2334 (1982).

27. D. Chernin and P. Sprangle, "Integer Resonances in the Modified Betatron," Particle Accelerators **12**, 101 (1982).

28. D. Chernin and P. Sprangle, "Transverse Beam Dynamics in the Modified Betatron," Particle Accelerators **12**, 85 (1982).

29. G. Barak and N. Rostoker, "Orbital Stability of the High-Current Betatron," Phys. Fluids **26**, 856 (1983).

30. W.M. Manheimer and J.M. Finn, "Self-Consistent Theory of Equilibrium and Acceleration of a High-Current Electron Ring in a Modified Betatron," Particle Accelerators **14**, 29 (1983).

31. C.A. Kapetanakos, P. Sprangle, D.P. Chernin, S.J. Marsh and I. Haber, "Equilibrium of a High-Current Electron Ring in a Modified Betatron Accelerator," Phys. Fluids **26**, 1634 (1983).

32. D. Chernin, "Mode Coupling in the Modified Betatron," Particle Accelerators **14**, 139 (1984).

33. P. Sprangle and D. Chernin, "Beam Current Limitations Due to Instabilities in Modified and Conventional Betatrons," Particle Accelerators **15**, 35 (1984).

34. H. Ishizuka, G. Lindley, B. Mandelbaum, A. Fisher and N. Rostoker, "Formation of a High-Current Electron Beam in Modified Betatron Fields," Phys. Rev. Lett. **53**, 266 (1984).

35. C.A. Kapetanakos, S.J. Marsh and P. Sprangle, "Dynamics of a High-Current Electron Ring in a Conventional Betatron Accelerator," Particle Accelerators **14**, 261 (1984).

36. H.S. Uhm, R.C. Davidson and J.J. Petillo, "Kinetic Stability Properties of an Intense Relativistic Electron Ring in a High-Current Betatron Accelerator," Phys. Fluids **28**, 2537 (1985).

37. D. Chernin and P. Sprangle, "Beam Current Limitations due to Instabilities in Modified and Conventional Betatrons," Particle Accelerators **18**, 35 (1985).

38. P. Sprangle and J.L. Vomvoridis, "Longitudinal and Transverse Instabilities in a High-Current Modified Betatron Electron Accelerator," Particle Accelerators **18**, 1 (1985).

39. B.B. Godfrey and T.P. Hughes, "Long-Wavelength Negative-Mass Instabilities in High-Current Betatrons," Phys. Fluids **28**, 669 (1985).

40. G.A. Roberts and N. Rostoker, "Adiabatic Beam Dynamics in a Modified Betatron," Phys. Fluids **28**, 1968 (1985).

41. J.J. Petillo and R.C. Davidson, "Kinetic Equilibrium and Stability Properties of High-Current Betatrons," Phys. Fluids **30**, 2477 (1987).

42. D.W. Kerst, "Acceleration of Electrons by Magnetic Induction," Phys. Rev. **58**, 841 (1940).

43. D.W. Kerst, G.D. Adams, H.W. Koch and C.S. Robinson, "Operation of a 300 MeV Betatron," Phys. Rev. **78**, 297 (1950).

44. C. Roberson, A. Mondelli and D. Chernin, "A High-Current Betatron with Stellarator Fields," Phys. Rev. Lett. **50**, 507 (1983).

45. C. Roberson, A. Mondelli and D. Chernin, "The Stellatron Accelerator," Particle Accelerators **17**, 79 (1985).

46. D. Chernin, "Beam Stability in a Stellatron," Phys. Fluids **29**, 556 (1986).

47. G. Roberts and N. Rostoker, "Effect of Quasiconfined Particles and $\ell = 2$ Stellarator Fields on the Negative-Mass Instability in a Modified Betatron," Phys. Fluids **29**, 333 (1986).

48. B. Mandelbaum, H. Ishizuka, A. Fisher and N. Rostoker, "Injection, Trapping and Acceleration of an Electron Beam in a Stellatron Accelerator," Phys. Fluids **31**, 916 (1988).

49. M.E. Rensink, "Thin E-Layer Equilibria," Phys. Fluids **16**, 443 (1973).

50. R.C. Davidson and S.M. Mahajan, "Intense Relativistic Nonneutral E-Layers-Equilibrium Theory," Phys. Fluids **17**, 2090 (1974).

51. R.C. Davidson and C.D. Striffler, "Vlasov Equilibria for Intense Hollow Relativistic Electron Beams," J. Plasma Phys. **12**, 353 (1974).

52. C.D. Striffler, C.A. Kapetanakos and R.C. Davidson, "Equilibrium Properties of a Rotating Nonneutral E-Layer in a Cusped Magnetic Field," Phys. Fluids **18**, 1374 (1975).

53. J.J. Bollinger and D.J. Wineland, "Strongly Coupled Nonneutral Ion Plasma," Phys. Rev. Lett. **53**, 348 (1984).

54. S.L. Gilberg, J.J. Bollinger and D.J. Wineland, "Shell Structure Phase of Magnetically Confined Strongly Coupled Plasmas," Phys. Rev. Lett. 60, 2022 (1988).

55. D.H.E. Dubin and T.M. O'Neil, "Computer Simulation of Ion Clouds in a Penning Trap," Phys. Rev. Lett. 60, 511 (1988).

56. L.D. Landau and E.M. Lifshitz, The Classical Theory of Fields (Pergamon, New York, 1971), p. 281.

57. E.F. Chrien, E.J. Valeo, R.M. Kulsrud and C.R. Oberman, "Propagation of Ion Beams Through a Tenuous Magnetized Plasma," Phys. Fluids 29, 1675 (1986).

58. J.D. Lawson, "Perveance and the Bennett Pinch Relation in Partially Neutralized Electron Beams," J. Electron. Control 5, 146 (1958).

59. J.D. Lawson, "On the Classification of Electron Streams," J. Nucl. Energy, Part C: Plasma Physics 1, 31 (1959).

60. G.J. Budker, "Relativistic Stabilized Electron Beam," in Proceedings of the 1956 CERN Symposium on High-Energy Accelerators and Pion Physics (CERN Scientific Information Service, Geneva, 1956), Volume 1, p. 68.

61. C.S. Gardner, "Bound on the Energy Available from a Plasma," Phys. Fluids 6, 839 (1963).

62. H.V. Wong, M.L. Sloan, J.R. Thompson and A.T. Drobot, "Stability of an Unneutralized Rigidly Rotating Electron Beam," Phys. Fluids 16, 902 (1973).

63. C.F. Driscoll, J.H. Malmberg and K.S. Fine, "Observation of Transport to Thermal Equilibrium in Pure Electron Plasmas," Phys. Rev. Lett. 60, 1290 (1988).

64. See, for example, R.C. Davidson, "Kinetic Waves and Instabilities in Uniform Plasma," in Handbook of Plasma Physics—Basic Plasma Physics, eds., M.N. Rosenbluth and R.Z. Sagdeev (North Holland, Amsterdam, 1983), Volume 1, pp. 521-585.

65. R.C. Davidson and B.H. Hui, "Influence of Self Fields on the Equilibrium and Stability Properties of Relativistic Electron Beam-Plasma Systems," Annals of Physics 94, 209 (1975).

66. L.D. Landau, "On the Vibrations of the Electronic Plasma," J. Phys. (USSR) 10, 25 (1946).

67. See, for example, I.B. Bernstein and S.K. Trehan, "Plasma Oscillations (I)," Nuclear Fusion 1, 3 (1960).

68. I.B. Bernstein, "Waves in a Plasma in a Magnetic Field," Phys. Rev. 109, 10 (1958).

69. R.A. Dory, G.E. Guest and E.G. Harris, "Unstable Electrostatic Plasma Waves Propagating Perpendicular to a Magnetic Field," Phys. Rev. Lett. 14, 131 (1965).

70. G. Benford, "Electron Beam Filamentation in Strong Magnetic Fields," Phys. Rev. Lett. 28, 1242 (1972).

71. R. Lee and M. Lampe "Electromagnetic Instabilities, Filamentation and Focusing of Relativistic Electron Beams," Phys. Rev. Lett. 31, 1390 (1973).

72. G. Benford, "Theory of Filamentation in Relativistic Electron Beams," Plasma Physics 15, 433 (1973).

73. R.C. Davidson, B.H. Hui and C.A. Kapetanakos, "Influence of Self Fields on the Filamentation Instability in Relativistic Beam-Plasma Systems," Phys. Fluids 18, 1040 (1975).

74. R.N. Sudan, "Plasma Electromagnetic Instabilities," Phys. Fluids 6, 57 (1963).

75. J.E. Scharer and A.W. Trivelpiece, "Cyclotron Wave Instabilities in a Plasma," Phys. Fluids 10, 591 (1967).

76. R.H. Levy, J.D. Daugherty and O. Buneman, "Ion Resonance Instability in Grossly Nonneutral Plasmas," Phys. Fluids 12, 2616 (1969).

77. R.C. Davidson and H.S. Uhm, "Influence of Finite Ion Larmor Radius Effects on the Ion Resonance Instability in a Nonneutral Plasma Column," Phys. Fluids 21, 60 (1978).

78. H.S. Uhm and R.C. Davidson, "Low-Frequency Flute Perturbations in Intense Nonneutral Electron and Ion Beams," Phys. Fluids 23, 1586 (1980).

79. J.J. Livingood, Principles of Cyclic Particle Accelerators (Van Nostrand, New York, 1961).

80. See, for example, R.W. Landau and V.K. Neil, "Negative-Mass Instability," Phys. Fluids 9, 2412 (1966).

The following references, while not cited directly in the text, are relevant to the general subject matter of this chapter.

D.J. Kaup, S.R. Choudhury and G.E. Thomas, "Second-Order Stability Analysis of the Vlasov-Poisson Equations in the Planar Magnetron," Phys. Fluids 31, 177 (1988).

R.R. Prasad and M. Krishnan, "Rigid Rotor Equilibria of Multifluid, Neutral Plasma Columns in Crossed Electric and Magnetic Fields," Phys. Fluids 30, 3496 (1987).

J.D. Crawford and T.M. O'Neil, "Nonlinear Collective Processes and the Confinement of a Pure-Electron Plasma," Phys. Fluids 30, 2076 (1987).

T.P. Hughes and B.B. Godfrey, "Electromagnetic Instability in a Quadrupole-Focusing Accelerator," Phys. Fluids **29**, 1698 (1986).

T.M. O'Neil and P.G. Hjorth, "Collisional Dynamics of a Strongly Magnetized Pure Electron Plasma," Phys. Fluids **28**, 3241 (1985).

P.D. Pedrow, J.B. Greenly, D.A. Hammer and R.N. Sudan, "Proton Ring Trapping in a Gated Magnetic Mirror," Appl. Phys. Lett. **47**, 225 (1985).

C.A. Kapetanakos, P. Sprangle, S.J. Marsh, D. Dialetis, C. Agritellis and A. Prakash, "Studies of a Rapid Electron Beam Accelerator (Rebatron)," Particle Accelerators **18**, 73 (1985).

G. Roberts and N. Rostoker, "Magnetic Damping of the $\ell = 1$ Diocotron Mode," Phys. Fluids **28**, 2547 (1985).

S.A. Prasad and T.M. O'Neil, "Vlasov Theory of Electrostatic Modes in a Finite Length Electron Column," Phys. Fluids **27**, 206 (1984).

D.L. Eggleston, T.M. O'Neil and J.H. Malmberg, "Collective Enhancement of Radial Transport in a Nonneutral Plasma," Phys. Rev. Lett. **53**, 982 (1984).

D. Chernin, A. Mondelli and C. Roberson, "A Bumpy-Torus Betatron," Phys. Fluids **27**, 2378 (1984).

C.F. Driscoll and J.H. Malmberg, "Length-Dependent Containment of a Pure Electron Plasma," Phys. Rev. Lett. **50**, 167 (1983).

S.A. Prasad and T.M. O'Neil, "Waves in a Pure Electron Plasma of Finite Length," Phys. Fluids **26**, 665 (1983).

P.L. Dreike, J.B. Greenly, D.A. Hammer and R.N. Sudan, "Formation and Dynamics of a Rotating Proton Ring in a Magnetic Mirror," Phys. Fluids **25**, 59 (1982).

T.M. O'Neil, "Centrifugal Separation of a Multispecies Pure Ion Plasma," Phys. Fluids **24**, 1447 (1981).

P.L. Dreike, J.B. Greenly, D.A. Hammer and R.N. Sudan, "Formation and Propagation of a Proton Ring," Phys. Rev. Lett. **46**, 539 (1981).

J. Chen and R.C. Davidson, "Thermal Equilibrium Properties of an Intense Ion Beam with Rotational and Axial Motion," Phys. Fluids **23**, 302 (1980).

H.S. Uhm and R.C. Davidson, "Influence of Axial Energy Spread on the Negative-Mass Instability in a Relativistic Nonneutral E-Layer," Phys. Fluids **21**, 265 (1978).

R.C. Davidson, S.M. Mahajan and M.J. Schwartz, "Synchrotron Radiation Spectrum for an Intense Relativistic Electron Ring," Phys. Fluids **17**, 1287 (1974).

R.C. Davidson and S.M. Mahajan, "Synchrotron Radiation Spectrum for an Intense Relativistic E-Layer," Phys. Fluids **17**, 2267 (1974).

M.L. Sloan and W.E. Drummond, "Autoresonant Accelerator Concept," Phys. Rev. Lett. **31**, 1234 (1973).

B.I. Bogema and R.C. Davidson, "Two-Rotating-Stream Instability in a Non-neutral Plasma," Phys. Fluids **14**, 1456 (1971).

T.H Stix, "Some Toroidal Equilibria for Plasma Under Magnetoelectric Confinement," Phys. Fluids **14**, 692 (1971).

T.H. Stix, "Stability of Cold Plasma Under Magnetoelectric Confinement," Phys. Fluids **14**, 702 (1971).

R.C. Davidson and N.A. Krall, "A Characteristic Instability of an Electron Gas of Uniform Density," Phys. Lett. **32A**, 187 (1970).

T.H. Stix "Negatively Charged Open-Ended Plasma to Strip and Confine Heavy Ions," Phys. Rev. Lett. **23**, 1093 (1969).

CHAPTER 5

MACROSCOPIC EQUILIBRIUM AND STABILITY PROPERTIES

In this chapter, use is made of the macroscopic fluid description developed in Sec. 2.3 to investigate the equilibrium and stability properties of a wide variety of nonneutral plasma systems. While a kinetic model (Chapter 4) provides a more complete physical description of collective processes, a macroscopic model based on the cold-fluid-Maxwell equations is usually more tractable in treating the important effects of finite geometry, particularly with regard to their influence on stability and wave propagation properties.[1-4] Following an analysis of the macroscopic equilibrium properties of a nonneutral plasma column confined by an applied magnetic field $B_0 \hat{e}_z$ (Secs. 5.1 and 5.2), the eigenvalue equation is derived for electrostatic perturbations about the general density profile $n_j^0(r)$ (Sec. 5.3). The dispersion relation is then obtained for perturbations about uniform-density plasma components, including the important influence of a conducting wall at radius $r = b$ (Sec. 5.4). This dispersion relation is analyzed for a wide range of stable oscillations and instabilities in nonneutral plasmas, ranging from the analogue of the Trivelpiece-Gould modes in a nonneutral plasma-filled waveguide (Sec. 5.5), to the ion resonance instability which occurs for flute perturbations ($k_z = 0$) about a nonneutral plasma column in which there is a relative rotation of the ion and electron components (Sec. 5.6). Electrostatic two-stream instabilities are investigated for a relativistic electron beam propagating with axial velocity $\beta_b c \hat{e}_z$ through a background plasma (Sec. 5.7). Following a derivation of the ordinary-mode electromagnetic dispersion relation, the stabilizing influence of intense self-magnetic fields on the filamentation instability is analyzed (Sec. 5.8). Finally, equilibrium and stability properties are investigated for a two-dimensional nonneutral ion plasma confined below the surface of liquid helium (Sec. 5.9). It is shown that the two-dimensional ion disk supports collective plasma oscillations, not unlike the collective modes characteristic of a three-dimensional nonneutral plasma.

5.1   Equilibrium Force Balance Equation

In this section, equilibrium properties ($\partial/\partial t = 0$) are investigated for a cold, multicomponent, nonneutral plasma column aligned parallel to a uniform applied magnetic field $B_0 \hat{e}_z$ (Fig. 5.1). The analysis is based on the macroscopic fluid description discussed in Sec. 2.3. The following simplifying assumptions are made:

(a) Equilibrium properties are uniform in the z-direction, with $\partial n_j^0(\mathbf{x})/\partial z = 0$ and $\partial \mathbf{V}_j^0(\mathbf{x})/\partial z = 0$, and there is no equilibrium electric field parallel to $B_0 \hat{e}_z$, i.e., $\mathbf{E}^0(\mathbf{x}) \cdot \hat{e}_z = 0$.

(b) The equilibrium radial density and velocity profiles are assumed to be azimuthally symmetric about the magnetic axis, i.e.,

$$n_j^0(\mathbf{x}) = n_j^0(r) , \quad \mathbf{V}_j^0(\mathbf{x}) = V_{\theta j}^0(r)\hat{e}_\theta + V_{zj}^0(r)\hat{e}_z , \qquad (5.1)$$

where r is the radial distance from the axis of symmetry, and $\hat{e}_\theta$ and $\hat{e}_z$ are unit vectors in the θ- and z-directions, respectively. The equilibrium continuity equation $\nabla \cdot [n_j^0(\mathbf{x})\mathbf{V}_j^0(\mathbf{x})] = 0$ is automatically satisfied for general profiles $n_j^0(r), V_{\theta j}^0(r)$ and $V_{zj}^0(r)$. The azimuthal current $J_\theta^0(r) = \sum_j e_j n_j^0(r) V_{\theta j}^0(r)$ generally induces an axial self-magnetic field $B_z^s(r)$, and the axial current $J_z^0(r) = \sum_j e_j n_j^0(r) V_{zj}^0(r)$ generally induces an azimuthal self-magnetic field $B_\theta^s(r)$.

(c) The plasma components are assumed to be sufficiently cold that the approximation $(\partial/\partial \mathbf{x}) \cdot \mathbf{P}_j^0(\mathbf{x}) \simeq 0$ is justified in the equilibrium force balance equation (2.35).

It is convenient to express

$$V_{zj}^0(r) = \beta_{zj}(r)c , \quad V_{\theta j}^0(r) = \omega_{rj}(r)r , \qquad (5.2)$$

in the subsequent analysis. Within the context of assumptions (a)–(c), the equilibrium field components are

$$\mathbf{E}^0(\mathbf{x}) = E_r(r)\hat{e}_r , \quad \mathbf{B}^0(\mathbf{x}) = [B_0 + B_z^s(r)]\,\hat{e}_z + B_\theta^s(r)\hat{e}_\theta , \qquad (5.3)$$

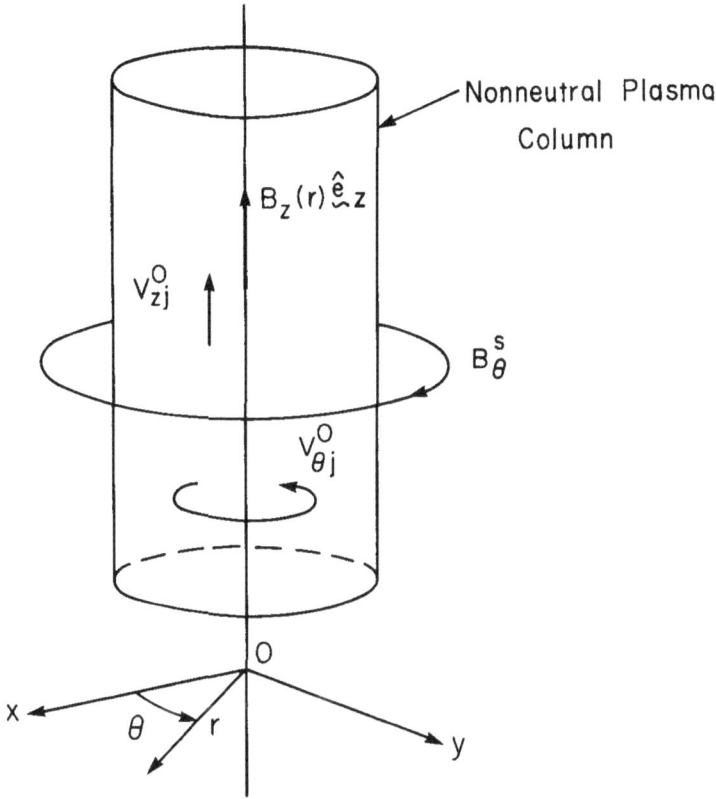

Figure 5.1. Schematic of a cylindrical nonneutral plasma column confined radially by an applied magnetic field $B_0 \hat{e}_z$. The lack of equilibrium charge neutrality produces a radial self-electric field $E_r(r)\hat{e}_r$, and the plasma current in the axial direction produces an azimuthal self-magnetic field $B_\theta^s(r)\hat{e}_\theta$. The azimuthal current associated with the equilibrium rotation of the plasma components generally produces a diamagnetic contribution $B_z^s(r)$ to the total axial magnetic field $B_z(r) = B_0 + B_z^s(r)$.

where the self fields are determined from the steady-state Maxwell equations

$$\frac{1}{r}\frac{\partial}{\partial r}\left[rE_r(r)\right] = \sum_j 4\pi e_j n_j^0(r) = 4\pi \rho^0(r) , \tag{5.4a}$$

$$\frac{1}{r}\frac{\partial}{\partial r}\left[rB_\theta^s(r)\right] = \sum_j 4\pi e_j n_j^0(r)\beta_{zj}(r) = \frac{4\pi}{c}J_z^0(r) , \tag{5.4b}$$

$$\frac{\partial}{\partial r} B_z^s(r) = -\sum_j \frac{4\pi e_j}{c} n_j^0(r) r \omega_{rj}(r) = -\frac{4\pi}{c} J_\theta^0(r) \ . \qquad (5.4c)$$

Integrating Maxwell's equations (5.4) gives the equilibrium radial electric field

$$E_r(r) = \frac{4\pi}{r} \sum_j e_j \int_0^r dr' \, r' n_j^0(r') \ , \qquad (5.5)$$

the equilibrium azimuthal self-magnetic field

$$B_\theta^s(r) = \frac{4\pi}{r} \sum_j e_j \int_0^r dr' \, r' n_j^0(r') \beta_{zj}(r') \ , \qquad (5.6)$$

and the equilibrium axial self-magnetic field

$$B_z^s(r) = \frac{4\pi}{c} \sum_j e_j \int_r^\infty dr' \, n_j^0(r') r' \omega_{rj}(r') \ . \qquad (5.7)$$

If the density $n_j^0(r)$ is equal to zero beyond some radius r_b, then $B_z^s(r \geq r_b)$ $= 0$ follows from Eq.(5.7).

In the present analysis, the axial and azimuthal motions of component j are allowed to be relativistic in the general case. The relativistic mass factor for a cold fluid element can be expressed as

$$\gamma_j^0(r) = \left[1 - \beta_{zj}^2(r) - \frac{r^2 \omega_{rj}^2(r)}{c^2} \right]^{-1/2} \ , \qquad (5.8)$$

and the azimuthal momentum $P_{\theta j}^0(r)$ can be expressed as $P_{\theta j}^0(r) = \gamma_j^0(r) \times m_j r \omega_{rj}(r)$, where $V_{\theta j}^0(r) = \omega_{rj}(r) r$. Making use of Eq.(2.35), radial force balance on the j'th component fluid element is given by

$$-\frac{\gamma_j^0 m_j V_{\theta j}^{0\,2}}{r} = e_j \left[E_r + \frac{1}{c} V_{\theta j}^0 (B_0 + B_z^s) - \frac{1}{c} V_{zj}^0 B_\theta^s \right] \qquad (5.9)$$

in the region where the equilibrium density $n_j^0(r)$ is nonzero. Substituting Eqs.(5.5)–(5.7) into Eq.(5.9) gives the radial force balance equation[1]

$$-\gamma_j^0(r)\omega_{rj}^2(r) = \epsilon_j\omega_{cj}\omega_{rj}(r)\left[1 + \frac{4\pi}{B_0 c}\sum_k e_k\int_r^\infty dr'\, n_k^0(r')r'\omega_{rk}(r')\right]$$

$$+ \frac{4\pi e_j}{m_j r^2}\sum_k e_k\int_0^r dr'\, r' n_k^0(r')[1 - \beta_{zj}(r)\beta_{zk}(r')]\,, \qquad (5.10)$$

where $\epsilon_j = \text{sgn}\, e_j$ denotes the sign of e_j, and $\omega_{cj} = |e_j|B_0/m_j c$ is the non-relativistic cyclotron frequency of component j in the applied magnetic field $B_0\hat{e}_z$.

For a cold, multicomponent nonneutral plasma, Eq.(5.10) is a statement of the radial balance of centrifugal, electric and magnetic forces on a fluid element. Note from Eq.(5.10) that there is considerable freedom in describing the equilibrium within the framework of a macroscopic cold-fluid model. In particular, any two of the equilibrium profiles $n_j^0(r)$, $\omega_{rj}(r)$ and $\beta_{zj}(r)$ can be prescribed arbitrarily, and the remaining profile calculated self-consistently from Eq.(5.10). Moreover, Eq.(5.10) is valid for arbitrary deviations from equilibrium charge neutrality and equilibrium current neutrality. For example, Eq.(5.10) can be applied to a one-component nonneutral plasma consisting only of electrons. It can also be applied to a wide variety of nonneutral beam-plasma admixtures, e.g., a relativistic electron beam propagating through a plasma background consisting of ions and electrons that provide partial charge and current neutralization.

5.2   Examples of Macroscopic Equilibria

In this section, use is made of the radial force balance equation (5.10) to investigate equilibrium properties for several examples of nonrelativistic and relativistic nonneutral plasmas and beam-plasma systems.

5.2.1 Nonrelativistic Nonneutral Plasma Column

We first consider Eq.(5.10) in the nonrelativistic regime where $\beta_{zj}^2(r)$, $r^2\omega_{rj}^2(r)/c^2 \ll 1$, and

$$|B_z^s(r)| \ll B_0 , \quad |\beta_{zj}(r)B_\theta^s(r)| \ll |E_r(r)| \tag{5.11}$$

over the cross section of the plasma column. The radial force balance equation (5.10) then reduces to[1]

$$\omega_{rj}^2(r) + \varepsilon_j\omega_{cj}\omega_{rj}(r) + \sum_k \frac{4\pi e_j e_k}{m_j r^2} \int_0^r dr'\, r'n_k^0(r') = 0, \tag{5.12}$$

where use has been made of $\gamma_j^0(r) \simeq 1$. Assuming that the density profile $n_j^0(r)$ is specified, Eq.(5.12) gives two equilibrium solutions for the angular velocity profile $\omega_{rj}(r)$. These are:

$$\omega_{rj}(r) = \omega_{rj}^\pm(r) \equiv -\frac{\varepsilon_j\omega_{cj}}{2}\left\{1 \pm \left(1 - \sum_k \frac{16\pi e_j e_k}{m_j\omega_{cj}^2 r^2}\int_0^r dr'r'n_k^0(r')\right)^{1/2}\right\}.$$

$$\tag{5.13}$$

For nonuniform density profile $n_j^0(r)$, note from Eq.(5.13) that there is generally a shear in the angular velocity profile, i.e., $\partial\omega_{rj}(r)/\partial r \neq 0$. In the special case where $n_j^0(r)$ is uniform out to some radius $r = r_b$,

$$n_j^0(r) = \begin{cases} \hat{n}_j = \text{const}, & 0 \leq r < r_b , \\ 0, & r > r_b , \end{cases} \tag{5.14}$$

Eq.(5.13) reduces to

$$\omega_{rj} = \omega_{rj}^\pm \equiv -\frac{\varepsilon_j\omega_{cj}}{2}\left\{1 \pm \left(1 - \sum_k \frac{8\pi e_j \hat{n}_k e_k}{m_j\omega_{cj}^2}\right)^{1/2}\right\}, \tag{5.15}$$

for $0 \leq r < r_b$. Note that $\omega_{rj}^\pm = \text{const}$. for the uniform density profile assumed in Eq.(5.14). Equation (5.15) represents the appropriate generalization of Eq.(3.6) to a multicomponent nonneutral plasma.[1] For the special case of a two-component nonneutral plasma consisting of electrons $(j = e)$ and singly-ionized ions $(j = i)$ that provide a partially neutralizing background with density $\hat{n}_i = f\hat{n}_e$ (where $f = \text{const.}$ = fractional charge neutralization), the equilibrium rotation velocities in Eq.(5.15) reduce to

$$\omega_{re} = \omega_{re}^{\pm} = \frac{\omega_{ce}}{2}\left\{1 \pm \left(1 - \frac{2\omega_{pe}^2}{\omega_{ce}^2}(1 - f)\right)^{1/2}\right\}, \qquad (5.16)$$

for the electrons, and

$$\omega_{ri} = \omega_{ri}^{\pm} = -\frac{\omega_{ci}}{2}\left\{1 \pm \left(1 + \frac{2\omega_{pi}^2}{\omega_{ci}^2}\frac{(1 - f)}{f}\right)^{1/2}\right\} \qquad (5.17)$$

for the ions. Here, $\omega_{pe}^2 = 4\pi\hat{n}_e e^2/m_e$ and $\omega_{pi}^2 = 4\pi\hat{n}_i e^2/m_i$. For complete charge neutralization ($f = 1$), Eqs.(5.16) and (5.17) reduce to the expected results, $\omega_{rj}^- = 0$ and $\omega_{rj}^+ = -\varepsilon_j\omega_{cj}$.

To summarize, Eqs.(5.12) and (5.13) can be used to investigate the equilibrium properties of multicomponent nonneutral plasmas for a wide variety of density profiles $n_j^0(r)$. Equations (5.12) and (5.13) are valid when the macroscopic fluid motion is cold and nonrelativistic, while allowing for arbitrary deviations from equilibrium charge neutrality. One of the most important consequences of the radial electric field $E_r(r)$ is to produce a differential rotation between plasma components. This can provide the free energy source to drive various cross-field streaming instabilities, such as the ion resonance instability (Sec. 5.6).

Problem 5.1   Consider a nonrelativistic nonneutral plasma column consisting only of electrons ($j = e$). Assume that the functional form of the angular velocity profile $\omega_{re}(r)$ is specified. Make use of the radial force balance equation (5.12) to show that

$$\omega_{pe}^2(r) = -\frac{1}{r}\frac{\partial}{\partial r}[r^2\omega_{re}^2(r) - r^2\omega_{ce}\omega_{re}(r)] \qquad (5.1.1)$$

where $\omega_{pe}^2(r) = 4\pi n_e^0(r)e^2/m_e$. Equation (5.1.1) determines the electron density profile $n_e^0(r)$ directly in terms of $\omega_{re}(r)$.

Problem 5.2   Assume that the electrons constituting the equilibrium in Problem 5.1 move on surfaces with constant canonical angular momentum, i.e.,

$$P_{\theta e}^0(r) = m_e[rV_{\theta e}^0(r) - r^2\omega_{ce}/2] = P_0 = \text{const.} \qquad (5.2.1)$$

Here, $\omega_{ce} = eB_0/m_e c$, $V_{\theta e}^0(r) = r\omega_{re}(r)$ is the azimuthal velocity of an electron fluid element, and the constant P_0 is independent of radius r. Such an equilibrium can occur when an electron layer is formed by an annular cathode ($P_0 = 0$ for a shielded cathode source). In the subsequent analysis, assume $P_0 = -m_e\omega_{ce}(r_b^-)^2/2 < 0$, where $r = r_b^-$ denotes the inner radius of the electron layer.

a. Show that the angular rotation velocity of the electron layer is

$$\omega_{re}(r) = \frac{1}{2}\omega_{ce}\left[1 - \left(\frac{r_b^-}{r}\right)^2\right] \tag{5.2.2}$$

for $r \geq r_b^-$.

b. Show that the corresponding equilibrium density profile is given self-consistently by

$$\omega_{pe}^2(r) = \frac{1}{2}\omega_{ce}^2\left[1 + \left(\frac{r_b^-}{r}\right)^4\right] \tag{5.2.3}$$

for $r \geq r_b^-$. For $P_0 = 0$ and $r_b^- = 0$, it follows from Eq.(5.2.3) that $2\omega_{pe}^2(r)/\omega_{ce}^2 = \text{const.}= 1$, which corresponds to a uniform-density plasma column at the Brillouin density limit defined in Eq.(3.8).

5.2.2   Relativistic Diamagnetic Equilibria

We now consider the radial force balance equation (5.10) in circumstances where the axial motion is nonrelativistic with

$$\beta_{zj}^2(r) \ll 1 - r^2\omega_{rj}^2(r)/c^2 , \tag{5.18}$$

but the azimuthal motion of the electrons is allowed to be relativistic. For present purposes, the nonneutral plasma is assumed to consist only of electrons ($j = e$) and no positive ions are present [$n_i^0(r) = 0$]. Neglecting the azimuthal self-magnetic field $B_\theta^s(r)$ in Eq.(5.10), but retaining the effects of the axial diamagnetic field $B_z^s(r)$, the radial force balance equation (5.10) can be expressed as[1]

$$-\gamma_e^0(r)\omega_{re}^2(r) = \frac{1}{r^2}\int_0^r dr'\,r'\omega_{pe}^2(r') - \omega_{re}(r)\left[\omega_{ce} - \frac{1}{c^2}\int_r^\infty dr'\,r'\omega_{re}(r')\omega_{pe}^2(r')\right]. \tag{5.19}$$

Here, $\omega_{ce} = eB_0/m_e c$ and $\omega_{pe}^2(r) = 4\pi n_e^0(r)e^2/m_e$, and the relativistic mass factor $\gamma_e^0(r)$ is defined by

$$\gamma_e^0(r) = \left[1 - \frac{r^2\omega_{re}^2(r)}{c^2}\right]^{-1/2}. \tag{5.20}$$

For specified electron density profile $\omega_{pe}^2(r)$, the radial force balance equation (5.19) can be used to calculate the angular velocity profile $\omega_{re}(r)$ self-consistently. On the other hand, if the profile for $\omega_{re}(r)$ is specified, then Eq.(5.19) can be viewed as an integral equation that determines $\omega_{pe}^2(r)$.

As a simple application of Eq.(5.19), consider the case where the density profile $n_e^0(r)$ extends over the interval $0 \leq r < r_b$, and the angular velocity profile is assumed to correspond to a rigid rotation with

$$\omega_{re}(r) = \omega_0 = \text{const.} \tag{5.21}$$

Substituting Eqs.(5.20) and (5.21) into Eq.(5.19) and solving for $\omega_{pe}^2(r) = 4\pi n_e^0(r)e^2/m_e$, some straightforward algebra gives[1]

$$n_e^0(r) = \frac{\hat{n}_e}{(1 - r^2\omega_0^2/c^2)^2}\left[1 + \frac{2\omega_0^2}{\omega_{pe}^2}\left(1 - \frac{(1 + r^2\omega_0^2/2c^2)}{(1 - r^2\omega_0^2/c^2)^{1/2}}\right)\right], \quad 0 \leq r < r_b. \tag{5.22}$$

Here, $r_b^2\omega_0^2/c^2 < 1$ is assumed, and $\omega_{pe}^2 \equiv 4\pi\hat{n}_e e^2/m_e$, where $\hat{n}_e = n_e^0(r = 0)$ is the on-axis value of the electron density. The total axial magnetic field $B_z(r) = B_0 + B_z^s(r)$ consistent with Eqs.(5.21), (5.22), $\partial B_z^s/\partial r = (4\pi e/c)n_e^0(r)\omega_0 r$, and $B_z(r = r_b) = B_0$ is given by (see also Problem 5.4)

$$B_z(r) = B_0\frac{(1 - r_b^2\omega_0^2/c^2)}{(1 - r^2\omega_0^2/c^2)}\left[1 + \frac{\omega_0}{\omega_{ce}}\frac{1}{(1 - r_b^2\omega_0^2/c^2)}\right.$$
$$\left. \times \left(\frac{r_b^2\omega_0^2/c^2}{(1 - r_b^2\omega_0^2/c^2)^{1/2}} - \frac{r^2\omega_0^2/c^2}{(1 - r^2\omega_0^2/c^2)^{1/2}}\right)\right], \tag{5.23}$$

where $\omega_{ce} = eB_0/m_e c$. Equation (5.23) is valid in the interval $0 \leq r < r_b$. Substituting Eqs.(5.22) and (5.23) into Eq.(5.19) and evaluating as $r \to 0$ gives the condition

$$-\omega_0^2 = \frac{1}{2}\omega_{pe}^2 - \omega_0\omega_{ce}\left[1 - \frac{r_b^2\omega_0^2}{c^2}\left(1 - \frac{\omega_0/\omega_{ce}}{(1 - r_b^2\omega_0^2/c^2)^{1/2}}\right)\right] . \quad (5.24)$$

Equation (5.24) relates ω_0 to the on-axis values of electron density (proportional to $\omega_{pe}^2 = 4\pi\hat{n}_e e^2/m_e$) and axial magnetic field $B_z(r = 0)$, which is proportional to the factor $\omega_{ce}[\cdots]$ in Eq.(5.24). In the limiting case of nonrelativistic electron flow with $r_b^2\omega_0^2/c^2 \ll 1$, Eqs.(5.22)–(5.24) reduce to the familiar results for a uniform equilibrium with $n_e^0(r) \simeq \hat{n}_e$, $B_z(r) \simeq B_0$, and $-\omega_0^2 \simeq \omega_{pe}^2/2 - \omega_0\omega_{ce}$, discussed in Sec. 3.1.

It is straightforward to show from Eqs.(5.22) and (5.24) that

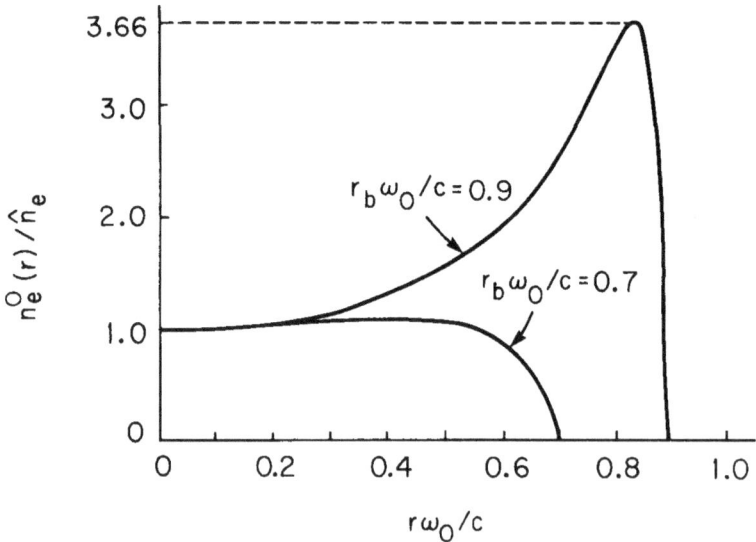

Figure 5.2. Plots of the normalized density profile $n_e^0(r)/\hat{n}_e$ versus $r\omega_0/c$ obtained from Eq.(5.22) for the two cases $r_b\omega_0/c = 0.7$ and $r_b\omega_0/c = 0.9$. Here, the equilibria satisfy $n_e^0(r = r_b) = 0$, so that ω_{pe}^2/ω_0^2 and ω_0/ω_{ce} are determined self-consistently in terms of $r_b\omega_0/c$ from Eqs.(5.26) and (5.27), respectively.

$$0 < \frac{\omega_0}{\omega_{ce}} < \frac{(1 - r_b^2\omega_0^2/c^2)^{3/2}}{(1 - r_b^2\omega_0^2/2c^2)} \tag{5.25}$$

is required to assure physically acceptable equilibria with $n_e^0(r) \geq 0$ over the interval $0 \leq r < r_b$. The detailed form of the density profile $n_e^0(r)$ depends on the parameter $\delta \equiv 2/3 + \omega_{pe}^2/3\omega_0^2$. For $2/3 < \delta < 1$, it is found that $n_e^0(r)$ decreases monotonically as a function of r. For $\delta > 1$, however, it can be shown that $n_e^0(r)$ increases away from the axis ($r = 0$), achieves a maximum at some radius $r = r_M < r_b$, and decreases monotonically to its value $n_e^0(r = r_b)$ at the boundary. From Eq.(5.22), the special case in which $n_e^0(r)$ approaches zero continuously as r approaches r_b requires

$$\frac{\omega_{pe}^2}{2\omega_0^2} = \frac{(1 + r_b^2\omega_0^2/2c^2)}{(1 - r_b^2\omega_0^2/c^2)^{1/2}} - 1 \,, \tag{5.26}$$

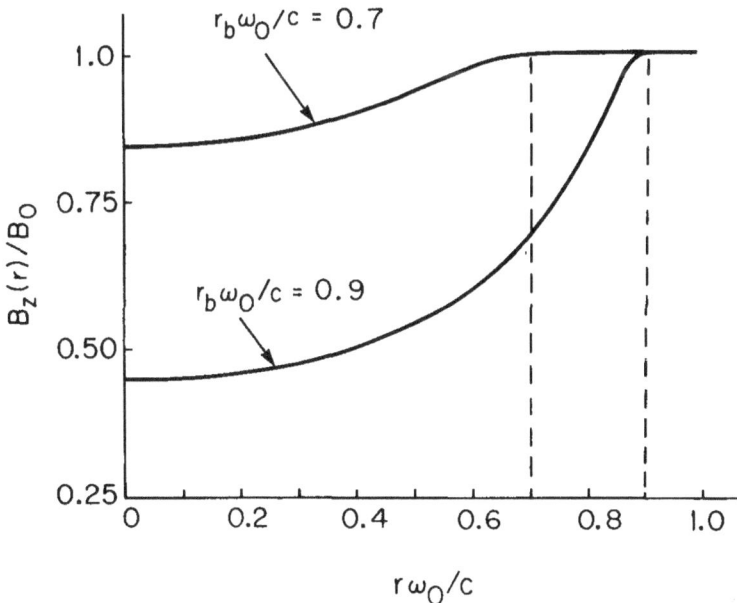

Figure 5.3. Plots of the normalized magnetic field profile $B_z(r)/B_0$ versus $r\omega_0/c$ obtained from Eq.(5.23) for the two cases $r_b\omega_0/c = 0.7$ and $r_b\omega_0/c = 0.9$, and parameters otherwise identical to Fig. 5.2.

which relates ω_0, r_b and $\omega_{pe}^2 = 4\pi\bar{n}_e e^2/m_e$ for rigid-rotor equilibria with $n_e^0(r = r_b) = 0$. Combining Eqs.(5.24) and (5.26) gives

$$\frac{\omega_0}{\omega_{ce}} = \frac{(1 - r_b^2\omega_0^2/c^2)^{3/2}}{(1 - r_b^2\omega_0^2/2c^2)} , \tag{5.27}$$

which corresponds to the upper limit of the inequality in Eq.(5.25). The density and magnetic field profiles in Eqs.(5.22) and (5.23) are illustrated in Figs. 5.2 and 5.3 for two such relativistic, diamagnetic equilibria with $r_b\omega_0/c = 0.7$ and $r_b\omega_0/c = 0.9$, and $n_e^0(r = r_b) = 0$.

Finally, for arbitrary angular velocity profile $\omega_{re}(r)$, it should be noted that a general formalism[5] can be developed to determine the profiles for $B_z(r)$ and $n_e^0(r)$ self-consistently from Maxwell's equations and the radial force balance Eq.(5.19) (Problem 5.4).

Problem 5.3   Consider a relativistic diamagnetic plasma column consisting only of electrons ($j = e$). Assume that the electron density profile $n_e^0(r)$ extends from $r = 0$ to some outer radius $r = r_b$. Make use of Eq.(5.7) and the radial force balance equation (5.19) to show that such a cold-fluid equilibrium model does not support field-reversed solutions with $B_z(r) = B_0 + B_z^s(r) < 0$ at any radius r in the interval $0 \leq r < r_b$. [Hint: Make use of the fact that the applied magnetic field satisfies $B_z(r = r_b) = B_0 > 0$.]

Problem 5.4   Consider a relativistic, diagmagnetic plasma column consisting only of electrons ($j = e$). Assume that $V_{ze}^0(r) = 0$ and $B_\theta^s(r) = 0$, and denote the normalized azimuthal flow velocity and relativistic mass factor by $\beta_\theta(r) = r\omega_{re}(r)/c$ and $\gamma_\theta(r) = [1 - \beta_\theta^2(r)]^{-1/2}$, respectively.

a. Show that the radial electric field $E_r(r)$ and axial magnetic field $B_z(r) = B_0 + B_z^s(r)$ are related by

$$\frac{\partial}{\partial r}B_z(r) = -\beta_\theta(r)\frac{1}{r}\frac{\partial}{\partial r}rE_r(r) \tag{5.4.1}$$

in the region where the electron density $n_e^0(r)$ is nonzero.

b. Operate on the radial force balance equation (5.9) with $r^{-1}(\partial/\partial r)[r \cdots]$ and show that $B_z(r)$ satisfies the differential equation[5]

$$\frac{\partial}{\partial r}B_z - \left[\frac{\gamma_\theta^2\beta_\theta}{r}\frac{\partial}{\partial r}(\beta_\theta r)\right]B_z = -\frac{mc^2}{e}\left[\frac{\gamma_\theta^2\beta_\theta}{r}\frac{\partial}{\partial r}(\gamma_\theta\beta_\theta^2)\right] . \tag{5.4.2}$$

For specified azimuthal velocity profile $\beta_\theta(r)$, Eq.(5.4.2) can be integrated to determine the axial magnetic field $B_z(r)$. Once the functional form of

$B_z(r)$ has been determined, the corresponding electron density profile can be calculated self-consistently from the Maxwell equation

$$n_e^0(r) = \frac{1}{4\pi e\beta_\theta(r)} \frac{\partial}{\partial r} B_z(r). \tag{5.4.3}$$

c. For $\beta_\theta(r) = r\omega_0/c$, integrate Eq.(5.4.2) from $r = 0$ to $r = r_b$, and obtain the expression for $B_z(r)$ in Eq.(5.23). Here, $\omega_0 = $ const. is the angular velocity of the (rigid-rotor) equilibrium, and $B_0 \equiv B_z(r = r_b)$ denotes the applied magnetic field in the region $r \geq r_b$.

5.2.3 Relativistic Beam-Plasma Equilibria

In this section, Eq.(5.10) is analyzed in circumstances where the axial motion is allowed to be relativistic and the $B_\theta^s(r)$ self-field contributions are retained in the equilibrium analysis. For simplicity, it is assumed that the axial velocity profile $V_{zj}^0(r) \equiv \beta_{zj}(r)c$ is uniform over the cross section of the plasma column and that the azimuthal motion of a fluid element is nonrelativistic, i.e.,

$$\beta_{zj}(r) \equiv \beta_j = \text{const.},$$
$$r^2\omega_{rj}^2(r)/c^2 \ll 1 - \beta_j^2 . \tag{5.28}$$

Defining $\gamma_j \equiv (1 - \beta_j^2)^{-1/2} = $ const., and neglecting $|B_z^s(r)| \ll B_0$, the equilibrium force balance equation (5.10) becomes[1]

$$\omega_{rj}^2(r) + \varepsilon_j\omega_{rj}(r)\frac{\omega_{cj}}{\gamma_j} + \sum_k \frac{4\pi e_j e_k}{\gamma_j m_j r^2}(1 - \beta_j\beta_k) \int_0^r dr'\, r' n_k^0(r') = 0 , \tag{5.29}$$

where use has been made of Eq.(5.28). Solving Eq.(5.29) for the rotation velocity $\omega_{rj}(r)$ gives

$$\omega_{rj}(r) = \omega_{rj}^\pm(r) \tag{5.30}$$

$$\equiv -\frac{\varepsilon_j\omega_{cj}}{2\gamma_j}\left\{1 \pm \left(1 - \sum_k \frac{16\pi e_j e_k \gamma_j}{m_j \omega_{cj}^2 r^2}(1 - \beta_j\beta_k) \int_0^r dr'\, r' n_k^0(r')\right)^{1/2}\right\} .$$

Note that $\partial\omega_{rj}(r)/\partial r$ is generally nonzero for nonuniform density profiles $n_j^0(r)$. Equation (5.30) represents a straightforward generalization of Eq.(5.13) to allow for relativistic axial motion.

As a simple application of Eq.(5.30), consider an intense electron beam ($j = b$) with relativistic axial motion propagating through a partially neutralizing plasma background ($j = e, i$). We assume uniform density profiles with $n_j^0(r) = \hat{n}_j$ for $0 \leq r < r_b (j = b, e, i)$, and define the fractional charge neutralization (f) and fractional current neutralization (f_M) provided by the background plasma according to

$$f = \frac{\displaystyle\sum_{k=e,i} \hat{n}_k e_k}{\hat{n}_b e} , \tag{5.31}$$

$$f_M = \frac{\displaystyle\sum_{k=e,i} \hat{n}_k e_k \beta_k}{\hat{n}_b e \beta_b} . \tag{5.32}$$

Here, $-e$ is the electron charge (for $j = b, e$). Note that $f = 1$ corresponds to $\sum_{k=b,e,i} \hat{n}_k e_k = 0$ and $E_r(r) = 0$, whereas $f_M = 1$ corresponds to $\sum_{k=b,e,i} \hat{n}_k e_k \beta_k = 0$ and $B_\theta^s(r) = 0$. For uniform density profiles, Eq.(5.30) readily gives for the beam electrons ($j = b$)

$$\omega_{rb} = \omega_{rb}^\pm \equiv \frac{\omega_{cb}}{2\gamma_b} \left\{ 1 \pm \left(1 - \frac{2\omega_{pb}^2 \gamma_b}{\omega_{cb}^2} \left[1 - f - \beta_b^2(1 - f_M) \right] \right)^{1/2} \right\},$$

$$\tag{5.33}$$

where $\gamma_b = (1 - \beta_b^2)^{-1/2}$, $\omega_{cb} = eB_0/m_e c$, and $\omega_{pb}^2 = 4\pi\hat{n}_b e^2/m_e$. The term proportional to $1 - f$ in Eq.(5.33) is associated with electric self-field effects and is *defocusing* whenever $f < 1$. On the other hand, the term proportional to $-\beta_b^2(1 - f_M)$ is associated with magnetic self-field effects and is *focusing* whenever $f_M < 1$. The equilibrium rotation velocity for the plasma ions ($j = i$) and plasma electrons ($j = e$) can be calculated from Eq.(5.30) in a similar manner.

The *net* self-field contribution to the beam rotation in Eq.(5.33) is focusing provided

$$\beta_b^2(1 - f_M) > 1 - f . \tag{5.34}$$

Whenever Eq.(5.34) is satisfied, radially confined beam equilibria exist even in the absence of an axial guide field $B_0\hat{e}_z$. Assuming Eq.(5.34) is satisfied and setting $B_0 = 0$ in Eq.(5.33), it readily follows that the beam rotation is given by

$$\omega_{rb} = \omega_{rb}^{\pm} \equiv \pm\frac{\omega_{pb}}{(2\gamma_b)^{1/2}}\left[\beta_b^2(1 - f_M) - (1 - f)\right]^{1/2}. \tag{5.35}$$

In the absence of a radial pressure gradient $(\partial P_b^0/\partial r = 0)$ and axial guide field $(B_0 = 0)$, the magnetically focused beam equilibrium described by Eq.(5.35) is necessarily rotating.

To summarize, within the context of Eq.(5.28), the radial force balance equation (5.29) and the solutions for $\omega_{rj}(r)$ in Eq.(5.30) can be used to investigate the equilibrium properties of cold, nonneutral beam-plasma systems for a wide variety of equilibrium density profiles.[1-4]

Problem 5.5   Consider a nonneutral beam-plasma column in which the motion parallel to $B_0\hat{e}_z$ is allowed to be relativistic. Assume that the axial flow velocity $\beta_{zj}(r)c \equiv \beta_j c = $ const. over the cross section of the plasma column, and that the fluid motion perpendicular to $B_0\hat{e}_z$ is nonrelativistic. Also assume that the equilibrium density profiles are uniform with $n_j^0(r) = \hat{n}_j = $ const. for $0 \le r < r_b$.

a. Show that the radial self-electric field $\mathbf{E}^0(\mathbf{x}) = E_r(r)\hat{e}_r$, and the azimuthal self-magnetic, $\mathbf{B}_0^s(\mathbf{x}) = B_\theta^s(r)\hat{e}_\theta$, can be expressed in Cartesian coordinates as

$$\mathbf{E}^0(\mathbf{x}) = 2\pi\sum_k \hat{n}_k e_k(x\hat{e}_x + y\hat{e}_y) , \tag{5.5.1}$$

and

$$\mathbf{B}_0^s(\mathbf{x}) = 2\pi\sum_k \hat{n}_k e_k \beta_k(-y\hat{e}_x + x\hat{e}_y) , \tag{5.5.2}$$

for $0 \le r < r_b$.

b. Consider the relativistic orbit equation

$$\frac{d}{dt}\mathbf{p}(t) = e_j\left[E^0[\mathbf{x}(t)] + \frac{1}{c}\mathbf{v}(t) \times (\mathbf{B}_0^s[\mathbf{x}(t)] + B_0\hat{e}_z)\right] \tag{5.5.3}$$

for a particle with charge e_j and rest mass m_j. Here $\mathbf{p}(t) = \gamma(t)m_j\mathbf{v}(t)$ is the particle momentum, and $\gamma(t) = [1 - \mathbf{v}^2(t)/c^2]^{-1/2}$ is the relativistic mass factor. For $v_\perp^2/c^2 \ll 1 - v_z^2/c^2$, approximate $v_z(t) \simeq v_z = $ const. and

$\gamma(t) \simeq (1 - v_z^2/c^2)^{-1/2} \equiv \gamma_z = \text{const.}$ in Eq.(5.5.3). Show from Eqs.(5.5.1)–(5.5.3) that the orbit equations for the motion perpendicular to $B_0 \hat{e}_z$ can be approximated by

$$\frac{d^2}{dt^2}x(t) = \frac{2\pi e_j}{\gamma_z m_j} \sum_k \hat{n}_k e_k \left(1 - \frac{v_z}{c}\beta_k\right) x(t) + \frac{1}{\gamma_z}\varepsilon_j \omega_{cj}\frac{d}{dt}y(t) \,, \quad (5.5.4)$$

$$\frac{d^2}{dt^2}y(t) = \frac{2\pi e_j}{\gamma_z m_j} \sum_k \hat{n}_k e_k \left(1 - \frac{v_z}{c}\beta_k\right) y(t) - \frac{1}{\gamma_z}\varepsilon_j \omega_{cj}\frac{d}{dt}x(t) \,, \quad (5.5.5)$$

where $\varepsilon_j = \text{sgn}\, e_j$ and $\omega_{cj} = |e_j| B_0/m_j c$. Equations (5.5.4) and (5.5.5) constitute the generalization of Eqs.(3.13a) and (3.13b) to a multicomponent nonneutral plasma with relativistic axial motion, but nonrelativistic motion perpendicular to $B_0 \hat{e}_z$. If the axial velocity spread of component j is small, then $v_z/c \simeq \beta_j$ in Eqs.(5.5.4) and (5.5.5).

5.2.4   Bennett Pinch Equilibrium

In this section, use is made of a macroscopic fluid description to construct a simple model of the Bennett pinch equilibrium[6-8] for a self-focused electron beam $(j = b)$ propagating through a partially neutralizing ion background $(j = i)$. The ions are taken to form a singly ionized background with density profile

$$n_i^0(r) = f n_b^0(r), \quad (5.36)$$

where $f = \text{const.} = $ fractional charge neutralization. It is also assumed that the ions are cold $(T_i = 0)$ and that the ion motion is nonrelativistic. On the other hand, the relativistic electron beam is assumed to have uniform axial velocity profile, $\beta_{zb}(r) \equiv \beta_b = \text{const.}$, and zero mean rotation velocity, $\omega_{rb}(r) = 0$, about the axis of symmetry. Allowing for uniform but finite transverse beam temperature T_b, the electron pressure profile is given by $P_b^0(r) = n_b^0(r)k_B T_b$, and radial force balance on an electron fluid element can be expressed as (for $V_{\theta b}^0 = 0$)

$$k_B T_b \frac{\partial}{\partial r} n_b^0(r) = -e n_b^0(r)[E_r(r) - \beta_b B_\theta^s(r)] \,. \quad (5.37)$$

Making use of Eqs.(5.5), (5.6), and (5.36), the radial force balance equation (5.37) can be expressed in the equivalent form

$$\frac{k_B T_b}{n_b^0(r)} \frac{\partial}{\partial r} n_b^0(r) = -\frac{4\pi e^2 [\beta_b^2 - (1 - f)]}{r} \int_0^r dr' \, r' n_b^0(r') \,. \tag{5.38}$$

For magnetically focused equilibria ($\beta_b^2 > 1 - f$), Eq.(5.38) can be solved for $n_b^0(r)$ to give the self-consistent density profile (Fig. 5.4)

$$n_b^0(r) = \frac{\hat{n}_b}{(1 + r^2/r_b^2)^2} \,, \tag{5.39}$$

where the effective beam radius r_b is defined by

$$r_b^2 = \frac{8\lambda_{Db}^2}{[\beta_b^2 - (1 - f)]} \,. \tag{5.40}$$

Here, $\lambda_{Db}^2 \equiv k_B T_b / 4\pi \hat{n}_b e^2$, and $\hat{n}_b = n_b^0(r = 0)$ is the on-axis beam density. Note from Eq.(5.39) that $\hat{n}_b$ is related to the number of electrons per unit axial length of the beam, $N_b = 2\pi \int_0^\infty dr \, r n_b^0(r)$, by $\hat{n}_b = N_b / \pi r_b^2$.

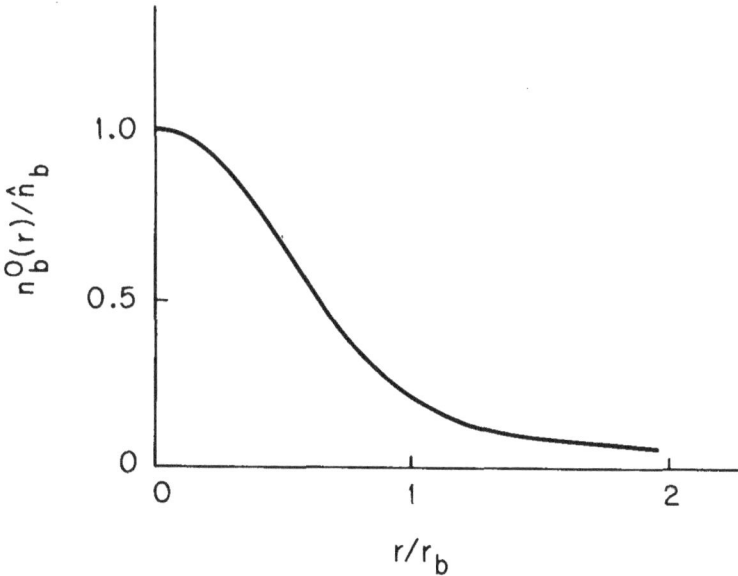

Figure 5.4. Plot of the normalized density profile $n_b^0(r)/\hat{n}_b$ versus r/r_b for the Bennett pinch equilibrium in Eq.(5.39).

Moreover, Eqs.(5.38) and (5.39) are valid whether or not there is an axial guide field $B_0 \hat{e}_z$, provided $V_{\theta b}^0(r) = 0$. In addition, it follows from Eq.(5.40) that the beam radius r_b can be large in units of λ_{Db} whenever β_b^2 is closely tuned to $(1 - f)$.

For the cold, nonrelativistic ions, radial force balance on a fluid element is given by Eq.(5.13) for $j = i$. Substituting Eqs.(5.36) and (5.39) into Eq.(5.13), it follows that

$$\omega_{ri}(r) = \omega_{ri}^{\pm}(r) = -\frac{\omega_{ci}}{2} \left\{ 1 \pm \left(1 + \frac{2\omega_{pi}^2}{\omega_{ci}^2} \frac{(1-f)}{f} \frac{1}{(1 + r^2/r_b^2)} \right)^{1/2} \right\},$$

(5.41a)

where $\omega_{pi}^2 = 4\pi f \hat{n}_b e^2/m_i$ is the on-axis ($r = 0$) ion plasma frequency-squared. If $B_0 = 0$, Eq.(5.41a) reduces to

$$\omega_{ri}(r) = \omega_{ri}^{\pm} = \mp\omega_{pi} \left(\frac{1-f}{2f} \right)^{1/2} \frac{1}{(1 + r^2/r_b^2)^{1/2}}, \qquad (5.41b)$$

which approaches zero as $r/r_b \to \infty$.

Problem 5.6   Generalize the Bennett pinch equilibrium analyzed in Sec. 5.2.4 to the case of an electron beam rotating with angular velocity $\omega_{rb} =$ const.[8] Assume nonrelativistic azimuthal motion with $r^2\omega_{rb}^2/c^2 \ll 1 - \beta_b^2 \equiv \gamma_b^{-2}$.

a. With assumptions otherwise identical to Sec. 5.2.4, show that radial force balance on a beam fluid element can be expressed as

$$-\gamma_b m_e \omega_{rb}^2 r n_b^0(r) + k_B T_b \frac{\partial}{\partial r} n_b^0(r) = -e n_b^0(r) \left[E_r(r) - \beta_b B_\theta^s(r) + \frac{1}{c}\omega_{rb} r B_0 \right].$$

(5.6.1)

b. Introduce the potentials $\phi_0(r)$ and $A_z^s(r)$, where $E_r(r) = -\partial\phi_0(r)/\partial r$ and $B_\theta^s(r) = -\partial A_z^s(r)/\partial r$. Make use of Maxwell's equations (5.4a) and (5.4b) to show that

$$\beta_b \phi_0(r) = (1 - f) A_z^s(r) . \qquad (5.6.2)$$

In the special case where $\beta_b^2 = (1 - f)$, note that Eq.(5.6.2) reduces to $\phi_0(r) = \beta_b A_z^s(r)$.

c. Integrate Eq.(5.6.1) with respect to r and show that the density profile for the beam electrons can be expressed as

$$n_b^0(r) = \hat{n}_b \exp\left\{-\left[\frac{m_e(\omega_{ce}\omega_{rb}-\gamma_b\omega_{rb}^2)r^2}{2k_BT_b} - \frac{e}{k_BT_b}\left(\frac{1-f-\beta_b^2}{1-f}\right)\phi_0(r)\right]\right\},$$

(5.6.3)

where $\hat{n}_b = n_b^0(r = 0)$ and $\omega_{ce} = eB_0/m_ec$. Equation (5.6.3) should be solved (numerically) in conjunction with Poisson's equation $r^{-1}(\partial/\partial r) \times [r\partial\phi_0/\partial r] = 4\pi e(1 - f)n_b^0(r)$ to determine the profiles for $\phi_0(r)$ and $n_b^0(r)$. Note that Eq.(5.6.3) generalizes the thermal equilibrium profile in Eq.(3.29) to the case of relativistic axial motion and a partially neutralizing ion background.

For the particular case in which $\beta_b^2 = 1 - f$ exactly, it follows that Eq.(5.6.3) gives the gaussian density profile[8]

$$n_b^0(r) = \hat{n}_b exp\left(-r^2/2r_b^2\right),$$

(5.6.4)

where $r_b^2 = k_BT_b/[m_e(\omega_{ce}\omega_{rb} - \gamma_b\omega_{rb}^2)]$.

To conclude Sec. 5.2, we reiterate that there is considerable latitude in describing the equilibrium properties of nonneutral plasmas within the framework of the macroscopic cold-fluid model based on the radial force balance equation (5.10) and the Maxwell equations (5.4a)–(5.4c). In particular, any two of the equilibrium profiles $n_j^0(r)$, $\omega_{rj}(r)$ and $\beta_{zj}(r)$ can be prescribed arbitrarily, and the remaining profile calculated self-consistently from Eq.(5.10). Of course, specific profile assumptions or constraint conditions should be guided by the particular experimental configuration that is being modeled. For the case of an intense relativistic electron beam, the constraint conditions are naturally guided by the source of beam electrons (e.g., cathode conditions and geometry).[9] For example, if the electrons are created with uniform energy and canonical angular momentum by a cylindrical cathode, then (for laminar electron flow) it is reasonable to assume that

$$H_e \equiv [\gamma_e^0(r) - 1]m_ec^2 - e\phi_0(r) = \text{const.},$$

(5.42a)

and

$$P_{\theta e} \equiv r[\gamma_e^0(r)m_eV_{\theta e}^0(r) - eA_\theta^0(r)/c] = \text{const.},$$

(5.42b)

over the radial cross section of the electron beam. Here, $A_\theta^0(r)$ is the vector potential for the axial magnetic field, i.e., $B_z(r) = r^{-1}(\partial/\partial r)[rA_\theta^0(r)]$. Constraint conditions such as those above will be used in Chapters 8 and 9, where we investigate some specialized aspects of intense beam propagation.

5.3   Electrostatic Eigenvalue Equation

Macroscopic studies of the propagation of (electrostatic) space-charge waves in a neutral plasma column[10,11] or a nonneutral plasma column[12] or electron beam[13-17] date back to the 1950's. In this section, the stability of nonrelativistic nonneutral plasma equilibria to small-amplitude electrostatic perturbations is investigated.[1-4] As in the analysis leading to Eq.(5.12), the azimuthal and axial motions of the plasma components are assumed to be nonrelativistic with $\beta_{zj}^2(r), r^2\omega_{rj}^2(r)/c^2 \ll 1$. In addition, the axial velocity of each plasma component is assumed to be independent of radius r, i.e.,

$$V_{zj}^0(r) \equiv V_j = \beta_j c = \text{const.}, \tag{5.43}$$

and the equilibrium self-magnetic field $B_\theta^s(r)$ is neglected in comparison with $E_r(r)$ in the radial force balance equation because $\beta_j^2 \ll 1$ is assumed in the present analysis. For general density profile $n_j^0(r)$, the equilibrium azimuthal velocity profile $V_{\theta j}^0(r) = \omega_{rj}(r)r$ is given by [Eqs.(5.12) and (5.13)]

$$V_{\theta j}^0(r) = \omega_{rj}^{\pm}(r)r \tag{5.44}$$

$$= -\frac{1}{2}\varepsilon_j\omega_{cj}r\left\{1 \pm \left(1 - \sum_k \frac{16\pi e_j e_k}{m_j\omega_{cj}^2 r^2}\int_o^r dr' r' n_k^0(r')\right)^{1/2}\right\},$$

which follows from radial force balance on a fluid element.

To examine the stability of various equilibrium configurations, each quantity of physical interest is expressed as its equilibrium value plus a perturbation. That is,

$$n_j(\mathbf{x}, t) = n_j^0(r) + \delta n_j(\mathbf{x}, t) ,$$

$$\mathbf{V}_j(\mathbf{x}, t) = V_{\theta j}^0(r)\hat{\mathbf{e}}_\theta + V_j\hat{\mathbf{e}}_z + \delta\mathbf{V}_j(\mathbf{x}, t) , \tag{5.45}$$

$$\mathbf{E}(\mathbf{x}, t) = E_r(r)\hat{\mathbf{e}}_r + \delta\mathbf{E}(\mathbf{x}, t),$$

$$\mathbf{B}(\mathbf{x}, t) = B_0\hat{\mathbf{e}}_z + \delta\mathbf{B}(\mathbf{x}, t),$$

where $\hat{\mathbf{e}}_r$, $\hat{\mathbf{e}}_\theta$, and $\hat{\mathbf{e}}_z$ are unit vectors in the r-, θ-, and z-directions, respectively. For small-amplitude perturbations, the evolution of $\delta n_j(\mathbf{x}, t)$, $\delta\mathbf{V}_j(\mathbf{x}, t)$, $\delta\mathbf{E}(\mathbf{x}, t)$ and $\delta\mathbf{B}(\mathbf{x}, t)$ is determined from Eqs.(2.41)–(2.46). In the electrostatic approximation, the perturbed magnetic field $\delta\mathbf{B}(\mathbf{x}, t)$ is assumed to remain negligibly small in Eqs.(2.42) and (2.43), and Eq.(2.43) is approximated by $\nabla \times \delta\mathbf{E}(\mathbf{x}, t) = 0$. This implies that the perturbed electric field $\delta\mathbf{E}(\mathbf{x}, t)$ can be expressed as the gradient of a scalar potential,

$$\delta\mathbf{E}(\mathbf{x}, t) = -\nabla\delta\phi(\mathbf{x}, t). \tag{5.46}$$

Since the motion of the plasma components is assumed to be nonrelativistic, the perturbed momentum of component j can be approximated in Eq.(2.42) by $\delta\mathbf{P}_j(\mathbf{x}, t) = m_j\delta\mathbf{V}_j(\mathbf{x}, t)$. Combining Eqs.(2.41), (2.42), (2.45), (5.45), and (5.46), the macroscopic fluid-Poisson equations for the perturbed quantities become

$$0 = \frac{\partial}{\partial t}\delta n_j(\mathbf{x}, t) + \nabla \cdot \left\{ n_j^0(r)\delta\mathbf{V}_j(\mathbf{x}, t) + \delta n_j(\mathbf{x}, t)\left[V_{\theta j}^0(r)\hat{\mathbf{e}}_\theta + V_j\hat{\mathbf{e}}_z\right] \right\}, \tag{5.47}$$

$$\frac{\partial}{\partial t}\delta\mathbf{V}_j(\mathbf{x}, t) + \left[V_{\theta j}^0(r)\hat{\mathbf{e}}_\theta + V_j\hat{\mathbf{e}}_z\right] \cdot \nabla\delta\mathbf{V}_j(\mathbf{x}, t) + \delta\mathbf{V}_j(\mathbf{x}, t) \cdot \nabla\left[V_{\theta j}^0(r)\hat{\mathbf{e}}_\theta\right]$$

$$= \frac{e_j}{m_j}\left(-\nabla\delta\phi(\mathbf{x}, t) + \frac{\delta\mathbf{V}_j(\mathbf{x}, t) \times B_0\hat{\mathbf{e}}_z}{c}\right), \tag{5.48}$$

$$\nabla^2\delta\phi(\mathbf{x}, t) = -4\pi\sum_j e_j\delta n_j(\mathbf{x}, t). \tag{5.49}$$

To determine the wave and stability properties characteristic of perturbations about equilibrium, a normal-mode approach is adopted. It is assumed that the time variation of perturbed quantities is of the form $\exp(-i\omega t)$, where the complex oscillation frequency ω is determined consistently from Eqs.(5.47)–(5.49). If $\text{Im}\,\omega > 0$, then the perturbations grow and the equilibrium configuration is *unstable*. In analyzing Eqs.(5.47)–(5.49), the perturbations are assumed to be spatially periodic in the z-direction with periodicity length L. The θ- and z-dependences of all perturbed quantities are Fourier decomposed according to

$$\delta\psi(r, \theta, z, t) = \sum_{\ell=-\infty}^{\infty} \sum_{k_z=-\infty}^{\infty} \delta\psi^\ell(r, k_z)\exp\{i(\ell\theta + k_z z - \omega t)\}, \tag{5.50}$$

where $k_z = 2\pi n/L$, and n is an integer. Substituting Eq.(5.50) into Eqs.(5.47)–(5.49), it can be shown that the Fourier amplitudes $\delta n_j^\ell(r, k_z)$, $\delta V_{rj}^\ell(r, k_z)$, etc., satisfy

$$-i\left(\omega - k_z V_j - \ell\omega_{rj}\right)\delta n_j^\ell + \frac{1}{r}\frac{\partial}{\partial r}\left(rn_j^0\delta V_{rj}^\ell\right) + \frac{i\ell n_j^0 \delta V_{\theta j}^\ell}{r} + ik_z n_j^0 \delta V_{zj}^\ell = 0,$$

(5.51)

$$-i\left(\omega - k_z V_j - \ell\omega_{rj}\right)\delta V_{rj}^\ell - \left(\varepsilon_j\omega_{cj} + 2\omega_{rj}\right)\delta V_{\theta j}^\ell = -\frac{e_j}{m_j}\frac{\partial}{\partial r}\delta\phi^\ell \quad (5.52)$$

$$-i\left(\omega - k_z V_j - \ell\omega_{rj}\right)\delta V_{\theta j}^\ell + \left[\varepsilon_j\omega_{cj} + \frac{1}{r}\frac{\partial}{\partial r}\left(r^2\omega_{rj}\right)\right]\delta V_{rj}^\ell = -\frac{e_j}{m_j}\frac{i\ell\delta\phi^\ell}{r},$$

(5.53)

$$-i\left(\omega - k_z V_j - \ell\omega_{rj}\right)\delta V_{zj}^\ell = -\frac{e_j}{m_j}ik_z\delta\phi^\ell,$$

(5.54)

$$\frac{1}{r}\frac{\partial}{\partial r}r\frac{\partial}{\partial r}\delta\phi^\ell - \frac{\ell^2}{r^2}\delta\phi^\ell - k_z^2\delta\phi^\ell = -4\pi\sum_j e_j\delta n_j^\ell,$$

(5.55)

where the (r, k_z) arguments of the perturbations have been suppressed. In Eqs.(5.51)–(5.55), $\varepsilon_j = \text{sgn}\, e_j$, $\omega_{cj} = |e_j|B_0/m_j c$, and the equilibrium angular velocity profile $\omega_{rj}(r) = V_{\theta j}^0(r)/r$ is related to the equilibrium density profile $n_j^0(r)$ by Eq.(5.44). The perturbations in density and mean fluid velocities in Eqs.(5.51)–(5.54) can be eliminated in favor of $\delta\phi^\ell(r, k_z)$. After some algebraic manipulation, Poisson's equation (5.55) for the perturbed electrostatic potential can be expressed in the form[1-3]

$$\frac{1}{r}\frac{\partial}{\partial r}\left[r\left(1 - \sum_j \frac{\omega_{pj}^2}{\nu_j^2}\right)\frac{\partial}{\partial r}\delta\phi^\ell\right] - \frac{\ell^2}{r^2}\left(1 - \sum_j \frac{\omega_{pj}^2}{\nu_j^2}\right)\delta\phi^\ell$$

$$-k_z^2\left(1 - \sum_j \frac{\omega_{pj}^2}{(\omega - k_z V_j - \ell\omega_{rj})^2}\right)\delta\phi^\ell$$

$$= -\frac{\ell\delta\phi^\ell}{r}\sum_j \frac{1}{(\omega - k_z V_j - \ell\omega_{rj})}\frac{\partial}{\partial r}\left[\frac{\omega_{pj}^2}{\nu_j^2}\left(\varepsilon_j\omega_{cj} + 2\omega_{rj}\right)\right]. \quad (5.56)$$

Here, $\omega_{pj}^2(r) \equiv 4\pi n_j^0(r)e_j^2/m_j$ is the local plasma frequency-squared, $\omega_{rj}(r) = \omega_{rj}^\pm(r)$ is the angular rotation velocity defined in Eq.(5.44), and $\nu_j^2(r,\omega)$ is defined by

$$\nu_j^2(r,\omega) \equiv \left(\omega - k_z V_j - \ell\omega_{rj}\right)^2 - \left(\varepsilon_j\omega_{cj} + 2\omega_{rj}\right)\left[\varepsilon_j\omega_{cj} + \frac{1}{r}\frac{\partial}{\partial r}\left(r^2\omega_{rj}\right)\right] .$$

(5.57)

The eigenvalue equation (5.56) is valid for arbitrary $\omega_{pj}^2(r)$ and $\omega_{rj}(r)$ consistent with Eq.(5.44). Operationally, the procedure is to solve Eq.(5.56) for $\delta\phi^\ell(r, k_z)$ and ω as an eigenvalue problem. The solution to Eq.(5.56) is accessible analytically for certain simple density profiles. As a first application of Eq.(5.56), the case where the density of each plasma component is uniform in the column interior is considered in Sec. 5.4. The stability of annular (hollow) density profiles is analyzed in Chapter 6 with particular emphasis on the diocotron instability.

Problem 5.7   Consider an electrically neutral plasma column satisfying $\sum_j n_j^0(r)e_j = 0$ and $V_{\theta j}^0(r) = 0$, which corresponds to no rotation of the plasma components. Make use of Eqs.(5.51)–(5.55) to derive the corresponding eigenvalue equation for $\delta\phi^\ell$. Show that the eigenvalue equation is identical in form to Eq.(5.56) supplemented by the algorithm

$$\omega - k_z V_z - \ell\omega_{rj} \rightarrow \omega - k_z V_z ,$$

$$\varepsilon_j\omega_{cj} + 2\omega_{rj} \rightarrow \varepsilon_j\omega_{cj} ,$$

$$\nu_j^2(r,\omega) \rightarrow \left(\omega - k_z V_z\right)^2 - \omega_{cj}^2 .$$

(5.7.1)

5.4   Dispersion Relation for a Uniform Density Plasma Column

As a first application of the electrostatic eigenvalue equation (5.56), consider perturbations about the uniform density profile (Fig. 5.5)

$$n_j^0(r) = \begin{cases} \hat{n}_j = \text{const.} , & 0 \leq r < r_b , \\ 0 , & r_b < r \leq b , \end{cases}$$

(5.58)

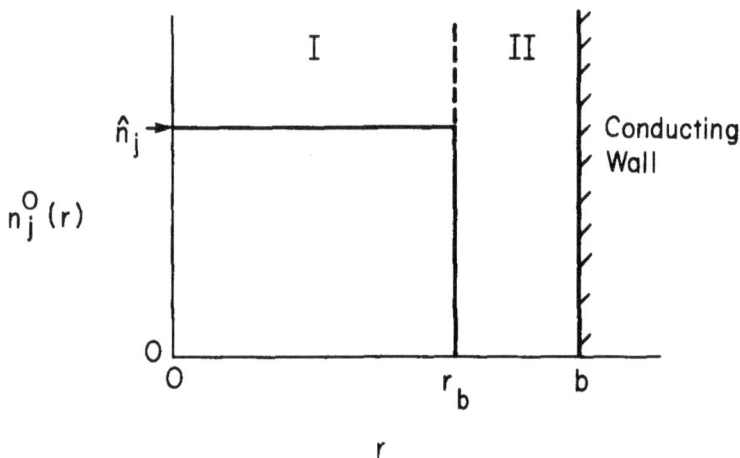

Figure 5.5.   Plot versus r of the uniform density profile $n_j^0(r)$ assumed in Eq.(5.58). A perfectly conducting wall is located at radius $r = b$.

where $r = b$ is the radial location of a perfectly conducting wall. The rotation velocity ω_{rj} is then given by [Eq.(5.15)]

$$\omega_{rj} = \omega_{rj}^{\pm} = -\frac{\varepsilon_j \omega_{cj}}{2}\left\{1 \pm \left(1 - \sum_k \frac{8\pi e_j \hat{n}_k e_k}{m_j \omega_{cj}^2}\right)^{1/2}\right\}\,, \qquad (5.59)$$

where $\omega_{cj} = |e_j|B_0/m_j c$ and $\varepsilon_j = \operatorname{sgn} e_j$. Because $\omega_{rj} = \text{const.}$, it follows from Eq.(5.57) that

$$\nu_j^2(r,\omega) = (\omega - k_z V_j - \ell\omega_{rj})^2 - \left(\omega_{rj}^+ - \omega_{rj}^-\right)^2\,, \qquad (5.60)$$

where use has been made of $(\varepsilon_j \omega_{cj} + 2\omega_{rj})^2 = (\omega_{rj}^+ - \omega_{rj}^-)^2$. Moreover, for the rectangular density profile in Eq.(5.58),

$$\frac{\partial}{\partial r}n_j^0(r) = -\hat{n}_j \delta(r - r_b)\,, \qquad (5.61)$$

and Eq.(5.56) reduces to[1-3]

$$\frac{1}{r}\frac{\partial}{\partial r}\left[r\left(1 - \sum_j \frac{\omega_{pj}^2(r)}{(\omega - k_z V_j - \ell\omega_{rj})^2 - (\omega_{rj}^+ - \omega_{rj}^-)^2}\right)\frac{\partial}{\partial r}\delta\phi^\ell\right]$$

$$-\frac{\ell^2}{r^2}\left[1 - \sum_j \frac{\omega_{pj}^2(r)}{(\omega - k_z V_j - \ell\omega_{rj})^2 - (\omega_{rj}^+ - \omega_{rj}^-)^2}\right]\delta\phi^\ell$$

$$-k_z^2\left[1 - \frac{\omega_{pj}^2(r)}{(\omega - k_z V_j - \ell\omega_{rj})^2}\right]\delta\phi^\ell \tag{5.62}$$

$$= \frac{\ell\delta\phi^\ell}{r}\sum_j \frac{(\varepsilon_j\omega_{cj} + 2\omega_{rj})}{(\omega - k_z V_j - \ell\omega_{rj})}\frac{\omega_{pj}^2}{(\omega - k_z V_j - \ell\omega_{rj})^2 - (\omega_{rj}^+ - \omega_{rj}^-)^2}\delta(r - r_b).$$

Here, $\omega_{pj}^2(r) = 4\pi n_j^0(r)e_j^2/m_j$ on the left-hand side of Eq.(5.62), and $\omega_{pj}^2 = 4\pi\hat{n}_j e_j^2/m_j$ on the right-hand side of Eq.(5.62). The term in Eq.(5.62) proportional to $\delta(r - r_b)$ corresponds to a surface-charge perturbation localized at $r = r_b$.

The eigenvalue equation (5.62) can be solved separately in the two regions in Fig. 5.5 corresponding to: Region I $(0 \le r < r_b)$, and Region II $(r_b < r \le b)$. Inside the plasma column $(0 \le r < r_b)$ Eq.(5.62) reduces to

$$\frac{1}{r}\frac{\partial}{\partial r}r\frac{\partial}{\partial r}\delta\phi_I^\ell - \frac{\ell^2}{r^2}\delta\phi_I^\ell + T^2\delta\phi_I^\ell = 0 , \tag{5.63}$$

whereas in the vacuum region $(r_b < r \le b)$ Eq.(5.62) reduces to

$$\frac{1}{r}\frac{\partial}{\partial r}r\frac{\partial}{\partial r}\delta\phi_{II}^\ell - \frac{\ell^2}{r^2}\delta\phi_{II}^\ell - k_z^2\delta\phi_{II}^\ell = 0 . \tag{5.64}$$

Here, the coefficient T^2 is defined by

$$T^2 \equiv -k_z^2\left[1 - \sum_j \frac{\omega_{pj}^2}{(\omega - k_z V_j - \ell\omega_{rj})^2}\right]$$

$$\times \left[1 - \sum_j \frac{\omega_{pj}^2}{(\omega - k_z V_j - \ell\omega_{rj})^2 - (\omega_{rj}^+ - \omega_{rj}^-)^2}\right]^{-1} , \tag{5.65}$$

where $\omega_{pj}^2 = 4\pi\hat{n}_j e_j^2/m_j = \text{const}$. The solutions to Eqs.(5.63) and (5.64) that remain finite at $r = 0$, vanish at $r = b$ (because $\delta E_\theta^\ell = 0$ at the conducting wall), and are continuous at the surface of the plasma column $(r = r_b)$ are given by

$$\delta\phi_I^\ell = AJ_\ell(Tr) , \quad 0 \le r < r_b , \qquad (5.66)$$

and

$$\delta\phi_{II}^\ell = AJ_\ell(Tr_b)\frac{[I_\ell(k_z r)K_\ell(k_z b) - K_\ell(k_z r)I_\ell(k_z b)]}{[I_\ell(k_z r_b)K_\ell(k_z b) - K_\ell(k_z r_b)I_\ell(k_z b)]} , \quad r_b < r \le b. \quad (5.67)$$

In Eqs.(5.66) and (5.67), A is a constant, $J_\ell(x)$ is the Bessel function of the first kind of order ℓ , and $I_\ell(x)$ and $K_\ell(x)$ are modified Bessel functions of order ℓ.

The remaining boundary condition relating $\delta\phi_I^\ell$ and $\delta\phi_{II}^\ell$ is obtained by multiplying the eigenvalue equation (5.62) by r, integrating across the surface of the plasma column from $r = r_b(1 - \varepsilon)$ to $r = r_b(1 + \varepsilon)$, and taking the limit $\varepsilon \to 0_+$. We obtain

$$r_b\left[\frac{\partial}{\partial r}\delta\phi_{II}^\ell\right]_{r=r_b} - r_b\left[1 - \sum_j \frac{\omega_{pj}^2}{(\omega - k_z V_j - \ell\omega_{rj})^2 - (\omega_{rj}^+ - \omega_{rj}^-)^2}\right]\left[\frac{\partial}{\partial r}\delta\phi_I^\ell\right]_{r=r_b}$$

$$= \ell[\delta\phi^\ell]_{r=r_b}\sum_j \frac{(\varepsilon_j\omega_{cj} + 2\omega_{rj})}{(\omega - k_z V_j - \ell\omega_{rj})}\frac{\omega_{pj}^2}{(\omega - k_z V_j - \ell\omega_{rj})^2 - (\omega_{rj}^+ - \omega_{rj}^-)^2} .$$

$$(5.68)$$

Equation (5.68) relates the discontinuity in the perturbed radial electric field at $r = r_b$ to the perturbed charge density at the surface of the plasma column. Substituting Eqs.(5.66) and (5.67) into Eq.(5.68), the condition for a nontrivial solution $(A \ne 0)$ can be expressed as[1-3]

$$\frac{1}{g_\ell} - \left[1 - \sum_j \frac{\omega_{pj}^2}{(\omega - k_z V_j - \ell\omega_{rj})^2 - (\omega_{rj}^+ - \omega_{rj}^-)^2}\right]Tr_b\frac{J_\ell'(Tr_b)}{J_\ell(Tr_b)}$$

$$= \ell\sum_j \frac{(\varepsilon_j\omega_{cj} + 2\omega_{rj})}{(\omega - k_z V_j - \ell\omega_{rj})}\frac{\omega_{pj}^2}{(\omega - k_z V_j - \ell\omega_{rj})^2 - (\omega_{rj}^+ - \omega_{rj}^-)^2} . \quad (5.69)$$

Here, the geometric factor g_ℓ is defined by

$$g_\ell = (k_z r_b)^{-1} \frac{K_\ell(k_z b) I_\ell(k_z r_b) - K_\ell(k_z r_b) I_\ell(k_z b)}{K_\ell(k_z b) I'_\ell(k_z r_b) - K'_\ell(k_z r_b) I_\ell(k_z b)} . \qquad (5.70)$$

The "prime" notation in Eqs.(5.69) and (5.70) denotes differentiation with respect to the complete argument, e.g., $J'_\ell(Tr_b) = [dJ_\ell(x)/dx]_{x=Tr_b}$.

Equation (5.69) is the *dispersion relation* for electrostatic waves in a cold, uniform-density, nonneutral plasma column. It relates the (complex) eigenfrequency ω to the axial wavenumber k_z, the azimuthal mode number ℓ, and properties of the equilibrium configuration (such as r_b/b, ω_{pj}^2, ω_{rj}, V_j, ω_{cj}, etc.). Generally speaking, the relative azimuthal rotation (Sec. 5.6) and/or the relative axial motion (Sec. 5.7) of the plasma components can provide the free energy to drive various instabilities which correspond to solutions to Eq.(5.69) with $\text{Im}\,\omega > 0$. Moreover, Eq.(5.69) can be simplified in various limiting regimes, which include: (a) $r_b \simeq b$ (nonneutral plasma-filled waveguide), (b) $k_z^2 b^2 \ll 1$ (long axial wavelengths), and (c) $r_b \ll b$ (limit of a thin plasma column). For example, for $k_z^2 b^2 \ll 1$, the geometric factor g_ℓ defined in Eq.(5.70) can be approximated by

$$g_\ell = \begin{cases} -\ln\left(\dfrac{b}{r_b}\right) , & \text{for } \ell = 0 , \\[3mm] -(\ell)^{-1}\dfrac{1 - (r_b/b)^{2\ell}}{1 + (r_b/b)^{2\ell}} , & \text{for } \ell \neq 0 . \end{cases} \qquad (5.71)$$

This, and other approximations will be used in Secs. 5.5–5.7 to investigate the electrostatic stability properties predicted by Eq.(5.69).

Problem 5.8 Consider a uniform-density, nonneutral plasma-filled waveguide with $r_b = b$ in Fig. 5.5. Show that the solution to the eigenvalue equation (5.62) is given by

$$\delta\phi^\ell = AJ_\ell(Tr) , \qquad 0 \leq r \leq b , \qquad (5.8.1)$$

and that the corresponding dispersion relation is

$$J_\ell(Tb) = 0 . \qquad (5.8.2)$$

Here, $A = \text{const.}$, and the coefficient T^2 is defined in Eq.(5.65).

5.5   Nonneutral Plasma-Filled Waveguide

5.5.1   Electrostatic Dispersion Relation

As a first application of the dispersion relation (5.69), consider a non-neutral plasma-filled waveguide in which the vacuum gap in Fig. 5.5 is sufficiently narrow $(b - r_b \ll b)$ that the approximation

$$r_b = b \tag{5.72}$$

can be made in Eq.(5.69). From Eqs.(5.70) and (5.72), it follows that the geometric factor $g_\ell \to 0$, and therefore the dispersion relation (5.69) reduces to[1]

$$J_\ell(Tb) = 0 \tag{5.73}$$

in the limit of a plasma-filled waveguide. Equation (5.73) also follows directly by solving the eigenvalue equation (5.62) for the case where $r_b = b$ (Problem 5.8). Equation (5.73) gives $T^2 b^2 = p_{\ell m}^2$, where $p_{\ell m}$ is the m'th zero of $J_\ell(p_{\ell m}) = 0$. Introducing the effective perpendicular wavenumber defined by $k_\perp^2 = p_{\ell m}^2/b^2$, the dispersion relation $T^2 = k_\perp^2$ can be expressed as

$$0 = 1 - \frac{k_\perp^2}{k^2} \sum_j \frac{\omega_{pj}^2}{(\omega - k_z V_j - \ell\omega_{rj})^2 - (\omega_{rj}^+ - \omega_{rj}^-)^2}$$

$$- \frac{k_z^2}{k^2} \sum_j \frac{\omega_{pj}^2}{(\omega - k_z V_j - \ell\omega_{rj})^2} , \tag{5.74}$$

where $k^2 \equiv k_z^2 + k_\perp^2$, $\omega_{pj}^2 = 4\pi\hat{n}_j e_j^2/m_j$, and use has been made of the definition of T^2 in Eq.(5.65). Here, the angular rotation velocity is $\omega_{rj} = \omega_{rj}^+$ or $\omega_{rj} = \omega_{rj}^-$, where $\omega_{rj}^\pm$ is defined in Eq.(5.59). Moreover, $(\omega_{rj}^+ - \omega_{rj}^-)^2$ can be expressed as

$$\left(\omega_{rj}^+ - \omega_{rj}^-\right)^2 = \omega_{cj}^2 \left(1 - \sum_k \frac{8\pi e_j \hat{n}_k e_k}{m_j \omega_{cj}^2}\right). \tag{5.75}$$

In the limit of a neutral plasma with $\sum_k \hat{n}_k e_k = 0$ and $\omega_{rj} = \omega_{rj}^- = 0$, it follows from Eq.(5.75) that $(\omega_{rj}^+ - \omega_{rj}^-)^2 = \omega_{cj}^2$. In this case, Eq.(5.74) reduces directly to the familiar dispersion relation for the Trivelpiece-Gould modes[10] in a charge-neutral plasma-filled waveguide. Conversely, the non-neutral dispersion relation (5.74) can be recovered by simply making the replacements

$$\omega \to \omega - \ell\omega_{rj} \ , \quad \omega_{cj}^2 \to \left(\omega_{rj}^+ - \omega_{rj}^-\right)^2 \tag{5.76}$$

in the corresponding neutral-plasma dispersion relation. The replacement $\omega \to \omega - \ell\omega_{rj}$ is expected since component j is rotating with angular velocity ω_{rj} in the nonneutral case. The replacement $\omega_{cj}^2 \to (\omega_{rj}^+ - \omega_{rj}^-)^2$ is expected since the particles gyrate with angular frequency $|\omega_{rj}^+ - \omega_{rj}^-|$ in the rotating frame (see the discussion in Sec. 3.2).

5.5.2   Trivelpiece-Gould Modes in a Pure Ion Plasma

For a nonneutral plasma with nonzero self-field parameter $s_j = (8\pi m_j c^2/e_j B_0^2)\sum_k \hat{n}_k e_k$, the detailed wave properties predicted by Eq.(5.74) can differ substantially from the case of a neutral plasma. For example, for a pure ion plasma ($\hat{n}_e = 0$), Eq.(5.74) can be solved to give

$$(\omega - \ell\omega_{ri} - k_z V_i)^2 \tag{5.77}$$

$$= \frac{1}{2}\left[\omega_{pi}^2 + (\omega_{ri}^+ - \omega_{ri}^-)^2\right]\left\{1 \pm \left[1 - \frac{k_z^2/k_\perp^2}{1 + k_z^2/k_\perp^2}\frac{4\omega_{pi}^2(\omega_{ri}^+ - \omega_{ri}^-)^2}{\left[\omega_{pi}^2 + (\omega_{ri}^+ - \omega_{ri}^-)^2\right]^2}\right]^{1/2}\right\},$$

where $\omega_{ri} = \omega_{ri}^-$ or $\omega_{ri} = \omega_{ri}^+$, and $(\omega_{ri}^+ - \omega_{ri}^-)^2 = \omega_{ci}^2(1 - 2\omega_{pi}^2/\omega_{ci}^2)$. For $k_z^2 \gg k_\perp^2$, the two solutions in Eq.(5.77) asymptote at ω_{pi}^2 and at $(\omega_{ri}^+ - \omega_{ri}^-)^2$. The analogue of the oscillations in Eq.(5.77) of course also exists for a pure electron plasma ($\hat{n}_i = 0$).[1] What is most striking about Eq.(5.77), however, is that the low-frequency oscillations (near the ion plasma frequency) in a pure ion plasma with $\hat{n}_e = 0$ do not resemble the plasma oscillations predicted by the cold-plasma dispersion relation (5.74) for the case of a charge-neutral plasma with $\hat{n}_e = Z_i\hat{n}_i$. For example, for $\sum_k \hat{n}_k e_k = 0$, and $\omega_{rj} = 0$ and $V_j = 0$, the plasma-oscillation branch in Eq.(5.74) gives $\omega^2 = \omega_{pe}^2 + \omega_{pi}^2 \simeq \omega_{pe}^2$ for $k_z^2/k_\perp^2 \gg 1$, which corresponds to (high-frequency) electron plasma oscillations. The solutions in Eq.(5.77) for a pure ion plasma are plotted versus $k_z/k_\perp$ in Fig. 5.6 for self-field parameter $s_i = 2\omega_{pi}^2/\omega_{ci}^2 = 1/2$, which corresponds to $(\omega_{ri}^+ - \omega_{ri}^-)^2 = 2\omega_{pi}^2$. For $k_z^2/k_\perp^2 \gg 1$, we note from Fig. 5.6 that the two solutions asymptote at ω_{pi}^2 and $(\omega_{ri}^+ - \omega_{ri}^-)^2$.

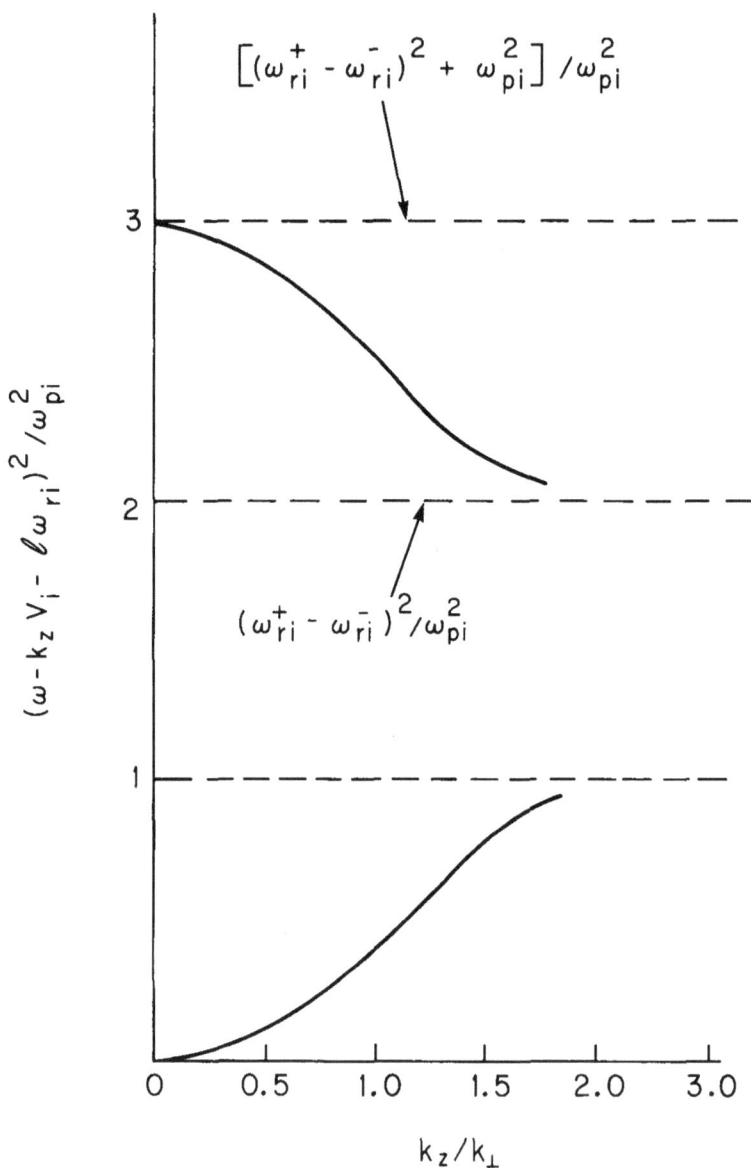

Figure 5.6. Plot of $(\omega - k_z V_i - \ell\omega_{ri})^2/\omega_{pi}^2$ versus $k_z/k_\perp$ obtained from Eq.(5.77) for $2\omega_{pi}^2/\omega_{ci}^2 = 1/2$ and $(\omega_{ri}^+ - \omega_{ri}^-)^2 = 2\omega_{pi}^2$.

The nonneutral analogue of the Trivelpiece-Gould modes[10] has been investigated experimentally for both pure electron plasmas[18] and pure ion plasmas.[19] We summarize here the experiments by Dimonte[19] on lithium ion plasmas with $\hat{n}_e = 0$. Figure 5.7 is a schematic of the experimental set-up. A heated, β-eucryptite source emits ^{7}Li$^+$ ions which form a 115-cm-long nonneutral plasma column confined radially by an axial magnetic field $B_0\hat{e}_z$. The grounded conducting wall in Fig. 5.7 is located at $b = 3.75$ cm. Ion plasma waves (corresponding to the lower branch in Fig. 5.6) are launched by an isolated section of the drift tube. Typical experimental results[19] are illustrated in Fig. 5.8 for magnetic field $B_0 = 1.5$ kG, central ion density $\hat{n}_i = 7.3 \times 10^6$ cm^{-3}, ion temperature $T_i = 1.2$ eV, and axial drift velocity $V_i = 0$. The radial density profile is approximately gaussian, $n_i^0(r) = \hat{n}_i \exp(-r^2/r_b^2)$, with characteristic column diameter $2r_b \simeq 3.2$ cm, which is much larger than the thermal ion Debye length $\lambda_{Di} = (k_B T_i/4\pi\hat{n}_i e^2)^{1/2} = 3$ mm. The (on-axis) value of $s_i = 2\omega_{pi}^2/\omega_{ci}^2$ is $s_i = 0.85$, with ion plasma frequency $\omega_{pi} = 1.35 \times 10^6$ s^{-1}. (This value of ω_{pi} is indicated by the arrow on the vertical axis in Fig. 5.8.) The measured values[19] of wave frequency ω versus normalized axial wavenumber $k_z b$ are indicated by the dots in Fig. 5.8 for wave excitations with azimuthal mode number $\ell = 0$. The solid curve

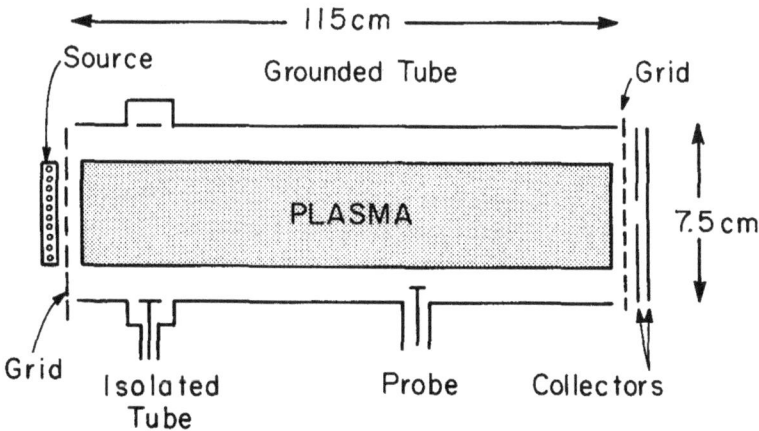

Figure 5.7. Schematic of experimental set-up used to investigate ion plasma waves in a ^{7}Li$^+$ plasma column [G. Dimonte, Phys. Rev. Lett. **46**, 26 (1981)].

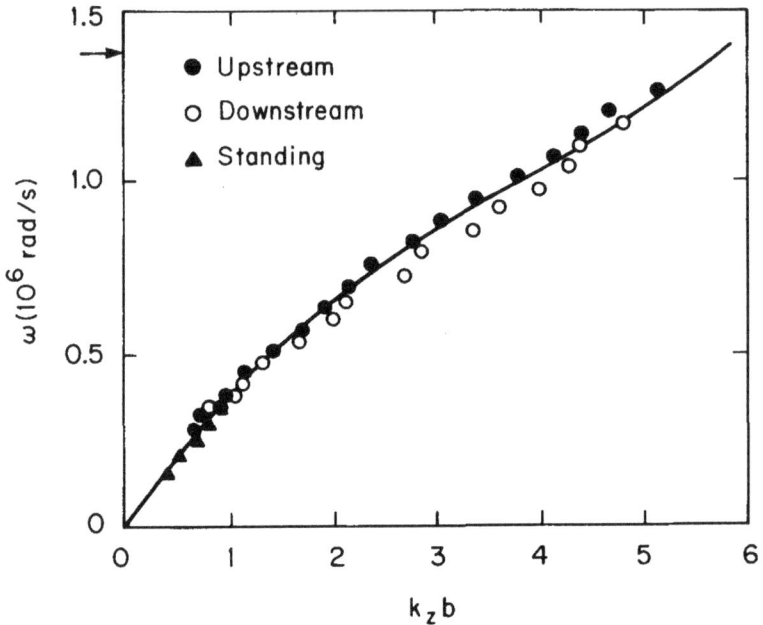

Figure 5.8.   The measured values of wave frequency ω versus $k_z b$ for $\ell = 0$ are indicated by the dots, and the solid curve corresponds to theory. The system parameters are $\hat{n}_i = 7.3 \times 10^6 \mathrm{cm}^{-3}$, $B_0 = 1.5$ kG, and the self-field parameter is $s_i = 2\omega_{pi}^2/\omega_{ci}^2 = 0.85$ [G. Dimonte, Phys. Rev. Lett. **46**, 26 (1981)].

in Fig. 5.8 is the theoretically-calculated dispersion curve,[19] taking into account the radial variation of the density profile $n_i^0(r)$. The agreement between theory and experiment is very good. While Fig. 5.8 corresponds to axial drift velocity $V_i = 0$, similar good agreement between theory and experiment is obtained when the grid in Fig. 5.7 is biased so that $V_i \neq 0$.

We reiterate that the low-frequency ion plasma oscillations characteristic of a pure ion plasma with $\hat{n}_e = 0$ (Figs. 5.6 and 5.8) do not resemble the plasma oscillations predicted by the cold-plasma dispersion relation (5.74) for the case of a charge-neutral plasma with $\hat{n}_e = Z_i \hat{n}_i$ (see Problem 5.9).

Problem 5.9   For comparison with the dispersion relation (5.77) for a pure ion plasma ($\hat{n}_e = 0$), consider the dispersion relation (5.74) for the case of a two-component ($j = e, i$) charge-neutral plasma with $\sum_k \hat{n}_k e_k = 0$, $\omega_{rj} = \omega_{rj}^- = 0$, and $V_j = 0$ (no drift motion along the magnetic field $B_0 \hat{e}_z$).

a. Show that Eq.(5.74) reduces to

$$0 = 1 - \frac{k_\perp^2}{k^2}\left(\frac{\omega_{pe}^2}{\omega^2 - \omega_{ce}^2} + \frac{\omega_{pi}^2}{\omega^2 - \omega_{ci}^2}\right) - \frac{k_z^2}{k^2}\frac{\omega_{pe}^2 + \omega_{pi}^2}{\omega^2} , \tag{5.9.1}$$

where $k^2 = k_z^2 + k_\perp^2$.

b. For nonzero k_z^2 and $k_\perp^2$, show that Eq.(5.9.1) supports six real solutions $(\pm\omega_1, \pm\omega_2, \pm\omega_3)$ located in the frequency intervals

$$0 < \omega_1^2 < \omega_{ci}^2 ,$$

$$\omega_{ci}^2 < \omega_2^2 < \omega_{ce}^2 ,$$

$$\omega_{ce}^2 < \omega_3^2 . \tag{5.9.2}$$

In Parts (c)-(e), assume $\omega_{ci}^2 < \omega_{pe}^2 < \omega_{ce}^2$.

c. For short axial wavelengths with $k_z^2/k_\perp^2 \gg 1$, show that the solutions to Eq.(5.9.1) asymptote at

$$\omega_1^2 = \omega_{ci}^2 ,$$

$$\omega_2^2 = \omega_{pe}^2 + \omega_{pi}^2 ,$$

$$\omega_3^2 = \omega_{ce}^2 . \tag{5.9.3}$$

d. For long axial wavelengths with $k_z^2/k_\perp^2 \ll 1$, show that the solutions to Eq.(5.9.1) reduce (approximately) to

$$\omega_1^2 = 0 ,$$

$$\omega_2^2 = \omega_{ci}^2 + \omega_{LH}^2 ,$$

$$\omega_3^2 = \omega_{ce}^2 + \omega_{pe}^2 \equiv \omega_{UH}^2 , \tag{5.9.4}$$

where $\omega_{LH}^2 \equiv \omega_{pi}^2/(1 + \omega_{pe}^2/\omega_{ce}^2)$, and use has been made of $m_e/m_i \ll 1$. Here ω_{UH} and ω_{LH} are known as the upper and lower hybrid frequencies, respectively, in a neutral plasma.[20,21]

e. Make use of Eqs.(5.9.2)-(5.9.4) to sketch ω^2 versus $k_z^2/k_\perp^2$ for the three solutions to Eq.(5.9.1).

5.5.3   Modified Two-Stream Instability

The dispersion relation (5.74) for a plasma-filled waveguide supports unstable solutions ($\text{Im}\,\omega > 0$) corresponding to the two-stream instability (Sec. 5.7) when there is a relative axial drift of the plasma components (different V_j for different components). This is true for both neutral

$(\sum_k \hat{n}_k e_k = 0)$ and nonneutral $(\sum_k \hat{n}_k e_k \neq 0)$ plasmas. An important property of Eq.(5.74) in the nonneutral case, however, is that the relative rotational motion (induced by the self-electric field) of the plasma components can also provide the free energy to drive a rotating two-stream instability.[1-3] To illustrate this point, consider Eq.(5.74) for the case of a two-component nonneutral plasma consisting of electrons and singly-ionized ions that provide partial charge neutralization ($\hat{n}_i = f\hat{n}_e$). It is further assumed that there is no axial drift of the plasma components ($V_e = 0 = V_i$), and that the electron and ion fluids are rotating with "slow" rotation velocities

$$\omega_{re} = \omega_{re}^- = \frac{1}{2}\omega_{ce}\left\{ 1 - \left(1 - \frac{2\omega_{pe}^2}{\omega_{ce}^2}(1-f) \right)^{1/2} \right\} , \qquad (5.78)$$

and

$$\omega_{ri} = \omega_{ri}^- = -\frac{1}{2}\omega_{ci}\left\{ 1 - \left(1 + \frac{2\omega_{pi}^2}{\omega_{ci}^2}\frac{1-f}{f} \right)^{1/2} \right\} , \qquad (5.79)$$

which follow from Eqs.(5.16) and (5.17). For a neutral plasma column with $f = 1$, note from Eqs.(5.78) and (5.79) that $\omega_{re}^- = 0 = \omega_{ri}^-$. For present purposes, it is also assumed that the perturbations have intermediate frequency satisfying

$$|\omega - \ell\omega_{re}^-|^2 \ll \left(\omega_{re}^+ - \omega_{re}^- \right)^2 ,$$

$$|\omega - \ell\omega_{ri}^-|^2 \gg \left(\omega_{ri}^+ - \omega_{ri}^- \right)^2 . \qquad (5.80)$$

The first inequality in Eq.(5.80) requires that the electron density be sufficiently far below the Brillouin density limit $[(2\omega_{pe}^2/\omega_{ce}^2)(1-f) < 1]$ that the electron fluid is strongly magnetized. Substituting Eq.(5.80) and $V_i = 0 = V_e$ into Eq.(5.74), we obtain the approximate dispersion relation

$$0 = 1 + \frac{k_\perp^2}{k^2}\frac{\omega_{pe}^2}{(\omega_{re}^+ - \omega_{re}^-)^2} - \frac{\omega_{pi}^2}{(\omega - \ell\omega_{ri}^-)^2} - \frac{k_z^2}{k^2}\frac{\omega_{pe}^2}{(\omega - \ell\omega_{re}^-)^2} , \qquad (5.81)$$

where use has been made of $k^2 = k_\perp^2 + k_z^2$.

Equation (5.81) has the classic form of the dispersion relation for the modified two-stream instability,[21] with the free energy source provided by the relative rotation $\omega_{re}^- - \omega_{ri}^-$ of the electron and ion components. It is

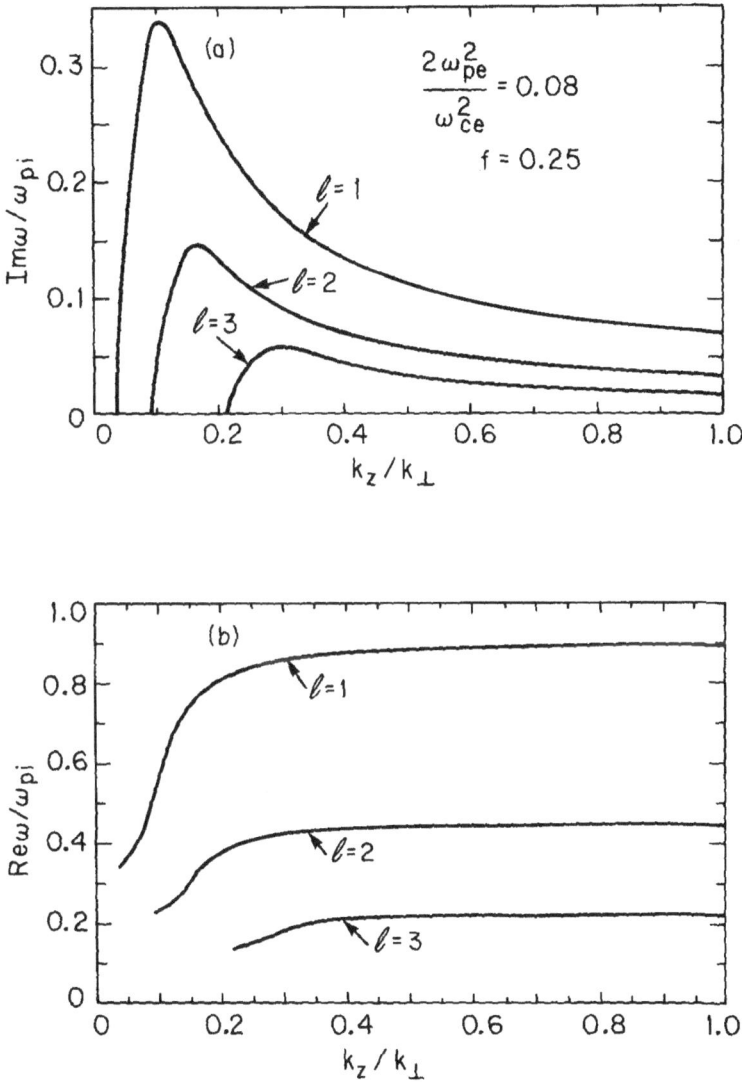

Figure 5.9. Plots versus $k_z/k_\perp$ of (a) the normalized growth rate $(\mathrm{Im}\,\omega)/\omega_{pi}$, and (b) the normalized real frequency $(\mathrm{Re}\,\omega)/\omega_{pi}$ of the modified two-stream instability determined from Eq.(5.83) for azimuthal mode numbers $\ell = 1, 2, 3$. Here, $2\omega_{pe}^2/\omega_{ce}^2 = 0.08$, $f = 0.25$, and $m_i/m_e = 1836$ are assumed, and $(\mathrm{Re}\,\omega)/\omega_{pi}$ is plotted only for the range of $k_z/k_\perp$ corresponding to instability.

convenient to introduce the effective lower hybrid frequency ω_{LH} defined by

$$\omega_{LH}^2 = \omega_{pi}^2 \left[1 + \frac{k_\perp^2}{k^2} \frac{\omega_{pe}^2}{(\omega_{re}^+ - \omega_{re}^-)^2} \right]^{-1} . \tag{5.82}$$

The dispersion relation (5.81) can then be expressed in the equivalent form

$$0 = 1 - \frac{\omega_{LH}^2}{(\omega - \ell\omega_{ri}^-)^2} - \frac{\mu\omega_{LH}^2}{(\omega - \ell\omega_{re}^-)^2} , \tag{5.83}$$

where $\mu \equiv (k_z^2/k^2)(m_i/fm_e)$. Note that μ can cover the range from $\mu \ll 1$ to $\mu \gg 1$, depending on the value of k_z^2/k^2. It can be shown from Eq.(5.83) that the necessary and sufficient condition for instability (Im $\omega > 0$) when $f \neq 0$ is given by (see Problem 5.13)

$$0 < \ell^2 \left(\omega_{re}^- - \omega_{ri}^- \right)^2 < \omega_{LH}^2 \left(1 + \mu^{1/3} \right)^3 . \tag{5.84}$$

For specified mode number ℓ , and specified values of f, $\omega_{pe}^2/\omega_{ce}^2$ and m_e/m_i, the right-most inequality in Eq.(5.84) determines the range of $\mu = (k_z^2/k^2)(m_i/fm_e)$ corresponding to instability. Evidently, Eq.(5.84) is easiest to satisfy when $\mu \gg 1$.

Typical numerical results obtained from Eqs.(5.83) and (5.84) are illustrated in Fig. 5.9 for the case of a low-density nonneutral plasma with $2\omega_{pe}^2/\omega_{ce}^2 = 0.08$ and fractional charge neutralization $f = 0.25$. Here, the normalized growth rate (Im $\omega)/\omega_{pi}$ and real oscillation frequency (Re $\omega)/\omega_{pi}$ are plotted versus $k_z/k_\perp$ for several values of azimuthal mode number ℓ. Note from Fig. 5.9 that there is a threshold value of $k_z/k_\perp$ for onset of the instability, and that maximum growth occurs for $k_z^2/k_\perp^2 \gg 1$. For specified system parameters, the inequality in Eq.(5.84) limits the number of ℓ-values corresponding to instability. For a low-density nonneutral plasma with $s_e = (2\omega_{pe}^2/\omega_{ce}^2)(1-f) \ll 1$ and $\omega_{re}^- \gg \omega_{ri}^-$, Eq.(5.84) gives $0 < \ell^2 < (8/s_e)[1/(1-f)]$ when $k_z^2/k_\perp^2 \gg 1$, which can be used to determine the range of unstable ℓ-values.

Problem 5.10   Consider the dispersion relation (5.81) and the criterion for the modified two-stream instability in Eq.(5.84) for the case of a low-density nonneutral plasma with $(2\omega_{pe}^2/\omega_{ce}^2)(1-f) \ll 1$ and $\omega_{re}^- \gg \omega_{ri}^-$.

a. For $k_z^2/k_\perp^2 \gg 1$, show that the range of unstable ℓ-values determined from Eq.(5.84) can be approximated by

$$0 < \ell^2 < \frac{8}{s_e(1-f)} \, , \tag{5.10.1}$$

where $s_e = (2\omega_{pe}^2/\omega_{ce}^2)(1-f)$ is the self-field parameter.

b. For $k_z^2/k_\perp^2 \gg 1$, show that the maximum growth rate of the instability, which occurs for the largest ℓ-values in Eq.(5.10.1), can be approximated by (see Problem 5.13)

$$(\mathrm{Im}\,\omega)_{\max} = \frac{3^{1/2}}{2} \left(\frac{fm_e}{m_i}\right)^{1/3} \omega_{pe} \, . \tag{5.10.2}$$

5.6 Stabililty Properties for Flute Perturbations

5.6.1 Electrostatic Dispersion Relation

In this section, the electrostatic dispersion relation (5.69) is considered for the case of flute perturbations with

$$k_z = 0 \, . \tag{5.85}$$

The plasma components have uniform density $\hat{n}_j$ out to radius $r = r_b < b$ (Fig. 5.5). For $k_z = 0$, Eq.(5.69) does not support normal-mode solutions when $\ell = 0$. For $\ell \neq 0$ and $k_z = 0$, use is made of Eq.(5.71) to simplify the geometric factor g_ℓ in Eq.(5.69). Taylor expanding $xJ_\ell'(x)/J_\ell(x) \simeq \ell[1 - x^2/2\ell(\ell+1) + \cdots]$ for $|x| \ll 1$, the dispersion relation (5.69) reduces to

$$\ell\frac{(r_b/b)^{2\ell}+1}{(r_b/b)^{2\ell}-1} - \ell\left[1 - \sum_j \frac{\omega_{pj}^2}{(\omega - \ell\omega_{rj})^2 - (\omega_{rj}^+ - \omega_{rj}^-)^2}\right]$$

$$= \ell\sum_j \frac{\omega_{pj}^2(\varepsilon_j\omega_{cj} + 2\omega_{rj})}{(\omega - \ell\omega_{rj})[(\omega - \ell\omega_{rj})^2 - (\omega_{rj}^+ - \omega_{rj}^-)^2]} \tag{5.86}$$

for $k_z = 0$ and $\ell \neq 0$. Here, $\omega_{pj}^2 = 4\pi\hat{n}_j e_j^2/m_j$, $\omega_{rj} = \omega_{rj}^\pm$ is the angular rotation velocity of component j, and $\omega_{rj}^\pm$ is defined in Eq.(5.59) for a uniform-density plasma column. Rearranging terms in Eq.(5.86), the dispersion relation can be expressed in the equivalent form[1]

$$0 = 1 - \sum_j \frac{\omega_{pj}^2 [1 - (r_b/b)^{2\ell}]}{2(\omega - \ell\omega_{rj})[(\omega - \ell\omega_{rj}) + (\varepsilon_j\omega_{cj} + 2\omega_{rj})]} . \tag{5.87}$$

Here,

$$\varepsilon_j\omega_{cj} + 2\omega_{rj} = \pm \left(\omega_{rj}^+ - \omega_{rj}^-\right) = \mp\varepsilon_j\omega_{cj} \left(1 - \sum_k \frac{8\pi e_j \hat{n}_k e_k}{m_j \omega_{cj}^2}\right)^{1/2} \tag{5.88}$$

follows directly from Eq.(5.59). The upper (lower) sign in Eq.(5.88) corresponds to $\omega_{rj} = \omega_{rj}^+ (\omega_{rj} = \omega_{rj}^-)$.

The dispersion relation (5.87) can be used to determine the complex eigenfrequency ω in terms of equilibrium properties. The detailed waves and instabilities predicted by Eq.(5.87) depend on the admixture of plasma components and the strength of the self-electric field $E_r(r)$ as measured by $\omega_{rj}^{\pm}$. An important feature of Eq.(5.87) is that the relative rotation of plasma components in a nonneutral plasma column can provide the free energy to drive rotating two-stream instabilities for flute perturbations with $k_z = 0$.[1] (See Sec. 5.5.3 for a discussion of rotating two-stream instabilities when $k_z \neq 0$ and $r_b = b$.) It should be noted that the dispersion relation (5.87) can also be derived by solving the eigenvalue equation (5.62) directly for the case $k_z = 0$ (Problem 5.11).

Problem 5.11   Consider the eigenvalue equation (5.62) for flute perturbations with $k_z = 0$.

a.   Show that the solution to Eq.(5.62) in the two regions in Fig. 5.5 can be expressed as

$$\delta\phi_{\rm I}^\ell = A \left(\frac{r}{r_b}\right)^\ell , \quad 0 \le r < r_b , \tag{5.11.1}$$

and

$$\delta\phi_{\rm II}^\ell = A \left[\left(\frac{r}{b}\right)^\ell - \left(\frac{b}{r}\right)^\ell\right] \left[\left(\frac{r_b}{b}\right)^\ell - \left(\frac{b}{r_b}\right)^\ell\right]^{-1} , \quad r_b < r \le b , \tag{5.11.2}$$

where $A = \text{const.}$ Here $\delta\phi_{\rm I}^\ell(r = 0) = 0 = \delta\phi_{\rm II}^\ell(r = b)$, and $\delta\phi^\ell$ is continuous at $r = r_b$.

b. Integrate Eq.(5.62) across the surface of the plasma column at $r = r_b$, and make use of Eqs.(5.11.1) and (5.11.2) to derive the dispersion relation (5.86) directly.

5.6.2   Stable Oscillations

As a first application of Eq.(5.87) corresponding to stable oscillations ($\text{Im}\,\omega = 0$), consider a neutral plasma with $\sum_k \hat{n}_k e_k = 0$ and $\omega_{rj} = \omega_{rj}^- = 0$ [Eq.(5.59)]. For a two-component neutral plasma consisting of electrons ($j = e$) and ions ($j = i$), assuming $\omega_{rj} = \omega_{rj}^- = 0$, the exact solutions to Eq.(5.87) are $\omega = 0$ and

$$\omega = \frac{1}{2}(\omega_{ce} - \omega_{ci}) \pm \frac{1}{2}\left\{ (\omega_{ce} - \omega_{ci})^2 + 4\omega_{ce}\omega_{ci} + 2\left(\omega_{pe}^2 + \omega_{pi}^2\right)\left[1 - \left(\frac{r_b}{b}\right)^{2\ell}\right]\right\}^{1/2}.$$

(5.89)

As a second application of Eq.(5.87) corresponding to stable oscillations, consider a fully nonneutral plasma column with no ions ($\hat{n}_i = 0$), and a single component of electrons rotating with angular velocity $\omega_{re} = \omega_{re}^+$ or $\omega_{re} = \omega_{re}^-$, where [see Eq. (3.6)]

$$\omega_{re}^{\pm} = \frac{\omega_{ce}}{2}\left\{ 1 \pm \left(1 - \frac{2\omega_{pe}^2}{\omega_{ce}^2}\right)^{1/2}\right\}.$$

(5.90)

In this case, the exact solutions to Eq.(5.87) reduce to

$$\omega = \ell\omega_{re} + \left(\frac{1}{2}\omega_{ce} - \omega_{re}\right) \pm \left\{\left[\frac{1}{2}\omega_{ce} - \omega_{re}\right]^2 + \frac{1}{2}\omega_{pe}^2\left[1 - \left(\frac{r_b}{b}\right)^{2\ell}\right]\right\}^{1/2}.$$

(5.91)

In the limit where the conducting wall is located at infinity, $b \to \infty$, it follows from Eqs.(5.90) and (5.91) that the normal-mode frequencies are

$$\omega = (\ell - 1)\omega_{re} + \omega_{ce}, \quad \text{and} \quad \omega = (\ell - 1)\omega_{re}.$$

(5.92)

5.6.3   Ion Resonance Instability

As an application of Eq.(5.87) corresponding to instability,[22-25] consider a two-component nonneutral plasma consisting of electrons and singly ionized ions. The ion and electron densities are related by $\hat{n}_i = f\hat{n}_e$, where $f = $ const. = fractional charge neutralization. It is assumed that the equilibrium consists of a single component of electrons rotating with angular velocity $\omega_{re} = \omega_{re}^-$, and a single component of ions rotating with angular velocity $\omega_{ri} = \omega_{ri}^-$, where ω_{re}^- and ω_{ri}^- are defined in Eqs.(5.16) and (5.17).

Because ω_{ri}^- and ω_{re}^- are generally different, it can be anticipated that Eq.(5.87) supports unstable solutions with $\text{Im}\,\omega > 0$, corresponding to a two-stream instability produced by the differential rotation of electrons and ions. For example, for the fundamental ($\ell = 1$) mode, it is straightforward to show that Eq.(5.87) can be expressed as

$$\frac{1}{1 - (r_b/b)^2} = \frac{\omega_{pe}^2}{2(\omega - \omega_{re}^-)(\omega - \omega_{re}^+)} + \frac{\omega_{pi}^2}{2(\omega - \omega_{ri}^-)(\omega - \omega_{ri}^+)} , \qquad (5.93)$$

where $\omega_{re}^\pm$ and $\omega_{ri}^\pm$ are defined in Eqs.(5.16) and (5.17). For a neutral plasma with $f = 1$, the solutions to Eq.(5.93) reduce to Eq.(5.89), and there is no instability. Furthermore, for a pure electron plasma with $f = 0$, the solutions to Eq.(5.93) reduce to Eq.(5.91) for $\ell = 1$, and there is no instability. For $f \neq 0$ and $f \neq 1$, however, Eq.(5.93) has one unstable solution with $\text{Im}\,\omega > 0$, at least in certain parameter regimes.

The $\ell = 1$ dispersion relation (5.93) can be solved numerically for the real oscillation frequency $\text{Re}\,\omega$ and growth rate $\text{Im}\,\omega$ of the unstable mode. Figure 5.10 illustrates the corresponding stability boundaries in the parameter space $(\omega_{pe}^2/\omega_{ce}^2, f)$ for hydrogen ions $(m_i/m_e = 1836)$ and values of r_b/b ranging from 0.25 to 0.75. For specified value of r_b/b, the region below the curve in Fig. 5.10 corresponds to stable oscillations with $\text{Im}\,\omega = 0$, whereas the region above the curve corresponds to instability $(\text{Im}\,\omega > 0)$. The shaded region in Fig. 5.10 corresponds to $(2\omega_{pe}^2/\omega_{ce}^2)(1 - f) > 1$ (or equivalently, $f < 1 - \omega_{ce}^2/2\omega_{pe}^2$), i.e., radially confined equilibria do not exist in this region because the (repulsive) space-charge force is too strong. For $r_b = b$ exactly, Eq.(5.93) supports only stable oscillations with $\text{Re}\,\omega = \omega_{ri}^-$, ω_{ri}^+, ω_{re}^- and ω_{re}^+.

The dispersion relation (5.93) can be solved numerically for the growth rate and real oscillation frequency of the ion resonance instability. Typi-

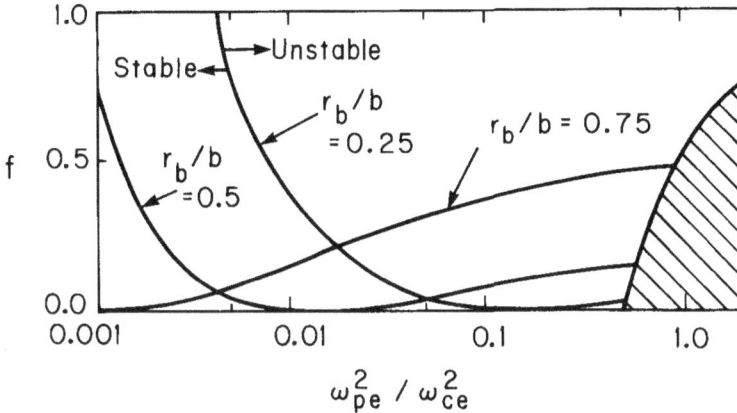

Figure 5.10. Plots of the $\ell = 1$ stability boundaries determined from Eq.(5.93) in the parameter space $(\omega_{pe}^2/\omega_{ce}^2, f)$ for $m_i/m_e = 1836$ and $r_b/b = 0.25$, 0.5, 0.75. For specified value of r_b/b, the region above the curve corresponds to the ion resonance instability (Im $\omega > 0$). Equilibria are forbidden in the shaded region where $(2\omega_{pe}^2/\omega_{ce}^2)(1-f) > 1$.

cal results are illustrated in Figs. 5.11 and 5.12 for $r_b/b = 0.5$. Here, Im ω and Re ω are plotted versus the fractional charge neutralization f for the two cases: $2\omega_{pe}^2/\omega_{ce}^2 = 1$ (Fig. 5.11) and $2\omega_{pe}^2/\omega_{ce}^2 = 0.02$ (Fig. 5.12). The growth rate and real frequency in Figs. 5.11 and 5.12 are normalized to the geometric-mean gyrofrequency $(\omega_{ce}\omega_{ci})^{1/2}$, and the stability results are presented for the fundamental $\ell = 1$ mode [Eq.(5.93)] as well as higher ℓ-values [Eq.(5.87)]. Evidently, the system is stable (Im $\omega = 0$) for a neutral plasma $(f = 1)$ or a one-component pure electron plasma $(f = 0)$. At intermediate values of f, however, the growth rate of the ion resonance instability can be substantial, particularly for moderate values of $2\omega_{pe}^2/\omega_{ce}^2$. In addition, for specified mode number ℓ, it is evident from Figs. 5.11 and 5.12 that there is a threshold value of f for onset of instability, which increases as ℓ is increased.[23]

Analytical investigations of the $\ell = 1$ dispersion relation (5.93) are simplified in the limit of a low-density nonneutral plasma with $(2\omega_{pe}^2/\omega_{ce}^2) \times (1 - f) \ll 1$. Making use of Eq.(5.16), $\omega_{re}^{\pm}$ can be expressed in the approximate forms $\omega_{re}^{+} = \omega_{ce}$, and

$$\omega_{re}^{-} = \frac{\omega_{pe}^2}{2\omega_{ce}}(1-f) \ . \tag{5.94}$$

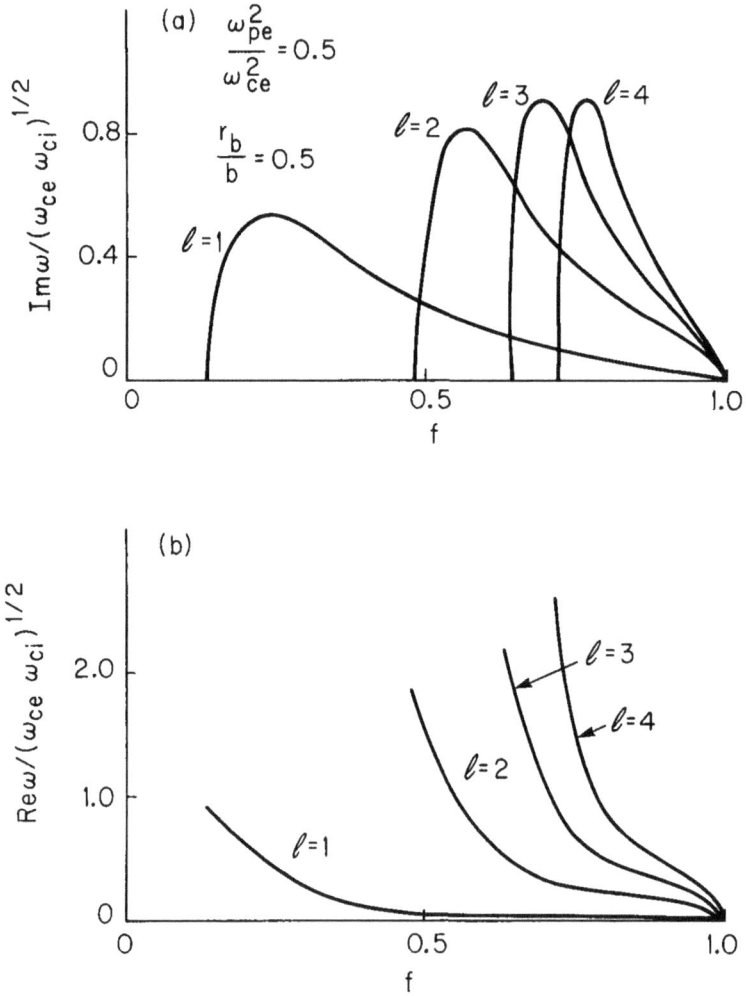

Figure 5.11. Plots versus the fractional charge neutralization f of (a) the normalized growth rate $(\mathrm{Im}\,\omega)/(\omega_{ce}\omega_{ci})^{1/2}$, and (b) the normalized real frequency $(\mathrm{Re}\,\omega)/(\omega_{ce}\omega_{ci})^{1/2}$ of the ion resonance instability determined from Eq.(5.87) for several values of the azimuthal mode number ℓ . Here, $2\omega_{pe}^2/\omega_{ce}^2 = 1$, $r_b/b = 0.5$ and $m_i/m_e = 1836$ are assumed, and $(\mathrm{Re}\,\omega)/(\omega_{ce}\omega_{ci})^{1/2}$ is plotted only for the range of f corresponding to instability.

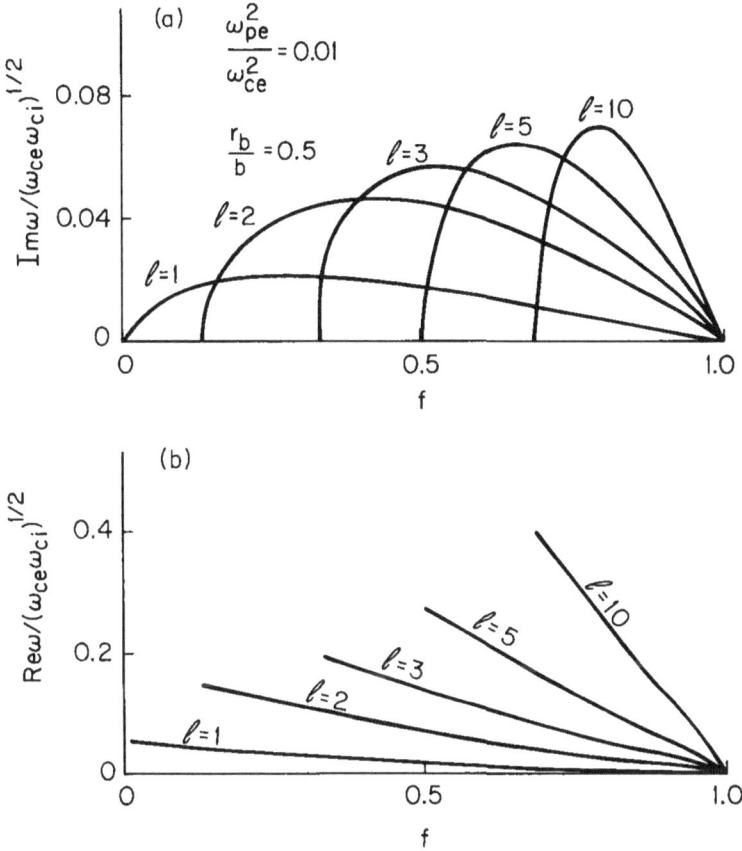

Figure 5.12.   Plots versus the fractional charge neutralization f of (a) the normalized growth rate $(\mathrm{Im}\,\omega)/(\omega_{ce}\omega_{ci})^{1/2}$, and (b) the normalized real frequency $(\mathrm{Re}\,\omega)/(\omega_{ce}\omega_{ci})^{1/2}$ of the ion resonance instability determined from Eq.(5.87) for several values of the azimuthal mode number ℓ. Here, $2\omega_{pe}^2/\omega_{ce}^2 = 0.02$, $r_b/b = 0.5$, and $m_i/m_e = 1836$ are assumed, and $(\mathrm{Re}\,\omega)/(\omega_{ce}\omega_{ci})^{1/2}$ is plotted only for the range of f corresponding to instability.

For low densities, note from Eq.(5.94) that ω_{re}^- is equal to the $E_r(r)\hat{e}_r \times B_0\hat{e}_z$ rotation velocity, i.e., $\omega_{re}^- \simeq -cE_r(r)/rB_0$ for $0 \leq r < r_b$. For frequencies well below the electron cyclotron frequency, $|\omega| \ll \omega_{ce}$, Eq.(5.93) reduces to[1]

$$\frac{1}{1-(r_b/b)^2} = -\frac{\omega_{pe}^2/2\omega_{ce}}{(\omega - \omega_{re}^-)} + \frac{\omega_{pi}^2}{2(\omega - \omega_{ri}^-)(\omega - \omega_{ri}^+)} \,, \tag{5.95}$$

where $\omega_{ri}^{\pm}$ and ω_{re}^- are defined in Eqs.(5.17) and (5.94). For a pure electron plasma with $f = 0$, the solution to Eq.(5.95) is

$$\omega = \omega_D \left(\frac{r_b}{b}\right)^2 ,\tag{5.96}$$

where $\omega_D \equiv \omega_{pe}^2/2\omega_{ce}$. If a low-density ion component is present ($0 < f \ll 1$), Eq.(5.95) has one unstable solution with $\mathrm{Im}\,\omega > 0$ provided $\omega_{ri}^- \approx \omega_D(r_b/b)^2$, i.e., provided there is a resonance between the electron diocotron wave and the azimuthal motion of the ions. For $f \ll 1$, it can be shown that instability exists provided $\omega_D(r_b/b)^2$ and ω_{ri}^- differ by an amount proportional to $\sqrt{f}$. Furthermore, for $f \ll 1$ and $\omega_D(r_b/b)^2$ very close to ω_{ri}^-, it can be shown that the growth rate is approximately[22]

$$\mathrm{Im}\,\omega = \omega_{pi} \left\{ \left(1 - \frac{r_b^2}{b^2}\right)^3 \left(\frac{b^2/2r_b^2}{2 - r_b^2/b^2}\right) \right\}^2 .\tag{5.97}$$

For the limiting case considered in Eq.(5.97), note that the growth rate is of order ω_{pi} times a geometric factor.

As a final point, the ion resonance instability, while modified by ion kinetic effects (Chapter 4),[24,25] also exists in high-current configurations such as the modified betatron[26] when a partially neutralizing ion background is present. Moreover, in circumstances where no ions are present ($\hat{n}_i = 0$) and there are two components of electrons rotating at different angular velocities, ω_{re}^- and ω_{re}^+, the dispersion relation (5.87) also supports a class of electron-electron cross-field instabilities with relatively large growth rates.

Problem 5.12   The cubic dispersion relation (5.95), valid for $\ell = 1$ and $(2\omega_{pe}^2/\omega_{ce}^2)(1 - f) \ll 1$, can be solved exactly to determine the growth rate and real oscillation frequency of the ion resonance instability.

a. Obtain a closed analytical expression for the threshold value of f for onset of the ion resonance instability when $f \ll 1$. That is, calculate f_{crit} such that $\mathrm{Im}\,\omega = 0$ for $f \leq f_{\mathrm{crit}}$, but $\mathrm{Im}\,\omega > 0$ for $f > f_{\mathrm{crit}}$.

b. Obtain a closed equation that determines the value of f that maximizes the growth rate of the ion resonance instability.

5.7   Two-Stream Instabilities in Relativistic Beam-Plasma Systems

5.7.1   Electrostatic Dispersion Relation

The electrostatic stability properties of relativistic beam-plasma systems are now examined in circumstances where the density profile $n_j^0(r)$ is uniform with $n_j^0(r) = \hat{n}_j = \text{const.}$, for $0 \leq r < r_b$, and $n_j^0(r) = 0$, for $r_b < r \leq b$ [Eq.(5.58) and Fig. 5.5]. It is also assumed that the axial velocity profile is uniform with $V_{zj}^0(r) = \beta_j c = \text{const.}$, and that the beam-plasma system is both charge neutralized and current neutralized with

$$\sum_k \hat{n}_k e_k = 0 \quad \text{and} \quad \sum_k \hat{n}_k e_k \beta_k = 0 \,, \tag{5.98}$$

which corresponds to zero equilibrium self fields, i.e., $E_r(r) = 0 = B_\theta^s(r)$. Assuming slow rotational equilibria with $\omega_{rj}(r) = \omega_{rj}^-(r)$, it follows from Eq.(5.30) that $\omega_{rj}(r) = 0$. That is, there is no equilibrium rotation of the individual plasma components. The counterstreaming axial motion of the beam-plasma components, however, can provide the free energy to drive the classical two-stream instability.[1,3,4,27−34]

An eigenvalue equation completely analogous to Eq.(5.56) can be derived in the relativistic case. For the density profile illustrated in Fig. 5.5, this eigenvalue equation can be solved exactly within the plasma ($0 \leq r < r_b$), in the vacuum region ($r_b < r \leq b$), and integrated across the boundary of the plasma column (at $r = r_b$) to give the dispersion relation that determines the complex oscillation frequency ω. Making use of the assumptions enumerated in the previous paragraph, but not presenting algebraic details, the resulting dispersion relation can be expressed as[1,27]

$$\frac{1}{g_\ell} - \left(1 - \sum_j \frac{\omega_{pj}^2/\gamma_j}{(\omega - k_z\beta_j c)^2 - \omega_{cj}^2/\gamma_j^2}\right) Tr_b \frac{J_\ell'(Tr_b)}{J_\ell(Tr_b)}$$

$$= \ell \sum_j \frac{\omega_{pj}^2 \varepsilon_j \omega_{cj}/\gamma_j^2}{(\omega - k_z\beta_j c)\left[(\omega - k_z\beta_j c)^2 - \omega_{cj}^2/\gamma_j^2\right]} \,. \tag{5.99}$$

In Eq.(5.99), $\varepsilon_j = \mathrm{sgn}\, e_j, \omega_{cj} = |e_j|B_0/m_j c$ is the nonrelativistic cyclotron frequency, $\omega_{pj}^2 = 4\pi \hat{n}_j e_j^2/m_j$ is the nonrelativistic plasma frequency-squared, and $\gamma_j = (1 - \beta_j^2)^{-1/2}$ is the relativistic mass factor associated with the axial motion. Moreover, the coefficient T^2 is defined by

$$T^2 = -k_z^2 \left(1 - \sum_j \frac{\omega_{pj}^2/\gamma_j^3}{(\omega - k_z \beta_j c)^2} \right) \left(1 - \sum_j \frac{\omega_{pj}^2/\gamma_j}{(\omega - k_z \beta_j c)^2 - \omega_{cj}^2/\gamma_j^2} \right)^{-1} ,$$

$$(5.100)$$

and the geometric factor g_ℓ is defined in Eq.(5.70).

The dispersion relation (5.99) can be used to investigate electrostatic stability properties for a wide range of beam-plasma parameters, values of r_b/b, etc. For present purposes, Eq.(5.99) is analyzed in limiting regimes that are analytically tractable and illustrate the essential features of the relativistic two-stream instability including the important influence of finite radial geometry.

We consider Eq.(5.99) for azimuthally symmetric $(\partial/\partial\theta = 0)$ perturbations and long axial wavelengths, i.e.,

$$\ell = 0 \quad \text{and} \quad k_z^2 b^2 \ll 1 . \tag{5.101}$$

From Eqs.(5.71) and (5.101), the geometric factor g_0 can be approximated by

$$g_0 = -\ln\left(\frac{b}{r_b}\right) . \tag{5.102}$$

It is also assumed that the perturbations have sufficiently high frequency in Eq.(5.99) that the ions can be treated as a fixed $(m_i \to \infty)$ neutralizing background. That is, only the beam electrons $(j = b)$ and the plasma electrons $(j = e)$ are treated as active plasma components in analyzing the dispersion relation (5.99). It is further assumed that the electrons are strongly magnetized with

$$\frac{\omega_{pj}^2}{\gamma_j} , \; |\omega - k_z \beta_j c|^2 \ll \frac{\omega_{cj}^2}{\gamma_j^2} \tag{5.103}$$

for $j = b, e$. Making use of Eqs.(5.101)–(5.103), the dispersion relation (5.99) can be approximated by

$$J_0(Tr_b) = g_0 T r_b J_0'(Tr_b) , \tag{5.104}$$

where

$$T^2 = -k_z^2 \left(1 - \sum_{j=b,e} \frac{\omega_{pj}^2/\gamma_j^3}{(\omega - k_z\beta_j c)^2} \right) , \qquad (5.105)$$

and $\omega_{pj}^2 = 4\pi\hat{n}_j e_j^2/m_j$. The approximate dispersion relation (5.104) is now analyzed for the case of a beam-plasma filled waveguide ($r_b = b$), and in the thin-beam limit ($r_b \ll b$).

5.7.2 Beam-Plasma Filled Waveguide

For $r_b = b$, it follows that $g_0 = \ln(b/r_b) = 0$ and Eq.(5.104) reduces to $J_0(Tb) = 0$, which has solutions $T^2 r_b^2 = p_{0m}^2$, where p_{0m} is the m'th zero of $J_0(x) = 0$. Introducing the effective perpendicular wavenumber defined by $k_\perp^2 \equiv p_{0m}^2/b^2$, it readily follows that Eq.(5.105) and $T^2 = k_\perp^2$ give the dispersion relation[1,27]

$$0 = 1 - \frac{k_z^2}{k^2} \sum_{j=b,e} \frac{\omega_{pj}^2/\gamma_j^3}{(\omega - k_z\beta_j c)^2} , \qquad (5.106)$$

where $k^2 \equiv k_z^2 + k_\perp^2$, and $k_\perp^2 = p_{0m}^2/b^2$.

For appropriate values of parameters, Eq.(5.106) supports solutions corresponding to an electrostatic two-stream instability between the beam electrons and the plasma electrons. Using standard techniques, it is straightforward to show that Eq.(5.106) supports one unstable solution ($\mathrm{Im}\,\omega > 0$) for k_z that satisfies (see Problem 5.13)

$$0 < k_z^2(\beta_b - \beta_e)^2 c^2 < \frac{\omega_{pb}^2}{\gamma_b^3} \frac{k_z^2}{k^2} \left[1 + \left(\frac{\hat{n}_e}{\hat{n}_b} \right)^{1/3} \frac{\gamma_b}{\gamma_e} \right]^3 . \qquad (5.107)$$

The right-most inequality in Eq.(5.107) can be expressed in the equivalent form

$$k_z^2(\beta_b - \beta_e)^2 c^2 < \frac{\omega_{pb}^2}{\gamma_b^3} \left[1 + \left(\frac{\hat{n}_e}{\hat{n}_b} \right)^{1/3} \frac{\gamma_b}{\gamma_e} \right]^3 - k_\perp^2 (\beta_b - \beta_e)^2 c^2 . \qquad (5.108)$$

A necessary condition for two-stream instability is that the right-hand side of Eq.(5.108) be positive, i.e.,

$$\omega_{pb}^2 > k_\perp^2 c^2 \gamma_b^3 (\beta_b - \beta_e)^2 \left[1 + \left(\frac{\hat{n}_e}{\hat{n}_b} \right)^{1/3} \frac{\gamma_b}{\gamma_e} \right]^{-3} , \qquad (5.109)$$

where $k_\perp^2 = p_{0m}^2/b^2$. For specified ω_{pb}, γ_b, etc., it is clear from Eq.(5.109) that the finite radial geometry (as measured by $k_\perp$) has a stabilizing influence on the two-stream instability. In terms of the magnitude of the beam current $I_b = |\hat{n}_b e \beta_b c| \pi r_b^2$, the necessary condition for instability in Eq.(5.109) can be expressed in the equivalent form

$$I_b > I_{\text{crit}} = I_A \frac{p_{0m}^2 \beta_b^2 \gamma_b^2 (1 - \beta_e/\beta_b)^2}{4 \left[1 + (\hat{n}_e/\hat{n}_b)^{1/3} (\gamma_b/\gamma_e) \right]^3} , \qquad (5.110)$$

where $I_A \equiv (m_e c^3/e) \beta_b \gamma_b \simeq 17000 \, \beta_b \gamma_b$ A is the Alfvén critical current.[35,36] It is clear from Eq.(5.110) that the stabilizing influence of finite radial geometry can lead to a substantial beam current threshold for the two-stream instability. In the extreme case of equidensity beam electrons and plasma electrons ($\hat{n}_b = \hat{n}_e$), the current neutrality condition gives $\beta_e = -\beta_b$ and $\gamma_e = \gamma_b$, assuming $\beta_i = 0$. Equation (5.110) then reduces to $I_b > I_{\text{crit}} = I_A p_{0m}^2 \beta_b^2 \gamma_b^2/8$. The lowest current threshold is for $m = 1$. Substituting $p_{01} \simeq 2.4$ gives $I_b > 0.72 \beta_b^2 \gamma_b^2 I_A$, or equivalently,

$$I_b > I_{\text{crit}} = 12.2 \beta_b^3 \gamma_b^3 \text{ kA} , \qquad (5.111)$$

for instability. In the equidensity case, in the limit where $I_b \gg I_{\text{crit}}$ (i.e., $k_\perp^2 c^2 \beta_b^2 \ll 2\omega_{pb}^2/\gamma_b^3$), it is straightforward to show that the maximum growth rate $\gamma = \text{Im}\,\omega$ obtained from Eq.(5.106) is given by

$$\gamma_{\max} = \frac{1}{2} \omega_{pb}/\gamma_b^{3/2} , \qquad (5.112)$$

and that the axial wavenumber at maximum growth is $[k_z^2]_{\max} = (3\omega_{pb}^2/4\gamma_b^3 \beta_b^2 c^2)$.

5.7.3   Thin-Beam Limit

In this section, the $\ell = 0$ dispersion relation (5.104) is analyzed in the limit of a thin beam with

$$r_b \ll b \,,$$
$$r_b^2 |T|^2 \ll 1 \,. \tag{5.113}$$

Expanding $[xJ_0'(x)/J_0(x)]_{x=Tr_b} \simeq -T^2 r_b^2/2$, Eq.(5.104) can be approximated by[1,27]

$$0 = 1 + \frac{1}{2} k_z^2 r_b^2 \ln(b/r_b) \left(1 - \sum_{j=b,e} \frac{\omega_{pj}^2/\gamma_j^3}{(\omega - k_z \beta_j c)^2} \right) \,, \tag{5.114}$$

where $\omega_{pj}^2 = 4\pi \hat{n}_j e_j^2/m_j$. Defining the effective perpendicular wavenumber $k_\perp$ by

$$k_\perp^2 = \frac{2}{r_b^2 \ln(b/r_b)} \,, \tag{5.115}$$

the dispersion relation (5.114) reduces to

$$0 = 1 - \frac{k_z^2}{k^2} \sum_{j=b,e} \frac{\omega_{pj}^2/\gamma_j^3}{(\omega - k_z \beta_j c)^2} \,, \tag{5.116}$$

where $k^2 = k_z^2 + k_\perp^2$, and $k_\perp^2$ is defined in Eq.(5.115). Apart from the definition of $k_\perp^2$, Eq.(5.116) is identical in form to Eq.(5.106). Therefore, the analogous current threshold condition necessary for onset of the two-stream instability can be expressed as

$$I_b > I_{\text{crit}} = I_A \frac{\beta_b^2 \gamma_b^2 (1 - \beta_e/\beta_b)^2}{2 \ln(b/r_b) \left[1 + (\hat{n}_e/\hat{n}_b)^{1/3} (\gamma_b/\gamma_e) \right]^3} \,, \tag{5.117}$$

where $I_A = (m_e c^3/e) \beta_b \gamma_b$ is the Alfvén critical current. Comparing Eqs.(5.106) and (5.117), the current threshold for instability tends to be smaller in the case of a thin beam with $r_b \ll b$.

Problem 5.13   The two-stream instability can be described by a generic dispersion relation of the form[37]

$$1 = R(\omega) \equiv \frac{\omega_{p1}^2}{\omega^2} + \frac{\omega_{p2}^2}{(\omega - k_z V)^2} \ . \tag{5.13.1}$$

a. Assume $k_z V > 0$ without loss of generality, and sketch $R(\omega)$ versus ω for real ω. Show that $R(\omega)$ has a single minimum in the interval $0 < \omega < k_z V$ located at $\omega = \omega_0$, where

$$\omega_0 = k_z V \frac{(\omega_{p1}^2)^{1/3}}{(\omega_{p1}^2)^{1/3} + (\omega_{p2}^2)^{1/3}} \tag{5.13.2}$$

b. Referring to Eqs.(5.13.1) and (5.13.2), show that the necessary and sufficient condition for instability (existence of a solution with $\text{Im}\,\omega > 0$) is given by $R(\omega_0) > 1$, or equivalently,

$$0 < k_z^2 V^2 < \left[(\omega_{p1}^2)^{1/3} + (\omega_{p2}^2)^{1/3} \right]^3 \ . \tag{5.13.3}$$

c. Make use of Eq.(5.13.1) to show that

$$\frac{\partial \omega}{\partial k_z} = V \frac{\omega_{p2}^2}{\omega_{p1}^2 (1 - k_z V/\omega)^3 + \omega_{p2}^2} \tag{5.13.4}$$

for complex ω. At maximum growth rate, show that the condition $(\partial/\partial k_z)(\text{Im}\,\omega) = 0$ requires that ω be of the form

$$\omega = \frac{k_z V}{[1 + \alpha \exp(in\pi/3)]} \tag{5.13.5}$$

for some real α and integer n.

d. Substitute Eq.(5.13.5) into the dispersion relation (5.13.1). Show that the conditions on α for which the real and imaginary parts of the resulting equation are satisfied are given by

$$\frac{k_z^2 V^2}{\omega_{p2}^2} = 1 + \frac{(1 + 2\alpha + 3\alpha^2)}{\alpha^3 (2 + \alpha)} \ , \tag{5.13.6}$$

$$\frac{\omega_{p1}^2}{\omega_{p2}^2} = \frac{(1 + 2\alpha)}{\alpha^3 (2 + \alpha)} \ . \tag{5.13.7}$$

Equations (5.13.6) and (5.13.7) determine the values of α and $k_z^2 V^2$ in terms of ω_{p1}^2 and ω_{p2}^2 corresponding to maximum growth.

e. Consider the case in which $\omega_{p1}^2/\omega_{p2}^2 = \varepsilon \ll 1$, and Eq.(5.13.7) gives $\alpha \simeq (2/\varepsilon)^{1/3}$. Show that the values of the growth rate $\text{Im}\,\omega$ and $k_z^2 V^2$ at maximum growth are given (approximately) by

$$(\text{Im}\,\omega)_{\max} = \frac{3^{1/2}}{2}\left(\frac{\omega_{p1}^2}{2\omega_{p2}^2}\right)^{1/3}\omega_{p2}\,. \tag{5.13.8}$$

$$k_z^2 V^2 = \omega_{p2}^2\left[1 + 3\left(\frac{\omega_{p1}^2}{2\omega_{p2}^2}\right)^{2/3}\right]\,. \tag{5.13.9}$$

5.8   Electromagnetic Filamentation Instability

In this section, an example is presented in which intense self-magnetic fields play an important role in determining the detailed stability behavior. In particular, consider a cylindrical relativistic electron beam ($j = b$) with uniform density $\hat{n}_b$ out to radius $r = r_b$ (Fig. 5.5) propagating with axial velocity $\beta_b c = \text{const.}$ parallel to a magnetic guide field $B_0\hat{e}_z$ through a cold, background plasma ($j = e, i$). The background plasma components have uniform density $\hat{n}_j$ and uniform axial velocity $\beta_j c$, and the rotational motion is assumed to be nonrelativistic with $r^2\omega_{rj}^2/c^2 \ll 1 - \beta_j^2$ for $j = b, e, i$. Paralleling the analysis in Sec. 5.2.3, the radial self-electric field $E_r(r)$ and azimuthal self-magnetic field $B_\theta^s(r)$ in the region $0 \leq r < r_b$ can be expressed as

$$E_r(r) = 2\pi\left(\sum_{k=b,e,i}\hat{n}_k e_k\right)r\,, \tag{5.118}$$

$$B_\theta^s(r) = 2\pi\left(\sum_{k=b,e,i}\hat{n}_k e_k \beta_k\right)r\,, \tag{5.119}$$

and the angular rotation velocity for component j is given by [Eq.(5.30)]

$$\omega_{rj} = \omega_{rj}^{\pm} = -\frac{\varepsilon_j\omega_{cj}}{2\gamma_j}\left\{1 \pm \left(1 - \sum_k\frac{8\pi e_j\gamma_j\hat{n}_k e_k}{m_j\omega_{cj}^2}(1 - \beta_j\beta_k)\right)^{1/2}\right\}\,, \tag{5.120}$$

where $\gamma_j = (1 - \beta_j^2)^{-1/2}$, and $\sum_k$ denotes summation over $k = b, e, i$.

The ordinary-mode filamentation instability[38-42] has been observed experimentally[43] in circumstances where a low-density electron beam is injected into a high-density preionized plasma, provided the instability criterion $4\pi\hat{n}_b\gamma_b m_e c^2\beta_b^2/B_0^2 = \gamma_b\omega_{pb}^2\beta_b^2/\omega_{cb}^2 > 1$ is satisfied. On the other hand, when a low-density electron beam is injected into a dense neutral gas and the instability criterion quoted above is satisfied, the beam is observed not to filament.[43] An essential difference between these two cases is the lack of current neutralization in the beam-plasma system formed when the beam is injected into neutral gas. When a low-density beam is injected into a dense preionized plasma, charge and current neutralization take place on a rapid time scale. However, when a low-density beam is injected into a dense neutral gas, there is avalanche breakdown of the gas, typically a few nanoseconds after charge neutralization occurs. Moreover, current neutralization is often incomplete.[43]

The absence of filamentation instability in circumstances where the beam-plasma system is not current neutralized strongly suggests that the azimuthal self-magnetic field has a stabilizing influence on the filamentation instability. To investigate this possibility, we examine stability properties of the relativistic beam-plasma system for perturbations with ordinary-mode electromagnetic polarization. Assuming azimuthally symmetric flute perturbations with $\partial/\partial\theta = 0$ and $\partial/\partial z = 0$, the perturbed electromagnetic field components are $\delta\mathbf{E} = \delta E_z(r,t)\hat{\mathbf{e}}_z$ and $\delta\mathbf{B} = \delta B_\theta(r,t)\hat{\mathbf{e}}_\theta$, where δB_θ and δE_z are related by

$$\frac{1}{c}\frac{\partial}{\partial t}\delta B_\theta(r,t) = \frac{\partial}{\partial r}\delta E_z(r,t) \ . \tag{5.121}$$

5.8.1 Ordinary-Mode Dispersion Relation

To investigate electromagnetic stability properties for $\partial/\partial\theta = 0$ and $\partial/\partial z = 0$, perturbed quantities are expressed as $\delta\psi(r,t) = \delta\psi(r)\exp(-i\omega t)$, where $\text{Im}\,\omega > 0$ corresponds to instability. For $\delta\mathbf{E} = \delta E_z(r,t)\hat{\mathbf{e}}_z$ and $\delta\mathbf{B} = \delta B_\theta(r,t)\hat{\mathbf{e}}_\theta$, the $\nabla\times\delta\mathbf{B}$ Maxwell equation reduces to

$$\left(\frac{1}{r}\frac{\partial}{\partial r}r\frac{\partial}{\partial r} + \frac{\omega^2}{c^2}\right)\delta E_z = -\frac{4\pi i\omega}{c^2}\left(\sum_j e_j n_j^0\delta V_{zj} + \sum_j e_j\beta_j c\delta n_j\right) ,$$

$$\tag{5.122}$$

where δV_{zj} and δn_j are determined self-consistently from the linearized continuity and force balance equations (2.41) and (2.42), and $\sum_j$ denotes summation over $j = b, e, i$. For cold beam-plasma components, Eqs.(2.41) and (2.42) become

$$- i\omega \delta n_j + \frac{1}{r}\frac{\partial}{\partial r}\left(r n_j^0 \delta V_{rj}\right) = 0 \,, \tag{5.123}$$

$$- i\omega \delta P_{rj} - \omega_{rj}\left(\delta P_{\theta j} + \gamma_j m_j \delta V_{\theta j}\right) - \varepsilon_j m_j \omega_{cj} \delta V_{\theta j}$$
$$= e_j \left(-\frac{1}{c}\delta V_{zj} B_\theta^s - \beta_j \delta B_\theta\right) \,, \tag{5.124}$$

$$- i\omega \delta P_{\theta j} + \omega_{rj}\left(\delta P_{rj} + \gamma_j m_j \delta V_{rj}\right) + \varepsilon_j m_j \omega_{cj} \delta V_{rj} = 0 \,, \tag{5.125}$$

$$- i\omega \delta P_{zj} = e_j \left(\delta E_z + \frac{1}{c}\delta V_{rj} B_\theta^s\right) \,, \tag{5.126}$$

where $\gamma_j = (1 - \beta_j^2)^{-1/2}$, $\omega_{cj} = |e_j|B_0/m_j c$ and $\varepsilon_j = \operatorname{sgn} e_j$. Here, the perturbed momentum of a fluid element is $\delta \mathbf{P}_j = \gamma_j m_j \delta \mathbf{V}_j + \delta\gamma_j m_j \mathbf{V}_j^0$, where $\delta\gamma_j = (\gamma_j^3/c^2)\mathbf{V}_j^0 \cdot \delta\mathbf{V}_j$ and $\mathbf{V}_j^0 = \omega_{rj} r \hat{\mathbf{e}}_\theta + \beta_j c \hat{\mathbf{e}}_z$ (see Problem 2.4). We make use of Eqs.(5.123)–(5.126) to express δV_{zj} and δn_j in terms of δE_z. After some straightforward algebra that makes use of $r^2 \omega_{rj}^2/c^2 \ll 1 - \beta_j^2$, the eigenvalue equation (5.122) can be expressed as

$$\frac{1}{r}\frac{\partial}{\partial r}\left[r\left(1 + \sum_{j=b,e,i}\frac{\beta_j^2 \omega_{pj}^2(r)/\gamma_j}{\omega^2 - (\omega_{rj}^+ - \omega_{rj}^-)^2}\right)\frac{\partial}{\partial r}\delta E_z\right]$$
$$+ \left(\frac{\omega^2}{c^2} - \sum_{j=b,e,i}\frac{\omega_{pj}^2(r)}{\gamma_j^3 c^2}\right)\delta E_z = 0 \,, \tag{5.127}$$

where $\omega_{pj}^2(r) = 4\pi n_j^0(r)e_j^2/m_j$, and $(\omega_{rj}^+ - \omega_{rj}^-)^2$ is determined from Eq.(5.120).

For slow-wave perturbations with $|\omega|^2 \ll \sum_j \omega_{pj}^2/\gamma_j^3$, the displacement current (the term proportional to ω^2/c^2) is neglected in Eq.(5.127). The solution to Eq.(5.127) that satisfies $\delta E_z(r = b) = 0$ at the conducting wall, with $\delta E_z(r)$ continuous at $r = r_b$, can be expressed as

$$\delta E_z(r) = \begin{cases} A J_0(Tr) \,, & 0 \leq r < r_b \,, \\[2mm] A J_0(Tr_b)\dfrac{\ln(r/b)}{\ln(r_b/b)} \,, & r_b < r \leq b \,. \end{cases} \tag{5.128}$$

Here, the coefficient T is defined by

$$T^2 = -\left(\sum_{j=b,e,i} \frac{\omega_{pj}^2}{\gamma_j^3 c^2}\right)\left[1 + \sum_{j=b,e,i} \frac{\beta_j^2 \omega_{pj}^2/\gamma_j}{\omega^2 - (\omega_{rj}^+ - \omega_{rj}^-)^2}\right]^{-1} , \qquad (5.129)$$

where $\omega_{pj}^2 = 4\pi\hat{n}_j e_j^2/m_j$. Enforcing continuity of $\delta B_\theta(r) = (ic/\omega) \times (\partial/\partial r)\delta E_z(r)$ at $r = r_b$ then gives the desired dispersion relation[38]

$$Tr_b\frac{J_0'(Tr_b)}{J_0(Tr_b)} = \frac{1}{\ln(r_b/b)} , \qquad (5.130)$$

where $J_0'(Tr_b) = [dJ_0(x)/dx]_{x=Tr_b}$.

The ordinary-mode dispersion relation (5.130) for slow-wave perturbations can be used to investigate detailed stability properties over a wide range of system parameters. For present purposes, we consider Eq.(5.130) for a sufficiently narrow vacuum gap $(b - r_b \ll b)$ that the dispersion relation can be approximated by

$$J_0(Tb) = 0 . \qquad (5.131)$$

Equation (5.131) implies $T^2 b^2 = p_{0n}^2$, where p_{0n} is the n'th zero of $J_0(p_{0n}) = 0$. Defining the effective perpendicular wavenumber by $k_\perp^2 = p_{0n}^2/b^2$, it follows that $T^2 b^2 = p_{0n}^2$ can be expressed in the equivalent form

$$1 + \sum_{j=b,e,i} \frac{\omega_{pj}^2}{\gamma_j^3 c^2 k_\perp^2} + \sum_{j=b,e,i} \frac{\beta_j^2 \omega_{pj}^2/\gamma_j}{\omega^2 - (\omega_{rj}^+ - \omega_{rj}^-)^2} = 0 . \qquad (5.132)$$

Equation (5.132) is used in Sec. 5.8.2 to calculate detailed properties of the filamentation instability.

Problem 5.14   Make use of Eqs.(5.123) and (5.126) to express δV_{zj} and δn_j in terms of δE_z. Assuming $r^2\omega_{rj}^2/c^2 \ll 1 - \beta_j^2$, show that the ordinary-mode eigenvalue equation (5.122) can be expressed in the form given in Eq.(5.127).

Problem 5.15   Consider the ordinary-mode eigenvalue equation (5.127) for the case of a nonrelativistic nonneutral plasma with $\beta_j = 0$ and $\gamma_j = 1$.

a.   Solve Eq.(5.127) for the case of a plasma-filled waveguide $(r_b = b)$. Show that

$$\delta E_z(r) = A J_0(Tr) , \quad 0 \le r \le b , \qquad (5.15.1)$$

where the coefficient T is defined by

$$T^2 = \left(\frac{\omega^2}{c^2} - \sum_j \frac{\omega_{pj}^2}{c^2} \right) . \tag{5.15.2}$$

Here, $A = \text{const.}$, and $\omega_{pj}^2 = 4\pi \hat{n}_j e_j^2 / m_j$.

b. Make use of the boundary condition $J_0(Tb) = 0$ to show that the normal-mode oscillation frequencies are given by

$$\omega^2 = c^2 k_\perp^2 + \sum_j \omega_{pj}^2 , \tag{5.15.3}$$

where $k_\perp^2 \equiv p_{0n}^2 / b^2$, and p_{0n} is the n'th zero of $J_0(p_{0n}) = 0$.

5.8.2   Filamentation Instability

In Eq.(5.132), the relative axial motion of the beam-plasma components (the term proportional to β_j^2) provides the free energy source to drive a Weibel-like electromagnetic instability[21] known as the filamentation instability.[38-42] For present purposes, we examine Eq.(5.132) for the case of a dense background plasma ($\hat{n}_e \gg \hat{n}_b$) and a sufficiently intense electron beam that $\omega_{pb}^2 \beta_b^2 \gg \omega_{pe}^2 \beta_e^2$. Eq.(5.132) then reduces to

$$1 + \frac{\omega_{pe}^2}{c^2 k_\perp^2} + \frac{\beta_b^2 \omega_{pb}^2 / \gamma_b}{\omega^2 - (\omega_{rb}^+ - \omega_{rb}^-)^2} = 0 , \tag{5.133}$$

where $\gamma_e \simeq 1$ is assumed. Assuming that the beam-plasma system is charge neutralized with $\sum_{k=b,e,i} \hat{n}_k e_k = 0$, but not generally current neutralized, it follows from Eq.(5.120) that

$$\left(\omega_{rb}^+ - \omega_{rb}^- \right)^2 = \frac{\omega_{cb}^2}{\gamma_b^2} \left[1 + \frac{2\gamma_b \omega_{pb}^2 \beta_b^2}{\omega_{cb}^2} (1 - f_M) \right] . \tag{5.134}$$

where $\omega_{pb}^2 = 4\pi \hat{n}_b e^2 / m_e$, $\omega_{cb} = eB_0 / m_e c$, and $f_M = \sum_{j=e,i} \hat{n}_j e_j \beta_j / (\hat{n}_b e \beta_b)$ is the fractional current neutralization defined in Eq.(5.32). Solving Eq.(5.133) for ω^2 gives[38]

$$\omega^2 = - \left[\frac{\beta_b^2 \omega_{pb}^2 / \gamma_b}{1 + \omega_{pe}^2 / c^2 k_\perp^2} - \left(\omega_{rb}^+ - \omega_{rb}^- \right)^2 \right] . \tag{5.135}$$

Equation (5.135) has one unstable solution ($\mathrm{Im}\,\omega > 0$) for short perturbation wavelengths ($c^2 k_\perp^2 \gg \omega_{pe}^2$) whenever the inequality

$$\frac{\omega_{pb}^2 \beta_b^2}{\gamma_b} > \frac{\omega_{cb}^2}{\gamma_b^2} \left[1 + \frac{2\gamma_b \omega_{pb}^2 \beta_b^2}{\omega_{cb}^2} (1 - f_M) \right] \qquad (5.136)$$

is satisfied. For complete current neutralization ($f_M = 1$), Eq.(5.136) reduces to $\gamma_b \omega_{pb}^2 \beta_b^2 / \omega_{cb}^2 > 1$, and the maximum growth rate is given by

$$(\mathrm{Im}\,\omega)_{\max} = \frac{\omega_{pb} \beta_b}{\gamma_b^{1/2}} \left(1 - \frac{\omega_{cb}^2}{\gamma_b \omega_{pb}^2 \beta_b^2} \right)^{1/2} \qquad (5.137)$$

for $c^2 k_\perp^2 \gg \omega_{pe}^2$. On the other hand, for $f_M < 1$, the growth rate is reduced from the value in Eq.(5.137) because of the stabilizing influence of the azimuthal self-magnetic field. Indeed, from Eq.(5.135), the instability is completely stabilized ($\mathrm{Im}\,\omega = 0$) whenever[38]

$$f_M < \frac{1}{2} + \frac{\omega_{cb}^2}{2\gamma_b \omega_{pb}^2 \beta_b^2} . \qquad (5.138)$$

The strong stabilizing influence of the azimuthal self-magnetic field on the filamentation instability is consistent with the experimental observations of Kapetanakos.[43]

5.9   Equilibrium and Collective Oscillation Properties of Two-Dimensional Nonneutral Plasma

Recent experiments[44,45] have demonstrated the existence of collective modes in a two-dimensional nonneutral ion layer residing below the surface of liquid helium. A typical cell geometry used in the experiments[44,45] is illustrated in Fig. 5.13. The positive ions form a two-dimensional charge disk held in position by electric potentials applied to the walls of the confining cell (shaped like a cylindrical pillbox), and by the vertical electric field resulting from the polarization of the liquid helium. The two horizontal electrodes A and B in Fig. 5.13 are a distance h apart, electrically insulated by a guard ring of radius b that forms the vertical wall of the cylindrical pillbox. The cell is partially immersed in liquid ^{4}He which fills the pillbox to a height $d \simeq h/2$. Helium ions are created by a field emission tip (below electrode B), and enter the cell through a screen grid.

The electric field due to the applied potential $V_{BA} > 0$ between the electrodes B and A forces the ions up towards the surface of the liquid helium. The ions also polarize the surrounding liquid helium (dielectric constant $= 1.06$), and this polarization exerts a downward force on the ions. At typical temperatures $T \sim 3 \times 10^{-2}$ K, the two forces create a potential well which confines the ions vertically in a thin layer with thickness less than 2×10^{-7} cm at a distance $\sim 4 \times 10^{-6}$ cm below the surface of the liquid.[44,45] A positive potential $V_G > 0$ applied to the guard ring provides radial confinement of the two-dimensional ion layer. Waves are excited in the experiments by imposing an oscillating potential $\tilde{\phi}_W$ on the guard ring at $r = b$ and are detected by measuring the current from a circular button located at the center of electrode A.

Prasad and Morales[46] have investigated the equilibrium properties and collective modes of oscillation in such a two-dimensional nonneutral ion plasma, including the influence of an applied axial magnetic field $B_0 \hat{e}_z$. In this section, we summarize several features of the equilibrium and wave properties for the case where $B_0 = 0$.

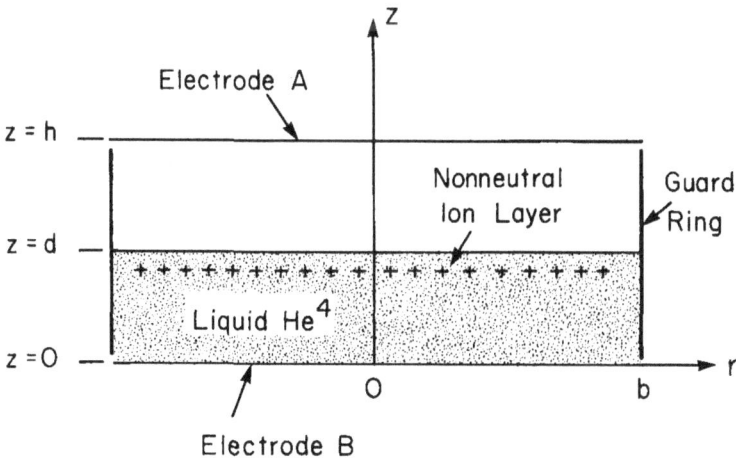

Figure 5.13. Typical experimental cell geometry used to create and study a two-dimensional nonneutral ion plasma located just below the surface of liquid helium [M.L. Ott-Rowland, et al., Phys. Rev. Lett. **49**, 1708 (1982)]. The positive ions form a circular charge disk held in position by electric potentials applied to the walls of the confining cell (shaped like a cylindrical pillbox), and by the vertical electric field resulting from the polarization of the liquid helium.

5.9.1   Two-Dimensional Disk Equilibrium

Under steady-state conditions $(\partial/\partial t = 0)$, it is assumed that the equilibrium ion density $\sigma_i^0(r)$ in the plane of the disk $(z = d)$ is distributed according to the Boltzmann factor $\exp(-e\phi_0/k_BT)$. Here, $+e$ is ion charge and T is the temperature of the plasma ions (assumed to be the same as the temperature of the liquid helium). Therefore, Poisson's equation for the electrostatic potential $\phi_0(r,z)$ can be expressed as

$$\left(\frac{1}{r}\frac{\partial}{\partial r}r\frac{\partial}{\partial r} + \frac{\partial^2}{\partial z^2}\right)\phi_0(r,z) \quad = -4\pi e\sigma_i^0(r)\delta(z-d) \qquad (5.139)$$

$$= -4\pi e\sigma_i^0(r=0)\exp\left\{-\frac{e}{k_BT}\left[\phi_0(r,z=d) - \phi_0(r=0,z=d)\right]\right\}\delta(z-d) \ .$$

Equation (5.139) is to be solved within the cell region $0 \leq r \leq b$ and $0 \leq z \leq h$ (Fig. 5.13) subject to the boundary conditions $\phi_0(r = b, z) = 0$, $\phi_0(r, z = 0) = V_{BA} - V_G$, and $\phi_0(r, z = h) = -V_G$. (For convenience, the zero of potential has been shifted by an amount $-V_G$ relative to the convention used at the beginning of Sec. 5.9.) We denote the n'th zero of $J_0(x) = 0$ by p_{0n}, and define $k_{0n} = p_{0n}/b$. It is straightforward to show that the solution to Eq.(5.139) for $0 \leq r \leq b$ can be expressed as[46]

$$\phi_0(r,z) = \sum_{n=1}^{\infty}\left[\alpha_n f_{0n}(z) + g_{0n}(z)\right]J_0(k_{0n}r) \ , \qquad (5.140)$$

where

$$f_{0n}(z) = \begin{cases} \dfrac{\sinh(k_{0n}z)}{\sinh(k_{0n}d)} \ , & 0 \leq z < d \ , \\[3mm] \dfrac{\sinh[k_{0n}(h - z)]}{\sinh[k_{0n}(h - d)]} \ , & d < z \leq h \ , \end{cases} \qquad (5.141)$$

and

$$g_{0n}(z) = -\frac{2\sinh[k_{0n}(z - d)]}{k_{0n}bJ_1(k_{0n}b)} \times \begin{cases} \dfrac{(V_{BA} - V_G)}{\sinh(k_{0n}d)} \ , & 0 \leq z < d \ , \\[3mm] \dfrac{V_G}{\sinh[k_{0n}(h - d)]} \ , & d < z \leq h \ . \end{cases}$$

$$(5.142)$$

The solution for $\phi_0(r, z)$ satisfies the correct boundary conditions at $z = 0$, $z = h$ and $r = b$, and is continuous across the ion layer at $z = d$.

Integrating Poisson's equation (5.139) from $z = d(1 - \varepsilon)$ to $z = d(1 + \varepsilon)$ and taking the limit $\varepsilon \to 0_+$ gives

$$\left[\frac{\partial}{\partial z} \phi_0 \right]_{z=d_+} - \left[\frac{\partial}{\partial z} \phi_0 \right]_{z=d_-} = -4\pi e \sigma_i^0(r) , \qquad (5.143)$$

which relates the discontinuity in axial electric field, $E_z^0 = -\partial\phi_0/\partial z$, to the equilibrium charge density $e\sigma_i^0(r)$ of the disk ions. Substituting Eqs.(5.140)–(5.142) into Eq.(5.143) and operating with $\int_0^b dr\, r J_0(k_{0n}r) \cdots$ gives[46]

$$\frac{1}{2} b^2 J_1^2(k_{0n}b) \left[\alpha_n \left(\frac{k_{0n} \sinh(k_{0n}h)}{\sinh(k_{0n}d) \sinh[k_{0n}(h-d)]} \right) \right.$$

$$\left. + \frac{2}{b J_1(k_{0n}b)} \left(\frac{V_G}{\sinh[k_{0n}(h-d)]} - \frac{(V_{BA} - V_G)}{\sinh(k_{0n}d)} \right) \right]$$

$$= 4\pi e \int_0^b dr\, r J_0(k_{0n}r) \sigma_i^0(r) \qquad (5.144)$$

$$= 4\pi e \int_0^b dr\, r J_0(k_{0n}r) \sigma_i^0(r=0) \exp\left\{ -\frac{e}{k_B T} \sum_{m=1}^{\infty} \alpha_m [J_0(k_{0m}r) - 1] \right\} .$$

Equation (5.144) is a nonlinear equation for the coefficients α_n, and hence for the ion density profile $\sigma_i^0(r)$.

Equation (5.144) has been solved numerically by successive iteration.[46] At low temperatures, the density profile $\sigma_i^0(r)$ is approximately uniform out to some radius $r = r_b < b$, where it decreases rapidly to zero in a thin boundary layer (Fig. 5.14). For a cell with short vertical height ($h \ll b$), the characteristic thickness of the boundary layer is of order the thermal Debye length λ_{Di} defined by

$$\lambda_{Di} = \left(\frac{k_B T h}{4\pi e^2 \sigma_i^0(r=0)} \right)^{1/2} . \qquad (5.145)$$

Figure 5.14 illustrates the equilibrium density profile $\sigma_i^0(r)$ determined numerically from Eqs.(5.139) and (5.144) for $h/b = 0.2$ and $\lambda_{Di}/h = 0.1, 0.2$, and for typical experimental values[44] $V_G = 11.6\, \sigma_i^0(r = 0)eh$, $V_{BA} = 9.9\, \sigma_i^0(r = 0)eh$, and $d = 0.467\, h$. Evidently, for the choice of parameters in Fig. 5.14, the ion density is uniform to high accuracy out to radius $r = r_b$, which is related to $\sigma_i^0(r = 0)$ and the total ion charge Q_i by $Q_i \simeq \sigma_i^0(r = 0)e\pi r_b^2$.

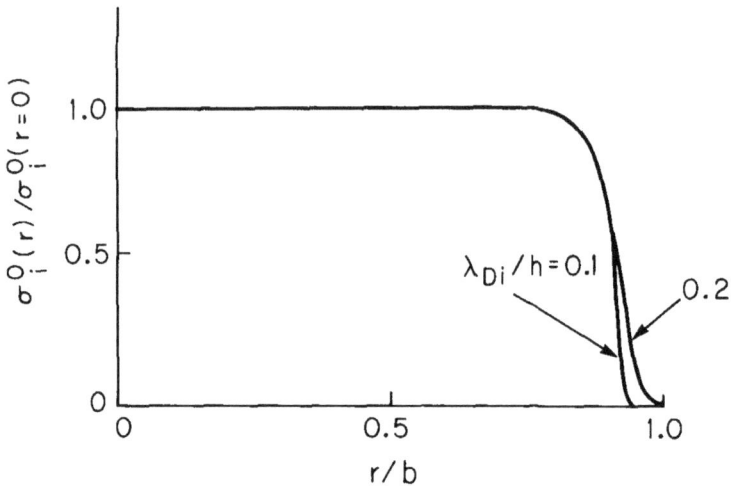

Figure 5.14. Plot versus r/b of the normalized ion density profile $\sigma_i^0(r)/\sigma_i^0(r = 0)$ determined numerically from Eqs.(5.139) and (5.144) for $h/b = 0.2$, $d/h = 0.467$, and applied voltages $V_G = 11.6\,\sigma_i^0(r = 0)eh$ and $V_{BA} = 9.9\,\sigma_i^0(r = 0)eh$. The two cases $\lambda_{Di}/h = 0.1, 0.2$ are presented in the figure [S.A. Prasad and G.J. Morales, Phys. Fluids **30**, 3475 (1987)].

5.9.2   Normal Modes of Oscillation

In this section, collective oscillations are investigated for electrostatic perturbations about the axisymmetric disk equilibrium described in Sec. 5.9.1. Perturbed quantities are expressed as $\delta\psi(r, \theta, z, t) = \sum_{\ell=-\infty}^{\infty} \delta\psi^\ell(r, z) \times \exp(i\ell\theta - i\omega t)$, and use is made of the linearized cold-fluid-Poisson equations (Sec. 2.3) to investigate electrostatic stability properties. It is assumed that the motion of the perturbed ion fluid occurs only in the plane $z = d$ with $\delta V_{zi} = 0$. The linearized continuity and force balance equations are given by

$$-i\omega\delta\sigma_i^\ell(r) + i\frac{\ell}{r}\sigma_i^0(r)\delta V_{\theta i}^\ell(r, z = d) + \frac{1}{r}\frac{\partial}{\partial r}\left[r\sigma_i^0(r)\delta V_{ri}^\ell(r, z = d)\right] = 0\,,$$

$$(5.146)$$

$$-i\omega\delta V_{ri}^\ell(r, z = d) = -\frac{e}{m_i}\frac{\partial}{\partial r}\delta\phi^\ell(r, z = d)\,,$$

$$(5.147)$$

$$-i\omega\delta V_{\theta i}^\ell(r, z = d) = -\frac{e}{m_i}\frac{i\ell}{r}\delta\phi^\ell(r, z = d)\,.$$

$$(5.148)$$

Here, the perturbed electrostatic potential is calculated self-consistently from the linearized Poisson equation

$$\left(\frac{1}{r}\frac{\partial}{\partial r}r\frac{\partial}{\partial r} - \frac{\ell^2}{r^2} + \frac{\partial^2}{\partial z^2}\right)\delta\phi^\ell(r,z) = -4\pi e\delta\sigma_i^\ell(r)\delta(z-d) . \tag{5.149}$$

Introducing the effective ion plasma frequency defined by

$$\omega_{pi}^2(r) = \frac{4\pi\sigma_i^0(r)e^2}{m_i h} , \tag{5.150}$$

it is straightforward to show from Eqs.(5.146)–(5.148) that

$$4\pi e\delta\sigma_i^\ell(r) = \frac{h\omega_{pi}^2(r)}{\omega^2}\frac{\ell^2}{r^2}\delta\phi^\ell(r,z=d) - \frac{1}{r}\frac{\partial}{\partial r}\left[\frac{h\omega_{pi}^2(r)}{\omega^2}r\frac{\partial}{\partial r}\delta\phi^\ell(r,z=d)\right] .$$

$$\tag{5.151}$$

The solution to Eq.(5.149) which satisfies $\delta\phi^\ell = 0$ at $r = b$, $z = 0$ and $z = h$, and is continuous across the ion layer at $z = d$ can be expressed as

$$\delta\phi^\ell(r,z) = \sum_{n=1}^{\infty}\beta_n\, f_{\ell n}(z)J_\ell(k_{\ell n}r) . \tag{5.152}$$

Here, $k_{\ell n}$ is the n'th zero of $J_\ell(k_{\ell n}b) = 0$, and $f_{\ell n}(z)$ is defined by

$$f_{\ell n}(z) = \begin{cases} \dfrac{\sinh(k_{\ell n}z)}{\sinh(k_{\ell n}d)} , & 0 \le z < d , \\[2ex] \dfrac{\sinh[k_{\ell n}(h-z)]}{\sinh[k_{\ell n}(h-d)]} , & d < z \le h . \end{cases} \tag{5.153}$$

Integrating Eq.(5.149) from $z = d(1-\varepsilon)$ to $z = d(1+\varepsilon)$ and taking the limit $\varepsilon \to 0_+$ gives

$$\left[\frac{\partial}{\partial z}\delta\phi^\ell\right]_{z=d_+} - \left[\frac{\partial}{\partial z}\delta\phi^\ell\right]_{z=d_-} = -4\pi e\delta\sigma_i^\ell(r) . \tag{5.154}$$

Substituting Eqs.(5.151)–(5.153) into Eq.(5.154) and operating with $\int_0^b dr\, rJ_\ell(k_{\ell m}r)\cdots$ gives the matrix dispersion equation[46]

$$\text{Det}\{D_{mn}(\omega)\} = 0 , \tag{5.155}$$

where $D_{mn}(\omega)$ is defined by

$$D_{mn}(\omega) = \int_0^b dr\, r\omega_{pi}^2(r) \left[k_{\ell m}k_{\ell n}J_\ell'(k_{\ell m}r)J_\ell'(k_{\ell n}r) + \frac{\ell^2}{r^2}J_\ell(k_{\ell m}r)J_\ell(k_{\ell n}r) \right]$$

$$- \omega^2 \frac{b}{2h} \frac{J_{\ell+1}^2(k_{\ell n}b)k_{\ell n}b\sinh(k_{\ell n}h)}{\sinh(k_{\ell n}d)\sinh[k_{\ell n}(h-d)]}\delta_{m,n} \,. \tag{5.156}$$

In Eq.(5.156), $J_\ell'(x)$ denotes $dJ_\ell(x)/dx$, and $\delta_{m,n}$ is the Kronecker delta with $\delta_{m,n} = 1$ for $m = n$, and $\delta_{m,n} = 0$ for $m \neq n$.

For specified density profile $\sigma_i^0(r)$, the eigenfrequencies can be determined numerically from Eqs.(5.155) and (5.156). Typical results[46] are illustrated in Fig. 5.15 for the case where $\sigma_i^0(r) = \sigma_i^0(r = 0) = $ const. over the interval $0 \leq r < r_b$, and the system parameters correspond to $r_b/b = 0.895$, $h/b = 0.2$, and $d/b = 0.093$. The eigenfrequencies in Fig. 5.15 are characterized by an azimuthal mode number ℓ and a radial mode number N. Note that the minimum eigenfrequency in Fig. 5.15 increases as the mode number ℓ is increased.

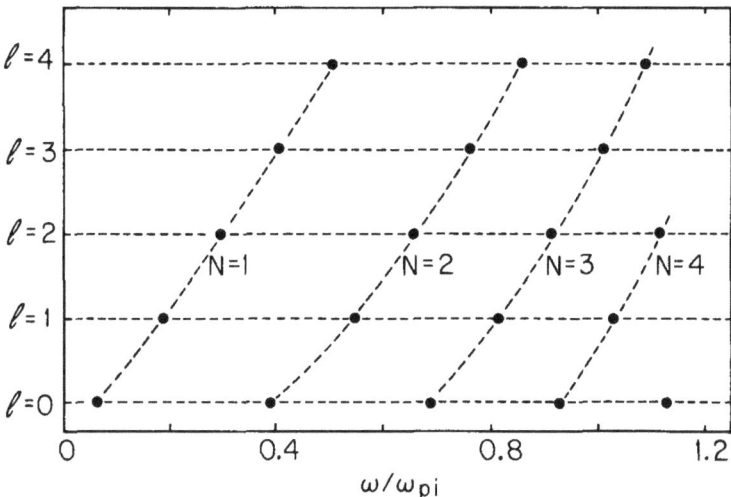

Figure 5.15. Eigenfrequencies for ion plasma oscillations in a uniform-density two-dimensional disk calculated numerically from Eqs.(5.155) and (5.156) for $r_b/b = 0.895$, $h/b = 0.2$, and $d/b = 0.093$. The modes presented in the figure are characterized by an azimuthal mode number ℓ and radial mode number N [S.A. Prasad and G.L. Morales, Phys. Fluids **30**, 3475 (1987)].

It is evident from Eq.(5.155) and Fig. 5.15 that a two-dimensional non-neutral ion plasma supports collective plasma oscillations over a wide frequency range. This property is not unlike that found for a three-dimensional nonneutral plasma. What is most striking in the present analysis, however, is that the wave perturbations are about a two-dimensional disk equilibrium. As indicated earlier, the normal-mode analysis can be extended[46] to include the influence of an applied axial magnetic field $B_0\hat{e}_z$. Assuming a nonrotating disk equilibrium with $V_{\theta i}^0(r, z = d) = 0$, the only modification to Eqs.(5.155) and (5.156) for the special case of azimuthally symmetric perturbations ($\ell = 0$) is the replacement

$$\omega^2 \rightarrow \omega^2 - \omega_{ci}^2 \,, \tag{5.157}$$

where $\omega_{ci} = eB_0/m_i c$ is the ion cyclotron frequency. Therefore, the solutions for ω^2 are upshifted by an amount ω_{ci}^2 relative to the unmagnetized case. In circumstances where an equilibrium rotation $V_{\theta i}^0(r, z = d)$ is induced, and there is an inverted population in the ion density profile $\sigma_i^0(r)$ as a function of radius r, it is reasonable to expect that the analogue of the diocotron instability (Chapter 6) exists for perturbations about a two-dimensional disk equilibrium.

Problem 5.16

a. Show that Eqs.(5.149) and (5.151) can be combined to give

$$\omega^2 \nabla^2 \left[\delta\phi^\ell(r, z) \exp(i\ell\theta)\right] = h\nabla_\perp \cdot \left\{\omega_{pi}^2(r)\nabla_\perp \left[\delta\phi^\ell(r, z) \exp(i\ell\theta)\right]\right\} \delta(z-d) \,,$$
$$\tag{5.16.1}$$

where $\omega_{pi}^2(r) = 4\pi\sigma_i^0(r)e^2/m_i h$, $\nabla_\perp = \hat{e}_r(\partial/\partial r) + \hat{e}_\theta r^{-1}(\partial/\partial\theta)$, and $\nabla^2 = r^{-1}(\partial/\partial r)(r\partial/\partial r) + r^{-2}\partial^2/\partial\theta^2 + \partial^2/\partial z^2$.

b. Multiply Eq.(5.16.1) by $\delta\phi^{\ell*}(r, z) \exp(-i\ell\theta)$ and integrate over the volume of the cell in Fig. 5.13, enforcing the boundary conditions $\delta\phi^\ell = 0$ at $r = b, z = 0$ and $z = h$. Show that

$$\omega^2 = h\frac{\int_0^{2\pi} d\theta \int_0^b dr\, r\omega_{pi}^2(r) \left|\nabla_\perp \left[\delta\phi^\ell(r, z = d) \exp(i\ell\theta)\right]\right|^2}{\int_0^{2\pi} d\theta \int_0^b dr\, r \int_0^h dz\, |\nabla \left[\delta\phi^\ell(r, z) \exp(i\ell\theta)\right]|^2} \,. \tag{5.16.2}$$

Equation (5.16.2) can be used to formulate a (Rayleigh-Ritz) variational principle to determine ω^2. The minimum value of the right-hand side of Eq.(5.16.2) gives the lowest eigenvalue ω^2, and the function $\delta\phi^\ell(r, z)$ that makes it a minimum is the corresponding eigenfunction.

Problem 5.17   Extend the normal-mode analysis in Sec. 5.9.2 to include the influence of an applied axial magnetic field $B_0 \hat{e}_z$. Assuming a nonrotating disk equilibrium with $V_{\theta i}^0(r, z = d) = 0$, and azimuthally symmetric perturbations with $\partial/\partial\theta = 0$, show that the only modification to Eqs.(5.155) and (5.156) is the replacement $\omega^2 \to \omega^2 - \omega_{ci}^2$, where $\omega_{ci} = eB_0/m_i c$ is the ion cyclotron frequency.

Chapter 5 References

1. R.C. Davidson, *Theory of Nonneutral Plasmas* (Addison-Wesley, Reading, Massachusetts, 1989), Chapter 2.

2. R.C. Davidson, "Waves and Instabilities in Nonneutral Plasmas," in *Nonneutral Plasma Physics*, eds., C.W. Roberson and C.F. Driscoll, AIP Conference Proceedings No. **175** (American Institute of Physics, New York, 1988), pp. 139-208.

3. R.C. Davidson, "Relativistic Electron Beam-Plasma Interaction with Intense Self Fields," in *Handbook of Plasma Physics - Basic Plasma Physics*, eds., M.N. Rosenbluth and R.Z. Sagdeev (North Holland, Amsterdam, 1984), Volume 2, pp. 729-819.

4. R.B. Miller, *Intense Charged Particle Beams* (Plenum, New York, 1982).

5. K.T. Tsang and R.C. Davidson, "Macroscopic Cold-Fluid Equilibrium Properties of Relativistic Nonneutral Electron Flow in a Cylindrical Diode," Phys. Rev. **A33** , 4284 (1986).

6. W.H. Bennett, "Magnetically Self-Focussing Streams," Phys. Rev. **45**, 890 (1934).

7. J.D. Lawson, "Perveance and the Bennett Pinch Relation in Partially Neutralized Electron Beams," J. Electron. Control **5**, 146 (1958).

8. R.C. Davidson and H.S. Uhm, "Thermal Equilibrium Properties of an Intense Relativistic Electron Beam," Phys. Fluids **22**, 1375 (1979).

9. M. Reiser, "Laminar-Flow Equilibria and Limiting Currents in Magnetically-Focused Relativistic Beams," Phys. Fluids **20**, 477 (1977).

10. A.W. Trivelpiece and R.W. Gould, "Plasma Waves in Cylindrical Plasma Columns," J. Appl. Phys. **30**, 1784 (1959).

11. A.W. Trivelpiece, *Slow-Wave Propagation in Plasma Waveguides* (San Francisco Press, San Francisco, California, 1967).

12. L.M. Linson, "Electrostatic Electron Oscillations in Cylindrical Nonneutral Plasmas," Phys. Fluids **14**, 805 (1971).

13. W.W. Rigrod and J.A. Lewis, "Wave Propagation Along a Magnetically-Focused Electron Beam," Bell System Tech. J. **33**, 399 (1954).

14. G.R. Brewer, "Some Effects of Magnetic Field Strength on Space-Charge Wave Propagation," Proc. IRE **44**, 896 (1956).

15. J. Labus, "Space-Charge Waves Along Magnetically Focused Electron Beam," Proc. IRE **45**, 854 (1957).

16. W.W. Rigrod, "Space-Charge Wave Harmonics and Noise Propagating in Rotating Electron Beams," Bell System Tech. J. **38**, 119 (1959).

17. H. Potzl, "Types of Waves in Magnetically Focused Electron Beams," Arch. Elek. Ubertragung 19, 367 (1965).

18. J.H. Malmberg and J.S. DeGrassie, "Properties of Nonneutral Plasmas," Phys. Rev. Lett. **35**, 577 (1975).

19. G. Dimonte, "Ion Langmuir Waves in a Nonneutral Plasma," Phys. Rev. Lett. **46**, 26 (1980).

20. T.H. Stix, *Theory of Plasma Waves* (McGraw Hill, New York, 1962).

21. R.C. Davidson, "Kinetic Waves and Instabilities in Uniform Plasma," in *Handbook of Plasma Physics - Basic Plasma Physics*, eds., M.N. Rosenbluth and R.Z. Sagdeev (North Holland, Amsterdam, 1983), Volume 1, pp. 521-585.

22. R.H. Levy, J.D. Daugherty and O. Buneman, "Ion Resonance Instability in Grossly Nonneutral Plasmas," Phys. Fluids **12**, 2616 (1969).

23. R.C. Davidson and H.S. Uhm, "Influence of Strong Self-Electric Fields on the Ion Resonance Instability in a Nonneutral Plasma Column," Phys. Fluids **20**, 1938 (1977).

24. R.C. Davidson and H.S. Uhm, "Influence of Finite Ion Larmor Radius Effects on the Ion Resonance Instability in a Nonneutral Plasma Column," Phys. Fluids **21**, 60 (1978).

25. H.S. Uhm and R.C. Davidson, "Influence of Finite Ion Larmor Radius and Equilibrium Self Electric Fields on the Ion Resonance Instability," Plasma Physics **20**, 579 (1978).

26. H.S. Uhm and R.C. Davidson, "Ion Resonance Instability in a Modified Betatron Accelerator," Phys. Fluids **25**, 2334 (1982).

27. L.S. Bogdankevich and A.A. Rukhadze, "Stability of Relativistic Electron Beams in a Plasma and the Problem of Critical Currents," Sov. Phys.-Usp. **14**, 163 (1971), and references therein.

28. A.A. Ivanov and L.I. Rudakov, "Intense Relativistic Electron Beam in a Plasma," Sov. Phys. JETP **31**, 715 (1970).

29. R.V. Lovelace and R.N. Sudan, "Plasma Heating by High-Current Relativistic Electron Beams," Phys. Rev. Lett. **27**, 1256 (1971).

30. L.E. Thode and R.N. Sudan, "Two-Stream Instability Heating of Plasmas by Relativistic Electron Beams," Phys. Rev. Lett. **30**, 732 (1973).

31. C.A. Kapetanakos and D.A. Hammer, "Plasma Heating by an Intense Relativistic Electron Beam," Appl. Phys. Lett. **23**, 17 (1973).

32. P.A. Miller and G.W. Kuswa, "Plasma Heating by an Intense Relativistic Electron Beam," Phys. Rev. Lett. **30** 958 (1973).

33. G.C. Goldenbaum, W.F. Dove, K.A. Gerber and B.G. Logan, "Plasma Heating by Intense Relativistic Electron Beams," Phys. Rev. Lett. **32**, 830 (1974).

34. D.A. Hammer, K.A. Gerber, W.F. Dove, G.C. Goldenbaum, B.G. Logan, K. Papadopoulos and A.W. Ali, "Intense Relativistic Electron Beam Interaction with a Cool Theta-Pinch Plasma," Phys. Fluids **21**, 483 (1978).

35. A. Alfvén, "On the Motion of Cosmic Rays in Interstellar Space," Phys. Rev. **55**, 425 (1939).

36. J.D. Lawson, "On the Classification of Electron Streams," J. Nuclear Energy, Part C: Plasma Phys. **1**, 31 (1959).

37. I.B. Bernstein and S.K. Trehan, "Plasma Oscillations (I)," Nuclear Fusion **1**, 3 (1960).

38. R.C. Davidson, B.H. Hui and C.A. Kapetanakos, "Influence of Self Fields on the Filamentation Instability in Relativistic Beam-Plasma Systems," Phys. Fluids **18**, 1040 (1975).

39. R.C. Davidson and B.H. Hui, "Influence of Self Fields on the Equilibrium and Stability Properties of Relativistic Electron Beam-Plasma Systems," Annals of Physics **94**, 209 (1975).

40. R. Lee and M. Lampe, "Electromagnetic Instabilities, Filamentation and Focusing of Relativistic Electron Beams," Phys. Rev. Lett. **31**, 1390 (1973).

41. G. Benford, "Electron Beam Filamentation in Strong Magnetic Fields," Phys. Rev. Lett. **28**, 1242 (1972).

42. G. Benford, "Theory of Filamentation in Relativistic Electron Beams," Plasma Physics **15**, 433 (1973).

43. C.A. Kapetanakos, "Filamentation of Relativistic Electron Beams Propagating in Dense Plasmas," Appl. Phys. Lett. **25**, 484 (1974).

44. S. Hannahs and G.A. Williams, "Plasma Wave Resonances in Positive Ions Under the Surface of Liquid Helium," Japanese Journal of Applied Physics **26-3**, 741 (1987).

45. M.L. Ott-Rowland, V. Kotsubo, J. Theobald and G.A. Williams, "Two-Dimensional Plasma Resonances in Positive Ions Under the Surface of Liquid Helium," Phys. Rev. Lett. **49**, 1708 (1982).

46. S.A. Prasad and G.J. Morales, "Equilibrium and Wave Properties of Two-Dimensional Ion Plasmas," Phys. Fluids **30**, 3475 (1987).

The following references, while not cited directly in the text, are relevant to the general subject matter of this chapter.

S.A. Prasad and G.J. Morales, "Magnetized Equilibrium of a Two-Dimensional Ion Plasma," Phys. Fluids **B1**, 1329 (1989).

S.A. Prasad and G.J. Morales, "Nonlinear Resonance of Two-Dimensional Ion Layers," Phys. Fluids **31**, 562 (1988).

W.R. Shanahan, "Filamentation Instability of a Radially Finite Relativistic Electron Beam," Phys. Fluids **29**, 1231 (1986).

G.J. Morales, "Effect of a DC Electric Field on the Trapping Dynamics of a Cold Electron Beam," Phys. Fluids **23**, 2472 (1980).

CHAPTER 6

THE DIOCOTRON INSTABILITY

The angular velocity shear was equal to zero ($\partial \omega_{rj}/\partial r = 0$) for the nonneutral plasma waves and instabilities investigated in Chapters 4 and 5. This restriction is now removed, and we examine several features of the classical diocotron ("slipping-stream") instability[1-4] in a low-density, pure electron plasma with $\omega_{pe}^2(r)/\omega_{ce}^2 \ll 1$ and $n_i^0(r) = 0$. The diocotron instability is one of the most ubiquitous instabilities in low-density nonneutral plasmas with shear in the flow velocity. For example, it can occur in propagating nonneutral electron beams and layers[5-11] and in low-voltage microwave generation devices such as magnetrons, traveling-wave tubes and ubitrons.[12] In this chapter, we describe the nonrelativistic guiding-center model used to investigate the evolution of the diocotron instability (Sec. 6.1), and make use of certain global conservation constraints to derive a sufficient condition for electrostatic stability of nonneutral electron flow in cylindrical geometry (Sec. 6.2). Linear stability properties are then investigated (Sec. 6.3), and the results are summarized for three experiments carried out for annular electron beams (Sec. 6.4). Then, for multimode wave excitation, a simple quasilinear model of the nonlinear evolution of the diocotron instability is presented (Sec. 6.5), including an investigation of the resonant diocotron instability (Sec. 6.6). Finally, for relativistic cross-field electron flow, it is shown that the diocotron instability can be completely stabilized by electromagnetic and relativistic effects (Sec. 6.7).

The analysis of the diocotron instability in Chapter 6 treats the nonneutral electron plasma as a macroscopic cold fluid with negligibly small pressure variations. To include thermal effects, one can make use of a kinetic description based on the Vlasov-Poisson equations (Chapter 4). Alternatively, for strongly magnetized electrons and negligibly small heat flow, a closed equation for the evolution of the pressure tensor can be developed (see Problem 2.3).

6.1   Nonrelativistic Guiding-Center Model

Consider a low-density nonneutral electron plasma in cylindrical geometry with

$$\frac{\omega_{pe}^2(r)}{\omega_{ce}^2} = \frac{4\pi n_e^0(r)m_e c^2}{B_0^2} \ll 1 \qquad (6.1)$$

and $n_i^0(r) = 0$. The electrons are confined radially by a uniform axial magnetic field $B_0 \hat{e}_z$, and cylindrical conducting walls are located at $r = a$ and $r = b$ (Fig. 6.1). The case where the central conductor is absent is treated simply by setting $a = 0$.

In the present analysis, a cold-fluid guiding-center model is adopted in which electron inertial effects are neglected ($m_e \rightarrow 0$), and the motion of

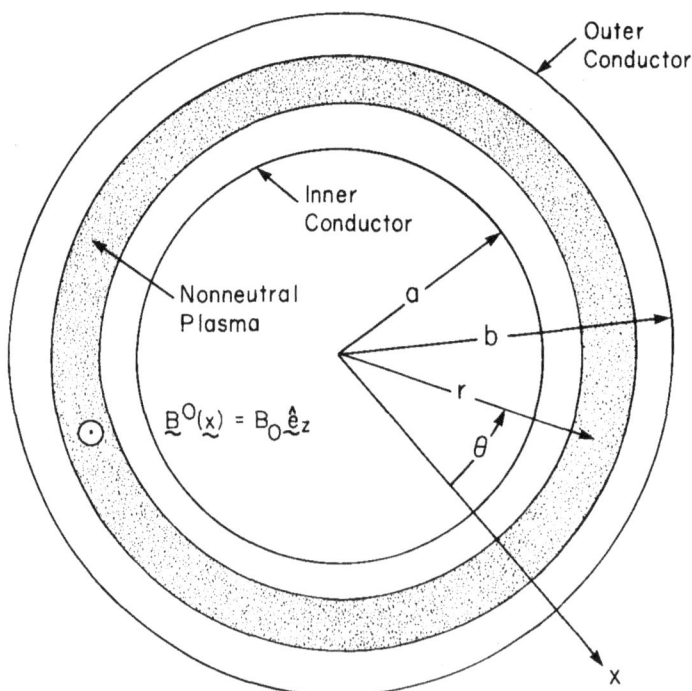

Figure 6.1.   A low-density nonneutral plasma is confined between cylindrical conducting walls at $r = a$ and $r = b$ by a uniform axial magnetic field $B_0 \hat{e}_z$.

a strongly magnetized electron fluid element is determined from

$$0 = -en_e(\mathbf{x}, t) \left[\mathbf{E}(\mathbf{x}, t) + \frac{1}{c} \mathbf{V}_e(\mathbf{x}, t) \times B_0 \hat{\mathbf{e}}_z \right] . \tag{6.2}$$

In the electrostatic approximation, $\mathbf{E}(\mathbf{x}, t) = -\nabla\phi(\mathbf{x}, t)$, and Eq.(6.2) gives

$$\mathbf{V}_e(x, t) = -\frac{c}{B_0} \nabla\phi(\mathbf{x}, t) \times \hat{\mathbf{e}}_z \tag{6.3}$$

for the perpendicular motion. In cylindrical geometry, Eq.(6.3) reduces to

$$V_{re}(r, \theta, t) = -\frac{c}{B_0 r} \frac{\partial}{\partial\theta} \phi(r, \theta, t) , \tag{6.4a}$$

$$V_{\theta e}(r, \theta, t) = \frac{c}{B_0} \frac{\partial}{\partial r} \phi(r, \theta, t) , \tag{6.4b}$$

where flute perturbations with $\partial/\partial z = 0$ have been assumed. The continuity equation (2.28) relates the evolution of the electron density $n_e(\mathbf{x}, t)$ to the flow velocity $\mathbf{V}_e(\mathbf{x}, t)$ by

$$\frac{\partial}{\partial t} n_e(\mathbf{x}, t) + \frac{\partial}{\partial \mathbf{x}} \cdot [n_e(\mathbf{x}, t) \mathbf{V}_e(\mathbf{x}, t)] = 0 . \tag{6.5}$$

Because $\nabla \cdot \mathbf{V}_e = 0$ follows from Eq.(6.3), Eq.(6.5) can be expressed as $\partial n_e/\partial t + \mathbf{V}_e \cdot \nabla n_e = 0$, or equivalently,

$$\left\{ \frac{\partial}{\partial t} - \frac{c}{B_0 r} \frac{\partial\phi}{\partial\theta} \frac{\partial}{\partial r} + \frac{c}{B_0 r} \frac{\partial\phi}{\partial r} \frac{\partial}{\partial\theta} \right\} n_e(r, \theta, t) = 0 . \tag{6.6}$$

Of course, Eq.(6.6) must be supplemented by Poisson's equation

$$\left(\frac{1}{r} \frac{\partial}{\partial r} r \frac{\partial}{\partial r} + \frac{1}{r^2} \frac{\partial^2}{\partial\theta^2} \right) \phi(r, \theta, t) = 4\pi e n_e(r, \theta, t) , \tag{6.7}$$

which relates self-consistently the electrostatic potential $\phi(r, \theta, t)$ to the electron density $n_e(r, \theta, t)$.

Equations (6.4), (6.6), and (6.7) constitute a *fully nonlinear* description of the system evolution in the cold-fluid guiding-center approximation with $m_e \rightarrow 0$. Although the ratio $\omega_{pe}^2(r)/\omega_{ce}^2 = 4\pi n_e^0(r) m_e c^2/B_0^2$ approaches zero in the limit of zero electron mass, note that the effective diocotron frequency defined by $\omega_D(r) = \omega_{pe}^2(r)/2\omega_{ce} = 2\pi n_e^0(r)ec/B_0$ remains finite as $m_e \to 0$. Assuming that the cylinders at $r = a$ and $r = b$ are perfectly conducting, it is required that

$$E_\theta(r,\theta,t) = -\frac{1}{r}\frac{\partial}{\partial\theta}\phi(r,\theta,t) = 0, \quad \text{at } r = a \text{ and } r = b , \tag{6.8}$$

which corresponds to zero tangential electric field at the conducting walls. Making use of $V_{re}(r,\theta,t) = -(c/B_0 r)(\partial/\partial\theta)\phi(r,\theta,t)$ [Eq.(6.4)], it follows from Eq.(6.8) that

$$V_{re}(r,\theta,t) = 0, \quad \text{at } r = a \text{ and } r = b \tag{6.9}$$

which corresponds to zero radial flow of the electron fluid at $r = a$ and $r = b$.

Before presenting a linear stability analysis of Eqs.(6.6) and (6.7) (Sec. 6.3), it is useful to develop a stability theorem based on global conservation constraints (Sec. 6.2).

6.2 Electrostatic Stability Theorem

In this section, a sufficient condition is derived for the equilibrium density profile $n_e^0(r)$ to be stable to electrostatic perturbations. In this regard, the macroscopic guiding-center model based on Eqs.(6.6) and (6.7) possesses certain global (spatially averaged) nonlinear conservation constraints that can be used to develop a stability theorem.[13]

Consider the quantity ΔU_G defined by

$$\Delta U_G = \int d^2x \left[G(n_e) - G(n_e^0) \right] , \tag{6.10}$$

where $G(n_e)$ is a (yet unspecified) smooth, differentiable function with $G(n_e \to 0) = 0$, and $\int d^2x \cdots = \int_0^{2\pi} d\theta \int_a^b dr\, r \cdots$ in cylindrical geometry. It follows from Eq.(6.5) that $\partial G(n_e)/\partial t = -\nabla \cdot [G(n_e)\mathbf{V}_e]$, so that

$$\frac{d}{dt}\Delta U_G = \int d^2x \frac{\partial}{\partial t}G(n_e)$$

$$= -\int_0^{2\pi} d\theta \int_a^b dr\, r \left[\frac{1}{r}\frac{\partial}{\partial r}(rV_{re}G) + \frac{1}{r}\frac{\partial}{\partial\theta}(V_{\theta e}G) \right] . \tag{6.11}$$

The $\partial/\partial\theta$ contribution in Eq.(6.11) integrates to zero by virtue of periodicity in the θ-direction. The $\partial/\partial r$ contribution in Eq.(6.11) integrates to zero because the radial flow velocity $V_{re}(r,\theta,t)$ is equal to zero at $r=a$ and $r=b$ [Eq.(6.9)]. From Eq.(6.11), we conclude that $(d/dt)\Delta U_G = 0$, or equivalently,

$$\Delta U_G = \int d^2x \left[G(n_e) - G(n_e^0)\right] = \text{const.} \qquad (6.12)$$

A special case of Eq.(6.12) is the conservation of total charge

$$\Delta U_q = -e \int d^2x \left(n_e - n_e^0\right) = \text{const.} = 0 , \qquad (6.13)$$

where the constant in Eq.(6.13) has been taken equal to zero. This is consistent for all times t provided zero net charge is introduced into the system by the initial density perturbation, i.e., provided $\int d^2x \delta n_e(r,\theta,t=0) = 0$.

A further global conservation constraint is related to the density-weighted average radial location of guiding centers. Defining

$$\Delta U_r = \int d^2x r^2 \left(n_e - n_e^0\right) , \qquad (6.14)$$

and making use of $\partial n_e/\partial t = -\nabla \cdot (n_e \mathbf{V}_e)$ gives

$$\frac{d}{dt}\Delta U_r = -\frac{2c}{B_0} \int_0^{2\pi} d\theta \int_a^b dr\, rn_e \frac{\partial}{\partial\theta}\phi . \qquad (6.15)$$

In obtaining Eq.(6.15), we have integrated by parts with respect to r and θ, and made use of periodicity in θ , as well as $V_{re}(r,\theta,t) = 0$ at $r=a$ and $r=b$. If the density factor $n_e(r,\theta,t)$ in Eq.(6.15) is eliminated in favor of $\phi(r,\theta,t)$ by means of Poisson's equation (6.7), then some straightforward algebra shows that the right-hand side of Eq.(6.15) is equal to zero (Problem 6.1). This gives $(d/dt)\Delta U_r = 0$, or equivalently,

$$\Delta U_r = \int d^2x r^2 \left(n_e - n_e^0\right) = \text{const.} \qquad (6.16)$$

Equation (6.16) is a statement of the conservation of the canonical angular momentum of an electron fluid element, $\int d^2x n_e(m_e rV_{\theta e} - eB_0 r^2/2c) = \text{const.}$, in the limit of zero electron mass $(m_e \to 0)$.

A sufficient condition for stability follows directly from Eqs.(6.12) and (6.16).[13] Defining an effective Helmholtz free energy function ΔF by $\Delta F = \Delta U_r + \Delta U_G$, it follows that

$$\Delta F = \int d^2x \left[r^2(n_e - n_e^0) + G(n_e) - G(n_e^0) \right] = \text{const.} \quad (6.17)$$

is an *exact* (nonlinear) conservation constraint. Expressing $\delta n_e(r, \theta, t) = n_e(r, \theta, t) - n_e^0(r)$, and Taylor expanding

$$G(n_e) = G(n_e^0) + G'(n_e^0)(\delta n_e) + \frac{1}{2}G''(n_e^0)(\delta n_e)^2 + \cdots , \quad (6.18)$$

it follows that Eq.(6.17) can be expressed as

$$\Delta F = \int d^2x \left\{ \left[r^2 + G'(n_e^0) \right] (\delta n_e) + \frac{1}{2}G''(n_e^0)(\delta n_e)^2 \right\} = \text{const.} , \quad (6.19)$$

correct to quadratic order in the density perturbation $\delta n_e(r, \theta, t)$.

The function $G(n_e^0)$ has been arbitrary up to this point. We now choose $G(n_e^0)$ to satisfy

$$G'(n_e^0) = -r^2 , \quad (6.20)$$

so that $G''(n_e^0) = -(\partial n_e^0/\partial r^2)^{-1}$. Equation (6.19) then becomes

$$\Delta F = \frac{1}{2} \int d^2x \frac{1}{[-\partial n_e^0/\partial r^2]} (\delta n_e)^2 = \text{const.} \quad (6.21)$$

It follows trivially from Eq.(6.21) that for monotonically decreasing density profiles with

$$\frac{1}{r}\frac{\partial}{\partial r}n_e^0(r) \le 0 , \quad \text{for } a \le r \le b , \quad (6.22)$$

the density perturbation $\delta n_e(r, \theta, t)$ cannot grow without bound, and the system is linearly stable.[13] That is to say, Eq.(6.22) is a *sufficient condition for stability* to small-amplitude electrostatic perturbations in the context of the cold-fluid guiding-center model based on Eqs.(6.6) and (6.7). Similarly, if $n_e^0(r)$ is an increasing function of r over the interval $a \le r \le b$, we find from Eq.(6.21) that the system is linearly stable.

We also conclude that a *necessary condition for instability* is that $\partial n_e^0/\partial r$ changes sign in the interval $a \le r \le b$. Alternatively, in terms of the equilibrium angular rotation velocity $\omega_{re}^- = -cE_r(r)/rB_0$, a necessary condition for instability is given by

$$\frac{\partial}{\partial r}\frac{1}{r}\frac{\partial}{\partial r}\left[r^2\omega_{re}^-(r)\right] \text{ changes sign in the interval } a \leq r \leq b. \quad (6.23)$$

Here, use has been made of the equilibrium Poisson equation, $(1/r) \times (\partial/\partial r)(rE_r) = -4\pi e n_e^0$, to eliminate $n_e^0(r)$ in favor of $\omega_{re}^-(r)$.

In conclusion, it should be pointed out that the sufficient condition for stability in Eq.(6.22) can also be derived directly from the electrostatic eigenvalue equation (6.28).[4,14] Moreover, the stability theorem can be extended to the case of relativistic electron flow and extraordinary-mode electromagnetic perturbations,[15] treating the electrons as a relativistic guiding-center fluid with $m_e \to 0$. Finally, in the preceding derivation of the stability theorem, it should be noted that $G(n_e^0)$ has been chosen to satisfy the condition $G'(n_e^0) = -r^2$ [Eq.(6.20)]. Strictly speaking, this is valid only if $n_e^0(r)$ is monotonic and there is not a finite range of r for which $\partial n_e^0(r)/\partial r = 0$. Otherwise, r is not a single-valued function of n_e^0.

Problem 6.1 Eliminate $n_e(r,\theta,t)$ in terms of $\phi(r,\theta,t)$ by means of Poisson's equation (6.7), and show that the expression for $d\Delta U_r/dt$ in Eq.(6.15) can be expressed as

$$\frac{d}{dt}\Delta U_r = -\frac{c}{2\pi e B_0}\int_0^{2\pi} d\theta \int_a^b dr\, r\frac{\partial\phi}{\partial\theta}\left(\frac{1}{r}\frac{\partial}{\partial r}r\frac{\partial\phi}{\partial r} + \frac{1}{r^2}\frac{\partial^2\phi}{\partial\theta^2}\right). \quad (6.1.1)$$

Integrate the right-hand side of Eq.(6.1.1) by parts with respect to θ and r, and make use of $\partial\phi/\partial\theta = 0$ at $r = a$ and $r = b$ to show that

$$\frac{d}{dt}\Delta U_r = 0. \quad (6.1.2)$$

6.3 Linear Stability Properties

6.3.1 Electrostatic Eigenvalue Equation

To investigate the linear stability properties associated with Eqs.(6.6) and (6.7), we assume small-amplitude perturbations about an azimuthally symmetric equilibrium $(\partial/\partial\theta = 0)$ characterized by the equilibrium density profile $n_e^0(r)$ and electrostatic potential $\phi_0(r)$ related by Poisson's equation

$$\frac{1}{r}\frac{\partial}{\partial r}r\frac{\partial}{\partial r}\phi_0(r) = 4\pi e n_e^0(r) \ . \tag{6.24}$$

The electron density and electrostatic potential are expressed as

$$n_e(r, \theta, t) = n_e^0(r) + \sum_{\ell=-\infty}^{\infty} \delta n_e^\ell(r)\exp(i\ell\theta - i\omega t) \ ,$$

$$\phi(r, \theta, t) = \phi_0(r) + \sum_{\ell=-\infty}^{\infty} \delta\phi^\ell(r)\exp(i\ell\theta - i\omega t) \ . \tag{6.25}$$

Here, ℓ is the azimuthal mode number of the perturbation, $\partial/\partial z = 0$ has been assumed, and ω is the (complex) oscillation frequency, with $\text{Im}\,\omega > 0$ corresponding to instability (temporal growth). For small-amplitude perturbations, the linearized continuity equation (6.6) gives

$$\left[\omega - \ell\omega_{re}^-(r)\right]\delta n_e^\ell(r) = -\frac{c\ell\delta\phi^\ell(r)}{B_0 r}\frac{\partial}{\partial r}n_e^0(r) \ , \tag{6.26}$$

where $\omega_{re}^-(r)$ is the equilibrium $E_r(r)\hat{e}_r \times B_0\hat{e}_z$ rotation velocity of the electron fluid defined by

$$\omega_{re}^-(r) = -\frac{cE_r(r)}{rB_0} = \frac{c}{rB_0}\frac{\partial}{\partial r}\phi_0(r) \ . \tag{6.27}$$

Substituting Eqs.(6.25) and (6.26) into Poisson's equation (6.7), we obtain the eigenvalue equation

$$\frac{1}{r}\frac{\partial}{\partial r}r\frac{\partial}{\partial r}\delta\phi^\ell(r) - \frac{\ell^2}{r^2}\delta\phi^\ell(r) = -\frac{\ell}{\omega_{ce}r}\frac{\delta\phi^\ell(r)}{\left[\omega - \ell\omega_{re}^-(r)\right]}\frac{\partial}{\partial r}\omega_{pe}^2(r) \ , \tag{6.28}$$

where $\omega_{ce} = eB_0/m_e c$ and $\omega_{pe}^2(r) = 4\pi n_e^0(r)e^2/m_e$, and the ratio $\omega_{pe}^2(r)/\omega_{ce} = 4\pi n_e^0(r)ec/B_0$ is independent of the electron mass m_e.

For specified equilibrium density profile $n_e^0(r)$, the eigenvalue equation (6.28) can be used to determine the eigenfunction $\delta\phi^\ell(r)$ and the complex eigenfrequency ω. Based on the discussion in Sec. 6.2, it is expected that an appropriate shear in the angular velocity profile $\omega_{re}^-(r)$, or equivalently, an equilibrium profile $n_e^0(r)$ with a sufficiently strong density maximum between $r = a$ and $r = b$ will provide the free energy to drive the diocotron instability.[1-4] It should be noted that Eq.(6.28) can also be obtained from the (more exact) eigenvalue equation (5.56). In particular, for $n_i^0(r) = 0$ and $\omega_{re}(r) \simeq \omega_{re}^-(r)$, it can be shown that Eq.(5.56) reduces to Eq.(6.28) in the limit of low-frequency flute perturbations about a tenuous nonneutral plasma satisfying

$$\left| \omega - \ell\omega_{re}^-(r) \right|^2 \ll \omega_{ce}^2 \, ,$$

$$\omega_{pe}^2(r) \ll \omega_{ce}^2 \, . \qquad (6.29)$$

The inequalities in Eq.(6.29) are intrinsic to the guiding-center model described in Sec. 6.1, where the electrons are treated as a strongly magnetized, massless fluid.

Problem 6.2   Consider the eigenvalue equation (5.56) derived for electrostatic perturbations with arbitrary frequency about general equilibrium density profile $n_j^0(r)$. For flute perturbations ($k_z = 0$) and zero ion density $\left[n_i^0(r) = 0 \right]$, make use of the inequalities in Eq.(6.29) to show that the general eigenvalue equation (5.56) reduces to Eq.(6.28) in the limit of low-frequency perturbations about a tenuous electron plasma rotating with angular velocity $\omega_{re}^-(r)$.

6.3.2   Diocotron Instability for an Annular Electron Layer

As a specific example, we solve the eigenvalue equation (6.28) for the choice of electron density profile $n_e^0(r)$ corresponding to the annular electron layer illustrated in Fig. 6.2. Here, $n_e^0(r)$ is specified by[3,4]

$$n_e^0(r) = \begin{cases} 0 \, , & a \le r < r_b^- \, , \\ \hat{n}_e = \text{const.} \, , & r_b^- < r < r_b^+ \, , \\ 0 \, , & r_b^+ < r \le b \, , \end{cases} \qquad (6.30)$$

where cylindrical conducting walls are located at $r = a$ and $r = b$, and r_b^- (r_b^+) denotes the inner (outer) radius of the electron layer. For present purposes, it is also assumed that the inner conductor at $r = a$ carries a charge Q per unit length, so that the radial electric field is $E_r(r = a) = 2Q/a$ at the surface of the conductor, and $E_r(r) = (2Q/a)(a/r)$ in the vacuum region $a \le r < r_b^-$. Solving Eq.(6.24) for $E_r(r) = -\partial\phi_0(r)/\partial r$

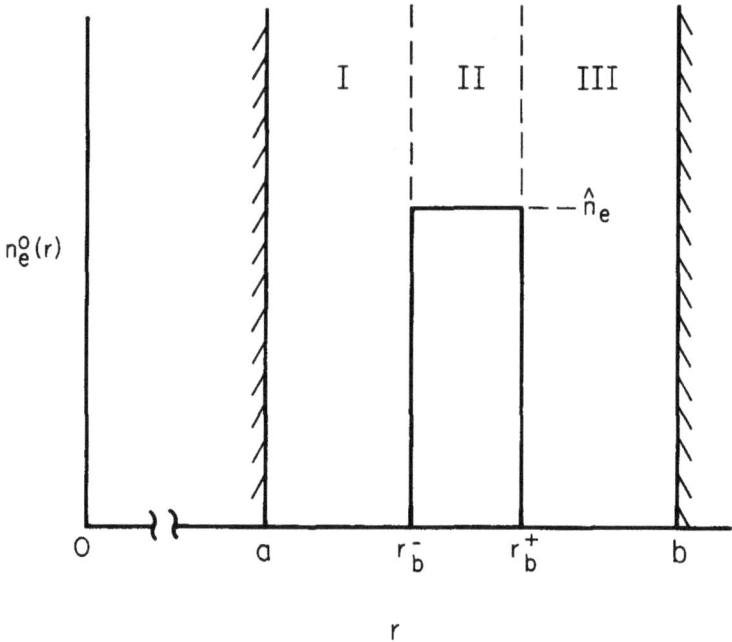

Figure 6.2. Annular electron density profile $n_e^0(r)$ assumed in Eq.(6.30). The inner conductor at $r = a$ carries a charge Q per unit length.

within the electron layer then gives for the angular rotation velocity $\omega_{re}^-(r) = -cE_r(r)/rB_0$,

$$\omega_{re}^-(r) = \omega_q \left(\frac{r_b^-}{r}\right)^2 + \omega_D \left[1 - \left(\frac{r_b^-}{r}\right)^2\right] , \quad r_b^- < r < r_b^+ . \quad (6.31)$$

Here, the constants ω_q and ω_D are defined by

$$\omega_q = -\frac{2Qc}{B_0(r_b^-)^2} \quad \text{and} \quad \omega_D = \frac{\omega_{pe}^2}{2\omega_{ce}} , \quad (6.32)$$

where $\omega_{pe}^2/2\omega_{ce} = 2\pi\hat{n}_e ec/B_0 = $ const. is the diocotron frequency. Note that ω_q is equal to the rotation velocity induced by the radial electric field at $r = r_b^-$ produced by the charge Q per unit length on the central conductor at $r = a$. On the other hand, $\omega_D = \omega_{pe}^2/2\omega_{ce} = $ const. is equal to the $E_r \hat{e}_r \times B_0 \hat{e}_z$ rotation velocity that would exist for a uniform-density nonneutral plasma column extending to the origin ($r_b^- = 0$) in the absence of a central conductor ($Q = 0$ and $a = 0$). For positive charge $Q > 0$, it follows from Eqs.(6.31) and (6.32) that $\omega_q < 0$, thereby *slowing* the natural sense of rotation of the electron layer due to self-field effects.

We now substitute Eq.(6.30) into Eq.(6.28) and make use of $\partial n_e^0(r)/\partial r = \hat{n}_e \delta(r - r_b^-) - \hat{n}_e \delta(r - r_b^+)$. The eigenvalue equation (6.28) can be expressed as

$$\frac{1}{r}\frac{\partial}{\partial r}r\frac{\partial}{\partial r}\delta\phi^\ell(r) - \frac{\ell^2}{r^2}\delta\phi^\ell(r) \tag{6.33}$$

$$= -\frac{\ell}{r}\frac{\omega_{pe}^2}{\omega_{ce}}\left\{ \frac{\delta\phi^\ell(r_b^-)}{[\omega - \ell\omega_{re}(r_b^-)]}\delta(r - r_b^-) - \frac{\delta\phi^\ell(r_b^+)}{[\omega - \ell\omega_{re}(r_b^+)]}\delta(r - r_b^+) \right\} ,$$

where $\omega_{pe}^2 = 4\pi\hat{n}_e e^2/m_e = \text{const.}$, $\omega_{ce} = eB_0/m_e c$, and $\omega_{re}^-(r)$ is defined in Eq.(6.31). The terms in Eq.(6.33) proportional to $\delta(r - r_b^-)$ and $\delta(r - r_b^+)$ correspond to surface-charge perturbations on the inner (r_b^-) and outer (r_b^+) surfaces, respectively, of the electron layer.

Except at $r = r_b^-$ and $r = r_b^+$, Eq.(6.33) reduces to

$$\frac{1}{r}\frac{\partial}{\partial r}r\frac{\partial}{\partial r}\delta\phi^\ell(r) - \frac{\ell^2}{r^2}\delta\phi^\ell(r) = 0 . \tag{6.34}$$

Equation (6.34) is solved separately in the three regions in Fig. 6.2 corresponding to: Region I ($a \le r < r_b^-$); Region II ($r_b^- < r < r_b^+$); and Region III ($r_b^+ < r \le b$). Enforcing $\delta\phi^\ell(r = a) = 0 = \delta\phi^\ell(r = b)$ so that the tangential electric field $\delta E_\theta^\ell = (-i\ell/r)\delta\phi^\ell$ vanishes at the two perfectly conducting walls, and enforcing continuity of $\delta\phi^\ell(r)$ at $r = r_b^-$ and $r = r_b^+$, we obtain

$$\delta\phi_I^\ell(r) = \left[B(r_b^-)^\ell + \frac{C}{(r_b^-)^\ell}\right]\left[\frac{(r/a)^{2\ell} - 1}{(r_b^-/a)^{2\ell} - 1}\right]\left(\frac{r_b^-}{r}\right)^\ell , \quad a \le r < r_b^- ,$$

$$\tag{6.35a}$$

$$\delta\phi_{II}^\ell(r) = \left(Br^\ell + \frac{C}{r^\ell}\right) , \quad r_b^- < r < r_b^+ , \tag{6.35b}$$

$$\delta\phi_{III}^\ell(r) = \left[B(r_b^+)^\ell + \frac{C}{(r_b^+)^\ell}\right]\left[\frac{1 - (r/b)^{2\ell}}{1 - (r_b^+/b)^{2\ell}}\right]\left(\frac{r_b^+}{r}\right)^\ell , \quad r_b^+ < r < b ,$$

$$\tag{6.35c}$$

where B and C are constants. Note that the solutions in Eq.(6.35) satisfy the requisite boundary conditions at $r = a$, r_b^-, r_b^+ and b. To eliminate the constants B and C and obtain a closed dispersion relation for the complex eigenfrequency ω, we integrate the eigenvalue equation (6.33) across the layer surfaces at $r = r_b^-$ and $r = r_b^+$. Multiplying Eq.(6.33) by r, and integrating with respect to r from $r_b^-(1 - \varepsilon)$ to $r_b^-(1 + \varepsilon)$ with $\varepsilon \to 0_+$, it is readily shown that

$$r_b^- \left[\frac{\partial}{\partial r} \delta\phi_{II}^\ell(r) \right]_{r=r_b^-} - r_b^- \left[\frac{\partial}{\partial r} \delta\phi_I^\ell(r) \right]_{r=r_b^-} = -\frac{\ell\omega_{pe}^2}{\omega_{ce}} \frac{\delta\phi^\ell(r_b^-)}{[\omega - \ell\omega_{re}^-(r_b^-)]} .$$

(6.36)

Similarly, integrating Eq.(6.33) across the outer surface of the layer at $r = r_b^+$, we obtain

$$r_b^+ \left[\frac{\partial}{\partial r} \delta\phi_{III}^\ell(r) \right]_{r=r_b^+} - r_b^+ \left[\frac{\partial}{\partial r} \delta\phi_{II}^\ell(r) \right]_{r=r_b^+} = \frac{\ell\omega_{pe}^2}{\omega_{ce}} \frac{\delta\phi^\ell(r_b^+)}{[\omega - \ell\omega_{re}^-(r_b^+)]} .$$

(6.37)

Equations (6.36) and (6.37) relate the discontinuity in radial electric field at the inner and outer surfaces of the electron layer to the perturbed charge at the two surfaces.

The expressions for $\omega_{re}^-(r)$ and $\delta\phi^\ell(r)$ in Eqs.(6.31) and (6.35) are substituted into Eqs.(6.36) and (6.37) to give two homogeneous linear equations relating the constants B and C. Setting the two-by-two determinant of the coefficients of B and C equal to zero gives the dispersion relation for the complex eigenfrequency ω. Some algebraic manipulation gives the quadratic equation[3,4]

$$(\omega/\omega_D)^2 - b_\ell (\omega/\omega_D) + c_\ell = 0 ,$$

(6.38)

where $\omega_D = \omega_{pe}^2/2\omega_{ce}$, and the coefficients b_ℓ and c_ℓ are defined by

$$b_\ell = \left(\ell \left\{ \left[1 - \left(\frac{r_b^-}{r_b^+} \right)^2 \right] + \frac{\omega_q}{\omega_D} \left[1 + \left(\frac{r_b^-}{r_b^+} \right)^2 \right] \right\} \left[1 - \left(\frac{a}{b} \right)^{2\ell} \right] \right.$$

$$\left. + \left[1 - \left(\frac{r_b^-}{r_b^+} \right)^{2\ell} \right] \left[\left(\frac{r_b^+}{b} \right)^{2\ell} - \left(\frac{a}{r_b^-} \right)^{2\ell} \right] \right) \left[1 - \left(\frac{a}{b} \right)^{2\ell} \right]^{-1} ,$$

(6.39)

and

$$c_\ell = \left(\ell^2 \frac{\omega_q}{\omega_D} \left[1 - \left(\frac{r_b^-}{r_b^+} \right)^2 + \frac{\omega_q}{\omega_D} \left(\frac{r_b^-}{r_b^+} \right)^2 \right] \left[1 - \left(\frac{a}{b} \right)^{2\ell} \right] \right.$$

$$- \ell \frac{\omega_q}{\omega_D} \left[1 - \left(\frac{a}{r_b^+} \right)^{2\ell} \right] \left[1 - \left(\frac{r_b^+}{b} \right)^{2\ell} \right] \tag{6.40}$$

$$+ \ell \left[1 - \left(\frac{r_b^-}{r_b^+} \right)^2 + \frac{\omega_q}{\omega_D} \left(\frac{r_b^-}{r_b^+} \right)^2 \right] \left[1 - \left(\frac{r_b^-}{b} \right)^{2\ell} \right] \left[1 - \left(\frac{a}{r_b^-} \right)^{2\ell} \right]$$

$$\left. - \left[1 - \left(\frac{r_b^+}{b} \right)^{2\ell} \right] \left[1 - \left(\frac{a}{r_b^-} \right)^{2\ell} \right] \left[1 - \left(\frac{r_b^-}{r_b^+} \right)^{2\ell} \right] \right) \left[1 - \left(\frac{a}{b} \right)^{2\ell} \right]^{-1} .$$

The solution to the dispersion relation (6.38) is

$$\omega = \frac{1}{2} \omega_D \left[b_\ell \pm (b_\ell^2 - 4 c_\ell)^{1/2} \right] . \tag{6.41}$$

If $b_\ell^2 \geq 4 c_\ell$, the oscillation frequencies defined in Eq.(6.41) are real ($\mathrm{Im}\,\omega = 0$), and the equilibrium configuration in Fig. 6.2 is stable. On the other hand, if $4 c_\ell > b_\ell^2$, then the oscillation frequencies in Eq.(6.41) are complex and form conjugate pairs with

$$\mathrm{Re}\,\omega = \frac{1}{2} b_\ell \omega_D ,$$

$$\mathrm{Im}\,\omega = \pm \frac{1}{2} \left(4 c_\ell - b_\ell^2 \right)^{1/2} \omega_D . \tag{6.42}$$

The solution with $\mathrm{Im}\,\omega > 0$ in Eq.(6.42) corresponds to instability. Note that $\mathrm{Re}\,\omega$ and $\mathrm{Im}\,\omega$ are both of order ω_D times a geometric factor.

Equations (6.38)–(6.41) can be used to investigate detailed stability properties of the equilibrium configuration in Eq.(6.30) and Fig. 6.2 over a wide range of system parameters

$$\frac{a}{b} , \quad \frac{r_b^-}{b} , \quad \frac{r_b^+}{b} \quad \text{and} \quad \frac{\omega_q}{\omega_D} .$$

Here, keep in mind that $a < r_b^- < r_b^+ < b$ (Fig. 6.2), and that ω_q is negative (positive) whenever the charge Q on the central conductor is positive (negative). Some algebraic manipulation shows that the necessary and sufficient condition for stability ($b_\ell^2 \geq 4c_\ell$) can be expressed as

$$
\left\{ -\ell \left(1 - \frac{\omega_q}{\omega_D} \right) \left[1 - \left(\frac{r_b^-}{r_b^+} \right)^2 \right] \left[1 - \left(\frac{a}{b} \right)^{2\ell} \right] \right.
$$

$$
\left. + 2 \left[1 + \left(\frac{a}{b} \right)^{2\ell} \right] - \left[1 + \left(\frac{r_b^-}{r_b^+} \right)^{2\ell} \right] \left[\left(\frac{a}{r_b^-} \right)^{2\ell} + \left(\frac{r_b^+}{b} \right)^{2\ell} \right] \right\}^2
$$

$$
\geq 4 \left(\frac{r_b^-}{r_b^+} \right)^{2\ell} \left[1 - \left(\frac{r_b^+}{b} \right)^{2\ell} \right]^2 \left[1 - \left(\frac{a}{r_b^-} \right)^{2\ell} \right]^2. \tag{6.43}
$$

The inequality in Eq.(6.43) assures stable oscillations with $\operatorname{Im}\omega = 0$. In the absence of a central conductor, ($Q = 0$ and $a = 0$), the condition for stability in Eq.(6.43) reduces to

$$
\left\{ -\ell \left[1 - \left(\frac{r_b^-}{r_b^+} \right)^2 \right] + 2 - \left[\left(\frac{r_b^+}{b} \right)^{2\ell} + \left(\frac{r_b^-}{b} \right)^{2\ell} \right] \right\}^2
$$

$$
\geq 4 \left(\frac{r_b^-}{r_b^+} \right)^{2\ell} \left[1 - \left(\frac{r_b^+}{b} \right)^{2\ell} \right]^2. \tag{6.44}
$$

Several important conclusions regarding the diocotron instability follow directly from examination of Eqs.(6.42)–(6.44):

(a) From Eq.(6.43), the system is stable ($\operatorname{Im}\omega = 0$) whenever $r_b^- = a$ or $r_b^+ = b$, i.e., whenever the inner or outer surface of the electron layer is in contact with the conducting wall. More generally, when $4c_\ell > b_\ell^2$, the instability growth rate is reduced when the surface of the layer is brought in closer proximity to the conducting wall (r_b^- approaches a, or r_b^+ approaches b).

(b) Increasing the ratio ω_q/ω_D to sufficiently large positive *or* negative values in Eq.(6.43) has a stabilizing influence for fixed geometric factors a/b, r_b^-/b and r_b^+/b. That is, charging the central conductor reduces the growth rate relative to the case where $Q = 0$.

(c) When $4c_\ell > b_\ell^2$, the number of unstable mode numbers increases as the layer thickness $r_b^+ - r_b^-$ is decreased.

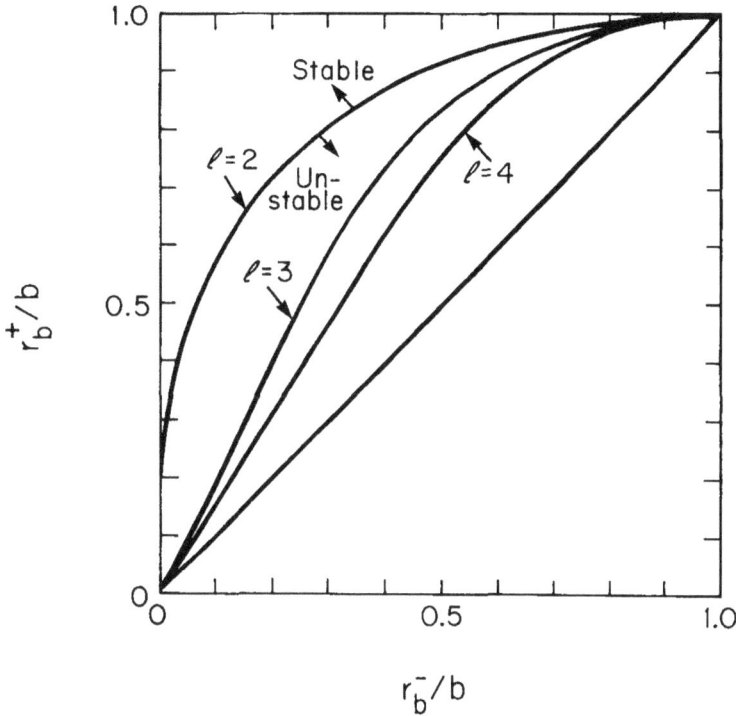

Figure 6.3. Stability-instability curves in $(r_b^-/b, r_b^+/b)$ parameter space calculated from Eq.(6.44) for mode numbers $\ell = 2, 3, 4$, in the absence of a central conductor ($Q = 0$ and $a = 0$). For specified ℓ, Eq.(6.38) has two stable solutions (Im $\omega = 0$) for values of r_b^-/b and r_b^+/b above the curve. For values of r_b^-/b and r_b^+/b below the curve and above the 45° line, Eq.(6.38) has one unstable solution with Im $\omega > 0$ [R.H. Levy, Phys. Fluids **8**, 1288 (1965)].

(d) In the absence of a central conductor ($Q = 0$ and $a = 0$), the condition for the fundamental $\ell = 1$ mode to be stable in Eq.(6.44) reduces to

$$\left[1 - \left(\frac{r_b^-}{r_b^+} \right)^2 \right]^2 \left[1 - \left(\frac{r_b^+}{b} \right)^2 \right]^2 \geq 0 . \tag{6.45}$$

It follows that the system is always stable for $\ell = 1$.

(e) In the special case of a uniform-density plasma column extending to the origin ($r_b^- = 0$, as well as $Q = 0$ and $a = 0$), the system is stable for all mode numbers ℓ, and the oscillation frequency can be expressed as (Problem 6.3)

$$\omega = \omega_D \left[\ell - 1 + (r_b^+/b)^{2\ell} \right] \ . \tag{6.46}$$

Equation (6.46) corresponds to stable oscillations excited on the surface of the plasma column.

In the absence of a central conductor ($Q = 0$ and $a = 0$), the inequality in Eq.(6.44) can be used to calculate stability-instability boundaries as functions of the geometric factors r_b^-/b and r_b^+/b, for various values of the harmonic number ℓ. Shown in Fig. 6.3 are the stability-instability boundaries for $\ell = 2, 3, 4$. Only the region above the 45° line in Fig. 6.3 has meaning because $r_b^+ > r_b^-$. For a given value of ℓ, the region of parameter space above the curve in Fig. 6.3 corresponds to stable oscillations (Im $\omega = 0$), whereas the region between the curve and the 45° line has one unstable solution (Im $\omega > 0$).[2] Evidently, for fixed value of r_b^-/b, a sufficiently large value of r_b^+/b assures stability. That is, proximity of a conducting wall has a stabilizing influence on the diocotron instability.

Problem 6.3   Consider a uniform-density nonneutral plasma column with no central conductor ($Q = 0$ and $a = 0$) and equilibrium density profile specified by

$$n_e^0(r) = \begin{cases} \hat{n}_e = \text{const.}, & 0 \leq r < r_b^+ \ , \\ 0 \ , & r_b^+ < r \leq b \ . \end{cases} \tag{6.3.1}$$

Here, $r = b$ denotes the radial location of a perfectly conducting cylindrical wall. Solve the electrostatic eigenvalue equation (6.28) in the two regions $0 \leq r < r_b^+$ and $r_b^+ < r \leq b$, and show that the (real) eigenfrequency ω is given by

$$\omega = \omega_D \left[\ell - 1 + (r_b^+/b)^{2\ell} \right] \ , \tag{6.3.2}$$

where $\omega_D = \omega_{pe}^2/2\omega_{ce}$. Sketch the radial form of the eigenfunction $\delta\phi^\ell(r)$ and show that it is highly peaked at the surface of the plasma column ($r = r_b^+$).

Problem 6.4   Consider the eigenvalue equation (6.33) for electrostatic perturbations about the annular electron density profile in Eq.(6.30) and Fig. 6.2.

a. Verify that the solutions for the eigenfunction $\delta\phi_\ell(r)$ in each of the three regions in Fig. 6.2 can be expressed in the form given in Eq.(6.35).

b. Substitute the solutions for $\delta\phi_\ell(r)$ into Eqs.(6.36) and (6.37) to obtain two homogeneous linear equations relating the constants B and C. Show that the condition for a nontrivial solution to these equations is given by the dispersion relation (6.38), where the constants b_ℓ and c_ℓ are defined in Eqs.(6.39) and (6.40).

c. Show that the necessary and sufficient condition for Eq.(6.38) to have stable solutions (Im $\omega = 0$) can be expressed in the form given in Eq.(6.43).

d. In the absence of a central conductor ($Q = 0$ and $a = 0$), show that Eq.(6.43) reduces to the inequality in Eq.(6.44).

6.3.3   Stable Oscillations for $\ell = 1$ and Arbitrary Density Profile

For perturbations with mode number $\ell = 1$, the electrostatic eigenvalue equation (6.28) can be expressed as

$$r \left[\omega + \frac{cE_r(r)}{rB_0} \right] \left\{ \frac{1}{r} \frac{\partial}{\partial r} r \frac{\partial}{\partial r} \delta\phi^1(r) - \frac{\delta\phi^1(r)}{r^2} \right\}$$

$$= \delta\phi^1(r) \frac{\partial}{\partial r} \left[\frac{1}{r} \frac{\partial}{\partial r} \frac{crE_r(r)}{B_0} \right] . \tag{6.47}$$

Here, use has been made of $\omega_{re}^-(r) = -cE_r(r)/rB_0$ and the equilibrium Poisson equation (6.24) to express

$$\frac{1}{\omega_{ce}} \frac{\partial}{\partial r} \omega_{pe}^2(r) = -\frac{\partial}{\partial r} \left[\frac{1}{r} \frac{\partial}{\partial r} \frac{crE_r(r)}{B_0} \right] . \tag{6.48}$$

For present purposes, we assume there is no internal conductor and allow for arbitrary (unspecified) density profile $n_e^0(r)$ in the interval $0 \leq r \leq b$, where $r = b$ is the radius of the outer conducting wall. The exact solution to Eq.(6.47) which is not singular at $r = 0$ can be expressed as[16]

$$\delta\phi^1(r) = B \left[r\omega + \frac{cE_r(r)}{B_0} \right] , \tag{6.49}$$

where B is a constant. Equation (6.49) can be verified by direct substitution into Eq.(6.47).

The boundary condition $\delta\phi^1(r = 0) = 0$ is automatically satisfied by Eq.(6.49) because $E_r(r = 0) = 0$. Enforcing $\delta\phi^1(r = b) = 0$ at the outer conductor determines the (real) oscillation frequency

$$\omega = -\frac{cE_r(r = b)}{bB_0} . \tag{6.50}$$

Integrating the equilibrium Poisson equation (6.34) from $r = 0$ to $r = b$ gives $E_r(r = b) = -4\pi e b^{-1} \int_0^b dr\, r n_e^0(r) = 2b^{-1}Q$, where $Q \equiv -2\pi e \int_0^b dr\, r \times n_e^0(r)$ is the total (negative) electric charge per unit length of the plasma column. The $\ell = 1$ oscillation frequency in Eq.(6.50) can then be expressed as

$$\omega = -\frac{2Qc}{b^2 B_0}. \tag{6.51}$$

Not only is the system stable for perturbations with mode number $\ell = 1$ and general density profile $n_e^0(r)$, a measurement of the $\ell = 1$ oscillation frequency can be used to infer the total charge Q per unit length that is confined [Eq.(6.51)].[16]

6.4 Experimental Results

In this section, we summarize the results of three experiments[9-11] on the diocotron instability, spanning more than three decades from 1956 to 1987.

We first consider the experimental results of Kyhl and Webster[9] (Fig. 6.4). A low-current (tens of μA), annular electron beam is emitted by a hollow cathode gun and collected by a moveable fluorescent-screen anode. No inner or outer conductors were present in the experiment.[9] The anode screen could be moved from $z = 0.5$ cm to $z = 8.6$ cm from the cathode, and the experiments were carried out for an 80 eV pulsed electron beam with duration of $10\,\mu$s, repeated every $305\,\mu$s. Figure 6.4 shows a series of photographs of the fluorescent-screen anode, taken at distances ranging from $z = 1$ cm [Fig. 6.4(a)] to $z = 8.7$ cm [Fig. 6.4(e)] from the cathode, for an axial guide field $B_0 = 580$ G. We note from Fig. 6.4 that the beam cross section distorts as it propagates axially. By $z = 7$ cm [Fig. 6.4(d)], the beam has broken up into a discrete set of vortex filaments, and the system has evolved to a coherent, nonlinear state. Evidently, the evolution of the diocotron instability in this configuration corresponds to spatial growth (as a function of z), rather than the case of temporal growth considered in Sec. 6.3. The visual display of experimental results[9] on the diocotron instability in Fig. 6.4 is among the most striking in the published literature.

As indicated earlier, the beam current and beam energy in the experiments by Kyhl and Webster[9] were very low (tens of μA and tens of eV, respectively). However, the diocotron instability can also develop

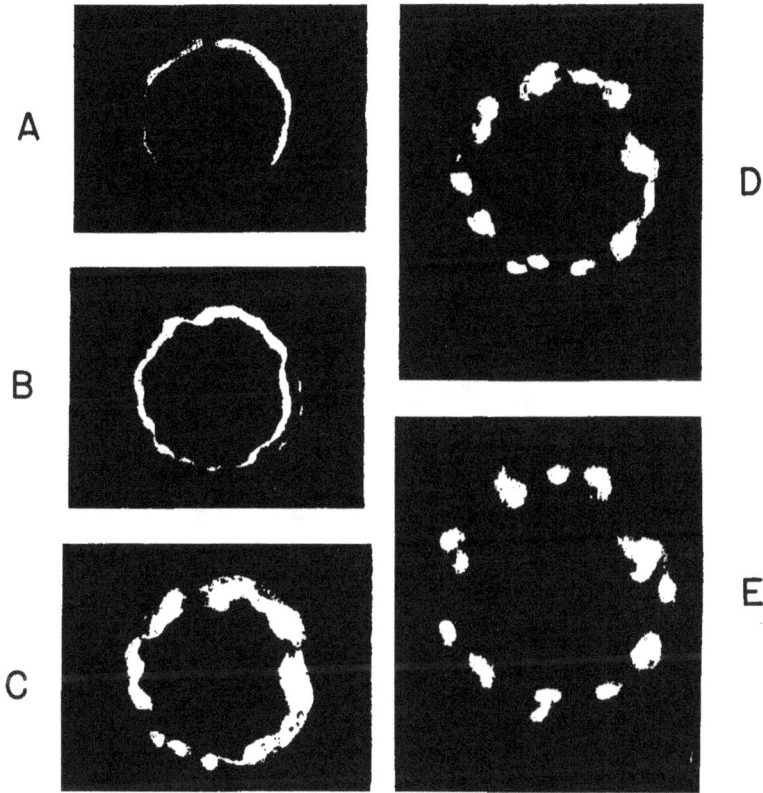

Figure 6.4. Photographs of the fluorescent-screen anode taken at distances of (a) 1 cm, (b) 3 cm, (c) 5 cm, (d) 7 cm, and (e) 8.7 cm from the cathode. The beam energy is $E_b = 80$ eV and the magnetic field is $B_0 = 580$ G [R.L. Kyhl and H.F. Webster, IRE Trans. Electron Devices **3**, 172 (1956)].

in intense annular electron beams at higher energies and currents. Typical experimental results obtained by Kapetanakos, et al.[10] are shown in Fig. 6.5 for a 400 keV, 40 kA, 50 ns-duration electron beam emitted by an annular cathode with outer radius 2.5 cm and thickness 0.3 cm. The axial guide field is $B_0 = 4$ kG, and the drift-tube pressure is about 0.2 torr. Figure 6.5 shows the damage patterns from four separate shots on Lucite plates located at distances of $z = 10, 35, 45$ and 52 cm, respectively, downstream from the anode foil. The linear theory of the diocotron instability, appropriately extended[10] to include relativistic beam dynamics and a partially neutralizing ion background, as well as spatial (rather than

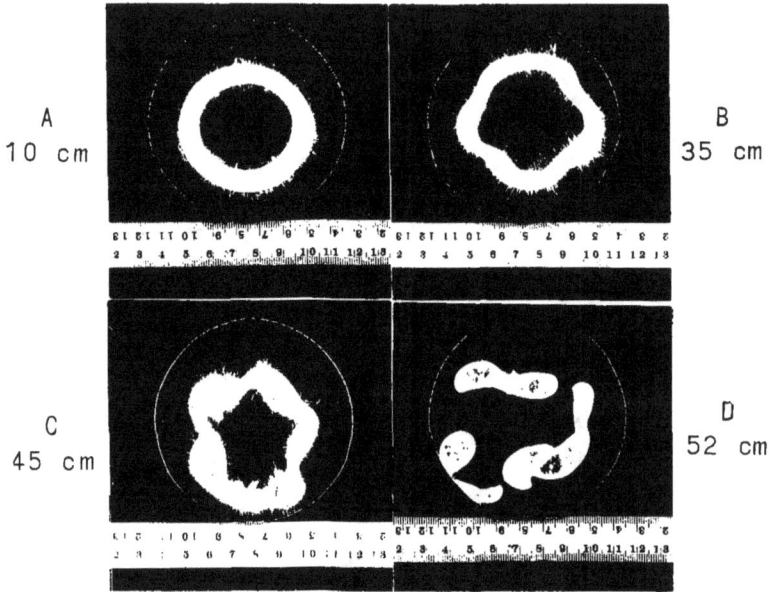

Figure 6.5. Lucite damage patterns at several distances from the anode using an annular cathode with outer radius 2.5 cm and thickness 0.3 cm. The beam current and energy are $I_b = 40$ kA and $E_b = 400$ keV, and the magnetic field is $B_0 = 4$ kG [C.A. Kapetanakos, et al., Phys. Rev. Lett. **30**, 1303 (1973)].

temporal) growth in the z-direction, explains several key features of the experimental results. For example, for the system parameters in Fig. 6.5, linear theory predicts that the layer is unstable only for azimuthal mode numbers $\ell = 5$ or $\ell = 6$, which is consistent with the beam distortion that occurs by $z = 35$ cm in Fig. 6.5. It is also found[10] that the distortion of the beam cross section is greatly reduced by bringing the outer conducting wall and outer boundary of the electron beam into closer proximity. (There was no inner conductor in the experiment.[10]) Furthermore, it is observed that the beam distortion increases for higher beam current and/or lower magnetic field (thereby increasing the value of $\omega_{pe}^2/2\omega_{ce}$). An important feature of the experiments[10] is that the angular velocity profile $\omega_{re}^-(r)$ is modified from Eq.(6.31) (with $\omega_q = 0$) to become

$$\omega_{re}^-(r) = \frac{\omega_{pe}^2}{2\omega_{ce}} \left(1 - f - \beta_{ze}^2\right) \left[1 - \left(\frac{r_b^-}{r}\right)^2\right] . \tag{6.52}$$

Here, f is the fractional charge neutralization provided by a partially neutralizing ion background with (assumed) density profile $n_i^0(r) = fn_e^0(r)$.
The β_{ze}^2 contribution in Eq.(6.52) is associated with the $V_{ze}^0 \hat{e}_z \times B_\theta^s(r)\hat{e}_\theta$
rotation of the beam induced by the axial electron current and azimuthal
self-magnetic field. Both effects *slow* the beam rotation relative to the
case where $f = 0$ and $\beta_{ze}^2 \ll 1$.

Generally speaking, despite the many orders-of-magnitude difference
between the beam currents and energies in the two experiments,[9,10] the

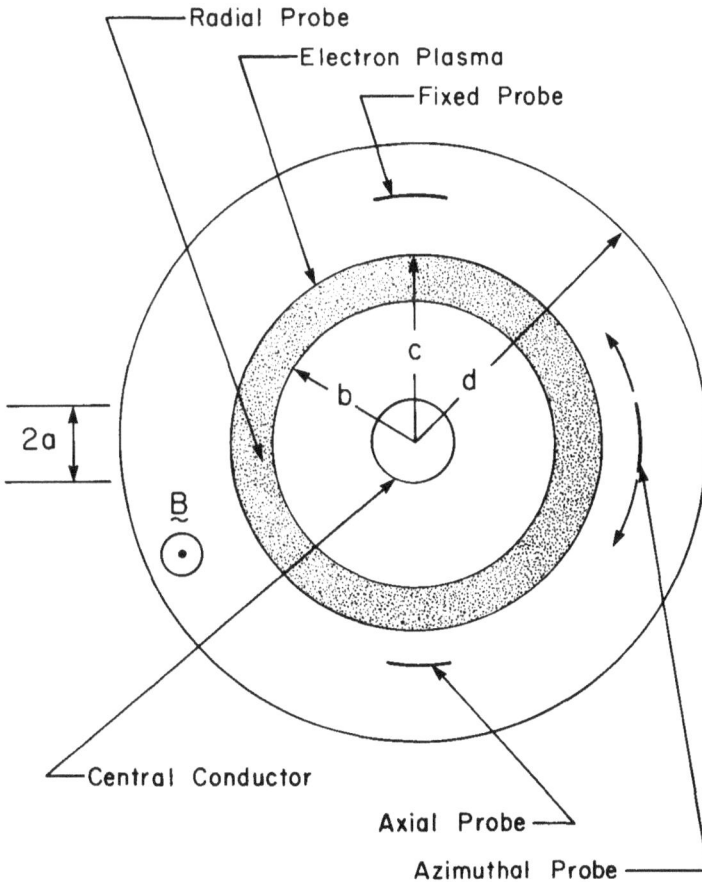

Figure 6.6. Cross section of the experimental set-up used to investigate the diocotron instability in an annular electron beam [G. Rosenthal,
G. Dimonte and A.Y. Wong, Phys. Fluids **30**, 3257 (1987)].

overall signature of nonlinear beam breakup by the diocoton instability is remarkably similar in Figs. 6.4 and 6.5. For the system parameters in Figs. 6.4 and 6.5, a few azimuthal mode numbers appear to dominate the (coherent) nonlinear evolution of the instability.

Finally, we summarize some aspects of experiments on the diocotron instability by Rosenthal, Dimonte and Wong.[11] The cross section of the experimental configuration[11] is illustrated in Fig. 6.6. An annular, non-neutral electron plasma of length $L = 126$ cm is created by a thermal cathode, and the average radius and thickness of the electron layer are 1.75 cm and 0.36 cm, respectively. The confining axial magnetic field B_0 can be varied up to 650 G, and typical plasma densities are in the range of 5×10^7 cm^{-3}, with $\omega_{pe}^2/\omega_{ce}^2 < 4.9 \times 10^{-3}$. For the experimental data shown in Fig. 6.7, the central conductor (at radius $a = 0.32$ cm) is at zero potential. Figure 6.7 shows the measured real frequency Reω of the diocotron mode, for wave excitations with $k_z = 0$ and azimuthal mode numbers $\ell = 3, 4, 5$, plotted versus $\hat{n}_e/B_0$, which is proportional

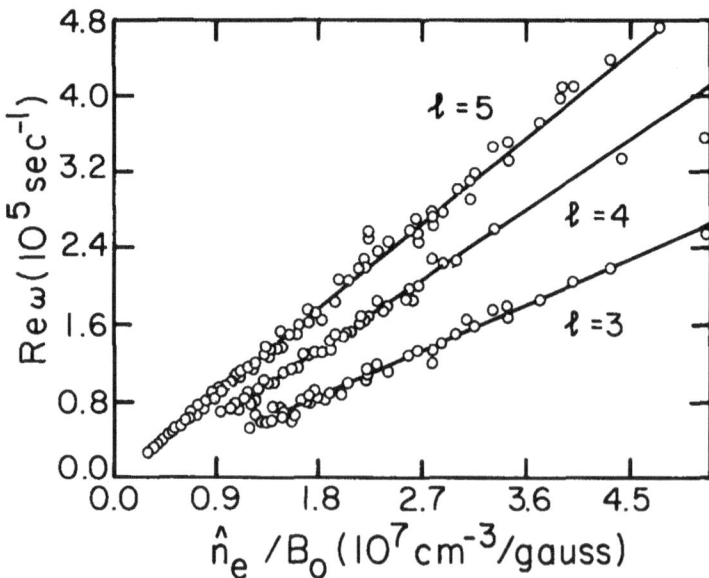

Figure 6.7. Measured values of the real frequency Reω of the diocotron mode are plotted versus $\hat{n}_e/B_0$ for wave excitations with $k_z = 0$ and azimuthal mode numbers $\ell = 3, 4, 5$ [G. Rosenthal, G. Dimonte and A.Y. Wong, Phys. Fluids **30**, 3257 (1987)].

to the diocotron frequency $\omega_D = \omega_{pe}^2/2\omega_{ce}$. The magnetic field and electron density in Fig. 6.7 vary over the range $130\,\text{G} < B_0 < 650\,\text{G}$, and $2 \times 10^7\,\text{cm}^{-3} < \hat{n}_e < 10^8\,\text{cm}^{-3}$. Moreover, the theoretical curves (solid lines) in Fig. 6.7 have been calculated[11] from Eq.(6.28) taking into account the bell-shaped nature of the density profile $n_e^0(r)$ measured in the experiments.[11] The agreement between theory and experiment in Fig. 6.7 is excellent, with $\text{Re}\,\omega$ scaling linearly with $\hat{n}_e/B_0$. Rosenthal, et al.[11] have also measured the saturation amplitude $\delta\phi_{\text{sat}}^\ell$ of the diocotron instability. For example, for azimuthal mode number $\ell = 5$ and magnetic field $B_0 > 300\,\text{G}$, it is found that $\delta\phi_{\text{sat}}^\ell$ scales linearly with electron density $\hat{n}_e$. On the other hand, for $B_0 < 200\,\text{G}$, it is found[11] that $\delta\phi_{\text{sat}}^\ell$ is independent of electron density. While nonlinear saturation mechanisms for the diocotron instability continue to be investigated theoretically, it is clear that such carefully controlled experiments will be required to provide the foundations for a definitive understanding.

Problem 6.5   Assume that the annular electron beam in the experiments by Kapetanakos, et al.[10] has uniform axial velocity $V_{ze}^0 = $ const. and uniform electron density profile specified by Eq.(6.30) with $a = 0$ (no central conductor).

a. Treat the electrons as a massless fluid ($m_e \to 0$) and show that radial force balance can be expressed as

$$0 = -en_e^0(r)\left[E_r(r) + \frac{\omega_{re}^-(r)r}{c}B_0 - \frac{V_{ze}^0 B_\theta^s(r)}{c}\right] , \qquad (6.5.1)$$

where $B_\theta^s(r)$ is the azimuthal self-magnetic field produced by the axial beam current.

b. Show that $B_\theta^s(r)$ within the electron beam can be expressed as

$$B_\theta^s(r) = -\frac{m_e}{2e}\beta_{ze}\omega_{pe}^2 r\left[1 - \left(\frac{r_b^-}{r}\right)^2\right] , \quad r_b^- < r < r_b^+ , \qquad (6.5.2)$$

where $\omega_{pe}^2 = 4\pi\hat{n}_e e^2/m_e$, and $\beta_{ze} = V_{ze}^0/c$.

c. For ion density profile $n_i^0(r) = fn_e^0(r)$, where $f = $ const. is the fractional charge neutralization, show that the radial electric field is given by

$$E_r(r) = -\frac{m_e}{2e}\omega_{pe}^2(1-f)r\left[1 - \left(\frac{r_b^-}{r}\right)^2\right] , \quad r_b^- < r < r_b^+ . \qquad (6.5.3)$$

d. Make use of Eqs.(6.5.1)–(6.5.3) to show that the angular rotation velocity of the electron beam is

$$\omega_{re}^-(r) = \frac{\omega_{pe}^2}{2\omega_{ce}} \left(1 - f - \beta_{ze}^2\right) \left[1 - \left(\frac{r_b^-}{r}\right)^2\right] , \quad r_b^- < r < r_b^+ , \quad (6.5.4)$$

where $\omega_{ce} = eB_0/m_e c$.

6.5   Quasilinear Evolution

When only a few unstable modes (ℓ-values) are excited, the evolution of the diocotron instability produces a coherent nonlinear distortion of the electron beam (see Figs. 6.4 and 6.5). On the other hand, for multimode excitation, it is reasonable to expect that the average macroscopic response of the system can be described by a quasilinear model, analogous to that developed in weak turbulence theories of velocity-space instabilities.[17] In this section, we describe such a quasilinear model of the diocotron instability.[18]

6.5.1   Theoretical Model

We make use of the macroscopic guiding-center model developed in Sec. 6.1 to describe the nonlinear evolution of sheared, nonrelativistic electron flow with velocity $\mathbf{V}_e = -(c/B_0)\nabla\phi \times \hat{\mathbf{e}}_z$. For simplicity, the analysis is carried out in the planar configuration illustrated in Fig. 6.8. Here, planar conductors are located at $x = 0$ and $x = d$, and the Cartesian components of flow velocity are

$$V_{xe}(x, y, t) = -\frac{c}{B_0} \frac{\partial}{\partial y} \phi(x, y, t) , \qquad (6.53a)$$

$$V_{ye}(x, y, t) = \frac{c}{B_0} \frac{\partial}{\partial x} \phi(x, y, t) . \qquad (6.53b)$$

Flute perturbations with $\partial/\partial z = 0$ are assumed, and the self-consistent

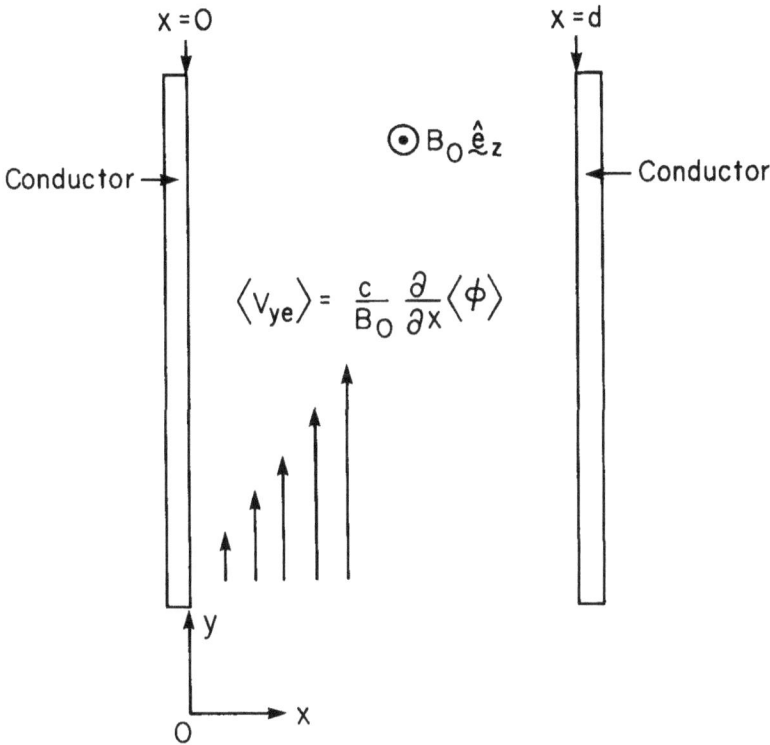

Figure 6.8. Schematic of cross-field electron flow between planar conductors located at $x = 0$ and $x = d$. The average electron flow is in the y-direction [Eq.(6.60)].

nonlinear evolution of the electron density $n_e(x, y, t)$ and electrostatic potential $\phi(x, y, t)$ are determined from the continuity-Poisson equations

$$\left\{ \frac{\partial}{\partial t} - \frac{c}{B_0} \frac{\partial \phi}{\partial y} \frac{\partial}{\partial x} + \frac{c}{B_0} \frac{\partial \phi}{\partial x} \frac{\partial}{\partial y} \right\} n_e(x, y, t) = 0, \qquad (6.54)$$

$$\left(\frac{\partial^2}{\partial x^2} + \frac{\partial^2}{\partial y^2} \right) \phi(x, y, t) = 4\pi e n_e(x, y, t) . \qquad (6.55)$$

For the planar configuration in Fig. 6.8, the coordinates x and y play a role similar to r and θ in cylindrical geometry [compare with Fig. 6.1 and Eqs.(6.6) and (6.7)].

The present analysis assumes that $\phi(x, y, t)$ and $n_e(x, y, t)$ are periodic in the y-direction with fundamental periodicity length L. Quantities are expressed as an average value (averaged over y) plus a perturbation, i.e.,

$$n_e(x, y, t) = \langle n_e \rangle (x, t) + \delta n_e(x, y, t) ,$$
$$\phi(x, y, t) = \langle \phi \rangle (x, t) + \delta\phi(x, y, t) . \tag{6.56}$$

Here, the average of $\psi(x, y, t) = \langle \psi \rangle (x, t) + \delta\psi(x, y, t)$ is defined by

$$\langle \psi \rangle (x, t) \equiv \frac{1}{L} \int_0^L dy\, \psi(x, y, t) . \tag{6.57}$$

Note that $\langle \delta\psi \rangle = 0$ and $(\partial/\partial y) \langle \psi \rangle = 0$ follow directly by definition. Similarly, it can be shown that $\langle \partial\psi/\partial y \rangle = 0$ and $\langle \chi \partial\psi/\partial y \rangle = -\langle \psi \partial\chi/\partial y \rangle$ for periodic functions ψ and χ in the y-direction. Substituting Eq.(6.56) into Eqs.(6.54) and (6.55) and taking the average, we find

$$\frac{\partial}{\partial t} \langle n_e \rangle (x, t) = \frac{c}{B_0} \frac{\partial}{\partial x} \left\langle \delta n_e(x, y, t) \frac{\partial}{\partial y} \delta\phi(x, y, t) \right\rangle , \tag{6.58}$$

and

$$\frac{\partial^2}{\partial x^2} \langle \phi \rangle (x, t) = 4\pi e \langle n_e \rangle (x, t) . \tag{6.59}$$

In obtaining Eq.(6.58) from Eq.(6.54), use has been made of

$$\left\langle \left(\frac{\partial \delta\phi}{\partial x} \right) \left(\frac{\partial \delta n_e}{\partial y} \right) \right\rangle = -\left\langle \delta n_e \left(\frac{\partial^2 \delta\phi}{\partial x \partial y} \right) \right\rangle .$$

Moreover, taking the average of Eq.(6.53) gives $\langle V_{xe} \rangle (x, t) = 0$ and

$$\langle V_{ye} \rangle (x, t) = V_E(x, t) \equiv \frac{c}{B_0} \frac{\partial}{\partial x} \langle \phi \rangle (x, t) . \tag{6.60}$$

We note that Eqs.(6.58)–(6.60) describe the (slow) nonlinear evolution of the average profiles $\langle n_e \rangle (x, t)$, $\langle \phi \rangle (x, t)$ and $\langle V_{ye} \rangle (x, t)$ in response to the (small-amplitude) perturbations δn_e and $\delta\phi$.

The equations describing the nonlinear evolution of the perturbations $\delta n_e(x, y, t)$ and $\delta\phi(x, y, t)$ are readily obtained by subtracting Eqs.(6.58) and (6.59) from Eqs.(6.54) and (6.55), respectively. This gives

$$\left(\frac{\partial}{\partial t} + V_E \frac{\partial}{\partial y}\right) \delta n_e - \frac{c}{B_0} \left(\frac{\partial}{\partial y} \delta\phi\right) \frac{\partial}{\partial x} \langle n_e \rangle \tag{6.61}$$

$$= \frac{c}{B_0} \left[\left(\frac{\partial}{\partial y}\delta\phi\right)\left(\frac{\partial}{\partial x}\delta n_e\right) - \left(\frac{\partial}{\partial x}\delta\phi\right)\left(\frac{\partial}{\partial y}\delta n_e\right) - \frac{\partial}{\partial x}\left\langle \delta n_e \frac{\partial}{\partial y}\delta\phi \right\rangle \right],$$

and

$$\left(\frac{\partial^2}{\partial x^2} + \frac{\partial^2}{\partial y^2}\right) \delta\phi = 4\pi e \delta n_e . \tag{6.62}$$

In Eq.(6.61), all terms explicitly bilinear in $\delta\phi\delta n_e$ have been transposed to the right-hand side.

Equations (6.58)–(6.62) constitute a closed description of the nonlinear evolution of the system, which is fully equivalent to the continuity-Poisson equations (6.54) and (6.55). In circumstances where the initial density profile $\langle n_e \rangle (x, 0)$ corresponds to linear instability, the perturbations $\delta\phi(x, y, t)$ and $\delta n_e(x, y, t)$ amplify, and the average density profile $\langle n_e \rangle (x, t)$ readjusts in response to the unstable field perturbations according to Eq.(6.58).

Problem 6.6   Consider the continuity-Poisson equations (6.54) and (6.55), and express $n_e = \langle n_e \rangle (x, t) + \delta n_e(x, y, t)$ and $\phi = \langle \phi \rangle (x, t) + \delta\phi(x, y, t)$. Here, averages are defined as in Eq.(6.57), and n_e and ϕ are assumed spatially periodic in the y-direction with fundamental periodicity length L.

a. Derive Eqs.(6.58) and (6.59) for the evolution of the average profiles $\langle n_e \rangle (x, t)$ and $\langle \phi \rangle (x, t)$.

b. Make use of Eqs.(6.54), (6.55), (6.58), and (6.59) to show that the perturbations $\delta n_e(x, y, t)$ and $\delta\phi(x, y, t)$ evolve according to Eqs.(6.61) and (6.62).

6.5.2   Quasilinear Kinetic Equations

A lowest-order *quasilinear* analysis of Eqs.(6.58)–(6.62) proceeds by neglecting all bilinear nonlinearities on the right-hand side of Eq.(6.61). The resulting equation for the evolution of $\delta n_e(x, y, t)$ is then solved in conjunction with Poisson's equation (6.62) for $\delta\phi(x, y, t)$, and the resulting expressions are substituted into Eq.(6.58) to determine the quasilinear response of the average density profile $\langle n_e \rangle$ (x, t) to the unstable field perturbations.[18]

With regard to Poisson's equation (6.62) for $\delta\phi$ and the continuity equation (6.61) for δn_e, it is convenient to Fourier decompose perturbed quantities with respect to their y-dependence. That is, we express

$$\delta\phi(x, y, t) = \sum_k \delta\phi_k(x, t) \exp(iky) ,$$

$$\delta n_e(x, y, t) = \sum_k \delta n_{ek}(x, t) \exp(iky) , \qquad (6.63)$$

where $k = 2\pi n/L$, L is the periodicity length in the y-direction, n is an integer, and the summation is from $n = -\infty$ to $n = +\infty$. Equation (6.62) then gives

$$\frac{\partial^2}{\partial x^2}\delta\phi_k - k^2\delta\phi_k = 4\pi e\delta n_{ek} , \qquad (6.64)$$

which relates $\delta\phi_k(x, t)$ and $\delta n_{ek}(x, t)$. At the quasilinear level of description, the right-hand side of Eq.(6.61) is approximated by zero, corresponding to the neglect of bilinear nonlinearities in the evolution of δn_e. Fourier decomposing Eq.(6.61) then gives

$$\left(\frac{\partial}{\partial t} + ikV_E\right)\delta n_{ek} = \frac{ikc}{B_0}\delta\phi_k\frac{\partial}{\partial x}\langle n_e\rangle , \qquad (6.65)$$

where $V_E(x, t)$ is given by

$$V_E(x, t) = \frac{c}{B_0}\frac{\partial}{\partial x}\langle\phi\rangle = \frac{4\pi ec}{B_0}\int_0^x dx' \langle n_e\rangle (x', t) . \qquad (6.66)$$

Here, use has been made of Eqs.(6.59) and (6.60), and $\langle E_x\rangle = -(\partial/\partial x)\langle\phi\rangle$ $= 0$ at $x = 0$ has been assumed, which corresponds to zero average normal electric field at the left-most conductor in Fig. 6.8. In Eqs.(6.65) and (6.66), the (slow) evolution of $\langle n_e\rangle$ (x, t) is calculated self-consistently in terms of $\delta\phi$ and δn_e from Eq.(6.58), which can be expressed in Fourier variables as

$$\frac{\partial}{\partial t}\langle n_e\rangle = \frac{c}{B_0}\frac{\partial}{\partial x}\sum_k \delta n_{ek}(-ik)\delta\phi_{-k} . \qquad (6.67)$$

Equations (6.64)–(6.67) constitute coupled nonlinear equations for the evolution of $\delta\phi_k$, δn_{ek} and $\langle n_e \rangle$ at the quasilinear level of description.

To analyze Eqs.(6.64) and (6.65), we consider amplifying ($\gamma_k > 0$) perturbations with time dependence of the form

$$\delta\phi_k(x,t) = \delta\hat{\phi}_k(x) \exp\left\{ \int_0^t dt'\, [-i\omega_k(t') + \gamma_k(t')] \right\},$$

$$\delta n_{ek}(x,t) = \delta\hat{n}_{ek}(x) \exp\left\{ \int_0^t dt'\, [-i\omega_k(t') + \gamma_k(t')] \right\}. \qquad (6.68)$$

In Eq.(6.68), the growth rate $\gamma_k(t)$ and oscillation frequency $\omega_k(t)$ are allowed to vary slowly in time in response to the slow evolution of $\langle n_e \rangle$ (x,t) in Eq.(6.67). Substituting Eq.(6.68) into Eq.(6.65) and solving for δn_{ek} gives

$$\delta n_{ek} = \frac{-(kc/B_0)\delta\phi_k}{\omega_k - kV_E + i\gamma_k} \frac{\partial}{\partial x} \langle n_e \rangle. \qquad (6.69)$$

Moreover, substituting Eq.(6.69) into Eq.(6.64), Poisson's equation for $\delta\phi_k$ becomes

$$\frac{\partial^2}{\partial x^2}\delta\phi_k - k^2\delta\phi_k = \frac{-(4\pi ekc/B_0)\delta\phi_k}{\omega_k - kV_E + i\gamma_k} \frac{\partial}{\partial x} \langle n_e \rangle, \qquad (6.70)$$

where $\gamma_k > 0$ is assumed. Note that Eqs.(6.69) and (6.70) are identical in form to the equations for δn_{ek} and $\delta\phi_k$ obtained in standard linear theory [compare with Eq.(6.28)], assuming perturbations about quasisteady equilibrium profiles $\langle n_e \rangle$ and V_E in planar geometry. The only difference in the present *quasilinear* analysis is that $\langle n_e \rangle$ (x,t) is allowed to vary slowly in time [Eq.(6.67)], which leads to a corresponding slow (adiabatic) variation of the growth rate $\gamma_k(t)$ and oscillation frequency $\omega_k(t)$ as calculated from the eigenvalue equation (6.70).

Substituting Eq.(6.69) into Eq.(6.67), the average density profile $\langle n_e \rangle$ (x,t) evolves according to

$$\frac{\partial}{\partial t} \langle n_e \rangle = \left(\frac{c}{B_0}\right)^2 \frac{\partial}{\partial x} \left[\sum_k \frac{ik^2 |\delta\phi_k|^2}{\omega_k - kV_E + i\gamma_k} \frac{\partial}{\partial x} \langle n_e \rangle \right]. \qquad (6.71)$$

In obtaining Eq.(6.71), use has been made of the conjugate symmetry, $\delta\phi_{-k}(x,t) = \delta\phi_k^*(x,t)$, which follows from Eq.(6.63) because $\delta\phi(x,y,t)$ is a real-valued function. Consistent with this conjugate symmetry, it can be shown from Eq.(6.68) that the oscillation frequency ω_k and growth rate γ_k satisfy the symmetries

$$\omega_{-k} = -\omega_k \quad \text{and} \quad \gamma_{-k} = \gamma_k \, , \qquad (6.72)$$

and the amplitude $\delta\hat{\phi}_k(x)$ satisfies $\delta\hat{\phi}_{-k} = \delta\hat{\phi}_k^*$. Making use of

$$\frac{1}{\omega_k - kV_E + i\gamma_k} = \frac{(\omega_k - kV_E) - i\gamma_k}{(\omega_k - kV_E)^2 + \gamma_k^2}$$

and the symmetries in Eq.(6.72) to eliminate the odd functions of k on the right-hand side of Eq.(6.71), we find that the quasilinear kinetic equation for $\langle n_e \rangle$ can be expressed as

$$\frac{\partial}{\partial t} \langle n_e \rangle = \frac{\partial}{\partial x} \left[D(x,t) \frac{\partial}{\partial x} \langle n_e \rangle \right] \, , \qquad (6.73)$$

where the diffusion coefficient $D(x,t)$ is defined by

$$D(x,t) = \left(\frac{c}{B_0} \right)^2 \sum_k \frac{k^2 |\delta\phi_k|^2 \gamma_k}{(\omega_k - kV_E)^2 + \gamma_k^2} \, , \qquad (6.74)$$

and $\gamma_k > 0$ is assumed. Moreover, from Eq.(6.68), the quantity $|\delta\phi_k|^2$ evolves according to the wave kinetic equation

$$\frac{\partial}{\partial t} |\delta\phi_k|^2 = 2\gamma_k |\delta\phi_k|^2 \, . \qquad (6.75)$$

To summarize, the quasilinear evolution of $\langle n_e \rangle$ and $|\delta\phi_k|^2$ is described by the coupled kinetic equations (6.73) and (6.75),[18] where the diffusion coefficient D is defined in Eq.(6.74). Moreover, the growth rate $\gamma_k(t)$ and oscillation frequency $\omega_k(t)$ are determined adiabatically from the linear eigenvalue equation (6.70) with $\langle n_e \rangle$ changing slowly according to the kinetic equation (6.73). Typically, if the initial profile $\langle n_e \rangle (x, t = 0)$ corresponds to instability with $\gamma_k(0) > 0$, the perturbations will amplify [Eq.(6.75)], and the density profile $\langle n_e \rangle (x,t)$ will readjust [Eq.(6.73)] in such a way as to reduce the growth rate $\gamma_k(t)$ and stabilize the instability [Eqs.(6.70) and (6.75)].

With regard to boundary conditions, the eigenvalue equation (6.70) is to be solved subject to

$$\delta\phi_k = 0 \, , \quad \text{at} \quad x = 0 \quad \text{and} \quad x = d \, , \qquad (6.76)$$

which assures that the tangential electric field, $\delta E_{yk} = -ik\delta\phi_k$, vanishes at the two conductors. Correspondingly, it follows from Eqs.(6.74) and (6.76) that the diffusion coefficient $D(x, t)$ is equal to zero at the two conductors, i.e.,

$$D = 0 \, , \quad \text{at} \quad x = 0 \quad \text{and} \quad x = d \, . \qquad (6.77)$$

The boundary conditions in Eqs.(6.76) and (6.77) play an important role in determining the detailed quasilinear evolution of the system.

6.5.3   Stabilization Process

To illustrate the general features of the quasilinear stabilization process, we consider the smooth initial density profile $\langle n_e \rangle \, (x, t = 0)$ corresponding to instability illustrated by the solid curve in Fig. 6.9(a). In addition to $\gamma_k(t = 0) > 0$, nonzero initial excitation of $|\delta\phi_k|^2$ is assumed. In Fig. 6.9(a), the density maximum at $t = 0$ is located at $x = x_m$. Moreover, from Eq.(6.66) and Fig. 6.9(a), the corresponding initial flow velocity profile $V_E(x, t = 0)$ has the form illustrated by the solid curve in Fig. 6.9(b), with inflection point $[V_E''(x, t = 0) = 0]$ located at $x = x_m$.

Several important features of the quasilinear evolution of the system follow directly from Eq.(6.70) and Eqs.(6.73)–(6.77).[18]

A.   Number Conservation:

First, number conservation readily follows upon integrating Eq.(6.73) from $x = 0$ to $x = d$. This gives

$$\frac{\partial}{\partial t} \int_0^d dx \, \langle n_e \rangle \, (x, t) = 0 \, , \qquad (6.78)$$

where use has been made of $D = 0$ at $x = 0$ and $x = d$ [Eq.(6.77)].

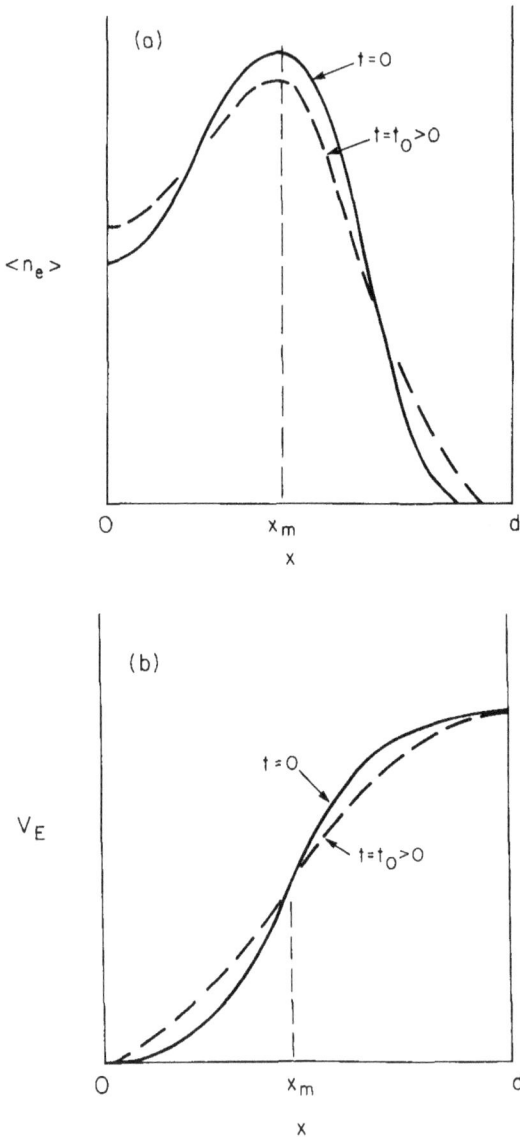

Figure 6.9. Quasilinear response of (a) the average density profile $\langle n_e \rangle (x,t)$ and (b) the average flow velocity $V_E(x,t) = \langle V_{ye} \rangle (x,t)$ to the amplifying field perturbations predicted by Eq.(6.70) and Eqs.(6.73)–(6.75) [R.C. Davidson, Phys. Fluids **28**, 1937 (1985)].

B.   Conservation of Average x-Location:

Second, the density-weighted, average x-location of the electrons is also conserved. Multiplying Eq.(6.73) by x and integrating from $x = 0$ to $x = d$ gives

$$\frac{\partial}{\partial t} \int_0^d dx\, x\, \langle n_e \rangle = -\int_0^d dx D \frac{\partial}{\partial x} \langle n_e \rangle \ , \tag{6.79}$$

where use has been made of Eq.(6.77). After some algebraic manipulation that makes use of the eigenvalue equation (6.70) (Problem 6.7), it can be shown that $\int_0^d dx D(\partial/\partial x) \langle n_e \rangle = 0$. Therefore, Eq.(6.79) reduces to

$$\frac{\partial}{\partial t} \int_0^d dx\, x\, \langle n_e \rangle = 0 \ . \tag{6.80}$$

Equation (6.80) is a statement that the density-weighted average x-location of the electrons is conserved, however complicated the quasilinear evolution of the system. The result in Eq.(6.80) is the planar analogue of the nonlinear conservation constraint (6.16) derived in cylindrical geometry in Sec. 6.2.

C.   Profile Evolution:

We consider Eqs.(6.73) and (6.74) for $\gamma_k > 0$, and integrate Eq.(6.73) from $x = 0$ to an arbitrary point $x(0 < x < d)$. Enforcing the boundary condition $D = 0$ at $x = 0$ [Eq.(6.77)], we obtain

$$\frac{\partial}{\partial t} \int_0^x dx\, \langle n_e \rangle\, (x, t) = D \frac{\partial}{\partial x} \langle n_e \rangle \ , \tag{6.81}$$

or equivalently, from Eq.(6.66),

$$\frac{\partial}{\partial t} V_E(x, t) = D \frac{\partial^2}{\partial x^2} V_E \ . \tag{6.82}$$

Comparing with Fig. 6.9, it follows from Eqs.(6.81) and (6.82) that

$$\frac{\partial}{\partial t} \int_0^x dx\, \langle n_e \rangle \Big|_{t=0} \gtrless 0 \ , \quad \text{for } x \lessgtr x_m \ , \tag{6.83}$$

and

$$\frac{\partial}{\partial t} V_E \bigg|_{t=0} \gtreqless 0 , \quad \text{for } x \lessgtr x_m , \qquad (6.84)$$

where $x = x_m$ corresponds to the density maximum in Fig. 6.9(a). That is, at a subsequent time $t_0 > 0$, the profiles for $\langle n_e \rangle$ and V_E have evolved to the form illustrated by the dashed curves in Fig. 6.9, corresponding to a weakening of the density gradients, and a partial fill-in of the density depression. Therefore, during the initial stages of instability, the quasilinear response of the system is in the direction of stabilization and reducing the growth rate [Eq.(6.70)]. As the potential perturbations continue to amplify for times $t > t_0$ [Eq.(6.75)], the density depression and velocity shear are further reduced in Fig. 6.9. This further decreases the instability growth rate γ_k, and the process continues until stabilization occurs.

To summarize, the quasilinear equations (6.70) and (6.73)–(6.75) can be used to calculate[18] the detailed nonlinear evolution of the diocotron instability for a wide range of initial profiles $\langle n_e \rangle (x, 0)$ corresponding to instability with $\gamma_k(0) > 0$. The evolution of $\langle n_e \rangle (x, t)$ and $|\delta \phi_k|^2$ is described by the kinetic equations (6.73) and (6.75), where the diffusion coefficient $D(x, t)$ is defined in Eq.(6.74). Moreover, the growth rate $\gamma_k(t)$ and real oscillation frequency $\omega_k(t)$ are determined adiabatically from the eigenvalue equation (6.70), where $\langle n_e \rangle (x, t)$ varies slowly according to the diffusion equation (6.73). Typically, if the initial profile $\langle n_e \rangle (x, 0)$ corresponds to instability with $\gamma_k(0) > 0$, then the potential perturbations amplify according to Eq.(6.75), thereby causing an increase in the diffusion coefficient $D(x, t)$, and a concomitant readjustment of the average density profile $\langle n_e \rangle (x, t)$ according to Eq.(6.73). The average density profile $\langle n_e \rangle (x, t)$ readjusts in such a way as to reduce the growth rate $\gamma_k(t)$ and stabilize the instability.

As a concluding remark, we reiterate that such a quasilinear model is expected to apply only in circumstances where there is multimode excitation (many unstable k-values) of the diocotron instability.

Problem 6.7 Consider the form of the quasilinear diffusion equation for $\langle n_e \rangle (x, t)$ given in Eq.(6.71).

a. Show from Eqs.(6.71) and (6.76) that

$$\frac{\partial}{\partial t} \int_0^x dx\, x\, \langle n_e \rangle = -\left(\frac{c}{B_0}\right)^2 \int_0^d dx \sum_k \frac{ik^2 |\delta \phi_k|^2}{\omega_k - kV_E + i\gamma_k} \frac{\partial}{\partial x} \langle n_e \rangle . \qquad (6.7.1)$$

b. Multiply the eigenvalue equation (6.70) by $(c/4\pi e B_0) k \delta\phi_{-k}$, integrate from $x = 0$ to $x = d$, and sum over k. Show that

$$-\left(\frac{c}{B_0}\right)^2 \int_0^d dx \sum_k \frac{k^2 |\delta\phi_k|^2}{\omega_k - kV_E + i\gamma_k} \frac{\partial}{\partial x} \langle n_e \rangle$$

$$= -\left(\frac{c}{4\pi e B_0}\right) \int_0^d dx \sum_k k \left(\left|\frac{\partial}{\partial x}\delta\phi_k\right|^2 + k^2 |\delta\phi_k|^2\right) . \quad (6.7.2)$$

To obtain Eq.(6.7.2), it is necessary to integrate by parts with respect to x, and make use of the conjugate symmetry $\delta\phi_{-k} = \delta\phi_k^*$.

c. Make use of Eqs.(6.7.1) and (6.7.2) and the fact that the summation $\sum_k \cdots$ in Eq.(6.7.2) is over an *odd* function of k to show that

$$\frac{\partial}{\partial t} \int_0^d dx \, x \, \langle n_e \rangle = 0 . \quad (6.7.3)$$

6.6   Resonant Diocotron Instability

Hollow density profiles (like those shown in Figs. 6.2 and 6.9) tend to give a strong version of the diocotron instability with growth rate comparable in magnitude to the real oscillation frequency (see Sec. 6.3). On the other hand, density profiles with a gentle density bump can result in a weak resonant version of the diocotron instability[14] characterized by relatively small growth rate with $|\gamma_k| \ll |\omega_k|$. Such a density profile is illustrated in Fig. 6.10. Here, $\langle n_e \rangle (x, 0)$ corresponds to a gentle density bump superimposed on a profile with uniform density $\hat{n}_e = $ const. extending from $x = 0$ to $x = x_b$. In the absence of the density bump, the profile in Fig. 6.10 is stable (because the layer is in contact with the conducting wall at $x = 0$). In the presence of the density bump, however, the profile $\langle n_e \rangle (x, 0)$ in Fig. 6.10 is subject to a weak resonant version of the diocotron instability. In this section, we make use of the formalism developed in Sec. 6.5 to describe the quasilinear evolution and stabilization of the resonant diocotron instability. While the initial profile shown in Fig. 6.10 constitutes one interesting application, the approach developed here is general and can be applied to a wide variety of profiles $\langle n_e \rangle (x, 0)$.

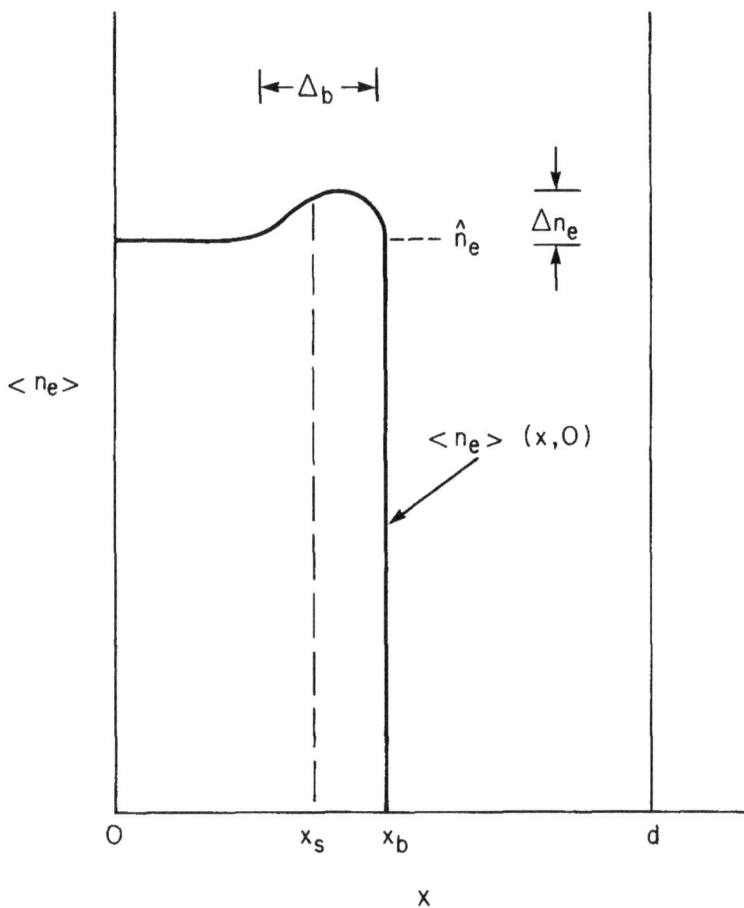

Figure 6.10. Schematic of electron density profile $\langle n_e \rangle (x, 0)$ with a gentle density bump. The resonance location x_s is determined from $\omega_k - kV_E(x_s) = 0$, where (ω_k, k) is the diocotron wave excited by the bulk electrons.

6.6.1 Instability Growth Rate

The eigenvalue equation (6.70) can be used to derive an effective dispersion relation for $\omega_k + i\gamma_k$ in circumstances where the functional form of

$\delta\hat{\phi}_k(x)$ is assumed to be known. Multiplying Eq.(6.70) by $\delta\hat{\phi}_{-k} = \delta\hat{\phi}_k^*$, integrating from $x = 0$ to $x = d$, and making use of the boundary conditions in Eq.(6.76), it can be shown that

$$0 = \varepsilon(k, \omega_k + i\gamma_k) \tag{6.85}$$

$$= \int_0^d dx \left\{ \left| \frac{\partial}{\partial x}\delta\hat{\phi}_k \right|^2 + k^2 \left| \delta\hat{\phi}_k \right|^2 - \frac{k\left|\delta\hat{\phi}_k\right|^2}{\omega_k - kV_E + i\gamma_k} \frac{4\pi ec}{B_0} \frac{\partial}{\partial x} \langle n_e \rangle \right\} .$$

For specified $\delta\hat{\phi}_k(x)$, Eq.(6.85) plays the role of a dispersion relation that determines $\omega_k + i\gamma_k$. For the case of weak resonant diocotron instability characterized by $|\gamma_k| \ll |\omega_k|$, the effective dispersion relation (6.85) can be further simplified. For small γ_k, we approximate

$$0 = \varepsilon(k, \omega_k + i\gamma_k) = \varepsilon_r(k, \omega_k) + i\left[\varepsilon_i(k, \omega_k) + \gamma_k \frac{\partial \varepsilon_r}{\partial \omega_k}\right] + \cdots , \tag{6.86}$$

and

$$\lim_{\gamma_k \to 0_+} \frac{1}{\omega_k - kV_E + i\gamma_k} = \frac{P}{\omega_k - kV_E} - i\pi\delta(\omega_k - kV_E) . \tag{6.87}$$

Here, P denotes Cauchy principal value, and ε_r and ε_i denote the real and imaginary parts of ε, respectively. Substituting Eqs.(6.85) and (6.87) into Eq.(6.86), and setting the real and imaginary parts equal to zero gives[18]

$$0 = \varepsilon_r(k, \omega_k) = \int_0^d dx \left\{ \left| \frac{\partial}{\partial x}\delta\hat{\phi}_k \right|^2 + k^2|\delta\hat{\phi}_k|^2 - \frac{kP|\delta\hat{\phi}_k|^2}{\omega_k - kV_E} \frac{4\pi ec}{B_0} \frac{\partial}{\partial x} \langle n_e \rangle \right\} , \tag{6.88}$$

and

$$\gamma_k = -\frac{\varepsilon_i}{\partial\varepsilon_r/\partial\omega_k} = -\pi \int_0^d dx |\delta\hat{\phi}_k|^2 \delta(\omega_k - kV_E)\frac{\partial}{\partial x} \langle n_e \rangle$$

$$\times \left[\int_0^d dx \frac{|\delta\hat{\phi}_k|^2 P}{(\omega_k - kV_E)^2} \frac{\partial}{\partial x} \langle n_e \rangle \right]^{-1} . \tag{6.89}$$

The resonant layer is denoted by $x = x_s(k)$ where

$$\omega_k - kV_E(x_s) = 0 . \tag{6.90}$$

It follows from Eq.(6.89) that the growth rate γ_k can be expressed as

$$\gamma_k = \pi \left[\frac{|\delta\hat{\phi}_k|^2}{|k\partial V_E/\partial x|} \frac{\partial}{\partial x} \langle n_e \rangle \right]_{x=x_s} \left[-\int_0^d dx \frac{|\delta\hat{\phi}_k|^2 P}{(\omega_k - kV_E)^2} \frac{\partial}{\partial x} \langle n_e \rangle \right]^{-1} .$$

(6.91)

For the case in which $\partial\varepsilon_r/\partial\omega_k < 0$, and therefore $[\cdots]^{-1} > 0$ in Eq.(6.91), it follows from Eq.(6.91) that $\gamma_k > 0$ (corresponding to instability) whenever the resonance location x_s falls in the region of positive density slope, i.e., whenever

$$\frac{\partial}{\partial x} \langle n_e \rangle \Big|_{x=x_s} > 0 , \qquad (6.92)$$

as illustrated in Fig. 6.10. In circumstances where the quasilinear response of the system is such that the density profile flattens in the vicinity of $x = x_s$ with $(\partial/\partial x) \langle n_e \rangle |_{x=x_s} \to 0$, it follows from Eq.(6.91) that $\gamma_k \to 0$, corresponding to marginal stability and saturation of the wave spectrum $|\delta\phi_k|^2$ [Eq.(6.75)].

Problem 6.8 Consider the profile illustrated in Fig. 6.10 in the absence of the density bump, so that the density is uniform and equal to $\hat{n}_e$ in the interval $0 \leq x < x_b$.

a. Show that the eigenvalue equation (6.70) for $\delta\hat{\phi}_k(x)$ reduces to

$$\frac{\partial^2}{\partial x^2}\delta\hat{\phi}_k - k^2\delta\hat{\phi}_k = \frac{k(\omega_{pe}^2/\omega_{ce})\delta\hat{\phi}_k}{\omega_k - kV_E(x) + i\gamma_k}\delta(x - x_b) , \quad 0 \leq x \leq d , \quad (6.8.1)$$

where $\omega_{pe}^2/\omega_{ce} = 4\pi\hat{n}_e ec/B_0$, and $V_E(x) = (\omega_{pe}^2/\omega_{ce})x$.

b. Solve Eq.(6.8.1) in the two regions in Fig. 6.10, and show that

$$\delta\hat{\phi}_k^I(x) = A\sinh(kx) , \qquad\qquad 0 \leq x < x_b ,$$

$$\delta\hat{\phi}_k^{II}(x) = A\sinh(kx_b)\frac{\sinh[k(d - x)]}{\sinh[k(d - x_b)]} , \quad x_b < x \leq d , \quad (6.8.2)$$

where A is a constant. Here, $\delta\hat{\phi}_k^I(x = 0) = 0 = \delta\hat{\phi}_k^{II}(x = d)$ has been enforced, as well as the continuity of $\delta\hat{\phi}_k(x)$ at $x = x_b$.

c. Integrate Eq.(6.8.1) across the layer surface at $x = x_b$. Show that the growth rate $\gamma_k = 0$, and the real oscillation frequency is given by

$$\omega_k - k V_E(x_b) = -\frac{(\omega_{pe}^2/\omega_{ce})}{\coth(k x_b) + \coth\left[k(d - x_b)\right]} \,, \qquad (6.8.3)$$

where $V_E(x_b) = (\omega_{pe}^2/\omega_{ce}) x_b$.

Problem 6.9   Consider the density profile in Fig. 6.10 including the gentle density bump.

a. Assume that the real oscillation frequency is given (approximately) by Eq.(6.8.3). Show that the resonance location $x_s(k)$ satisfying $\omega_k - k V_E(x_s)$ $= 0$ is determined from

$$k(x_b - x_s) = \frac{1}{\coth(k x_b) + \coth\left[k(d - x_s)\right]} \,. \qquad (6.9.1)$$

b. Consider the long-wavelength and short-wavelength limits of Eq.(6.9.1), ranging from $k^2(d - x_b)^2 \ll 1$ and $k^2 x_b^2 \ll 1$ to $k^2(d - x_b)^2 \gg 1$ and $k^2 x_b^2 \gg 1$. Show that the resonance location x_s spans the interval

$$\frac{x_b}{d} x_b < x_s < x_b \,, \qquad (6.9.2)$$

where the lower (upper) limit in Eq.(6.9.2) corresponds to long (short) wavelengths.

c. Make use of the expression for the growth (or damping) rate in Eq.(6.91) to show that γ_k can be expressed as

$$\gamma_k = \pi \frac{[\omega_k - k V_E(x_b)]^2}{|k V_E(x_b)|} \frac{\left|\delta\hat{\phi}_k^I\right|^2_{x=x_s}}{\left|\delta\hat{\phi}_k^{II}\right|^2_{x=x_b}} \frac{x_b}{\hat{n}_e} \left[\frac{\partial}{\partial x} \langle n_e\rangle\right]_{x=x_s} \,, \qquad (6.9.3)$$

where $V_E(x_b) = (\omega_{pe}^2/\omega_{ce}) x_b$. Here, the eigenfunctions have been approximated by Eq.(6.8.2), and the principal-value contribution in Eq.(6.91) has been evaluated by integrating over the bulk electrons with $(\partial/\partial x)\langle n_e\rangle \simeq -\hat{n}_e \delta(x - x_b)$. Evidently, the region of positive (negative) slope in Fig. 6.10 corresponds to growth (damping).

6.6.2   Quasilinear Stabilization Process

We consider Eqs.(6.74) and (6.75) in the limit of a continuous k-spectrum with

$$\sum_k \frac{k^2 |\delta\phi_k|^2}{8\pi} \cdots \rightarrow \int dk \mathcal{E}_k \cdots \,.$$

Here, $\mathcal{E}_k = k^2 |\delta\phi_k|^2 / 8\pi$ is the spectral energy density associated with the δE_y electric field perturbations. In the continuum limit, Eqs.(6.74) and (6.75) become

$$D(x,t) = \frac{8\pi c^2}{B_0^2} \int dk \frac{\gamma_k \mathcal{E}_k}{(\omega_k - kV_E)^2 + \gamma_k^2}, \qquad (6.93)$$

and

$$\frac{\partial}{\partial t}\mathcal{E}_k = 2\gamma_k \mathcal{E}_k, \qquad (6.94)$$

where γ_k is the growth rate.

For the case of weak resonant diocotron instability, the expression for the diffusion coefficient $D(x,t)$ in Eq.(6.93) can be further simplified. In particular, taking $\gamma_k \to 0_+$ in Eq.(6.93) in the *resonant* region of x-space where $\omega_k - kV_E(x) = 0$, it follows that D can be approximated by

$$D_r(x,t) = \frac{8\pi^2 c^2}{B_0^2} \int dk \mathcal{E}_k \delta(\omega_k - kV_E). \qquad (6.95)$$

On the other hand, in the *nonresonant* region of x-space where $(\omega_k - kV_E)^2 \gg \gamma_k^2$, it follows from Eq.(6.93) that D can be approximated by

$$D_{nr}(x,t) = \frac{8\pi c^2}{B_0^2} \int dk \frac{\gamma_k \mathcal{E}_k}{(\omega_k - kV_E)^2}. \qquad (6.96)$$

The approximate forms of D_r and D_{nr} in Eqs.(6.95) and (6.96) are calculated to the same accuracy as Eqs.(6.88) and (6.91) for ω_k and γ_k. The accuracy in Eqs.(6.95) and (6.96) is also required to ensure that the conservation relation $(\partial/\partial t) \int_0^d dx\, x \langle n_e \rangle (x,t) = 0$ is satisfied [see Eq.(6.80)].

In the resonant region of x-space satisfying $\omega_k - kV_E(x) = 0$, the quasilinear diffusion equation (6.73) for $\langle n_e \rangle (x,t)$ can be expressed as

$$\frac{\partial}{\partial t} \langle n_e \rangle = \frac{\partial}{\partial x} \left[D_r(x,t) \frac{\partial}{\partial x} \langle n_e \rangle \right] \qquad (6.97)$$

where $D_r(x,t)$ is defined in Eq.(6.95). A nonlinear entropy theorem describing the stabilization process in the resonant region of x-space follows directly from Eqs.(6.95) and (6.97). Multiplying Eq.(6.97) by $\langle n_e \rangle$ and integrating with respect to x gives

$$\frac{\partial}{\partial t} \int dx \, \langle n_e \rangle^2 = - \int dx D_r(x,t) \left(\frac{\partial}{\partial x} \langle n_e \rangle \right)^2 \qquad (6.98)$$

$$= -\frac{16\pi^2 c^2}{B_0^2} \int dx \int_0^\infty dk \delta(\omega_k - kV_E) \mathcal{E}_k(x,t) \left(\frac{\partial}{\partial x} \langle n_e \rangle \right)^2 \le 0 \, .$$

Analogous to the quasilinear stabilization of the one-dimensional bump-in-tail instability in velocity space,[17] the time-asymptotic $(t \to \infty)$ solution inferred from Eq.(6.98) necessarily satisfies

$$\frac{\partial}{\partial x} \langle n_e \rangle (x, t \to \infty) \bigg|_{x=x_s} = 0 \qquad (6.99)$$

in the resonant region where $\omega_k - kV_E(x_s) = 0$. We conclude from Eqs.(6.91) and (6.99) that

$$\gamma_k(t \to \infty) = 0 \, , \qquad (6.100)$$

corresponding to plateau formation and quasilinear stabilization of the instability. From Eq.(6.100) and $(\partial/\partial t)\mathcal{E}_k = 2\gamma_k\mathcal{E}_k$, we also conclude that the spectral energy density saturates at a steady asymptotic level $\mathcal{E}_k(x, \infty)$.

A detailed analysis of the quasilinear stabilization of the resonant diocotron instability can be carried out for the density profile with the gentle bump in Fig. 6.10.[18] In this regard, use is made of the expression for the growth rate $\gamma_k(t)$ in Eq.(6.91), the wave kinetic equation (6.94) for the evolution of $\mathcal{E}_k(x,t)$, the particle kinetic equation (6.97) for the evolution of $\langle n_e \rangle (x,t)$ in the resonant region of x-space, the particle kinetic equation for the evolution of $\langle n_e \rangle (x,t)$ in the nonresonant region of x-space [approximate $D(x,t)$ by $D_{nr}(x,t)$ in Eq.(6.73)], and the eigenvalue equation (6.70) for determination of $\delta\hat{\phi}_k(x)$ and ω_k. Without presenting algebraic details,[18] the qualitative features of the stabilization process correspond to a slight outward expansion of the bulk density profile in Fig. 6.10, together with an inward motion and flattening of the density bump corresponding to $\gamma_k(t \to \infty) = 0$. An estimate of the saturated energy in the electric field perturbations gives

$$\frac{\langle \delta E_y^2(x_b, \infty) \rangle}{8\pi} \approx \frac{1}{6} \frac{\Delta n_e}{\hat{n}_e} \frac{\Delta_b^2}{x_b^2} \frac{[\langle E_x \rangle (x_b)]^2}{8\pi} . \tag{6.101}$$

Here, Δn_e and Δ_b measure the height and width of the initial density bump in Fig. 6.10, and $\langle E_x \rangle (x_b) \simeq -4\pi e\hat{n}_e x_b$ is the x-component of the average electric field at the surface of the electron layer $(x = x_b)$ calculated from $(\partial/\partial x) \langle E_x \rangle = -4\pi e \langle n_e \rangle$. It is evident from Eq.(6.101) that the amplitude of the perturbed field can saturate at a substantial level, even for a moderately small density bump as measured by $(\Delta n_e)/\hat{n}_e$.

6.7   Influence of Electromagnetic and Relativistic Effects

In this section, we extend the linear theory of the diocotron instability in Sec. 6.3 to the case in which the cross-field electron flow is relativistic, and electromagnetic effects are retained in the stability analysis. The analysis[19,20] makes use of a macroscopic model based on the cold-fluid-Maxwell equations to investigate the extraordinary-mode stability properties of a relativistic, low-density electron layer in the planar diode configuration illustrated in Fig. 6.8. Here, planar conductors are located at $x = 0$ and $x = d$; $V_{ye}^0(x) = -cE_x(x)/B_z(x)$ is the equilibrium flow velocity in the y-direction; and $E_x(x)$ and $B_z(x)$ are the equilibrium field components determined self-consistently from the steady-state $(\partial/\partial t = 0)$ Maxwell equations

$$\frac{\partial}{\partial x} E_x(x) = -4\pi e n_e^0(x) , \tag{6.102}$$

$$\frac{\partial}{\partial x} B_z(x) = \frac{4\pi e}{c} n_e^0(x) V_{ye}^0(x) , \tag{6.103}$$

where $n_e^0(x)$ is the equilibrium density profile of the layer electrons. Note that the diamagnetic variation of $B_z(x)$ is retained in Eq.(6.103), an effect which is important when the flow is relativistic. For simplicity, the present analysis also assumes $E_x(x = 0) = 0$ at the left-most conductor in Fig. 6.8.

The stability analysis makes use of the extraordinary-mode eigenvalue equation (8.101) derived for flute perturbations with polarization

$$\delta\mathbf{E}(\mathbf{x},t) = \delta E_x(x,y,t)\hat{\mathbf{e}}_x + \delta E_y(x,y,t)\hat{\mathbf{e}}_y ,$$

$$\delta\mathbf{B}(\mathbf{x},t) = \delta B_z(x,y,t)\hat{\mathbf{e}}_z . \tag{6.104}$$

For purposes of describing the diocotron instability, we specialize to the case of low-frequency perturbations about a tenuous electron layer satisfying

$$\omega_{pe}^2(x) \ll \omega_{ce}^2(x)/\gamma_e^0(x) ,$$

$$\left|\omega - kV_{ye}^0(x)\right|^2 \ll \omega_{ce}^2(x)/\gamma_e^{02}(x) . \tag{6.105}$$

Here, $\omega_{pe}(x) = \left[4\pi n_e^0(x)e^2/m\right]^{1/2}$ and $\omega_{ce}(x) = eB_z(x)/m_ec$ are the nonrelativitic electron plasma frequency and cyclotron frequency, respectively; ω is the complex oscillation frequency; k is the perturbation wavenumber in the y-direction; and

$$\gamma_e^0(x) = \left[1 - \left(\frac{E_x}{B_z}\right)^2\right]^{-1/2}$$

is the relativistic mass factor of an electron fluid element. From Eqs.(6.102) and (6.103), it can be shown that $B_z(x)/\gamma_e^0(x) = $ const. across the radial extent of the electron layer (Sec. 8.3.3). Within the context of Eq.(6.105), the electrons are treated as a massless ($m_e \to 0$) guiding-center fluid in which $\omega_{pe}^2(x)/\omega_{ce}^2(x) \to 0$, but the ratio

$$\frac{\omega_{pe}^2(x)}{\omega_{ce}(x)} = \frac{4\pi n_e^0(x)ec}{B_z(x)} \equiv c\kappa(x) \tag{6.106}$$

remains finite.

Perturbed quantities are expressed as $\delta E_y(x,y,t) = \sum_k \delta E_{yk}(x) \times \exp(iky - i\omega t)$, etc., where $\mathrm{Im}\,\omega > 0$ corresponds to instability. Introducing the effective potential amplitude $\Phi_k(x)$ defined by $\Phi_k(x) \equiv (i/k)\delta E_{yk}(x)$, and making use of Eq.(6.105) to simplify the extraordinary-mode eigenvalue equation (8.101) derived for arbitrary ω and $\omega_{pe}^2/\omega_{ce}^2$, we obtain

$$
\frac{\partial^2}{\partial x^2} \Phi_k(x) - k^2 \left[1 - \frac{\omega^2}{c^2 k^2} + \frac{\omega_{pe}^4(x)}{c^2 k^2 \omega_{ce}^2(x)} \right] \Phi_k(x)
$$

$$
= - \frac{4\pi e c k \Phi_k(x)}{\left[\omega - k V_{ye}^0(x) \right]} \left[1 - \frac{V_{ye}^0(x)}{c} \frac{\omega}{ck} \right] \frac{\partial}{\partial x} \left[\frac{n_e^0(x)}{B_z(x)} \right] . \quad (6.107)
$$

Although a tenuous electron layer and $m_e \to 0$ have been assumed, the eigenvalue equation (6.107) is fully electromagnetic and valid for relativistic flow velocities. In this regard, Eq.(6.107) can be used to investigate the low-frequency stability properties of a low-density electron layer for a broad range of equilibrium profiles consistent with the steady-state Maxwell equations (6.102) and (6.103). For nonrelativistic electron flow with $V_{ye}^{02}/c^2 \ll 1$ and $B_z(x) \simeq B_0 = $ const., and slow-wave perturbations with $|\omega^2/c^2 k^2| \ll 1$, we note that Eq.(6.107) reduces to the electrostatic eigenvalue equation (6.70), or equivalently, the planar limit of Eq.(6.28).

As a particular example that is analytically tractable, we consider perturbations about the class of equilibrium profiles where $n_e^0(x)/\gamma_e^0(x)$ has the simple rectangular shape (Fig. 6.11)[20]

$$
\frac{n_e^0(x)}{\gamma_e^0(x)} = \begin{cases} 0 , & 0 \leq x < x_b^- , \\ \dfrac{\hat{n}_e}{\hat{\gamma}_e} = \text{const.} , & x_b^- < x < x_b^+ , \\ 0 , & x_b^+ < x \leq d . \end{cases} \quad (6.108)
$$

In Eq.(6.108), $\hat{n}_e \equiv n_e^0(x = x_b^+)$ and $\hat{\gamma}_e \equiv \gamma_e^0(x = x_b^+)$ denote the electron density and energy, respectively, at the outer edge ($x = x_b^+$) of the electron layer. Moreover, Eq.(6.108) generally allows for the inner edge of the electron layer ($x = x_b^-$) to be detached from the conducting wall at $x = 0$, although the case $x_b^- = 0$ is not excluded. A configuration like that illustrated in Fig. 6.11 or Eq.(6.108) could occur in a high-voltage diode if the magnetically-insulated electron sheath becomes detached from the cathode (at $x = 0$).

For $n_e^0(x)/\gamma_e^0(x)$ specified by Eq.(6.108) it can be shown from the steady-state Maxwell equations (6.102) and (6.103) that the self-consistent equilibrium profiles for the magnetic field $B_z(x)$, the electric field $E_x(x)$, the

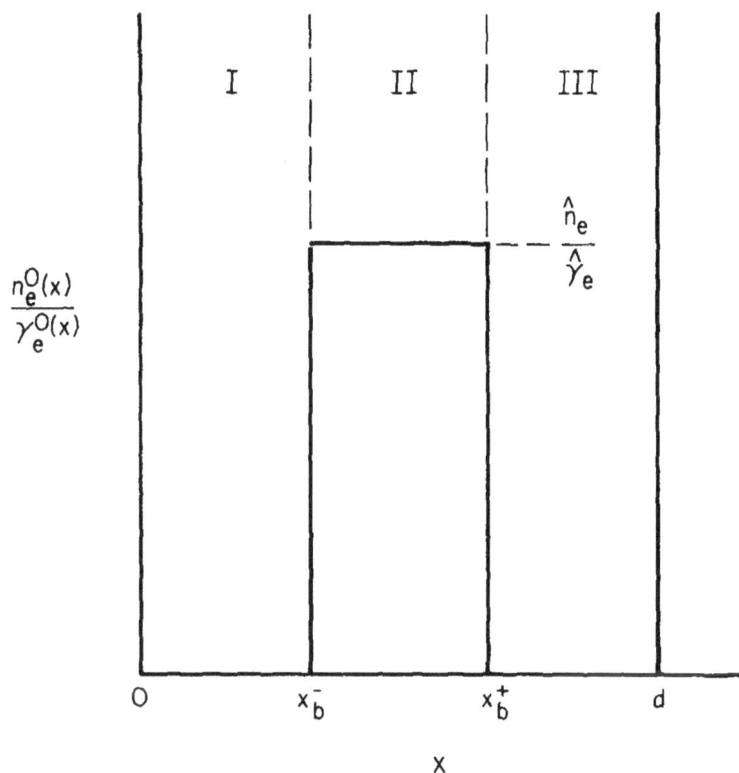

Figure 6.11.   Plot of $n_e^0(x)/\gamma_e^0(x)$ versus x for the rectangular profile in Eq.(6.108).

relativistic mass factor $\gamma_e^0(x)$, and the flow velocity $V_{ye}^0(x)$ can be expressed as

$$B_z(x) = B_0 \frac{\cosh\left[\kappa(x - x_b^-)\right]}{\cosh\left[\kappa(x_b^+ - x_b^-)\right]} \ , \tag{6.109a}$$

$$E_x(x) = -B_0 \frac{\sinh\left[\kappa(x - x_b^-)\right]}{\cosh\left[\kappa(x_b^+ - x_b^-)\right]} \ , \tag{6.109b}$$

$$\gamma_e^0(x) = \cosh\left[\kappa(x - x_b^-)\right] \ , \tag{6.109c}$$

$$V_{ye}^0(x) = c\tanh\left[\kappa(x - x_b^-)\right] \ , \tag{6.109d}$$

within the electron layer $(x_b^- < x < x_b^+)$. Here, $B_0 \equiv B_z(x = x_b^+)$ is the applied magnetic field in the outer vacuum region $(x_b^+ < x \le d)$ in Fig. 6.11. Moreover, the constant κ is defined by

$$c\kappa = \frac{\omega_{pe}^2(x = x_b^+)}{\omega_{ce}(x = x_b^+)} = \frac{4\pi\hat{n}_e ec}{B_0} \tag{6.110}$$

where $\hat{n}_e$ and B_0 are the density and magnetic field at the outer edge of the electron layer $(x = x_b^+)$. Indeed, because $n_e^0(x)/B_z(x) = \text{const.}$ follows from Eqs.(6.108) and (6.109), it is readily shown that

$$c\kappa(x) \equiv \frac{\omega_{pe}^2(x)}{\omega_{ce}(x)} = c\kappa = \text{const.} \tag{6.111}$$

throughout the cross section of the electron layer $(x_b^- < x < x_b^+)$. Here, κ is the constant defined in Eq.(6.110).

Making use of Eq.(6.111), it follows that $4\pi ec(\partial/\partial x)\left[n_e^0(x)/B_z(x)\right] = c\kappa\left[\delta(x - x_b^-) - \delta(x - x_b^+)\right]$ for the choice of equilibrium profiles in Eqs.(6.108) and (6.109). Therefore, the eigenvalue equation (6.107) can be expressed as[20]

$$\frac{\partial^2}{\partial x^2}\Phi_k(x) - \kappa^2(x,\omega)\Phi_k(x) \tag{6.112}$$

$$= -\frac{ck}{\omega}\kappa\Phi_k(x_b^-)\delta(x - x_b^-) + \frac{ck\kappa\Phi_k(x_b^+)}{\omega - kV_d}\left(1 - \frac{V_d}{c}\frac{\omega}{ck}\right)\delta(x - x_b^+) .$$

Here, $V_d \equiv V_{ye}^0(x = x_b^+) = c\tanh\left[\kappa(x_b^+ - x_b^-)\right]$ is the flow velocity at the outer edge of the electron layer, and $\kappa^2(x,\omega)$ is defined by

$$\kappa^2(x,\omega) = \begin{cases} \kappa_v^2(\omega) = k^2\left(1 - \dfrac{\omega^2}{c^2k^2}\right) , & 0 \le x < x_b^- , \\[3mm] \kappa_b^2(\omega) = k^2\left(1 - \dfrac{\omega^2}{c^2k^2} + \dfrac{\kappa^2}{k^2}\right) , & x_b^- < x < x_b^+ , \quad (6.113) \\[3mm] \kappa_v^2(\omega) = k^2\left(1 - \dfrac{\omega^2}{c^2k^2}\right) , & x_b^+ < x \le d . \end{cases}$$

Note that the right-hand side of Eq.(6.112) corresponds to surface perturbations on the outer $(x = x_b^+)$ and inner $(x = x_b^-)$ surfaces of the electron layer. Note also from Eq.(6.113) that $\kappa^2(x, \omega)$ is constant (independent of x) in each of the three regions. It is clear from the eigenvalue equation (6.112) that the relativistic flow parameter

$$\theta = \kappa(x_b^+ - x_b^-) = \frac{4\pi \hat{n}_e e(x_b^+ - x_b^-)}{B_0} \tag{6.114}$$

will play an important role in determining the detailed stability properties. For example, from Eq.(6.109), $V_d = V_{ye}^0(x = x_b^+)$ can be expressed as

$$V_d = c \tanh \theta . \tag{6.115}$$

The right-hand side of the eigenvalue equation (6.112) vanishes except at the inner and outer surfaces of the electron layer at $x = x_b^-$ and $x = x_b^+$. We therefore solve Eq.(6.112) separately in each of the three regions in Fig. 6.11, enforcing the continuity of $\Phi_k(x)$ at $x = x_b^-$ and $x = x_b^+$, and setting $\Phi_k(x = 0) = 0 = \Phi_k(x = d)$, which corresponds to zero tangential electric field, $\delta E_{yk} = -ik\Phi_k = 0$, at the conducting walls. The solutions to Eq.(6.112) in the three regions are given by

$$\Phi_k^I(x) = B \frac{\sinh\left[\kappa_v(\omega)x\right]}{\sinh\left[\kappa_v(\omega)x_b^-\right]} , \quad 0 \le x < x_b^- , \tag{6.116a}$$

$$\Phi_k^{II}(x) = B \frac{\sinh\left[\kappa_b(\omega)(x_b^+ - x)\right]}{\sinh\left[\kappa_b(\omega)(x_b^+ - x_b^-)\right]} + C \frac{\sinh\left[\kappa_b(\omega)(x - x_b^-)\right]}{\sinh\left[\kappa_b(\omega)(x_b^+ - x_b^-)\right]} ,$$
$$x_b^- < x < x_b^+ , \tag{6.116b}$$

$$\Phi_k^{III}(x) = C \frac{\sinh\left[\kappa_v(\omega)(d - x)\right]}{\sinh\left[\kappa_v(\omega)(d - x_b^+)\right]} , \quad x_b^+ < x \le d . \tag{6.116c}$$

The remaining boundary conditions are obtained by integrating the eigenvalue equation across the layer surfaces at $x = x_b^-$ and $x = x_b^+$. This gives

$$\left[\frac{\partial}{\partial x}\Phi_k^{II}\right]_{x=x_b^-} - \left[\frac{\partial}{\partial x}\Phi_k^I\right]_{x=x_b^-} = -\frac{ck\kappa}{\omega}\Phi_k(x_b^-) , \tag{6.117}$$

and

$$\left[\frac{\partial}{\partial x}\Phi_k^{III}\right]_{x=x_b^+} - \left[\frac{\partial}{\partial x}\Phi_k^{II}\right]_{x=x_b^+} = \frac{ck\kappa\Phi_k(x_b^+)}{\omega - kV_d}\left(1 - \frac{V_d}{c}\frac{\omega}{ck}\right) , \tag{6.118}$$

where $V_d = c \tanh \theta$.

Substituting Eq.(6.116) into Eqs.(6.117) and (6.118), the constants B and C can be eliminated to obtain a closed dispersion relation for the complex eigenfrequency ω. After some algebraic manipulation, we obtain[20]

$$\left\{ \kappa_v(\omega) \coth \left[\kappa_v(\omega)(x_b^+ - d) \right] - \kappa_b(\omega) \coth \left[\kappa_b(\omega)(x_b^+ - x_b^-) \right] - \kappa_v(\omega) f^+(\omega) \right\}$$

$$\times \left\{ \kappa_v(\omega) \coth \left[\kappa_v(\omega) x_b^- \right] + \kappa_b(\omega) \coth \left[\kappa_b(\omega)(x_b^+ - x_b^-) \right] - \kappa_v(\omega) f^-(\omega) \right\}$$

$$= - \frac{\kappa_b^2(\omega)}{\sinh^2 \left[\kappa_b(\omega)(x_b^+ - x_b^-) \right]} \, . \tag{6.119}$$

Here, $\kappa_v(\omega)$ and $\kappa_b(\omega)$ are the dielectric functions defined in Eq.(6.113), and $f^-(\omega)$ and $f^+(\omega)$ are the coupling coefficients at the surfaces of the electron layer defined by

$$f^-(\omega) = \frac{k}{\kappa_v(\omega)} \frac{c\kappa}{\omega} \, ,$$

$$f^+(\omega) = \frac{k}{\kappa_v(\omega)} \frac{c\kappa}{\omega - kV_d} \left(1 - \frac{V_d}{c} \frac{\omega}{ck} \right) \, . \tag{6.120}$$

The dispersion relation (6.119) is a transcendental equation that determines the complex oscillation frequency ω in terms of the wavenumber k and equilibrium layer parameters such as $\kappa, V_d, x_b^+ - x_b^-$, etc. In Eq.(6.119), we note that perturbations on the outer surface of the electron layer (the first factor on the left-hand side) are coupled to perturbations on the inner surface of the layer (the second factor on the left-hand side) through body-wave perturbations within the layer (the right-hand side). The dispersion relation (6.119) is fully electromagnetic and valid for relativistic electron flow in the limit where the electrons are treated as a massless, guiding-center fluid. In this regard, Eq.(6.119) can be used to investigate detailed stability properties for low-frequency perturbations about a tenuous electron layer [Eq.(6.105)] for the class of equilibrium profiles described by Fig. 6.11 and Eqs.(6.108) and (6.109).

As a limiting case that is analytically tractable, we first assume long-wavelength perturbations $k^2 d^2 \ll 1$. For $\left| \omega^2/c^2 k^2 \right| \lesssim 1$, it readily follows from Eq.(6.113) that $\left| \kappa_v(x_b^+ - d) \right| \ll 1$, $\left| \kappa_v x_b^- \right| \ll 1$, and $\left| \kappa_b(x_b^+ - x_b^-) \right| \simeq \kappa(x_b^+ - x_b^-) = \theta$, when $k^2 d^2 \ll 1$. Examination of Eq.(6.119) in the long-wavelength regime shows that the dispersion relation can be approximated by the quadratic equation $\omega^2 - b\omega + c = 0$, where the coefficients b and c are independent of ω, but depend on $\theta = \kappa(x_b^+ - x_b^-)/c$ and the dimensionless geometric factors

$$\Delta_i = \frac{x_b^-}{d} ,$$

$$\Delta_b = \frac{x_b^+ - x_b^-}{d} , \qquad (6.121)$$

$$\Delta_0 = \frac{d - x_b^+}{d} .$$

It is found that the necessary and sufficient condition for instability $(4c - b^2 > 0)$ in the long wavelength regime $(k^2 d^2 \ll 1)$ can be expressed as[20]

$$g \equiv \frac{2(\Delta_0 \Delta_i)^{1/2}}{\Delta_b} > \frac{\sinh \theta}{\theta} . \qquad (6.122)$$

This is illustrated schematically in Fig. 6.12, where the solid curve corresponds to $2(\Delta_0 \Delta_i)^{1/2}/\Delta_b = (\sinh \theta)/\theta$, which separates the unstable and stable regions. Moreover, for $k^2 d^2 \ll 1$, whenever the inequality in Eq.(6.122) is satisfied, the corresponding normalized growth rate $\text{Im}\,\omega/c\,|k|$ and real oscillation frequency are given by[20]

$$\frac{\text{Im}\,\omega}{c\,|k|} = \frac{1}{2}\theta \left(g^2 - \frac{\sinh^2 \theta}{\theta^2} \right)^{1/2} \left[\cosh \theta + \frac{\theta}{\Delta_b \sinh \theta} (\Delta_i \cosh^2 \theta + \Delta_0) \right]^{-1} ,$$

$$(6.123)$$

and

$$\frac{\text{Re}\,\omega}{ck} = \frac{1}{2} \left(\sinh \theta + \frac{2\Delta_i}{\Delta_b} \theta \cosh \theta \right) \left[\cosh \theta + \frac{\theta}{\Delta_b \sinh \theta} (\Delta_i \cosh^2 \theta + \Delta_0) \right]^{-1} ,$$

$$(6.124)$$

where $g = 2(\Delta_0 \Delta_i)^{1/2}/\Delta_b$. It readily follows that $g = 0$ for $\Delta_i = 0$ or for $\Delta_0 = 0$. Because $g > (\sinh \theta)/\theta$ is a necessary and sufficient condition for instability $(\text{Im}\,\omega > 0)$, we therefore conclude that the system is stable $(\text{Im}\,\omega = 0)$ whenever the electron layer is in contact with the cathode $(\Delta_i = 0)$ or the anode $(\Delta_0 = 0)$, as expected. In the nonrelativistic, electrostatic regime $(\theta \ll 1)$, we further note that Eqs.(6.123) and (6.124) reduce to $\text{Im}\,\omega/c\,|k| = (\theta/2) \left[4\Delta_i \Delta_0 - \Delta_b^2 \right]^{1/2}$ and $\text{Re}\,\omega/ck = (\theta/2)(\Delta_b + 2\Delta_i)$.

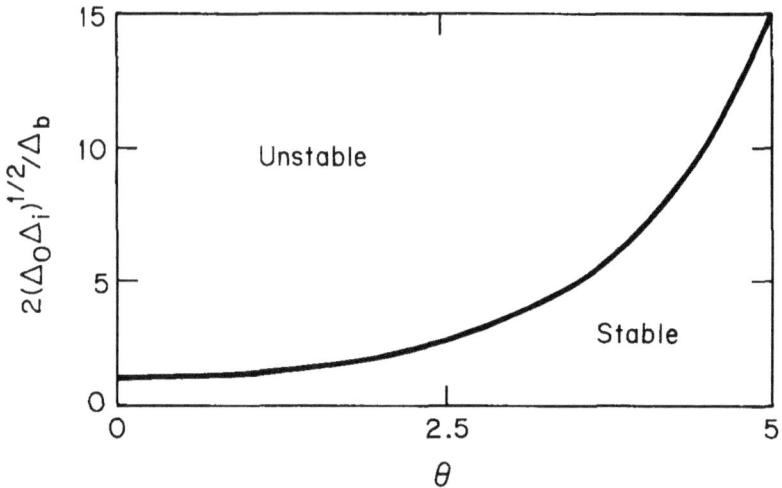

Figure 6.12. For long-wavelength perturbations $(k^2d^2 \ll 1)$, the diocotron instability is stabilized by relativistic and electromagnetic effects whenever $2(\Delta_0\Delta_i)^{1/2}/\Delta_b < (\sinh\theta)/\theta$ [R.C. Davidson, K.T. Tsang and H.S. Uhm, Phys. Fluids **31**, 1727 (1988)].

Removing the restriction $k^2d^2 \ll 1$, the full dispersion relation (6.119) has been solved numerically for $\text{Im}\,\omega/c\,|k|$ and $\text{Re}\,\omega/ck$.[19,20] Typical results are illustrated in Fig. 6.13 for $\Delta_i = x_b^-/d = 0.3$ and $\Delta_0 = (d - x_b^+)/d = 0.6$. As a general remark, for increasing values of kd, it is found that there is a concomitant decrease in both the maximum normalized growth rate and the range of θ corresponding to instability. Moreover, the long-wavelength *stability* criterion, $2(\Delta_0\Delta_i)^{1/2}/\Delta_b < (\sinh\theta)/\theta$, is found to be a sufficient condition for the diocotron instability to be stabilized at all perturbation wavelengths.[20]

The analysis in Sec. 6.7 represents a rather striking example in which an instability that is highly ubiquitous in the nonrelativistic, electrostatic regime, can be completely stabilized by relativistic and electromagnetic effects.

As a final point, for moderate values of $\omega_{pe}^2(x)/\omega_{ce}^2(x)$, it should be emphasized that the full extraordinary-mode eigenvalue equation (8.101) gives rise to an instability known as the magnetron instability,[2,15,21–24] which is of considerable importance for intense nonneutral electron flow in crossed-field microwave devices[12,25] and in magnetically insulated high-voltage diodes.[26,27] The magnetron instability has been extensively investigated for relativistic electron flow in planar diode geometry (Chapter 8).

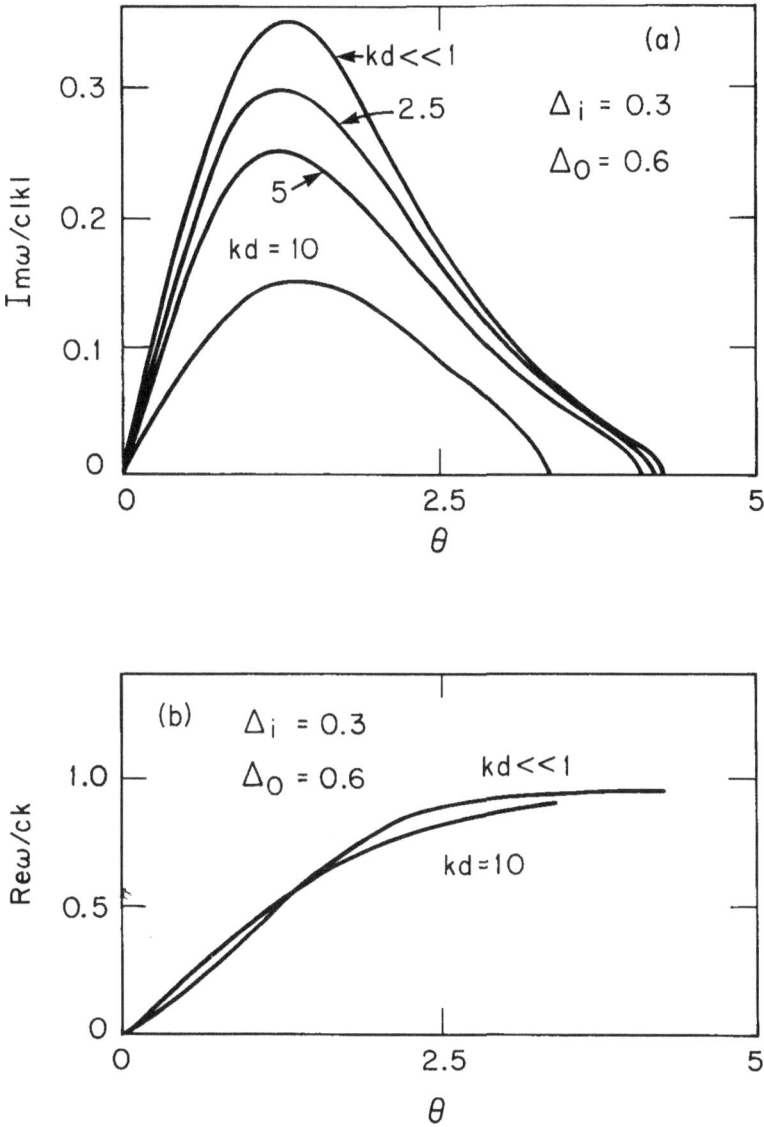

Figure 6.13.   Plots of (a) $\mathrm{Im}\,\omega/c\,|k|$ and (b) $\mathrm{Re}\,\omega/ck$ versus the relativistic flow parameter θ obtained numerically from Eq.(6.119) for $\Delta_i = 0.3$, $\Delta_0 = 0.6$, and several values of kd [R.C. Davidson, K.T. Tsang and H.S. Uhm, Phys. Fluids **31**, 1727 (1988)].

This instability, and more generally speaking, the stability of both electron and ion flow[21-24,28] in high-voltage diodes, play an important role in determining the operating characteristics of high-power diodes with applications that include particle-beam fusion.[29]

Problem 6.10   Consider the equilibrium profiles specified by Eqs.(6.108) and (6.109) within the electron layer $(x_b^- < x < x_b^+)$.

a. Show from Eqs.(6.108) and (6.109) that

$$\frac{\omega_{pe}^2(x)}{\omega_{ce}(x)} = \frac{4\pi\hat{n}_e ec}{B_0} \equiv c\kappa = \text{const.} \tag{6.10.1}$$

b. Verify by direct substitution that the equilibrium profiles in Eqs.(6.108) and (6.109) solve Maxwell's equations (6.102) and (6.103) for $E_x(x)$ and $B_z(x)$.

Problem 6.11   Consider the eigenvalue equation (6.112) in the nonrelativistic, electrostatic regime with $V_d^2/c^2 \ll 1$, $\left|\omega^2/c^2k^2\right| \ll 1$ and $\kappa^2/k^2 \ll 1$.

a. Solve Eq.(6.112) in each of the three regions in Fig. 6.11 and determine the eigenfunction $\Phi_k(x)$ and complex eigenfrequency ω .

b. Assuming $k^2d^2 \ll 1$, show that the real oscillation frequency and growth rate of the unstable branch are given by

$$\text{Re}\,\omega = \frac{1}{2}\Delta_b kd\frac{\omega_{pe}^2}{\omega_{ce}}\left(\Delta_b + 2\Delta_i\right) , \tag{6.11.1}$$

and

$$\text{Im}\,\omega = \frac{1}{2}\Delta_b |k|\, d\frac{\omega_{pe}^2}{\omega_{ce}}\left[4\Delta_i\Delta_0 - \Delta_b^2\right]^{1/2} , \tag{6.11.2}$$

when $4\Delta_i\Delta_0 > \Delta_b^2$. Here $\omega_{pe}^2/\omega_{ce} = 4\pi\hat{n}_e ec/B_0 \equiv c\kappa$, and Δ_i, Δ_b and Δ_0 are the geometric factors defined in Eq.(6.121).

c. The system is stable $(\text{Im}\,\omega = 0)$ whenever $4\Delta_i\Delta_0 < \Delta_b^2$. What is the corresponding expression for $\text{Re}\,\omega$?

Chapter 6 References

1. C.C. MacFarlane and H.G. Hay, "Wave Propagation in a Slipping Stream of Electrons; Small-Amplitude Theory," Proc. Phys. Soc. (London) **63B**, 409 (1950).

2. O. Buneman, R.H. Levy and L.M. Linson, "Stability of Crossed-Field Electron Beams," J. Appl. Phys. **37**, 3203 (1966).

3. R.H. Levy, "Diocotron Instability in Cylindrical Plasma," Phys. Fluids 8, 1288 (1965).

4. R.C. Davidson, *Theory of Nonneutral Plasmas* (Addison-Wesley, Reading, Massachusetts, 1989), Chapter 2, Sec. 2.10.

5. H.F. Webster, "Breakup of Hollow Electron Beams," J. Appl. Phys. 26, 1386 (1955).

6. C.C. Cutler, "Instability in Hollow and Strip Electron Beams," J. Appl. Phys. 27, 1028 (1956).

7. J.R. Pierce, "Instability of Hollow Beams," IRE Trans. Electron Devices ED-3, 183 (1956).

8. O. Buneman, "Ribbon Beams," J. Electron. Control 3, 507 (1957).

9. R.L. Kyhl and H.F. Webster, "Breakup of Hollow Cylindrical Electron Beams," IRE Trans. Electron Devices ED-3, 172 (1956).

10. C.A. Kapetanakos, D.A. Hammer, C. Striffler and R.C. Davidson, "Destructive Instabilities in Hollow Intense Relativistic Electron Beams," Phys. Rev. Lett. 30, 1303 (1973).

11. G. Rosenthal, G. Dimonte and A.Y. Wong, "Stabilization of the Diocotron Instability in an Annular Plasma," Phys. Fluids 30, 3257 (1987).

12. Y.Y. Lau, "Theory of Crossed-Field Devices and a Comparative Study of Other Radiation Sources," in *High-Power Microwave Sources*, eds. V. Granatstein and I. Alexeff (Artech House, Boston, Massachusetts, 1987), p. 309.

13. R.C. Davidson, "Macroscopic Guiding-Center Stability Theorem for Nonrelativistic Nonneutral Electron Flow," Phys. Fluids 27, 1804 (1984).

14. R.J. Briggs, J.D. Daugherty, and R.H. Levy, "Role of Landau Damping in Cross-Field Electron Beams and Inviscid Shear Flow," Phys. Fluids 13, 421 (1970).

15. R.C. Davidson, K.T. Tsang and J.A. Swegle, "Macroscopic Extraordinary-Mode Stability Properties of Relativistic Nonneutral Electron Flow in a Planar Diode with Applied Magnetic Field," Phys. Fluids 27, 2332 (1984).

16. R.H. Levy, "Two New Results in Cylindrical Diocotron Theory," Phys. Fluids 11, 920 (1968).

17. R.C. Davidson, *Methods in Nonlinear Plasma Theory* (Academic Press, New York, 1972).

18. R.C. Davidson, "Quasilinear Theory of the Diocotron Instability for Nonrelativistic Nonneutral Electron Flow in Planar Geometry," Phys. Fluids 28, 1937 (1985).

19. R.C. Davidson, K.T. Tsang and H.S. Uhm, "Stabilization of Diocotron Instability by Relativistic and Electromagnetic Effects for Intense Nonneutral Electron Flow," Physics Letters A **125**, 61 (1987).

20. R.C. Davidson, K.T. Tsang and H.S. Uhm, "Diocotron Instability for Intense Relativistic Nonneutral Electron Flow in Planar Geometry," Phys. Fluids **31**, 1727 (1988).

21. J. Swegle and E. Ott, "Linear Waves and Instabilities in Magnetically Insulated Gaps," Phys. Fluids **24**, 1821 (1981).

22. D. Chernin and Y.Y. Lau, "Stability of Laminar Electron Layers," Phys. Fluids **27**, 2319 (1984).

23. R.C. Davidson and K.T. Tsang, "Influence of Profile Shape on the Extraordinary-Mode Stability Properties of Relativistic Nonneutral Electron Flow in a Planar Diode with Applied Magnetic Field," Phys. Fluids **28**, 1169 (1985).

24. C.L. Chang, E. Ott and T.M. Antonsen, Jr., "Linear Stability of Obliquely Propagating Electromagnetic Waves in Magnetically Insulated Gaps," Phys. Fluids **29**, 3851 (1986).

25. A. Palevsky and G. Bekefi, "Microwave Emission from Pulsed, Relativistic Electron Beam Diodes II: The Multiresonator Magnetron," Phys. Fluids **22**, 986 (1979).

26. R.B. Miller, *Intense Charged Particle Beams* (Plenum, New York, 1982).

27. M.P. Desjarlais, "Impedance Characteristics of Applied-B Ion Diodes," Phys. Rev. Lett. **59**, 2295 (1987).

28. C.L. Chang, D.P. Chernin, A.T. Drobot, E. Ott and T.M. Antonsen, Jr., "Electromagnetic Stability of High-Power Ion Diodes," Phys. Fluids **29**, 1528 (1986).

29. J.P. VanDevender and D.L. Cook, "Inertial Confinement Fusion with Light Ion Beams," Science **232**, 831 (1986).

The following references, while not cited directly in the text, are relevant to the general subject matter of this chapter.

R.C. Davidson, H.-W. Chan, C. Chen and S. Lund, "Large-Amplitude Coherent Structures in Nonneutral Plasmas with Circulating Electron Flow," in AIP Conference Proceedings on Nonlinear and Relativistic Effects in Plasmas (American Institute of Physics, New York, 1990), in press.

C.F. Driscoll and K.S. Fine, "Experiments on Vortex Dynamics in Pure Electron Plasmas," in AIP Conference Proceedings on Nonlinear and Relativistic Effects in Plasmas (American Institute of Physics, New York, 1990), in press.

G. Rosenthal and A.Y Wong, "Localized Density Clumps Generated in a Magnetized Nonneutral Plasma," submitted for publication (1990).

K.S. Fine, C.F. Driscoll and J.H. Malmberg, "Measurements of a Nonlinear Diocotron Mode in Pure Electron Plasmas," Phys. Rev. Lett. 63, 2232 (1989).

T. Chiueh, "Expansion of Shear Flow and Vortex Merger in a Two-Dimensional Turbulent Mixing Layer," Phys. Fluids B1, 1163 (1989).

C.F. Driscoll, J.H. Malmberg, K.S. Fine, R.A. Smith, X.-P. Huang and R.W. Gould, "Growth and Decay of Turbulent Vortex Structures in Pure Electron Plasmas," in *Plasma Physics and Controlled Nuclear Fusion Research 1988* (IAEA, Vienna, 1989) Vol. 3, p. 507.

J.H. Malmberg, C.F. Driscoll, B. Beck, D.L. Eggleson, J. Fajans, K.S. Fine, X.-P. Huang and A.W. Hyatt, "Experiments with Pure Electron Plasma," eds., C.W. Roberson and C.F. Driscoll, AIP Conference Proceedings No. 175 (American Institute of Physics, New York, 1988), p. 28.

D.J. Kaup, S.R. Choudhury, and G.E. Thomas, "Full Second-Order Cold-Fluid Theory of the Diocotron and Magnetron Resonances," Phys. Rev. A38, 1402 (1988).

K.W. Chen, S.H. Kim and M.C. McKinley, "Low-Frequency Azimuthally Propagating Diocotron Waves in a Nonneutral Electron Beam Column," Phys. Fluids 30, 3306 (1987).

S.A. Prasad and J.H. Malmberg, "A Nonlinear Diocotron Mode," Phys. Fluids 29, 2196 (1986).

W.D. White, J.H. Malmberg and C.F. Driscoll, "Resistive Wall Destabilization of Diocotron Waves," Phys. Rev. Lett. 49, 1822 (1982).

J.D. Daugherty, J.E. Eninger and G.S. Janes, "Experiments on the Injection and Containment of Electron Clouds in a Toroidal Apparatus," Phys. Fluids 12, 2677 (1969).

A. Nocentini, H.L. Berk, and R.N. Sudan, "Kinetic Theory of the Diocotron Instability," J. Plasma Phys. 2, 311 (1968).

R.H. Levy, "Electrostatic Equilibrium of Electron Clouds in Magnetic Fields," Phys. Fluids 11, 772 (1968).

J.D. Daugherty and R.H. Levy, "Equilibrium of Electron Clouds in Toroidal Magnetic Fields," Phys. Fluids 10, 155 (1967).

W. Knauer, "Diocotron Instability in Plasmas and Gas Discharges," J. Appl. Phys. 37, 602 (1966).

V.K. Neil and W. Heckrotte, "Relation Between Diocotron and Negative-Mass Instabilities," J. Appl. Phys. 36, 2761 (1965).

CHAPTER 7

COHERENT ELECTROMAGNETIC WAVE GENERATION BY THE CYCLOTRON MASER AND FREE ELECTRON LASER INSTABILITIES

7.1 Introduction

The stimulated emission of radiation by the (unstable) interaction of energetic electrons with a static magnetic field structure forms the basis for many practical devices used to generate coherent electromagnetic waves. Examples include klystrons, ubitrons, cyclotron autoresonance masers and gyrotrons,[1-4] magnetrons,[5-8] and free electron lasers,[9-12] to mention a few. Indeed, an entire treatise could easily be devoted to the study of coherent electromagnetic wave generation by free electron devices. Therefore, in this book, we describe the basic instability mechanisms for only a few prominent configurations used for coherent radiation generation. In particular, the magnetron instability, which occurs for magnetically insulated electron flow in high-voltage diodes, is discussed in Chapter 8 with particular emphasis on applications to relativistic magnetrons.[5-8] In the present chapter, we limit discussion to the cyclotron maser instability (Secs. 7.2–7.4) and the free electron laser instability (Secs. 7.5–7.12). The cyclotron maser instability is the basic instability mechanism used to generate coherent radiation in gyrotrons and cyclotron autoresonance masers.[1-4]

As background, the cyclotron maser instability was first investigated theoretically in the 1950s by Twiss,[13] Schneider,[14] and Gapanov,[15] and measured experimentally in the 1960s by Hirshfield and Wachtel,[16] and Gapanov, Petelin and Yulpatov.[17] The cyclotron maser instability occurs when a (moderately) relativistic electron beam with an inverted population in perpendicular momentum $p_\perp$ interacts with a uniform magnetic field $\mathbf{B}^0(\mathbf{x}) = B_0\hat{\mathbf{e}}_z$.[18-22] For a right-circularly-polarized electromagnetic wave (ω, k_z) propagating parallel to $B_0 \hat{e}_z$, there is a strong interaction between the wave field and the electron gyromotion when the cyclotron resonance condition

$$\omega - k_z V_{ze} = \frac{\omega_{ce}}{\hat{\gamma}_e} \qquad (7.1)$$

is satisfied [see Eq.(7.11)]. Here, V_{ze} is the axial velocity of the electron beam, $\omega_{ce} = eB_0/m_e c$ is the nonrelativistic electron cyclotron frequency, and $\hat{\gamma}_e = (1 - V_{\perp e}^2/c^2 - V_{ze}^2/c^2)^{-1/2}$ is the relativistic mass factor. The inverted population in perpendicular momentum leads to an (unstable) coupling with the (fast) electromagnetic wave propagating along $B_0 \hat{e}_z$, and there is concomitant growth of the wave field. Gyrotrons, which are based on the cyclotron maser instability mechanism, operate at relatively small values of $k_z V_{ze}$ and have proven to be efficient generators of coherent electromagnetic radiation in the centimeter and millimeter wavelength range.[1,2] It is clear from Eq.(7.1), however, that the ability to produce even higher frequencies (shorter wavelengths) is set by the technological limits on producing high magnetic fields (as measured by $\omega_{ce}/\hat{\gamma}_e$) over a sufficiently large interaction region.

Early research on free electron lasers dates back to the 1950s and 1960s with the work of Motz[23,24] on the interaction of an electron beam with an undulating magnetic field, and the work of Phillips[25,26] on high-power ubitrons. The major growth of research on free electron lasers in the 1970s and 1980s was stimulated by the classic investigations of Madey,[27] Elias, et al.,[28] and Deacon, et al.,[29] which led to the first successful experimental operation of a free electron laser oscillator.[29] In contrast to the cyclotron maser instability, the free electron laser instability[30-37] occurs when a relativistic electron beam with narrow energy spread propagates in the z-direction and interacts with a transverse (helical, say) wiggler magnetic field $\mathbf{B}_w^0(\mathbf{x}) = -(m_e c^2 k_0/e) a_w [\cos(k_0 z) \hat{e}_x + \sin(k_0 z) \hat{e}_y]$. Here, $a_w = eB_w/m_e c^2 k_0$ is the normalized field amplitude, and $\lambda_w = 2\pi/k_0$ is the wiggler wavelength. The transverse wiggler field causes an oscillatory modulation of the x- and y-motion of the electrons with perpendicular velocity $V_{\perp e} = c a_w/\hat{\gamma}_e$, and there is a corresponding strong interaction between the electrons and an electromagnetic wave (ω, k_z) propagating parallel to $\hat{e}_z$ when the resonance condition

$$\omega - k_z V_{ze} = k_0 V_{ze} \qquad (7.2)$$

is satisfied [see Eq.(7.62)]. In a frame of reference moving with axial velocity V_{ze}, the frequency of the electromagnetic wave is $\omega' = \gamma_z \times (\omega - k_z V_{ze})$, and the frequency of the (backward-traveling) wiggler field is $\omega_0' = \gamma_z k_0 V_{ze}$, where $\gamma_z = (1 - V_{ze}^2/c^2)^{-1/2}$. Therefore, the resonance condition in Eq.(7.2) can be viewed quantum mechanically as energy conservation ($\hbar\omega' = \hbar\omega_0'$) in which the recoil energy of the electrons in the beam frame is neglected. The free electron laser emission process can be interpreted as stimulated Compton scattering[38-40] in which the detailed growth properties are strongly influenced by the form of the beam distribution function f_e^0 and the charge and current bunching of the electrons.

For a sufficiently tenuous electron beam, we solve the resonance condition $\omega - k_z V_{ze} = k_0 V_{ze}$ in conjunction with the vacuum dispersion relation $\omega = ck_z$ for a (forward-traveling) light wave. Denoting the solutions by $(\hat{\omega}, \hat{k}_z)$, this gives $c\hat{k}_z = k_0 V_{ze}/(1 - V_{ze}/c)$, which is the wavenumber at which radiation emission is expected to be a maximum. Making use of $\hat{\gamma}_e = (1 - V_{ze}^2/c^2 - V_{\perp e}^2/c^2)^{-1/2}$ and $V_{\perp e} = c\,a_w/\hat{\gamma}_e$, the above expression for $\hat{k}_z$ is readily expressed as [Eq.(7.64)]

$$\hat{k}_z = \frac{\hat{\gamma}_e^2(1 + \beta_{ze})\beta_{ze}}{(1 + a_w^2)}k_0 \, ,$$

where $\beta_{ze} = V_{ze}/c$. Depending on the value of $\hat{\gamma}_e$, note that $\hat{k}_z$ can be upshifted by a significant amount relative to the wiggler wavenumber k_0. Indeed, two of the most attractive features of free electron lasers are frequency tunability (by varying $\hat{\gamma}_e$ and/or a_w), and the accessibility of regions of the electromagnetic spectrum which are unattainable with conventional lasers and microwave sources. Free electron lasers have been used to generate coherent radiation at wavelengths ranging from the visible,[41] to the infrared,[29,42-44] to the microwave[45-51] region of the electromagnetic spectrum.

7.2   Cyclotron Maser Instability

The classical cyclotron maser instability[18-22] in a uniform plasma immersed in a magnetic field $B_0\hat{e}_z$ is a transverse electromagnetic instability driven by an inverted population in the perpendicular momentum dependence of the electron distribution function $F_e(p_\perp^2, p_z)$. This instability has a wide range of applicability to astrophysical and space plasmas,[52-57] including Jupiter's decametric radio emission and Earth's auroral kilometric radiation, and to coherent radiation generation by relativistic electron beams in microwave devices such as gyrotrons[1-3] and cyclotron autoresonance masers (CARMS).[4,58-63] To illustrate the basic features of the cyclotron maser instability, we first investigate stability properties for a tenuous electron beam ($s_e = \hat{\gamma}_e \omega_{pe}^2 / \omega_{ce}^2 \ll 1$) with uniform density, assuming electromagnetic wave propagation parallel to the guide field $B_0 \hat{e}_z$ (Sec. 7.2.1). The influence of self-electric and self-magnetic fields on stability behavior is then examined, neglecting the effects of finite radial geometry (Sec. 7.2.2). For a tenuous, annular electron beam in a cylindrical waveguide, the cyclotron maser instability is investigated including the radial dependence of the equilibrium profiles and the electromagnetic wave field (Sec. 7.3). Finally, some experimental results on high-power microwave generation by step-tunable gyrotrons are summarized (Sec. 7.4).

7.2.1   Stability Properties for a Tenuous Electron Beam

To illustrate the basic features of the cyclotron maser instability, we consider a relativistic electron beam with uniform density ($\hat{n}_e$) propagating parallel to a uniform guide field $B_0 \hat{e}_z$. It is assumed that the electron density is sufficiently low ($s_e = \hat{\gamma}_e \omega_{pe}^2 / \omega_{ce}^2 \ll 1$) that the equilibrium self fields have a negligible influence on the particle trajectories. Expressing the equilibrium distribution function as $f_e^0(\mathbf{x}, \mathbf{p}) = \hat{n}_e F_e(p_\perp^2, p_z)$, where $p_\perp = (p_x^2 + p_y^2)^{1/2}$ is the perpendicular momentum, the dispersion relation for transverse electromagnetic waves propagating parallel to $B_0 \hat{e}_z$ can be expressed as (Problem 7.1)[22]

$$
0 = D_\pm^T(k_z, \omega) = \omega^2 - c^2 k_z^2 + \omega_{pe}^2 \int \frac{d^3 p}{\gamma} \frac{p_\perp / 2}{(\gamma \omega - k_z p_z / m_e \pm \omega_{ce})}
$$

$$
\times \left[\left(\gamma \omega - \frac{k_z p_z}{m_e} \right) \frac{\partial}{\partial p_\perp} + \frac{k_z p_\perp}{m_e} \frac{\partial}{\partial p_z} \right] F_e(p_\perp^2, p_z) . \tag{7.3}
$$

In obtaining Eq.(7.3) from the linearized Vlasov-Maxwell equations, perturbed quantities are assumed to vary as $\delta\psi(z,t) = \sum_{k_z = -\infty}^{\infty} \delta\psi(k_z) \times exp(ik_z z - i\omega t)$, where $k_z = 2\pi m / L$, m is an integer, and L is the fundamental periodicity length in the z-direction. In Eq.(7.3) ω is the complex oscillation frequency with $\text{Im}\,\omega > 0$ corresponding to instability (temporal growth), k_z is the axial wavenumber, $\gamma = (1 + \mathbf{p}^2 / m_e^2 c^2)^{1/2}$ is the relativistic mass factor, and $\omega_{ce} = eB_0 / m_e c$ and $\omega_{pe} = (4\pi \hat{n}_e e^2 / m_e)^{1/2}$ are the nonrelativistic electron cyclotron frequency and plasma frequency, respectively. Moreover, the electromagnetic wave polarization is such that $\delta\mathbf{E} \cdot \hat{e}_z = 0 = \delta\mathbf{B} \cdot \hat{e}_z$ (see also Fig. 4.13), and the lower and upper signs in Eq.(7.3) correspond to waves with right-hand circular polarization $[\delta E_x(k_z) = -i\delta E_y(k_z)]$ and left-hand circular polarization $[\delta E_x(k_z) = +i\delta E_y(k_z)]$, respectively. Expressing $\int d^3 p \cdots = 2\pi \int_0^\infty dp_\perp p_\perp \int_{-\infty}^\infty dp_z \cdots$

in the dispersion relation (7.3) and integrating by parts with respect to $p_\perp$ and p_z gives

$$0 = D_\pm^T(k_z, \omega) = \omega^2 - c^2 k_z^2 - \omega_{pe}^2 \int \frac{d^3 p}{\gamma} F_e(p_\perp^2, p_z)$$

$$\times \left[\frac{\gamma\omega - k_z p_z/m_e}{\gamma\omega - k_z p_z/m_e \pm \omega_{ce}} - \frac{p_\perp^2}{2m_e^2 c^2} \frac{\omega^2 - c^2 k_z^2}{(\gamma\omega - k_z p_z/m_e \pm \omega_{ce})^2} \right] , \quad (7.4)$$

which is fully equivalent to Eq.(7.3).

If $F_e(p_\perp^2, p_z)$ is an isotropic function of $p_\perp^2 + p_z^2$, and decreases monotonically with $\partial F_e/\partial(p_\perp^2 + p_z^2) \leq 0$, then the solutions to Eq.(7.4) [or Eq.(7.3)] correspond to stable oscillations. On the other hand, if $F_e(p_\perp^2, p_z)$ has the nonthermal feature corresponding to an inverted population in perpendicular momentum $p_\perp$, then there is a free energy source to drive the cyclotron maser instability. As an example corresponding to strong instability, consider the equilibrium distribution function specified by

$$F_e(p_\perp^2, p_z) = \frac{1}{2\pi p_\perp} \delta\left(p_\perp - \hat{\gamma}_e m_e V_{\perp e}\right) \delta\left(p_z - \hat{\gamma}_e m_e V_{ze}\right) , \quad (7.5)$$

where $\hat{\gamma}_e$, $V_{\perp e}$ and V_{ze} are constants. Equation (7.5) corresponds to a uniform electron beam propagating parallel to $B_0 \hat{e}_z$ with axial momentum $\hat{\gamma}_e m_e V_{ze}$. The $p_\perp$-dependence in Eq.(7.5) is inverted (Fig. 7.1), and all electrons have perpendicular momentum $p_\perp = \hat{\gamma}_e m_e V_{\perp e}$ and total energy $\gamma = (1 + \hat{\gamma}_e^2 V_{\perp e}^2/c^2 + \hat{\gamma}_e^2 V_{ze}^2/c^2)^{1/2}$. Without loss of generality, we make the identification $\hat{\gamma}_e = (1 + \hat{\gamma}_e^2 V_{\perp e}^2/c^2 + \hat{\gamma}_e^2 V_{ze}^2/c^2)^{1/2}$, or equivalently

$$\hat{\gamma}_e = \left(1 - V_{\perp e}^2/c^2 - V_{ze}^2/c^2\right)^{-1/2} . \quad (7.6)$$

Substituting Eq.(7.5) into Eq.(7.4) and rearranging terms gives the dispersion relation[18,22]

$$(\omega^2 - c^2 k_z^2) \left[1 + \frac{1}{2} \frac{V_{\perp e}^2}{c^2} \frac{\omega_{pe}^2/\hat{\gamma}_e}{(\omega - k_z V_{ze} \pm \omega_{ce}/\hat{\gamma}_e)^2} \right]$$

$$= \frac{\omega_{pe}^2}{\hat{\gamma}_e} \left[\frac{\omega - k_z V_{ze}}{\omega - k_z V_{ze} \pm \omega_{ce}/\hat{\gamma}_e} \right] . \quad (7.7)$$

In the remainder of this section, only the lower $(-)$ sign in Eq.(7.7) is considered, which corresponds to waves with right-circular polarization.

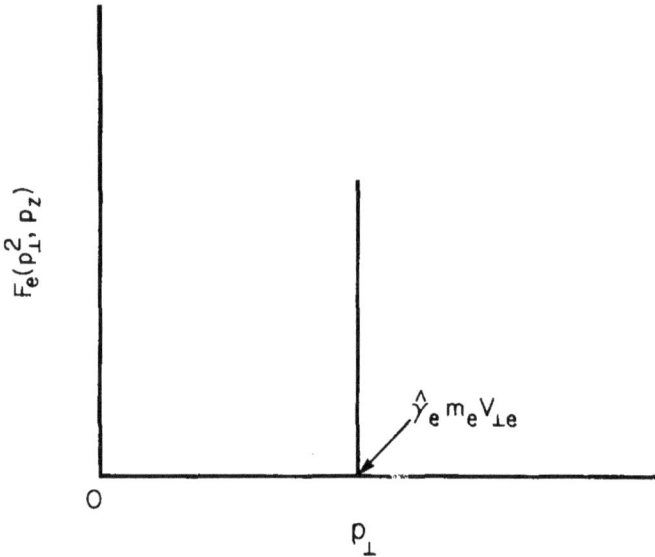

Figure 7.1. The distribution function $F_e(p_\perp^2, p_z)$ in Eq.(7.5) has an inverted population in perpendicular momentum $p_\perp$ and is subject to the cyclotron maser instability.

If $V_{\perp e} = 0$, there is no free energy source to drive instability, and Eq.(7.7) gives stable oscillations with $\text{Im}\,\omega = 0$. This follows upon expressing Eq.(7.7) in the form

$$\left(\omega^2 - c^2 k_z^2 - \frac{\omega_{pe}^2}{\hat{\gamma}_e}\right)\left(\omega - k_z V_{ze} - \frac{\omega_{ce}}{\hat{\gamma}_e}\right) = \frac{\omega_{pe}^2}{\hat{\gamma}_e}\frac{\omega_{ce}}{\hat{\gamma}_e} \qquad (7.8)$$

for $V_{\perp e} = 0$ and $\hat{\gamma}_e = (1 - V_{ze}^2/c^2)^{-1/2}$. For large values of k_z, the fast-wave branch ($\omega^2 > c^2 k_z^2$) in Eq.(7.8) asymptotes at $\omega^2 = c^2 k_z^2 + \omega_{pe}^2/\hat{\gamma}_e$, whereas the slow-wave whistler branch ($\omega^2 < c^2 k_z^2$) asymptotes at $\omega = k_z V_{ze} + \omega_{ce}/\hat{\gamma}_e$. This is illustrated in Fig. 7.2, where $(\omega - k_z V_{ze})/\hat{\omega}_{ce}$ is plotted versus $ck_z/\hat{\omega}_{ce}$ for $V_{ze} = 0.866c$, $\hat{\gamma}_e = 2$ and $\hat{\gamma}_e\omega_{pe}^2/\omega_{ce}^2 = 0.1$. Here, $\hat{\omega}_{ce}$ is the relativistic cyclotron frequency defined by $\hat{\omega}_{ce} = \omega_{ce}/\hat{\gamma}_e$.

The results are very different in circumstances where $V_{\perp e} \neq 0$ in the dispersion relation (7.7). Indeed, it is found that the slow-wave branch exhibits the electron whistler instability[22] for sufficiently large k_z, whereas

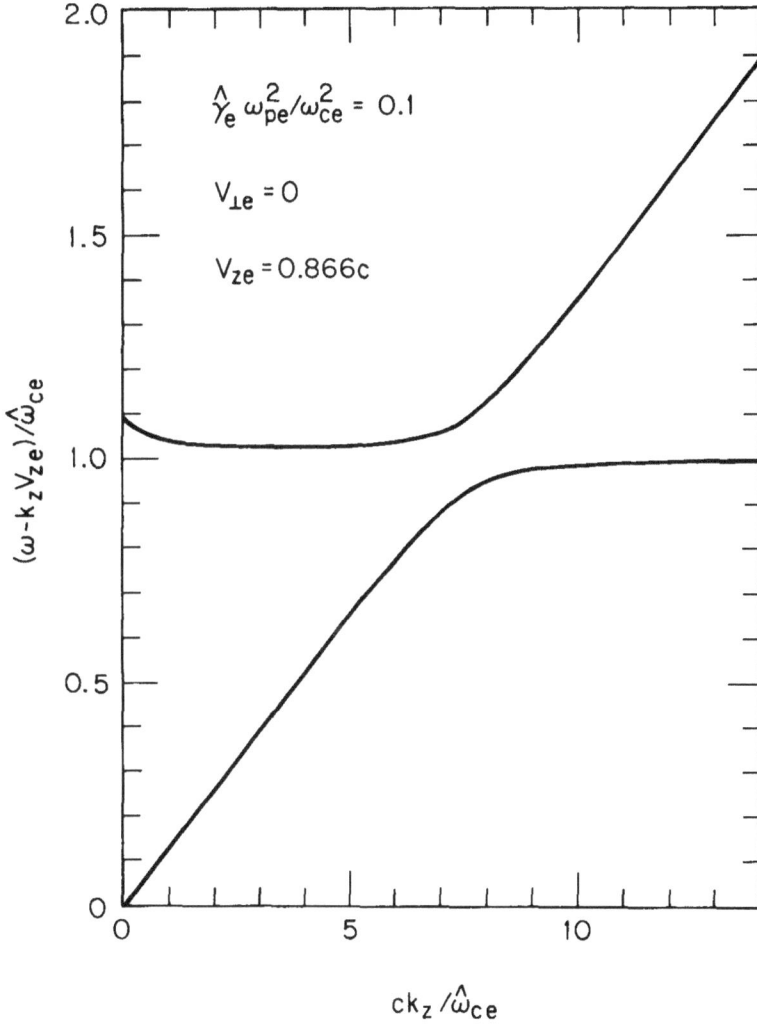

Figure 7.2. Plot versus $ck_z/\hat{\omega}_{ce}$ of the normalized real oscillation frequency $(\omega - k_z V_{ze})/\hat{\omega}_{ce}$ obtained from the dispersion relation (7.8) for $V_{\perp e} = 0$, $V_{ze} = 0.866\,c$, $\hat{\gamma}_e = 2$ and $\hat{\gamma}_e \omega_{pe}^2/\omega_{ce}^2 = 0.1$. For $V_{\perp e} \neq 0$, the upper (fast-wave) branch in the figure is subject to the cyclotron maser instability.

the fast-wave branch exhibits the cyclotron maser instability[22] for sufficiently long wavelengths. For small values of $\hat{\gamma}_e \omega_{pe}^2 / \omega_{ce}^2 \ll 1$, the demarcation between the two instabilities in k_z-space is determined (approximately) from the condition that the light line, $\hat{\omega} = c\hat{k}_z$, intersects the Doppler-shifted cyclotron resonance condition, $\hat{\omega} = \hat{k}_z V_{ze} + \omega_{ce}/\hat{\gamma}_e$. Solving for $\hat{k}_z$ gives

$$c\hat{k}_z = \frac{\omega_{ce}/\hat{\gamma}_e}{1 - V_{ze}/c} , \qquad (7.9)$$

where $\hat{\gamma}_e$ is defined in Eq.(7.6). Therefore, the subsequent analysis of the cyclotron maser instability examines the region of k_z-space where $k_z < \hat{k}_z$. Note from Eq.(7.9) that the relativistic axial motion of the electron beam can upshift the value of $\hat{k}_z$ by a significant amount because of the factor $(1 - V_{ze}/c)^{-1}$. For fixed value of $\omega_{ce}/\hat{\gamma}_e$, it is precisely this effect that enables CARMs to generate radiation at a higher frequency than gyrotrons (where V_{ze}/c is typically smaller). The basic instability mechanism (the cyclotron maser instability) is similar in the two cases.[18−22,58−63]

We now examine the dispersion relation (7.7) in the region where $k_z < \hat{k}_z$. For $V_{\perp e} \neq 0$ and $\hat{\gamma}_e \omega_{pe}^2 / \omega_{ce}^2 \ll 1$, the real oscillation frequency and growth rate of the cyclotron maser instability are obtained to good accuracy in the region of maximum growth by balancing the two terms in square brackets on the left-hand side of Eq.(7.7). For right-circular polarization this gives

$$\left(\omega - k_z V_{ze} - \frac{\omega_{ce}}{\hat{\gamma}_e} \right)^2 = -\frac{1}{2} \frac{V_{\perp e}^2}{c^2} \frac{\omega_{pe}^2}{\hat{\gamma}_e} . \qquad (7.10)$$

Solving Eq.(7.10) for $\mathrm{Re}\,\omega$ and $\mathrm{Im}\,\omega$, we obtain

$$\mathrm{Re}\,\omega = k_z V_{ze} + \frac{\omega_{ce}}{\hat{\gamma}_e} ,$$

$$\mathrm{Im}\,\omega = \left(\frac{1}{2} \frac{V_{\perp e}^2}{c^2} \frac{\omega_{pe}^2}{\hat{\gamma}_e} \right)^{1/2} \equiv \Gamma_e , \qquad (7.11)$$

for the unstable branch with $\mathrm{Im}\,\omega > 0$. Note from Eq.(7.11) that $(\mathrm{Im}\,\omega)/(\omega_{ce}/\hat{\gamma}_e) = (V_{\perp e}^2 \hat{\gamma}_e \omega_{pe}^2 / 2c^2 \omega_{ce}^2)^{1/2} \ll 1$ for a tenuous electron beam with $\hat{\gamma}_e \omega_{pe}^2 / \omega_{ce}^2 \ll 1$, even when the perpendicular electron motion is highly relativistic (as measured by $V_{\perp e}/c$).

The full dispersion relation (7.7) has been solved numerically over a wide range of system parameters.[22,62] Typical results for the cyclotron maser instability are presented in Fig. 7.3 for the case where $V_{\perp e} = 0.405c$,

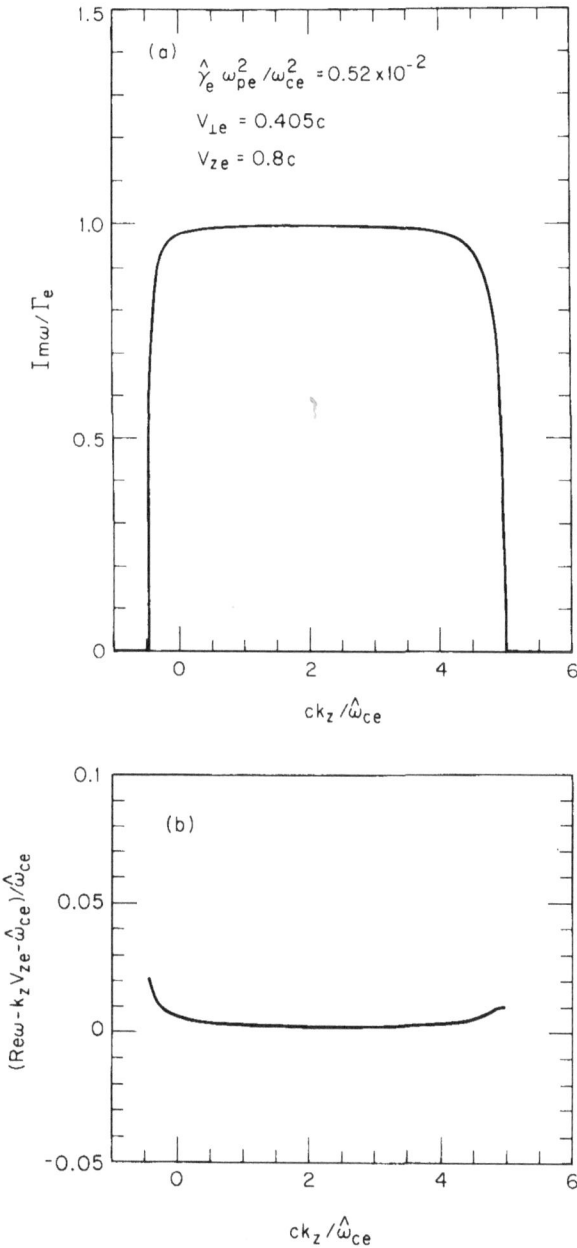

Figure 7.3. Linear growth properties of the cyclotron maser instability. The figures show plots versus $ck_z/\hat{\omega}_{ce}$ of (a) the normalized growth rate $\mathrm{Im}\,\omega/\Gamma_e$ and (b) the normalized real oscillation frequency $(\mathrm{Re}\,\omega - k_z V_{ze} - \hat{\omega}_{ce})/\hat{\omega}_{ce}$ obtained from the dispersion relation (7.7). The choice of system parameters corresponds to $\hat{\gamma}_e \omega_{pe}^2/\omega_{ce}^2 = 0.52 \times 10^{-2}$, $V_{\perp e} = 0.405\,c$, $V_{ze} = 0.8\,c$ and $\hat{\gamma}_e = 2.26$.

$V_{ze} = 0.800c$, $\hat{\gamma}_e = 2.26$, and $\hat{\gamma}_e \omega_{pe}^2/\omega_{ce}^2 = 0.52 \times 10^{-2}$. Here, the normalized growth rate $(\mathrm{Im}\,\omega)/\Gamma_e$ and the shifted real oscillation frequency $(\mathrm{Re}\,\omega - k_z V_{ze} - \hat{\omega}_{ce})/\hat{\omega}_{ce}$ are plotted versus $ck_z/\hat{\omega}_{ce}$ over the unstable range of wavenumbers in Figs. 7.3(a) and 7.3(b), respectively. (As before, $\hat{\omega}_{ce}$ is defined by $\hat{\omega}_{ce} = \omega_{ce}/\hat{\gamma}_e$.) It is evident from Fig. 7.3(a) that Eq.(7.11) provides an excellent estimate of the linear growth rate of the cyclotron maser instability over most of the unstable range of k_z. Moreover, it follows from Eq.(7.9) that $c\hat{k}_z/\hat{\omega}_{ce} = 5$ for $V_{ze} = 0.8c$, which is in excellent agreement with the abrupt decrease to zero of the growth rate at the upper end of the range in Fig. 7.3(a).

To summarize, the essential features of the cyclotron maser instability are contained in the unstable fast-wave solutions to Eq.(7.7) for sufficiently long axial wavelengths $(k_z < \hat{k}_z)$. While detailed stability behavior is modified by self-field effects[64] (Sec. 7.2.2), finite radial geometry (Sec. 7.3),[65-68] wall impedance,[69] and/or a spread in momentum of the beam electrons,[70] the basic instability mechanism is still associated with an inverted population in perpendicular momentum $p_\perp$. Furthermore, a relativistic treatment is necessary because the term driving the instability in Eq.(7.7) is proportional to $V_{\perp e}^2/c^2$, which is neglected in the nonrelativistic limit.

Problem 7.1   Consider transverse electromagnetic wave perturbations with $\delta\mathbf{E} \cdot \hat{e}_z = 0 = \delta\mathbf{B} \cdot \hat{e}_z$ propagating parallel to a uniform magnetic field $B_0\hat{e}_z$ through a uniform electron plasma with equilibrium distribution function $f_e^0(\mathbf{x}, \mathbf{p}) = \hat{n}_e F_e(p_\perp^2, p_z)$. Express perturbed quantities as

$$\delta\psi(z, t) = \sum_{k_z=-\infty}^{\infty} \delta\psi(k_z)exp(ik_z z - i\omega t) \,, \qquad (7.1.1)$$

where $\mathrm{Im}\,\omega > 0$ corresponds to instability.

a. Show that the linearized Maxwell equations can be expressed as

$$\left(\omega^2 - c^2 k_z^2\right)[\delta E_x(k_z) \pm i\delta E_y(k_z)] = 4\pi e i\omega \int d^3p \left(v_x \pm iv_y\right)\delta f_e(k_z, \mathbf{p}) \,.$$
$$(7.1.2)$$

b. Neglect equilibrium self-field effects and make use of the method of characteristics (Sec. 2.2) to integrate the linearized Vlasov equation. Show that the amplitude of the perturbed distribution function $\delta f_e(k_z, \mathbf{p})$ can be expressed as

$$\delta f_e(k_z, \mathbf{p}) = \frac{e\hat{n}_e}{\omega} \int_{-\infty}^{t} dt' exp\left[ik_z(z' - z) - i\omega(t' - t)\right]$$

$$\times \left[v_x'(t')\delta E_x(k_z) + v_y'(t')\delta E_y(k_z)\right]$$

$$\times \left[\frac{\gamma m\omega}{p_\perp} \frac{\partial}{\partial p_\perp} + k_z \left(\frac{\partial}{\partial p_z} - \frac{p_z}{p_\perp} \frac{\partial}{\partial p_\perp}\right)\right] F_e(p_\perp^2, p_z), \quad (7.1.3)$$

where

$$z'(t') = z + \frac{p_z}{\gamma m_e}(t' - t),$$

$$v_x'(t') = \frac{p_\perp}{\gamma m_e} \cos\left[\phi + \frac{\omega_{ce}}{\gamma}(t' - t)\right],$$

$$v_y'(t') = \frac{p_\perp}{\gamma m_e} \sin\left[\phi + \frac{\omega_{ce}}{\gamma}(t' - t)\right]. \quad (7.1.4)$$

Here, $v_x'(t' = t) = v_x = (p_\perp/\gamma m_e)\cos\phi$, $v_y'(t' = t) = v_y = (p_\perp/\gamma m_e)\sin\phi$, $\gamma = (1 + p_\perp^2/m_e^2 c^2 + p_z^2/m_e^2 c^2)^{1/2}$, and ϕ is the phase angle of the perpendicular momentum $p_\perp$.

c. Substitute Eqs.(7.1.3)–(7.1.4) into Eq.(7.1.2) and integrate over the phase angle ϕ with $\int d^3p \cdots = \int_0^{2\pi} d\phi \int_0^{\infty} dp_\perp\, p_\perp \int_{-\infty}^{\infty} dp_z \cdots$. Show that

$$D_\pm^T(k_z, \omega)\left[\delta E_x(k_z) \pm i\delta E_y(k_z)\right] = 0, \quad (7.1.5)$$

where the dielectric function $D_\pm^T(k_z, \omega)$ is defined in Eq.(7.3).

7.2.2   Influence of Intense Equilibrium Self Fields

For a relativistic electron beam with uniform density $(\hat{n}_e)$ and uniform axial velocity $(V_{ze} = \beta_{ze}c)$ propagating parallel to a uniform guide field $B_0\hat{e}_z$, the radial self-electric field and azimuthal self-magnetic field can be expressed for $0 \leq r < r_b$ as (see Sec. 5.8)

$$E_r(r) = -\frac{m_e}{2e}\omega_{pe}^2 r,$$

$$B_\theta^s(r) = -\frac{m_e}{2e}\omega_{pe}^2 \beta_{ze} r. \quad (7.12)$$

Here, r is the radial distance from the axis of symmetry, and r_b is the beam radius. In this section, we examine the influence of the self fields in

Eq.(7.12) on the cyclotron maser instability[64] analyzed in Sec. 7.2.1 for the case of a tenuous electron beam with $\hat{\gamma}_e \omega_{pe}^2 / \omega_{ce}^2 \ll 1$. The single-particle trajectories of a relativistic electron moving in the crossed electric and magnetic fields, $E_r(r)\hat{e}_r$ and $B_0\hat{e}_z + B_\theta^s(r)\hat{e}_\theta$, are readily calculated for the case where $\nu/\hat{\gamma}_e = \omega_{pe}^2 r_b^2 / 4\hat{\gamma}_e c^2 \ll 1$. For an electron with energy $\gamma m_e c^2$ and axial velocity $v_z \simeq$ const., it is found that the electron motion in the plane perpendicular to $B_0\hat{e}_z$ is biharmonic, with oscillatory components at the frequencies $\omega_e^+(\gamma, v_z)$ and $\omega_e^-(\gamma, v_z)$ defined by (Problem 7.2)

$$
\omega_e^\pm(\gamma, v_z) = \frac{\omega_{ce}}{2\gamma} \left\{ 1 \pm \left[1 - \frac{2\gamma\omega_{pe}^2}{\omega_{ce}^2} \left(1 - \beta_{ze}\frac{v_z}{c} \right) \right]^{1/2} \right\}, \qquad (7.13)
$$

where $\beta_{ze} = V_{ze}/c$, $\omega_{ce} = eB_0/m_e c$, and $\omega_{pe}^2 = 4\pi\hat{n}_e e^2/m_e$. The contribution proportional to $(2\gamma\omega_{pe}^2/\omega_{ce}^2)\beta_{ze}v_z/c$ in Eq.(7.13) arises from the (focusing) effects of the azimuthal self-magnetic field $B_\theta^s(r)$. The term proportional to $-2\gamma\omega_{pe}^2/\omega_{ce}^2$ in Eq.(7.13) arises from the (defocusing) effects of the radial self-electric field $E_r(r)$. If, as in Eq.(7.5), the electrons composing the electron beam have negligible momentum and energy spreads about $p_z = \hat{\gamma}_e m_e V_{ze}$, $p_\perp = \hat{\gamma}_e m_e V_{\perp e}$ and $\gamma = \hat{\gamma}_e = (1 - V_{\perp e}^2/c^2 - V_{ze}^2/c^2)^{-1/2}$, then the expressions for $\omega_e^\pm(\gamma, v_z)$ in Eq.(7.13) reduce to the laminar rotation velocities $\omega_{re}^\pm$ defined by [see also Eq.(5.120)]

$$
\omega_{re}^\pm \equiv \omega_e^\pm(\hat{\gamma}_e, \beta_{ze}c) = \frac{\omega_{ce}}{2\hat{\gamma}_e} \left\{ 1 \pm \left[1 - \frac{2\hat{\gamma}_e\omega_{pe}^2}{\omega_{ce}^2} \left(1 - \beta_{ze}^2 \right) \right]^{1/2} \right\}. \qquad (7.14)
$$

In the nonrelativistic limit with $\beta_{ze}^2 \ll 1$ and $\hat{\gamma}_e \to 1$ Eq.(7.14) reduces to the familiar result $\omega_{re}^\pm = (\omega_{ce}/2)\{1 \pm (1 - 2\omega_{pe}^2/\omega_{ce}^2)^{1/2}\}$, obtained in Eq.(3.6).

The dispersion relation (7.3) and analysis of the cyclotron maser instability in Sec. 7.2.1 assumed a tenuous electron beam with negligibly small self fields, which is equivalent to the approximation $(2\gamma\omega_{pe}^2/\omega_{ce}^2) \times (1 - \beta_{ze}v_z/c) \ll 1$ in Eq.(7.13). In this case, Eq.(7.13) can be approximated by $\omega_e^+(\gamma, v_z) = \omega_{ce}/\gamma$ and $\omega_e^-(\gamma, v_z) = 0$ and the electron motion perpendicular to $B_0\hat{e}_z$ corresponds to circular gyrations at the relativistic cyclotron frequency ω_{ce}/γ. For increasing values of $(2\gamma\omega_{pe}^2/\omega_{ce}^2)(1 - \beta_{ze}v_z/c)$, however, both oscillatory components of the perpendicular motion at frequencies $\omega_e^+(\gamma, v_z)$ and $\omega_e^-(\gamma, v_z)$ play an important role in determining the stability behavior. The corresponding dispersion relation including self-field effects has been derived[64] for the case of electromagnetic waves propagating parallel to $B_0\hat{e}_z$, neglecting the influence of finite radial geometry as in Sec. 7.2.1. For the choice of distribution function $F_e(p_\perp^2, p_z)$ in Eq.(7.5), which has an inverted population in perpendicular momentum $p_\perp$, the dispersion relation for electromagnetic waves with right-circular polarization can be expressed as

$$
\omega^2 - c^2 k_z^2 = \frac{\omega_{pe}^2}{\hat{\gamma}_e} \frac{(\omega - k_z V_{ze})^2}{(\omega - k_z V_{ze} - \omega_{re}^+)(\omega - k_z V_{ze} - \omega_{re}^-)} - \frac{1}{2} \frac{V_{\perp e}^2}{c^2} \frac{\omega_{pe}^2}{\hat{\gamma}_e}
$$

$$
\times \frac{(\omega^2 - c^2 k_z^2) \left[(\omega - k_z V_{ze})^2 - \left(\frac{\omega_{pe}^2}{2\hat{\gamma}_e}\right)(1 - \beta_{ze}^2) \right] + (\omega - k_z V_{ze})^2 \left(\frac{\omega_{pe}^2}{2\hat{\gamma}_e}\right)}{(\omega - k_z V_{ze} - \omega_{re}^+)^2 (\omega - k_z V_{ze} - \omega_{re}^-)^2} .
$$

$$
(7.15)
$$

Here, $\omega_{pe}^2 = 4\pi \hat{n}_e e^2 / m_e$, $\hat{\gamma}_e = (1 - V_{\perp e}^2/c^2 - V_{ze}^2/c^2)^{-1/2}$, and $\omega_{re}^\pm$ is defined in Eq.(7.14).

As a check on Eq.(7.15), consider the limit of a tenuous electron beam with $(2\hat{\gamma}_e \omega_{pe}^2 / \omega_{ce}^2)(1 - \beta_{ze}^2) \to 0$. In this limiting case, $\omega_{re}^+ \to \omega_{ce}/\hat{\gamma}_e$, $\omega_{re}^- \to 0$, and the terms quadratic in the electron density [proportional to $(V_{\perp e}^2/c^2)(\omega_{pe}^2/\hat{\gamma}_e)^2$] are neglected on the right-hand side of Eq.(7.15). It follows directly by inspection that Eq.(7.15) reduces to Eq.(7.7), as expected.

For increasing values of $(2\hat{\gamma}_e \omega_{pe}^2 / \omega_{ce}^2)(1 - \beta_{ze}^2)$, however, the stability behavior predicted by Eq.(7.15) is modified by self-field effects. Not only is the high-frequency (cyclotron maser) branch, which is associated with the resonance $\omega - k_z V_{ze} \approx \omega_{re}^+$ in Eq.(7.15), affected by self fields, it is found that a new low-frequency mode develops, which is associated with the resonance $\omega - k_z V_{ze} \approx \omega_{re}^-$ in Eq.(7.15). This low-frequency mode, which is *absent* in the limit $(2\hat{\gamma}_e \omega_{pe}^2 / \omega_{ce}^2)(1 - \beta_{ze}^2) \to 0$ is found to be unstable ($\text{Im}\,\omega > 0$) with somewhat lower growth rate than the cyclotron maser instability.[64] The existence of such a low-frequency mode is consistent with the spontaneous emission calculation in Sec. 3.5, which shows the presence of a low-frequency excitation associated entirely with self-field effects.

For present purposes, we consider only the stability properties associated with the cyclotron maser branch in Eq.(7.15). Typical numerical solutions to the dispersion relation (7.15) are illustrated in Fig. 7.4 for the cyclotron maser instability.[64] Here, $(\text{Re}\,\omega - k_z V_{ze})/\hat{\omega}_{ce}$ and $(\text{Im}\,\omega)/\hat{\omega}_{ce}$ are plotted versus $ck_z/\hat{\omega}_{ce}$ for $V_{\perp e} = 0.5c$, $V_{ze} = \beta_{ze} c = 0.2c$, $\hat{\gamma}_e = 1.19$, $\hat{\omega}_{ce} = \omega_{ce}/\hat{\gamma}_e$, and values of $s_e = \hat{\gamma}_e \omega_{pe}^2 / \omega_{ce}^2$ ranging from 0.1 to 0.5. [Strictly speaking, the self-field parameter occurring in the definition of

$\omega_{re}^{\pm}$ in Eq.(7.14) is $(2\hat{\gamma}_e\omega_{pe}^2/\omega_{ce}^2)(1 - \beta_{ze}^2)$, which differs by a factor of $2(1 - \beta_{ze}^2)$ from the definition of s_e used here.] Note from Fig. 7.4(a) that $(\operatorname{Re}\omega - k_z V_{ze}) \simeq \hat{\omega}_{ce} = \omega_{ce}/\hat{\gamma}_e$ is approximately constant for the entire range of the parameter $s_e = \hat{\gamma}_e\omega_{pe}^2/\omega_{ce}^2$ considered in the figure. Moreover, the normalized growth rate $(\operatorname{Im}\omega)/\hat{\omega}_{ce}$ increases for values of s_e up to 0.3 in Fig. 7.4(b), and then *decreases* as s_e is further increased. The growth rate and real frequency are plotted in Fig. 7.4 for negative and positive values of k_z. It is evident from Fig. 7.4(b) that the instability bandwidth decreases as the value of s_e is increased.

An important empirical result obtained by solving Eq.(7.15) for the cyclotron maser mode, even for moderate values of s_e, is that the wavenumber corresponding to maximum growth rate (denote by k_{z0}) is obtained to good accuracy from the solution to the simultaneous equations[64]

$$\omega - k_z V_{ze} = \omega_{ce}/\hat{\gamma}_e \ ,$$

$$\omega\frac{\partial\omega}{\partial k_z} = c^2 k_z \ , \tag{7.16}$$

when $V_g = \partial\omega/\partial k_z = V_{ze}$. Solving Eq.(7.16) readily gives

$$ck_{z0} = \frac{\beta_{ze}\omega_{ce}/\hat{\gamma}_e}{1 - \beta_{ze}^2} = \frac{\gamma_z^2\beta_{ze}\omega_{ce}}{\hat{\gamma}_e} \ , \tag{7.17}$$

where $\beta_{ze} = V_{ze}/c$ and $\gamma_z = (1 - \beta_{ze}^2)^{-1/2}$. For $\beta_{ze} = 0.2$ and $\gamma_z = 1.02$, Eq.(7.17) gives $ck_{z0}/(\omega_{ce}/\hat{\gamma}_e) \simeq 0.21$, which is in excellent agreement with Fig. 7.4(b). The expression for k_{z0} in Eq.(7.17) is identical to Eq.(7.44), calculated for the cyclotron maser instability in a cylindrical waveguide operating above cut-off.

It is convenient to transform the dispersion relation (7.15) to a frame of reference moving with axial velocity V_{ze} (the beam velocity) relative to the laboratory frame. The frequency and wavenumber (ω', k_z') in the moving frame are related to (ω, k_z) in the laboratory frame by

$$\omega' = \gamma_z\left(\omega - k_z V_{ze}\right) \ ,$$

$$k_z' = \gamma_z\left(k_z - \omega V_{ze}/c^2\right) , \tag{7.18}$$

and the inverse Lorentz transformation is given by

$$\omega = \gamma_z\left(\omega' + k_z' V_{ze}\right) ,$$

$$k_z = \gamma_z\left(k_z' + \omega' V_{ze}/c^2\right) , \tag{7.19}$$

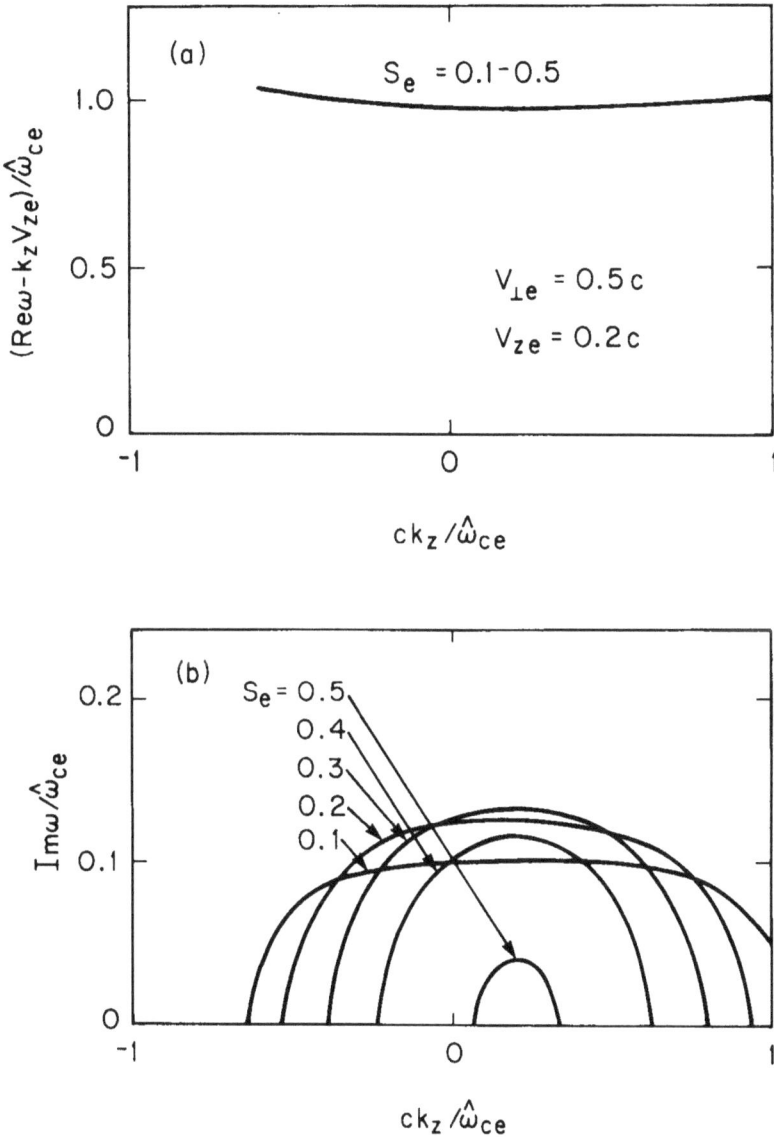

Figure 7.4. Linear growth properties of the cyclotron maser instability. The figures show plots versus $ck_z/\hat{\omega}_{ce}$ of (a) the normalized real frequency $(\text{Re}\,\omega - k_z V_{ze})/\hat{\omega}_{ce}$ and (b) the normalized growth rate $\text{Im}\,\omega/\hat{\omega}_{ce}$ obtained from the dispersion relation (7.15). The choice of system parameters corresponds to $V_{\perp e} = 0.5\,c, V_{ze} = 0.2\,c$ and $\hat{\gamma}_e = 1.19$, and values of the self-field parameter $s_e = \hat{\gamma}_e \omega_{pe}^2/\omega_{ce}^2$ ranging from 0.1 to 0.5 [H.S. Uhm and R.C. Davidson, Phys. Fluids **29**, 2713 (1986)].

where $\gamma_z = (1 - \beta_{ze}^2)^{-1/2}$. Substituting Eq.(7.19) into Eq.(7.15), the dispersion relation in the beam frame can be expressed as (Problem 7.3)[64]

$$
\omega'^2 - c^2 k_z'^2 = \frac{\omega_{pe}^2}{\hat{\gamma}_e} \frac{\omega'^2}{\left[\omega'^2 - \left(\dfrac{\gamma_z \omega_{ce}}{\hat{\gamma}_e} \right) \omega' + \dfrac{\omega_{pe}^2}{2\hat{\gamma}_e} \right]}
$$

$$
- \frac{\gamma_z^2}{2} \frac{V_{\perp e}^2}{c^2} \frac{\omega_{pe}^2}{\hat{\gamma}_e} \frac{\left[(\omega'^2 - c^2 k_z'^2) \left(\omega'^2 - \dfrac{\omega_{pe}^2}{2\hat{\gamma}_e} \right) + \dfrac{\omega'^2 \omega_{pe}^2}{2\hat{\gamma}_e} \right]}{\left[\omega'^2 - \left(\dfrac{\gamma_z \omega_{ce}}{\hat{\gamma}_e} \right) \omega' + \dfrac{\omega_{pe}^2}{2\hat{\gamma}_e} \right]^2}. \tag{7.20}
$$

Comparing Eqs.(7.16)–(7.18), it is evident that maximum growth occurs for $k_z' \simeq 0$ and $\omega' \simeq \gamma_z \omega_{ce}/\hat{\gamma}_e$ in the beam frame. Setting $k_z' = 0$ in Eq.(7.20) gives (exactly)

$$
0 = \left[\omega'^2 - \left(\frac{\gamma_z \omega_{ce}}{\hat{\gamma}_e} \right) \omega' + \frac{\omega_{pe}^2}{2\hat{\gamma}_e} \right]^2 - \frac{\omega_{pe}^2}{\hat{\gamma}_e} \left[\omega'^2 - \left(\frac{\gamma_z \omega_{ce}}{\hat{\gamma}_e} \right) \omega' + \frac{\omega_{pe}^2}{2\hat{\gamma}_e} \right]
$$

$$
+ \frac{\gamma_z^2}{2} \frac{V_{\perp e}^2}{c^2} \frac{\omega_{pe}^2}{\hat{\gamma}_e} \omega'^2 , \tag{7.21}
$$

which can be used to determine ω' at maximum growth. Approximating $\omega'^2 \simeq \gamma_z^2 \omega_{ce}^2 / \hat{\gamma}_e^2$ in the term proportional to $V_{\perp e}^2/c^2$ in Eq.(7.21) readily gives

$$
\omega'^2 - \gamma_z \frac{\omega_{ce}}{\hat{\gamma}_e} \omega' \mp \frac{i}{2} \frac{\omega_{ce}^2}{\hat{\gamma}_e^2} \left[s_e \left(\frac{2\gamma_z^4 V_{\perp e}^2}{c^2} - s_e \right) \right]^{1/2} = 0 \tag{7.22}
$$

for $s_e = \hat{\gamma}_e \omega_{pe}^2 / \omega_{ce}^2 < 2\gamma_z^4 V_{\perp e}^2/c^2$. Treating the term $[\cdots]^{1/2}$ as small in Eq.(7.22), and substituting $\omega' = \gamma_z(\omega - k_{z0} V_{ze})$ where $ck_{z0} = \gamma_z^2 \beta_{ze} \omega_{ce}/\hat{\gamma}_e$ [Eqs.(7.17)–(7.18)], we readily obtain the solutions

$$
\mathrm{Re}\,\omega = k_{z0} V_{ze} + \frac{\omega_{ce}}{\hat{\gamma}_e} = \gamma_z^2 \frac{\omega_{ce}}{\hat{\gamma}_e} , \tag{7.23}
$$

and

$$
\mathrm{Im}\,\omega = \frac{1}{2} \frac{\omega_{ce}}{\gamma_z^2 \hat{\gamma}_e} \left[s_e \left(\frac{2\gamma_z^4 V_{\perp e}^2}{c^2} - s_e \right) \right]^{1/2} . \tag{7.24}
$$

For $s_e = \hat{\gamma}_e \omega_{pe}^2 / \omega_{ce}^2 \ll 2\gamma_z^4 V_{\perp e}^2 / c^2$, the maximum growth rate in Eq.(7.24) reduces to the familiar result $\mathrm{Im}\,\omega = (V_{\perp e}^2 \omega_{pe}^2 / 2c^2 \hat{\gamma}_e)^{1/2}$ calculated for a tenuous electron beam in Sec. 7.2.1 [see Eq.(7.11)]. As the value of s_e is increased, however, the growth rate in Eq.(7.24) passes through the maximum value

$$(\mathrm{Im}\,\omega)_{\max} = \frac{1}{2}\gamma_z^2 \frac{V_{\perp e}^2}{c^2} \frac{\omega_{ce}}{\hat{\gamma}_e} , \qquad (7.25)$$

which is attained for normalized beam density $s_e = \hat{\gamma}_e \omega_{pe}^2 / \omega_{ce}^2$ satisfying

$$s_e = s_e^m \equiv \gamma_z^4 \frac{V_{\perp e}^2}{c^2} . \qquad (7.26)$$

Moreover, for $s_e > s_e^m$, the growth rate in Eq.(7.24) decreases to zero as s_e approaches

$$s_e = s_e^0 \equiv 2\gamma_z^4 \frac{V_{\perp e}^2}{c^2} . \qquad (7.27)$$

This behavior is illustrated in Fig. 7.5, where $(\mathrm{Im}\,\omega)/\hat{\omega}_{ce}$ is plotted versus $s_e = \hat{\gamma}_e \omega_{pe}^2 / \omega_{ce}^2$ for $k_z = k_{z0}$ and the choice of system parameters $V_{\perp e} = 0.5c$, $V_{ze} = \beta_{ze} c = 0.2c$, and $\hat{\gamma}_e = 1.19$. The solid curve in Fig. 7.5 is obtained from the numerical solution[64] to the full dispersion relation (7.15), whereas the dashed curve is obtained from the analytical solution in Eq.(7.24). Evidently, Eq.(7.24) provides an excellent estimate of the growth rate, at least for system parameters where $2\gamma_z^4 V_{\perp e}^2 / c^2$ is small in comparison with unity. Introducing the magnitude of the axial beam current I defined by $I = \pi r_b^2 \hat{n}_e |-e\beta_{ze} c|$, the threshold condition $(s_e > s_e^0)$ for complete stabilization of the cyclotron maser instability can be expressed in the equivalent form $I > I_{cr}$, where the critical current is defined by

$$I_{cr} = \frac{1}{2} \left(\frac{\omega_{ce}^2 r_b^2}{\hat{\gamma}_e c^2} \right) \left(\frac{m_e c^3}{e} \right) \gamma_z^4 \frac{V_{\perp e}^2}{c^2} \beta_{ze} , \qquad (7.28)$$

and $m_e c^3 / e \simeq 17,000$ A.

It is also evident from Fig. 7.4(b) that the bandwidth of the cyclotron maser instability in k_z-space decreases as s_e is increased. Denoting the range of k_z-values corresponding to instability $(\mathrm{Im}\,\omega > 0)$ by $k_{z0} - \Delta k_z < k_z < k_{z0} + \Delta k_z$, it can be shown[64] from Eq.(7.20) for $2\gamma_z^4 V_{\perp e}^2 / c^2 < 1$ that the total normalized bandwidth is given by approximately

$$2\left(\frac{c\Delta k_z}{\hat{\omega}_{ce}} \right) = 2\gamma_z^5 \left[1 - \left(\frac{s_e}{s_e^0} \right)^{1/2} \right]^{1/2} . \qquad (7.29)$$

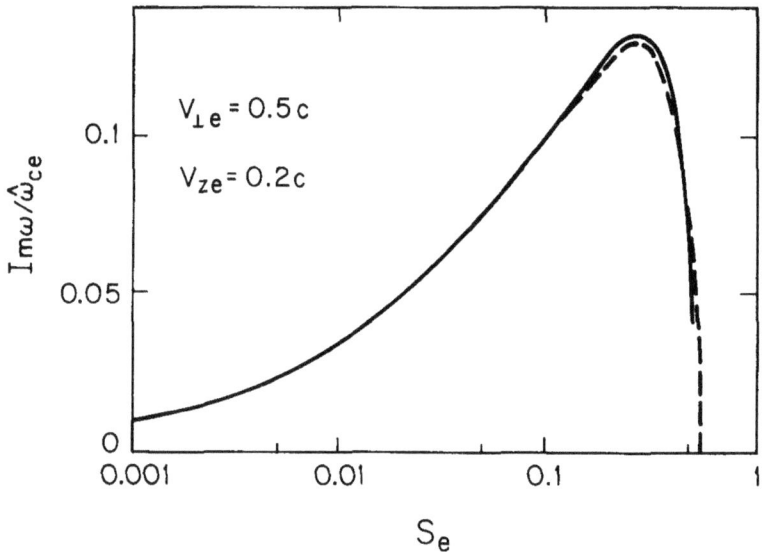

Figure 7.5. Linear growth properties of the cyclotron maser instability. The figure shows plots versus $s_e = \hat{\gamma}_e \omega_{pe}^2 / \omega_{ce}^2$ of the normalized growth rate $\text{Im}\,\omega/\hat{\omega}_{ce}$ for $k_z = k_{z0}$ obtained numerically from the full dispersion relation (7.15) (solid curve) and from the analytical estimate in Eq.(7.24) (dashed curve). The choice of system parameters corresponds to $V_{\perp e} = 0.5\,c$, $V_{ze} = 0.2\,c$ and $\hat{\gamma}_e = 1.19$ [H.S. Uhm and R.C. Davidson, Phys. Fluids **29**, 2713 (1986)].

Here, $\hat{\omega}_{ce} = \omega_{ce}/\hat{\gamma}_e$, $\gamma_z = (1 - \beta_{ze}^2)^{-1/2}$, and $s_e^0 = 2\gamma_z^4 V_{\perp e}^2 / c^2$ is the density threshold defined in Eq.(7.27) for complete stabilization of the cyclotron maser instability. Note from Eq.(7.29) that the bandwidth is relatively broad for a tenuous electron beam with $s_e \ll s_e^0$, and decreases to zero as s_e approaches s_e^0, as does the maximum growth rate in Eq.(7.24).

To summarize, as the parameter $s_e = \hat{\gamma}_e \omega_{pe}^2 / \omega_{ce}^2$ is increased to sufficiently large values, properties of the cyclotron maser instability are altered significantly, at least for the case of a uniform electron beam assumed in the present analysis. Instability ceases when s_e is increased to the value $s_e^0 = 2\gamma_z^4 V_{\perp e}^2 / c^2$. For $s_e \ll s_e^0$, however, the tenuous-beam assumption made in Sec. 7.2.1 is a good approximation for describing detailed stability behavior.

Problem 7.2   Consider the motion of an individual test electron in the crossed equilibrium electric and magnetic fields, $\mathbf{E}^0(\mathbf{x}) = E_r(r)\hat{\mathbf{e}}_r$ and $\mathbf{B}^0(\mathbf{x}) = B_0\hat{\mathbf{e}}_z + B_\theta^s(r)\hat{\mathbf{e}}_\theta$, where $E_r(r)$ and $B_\theta^s(r)$ are defined in Eq.(7.12). Here, $\hat{\mathbf{e}}_r$ and

$\hat{e}_\theta$ are unit vectors in the radial and azimuthal directions in cylindrical polar coordinates.

a. Show that the equations of motion of the test electron can be expressed in Cartesian coordinates as

$$\frac{d}{dt}p_x(t) = \frac{1}{2}m_e\omega_{pe}^2 x(t)\left(1 - \frac{1}{c}v_z(t)\beta_{ze}\right) - m_e\omega_{ce}v_y(t) , \quad (7.2.1)$$

$$\frac{d}{dt}p_y(t) = \frac{1}{2}m_e\omega_{pe}^2 y(t)\left(1 - \frac{1}{c}v_z(t)\beta_{ze}\right) + m_e\omega_{ce}v_x(t) , \quad (7.2.2)$$

$$\frac{d}{dt}p_z(t) = \frac{1}{2}m_e\frac{\omega_{pe}^2\beta_{ze}}{c}\left[x(t)v_x(t) + y(t)v_y(t)\right] , \quad (7.2.3)$$

within the electron beam ($0 \le r < r_b$). Here, $\mathbf{v}(t) = d\mathbf{x}(t)/dt$, $\mathbf{p}(t) = \gamma(t)m_e\mathbf{v}(t)$, $\gamma(t) = [1 + \mathbf{p}^2(t)/m_e^2c^2]^{1/2}$, and the notation is otherwise the same as in Sec. 7.2.1.

b. Show that Eqs.(7.2.1)–(7.2.3) possess the single-particle constants of the motion, H and P_z, where

$$\frac{d}{dt}H = \frac{d}{dt}\left[\gamma(t)m_ec^2 - \frac{1}{4}m_ec^2\frac{\omega_{pe}^2[x^2(t) + y^2(t)]}{c^2}\right] = 0 , \quad (7.2.4)$$

$$\frac{d}{dt}P_z = \frac{d}{dt}\left[p_z(t) - \frac{1}{4}m_e\beta_{ze}c\frac{\omega_{pe}^2[x^2(t) + y^2(t)]}{c^2}\right] = 0 . \quad (7.2.5)$$

Equations (7.2.4) and (7.2.5) correspond to the conservation of total energy and axial canonical momentum.

c. For $\nu/\hat{\gamma}_e = \omega_{pe}^2 r_b^2/4\hat{\gamma}_e c^2 \ll 1$, it follows from Eqs.(7.2.4) and (7.2.5) that $\gamma(t) \simeq \gamma = $ const. and $p_z(t) \simeq p_z = \gamma m_e v_z = $ const. to lowest order. Treating γ and v_z as constants, show from Eqs.(7.2.1) and (7.2.2) that the perpendicular orbit $x(t) + iy(t)$ satisfies

$$\left[\frac{d^2}{dt^2} - i\frac{\omega_{ce}}{\gamma}\frac{d}{dt} - \frac{1}{2}\frac{\omega_{pe}^2}{\gamma}\left(1 - \frac{1}{c}v_z\beta_{ze}\right)\right][x(t) + iy(t)] = 0 . \quad (7.2.6)$$

d. Equation (7.2.6) is the generalization of Eq.(3.13) for a uniform-density relativistic electron beam with $\nu/\hat{\gamma}_e \ll 1$. Show that the solutions to Eq.(7.2.6) for $x(t)$ and $y(t)$ are biharmonic, with oscillatory components at the frequencies $\omega_e^\pm(\gamma, v_z)$ defined in Eq.(7.13).

Problem 7.3   The electromagnetic dispersion relation (7.15) includes the full influence of the equilibrium self-electric and self-magnetic fields. Make use of the Lorentz transformation in Eq.(7.19) to show that the dispersion relation (7.15) can be expressed in the form given in Eq.(7.20). Equation (7.20) relates the frequency ω' and wavenumber k_z' in a frame of reference moving with velocity $V_{ze}\hat{e}_z$ (the axial velocity of the electron beam in the laboratory frame).

Problem 7.4   Unlike Eq.(7.15), the electromagnetic dispersion relation (7.7) includes no effects of the equilibrium self-electric and self-magnetic fields. (That is, a tenuous electron beam with $\hat{\gamma}_e \omega_{pe}^2/\omega_{ce}^2 \ll 1$ has been assumed.)

a. Make use of the Lorentz transformation in Eq.(7.19) to show that the dispersion relation (7.7) can be expressed as

$$\left(\omega'^2 - c^2 k_z'^2\right)\left[1 + \frac{\gamma_z^2}{2}\frac{V_{\perp e}^2}{c^2}\frac{\omega_{pe}^2/\hat{\gamma}_e}{\left(\omega' - \gamma_z \omega_{ce}/\hat{\gamma}_e\right)^2}\right] = \frac{\omega_{pe}^2}{\hat{\gamma}_e}\frac{\omega'}{\left(\omega' - \gamma_z \omega_{ce}/\hat{\gamma}_e\right)}$$

$$(7.4.1)$$

for electromagnetic waves with right-circular polarization. Here, γ_z is defined by $\gamma_z = (1 - V_{ze}^2/c^2)^{-1/2}$, and Eq.(7.4.1) relates to ω' and k_z' in a frame of reference moving with the axial velocity $V_{ze}\hat{e}_z$ of the electron beam.

b. The maximum growth rate of the cyclotron maser instability occurs for $k_z' = 0$ in the moving frame. If $2\gamma_z^4 V_{\perp e}^2/c^2$ is treated as a small parameter, then $\omega' \approx \gamma_z \omega_{ce}/\hat{\gamma}_e$ is also a good approximation near maximum growth. For $k_z' = 0$ and $\omega' \approx \gamma_z \omega_{ce}/\hat{\gamma}_e$, show that Eq.(7.4.1) can be approximated by

$$\left(\omega' - \frac{\gamma_z \omega_{ce}}{\hat{\gamma}_e}\right)^2 - \frac{\omega_{pe}^2}{\gamma_z \omega_{ce}}\left(\omega' - \frac{\gamma_z \omega_{ce}}{\hat{\gamma}_e}\right) + \frac{1}{2}\gamma_z^2\frac{V_{\perp e}^2}{c^2}\frac{\omega_{pe}^2}{\hat{\gamma}_e} = 0, \quad (7.4.2)$$

or in laboratory frame variables by

$$\left(\omega - k_z V_{ze} - \frac{\omega_{ce}}{\hat{\gamma}_e}\right)^2 - \frac{\omega_{pe}^2}{\gamma_z^2 \omega_{ce}}\left(\omega - k_z V_{ze} - \frac{\omega_{ce}}{\hat{\gamma}_e}\right) + \frac{1}{2}\frac{V_{\perp e}^2}{c^2}\frac{\omega_{pe}^2}{\hat{\gamma}_e} = 0.$$

$$(7.4.3)$$

c. Show that the necessary and sufficient condition for Eq.(7.4.3) to have stable oscillatory solutions (Im $\omega = 0$) is

$$s_e > 2\gamma_z^4\frac{V_{\perp e}^2}{c^2}, \quad (7.4.4)$$

where $s_e = \hat{\gamma}_e \omega_{pe}^2/\omega_{ce}^2$. Quite remarkably, the density threshold for stabilization of the cyclotron maser instability in Eq.(7.4.4) [calculated from Eq.(7.4.1), which neglects self-field effects] is identical to that in Eq.(7.27) [calculated from Eq.(7.20), which includes self-field effects]. Given the dissimilar algebraic features of the two dispersion relations [compare Eqs.(7.20) and (7.4.1)], this should be viewed as a somewhat fortuitous coincidence, tempered by the assumption that $2\gamma_z^4 V_{\perp e}^2/c^2$ is less than unity.

7.3 Cyclotron Maser Instability for an Annular Electron Beam

In this section, again assuming a tenuous electron beam ($\hat{\gamma}\omega_{pe}^2/\omega_{ce}^2 \ll 1$), we make use of the linearized Vlasov-Maxwell equations to investigate properties of the cyclotron maser instability for a relativistic electron beam propagating parallel to a uniform guide field $B_0\hat{e}_z$ in a perfectly conducting cylindrical waveguide with radius $r = b$ (Fig. 7.6). Neglecting equilibrium self-field effects, but including the full influence of finite radial geometry, the Vlasov-Maxwell equations have been used to investigate detailed stability behavior for transverse electric (TE) and transverse magnetic (TM) excitations of the cyclotron maser instability, including the effects of general magnetic harmonic number, spreads in energy and axial momentum of the beam electrons, as well as stability properties for both annular and solid electron beams.[66-70] For present purposes, we specialize to the case of an annular electron beam with average radius r_0 and radial thickness $2a$ (Fig. 7.6). The analysis neglects the influence of equilibrium self fields and assumes

$$\frac{\nu}{\hat{\gamma}_e} = \frac{N_e}{\hat{\gamma}_e}\frac{e^2}{m_e c^2} \ll 1 \, . \tag{7.30}$$

Here, ν is Budker's parameter, and $N_e = 2\pi \int_0^b dr \, r n_e^0(r)$ is the number of electrons per unit axial length of the electron beam.

For the cylindrically symmetric equilibrium configuration in Fig. 7.6, the single-particle constants of the motion, neglecting self-field effects, are given by

$$H = \left(m_e^2 c^4 + c^2 \mathbf{p}^2\right)^{1/2} - m_e c^2 = (\gamma - 1)m_e c^2 \, , \tag{7.31a}$$

$$P_\theta = r\left(p_\theta - \frac{e}{2c}B_0 r\right) \, , \tag{7.31b}$$

$$p_z = \gamma m_e v_z \, . \tag{7.31c}$$

Assuming zero spreads in energy and axial momentum, we consider the self-consistent Vlasov equilibrium $f_e^0(r, \mathbf{p})$ for an annular electron beam specified by

$$f_e^0(r, \mathbf{p}) = \frac{N_e \omega_{ce}}{4\pi^2 \hat{\gamma}_e m_e c^2}\delta\left(\gamma - \hat{\gamma}_e\right)\delta\left(p_z - \hat{\gamma}_e m_e V_{ze}\right)\delta\left(P_\theta - P_0\right) \, . \tag{7.32}$$

Conducting Wall

Annular
Electron Beam

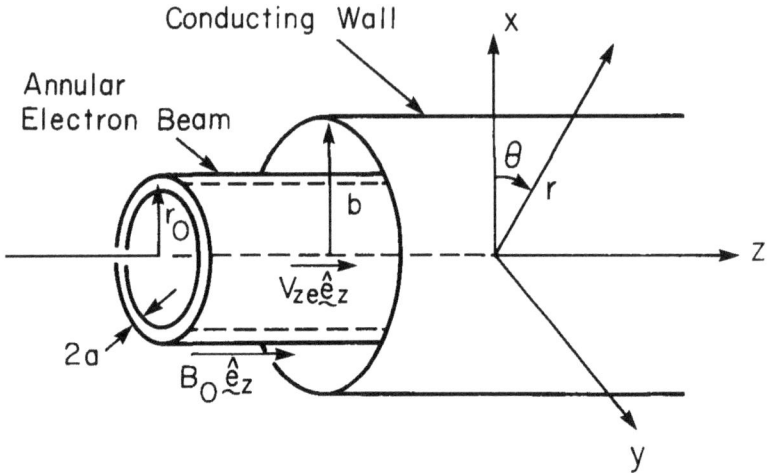

Figure 7.6. Schematic of an annular electron beam propagating parallel to a uniform guide field $B_0\hat{e}_z$ in a perfectly conducting cylindrical waveguide with radius $r = b$.

Here, γ, P_θ and p_z are defined in Eq.(7.31), $N_e = 2\pi \int_0^b dr\, r \int d^3p\, f_e^0$ is the number of electrons per unit axial length, $\omega_{ce} = eB_0/m_ec$ is the nonrelativistic cyclotron frequency, $\hat{\gamma}_e m_e V_{ze} = $ const. is the axial momentum of the beam electrons, and the constants $\hat{\gamma}_e$ and P_0 are defined by

$$\hat{\gamma}_e = \left(1 - \frac{V_{\perp e}^2}{c^2} - \frac{V_{ze}^2}{c^2}\right)^{-1/2},$$

$$P_0 = -\tfrac{1}{2}m_e\omega_{ce}\left(r_0^2 - \frac{\hat{\gamma}_e^2 V_{\perp e}^2}{\omega_{ce}^2}\right), \tag{7.33}$$

where r_0 and $V_{\perp e}$ are constants. For $\hat{\gamma}_e V_{\perp e}/\omega_{ce} \ll r_0$, it is readily shown from Eqs.(7.32) and (7.33) that the equilibrium density profile $n_e^0(r) = \int d^3p\, f_e^0(r, \mathbf{p})$ corresponds to a thin electron layer ($2a \ll r_0$) extending from radius $r = r_0 - a$ to radius $r = r_0 + a$, where a is defined by (Problem 7.5)

$$a = \frac{\hat{\gamma}_e V_{\perp e}}{\omega_{ce}}. \tag{7.34}$$

In terms of the average density $\bar{n}_e = (2ar_0)^{-1}\int_{r_0-a}^{r_0+a} dr\, rn_e^0(r)$, the inequality in Eq.(7.30) can be expressed in the equivalent form

$$\frac{\nu}{\hat{\gamma}_e} = \frac{\bar{\omega}_{pe}^2 a r_0}{\hat{\gamma}_e c^2} \ll 1 , \qquad (7.35)$$

where $\bar{\omega}_{pe}^2 = 4\pi \bar{n}_e e^2 / m_e$.

Assuming azimuthally symmetric TE-mode excitations with $\partial/\partial\theta = 0$, the perturbed electromagnetic field components are

$$\delta\mathbf{E}(\mathbf{x}, t) = \delta E_\theta(r, z, t)\hat{\mathbf{e}}_\theta ,$$

$$\delta\mathbf{B}(\mathbf{x}, t) = \delta B_r(r, z, t)\hat{\mathbf{e}}_r + \delta B_z(r, z, t)\hat{\mathbf{e}}_z , \qquad (7.36)$$

where $\delta\mathbf{E}$ and $\delta\mathbf{B}$ are related through Maxwell's equations to the perturbed beam current $\delta\mathbf{J}_e(\mathbf{x}, t) = -e \int d^3p \, \mathbf{v}\delta f_e(r, z, \mathbf{p}, t)$. We express the perturbed quantities as $\delta E_\theta(r, z, t) = \sum_{k_z = -\infty}^{\infty} \delta E_\theta(r, k_z)exp(ik_z z - i\omega t)$, etc., where ω is the (complex) oscillation frequency, $k_z = 2\pi m/L$ is the axial wavenumber, m is an integer, and L is the fundamental periodicity length in the z-direction. In the absence of an electron beam ($\nu \to 0$), the radial dependence of the vacuum eigenfunction is given by

$$\delta E_\theta(r, k_z) = a(k_z)J_1(k_{0n}r) , \qquad (7.37)$$

where $J_n(x)$ is the Bessel function of the first kind of order n. Here, k_{0n} is the n'th zero of $J_1(k_{0n}b) = 0$, which corresponds to $[\delta E_\theta]_{r=b} = 0$ at the perfectly conducting wall, as required. Furthermore, in the absence of the electron beam, the normal modes of oscillation in the vacuum waveguide are given by $\omega^2 = c^2 k_z^2 + c^2 k_{0n}^2$.

For a tenuous, annular electron beam with $\nu/\hat{\gamma}_e \ll 1$, and $2a \ll r_0$, the linearized Vlasov equation can be solved for the perturbed distribution function $\delta f_e(r, z, \mathbf{p}, t)$. Assuming azimuthally symmetric TE_{0n}-mode perturbations about the equilibrium distribution function $f_e^0(r, \mathbf{p})$ in Eq.(7.32), the resulting dispersion relation for the fundamental mode can be expressed as[70]

$$\omega^2 - c^2 k_z^2 - c^2 k_{0n}^2 = \frac{4\nu c^2}{\hat{\gamma}_e[bJ_2(k_{0n}b)]^2} \qquad (7.38)$$

$$\times \left[Q_1 \frac{(\omega - k_z V_{ze})}{(\omega - k_z V_{ze} - \omega_{ce}/\hat{\gamma}_e)} - H_1 \frac{V_{\perp e}^2}{c^2} \frac{(\omega^2 - c^2 k_z^2)}{(\omega - k_z V_{ze} - \omega_{ce}/\hat{\gamma}_e)^2} \right] .$$

Here, the geometric factors H_1 and Q_1 occurring in Eq.(7.38) are defined by[70]

$$H_1 = [J_1(k_{0n}r_0)J_1'(k_{0n}a)]^2 \ ,$$

$$Q_1 = 2H_1 + 2(k_{0n}a) [J_1(k_{0n}r_0)]^2 J_1'(k_{0n}a)J_1''(k_{0n}a) \ , \qquad (7.39)$$

where $J_n'(x)$ denotes $dJ_n(x)/dx$.

It is instructive to compare the dispersion relation (7.38) for an annular electron beam in a circular waveguide, with the dispersion relation (7.7) for a uniform-density beam neglecting the effects of radial geometry. First, the vacuum wave contribution $\omega^2 - c^2k_z^2$ is replaced by $\omega^2 - c^2k_z^2 - c^2k_{0n}^2$ on the left-hand side of Eq.(7.38), where the term, $-c^2k_{0n}^2$, is associated with the radial mode structure of the vacuum eigenfunction. Second, in the remaining terms in the dispersion relation (7.7), the (uniform) beam density enters only through the factor $\omega_{pe}^2/\hat{\gamma}_e$. In Eq.(7.38), however, the radial geometry of the electron beam and the cylindrical waveguide enters the dispersion relation in a strong manner through the factors $Q_1(4\nu/\hat{\gamma}_e)/[bJ_2(k_{0n}b)]^2$ and $H_1(4\nu/\hat{\gamma}_e)/[bJ_2(k_{0n}b)]^2$. The term in Eq.(7.38) proportional to $H_1V_{\perp e}^2/c^2$ drives the cyclotron maser instability. Referring to the definition of H_1 in Eq.(7.39a), it is evident that this term has the largest effect when the beam radius r_0 is positioned so that $[J_1(k_{0n}r_0)]^2$ is a maximum, which corresponds to a maximum in the vacuum electric field δE_θ defined in Eq.(7.37). While the precise ω-dependence on the right-hand side of Eq.(7.38) will differ somewhat for different choices of equilibrium distribution function $f_e^0(r, \mathbf{p})$, the dispersion relation (7.38) describes the essential features of the cyclotron maser instability including the effects of finite radial geometry.

We now analyze the dispersion relation (7.38) treating $\nu/\hat{\gamma}_e \ll 1$ as a small parameter. Close to cyclotron resonance $(\omega - k_z V_{ze} - \omega_{ce}/\hat{\gamma}_e \simeq 0)$, the term proportional to $H_1V_{\perp e}^2/c^2$ dominates the term proportional to Q_1 on the right-hand side of Eq.(7.38). Furthermore, close to the vacuum waveguide solution, $\omega^2 - c^2k_z^2 - c^2k_{0n}^2 \simeq 0$, the factor $H_1(\omega^2 - c^2k_z^2)$ can be replaced by $H_1c^2k_{0n}^2$ on the right-hand side of Eq.(7.38). The dispersion relation (7.38) can then be approximated by

$$\left(\omega^2 - c^2k_z^2 - c^2k_{0n}^2\right)\left(\omega - k_z V_{ze} - \frac{\omega_{ce}}{\hat{\gamma}_e}\right)^2 = -\frac{4\nu c^2 H_1 c^2 k_{0n}^2}{\hat{\gamma}_e[bJ_2(k_{0n}b)]^2}\frac{V_{\perp e}^2}{c^2} \ , \tag{7.40}$$

where H_1 is defined in Eq.(7.39). Equation (7.40) is a fourth-order algebraic equation for the (complex) oscillation frequency ω which can be

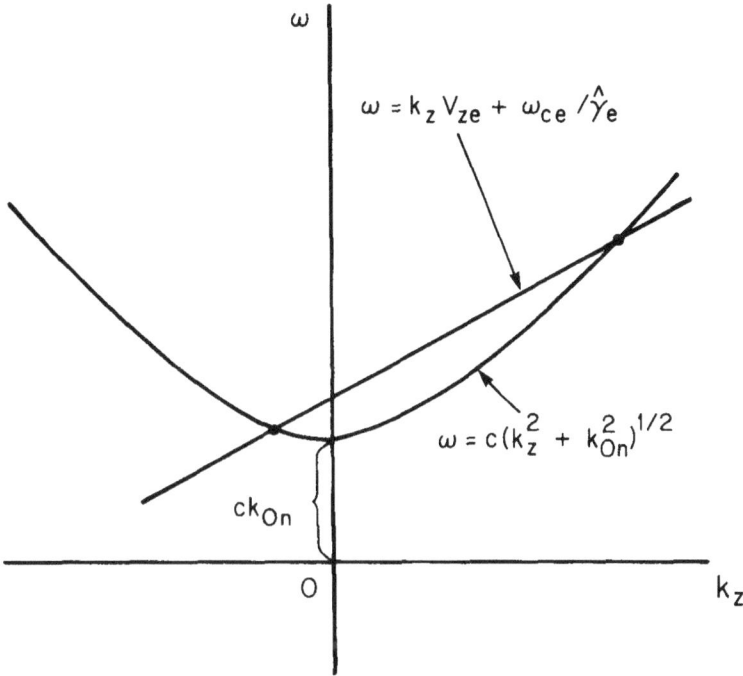

Figure 7.7. Plots showing the intersection of the two curves $\omega = k_z V_{ze} + \omega_{ce}/\hat{\gamma}_e$ and $\omega = c(k_z^2 + k_{0n}^2)^{1/2}$. For $\gamma_z \omega_{ce}/\hat{\gamma}_e = ck_{0n}$, the two curves are tangent at the point $(\omega_0, k_{z0}) = (\gamma_z^2 \omega_{ce}/\hat{\gamma}_e, \gamma_z^2 \beta_{ze} \omega_{ce}/c\hat{\gamma}_e)$.

solved over a wide range of system parameters. Of particular interest are solutions near the frequency and wavenumber (ω_0, k_{z0}) corresponding to the interaction of the cyclotron resonance mode with the vacuum waveguide mode (Fig. 7.7), i.e.,

$$\omega_0 = k_{z0} V_{ze} + \frac{\omega_{ce}}{\hat{\gamma}_e} \, ,$$

$$\omega_0 = c \left(k_{z0}^2 + k_{0n}^2 \right)^{1/2} \, . \tag{7.41}$$

Solving Eq.(7.41) for k_{z0} gives

$$ck_{z0} = \gamma_z^2 \beta_{ze} \frac{\omega_{ce}}{\hat{\gamma}_e} \pm \gamma_z \left(\gamma_z^2 \frac{\omega_{ce}^2}{\hat{\gamma}_e^2} - c^2 k_{0n}^2 \right)^{1/2} \, , \tag{7.42}$$

which requires $\gamma_z \omega_{ce}/\hat{\gamma}_e \geq ck_{0n}$ for intersection to occur. Here, β_{ze} and γ_z are defined by $\beta_{ze} = V_{ze}/c$ and $\gamma_z = (1 - \beta_{ze}^2)^{-1/2}$. The strongest coupling between the two modes actually occurs when the straight line $\omega = k_z V_{ze} + \omega_{ce}/\hat{\gamma}_e$ is exactly tangential to the curve $\omega = c(k_z^2 + k_{0n}^2)^{1/2}$ in Fig. 7.7, which requires $\omega \partial\omega/\partial k_z = c^2 k_z$ where $\partial\omega/\partial k_z = V_{ze}$. This condition imposes the requirement that the value of magnetic field be chosen such that

$$\gamma_z \frac{\omega_{ce}}{\hat{\gamma}_e} = ck_{0n} , \qquad (7.43)$$

which assures only one intersection point in Eq.(7.42) and Fig. 7.7.

For present purposes, it is assumed that Eq.(7.43) is satisfied, in which case ω_0 and k_{z0} are given by

$$\omega_0 = \frac{\gamma_z^2 \omega_{ce}}{\hat{\gamma}_e} ,$$

$$ck_{z0} = \frac{\gamma_z^2 \beta_{ze} \omega_{ce}}{\hat{\gamma}_e} . \qquad (7.44)$$

Note that the definition of k_{z0} in Eq.(7.44) is identical to the definition in Eq.(7.17), which was found to be the wavenumber for maximum growth of the cyclotron maser instability associated with the dispersion relation (7.15). In the analysis of Eq.(7.40), we express

$$\omega = \omega_0 + \delta\omega ,$$

$$k_z = k_{z0} + \delta k_z , \qquad (7.45)$$

treating $\delta\omega$ and δk_z as small quantities. To leading order, the dispersion relation (7.40) can be expressed as

$$\left(\frac{\delta\omega}{\hat{\omega}_{ce}} - \frac{V_{ze}}{c} \frac{c \delta k_z}{\hat{\omega}_{ce}} \right)^3 = -2 \frac{\nu}{\hat{\gamma}_e} \frac{V_{\perp e}^2}{c^2} \gamma_z^2 g_1 , \qquad (7.46)$$

where $\hat{\omega}_{ce} = \omega_{ce}/\hat{\gamma}_e$ is the relativistic cyclotron frequency, and g_1 is the geometric factor defined by

$$g_1 = \frac{H_1}{[k_{0n} b J_2(k_{0n} b)]^2} . \qquad (7.47)$$

Solving Eq.(7.46) for $\delta\omega$ gives

$$\mathrm{Im}\,\delta\omega = \frac{\sqrt{3}}{2}\left(2\frac{\nu}{\hat{\gamma}_e}\frac{V_{\perp e}^2}{c^2}\gamma_z^2 g_1\right)^{1/3}\frac{\omega_{ce}}{\hat{\gamma}_e},$$

$$\mathrm{Re}\,\delta\omega = V_{ze}\delta k_z + \frac{1}{2}\left(2\frac{\nu}{\hat{\gamma}_e}\frac{V_{\perp e}^2}{c^2}\gamma_z^2 g_1\right)^{1/3}\frac{\omega_{ce}}{\hat{\gamma}_e}, \qquad (7.48)$$

for the unstable branch with $\mathrm{Im}\,\delta\omega > 0$.

In addition to the strong geometric dependence through the factor g_1, it is evident that the cyclotron maser growth rate in Eq.(7.48) scales as $(\bar{n}_e V_{\perp e}^2/c^2)^{1/3}$ rather than $(\bar{n}_e V_{\perp e}^2/c^2)^{1/2}$, as found in Eq.(7.11) for the case of a uniform-density beam neglecting radial variations. This difference can be traced to the fact that the vacuum contribution to the dispersion relation (7.38) is proportional to $\omega^2 - c^2 k_z^2 - c^2 k_{0n}^2$, rather than proportional to $\omega^2 - c^2 k_z^2$ as in Eq.(7.7). Stabilization of the cyclotron maser instability at sufficiently large values of $|\delta k_z| = |k_z - k_{z0}|$ requires an analysis of the more accurate dispersion relation (7.40). A detailed examination of Eq.(7.40) shows that the instability is stabilized ($\mathrm{Im}\,\delta\omega = 0$) whenever $c^2(\delta k_z)^2$ exceeds the critical value defined by (Problem 7.6)

$$c^2(\delta k_z)_{cr}^2 = 2\gamma_z^4\frac{\omega_{ce}^2}{\hat{\gamma}_e^2}\left(\frac{27}{2}\frac{\nu}{\hat{\gamma}_e}\frac{V_{\perp e}^2}{c^2}\gamma_z^2 g_1\right)^{1/3}. \qquad (7.49)$$

In the derivation of the dispersion relation (7.38), it should be noted that no assumption has been made regarding the relative size of $V_{\perp e}/c$ and $\beta_{ze} = V_{ze}/c$. Therefore, Eq.(7.38) can be used both for cyclotron autoresonance maser (CARM) applications (moderate values of β_{ze}) and for gyrotron applications (typically, small values of β_{ze}). For the limiting case where $\beta_{ze}^2 \ll 1$ and $\gamma_z \simeq 1$, we note from Eqs.(7.43) and (7.44) that $c^2 k_{z0}^2 \simeq \beta_{ze}^2\omega_{ce}^2/\hat{\gamma}_e^2 \ll \omega_{ce}^2/\hat{\gamma}_e^2 \simeq c^2 k_{0n}^2$, which corresponds to operation near the cut-off frequency with $k_{0n}^2 \gg k_{z0}^2$ and $\omega_0 \simeq ck_{0n}$. In addition, with the appropriate generalization of Eq.(7.38), it should be pointed out that the cyclotron maser instability can be stabilized by relatively modest values of energy spread $(\Delta\gamma/\hat{\gamma}_e)$ and/or axial momentum spread $(\Delta p_z/\hat{\gamma}_e m_e V_{ze})$.[70] Moreover, the derivation of the dispersion relation (7.38) can be extended to the case of $\mathrm{TE}_{\ell n}$-mode perturbations with general azimuthal mode number ℓ.

In this book, it is not intended to treat the nonlinear evolution and saturation of the cyclotron maser instability. However, it should be pointed out that both coherent[71,72] and quasilinear (for multimode-excitation)[73] theories of the nonlinear evolution and saturation of the cyclotron maser instability have been developed. In addition, the nonlinear Vlasov-Maxwell equations have been used to obtain a bound[74] on the unstable field energy that can develop for a given input distribution of beam electrons.

Problem 7.5   Consider the formal expression for the density profile $n_e^0(r) = \int d^3p\, f_e^0(r, \mathbf{p})$ for the choice of equilibrium distribution function in Eq.(7.32).

a. Carry out the integrations over p_z and p_θ, and show that the envelope of turning points in the radial orbit ($r = r_b$ when $p_r = 0$) is determined from

$$
\left[1 + \hat{\gamma}_e^2 \frac{V_{ze}^2}{c^2} + \frac{1}{4} \frac{\omega_{ce}^2}{r_b^2 c^2} \left(r_b^2 - r_0^2 + \hat{\gamma}_e^2 \frac{V_{\perp e}^2}{\omega_{ce}^2} \right)^2 \right]^{1/2}
$$

$$
= \hat{\gamma}_e \equiv \left(1 + \frac{\hat{\gamma}_e^2 V_{ze}^2}{c^2} + \frac{\hat{\gamma}_e^2 V_{\perp e}^2}{c^2} \right)^{1/2} . \qquad (7.5.1)
$$

b. For a thin electron layer with $a \ll r_0$, show that the inner and outer boundaries of the electron layer determined from Eq.(7.5.1) are given by $r_b = r_b^{\pm} = r_0 \pm a$, where a is defined by

$$
a = \hat{\gamma}_e \frac{V_{\perp e}}{\omega_{ce}} . \qquad (7.5.2)
$$

Problem 7.6   Analyze the dispersion relation (7.40) retaining higher-order contributions in $\delta\omega = \omega - \omega_0$ and $\delta k_z = k_z - k_{z0}$, where ω_0 and k_{z0} are defined in Eq.(7.44).

a. Show that Eq.(7.40) can be expressed as

$$
\left[\delta\Omega + \frac{\beta_{ze} c \delta k_z}{\gamma_z^2 \hat{\omega}_{ce}} \delta\Omega + \frac{1}{2\gamma_z^2}(\delta\Omega)^2 - \frac{c^2(\delta k_z)^2}{2\gamma_z^4 \hat{\omega}_{ce}^2} \right] (\delta\Omega)^2 = -2 \frac{\nu}{\hat{\gamma}_e} \frac{V_{\perp e}^2}{c^2} \gamma_z^2 g_1 ,
$$
$$
(7.6.1)
$$

where the geometric factor g_1 is defined in Eq.(7.47), $\delta\Omega \equiv \delta\omega/\hat{\omega}_{ce} - \beta_{ze} c \delta k_z / \hat{\omega}_{ce}$, and $\hat{\omega}_{ce} = \omega_{ce}/\hat{\gamma}_e$. If the left-hand side of Eq.(7.6.1) is approximated by $(\delta\Omega)^3$, then Eq.(7.6.1) reduces to Eq.(7.46), which was analyzed in Sec. 7.3.

b. In solving Eq.(7.6.1) for $\delta\Omega$, it is valid to approximate $|\delta\Omega| \ll 1$ and $|c\delta k_z/\gamma_z^2 \hat{\omega}_{ce}| \ll 1$ for the unstable cyclotron maser branch. (Here, $\nu/\hat{\gamma}_e \ll 1$ is assumed.) In this case, Eq.(7.6.1) can be approximated by

$$
\left[\delta\Omega - (\delta K_z)^2 \right] (\delta\Omega)^2 = -2 \frac{\nu}{\hat{\gamma}_e} \frac{V_{\perp e}^2}{c^2} \gamma_z^2 g_1 , \qquad (7.6.2)
$$

where $\delta K_z \equiv c\delta k_z/\sqrt{2}\gamma_z^2\hat{\omega}_{ce}$. Show that the solutions to Eq.(7.6.2) corresond to stable oscillations ($\text{Im}\,\delta\Omega = 0$) provided the inequality

$$(\delta K_z)^2 > \left(\frac{27}{2}\frac{\nu}{\hat{\gamma}_e}\frac{V_{\perp e}^2}{c^2}\gamma_z^2 g_1\right)^{1/3} \qquad (7.6.3)$$

is satisfied. Equation (7.6.3) is identical to the estimate in Eq.(7.49).

7.4 High-Power Step-Tunable Gyrotron Experiments

The technical progress in the development of gyrotron sources of coherent electromagnetic radiation has been substantial, both with regard to the achievement of high power levels[1-3] and the achievement of step-tunable operation at high frequencies.[1] In this section, we give a brief summary of the experiments by Kreischer and Temkin[1] in single-mode operation of a gyrotron oscillator at power levels of more than 0.5 MW and frequencies up to 243 GHz. Figure 7.8 shows a schematic of the experimental configuration.[1] The gyrotron operates at 4 Hz repetition rate with 3-μs pulses. The magnetron injection gun produces an annular electron beam with $V_{\perp e}/V_{ze} = 1.93$ and a spread in $V_{\perp e}$ of 4% at beam energy and current of 80 keV and 35 A. The magnetic field in the cavity is produced by a Bitter magnet which produces fields up to $B_0 = 97$ kG. The cavity consists of

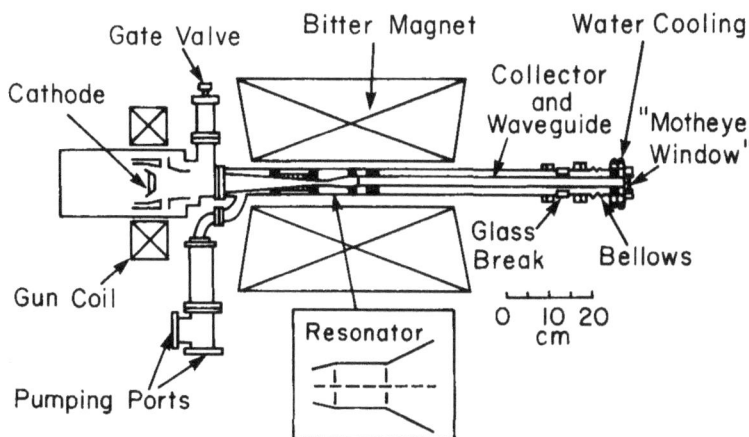

Figure 7.8. Schematic diagram of high-power step-tunable gyrotron configuration [K.E. Kreischer and R.J. Temkin, Phys. Rev. Lett. **59**, 547 (1987)].

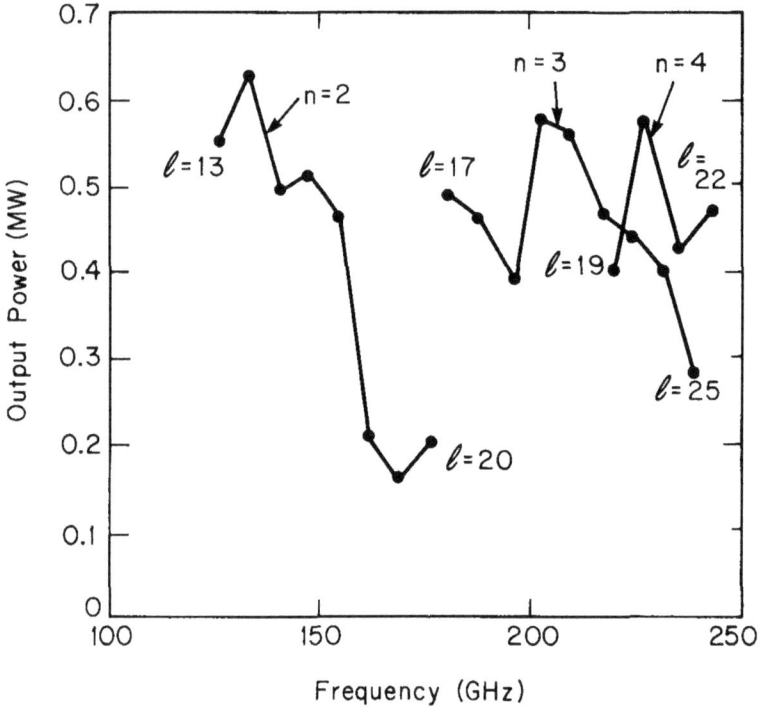

Frequency (GHz)

Figure 7.9. Measured microwave power versus frequency in step-tunable gyrotron experiments operating at $80\,\mathrm{kV}$ and $35\,\mathrm{A}$ in the $\mathrm{TE}_{\ell n}$ mode. Here, ℓ is the azimuthal mode number and n is the radial mode number [K.E. Kreischer and R.J. Temkin, Phys. Rev. Lett. 59, 547 (1987)].

a straight cylinder of length $L = 1.28\,\mathrm{cm}$ and radius $b = 0.75\,\mathrm{cm}$, and is terminated at both ends by linear tapers. The cyclotron maser radiation is transmitted through a waveguide to a "motheye" window, and broadcasts into a shielded box where measurements of the power, frequency and radiation pattern are made.

The waveguide mode excited in the experiments[1] is a $\mathrm{TE}_{\ell n}$ mode with azimuthal electric field $\delta E_\theta^\ell = a_\ell J_\ell'(k_{\ell n}r)$, where $k_{\ell n}$ is the n'th zero of $J_\ell'(k_{\ell n}b) = 0$. The oscillation frequency is determined from $\omega_0^2 = c_\ell^2 k_{\ell n}^2 + c^2 k_{z0}^2$ and $\omega_0 = k_{z0}V_{ze} + \omega_{ce}/\hat{\gamma}_e$, where $\omega_{ce} = eB_0/m_e c$ and $\hat{\gamma}_e = (1 - V_{\perp e}^2/c^2 - V_{ze}^2/c^2)^{-1/2}$. Moreover, in the experiments,[1] $k_{z0} = m\pi/L$ where $m = 1$. For $\beta_{ze}^2 \ll 1$ and operation near cut-off, it follows that $k_{\ell n}^2 \gg k_{z0}^2$ and $\omega_0 \simeq ck_{\ell n} \simeq \omega_{ce}/\hat{\gamma}_e$. For given values of $\hat{\gamma}_e$ and b, a mode represented by $k_{\ell n}$ and oscillating at frequency ω_0 is excited only over a narrow range of B_0. Therefore, by varying the magnetic field B_0, a sequence of discrete modes is excited in the experiments. Moreover, the value of the beam radius r_0 is adjusted so that the value of r_0/b gives strong coupling for a particular $\text{TE}_{\ell n}$ excitation. For example, it is found[1] that positioning the beam radius so that r_0/b is in the interval from 0.69 to 0.72 ensures strong excitation of the $(\ell, n) = (15, 2)$ mode, and suppresses neighboring parasitic modes.

The power output and step tunability of the gyrotron are illustrated in Fig. 7.9 for magnetic field B_0 ranging from 48 kG to 97 kG. The dominant $\text{TE}_{\ell n}$ modes observed in the experiments correspond to radial mode numbers $n = 2$, 3, and 4. The $n = 2$ sequence in Fig. 7.9 corresponds to azimuthal mode numbers ranging from $\ell = 13$ to $\ell = 20$, with a maximum power of 645 kW occurring for the $(\ell, n) = (14, 2)$ mode at 134 GHz. The highest-order mode in Fig. 7.9 corresponds to $(\ell, n) = (22, 4)$, where 470 kW is generated at 243 GHz in a single mode.

As a final point, mode stability and the suppression of nearby competing modes are observed to persist, even for operation in high-order cavity modes.[1] These encouraging results increase the prospects of developing megawatt-level cw gyrotrons in the 200–300 GHz range.

7.5   Free Electron Laser Instability

7.5.1   Introduction

For several decades, physicists have aspired to develop a source of coherent electromagnetic radiation with broad tunability in frequency. Ingenious techniques, such as the dye laser, have relied on active atomic or molecular media, and therefore have been limited in tuning range by the details of atomic structure. Several authors have realized that the constraints associated with atomic structure would not apply to laser sources based on the stimulated radiation generated by energetic free electrons interacting with a spatially periodic transverse magnetic field (so-called "wiggler" magnetic field). As an example of a free electron laser,[9-12] the oscillator configuration used in the classic experiments by Deacon, et al.,[29] is illustrated in Fig. 7.10. Here, a superconducting helix generates a periodic wiggler field (wavelength $\lambda_w = 3.2$ cm, and amplitude $B_w = 2.4$ kG), and a high-quality 43 MeV electron beam propagates along the axis of the helix. A stimulating electromagnetic signal (produced by a CO_2 laser in the experiments[29]) interacts with the beam electrons and is amplified by several orders-of-magnitude through the free electron laser interaction process (Secs. 7.5–7.8). Feedback is provided by the two mirrors shown at the ends of the interaction region in Fig. 7.10.

Since the experiments[29] by Deacon, et al. (see Sec. 7.5.3), the free electron laser interaction process has been used to generate coherent radiation at wavelengths ranging from the visible,[41] to the infrared,[42-44] to the microwave[45-51] region of the electromagnetic spectrum. Indeed, experimental investigations have been very successful over a wide range of electron beam energy (150 keV to 160 MeV) and current (tens of milliamperes to the multi-kiloampere range). Theoretical investigations of the free electron laser instability have included fundamental studies of stability and orbit behavior in field configurations with[75-80] and without[30-37, 81-87] an axial magnetic guide field; investigations of nonlinear effects and saturation mechanisms;[88-93] analyses of novel wiggler configurations, such as the longitudinal wiggler[94-99] and circular (cross-field)[100-102] free electron lasers; investigations of short-wavelength, static and electromagnetic (e.g., gyrotron-powered)[103] wiggler configurations, which reduce the wiggler wavelength λ_w, and therefore the accelerator voltage required for given output laser wavelength $\lambda_z = 2\pi/k_z$; studies of free electron laser radiation generation by a periodic dielectric medium;[104] and the development of advanced accelerator concepts, such as the two-beam accelerator.[105] Experimental investigations of advanced concepts have included measurements[106-109] of optical guiding[110, 111] of the radiation channel by beam dielectric effects, and coherent radiation generation in circular free electron laser configurations[112-114] in which a rotating annular electron beam interacts with a transverse (radial) wiggler field.

In the short space of this chapter, it is not possible to give a thorough delineation of the many technical advances in free electron lasers.[9-12]

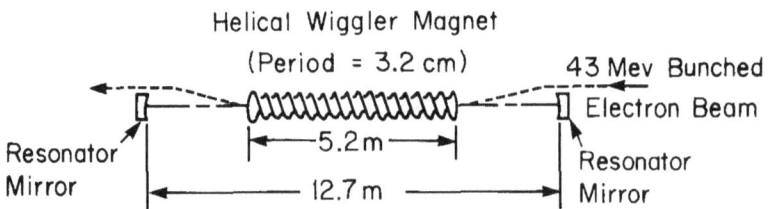

Helical Wiggler Magnet
(Period = 3.2 cm)   43 Mev Bunched
Electron Beam
Resonator Mirror ├──5.2 m──┤ Resonator Mirror
├──────── 12.7 m ────────┤

Figure 7.10. Schematic diagram of a free electron laser oscillator in which a 43 MeV electron beam interacts with a helical wiggler magnetic field ($B_w = 2.4$ kG and $\lambda_w = 3.2$ cm). The wavelength of the stimulated radiation is 3.417 μm [D.A.G. Deacon, et al., Phys. Rev. Lett. **38**, 892 (1977)].

Therefore, particular emphasis in the present treatment is placed on describing fundamental concepts which are important in the understanding and operation of free electron lasers. Following a discussion of the spontaneous emission from a test electron in a helical wiggler magnetic field (Sec. 7.5.2), the Vlasov-Maxwell equations are used to derive the kinetic dispersion relation which describes the stimulated emission of radiation by the free electron laser instability in the exponential-gain regime (Sec. 7.6). Linear growth properties are analyzed in detail for a cold electron beam (Sec. 7.7) and in the warm-beam Compton regime (Sec. 7.8). In this regard, the linear stability analysis in Secs. 7.6–7.8 is carried out for a free electron laser oscillator (temporal growth), but is readily extended to the case of a free electron laser amplifier (spatial growth). The nonlinear evolution of the free electron laser instability is investigated for multimode excitation (Sec. 7.9) in circumstances where the spectrum of (growing) waves is sufficiently broad that a quasilinear description is valid. Then, for the case where a single, coherent electromagnetic wave has evolved to a saturated state, the sideband instability associated with the motion of trapped electrons in the ponderomotive potential is investigated (Sec. 7.10). The analysis in Secs. 7.5–7.10 is carried out for a helical wiggler magnetic field. For the case of a planar wiggler magnetic field, however, there can be harmonic radiation generation due to the axial modulation of the electron orbits. In this regard, linear growth properties for a planar wiggler free electron laser are examined in the Compton-regime approximation (Sec. 7.11). Finally, brief summaries of several experiments are presented which illustrate fundamental aspects of free electron laser physics. These include: free electron laser oscillator operation at 3.417 μm (Sec. 7.5.3); tunable free electron laser operation in the Raman regime (Sec. 7.7.3); experimental evidence of the sideband instability (Sec. 7.10.4); and high-gain free electron laser amplifier experiments in the Compton regime (Sec. 7.12).

7.5.2   Spontaneous Emission from a Test Electron in a Helical Wiggler Magnetic Field

To illustrate the basic frequency-upshift mechanism in a free electron laser, we consider the spontaneous emission of electromagnetic radiation from a relativistic test electron moving with axial momentum $p_z = \gamma m_e v_z$ through a transverse helical wiggler magnetic field

$$\mathbf{B}_w^0(\mathbf{x}) = -B_w \left[\cos(k_0 z)\hat{\mathbf{e}}_x + \sin(k_0 z)\hat{\mathbf{e}}_y\right]. \tag{7.50}$$

Here, B_w = const. is the amplitude of the wiggler field, and λ_w = $2\pi/k_0$ = const. is the wiggler wavelength. Equation (7.50) is a good approximation to the magnetic field near the magnetic axis of a straight cylinder with bifilar helical current windings[115] with constant pitch and current. The vector potential $\mathbf{A}_w^0(\mathbf{x})$ consistent with Eq.(7.50) and $\nabla \times \mathbf{A}_w^0 = \mathbf{B}_w^0$ is given by

$$\mathbf{A}_w^0(\mathbf{x}) = \left(\frac{B_w}{k_0}\right)[\cos(k_0 z)\hat{\mathbf{e}}_x + \sin(k_0 z)\hat{\mathbf{e}}_y] \ . \tag{7.51}$$

For electron motion in the static magnetic field $\mathbf{B}_w^0(\mathbf{x})$, the energy $\gamma = (1 + \mathbf{p}^2/m_e^2 c^2)^{1/2}$ is of course conserved, and the transverse orbits, $x'(t')$ and $y'(t')$, are determined from

$$\gamma m_e \frac{d}{dt'}v_x'(t') = -\frac{e}{c}B_w v_z'(t')\sin[k_0 z'(t')] = \frac{eB_w}{ck_0}\frac{d}{dt'}\cos[k_0 z'(t')] \ , \tag{7.52a}$$

$$\gamma m_e \frac{d}{dt'}v_y'(t') = \frac{e}{c}B_w v_z'(t')\cos[k_0 z'(t')] = \frac{eB_w}{ck_0}\frac{d}{dt'}\sin[k_0 z'(t')] \ , \tag{7.52b}$$

where "primed" notation has been used to denote the time-dependent orbits in the laboratory frame. Assuming zero transverse motion of the test electron in the absence of the wiggler field ($B_w = 0$), the solutions to Eqs.(7.52a) and (7.52b) can be expressed as

$$v_x'(t') = \frac{ca_w}{\gamma}\cos[k_0 z'(t')] \ , \tag{7.53a}$$

$$v_y'(t') = \frac{ca_w}{\gamma}\sin[k_0 z'(t')] \ , \tag{7.53b}$$

where

$$a_w = \frac{eB_w}{m_e c^2 k_0} \tag{7.54}$$

is the normalized amplitude of the wiggler magnetic field. (For example, $a_w = 0.59$ when $B_w = 1\,\text{kG}$ and $\lambda_w = 2\pi\,\text{cm} = 6.28\,\text{cm}$.) Equations (7.53a) and (7.53b) correspond to zero transverse canonical momentum, $P_x \equiv \gamma m_e v_x' - (e/c)A_{xw}^0(z') = 0 = \gamma m_e v_y' - (e/c)A_{yw}^0(z') \equiv P_y$. From Eqs.(7.50) and (7.53) it also follows that

$$\gamma m_e \frac{dv_z'}{dt'} = -\frac{e}{c}\left[v_x' B_{yw}^0(z') - v_y' B_{xw}^0(z')\right] = 0 \ , \tag{7.55}$$

so that $v'_z(t') = v_z = \cos t$. That is, for the constant-amplitude helical wiggler field in Eq.(7.50), the axial motion of the electron is free-streaming with

$$z'(t') = z + v_z(t' - t) ,\qquad (7.56)$$

where $z'(t' = t) = z$ and $dz'/dt' = v_z$. On the other hand, the wiggler field induces an oscillatory modulation of the transverse orbits according to Eq.(7.53). Calculating the electron energy for $p_z = \gamma m_e v_z = \text{const.}$, and $v'_x(t')$ and $v'_y(t')$ specified by Eq.(7.53), it is readily shown that

$$\gamma = \left(1 + \frac{p_z^2}{m_e^2 c^2} + a_w^2\right)^{1/2} ,\qquad (7.57)$$

where the wiggler-induced contribution is incorporated in the (constant) factor a_w^2.

For electromagnetic radiation emitted in the forward direction (along $\hat{e}_z$), it is readily shown from Eqs.(3.40) and (7.56) that the spontaneous emission from an individual test electron is given in the far-field region by

$$\eta(\omega) = \frac{1}{T}\frac{d^2 I}{d\omega d\Omega} \qquad (7.58)$$

$$= \frac{e^2 \omega^2}{4\pi^2 c^3 T}\left|\int_0^T d\tau \left[v'_x(\tau)\hat{e}_x + v'_y(\tau)\hat{e}_y\right]\exp\left(ik_z z - i\omega\tau + ik_z v_z \tau\right)\right|^2 .$$

Here, ω is the frequency and k_z is the wavenumber of the emitted radiation; $d^2 I/d\omega d\Omega$ is the energy radiated per unit frequency interval per unit solid angle;[116,117] $z'(\tau) = z + v_z\tau$ is the free-streaming orbit along $\hat{e}_z$, where $\tau = t' - t$ is the shifted time variable; $T = L/v_z$ is the length of time that the electron is in the interaction region (of length L); and $v'_x(\tau)$ and $v'_y(\tau)$ are the components of the wiggler-induced transverse velocity defined in Eq.(7.53). Substituting Eq.(7.56) into Eq.(7.53), it is evident that the transverse velocity components undergo a temporal modulation proportional to $\exp(\pm i k_0 v_z \tau)$. This modulation beats with the radiating wave contribution in the integrand of Eq.(7.58) to give a τ-dependence proportional to

$$\exp\left\{-i\left[\omega - (k_z \pm k_0)v_z\right]\tau\right\} .\qquad (7.59)$$

Correspondingly, the spontaneous emission spectrum $\eta(\omega)$ is strongly peaked for ω in the vicinity of $\omega = (k_z + k_0)v_z$ (upshifted emission peak) as well as $\omega = (k_z - k_0)v_z$ (downshifted emission peak). For present purposes, we focus on the upshifted portion of the emission spectrum $\eta(\omega)$ peaked near $\omega = (k_z + k_0)v_z$, and neglect the oscillatory contributions in the integrand of Eq.(7.58) proportional to $\exp\{-i[\omega - (k_z - k_0)v_z]\tau\}$. Some straightforward algebra that makes use of Eqs.(7.53), (7.56), and (7.58) gives

$$\eta(\omega) = \frac{a_w^2 e^2 \omega^2 T}{\gamma^2 8\pi^2 c} \frac{\sin^2[(\omega - k_z v_z - k_0 v_z)T/2]}{[(\omega - k_z v_z - k_0 v_z)T/2]^2} . \tag{7.60}$$

As expected, Eq.(7.60) has a strong maximum at $\omega = (k_z + k_0)v_z$, and $\eta(\omega)$ is proportional to $v_{\perp w}^2 = c^2 a_w^2/\gamma^2$ induced by the wiggler field.

Equation (7.60) has been derived classically by calculating the radiation emitted in the forward direction by a charge oscillating in the wiggler field. The emission process can also be interpreted quantum mechanically as stimulated Compton scattering.[38,39] In a frame of reference moving with the axial velocity v_z of the electron, the wiggler magnetic field appears as a backward-traveling electromagnetic wave with frequency $\omega_0' = \gamma_z k_0 v_z$ and wavenumber $k_{z0}' = \gamma_z k_0$, where $\gamma_z = (1 - v_z^2/c^2)^{-1/2}$. On the other hand, the emitted electromagnetic wave (ω, k_z) has frequency $\omega' = \gamma_z(\omega - k_z v_z)$ and wavenumber $k_z' = \gamma_z(k_z - \omega v_z/c^2)$ in the moving frame. Neglecting the recoil energy of the electron in the interaction process, energy conservation in the moving frame is given approximately by

$$\hbar\omega' = \hbar\omega_0' , \tag{7.61}$$

or equivalently $\omega - k_z v_z = k_0 v_z$, which corresponds to the location of the emission maximum in Eq.(7.60).

For a collection of electrons making up an electron beam, the electromagnetic radiation generated by the interaction of the electrons with the wiggler field in Eq.(7.50) can be affected significantly by collective processes (e.g., charge and current bunching of the beam electrons, and the detailed form of the beam distribution function f_e^0). Nonetheless, the essential signature of the emission process evident in Eq.(7.60), namely, radiation generation which is strongly peaked for $\omega \simeq (k_z + k_0)v_z$, remains intact in analyses of the free electron laser instability based on the Vlasov-Maxwell equations including collective effects (Secs. 7.6–7.9). We assume that the beam electrons have narrow velocity spread $(\Delta v_z/V_{ze} \ll 1)$ about an average axial velocity V_{ze}, and that the electron beam is sufficiently tenuous that the (forward-traveling) electromagnetic wave has frequency $\omega \simeq ck_z$. Then the radiation spectrum will be strongly peaked for (upshifted) frequency and wavenumber (ω, k_z) in the vicinity of $(\hat{\omega}, \hat{k}_z)$ satisfying the simultaneous resonance conditions

$$\hat{\omega} = (\hat{k}_z + k_0)V_{ze} \ ,$$

$$\hat{\omega} = c\hat{k}_z \ . \tag{7.62}$$

Solving Eq.(7.62) for $\hat{k}_z$ gives

$$\hat{k}_z = \frac{\beta_{ze}}{(1 - \beta_{ze})}k_0 \ , \tag{7.63}$$

where $\beta_{ze} = V_{ze}/c$ is the normalized axial velocity of the electron beam. From Eq.(7.57), the average electron energy is $\hat{\gamma}_e = (1 + \hat{\gamma}_e^2\beta_{ze}^2 + a_w^2)^{1/2}$, which gives $1 - \beta_{ze}^2 = (1 + a_w^2)/\hat{\gamma}_e^2$. Eliminating $1 - \beta_{ze}$ in Eq.(7.63), we obtain

$$\hat{k}_z = \frac{\hat{\gamma}_e^2(1 + \beta_{ze})\beta_{ze}}{(1 + a_w^2)}k_0 \tag{7.64}$$

for the upshifted peak in the radiation spectrum. For sufficiently large $\hat{\gamma}_e$, it is clear from Eq.(7.64) that $\hat{k}_z$ exceeds k_0 by a large amount, corresponding to electromagnetic wave generation at wavelengths $\hat{\lambda}_z = 2\pi/\hat{k}_z$ much shorter than the wiggler wavelength $\lambda_w = 2\pi/k_0$. For example, if $a_w = 1$, $\lambda_w = 2\pi\,\text{cm} = 6.28\,\text{cm}$, and $\hat{\gamma}_e = 50$, then $\hat{\beta}_{ze} \simeq 1$ and $\hat{\lambda}_z = \lambda_w/2500 = 25\,\mu\text{m}$. Equation (7.64) also illustrates the tunability of the free electron laser interaction process. For example, the wavelength $\hat{\lambda}_z = 2\pi/\hat{k}_z$ of the radiation generated can be varied by changing the beam energy $\hat{\gamma}_e$ and/or the amplitude a_w of the wiggler field.

Problem 7.7   Make use of Eqs.(7.53) and (7.58) to calculate the spontaneous emission $\eta(\omega)$ in the forward direction (along $\hat{e}_z$) from a test electron moving in the helical wiggler magnetic field in Eq.(7.50). Include the emission contributions at both $\omega = (k_z + k_0)v_z$ and $\omega = (k_z - k_0)v_z$. In the vicinity of the upshifted emission peak, show that $\eta(\omega)$ can be approximated by Eq.(7.60).

7.5.3   Free Electron Laser Oscillator Experiments

In the classic free electron oscillator experiments by Deacon, et al.,[29] both the spontaneous emission spectrum and the stimulated free electron laser

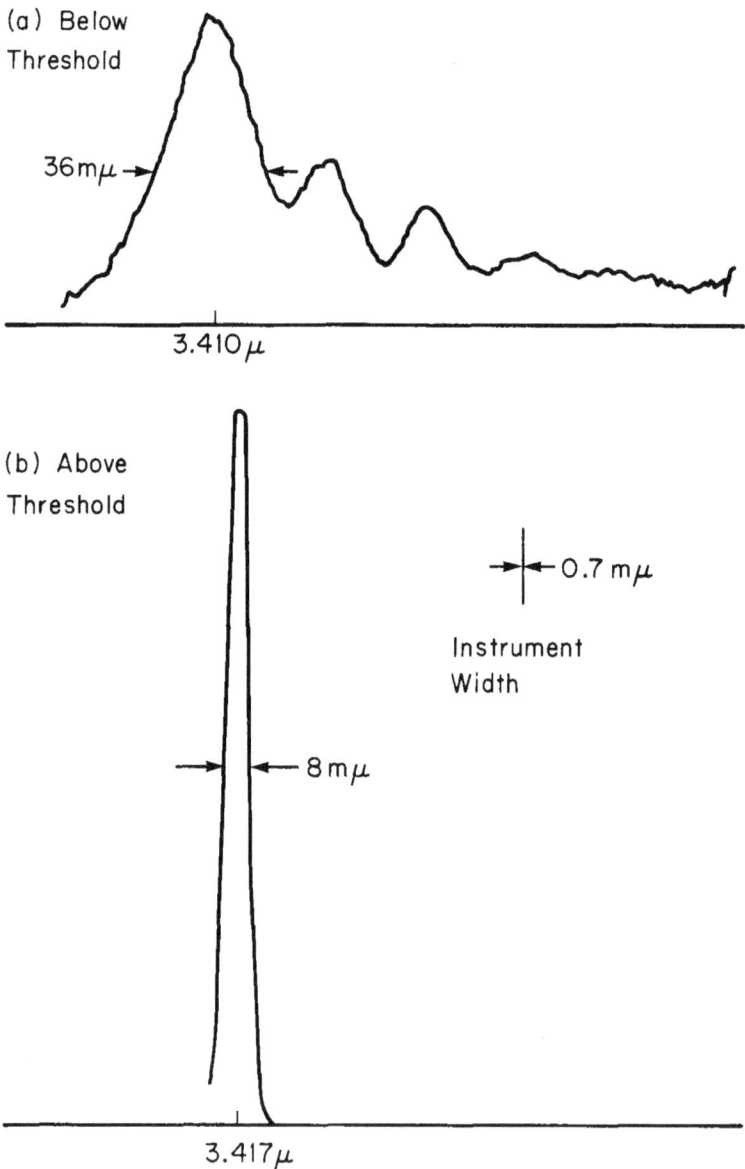

(a) Below
Threshold

36mμ →| ←

3.410 μ

(b) Above
Threshold

→| ←0.7 mμ

Instrument
Width

→| ← 8mμ

3.417μ

Figure 7.11. Measured power spectra for (a) the spontaneous radiation emitted by the electron beam, and (b) the free electron laser radiation above threshold for the experimental configuration in Fig. 7.10. The oscillator power in Fig. 7.11(b) is a factor of 10^8 larger than the spontaneous radiation [D.A.G. Deacon, et al., Phys. Rev. Lett. **38**, 892 (1977)].

radiation were measured (Fig. 7.11). The experimental configuration and the measured emission spectrum are illustrated in Figs. 7.10 and 7.11, respectively. A superconducting helix generates a periodic wiggler field with wavelength $\lambda_w = 2\pi/k_0 = 3.2\,\mathrm{cm}$ and amplitude $B_w = 2.4\,\mathrm{kG}$ (Fig. 7.10). A high-quality electron beam ($\Delta\gamma/\hat{\gamma}_e \simeq 0.05\%$) with kinetic energy $(\hat{\gamma}_e - 1)m_e c^2 = 43.5\,\mathrm{MeV}$ propagates along the axis of the helix, and CO_2 laser radiation[28] which interacts with the beam electrons is amplified through the free electron laser interaction process [Fig. 7.11(b)]. Feedback is provided by the two mirrors shown at the ends of the interaction region in Fig. 7.10. In the absence of the stimulating CO_2 laser signal, the measured spectrum[29] of the spontaneous radiation generated by the low-current electron beam (average current $\simeq 130\,\mu\mathrm{A}$, and peak current $\simeq 2.6\mathrm{A}$) is shown in Fig. 7.11(a). Above threshold, however, the free electron laser oscillator power increases by a factor of 10^8 over the spontaneous radiation, and the stimulated emission is strongly peaked at $3.417\,\mu\mathrm{m}$ as shown in Fig. 7.11(b). This was the first experiment[29] to demonstrate free electron laser oscillator operation in the Compton regime (see Sec. 7.7.5), albeit at modest power levels (peak power $\simeq 7\,\mathrm{kW}$).

7.6 Kinetic Description of the Free Electron Laser Instability

7.6.1 Theoretical Model

In this section, we make use of the Vlasov-Maxwell equations (Sec. 2.2) to develop a kinetic description of the free electron laser instability for a relativistic electron beam propagating in the z-direction through the constant-amplitude helical wiggler magnetic field specified by Eq.(7.50). The beam charge and current densities, $-e\hat{n}_e$ and $-e\hat{n}_e\beta_{ze}c$, are assumed to be sufficiently low that the equilibrium self fields, $\mathbf{E}_0^s(\mathbf{x})$ and $\mathbf{B}_0^s(\mathbf{x})$, can be neglected in describing the linear and nonlinear evolution of the system. Moreover, for short-wavelength perturbations, the effects of finite radial geometry are neglected. That is, we approximate

$$\frac{\partial}{\partial x} = 0 = \frac{\partial}{\partial y}$$

in the Vlasov-Maxwell equations and treat the beam and radiation field as spatially uniform perpendicular to the propagation direction.

Introducing the perturbed potentials, $\delta\phi(z,t)$ and $\delta\mathbf{A}(\mathbf{x},t) = \delta A_x(z,t)\hat{\mathbf{e}}_x$ $+ \delta A_y(z,t)\hat{\mathbf{e}}_y$, the electromagnetic fields can be expressed in the Coulomb gauge as

$$\delta\mathbf{E}(\mathbf{x},t) = -\frac{1}{c}\frac{\partial}{\partial t}\delta A_x(z,t)\hat{\mathbf{e}}_x - \frac{1}{c}\frac{\partial}{\partial t}\delta A_y(z,t)\hat{\mathbf{e}}_y - \frac{\partial}{\partial z}\delta\phi(z,t)\hat{\mathbf{e}}_z \,, \quad (7.65a)$$

$$\delta\mathbf{B}(\mathbf{x},t) = -\frac{\partial}{\partial z}\delta A_y(z,t)\hat{\mathbf{e}}_x + \frac{\partial}{\partial z}\delta A_x(z,t)\hat{\mathbf{e}}_y \,, \quad (7.65b)$$

where $\delta\mathbf{B} = \nabla \times \delta\mathbf{A}$ and $\delta\mathbf{E} = -(1/c)(\partial/\partial t)\delta\mathbf{A} - \nabla\delta\phi$. In the present geometry, there are two exact single-particle constants of the motion in the combined equilibrium and perturbed field configuration. These are the canonical momenta, P_x and P_y, transverse to the beam propagation direction, i.e.,

$$P_x = p_x - \frac{e}{c}A_{xw}^0(z) - \frac{e}{c}\delta A_x(z,t) = \text{const.} \,, \quad (7.66a)$$

$$P_y = p_y - \frac{e}{c}A_{yw}^0(z) - \frac{e}{c}\delta A_y(z,t) = \text{const.} \,, \quad (7.66b)$$

where $A_{xw}^0(z) = (B_w/k_0)\cos(k_0 z)$, $A_{yw}^0(z) = (B_w/k_0)\sin(k_0 z)$ [Eq.(7.51)], and p_x and p_y are the transverse mechanical momenta. The potential perturbations $\delta\phi(z,t)$, $\delta A_x(z,t)$ and $\delta A_y(z,t)$ are determined self-consistently from the Maxwell equations

$$\left(\frac{1}{c^2}\frac{\partial^2}{\partial t^2} - \frac{\partial^2}{\partial z^2}\right)\delta A_x = -\frac{4\pi e}{c}\int d^3p\, v_x \left(f_e - f_e^0\right) \,, \quad (7.67)$$

$$\left(\frac{1}{c^2}\frac{\partial^2}{\partial t^2} - \frac{\partial^2}{\partial z^2}\right)\delta A_y = -\frac{4\pi e}{c}\int d^3p\, v_y \left(f_e - f_e^0\right) \,, \quad (7.68)$$

$$\frac{\partial^2}{\partial z^2}\delta\phi = 4\pi e\int d^3p \left(f_e - f_e^0\right) \,, \quad (7.69)$$

where $f_e^0(z,\mathbf{p})$ is the equilibrium ($\partial/\partial t = 0$) distribution function, and $f_e(z,\mathbf{p},t)$ solves the nonlinear Vlasov equation

$$\left[\frac{\partial}{\partial t} + v_z\frac{\partial}{\partial z} - e\left(\delta\mathbf{E} + \frac{\mathbf{v} \times (\mathbf{B}_w^0 + \delta\mathbf{B})}{c}\right) \cdot \frac{\partial}{\partial\mathbf{p}}\right] f_e(z,\mathbf{p},t) = 0 \,. \quad (7.70)$$

In Eqs.(7.67)–(7.70) the particle velocity $\mathbf{v}$ and momentum $\mathbf{p}$ are related by $m_e\mathbf{v} = \mathbf{p}/(1 + \mathbf{p}^2/m_e^2 c^2)^{1/2}$.

For present purposes, we examine the class of exact solutions to Eq.(7.70) of the form[36,37]

$$f_e(z, \mathbf{p}, t) = \hat{n}_e \delta(P_x) \delta(P_y) F_e(z, p_z, t) , \qquad (7.71)$$

where $\hat{n}_e$ = const. is the average density, and P_x and P_y are the exact constants of the motion defined in Eq.(7.66). Note from Eq.(7.71) that the effective transverse motion of the beam electrons is "cold." Substituting Eq.(7.71) into Eq.(7.70) and making use of Eq.(7.66) we find that $F_e(z, p_z, t)$ evolves according to the one-dimensional nonlinear Vlasov equation[36]

$$\left[\frac{\partial}{\partial t} + v_z \frac{\partial}{\partial z} - \frac{\partial}{\partial z} \hat{H}(z, p_z, t) \frac{\partial}{\partial p_z} \right] F_e(z, p_z, t) = 0 , \qquad (7.72)$$

where $\hat{H}(z, p_z, t)$ is defined by

$$\hat{H}(z, p_z, t) = \gamma_T(z, p_z, t) m_e c^2 - e \delta\phi(z, t) . \qquad (7.73)$$

In Eq.(7.73),

$$\gamma_T(z, p_z, t) = \left[1 + \frac{p_z^2}{m_e^2 c^2} + \frac{e^2}{m_e^2 c^4} \left(A_{xw}^0 + \delta A_x \right)^2 + \frac{e^2}{m_e^2 c^4} \left(A_{yw}^0 + \delta A_y \right)^2 \right]^{1/2} \qquad (7.74)$$

is the particle energy for $P_x = 0 = P_y$. Moreover, substituting Eq.(7.71) into Eqs.(7.67)–(7.69), the potential perturbations $\delta A_x(z, t)$, $\delta A_y(z, t)$ and $\delta\phi(z, t)$ evolve nonlinearly according to[36]

$$\left(\frac{1}{c^2} \frac{\partial^2}{\partial t^2} - \frac{\partial^2}{\partial z^2} \right) \delta A_x = -\frac{\omega_{pe}^2}{c^2} \left[\left(A_{xw}^0 + \delta A_x \right) \int \frac{dp_z}{\gamma_T} F_e - A_{xw}^0 \int \frac{dp_z}{\gamma} F_e^0 \right] , \qquad (7.75)$$

$$\left(\frac{1}{c^2} \frac{\partial^2}{\partial t^2} - \frac{\partial^2}{\partial z^2} \right) \delta A_y = -\frac{\omega_{pe}^2}{c^2} \left[\left(A_{yw}^0 + \delta A_y \right) \int \frac{dp_z}{\gamma_T} F_e - A_{yw}^0 \int \frac{dp_z}{\gamma} F_e^0 \right] , \qquad (7.76)$$

$$\frac{\partial^2}{\partial z^2} \delta\phi = 4\pi e \hat{n}_e \int dp_z \left(F_e - F_e^0 \right) . \qquad (7.77)$$

Here, $\omega_{pe}^2 = 4\pi \hat{n}_e e^2 / m_e$ is the nonrelativistic plasma frequency-squared; $F_e^0(p_z)$ is the equilibrium distribution function for $\delta A_x = 0 = \delta A_y$ and $\partial/\partial t = 0$; $\gamma_T(z, p_z, t)$ is defined in Eq.(7.74); and $F_e(z, p_z, t)$ solves the

nonlinear Vlasov equation (7.72). Moreover, $\gamma = (1 + p_z^2/m_e^2 c^2 + a_w^2)^{1/2}$
is the electron energy in the absence of perturbations ($\delta A_x = 0 = \delta A_y$),
where use has been made of $(e^2/m_e^2 c^4)(A_{xw}^{02} + A_{yw}^{02}) = e^2 B_w^2/m_e^2 c^4 k_0^2 \equiv a_w^2$.
Within the context of the present model, Eqs.(7.72) and (7.75)–(7.77)
describe the exact nonlinear evolution of the system for perturbations
about the general beam equilibrium $F_e^0(p_z)$. In this regard, note from
Eqs.(7.72) and (7.73) that the axial force F_z on an electron in the phase
space (z, p_z) is given by

$$F_z = -\frac{\partial}{\partial z}\hat{H} \tag{7.78}$$

$$= -\frac{e^2}{2\gamma_T m_e c^2}\frac{\partial}{\partial z}\left[2A_{xw}^0\delta A_x + 2A_{yw}^0\delta A_y + (\delta A_x)^2 + (\delta A_y)^2\right] + e\frac{\partial}{\partial z}\delta\phi \ .$$

That is, the effective ponderomotive potential is proportional to $2A_{xw}^0\delta A_x + 2A_{yw}^0\delta A_y + (\delta A_x)^2 + (\delta A_y)^2$.

Problem 7.8   For the class of exact solutions $f_e(z, \mathbf{p}, t)$ in Eq.(7.71) to the
nonlinear Vlasov equation, integrate Eq.(7.70) over the transverse mechanical momenta p_x and p_y and show that the resulting (one-dimensional) Vlasov
equation for $F_e(z, p_z, t)$ is given by Eq.(7.72). Also show that the Maxwell
equations (7.67)–(7.69) reduce to Eqs.(7.75)–(7.77) for the class of distribution
functions in Eq.(7.71).

7.6.2   Kinetic Dispersion Relation

To determine the linear growth properties of the free electron laser instability, we analyze the Vlasov-Maxwell equations (7.72) and (7.75)–(7.77),
retaining only linear terms in the perturbations $\delta A_x(z, t)$, $\delta A_y(z, t)$,
$\delta\phi(z, t)$ and $\delta F_e(z, p_z, t) = F_e(z, p_z, t) - F_e^0(p_z)$. In this regard, it is
convenient to introduce the dimensionless potentials $\delta\Phi(z, t)$, $\delta A_\pm(z, t)$
and $A_w^\pm(z)$ defined by

$$\delta\Phi = \frac{e}{m_e c^2}\delta\phi \ , \tag{7.79a}$$

$$\delta A_\pm = \frac{e}{m_e c^2}(\delta A_x \pm i\delta A_y) \ , \tag{7.79b}$$

$$A_w^\pm = \frac{e}{m_e c^2}(A_{xw}^0 \pm iA_{yw}^0) = a_w \exp(\pm ik_0 z) \ , \tag{7.79c}$$

where $a_w = eB_w/m_e c^2 k_0$. From Eqs.(7.74) and (7.78), the perturbed force δF_z and the inverse relativistic mass factor γ_T^{-1} can be approximated by

$$\delta F_z = -\frac{m_e c^2 a_w}{2\gamma_T} \frac{\partial}{\partial z} \left[\exp(ik_0 z)\delta A_- + \exp(-ik_0 z)\delta A_+\right] + m_e c^2 \frac{\partial}{\partial z}\delta\Phi ,$$
(7.80)

and

$$\frac{1}{\gamma_T} = \frac{1}{\gamma} - \frac{a_w}{2\gamma^3} \left[\exp(ik_0 z)\delta A_- + \exp(-ik_0 z)\delta A_+\right] , \qquad (7.81)$$

where $\gamma = (1 + p_z^2/m_e^2 c^2 + a_w^2)^{1/2}$. Making use of Eqs.(7.80) and (7.81) in Eqs.(7.70) and (7.75)–(7.77), the linearized Vlasov-Maxwell equations can be expressed as[36]

$$\left(\frac{\partial}{\partial t} + v_z \frac{\partial}{\partial z}\right)\delta F_e = m_e c^2 \left\{\frac{a_w}{2\gamma}\frac{\partial}{\partial z}\left[\exp(ik_0 z)\delta A_- + \exp(-ik_0 z)\delta A_+\right]\right.$$
$$\left. - \frac{\partial}{\partial z}\delta\Phi\right\}\frac{\partial}{\partial p_z}F_e^0(p_z) , \qquad (7.82)$$

$$\frac{\partial^2}{\partial z^2}\delta\Phi = \frac{\omega_{pe}^2}{c^2}\int dp_z\, \delta F_e , \qquad (7.83)$$

$$\left(c^2\frac{\partial^2}{\partial z^2} - \frac{\partial^2}{\partial t^2} - \omega_{pe}^2\int\frac{dp_z}{\gamma}F_e^0\right)\delta A_+ = \omega_{pe}^2\left\{a_w \exp(ik_0 z)\int\frac{dp_z}{\gamma}\delta F_e\right.$$
$$\left. - \frac{1}{2}a_w^2\exp(ik_0 z)\left[\exp(ik_0 z)\delta A_- + \exp(-ik_0 z)\delta A_+\right]\int\frac{dp_z}{\gamma^3}F_e^0\right\}, (7.84)$$

and

$$\left(c^2\frac{\partial^2}{\partial z^2} - \frac{\partial^2}{\partial t^2} - \omega_{pe}^2\int\frac{dp_z}{\gamma}F_e^0\right)\delta A_- = \omega_{pe}^2\left\{a_w \exp(-ik_0 z)\int\frac{dp_z}{\gamma}\delta F_e\right.$$
$$\left. - \frac{1}{2}a_w^2\exp(-ik_0 z)\left[\exp(ik_0 z)\delta A_- + \exp(-ik_0 z)\delta A_+\right]\int\frac{dp_z}{\gamma^3}F_e^0\right\}. (7.85)$$

In the limit of zero wiggler amplitude ($a_w \to 0$), Eqs.(7.82)–(7.85) give the usual uncoupled electromagnetic and electrostatic dispersion relations for perturbations about a one-dimensional, relativistic electron plasma. On the other hand, for $a_w \neq 0$, the wiggler-induced currents in Eqs.(7.84) and (7.85) can lead to the free electron laser instability, depending on the detailed form of the equilibrium distribution function $F_e^0(p_z)$. Moreover, the longitudinal ($\delta\Phi$) and transverse ($\delta A_\pm$) perturbations in

Eqs.(7.82)–(7.85) are inexorably coupled through the wiggler field, and only in the tenuous-beam limit (sufficiently small ω_{pe}/ck_0) do the longitudinal and transverse modes decouple (Compton-regime approximation discussed later in Sec. 7.6.2).

For the case of temporal growth, we examine solutions to Eqs.(7.82)–(7.85) of the form

$$\delta A_-(z,t) = \sum_{k_z=-\infty}^{\infty} \delta A_-(k_z) \exp(ik_z z - i\omega t) , \qquad (7.86)$$

etc., where $\mathrm{Im}\,\omega > 0$ corresponds to instability. Here, $k_z = 2\pi n/L$, where n is an integer, and L is the fundamental periodicity length in the z-direction. Inspection of Eqs.(7.82)–(7.85) shows that the Fourier amplitude $\delta A_-(k_z)$ couples directly to the amplitudes $\delta F_e(k_z + k_0, p_z)$, $\delta\Phi(k_z + k_0)$ and $\delta A_+(k_z + 2k_0)$. Solving Eq.(7.82) readily gives for the amplitude of the perturbed distribution function

$$\delta F_e(k_z + k_0, p_z) = m_e c^2 \frac{(k_z + k_0)\partial F_e^0/\partial p_z}{[\omega - (k_z + k_0)v_z]} \qquad (7.87)$$

$$\times \left\{ \delta\Phi(k_z + k_0) - \frac{a_w}{2\gamma}[\delta A_-(k_z) + \delta A_+(k_z + 2k_0)] \right\} ,$$

where $v_z = p_z/\gamma m_e$ and

$$\gamma = \left(1 + \frac{p_z^2}{m_e^2 c^2} + a_w^2 \right)^{1/2} .$$

From Eqs.(7.83)–(7.85) for the field perturbations, it is evident that integrals of the form $\omega_{pe}^2 \int dp_z\, \delta F_e$ and $\omega_{pe}^2 \int dp_z\, \delta F_e/\gamma$ are required. Therefore, comparing with Eq.(7.87), it is convenient to introduce the effective susceptibility coefficients $\chi^{(n)}(k_z + k_0, \omega)$ defined by

$$\chi^{(n)}(k_z + k_0, \omega) = m_e c^2 \omega_{pe}^2 \int \frac{dp_z}{\gamma^n} \frac{(k_z + k_0)\partial F_e^0/\partial p_z}{\omega - (k_z + k_0)v_z} , \qquad (7.88)$$

where $n = 0, 1, 2$, and $\mathrm{Im}\,\omega > 0$ is assumed. For future reference, it is also convenient to introduce the longitudinal and transverse dielectric functions defined by

$$D^L(k_z + k_0, \omega) = 1 + \frac{m_e \omega_{pe}^2}{(k_z + k_0)^2} \int dp_z \frac{(k_z + k_0)\partial F_e^0/\partial p_z}{\omega - (k_z + k_0)v_z} \,, \quad (7.89)$$

$$D^T(k_z, \omega) = \omega^2 - c^2 k_z^2 - \omega_{pe}^2 \int \frac{dp_z}{\gamma} F_e^0(p_z) \,, \quad (7.90)$$

$$D^T(k_z + 2k_0, \omega) = \omega^2 - c^2(k_z + 2k_0)^2 - \omega_{pe}^2 \int \frac{dp_z}{\gamma} F_e^0(p_z) \,, \quad (7.91)$$

We substitute Eqs.(7.86) and (7.87) into the field equations (7.83)–(7.85). Some straightforward algebra gives the coupled equations[36]

$$\frac{1}{2} a_w^2 \left[\alpha_3 \omega_{pe}^2 + \chi^{(2)}(k_z + k_0, \omega)\right] \delta A_-(k_z)$$

$$+ \left\{ D^T(k_z + 2k_0, \omega) + \frac{1}{2} a_w^2 \left[\alpha_3 \omega_{pe}^2 + \chi^{(2)}(k_z + k_0, \omega)\right]\right\} \delta A_+(k_z + 2k_0)$$

$$- a_w \chi^{(1)}(k_z + k_0, \omega)\delta\Phi(k_z + k_0) = 0 \,, \quad (7.92)$$

$$\left\{ D^T(k_z, \omega) + \frac{1}{2} a_w^2 \left[\alpha_3 \omega_{pe}^2 + \chi^{(2)}(k_z + k_0, \omega)\right]\right\} \delta A_-(k_z)$$

$$+ \frac{1}{2} a_w^2 \left[\alpha_3 \omega_{pe}^2 + \chi^{(2)}(k_z + k_0, \omega)\right] \delta A_+(k_z + 2k_0)$$

$$- a_w \chi^{(1)}(k_z + k_0, \omega)\delta\Phi(k_z + k_0) = 0 \,, \quad (7.93)$$

$$- a_w \chi^{(1)}(k_z + k_0, \omega)\delta A_-(k_z) - a_w \chi^{(1)}(k_z + k_0, \omega)\delta A_+(k_z + 2k_0)$$

$$+ 2c^2(k_z + k_0)^2 D^L(k_z + k_0, \omega)\delta\Phi(k_z + k_0) = 0 \,, \quad (7.94)$$

where $\alpha_3 \equiv \int dp_z \, F_e^0(p_z)/\gamma^3$. Here $\chi^{(2)}(k_z + k_0, \omega)$, $\chi^{(1)}(k_z + k_0, \omega)$, $D^T(k_z, \omega)$, $D^T(k_z + 2k_0, \omega)$, and $D^L(k_z + k_0, \omega)$ are defined in Eqs.(7.88)–(7.91). The condition for a nontrivial solution to Eqs.(7.92)–(7.94) is that the three-by-three determinant of the coefficients of the amplitudes

$\delta A_-(k_z)$, $\delta A_+(k_z + 2k_0)$ and $\delta\Phi(k_z + k_0)$ vanish. This gives the dispersion relation[36]

$$c^2(k_z + k_0)^2 D^L(k_z + k_0, \omega) D^T(k_z, \omega) D^T(k_z + 2k_0, \omega)$$

$$= \frac{1}{2} a_w^2 \left[D^T(k_z, \omega) + D^T(k_z + 2k_0, \omega) \right] \left\{ \left[\chi^{(1)}(k_z + k_0, \omega) \right]^2 \right.$$

$$\left. - c^2(k_z + k_0)^2 D^L(k_z + k_0, \omega) \left[\alpha_3 \omega_{pe}^2 + \chi^{(2)}(k_z + k_0, \omega) \right] \right\} \ . \quad (7.95)$$

The dispersion relation (7.95) is an exact consequence of the linearized Vlasov-Maxwell equations (7.82)–(7.85), and can be used to determine the detailed stability behavior for perturbations about general equilibrium distribution function $F_e^0(p_z)$. Although equilibrium self fields have been neglected in the derivation of Eq.(7.95), no *a priori* restriction has been made to small wiggler amplitude (as measured by a_w) or to low beam density (as measured by $\omega_{pe}^2/c^2 k_0^2$). Because Eq.(7.95) contains no approximation apart from the linearized approximation, we refer to Eq.(7.95) as the full dispersion relation (FDR). For $a_w = 0$, note that the right-hand side of Eq.(7.95) is equal to zero, and the longitudinal and transverse modes in Eq.(7.95) are completely decoupled. For $a_w \neq 0$, however, the longitudinal and transverse polarizations are coupled through the wiggler field. Some straightforward algebra that makes use of Eqs.(7.92)–(7.94) and the dispersion relation (7.95) gives the relative amplitudes

$$\frac{\delta\Phi(k_z + k_0)}{\delta A_-(k_z)} = \frac{a_w}{2c^2(k_z + k_0)^2} \quad (7.96)$$

$$\times \chi^{(1)}(k_z + k_0, \omega) \frac{[D^T(k_z + 2k_0, \omega) + D^T(k_z, \omega)]}{D^L(k_z + k_0, \omega) D^T(k_z + 2k_0, \omega)} \ ,$$

$$\frac{\delta A_+(k_z + 2k_0)}{\delta A_-(k_z)} = \frac{D^T(k_z, \omega)}{D^T(k_z + 2k_0, \omega)} \ , \quad (7.97)$$

where ω and k_z are related by Eq.(7.95). Because $\chi^{(1)}(k_z + k_0, \omega)$ is proportional to ω_{pe}^2 [Eq.(7.88)], the longitudinal excitation $\delta\Phi(k_z + k_0)$ in Eq.(7.96) is expected to be negligibly small for sufficiently low beam density.

For sufficiently low beam density, it is often customary to approximate $\delta\Phi = 0$ at the outset in the analysis of the linearized Vlasov-Maxwell equations (7.82)–(7.85). This is the so-called *Compton-regime* approximation.

The resulting dispersion relation is obtained by setting the two-by-two determinant of the coefficients of $\delta A_-(k_z)$ and $\delta A_+(k_z + 2k_0)$ in Eqs.(7.92) and (7.93) equal to zero, which gives

$$D^T(k_z, \omega)D^T(k_z + 2k_0, \omega) \qquad (7.98)$$

$$= -\frac{1}{2}a_w^2 \left[D^T(k_z, \omega) + D^T(k_z + 2k_0, \omega) \right] \left[\alpha_3 \omega_{pe}^2 + \chi^{(2)}(k_z + k_0, \omega) \right] .$$

We refer to Eq.(7.98) as the Compton-regime dispersion relation (CDR). Equation (7.98) also follows directly from the full dispersion relation whenever the $[\chi^{(1)}]^2$ contribution can be neglected on the right-hand side of Eq.(7.95). Strictly speaking, the range of validity of the Compton-regime dispersion relation (7.98) depends on the values of a_w, $\omega_{pe}^2/c^2k_0^2$, and $\hat{\gamma}_e$. Indeed, the range of validity of Eq.(7.98) can be ascertained only by a detailed comparison of the stability results with the full dispersion relation (7.95), which makes no *a priori* approximation that the longitudinal potential $\delta\Phi$ is negligibly small.

Before proceeding with an analysis of Eqs.(7.95) and (7.98), we reiterate that the kinetic dispersion relation has been derived for general choice of beam distribution function $F_e^0(p_z)$. Although the derivation in Secs. 7.6.1 and 7.6.2 has assumed a helical wiggler field [Eq.(7.50)] and neglected transverse spatial variations $(\partial/\partial x = 0 = \partial/\partial y)$, the kinetic treatment of the free electron laser instability can be extended to the case of a planar magnetic wiggler (Sec. 7.11),[118,119] and to include the effects of finite radial geometry for both annular[82] and solid[83] electron beams.

Problem 7.9   Consider the linearized Vlasov-Maxwell equations (7.82)–(7.85) for perturbations about the equilibrium distribution function $F_e^0(p_z)$.

a. Express perturbed quantities in the form shown in Eq.(7.86) where $\text{Im}\,\omega > 0$. Derive the coupled equations (7.92)–(7.94) relating the amplitudes $\delta A_-(k_z)$, $\delta A_+(k_z + 2k_0)$ and $\delta\Phi(k_z + k_0)$.

b. Set the three-by-three determinant of the dispersion matrix represented by Eqs.(7.92)–(7.94) equal to zero and show that the resulting dispersion relation for the free electron laser instability is given (exactly) by Eq.(7.95).

c. Make use of Eqs.(7.92)–(7.95) to derive the polarization conditions relating the wave amplitudes in Eqs.(7.96) and (7.97).

Problem 7.10   Repeat the calculation in Problem 7.9 making the *Compton-regime* approximation $(\delta\Phi \simeq 0)$ *ab initio*. Show that the resulting dispersion relation for the free electron laser instability is given by Eq.(7.98).

7.7   Cold-Beam Stability Properties

7.7.1   Linear Dispersion Relation

A cold electron beam with negligibly small momentum spread ($\Delta p_z \simeq 0$) has the largest free energy to drive the free electron laser instability. Therefore, as a reference case corresponding to strong instability, we first consider a cold electron beam with distribution function

$$F_e^0(p_z) = \delta(p_z - \hat{\gamma}_e m_e V_{ze}) \ . \tag{7.99}$$

Here, $V_{ze} = $ const. is the axial velocity of the beam electrons, the normalization of $F_e^0(p_z)$ is $\int dp_z \, F_e^0(p_z) = 1$, and

$$\hat{\gamma}_e = \left(1 + \frac{\hat{\gamma}_e^2 V_{ze}^2}{c^2} + a_w^2 \right)^{1/2}$$

is the total electron energy. Defining $\beta_{ze} = V_{ze}/c$ and $\gamma_z = (1 - \beta_{ze}^2)^{-1/2}$, it follows that $\hat{\gamma}_e$ and γ_z are related by

$$\hat{\gamma}_e^2 = \gamma_z^2(1 + a_w^2) \ . \tag{7.100}$$

To simplify the expressions for $\chi^{(n)}(k_z + k_0, \omega)$ occurring in the dispersion relation (7.95), we make use of $(\partial/\partial p_z)(p_z/\gamma m_e) = (\gamma m_e)^{-1} \times (1 - v_z^2/c^2)$ and $(\partial/\partial p_z)(1/\gamma^n) = -(n/\gamma^{n+1})(v_z/m_e c^2)$, where $v_z = p_z/\gamma m_e$ and $\gamma = (1 + p_z^2/m_e^2 c^2 + a_w^2)^{1/2}$. Substituting Eq.(7.99) into Eq.(7.88) then gives the cold-beam susceptibilities

$$\chi^{(0)}(k_z + k_0, \omega) = -\frac{\omega_{pe}^2}{\hat{\gamma}_e \gamma_z^2} \frac{c^2(k_z + k_0)^2}{[\omega - (k_z + k_0)V_{ze}]^2} \ , \tag{7.101}$$

$$\chi^{(1)}(k_z + k_0, \omega) = \frac{\omega_{pe}^2}{\hat{\gamma}_e^2} \frac{[\omega(k_z + k_0)V_{ze} - c^2(k_z + k_0)^2]}{[\omega - (k_z + k_0)V_{ze}]^2} \ , \tag{7.102}$$

$$\chi^{(2)}(k_z + k_0, \omega) = \frac{\omega_{pe}^2}{\hat{\gamma}_e^3} \frac{[2\omega(k_z + k_0)V_{ze} - c^2(k_z + k_0)^2(1 + \beta_{ze}^2)]}{[\omega - (k_z + k_0)V_{ze}]^2} \ . \tag{7.103}$$

Moreover, $\int dp_z \, F_e^0/\gamma = 1/\hat{\gamma}_e$ and $\alpha_3 = \int dp_z \, F_e^0/\gamma^3 = 1/\hat{\gamma}_e^3$ for the choice of distribution function in Eq.(7.99). Substituting Eqs.(7.101)–(7.103) into Eq.(7.95) and combining terms, some algebraic manipulation gives the full dispersion relation (FDR)

$$\left\{ [\omega - (k_z + k_0) V_{ze}]^2 - \frac{\omega_{pe}^2}{\hat{\gamma}_e \gamma_z^2} \right\} \left[\omega^2 - c^2 k_z^2 - \frac{\omega_{pe}^2}{\hat{\gamma}_e} \right] \left[\omega^2 - c^2 (k_z + 2k_0)^2 - \frac{\omega_{pe}^2}{\hat{\gamma}_e} \right]$$

$$= -\frac{a_w^2}{\hat{\gamma}_e^3} \omega_{pe}^2 \left[\omega^2 - c^2 (k_z + k_0)^2 - \frac{\omega_{pe}^2}{\hat{\gamma}_e} \right] \left[\omega^2 - c^2 (k_z + k_0)^2 - c^2 k_0^2 - \frac{\omega_{pe}^2}{\hat{\gamma}_e} \right]$$

$$(7.104)$$

for a cold electron beam with equilibrium distribution function specified by Eq.(7.99).[85]

Similarly, substituting Eq.(7.103) into Eq.(7.98), we obtain the Compton-regime disperson relation (CDR)

$$\left[\omega - (k_z + k_0) V_{ze} \right]^2 \left[\omega^2 - c^2 k_z^2 - \frac{\omega_{pe}^2}{\hat{\gamma}_e} \right] \left[\omega^2 - c^2 (k_z + 2k_0)^2 - \frac{\omega_{pe}^2}{\hat{\gamma}_e} \right]$$

$$= -\frac{a_w^2}{\hat{\gamma}_e^3} \omega_{pe}^2 \left[\omega^2 - c^2 (k_z + k_0)^2 \right] \left[\omega^2 - c^2 (k_z + k_0)^2 - c^2 k_0^2 - \frac{\omega_{pe}^2}{\hat{\gamma}_e} \right] \quad (7.105)$$

for a cold electron beam.[85] Comparing the various factors in Eqs.(7.104) and (7.105), it is evident that the inequalities

$$\frac{\omega_{pe}^2}{\hat{\gamma}_e \gamma_z^2} \ll |\omega - (k_z + k_0) V_{ze}|^2 \,,$$

$$\frac{\omega_{pe}^2}{\hat{\gamma}_e} \ll |\omega^2 - c^2 (k_z + k_0)^2| \,, \quad (7.106)$$

are required for validity of the Compton-regime dispersion relation (7.105). Whenever Eq.(7.106) is satisfied, the full dispersion relation (7.104) can be approximated by Eq.(7.105).

Davies, et al.,[85] have carried out a thorough analytical and numerical investigation of the cold-beam dispersion relations (7.104) and (7.105). For present purposes, several important features of the stability behavior are summarized in Secs. 7.7.2–7.7.6.

Problem 7.11   Consider a cold electron beam with equilibrium distribution function $F_e^0(p_z)$ specified by Eq.(7.100).

a. Make use of Eqs.(7.88) and (7.100) to show that the cold-beam susceptibilities are given by Eqs.(7.101)–(7.103).

b. Show that the full dispersion relation (7.95) can be expressed in the form
given in Eq.(7.104) for a cold electron beam.

c. Show that the Compton-regime dispersion relation (7.98) can be expressed
in the form given in Eq.(7.105) for a cold electron beam.

7.7.2   Conditions for Free Electron Laser Interaction

Consider first the full dispersion relation (7.104) for nonzero wiggler
amplitude ($a_w \neq 0$). The free electron laser instability occurs when
the negative-energy longitudinal space-charge wave $\omega_L = (k_z + k_0)V_{ze} -
\omega_{pe}/\hat{\gamma}_e^{1/2}\gamma_z$ on the left-hand side of Eq.(7.104) couples with the transverse
electromagnetic wave $\omega_T = (c^2 k_z^2 + \omega_{pe}^2/\hat{\gamma}_e)^{1/2}$ on the left-hand side
of Eq.(7.104) through the wiggler magnetic field a_w on the right-hand
side of Eq.(7.104). This interaction is the strongest for frequency and
wavenumber (ω, k_z) in the vicinity of $(\hat{\omega}, \hat{k}_z)$ satisfying the simultaneous
resonance conditions (Fig. 7.12)

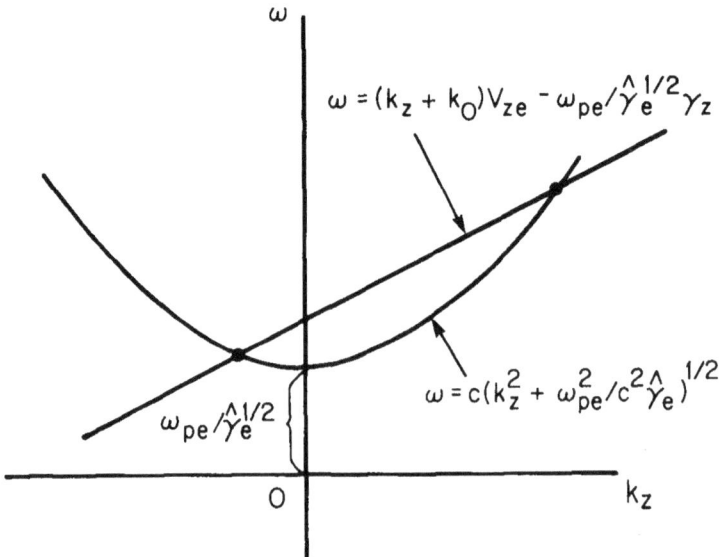

Figure 7.12. Plots illustrating the intersection of the two curves
$\omega = (k_z + k_0)V_{ze} - \omega_{pe}/\hat{\gamma}_e^{1/2}\gamma_z$ and $\omega = c(k_z^2 + \omega_{pe}^2/c^2\hat{\gamma}_e)^{1/2}$. For intersection to occur, $2\omega_{pe}/ck_0 < \beta_{ze}\hat{\gamma}_e^{1/2}\gamma_z$ is required.

$$\hat{\omega} = (\hat{k}_z + k_0)V_{ze} - \frac{\omega_{pe}}{\hat{\gamma}_e^{1/2}\gamma_z} ,$$

$$\hat{\omega} = \left(c^2\hat{k}_z^2 + \frac{\omega_{pe}^2}{\hat{\gamma}_e} \right)^{1/2} . \tag{7.107}$$

We denote the upshifted and downshifted wavenumbers satisfying Eq.(7.107) by $\hat{k}_z^+$ and $\hat{k}_z^-$, respectively. Some straightforward algebra that makes use of $1/\gamma_z^2 = 1 - \beta_{ze}^2$ gives

$$\hat{k}_z^{\pm} = \frac{k_0}{(1-\beta_{ze}^2)}\left[\beta_{ze}^2 - \beta_{ze}\frac{\omega_{pe}}{\hat{\gamma}_e^{1/2}\gamma_z ck_0} \pm \left(\beta_{ze}^2 - 2\beta_{ze}\frac{\omega_{pe}}{\hat{\gamma}_e^{1/2}\gamma_z ck_0} \right)^{1/2} \right] . \tag{7.108}$$

For a tenuous electron beam with $\omega_{pe}/ck_0 \ll \hat{\gamma}_e^{1/2}\gamma_z\beta_{ze}$, the upshifted wavenumber in Eq.(7.108) reduces to $\hat{k}_z^+ = k_0\beta_{ze}/(1-\beta_{ze})$, which is identical to the estimate in Eq.(7.63), whereas the downshifted wavenumber reduces to $\hat{k}_z^- = -k_0\beta_{ze}/(1+\beta_{ze})$. Note that intersection of the two curves in Eq.(7.107) and Fig. 7.12 occurs only if $\omega_{pe}/ck_0 < \beta_{ze}\hat{\gamma}_e^{1/2}\gamma_z/2$. When this inequality is satisfied, the free electron laser growth rate curve ($\text{Im}\,\omega$ versus k_z) calculated from Eq.(7.104) is found[85] to have two distinct maxima at the downshifted wavenumber $\hat{k}_z^-$ and the upshifted wave-number $\hat{k}_z^+$. As ω_{pe}/ck_0 is increased to the value $\omega_{pe}/ck_0 = \beta_{ze}\hat{\gamma}_e^{1/2}\gamma_z/2$, the two wavenumbers $\hat{k}_z^+$ and $\hat{k}_z^-$ coalesce in Eq.(7.108), and for $\omega_{pe}/ck_0 > (\beta_{ze}/2)\hat{\gamma}_e^{1/2}\gamma_z$ the dispersion relation (7.104) gives a growth rate curve for $\text{Im}\,\omega$ which has a single maximum as a function of k_z.

7.7.3   Tunable Free Electron Laser Experiments

Before proceeding with an analysis of the full dispersion relation (7.104) for a cold electron beam, it is useful to summarize briefly the experimental measurements of Fajans, et al.,[47] for a helical wiggler free electron laser operating in the Raman regime. In the experiments,[47] narrow-band emission ($\Delta\omega/\hat{\omega} \approx 2\%$) is measured, from a tunable ($7\,\text{GHz} < \hat{\omega}/2\pi < 18\,\text{GHz}$) free electron laser operating in a TE_{11} waveguide mode at power levels up to $100\,\text{kW}$ and efficiencies of radiation generation up to 12%. Although the experiment operates in the Raman regime where longitudinal perturbations play an important role ($\delta\phi \neq 0$), the waveguide cut-off frequency ($f_c \equiv ck_\perp/2\pi$) satisfies $c^2k_\perp^2 \gg \omega_{pe}^2/\hat{\gamma}_e$ so that the resonance conditions for free electron laser interaction can be approximated by $\hat{\omega} = (\hat{k}_z + k_0)V_{ze}$

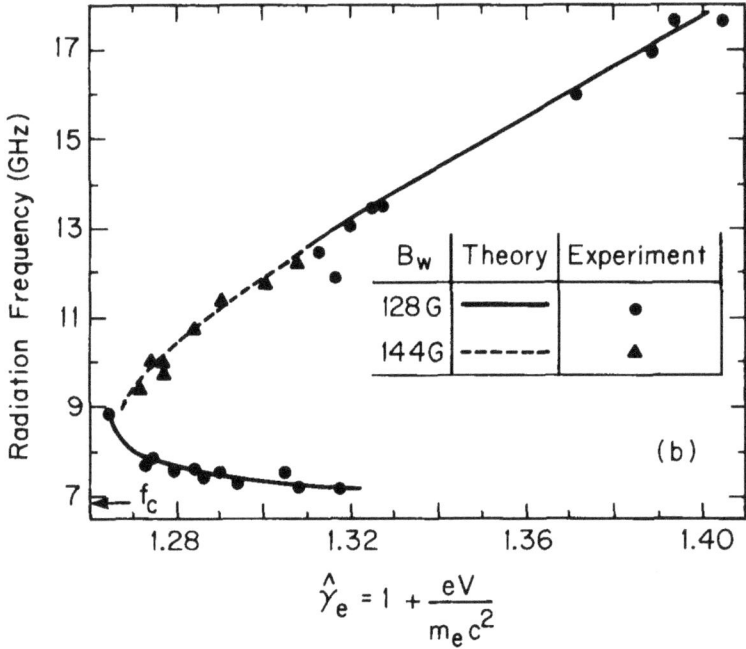

Figure 7.13. Experimental demonstration of free electron laser tunability for radiation frequency ranging from 7 GHz to 18 GHz and beam kinetic energy $(\hat{\gamma}_e - 1)m_e c^2 = eV$ ranging from 135 keV to 200 keV. The solid and dashed curves correspond to the theoretical prediction in Eq.(7.109) [J. Fajans, G. Bekefi, Y.Z. Yin and B. Lax, Phys. Rev. Lett. **53**, 246 (1984)].

and $\hat{\omega} = c(\hat{k}_z^2 + k_\perp^2)^{1/2}$, where the perpendicular wavenumber $k_\perp$ is determined from the radial mode structure. This gives for the upshifted $(\hat{\omega}^+)$ and downshifted $(\hat{\omega}^-)$ frequencies

$$\hat{\omega}^\pm = \gamma_z^2 \beta_{ze} c k_0 \left\{ 1 \pm \beta_{ze} \left[1 - \left(\frac{k_\perp}{\gamma_z \beta_{ze} k_0} \right)^2 \right]^{1/2} \right\} , \qquad (7.109)$$

where $\gamma_z = (1 - \beta_{ze}^2)^{-1/2}$. In the experiments,[47] the beam current is in the range of 5A, and a uniform axial magnetic field $B_0 \hat{e}_z$ $(0.7 \, \text{kG} < B_0 < 7 \, \text{kG})$ is used to guide the electron beam. Moreover, the wiggler wavelength $\lambda_w = 3.3 \, \text{cm}$, and the amplitude of the wiggler field can be varied up to $B_w = 1.5 \, \text{kG}$. Away from cyclotron resonance $(\omega_{ce} = eB_0/m_e c \neq \hat{\gamma}_e \beta_{ze} c k_0)$, the perpendicular velocity induced by the wiggler magnetic field is given by $V_{\perp e} = c|a_w \beta_{ze} c k_0/(\hat{\gamma}_e \beta_{ze} c k_0 - \omega_{ce})|$, where

$V_{1e} = c(1 - \beta_{ze}^2 - 1/\hat{\gamma}_e^2)^{1/2}$. Without presenting experimental details,[47] Fajans, et al., have investigated free electron laser tunability for radiation frequencies ranging from 7 GHz to 18 GHz, and beam kinetic energy $(\hat{\gamma}_e - 1)m_e c^2 = eV$ ranging from 135 keV to 200 keV (here, V is the accelerator voltage). The experimental results[47] are presented in Fig. 7.13 for the case in which a low-power ($\sim$ 100 mW) monochromatic electromagnetic signal of known frequency $f = \omega/2\pi$ is injected into the helical wiggler field, thereby stimulating the free electron laser emission process. The curves in Fig. 7.13 correspond to the theoretical prediction in Eq.(7.109) for $B_w = 128$ G (solid curve) and $B_w = 144$ G (dashed curve), and the axial guide field is $B_0 = 1.62$ kG. By varying the injected frequency in successive experimental shots, the frequency-voltage characteristics of the free electron laser instability can be measured directly (the dots and triangles in Fig. 7.13). Evidently, the agreement between theory and experiment is excellent, and above cut-off ($\hat{\omega} > 2\pi f_c = ck_\perp$) there are generally *two* emission frequencies corresponding to the frequencies $\hat{\omega}^-$ and $\hat{\omega}^+$ defined in Eq.(7.109).

7.7.4   Analysis of the Full Dispersion Relation

We now return to an analysis of the full dispersion relation (7.104) for a cold electron beam. Equation (7.104) has been solved[85] numerically over a wide range of dimensionless system parameters a_w, ω_{pe}/ck_0, $\hat{\gamma}_e$ and γ_z. Here, $\hat{\gamma}_e$ and $\gamma_z = (1 - \beta_{ze}^2)^{-1/2}$ are related by $\hat{\gamma}_e^2 = \gamma_z^2(1 + a_w^2)$ [Eq.(7.100)], which can also be expressed as

$$\beta_{ze}^2 = 1 - \frac{1 + a_w^2}{\hat{\gamma}_e^2} . \tag{7.110}$$

Typical results[85] obtained from Eq.(7.104) are illustrated in Fig. 7.14 for the upshifted growth rate curve. Here, the normalized growth rate Im ω/ck_0 is plotted versus k_z/k_0 for the choice of system parameters $\hat{\gamma}_e = 50$, $a_w = 0.75$ and $\omega_{pe}/ck_0 = 0.042$. The corresponding values of γ_z and β_{ze} calculated from Eq.(7.100) [or Eq.(7.110)] are $\gamma_z = 40$ and $\beta_{ze} \simeq 3199/3200$. As expected, the maximum growth rate of the free electron laser instability in Fig. 7.14 occurs for $k_z = \hat{k}_z^+ \simeq 3200 k_0$, where $\hat{k}_z^+$ is estimated from Eq.(7.108) [For the choice of system parameters in Fig. 7.14, the dimensionless quantity $\omega_{pe}/ck_0\hat{\gamma}_e^{1/2}\gamma_z = 1.5 \times 10^{-4}$ is very small, so that $\hat{k}_z^+ \simeq k_0\beta_{ze}/(1 - \beta_{ze})$ is a good approximation for the upshifted maximum.] Although the full width at half-maximum in Fig. 7.14

is relatively narrow $(\Delta k_z/\hat{k}_z^+ \simeq 1\%)$, the long tail on the growth rate curve in Fig. 7.14 gives a total bandwidth of $\Delta k_z/\hat{k}_z^+ \sim 25\%$. For the choice of system parameters in Fig. 7.14, the real oscillation frequency is given to good accuracy by $\mathrm{Re}\,\omega \simeq (k_z + k_0)V_{ze}$ over the entire range of instability where $\mathrm{Im}\,\omega \geq 0$. As a general remark, the inclusion of an axial momentum spread[86] in the distribution function $F_e^0(p_z)$ decreases the growth rate of the instability relative to the cold-beam results in Fig. 7.14.

An important property of the full dispersion relation (7.104) is that the free electron laser instability is completely stabilized for sufficiently large $k_z \gg \hat{k}_z^+$. This follows directly by taking the $k_z \to \infty$ limit in Eq.(7.104), where the dispersion relation for the longitudinal space-charge branch reduces to

$$[\omega - (k_z + k_0)V_{ze}]^2 = \frac{\omega_{pe}^2}{\hat{\gamma}_e\gamma_z^2} - a_w^2\frac{\omega_{pe}^2}{\hat{\gamma}_e^3} = \frac{\omega_{pe}^2}{\hat{\gamma}_e^3} \ . \tag{7.111}$$

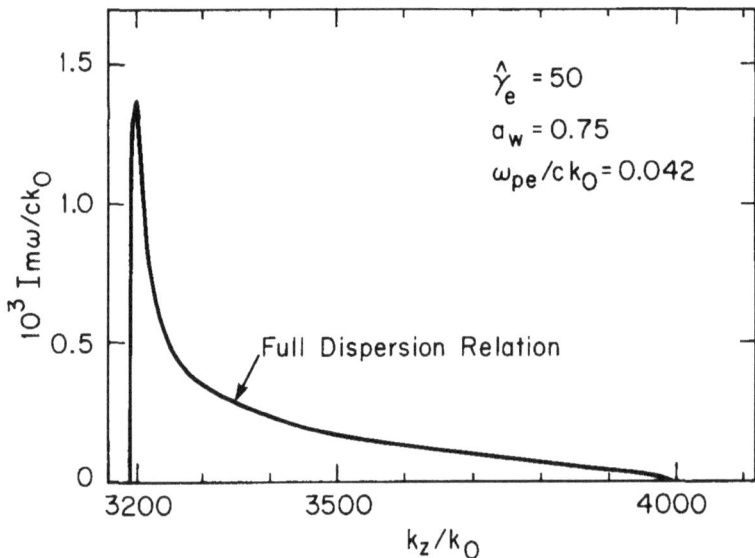

Figure 7.14.   Linear growth properties of the free electron laser instability. The figure shows a plot of the normalized growth rate $Im\omega/ck_0$ versus k_z/k_0 obtained for the upshifted growth rate curve from the full dispersion relation (7.104) for a cold electron beam. The choice of system parameters corresponds to $\hat{\gamma}_e = 50$, $a_w = 0.75$ and $\omega_{pe}/ck_0 = 0.042$ [J.A. Davies, R.C. Davidson and G.L. Johnston, J. Plasma Phys. **33**, 387 (1985)].

Here, use has been made of $\hat{\gamma}_e^2 = \gamma_z^2(1 + a_w^2)$, and Eq.(7.111) gives stable oscillations with $\text{Im}\,\omega = 0$ and $\text{Re}\,\omega = (k_z + k_0)V_{ze} \pm \omega_{pe}/\hat{\gamma}_e^{3/2}$ for $k_z \to \infty$.

7.7.5 Compton-Regime Approximation

We now consider the Compton-regime dispersion relation (7.105), which is expected to be valid for a sufficiently tenuous electron beam [Eq.(7.106)]. While Eq.(7.105) can give a good estimate of the magnitude and location of the maximum growth rate of the free electron laser instability, the detailed shape of the $\text{Im}\,\omega$ versus k_z curve differs from that calculated from the full dispersion relation (7.104).[85] For example, for large $k_z \gg \hat{k}_z^+$, Eq.(7.105) gives

$$[\omega - (k_z + k_0)V_{ze}]^2 = -a_w^2 \frac{\omega_{pe}^2}{\hat{\gamma}_e^3} . \tag{7.112}$$

Equation (7.112) has one unstable solution with $\text{Im}\,\omega = a_w\omega_{pe}/\hat{\gamma}_e^{3/2}$ and $\text{Re}\,\omega = (k_z + k_0)V_{ze}$ as $k_z \to \infty$. Comparing Eq.(7.112) with Eq.(7.111) (where $Im\omega = 0$), it is evident that the Compton-regime dispersion relation (7.105) breaks down at short wavelengths with $k_z \gg \hat{k}_z^+$.

This is also evident from Fig. 7.15 where $\text{Im}\,\omega/ck_0$ is plotted versus k_z/k_0 (for the upshifted peak) for the choice of system parameters $\hat{\gamma}_e = 2$, $a_w = 1$, $\omega_{pe}/ck_0 = 1.414 \times 10^{-2}$, $\gamma_z = 1.414$ and $\beta_{ze} = 0.707$. The solid curve in Fig. 7.15 has been obtained from the full dispersion relation (7.104), whereas the dashed curve has been obtained from the Compton-regime dispersion relation (7.105). Evidently, the magnitude and location of the maximum growth rates calculated from Eqs.(7.104) and (7.105) are in very good agreement. However, as expected from Eq.(7.112), the Compton-regime dispersion relation (7.105) (incorrectly) gives nonzero growth rate at short wavelengths with $k_z \gg \hat{k}_z^+$.

Analytical estimates of the maximum growth rate are readily obtained from the Compton-regime dispersion relation (7.105). For $\omega \simeq (k_z + k_0)V_{ze}$ and $\omega^2 \simeq c^2 k_z^2 + \omega_{pe}^2/\hat{\gamma}_e$, it is readily shown that Eq.(7.105) can be approximated by

$$[\omega - (k_z + k_0)V_{ze}]^2 \left(\omega^2 - c^2 k_z^2 - \frac{\omega_{pe}^2}{\hat{\gamma}_e} \right) = \frac{1}{2} \frac{a_w^2 \omega_{pe}^2}{\hat{\gamma}_e^3 \gamma_z^2} c^2 (k_z + k_0)^2 . \tag{7.113}$$

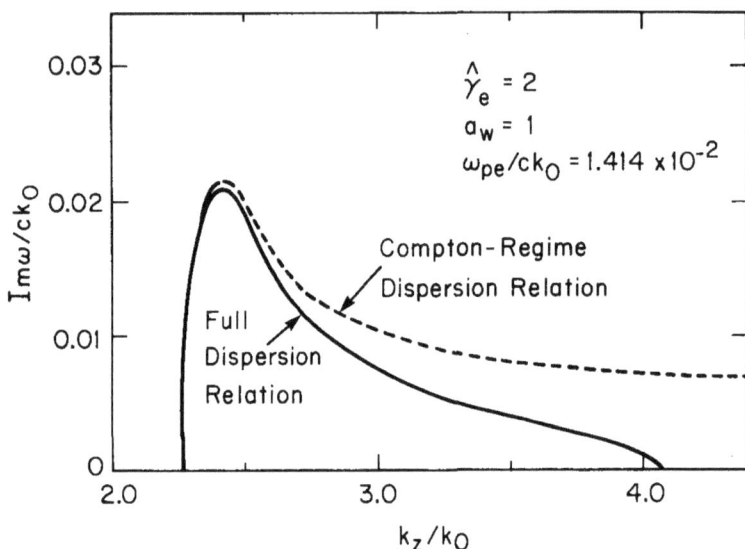

Figure 7.15. Linear growth properties of the free electron laser instability. The figure shows plots of the normalized growth rate $Im\omega/ck_0$ versus k_z/k_0 obtained for a cold electron beam from the full dispersion relation (7.104) (solid curve) and the Compton-regime dispersion relation (7.105) (dashed curve). The choice of system parameters corresponds to $\hat{\gamma}_e = 2$, $a_w = 1$ and $\omega_{pe}/ck_0 = 1.414 \times 10^{-2}$ [J.A. Davies, R.C. Davidson and G.L. Johnston, J. Plasma Phys. **33**, 387 (1985)].

Approximating $k_z = \hat{k}_z^+ \simeq k_0\beta_{ze}/(1-\beta_{ze})$ at maximum growth, we obtain (Problem 7.12)

$$\omega - (k_z + k_0)V_{ze} = \left[\frac{a_w^2\omega_{pe}^2ck_0}{4\hat{\gamma}_e^3}\frac{(1+\beta_{ze})}{\beta_{ze}} \right]^{1/3} \left(-\frac{1}{2} + i\frac{\sqrt{3}}{2} \right) \quad (7.114)$$

for the unstable solution ($Im\,\omega > 0$) to Eq.(7.113). For the Compton-regime approximation to be valid, the inequality $|\omega - (k_z + k_0)V_{ze}|^2 \gg \omega_{pe}^2/\hat{\gamma}_e\gamma_z^2$ must be satisfied at maximum growth [Eq.(7.106)]. Making use of Eq.(7.114), this criterion can be expressed as

$$\frac{\omega_{pe}}{ck_0} \ll \frac{1}{4}\frac{a_w^2\hat{\gamma}_e^{3/2}}{(1+a_w^2)^{3/2}}\frac{(1+\beta_{ze})}{\beta_{ze}} , \quad (7.115)$$

where $\beta_{ze}^2 = 1 - (1+a_w^2)/\hat{\gamma}_e^2$, and use has been made of $\hat{\gamma}_e^2 = \gamma_z^2(1+a_w^2)$.

For a highly relativistic electron beam with $\hat{\gamma}_e^2 \gg 1 + a_w^2$ and $\beta_{ze} \simeq 1$, the inequality in Eq.(7.115) is relatively easy to satisfy, even at moderate beam density.

A detailed numerical analysis of the full dispersion relation (7.104) and the Compton-regime dispersion relation (7.105) has been carried out[85] over a wide range of system parameters $(1 + a_w^2)/\hat{\gamma}_e^2$ and ω_{pe}/ck_0. It is found[85] that the real oscillation frequency and the maximum growth rate of the upshifted peak determined from the Compton-regime dispersion relation (7.105) are accurate to within 10% provided

$$\frac{\omega_{pe}}{ck_0} \leq \left[\frac{\omega_{pe}}{ck_0}\right]_{\text{cr}} \equiv \frac{1}{25} \frac{a_w^2 \hat{\gamma}_e^{3/2}}{(1 + a_w^2)^{3/2}} \frac{(1 + \beta_{ze})}{\beta_{ze}}. \tag{7.116}$$

[The accuracy is within 5% if the factor of 1/25 in Eq.(7.116) is replaced by 1/70.] In terms of the magnitude of the axial electron current $I_b = |-e|\hat{n}_e \pi r_b^2 \beta_{ze} c$, where r_b is the radius of the electron beam, the inequality in Eq.(7.116) can be expressed in the equivalent form

$$I_b \leq I_{\text{cr}} \equiv \frac{m_e c^3}{e} \frac{k_0^2 r_b^2}{2500} \frac{a_w^4 \hat{\gamma}_e^3}{(1 + a_w^2)^3} \frac{(1 + \beta_{ze})^2}{\beta_{ze}}. \tag{7.117}$$

For example, if $k_0 r_b = 1/2$, $a_w = 1$ and $\hat{\gamma}_e = 10$, then Eq.(7.117) gives $I_{\text{cr}} \simeq 850\,\text{A}$.

To summarize, for a cold, tenuous electron beam satisfying Eq.(7.115), the Compton-regime dispersion relation (7.105) [or the approximate form in Eq.(7.113)] provides an excellent approximation to the full dispersion relation (7.104) in the region near maximum growth of the upshifted peak in the growth rate curve. Furthermore, whenever Eq.(7.115) is satisfied, the longitudinal electric field is negligibly small ($\delta\Phi \simeq 0$), and the polarization is predominantly transverse electromagnetic. Finally, Eqs.(7.116) and (7.117) place practical limits on the range of applicability of the Compton-regime approximation.

Problem 7.12   Consider the cold-beam Compton-regime dispersion relation (7.113).

a. Neglect $\omega_{pe}^2/\hat{\gamma}_e$ in comparison with $\omega^2 - c^2 k_z^2$ on the left-hand side of Eq.(7.113), and express $k_z = \hat{k}_z^+ + \delta k_z$ and $\omega = c\hat{k}_z^+ + \delta\omega$, where $\hat{k}_z^+ = k_0 \beta_{ze}/(1 - \beta_{ze})$ is the solution to the simultaneous resonance conditions,

$\hat{\omega} = c\hat{k}_z$ and $\hat{\omega} = (\hat{k}_z + k_0)V_{ze}$. For $|\delta\omega/c\hat{k}_z^+| \ll 1$ and $\delta k_z = 0$, show that the dispersion relation (7.113) can be approximated by

$$(\delta\omega)^3 = \frac{a_w^2 \omega_{pe}^2 ck_0}{4\hat{\gamma}_e^3} \frac{(1 + \beta_{ze})}{\beta_{ze}}, \tag{7.12.1}$$

where $c\beta_{ze} = V_{ze}$.

b. Make use of Eq.(7.12.1) to show that the wiggler-induced growth rate and the real frequency shift of the free electron laser instability are given by Eq.(7.114).

7.7.6   Raman-Regime Approximation

When the beam density is sufficiently high that the inequality in Eq.(7.115) is violated, the longitudinal electric field $[\delta E_z = -(\partial/\partial z)\delta\phi]$ plays an important role in determining detailed properties of the free electron laser instability. In this case, a more accurate analysis of the full dispersion relation (7.104) is required which includes the effects of the longitudinal plasma oscillations occurring in the dielectric factor $[\omega - (k_z + k_0)V_{ze}]^2 - \omega_{pe}^2/\hat{\gamma}_e\gamma_z^2$. In the vicinity of the growth rate maximum, we approximate

$$\left[\omega - (k_z + k_0)V_{ze}\right]^2 - \frac{\omega_{pe}^2}{\hat{\gamma}_e\gamma_z^2} = -\left(\frac{2\omega_{pe}}{\hat{\gamma}_e^{1/2}\gamma_z}\right)\left[\omega - (k_z + k_0)V_{ze} + \frac{\omega_{pe}}{\hat{\gamma}_e^{1/2}\gamma_z}\right],$$

$$\omega^2 - c^2k_z^2 - \frac{\omega_{pe}^2}{\hat{\gamma}_e} = 2\left(c^2k_z^2 + \frac{\omega_{pe}^2}{\hat{\gamma}_e}\right)^{1/2}\left[\omega - \left(c^2k_z^2 + \frac{\omega_{pe}^2}{\hat{\gamma}_e}\right)^{1/2}\right], \tag{7.118}$$

in the full dispersion relation (7.104). The remaining factors of ω^2 in Eq.(7.104) are also evaluated with $\omega^2 = c^2k_z^2 + \omega_{pe}^2/\hat{\gamma}_e$. The full dispersion relation can then be approximated by

$$\left[\omega - (k_z + k_0)V_{ze} + \frac{\omega_{pe}}{\hat{\gamma}_e^{1/2}\gamma_z}\right]\left[\omega - \left(c^2k_z^2 + \frac{\omega_{pe}^2}{\hat{\gamma}_e}\right)^{1/2}\right]$$

$$= -\frac{a_w^2\gamma_z}{4\hat{\gamma}_e^{5/2}}\frac{\omega_{pe}ck_0(2ck_z + ck_0)}{(c^2k_z^2 + \omega_{pe}^2/\hat{\gamma}_e)^{1/2}} \equiv -R. \tag{7.119}$$

Equation (7.119) is often referred to as the *Raman-regime* dispersion relation. Note that the two factors on the left-hand side of Eq.(7.119) vanish identically when the simultaneous resonance conditions in Eq.(7.107) are satisfied.

We introduce the effective frequency mismatch $c\Delta K_z$ defined by

$$c\Delta K_z = \left(c^2 k_z^2 + \frac{\omega_{pe}^2}{\hat{\gamma}_e} \right)^{1/2} - (k_z + k_0)V_{ze} + \frac{\omega_{pe}}{\hat{\gamma}_e^{1/2}\gamma_z} \,. \qquad (7.120)$$

Solving the quadratic equation (7.119) gives for the unstable branch $(\text{Im}\,\omega > 0)$

$$\text{Re}\,\omega = (k_z + k_0)V_{ze} - \frac{\omega_{pe}}{\hat{\gamma}_e^{1/2}\gamma_z} + \frac{1}{2}c\Delta K_z \,,$$

$$\text{Im}\,\omega = \frac{1}{2}\left(4R - c^2 \Delta K_z^2 \right)^{1/2} \,. \qquad (7.121)$$

Here, R is defined in Eq.(7.119), and $c^2\Delta K_z^2 < 4R$ is required for instability. Evidently, maximum growth in Eq.(7.121) occurs for $\Delta K_z = 0$, in

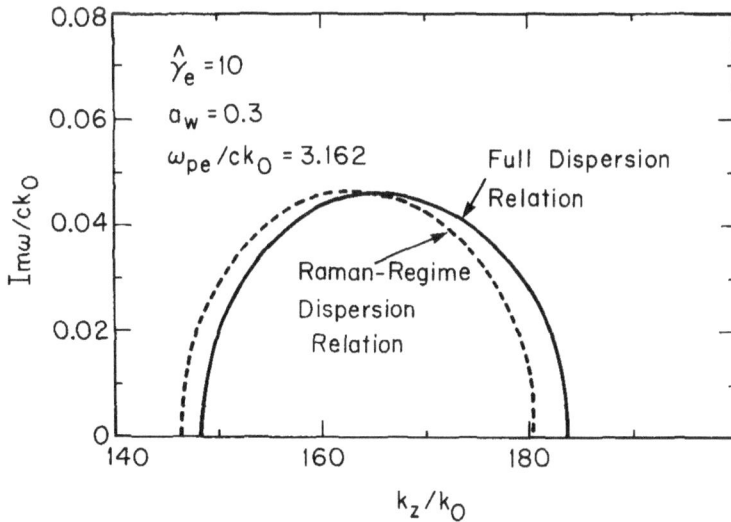

Figure 7.16. Linear growth properties of the free electron laser instability. The figure shows plots of the normalized growth rate $Im\omega/ck_0$ versus k_z/k_0 obtained for a cold electron beam from the full dispersion relation (7.104) (solid curve) and the Raman-regime dispersion relation (7.119) (dashed curve). The choice of system parameters corresponds to $\hat{\gamma}_e = 10$, $a_w = 0.3$ and $\omega_{pe}/ck_0 = 3.162$ [J.A. Davies, R.C. Davidson and G.L. Johnston, J. Plasma Phys. **33**, 387 (1985)].

which case $\text{Im}\,\omega = R^{1/2}$ and $k_z = \hat{k}_z^+$ or $k_z = \hat{k}_z^-$ [Eq.(7.108)]. Furthermore, the instability is stabilized ($\text{Im}\,\omega = 0$) whenever $c^2\Delta K_z^2 > 4R$.

Typical stability results[85] in the Raman regime are illustrated in Fig. 7.16 for the choice of system parameters $\hat{\gamma}_e = 10$, $a_w = 0.3$; $\omega_{pe}/ck_0 = 3.162$, and $\gamma_z = 9.578$. Figure 7.16 shows plots of the normalized growth rate $\text{Im}\,\omega/ck_0$ versus k_z/k_0 obtained numerically for the upshifted peak from the full dispersion relation (7.104) (solid curve) and the Raman-regime dispersion relation (7.119) (dashed curve). For the choice of system parameters in Fig. 7.16 the maximum growth rate, the bandwidth, and the shape of the growth rate curve calculated from the approximate dispersion relation (7.119) are in excellent agreement with Eq.(7.104). Furthermore, the value of $\hat{k}_z^+$ calculated from Eq.(7.108) is $\hat{k}_z^+ \simeq 164\,k_0$.

7.8 Warm-Beam Compton-Regime Stability Properties

Assuming sufficiently low beam density [see Eq.(7.115)], we now make use of the Compton-regime dispersion relation (7.98) to investigate the influence of an axial momentum spread Δp_z on detailed stability behavior. For $D^T(k_z,\omega) \simeq 0$ and $D^T(k_z+2k_0,\omega) \neq 0$, the dispersion relation (7.98) can be approximated by

$$D^T(k_z,\omega) = -\tfrac{1}{2}a_w^2\left[\alpha_3\omega_{pe}^2 + \chi^{(2)}(k_z + k_0,\omega)\right] . \qquad (7.122)$$

Here, $\chi^{(2)}(k_z + k_0,\omega)$ and $D^T(k_z,\omega)$ are defined in Eqs.(7.88) and (7.90), respectively, and the Compton-regime dispersion relation (7.122) can be expressed as

$$0 = D^c(k_z,\omega) \equiv \omega^2 - c^2k_z^2 - \omega_{pe}^2\int\frac{dp_z}{\gamma}\left(1 - \frac{a_w^2}{2\gamma^2}\right)F_e^0(p_z)$$
$$+ \tfrac{1}{2}a_w^2m_ec^2\omega_{pe}^2\int\frac{dp_z}{\gamma^2}\frac{(k_z + k_0)\partial F_e^0/\partial p_z}{\omega - (k_z + k_0)v_z} . \qquad (7.123)$$

In Eq.(7.123), $\omega_{pe}^2 = 4\pi\hat{n}_ee^2/m_e$, $v_z = p_z/\gamma m_e$, $\gamma = (1+p_z^2/m_e^2c^2+a_w^2)^{1/2}$, and $\text{Im}\,\omega > 0$ corresponds to instability (temporal growth).

The dispersion relation (7.123) describes the (unstable) interaction of the transverse electromagnetic wave ($\omega^2 \approx c^2k_z^2$) with the wiggler-induced current perturbations in the electron beam. For a cold electron beam with negligible momentum spread ($\Delta p_z = 0$), the electron distribution function is $F_e^0(p_z) = \delta(p_z - \hat{\gamma}_e m_e V_{ze})$, and the susceptibility factor $\chi^{(2)}(k_z + k_0, \omega)$ is given in Eq.(7.103). For the upshifted growth rate curve, the corresponding maximum growth rate calculated from Eq.(7.123) can be expressed as [see Eq.(7.114)]

$$[\text{Im}\,\omega]_{max} = ck_0 \frac{\sqrt{3}}{2} \left[\frac{a_w^2 \omega_{pe}^2}{4\hat{\gamma}_e^3 c^2 k_0^2} \frac{(1 + \beta_{ze})}{\beta_{ze}} \right]^{1/3} \tag{7.124}$$

for $\Delta p_z = 0$. Here, $\hat{\gamma}_e^2 = \gamma_z^2(1 + a_w^2)$, $\gamma_z = (1 - \beta_{ze}^2)^{-1/2}$ and $\beta_{ze}^2 = 1 - (1 + a_w^2)/\hat{\gamma}_e^2$. Strictly speaking, Eq.(7.124) is also valid when $\Delta p_z \neq 0$ provided the momentum spread is sufficiently small that

$$|(k_z + k_0)\Delta v_z| \ll [\text{Im}\,\omega]_{max} . \tag{7.125}$$

In Eq.(7.125), we estimate $k_z = \hat{k}_z^+ \simeq k_0\beta_{ze}/(1 - \beta_{ze})$ at maximum growth, and the axial velocity spread by $\Delta v_z = \Delta p_z/\gamma_z^2 \hat{\gamma}_e m_e$. Equation (7.125) can then be expressed as

$$\frac{\Delta p_z}{\hat{\gamma}_e m_e V_{ze}} \ll \frac{1}{\beta_{ze}(1 + \beta_{ze})} \frac{\sqrt{3}}{2} \left[\frac{a_w^2 \omega_{pe}^2}{4\hat{\gamma}_e^3 c^2 k_0^2} \frac{(1 + \beta_{ze})}{\beta_{ze}} \right]^{1/3} . \tag{7.126}$$

Equation (7.126) can also be expressed in terms of the energy spread $\Delta\gamma$ by making use of the estimate $\Delta\gamma/\hat{\gamma}_e = \beta_{ze}^2 \Delta p_z/\hat{\gamma}_e m_e V_{ze}$. For example, if $\hat{\gamma}_e = 50$, $a_w = 0.75$ and $\omega_{pe}/ck_0 = 0.42$, then the right-hand side of Eq.(7.126) reduces to 0.23×10^{-2}. In this case, the fractional momentum spread must be considerably less than 0.23% for the cold-beam Compton dispersion relation to provide a valid description of the free electron laser instability.

As the momentum spread is increased to sufficiently large values, it is evident from the kinetic dispersion relation (7.123) that the "ponderomotive" wave with phase velocity $\omega/(k_z + k_0)$ interacts resonantly with the distribution function $F_e^0(p_z)$ (Fig. 7.17). The resulting growth rate $\text{Im}\,\omega$ is reduced relative to the cold-beam value in Eq.(7.124). Indeed, for large enough Δp_z (but $\Delta p_z/\hat{\gamma}_e m_e V_{ze}$ still in the several percent range), the growth rate is reduced substantially relative to the value in Eq.(7.124). Of course, for specified distribution function $F_e^0(p_z)$, Eq.(7.123) can be solved numerically for the real oscillation frequency $\omega_r = \text{Re}\,\omega$ and the growth rate $\omega_i = \text{Im}\,\omega$. A limiting case where the stabilizing influence of axial momentum spread can be calculated analytically is the so-called

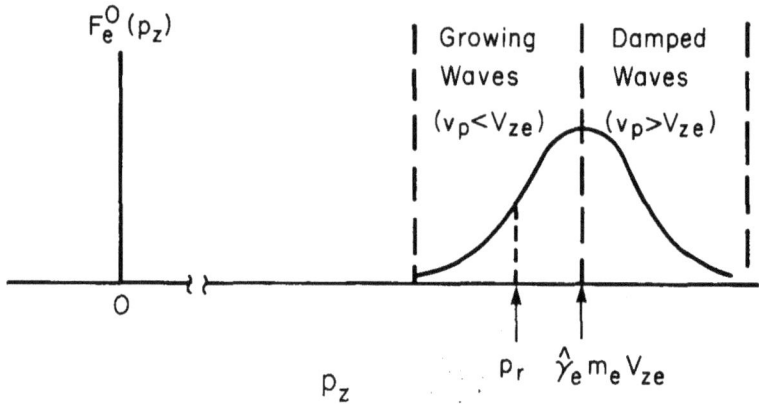

Figure 7.17. Illustrative plot of the beam distribution function $F_e^0(p_z)$ versus p_z in the warm-beam Compton regime. Instability exists ($\operatorname{Im}\omega > 0$) when the resonant momentum p_r satisfies $p_r < \hat{\gamma}_e m_e V_{ze}$. Here, $p_r = \gamma_p m_e v_p (1 + a_w^2)^{1/2}$ corresponds to the resonant velocity $v_z = v_p = \omega_r/(k_z + k_0)$ [Eqs.(7.134) and (7.135)].

warm-beam Compton-regime approximation where the axial momentum spread is sufficiently large that

$$|(k_z + k_0)\Delta v_z| \gg |\operatorname{Im}\omega| . \qquad (7.127)$$

[For typical system parameters, Eq.(7.127) requires a fractional momentum spread in the few-to-several percent range.] Expressing $\omega = \omega_r + i\omega_i$, the Compton-regime dispersion relation

$$D^c(k_z, \omega_r + i\omega_i) = 0$$

is expanded for weak resonant growth ($|\omega_i| \ll |\omega_r|$) according to[93]

$$0 = D_r^c(k_z, \omega_r) + i\left[\omega_i \frac{\partial D_r^c}{\partial \omega_r} + D_i^c(k_z, \omega_r)\right] + \cdots , \qquad (7.128)$$

where $D_r^c = \operatorname{Re} D^c$ and $D_i^c = \operatorname{Im} D^c$. Use is also made of

$$\lim_{\omega_i \to 0_+} \frac{1}{\omega_r - (k_z + k_0)v_z + i\omega_i} = \frac{P}{\omega_r - (k_z + k_0)v_z} - i\pi\delta\left[\omega_r - (k_z + k_0)v_z\right],$$

$$(7.129)$$

where P denotes Cauchy principal value. Substituting Eq.(7.129) into Eq.(7.123), and setting the real and imaginary parts of Eq.(7.128) equal to zero gives

$$0 = D_r^c(k_z, \omega_r) = \omega_r^2 - c^2 k_z^2 - \omega_{pe}^2 \int \frac{dp_z}{\gamma} \left(1 - \frac{a_w^2}{2\gamma^2}\right) F_e^0(p_z)$$

$$+ \frac{1}{2} a_w^2 m_e c^2 \omega_{pe}^2 \int \frac{dp_z}{\gamma^2} \frac{P(k_z + k_0)\partial F_e^0/\partial p_z}{\omega_r - (k_z + k_0)v_z} , \qquad (7.130)$$

and

$$\omega_i = -\frac{D_i^c(k_z, \omega_r)}{\partial D_r^c(k_z, \omega_r)/\partial \omega_r} . \qquad (7.131)$$

Here, $D_i^c(k_z, \omega_r)$ is defined by

$$D_i^c(k_z, \omega_r) = -\frac{\pi}{2} a_w^2 m_e c^2 \omega_{pe}^2 \int \frac{dp_z}{\gamma^2} \delta\left[\omega_r - (k_z + k_0)v_z\right](k_z + k_0)\frac{\partial F_e^0}{\partial p_z} . \qquad (7.132)$$

Equations (7.130) and (7.131) can be used to determine the real oscillation frequency $\omega_r = \text{Re}\,\omega$ and the growth rate $\omega_i = \text{Im}\,\omega$ for the case of weak resonant growth with $|\omega_i| \ll |\omega_r|$, $|(k_z + k_0)\Delta v_z|$. Note that the functional form of $F_e^0(p_z)$ has not yet been specified, and therefore Eqs.(7.130) and (7.131) have a wide range of applicability. Making use of $v_z = p_z/\gamma m_e$, $\gamma = (1 + p_z^2/m_e^2 c^2 + a_w^2)^{1/2}$, and $\partial v_z/\partial p_z = (1 + a_w^2)/\gamma^3 m_e$, the integration over p_z in Eq.(7.132) can be carried out exactly to give

$$D_i^c(k_z, \omega_r) = -\frac{\pi}{2} \frac{a_w^2 m_e^2 c^2 \omega_{pe}^2}{(1 + a_w^2)} \frac{(k_z + k_0)}{|k_z + k_0|} \left[\gamma \frac{\partial}{\partial p_z} F_e^0(p_z)\right]_{v_z = v_p} , \qquad (7.133)$$

where $v_p = \omega_r/(k_z + k_0)$ is the ponderomotive velocity.[93] For small ω_{pe}^2, we approximate $\partial D_r^c/\partial \omega_r = 2\omega_r$, and the growth rate in Eq.(7.131) reduces to

$$\text{Im}\,\omega = \frac{\pi}{4} \frac{a_w^2 m_e^2 c^2 \omega_{pe}^2}{(1 + a_w^2)} \frac{1}{|k_z + k_0|} \left[\frac{\gamma}{v_z} \frac{\partial}{\partial p_z} F_e^0(p_z)\right]_{v_z = v_p} . \qquad (7.134)$$

At the resonant velocity $v_z = v_p \equiv \omega_r/(k_z + k_0)$, we make use of $v_z = p_z/\gamma m_e$ and $\gamma = (1 + p_z^2/m_e^2 c^2 + a_w^2)^{1/2}$ to show that the resonant momentum $(p_z = p_r)$ and the resonant energy $(\gamma = \gamma_r)$ can be expressed as

$$p_r = \gamma_p m_e v_p (1 + a_w^2)^{1/2} \, ,$$
$$\gamma_r = \gamma_p (1 + a_w^2)^{1/2} \, , \qquad (7.135)$$

where $\gamma_p \equiv (1 - v_p^2/c^2)^{-1/2}$. It is evident from Eq.(7.134) that instability exists $(\mathrm{Im}\,\omega > 0)$ in the region of positive slope in Fig. 7.17 where $p_r < \hat{\gamma}_e m_e V_{ze}$ (or equivalently $v_p < V_{ze}$), whereas the waves are damped $(\mathrm{Im}\,\omega < 0)$ for $p_r > \hat{\gamma}_e m_e V_{ze}$.

As a simple application of Eq.(7.134), consider the Gaussian distribution function

$$F_e^0(p_z) = \frac{1}{\sqrt{\pi}\Delta} \exp\left[-\frac{(p_z - \hat{\gamma}_e m_e V_{ze})^2}{\Delta^2}\right] \, , \qquad (7.136)$$

where Δ is the axial momentum spread. Substituting Eq.(7.136) into Eq.(7.134) gives the growth rate

$$\frac{\mathrm{Im}\,\omega}{ck_0} = \frac{\pi}{4} \frac{a_w^2 m_e^2 c^2}{(1 + a_w^2)^{1/2}} \frac{\omega_{pe}^2}{c^2 k_0^2} \frac{\gamma_p ck_0}{\omega_r} \frac{2}{\sqrt{\pi}} \frac{(\hat{\gamma}_e m_e V_{ze} - p_r)}{\Delta^3}$$
$$\times \exp\left[-\frac{(p_r - \hat{\gamma}_e m_e V_{ze})^2}{\Delta^2}\right] \, , \qquad (7.137)$$

where $p_r = \gamma_p m_e v_p (1 + a_w^2)^{1/2}$. The maximum growth rate in Eq.(7.137) occurs for $\hat{\gamma}_e m_e V_{ze} - p_r = \Delta/\sqrt{2}$. If we further approximate $\gamma_p \simeq \gamma_z = \hat{\gamma}_e/(1 + a_w^2)^{1/2}$ in the coefficient in Eq.(7.137), the maximum growth rate in the warm-beam Compton regime is given by

$$[\mathrm{Im}\,\omega]_{\max} = ck_0 \left(\frac{\pi}{8}\right)^{1/2} \frac{a_w^2}{(1 + a_w^2)} \frac{\omega_{pe}^2}{c^2 k_0^2} \frac{ck_0}{\omega_r} \frac{\hat{\gamma}_e m_e^2 c^2}{\Delta^2} \exp(-0.5) \, , \quad (7.138)$$

where $\omega_r \simeq ck_0 \beta_{ze}/(1 - \beta_{ze})$. In the region of validity of Eq.(7.138), the maximum growth rate is substantially less than the cold-beam result in Eq.(7.124).

Typical numerical results obtained from Eqs.(7.134) and (7.137) are illustrated in Fig. 7.18.[93] Here, $\mathrm{Im}\,\omega/ck_0$ is plotted versus k_z/k_0 for the choice of system parameters $\hat{\gamma}_e = 47.1, \Delta/\hat{\gamma}_e m_e V_{ze} = 1.414 \times 10^{-2}$, $\omega_{pe}/ck_0 = 0.646$, and two values of the normalized wiggler amplitude corresponding to $a_w = 0.823$ and $a_w = 0.718$. Several points are noteworthy from Fig. 7.18. First, the momentum spread in the beam electrons causes a significant reduction in growth rate relative to the cold-beam estimate in Eq.(7.124) (where $\Delta = 0$). For example, for $a_w = 0.718$, Eq.(7.124) gives $[\text{Im}\,\omega]_{\max} = 8.6 \times 10^{-3}ck_0$, whereas Eq.(7.137) and Fig. 7.18 give $[\text{Im}\,\omega]_{\max} = 2 \times 10^{-3}ck_0$. Second, unlike the cold-beam Compton-regime dispersion relation, the waves are Landau damped at sufficiently short wavelengths (compare Figs. 7.15 and 7.18 for large k_z). Third, the approximate locations of the growth rate maxima in Fig. 7.18 are given by $k_z \simeq \hat{k}_z^+ = k_0\beta_{ze}/(1 - \beta_{ze})$, which can be expressed as [see Eq.(7.64)]

$$\hat{k}_z^+ = \frac{\hat{\gamma}_e^2(1 + \beta_{ze})\beta_{ze}}{(1 + a_w^2)}k_0 \simeq \frac{2\hat{\gamma}_e^2}{(1 + a_w^2)}k_0 \tag{7.139}$$

for $\beta_{ze} \simeq 1$. For fixed $\hat{\gamma}_e = 47.1$, Eq.(7.139) gives $\hat{k}_z^+/k_0 = 2646$ when $a_w = 0.823$, and $\hat{k}_z^+/k_0 = 2928$ when $a_w = 0.718$. Because $\hat{k}_z^+ = k_0\beta_{ze}/(1 - \beta_{ze})$ corresponds to $v_p = V_{ze}$ and $\omega_r = ck_z$, we note from Figs. 7.17 and 7.18 that Eq.(7.139) actually gives a good estimate

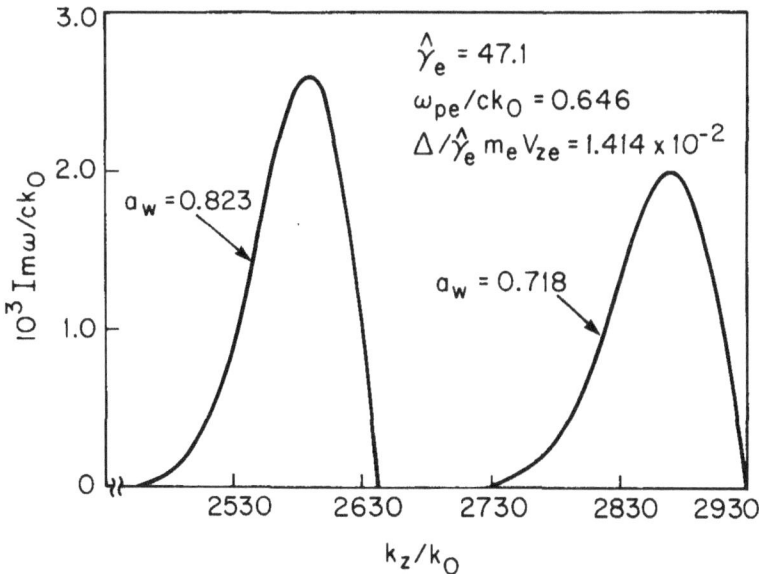

Figure 7.18.   Linear growth properties of the free electron laser instability. The figure shows plots of the normalized growth rate $Im\omega/ck_0$ versus k_z/k_0 obtained from Eqs.(7.134) and (7.137) in the warm-beam Compton regime. The choice of system parameters corresponds to $\hat{\gamma}_e = 47.1$, $\omega_{pe}/ck_0 = 0.646$ and $\Delta/\hat{\gamma}_e m_e V_{ze} = 1.414 \times 10^{-2}$, and the two values of normalized wiggler amplitude $a_w = 0.823$ and $a_w = 0.718$ [A. Dimos and R.C. Davidson, Phys. Fluids **28**, 677 (1985)].

of the marginal stability point (where $\text{Im}\,\omega = 0$) at the right-hand edge of the growth rate curves in Fig. 7.18. Finally, the bandwidth of the instability in k_z-space can be estimated from $\Delta v_z = \Delta v_p$, where $\Delta v_z = [(1 + a_w^2)/\hat{\gamma}_e^3 m_e]\Delta p_z$ and $v_p = \omega_r/(k_z + k_0) \simeq c k_z/(k_z + k_0)$. This gives $\Delta v_p \simeq c(1 - \beta_{ze})(\Delta k_z)/(k_z + k_0)$. Making use of $k_z \simeq \hat{k}_z^+ = k_0\beta_{ze}/(1 - \beta_{ze})$, the normalized bandwidth can be expressed as

$$\frac{\Delta k_z}{\hat{k}_z^+} = (1 + \beta_{ze})\frac{\Delta p_z}{\hat{\gamma}_e m_e V_{ze}} \tag{7.140}$$

in the warm-beam Compton regime.

To summarize, the kinetic version of the Compton-regime dispersion relation in Eq.(7.123) can be used to investigate detailed stability properties for a helical-wiggler free electron laser in circumstances where the beam density is sufficiently low [Eq.(7.115)]. For specified distribution function $F_e^0(p_z)$, the dispersion relation (7.123) can be solved numerically over a wide range of system parameters. In circumstances where the axial momentum spread is sufficiently large [Eq.(7.127)], the growth rate in the warm-beam Compton regime is given by Eq.(7.134).

Problem 7.13   Consider the Compton-regime dispersion relation (7.123) for an equilibrium distribution function $F_e^0(p_z)$ with small (but finite) momentum spread $\Delta \ll \hat{\gamma}_e m_e V_{ze}$. Assume that $F_e^0(p_z)$ is symmetric about the average momentum $\langle p_z \rangle = \hat{\gamma}_e m_e V_{ze}$.

a.   Taylor expand the energy $\gamma = (1 + a_w^2 + p_z^2/m_e^2 c^2)^{1/2}$ and axial velocity $v_z = p_z/\gamma m_e$ about $p_z = \hat{\gamma}_e m_e V_{ze}$ and show that

$$\frac{1}{\gamma^2} = \frac{1}{\hat{\gamma}_e^2} - \frac{2\beta_{ze}}{\hat{\gamma}_e^2}\frac{(p_z - \hat{\gamma}_e m_e V_{ze})}{\hat{\gamma}_e m_e c} + \cdots , \tag{7.13.1}$$

$$v_z = V_{ze} + \frac{(p_z - \hat{\gamma}_e m_e V_{ze})}{\gamma_z^2 \hat{\gamma}_e m_e} + \cdots . \tag{7.13.2}$$

Here, $\beta_{ze} = V_{ze}/c$, and $\hat{\gamma}_e$ and $\gamma_z = (1 - \beta_{ze}^2)^{-1/2}$ are related by $\hat{\gamma}_e^2 = \gamma_z^2(1 + a_w^2)$.

b.   For small momentum spread, make use of Eqs.(7.13.1) and (7.13.2) to show that the kinetic susceptibility occurring in Eq.(7.123) can be aproximated by

$$m_e c^2 \omega_{pe}^2 \int \frac{dp_z}{\gamma^2}\frac{(k_z + k_0)\partial F_e^0/\partial p_z}{\omega - (k_z + k_0)v_z} = \frac{m_e c^2 \omega_{pe}^2}{\hat{\gamma}_e^2}\left(1 - 2\beta_{ze}\gamma_z^2\frac{[\omega - (k_z + k_0)V_{ze}]}{c(k_z + k_0)}\right)$$

$$\times \int d\tilde{p}_z \frac{(k_z + k_0)\partial F_e^0/\partial \tilde{p}_z}{\omega - (k_z + k_0)V_{ze} - [(k_z + k_0)/\gamma_z^2\hat{\gamma}_e m_e]\tilde{p}_z} , \tag{7.13.3}$$

where $\tilde{p}_z = p_z - \hat{\gamma}_e m_e V_{ze}$ is the shifted momentum variable.

c. Simplify Eq.(7.13.3) for the choice of Gaussian distribution function

$$F_e^0(p_z) = \frac{1}{\sqrt{\pi}\Delta} \exp\left(-\frac{\tilde{p}_z^2}{\Delta^2}\right) , \qquad (7.13.4)$$

where Δ is the momentum spread. Introduce the plasma dispersion function $Z(\eta)$ defined by

$$Z(\eta) = \frac{1}{\sqrt{\pi}} \int_{-\infty}^{\infty} dx \, \frac{\exp(-x^2)}{x - \eta} , \qquad (7.13.5)$$

where η is the dimensionless parameter

$$\eta = \gamma_z^2 \frac{[\omega - (k_z + k_0)V_{ze}]}{c(k_z + k_0)} \frac{\hat{\gamma}_e m_e c}{\Delta} . \qquad (7.13.6)$$

Show that the kinetic susceptibility in Eq.(7.13.3) can be expressed as

$$m_e c^2 \omega_{pe}^2 \int \frac{dp_z}{\gamma^2} \frac{(k_z + k_0)\partial F_e^0/\partial p_z}{\omega - (k_z + k_0)v_z}$$

$$= \frac{2\gamma_z^2 m_e^2 c^2}{\Delta^2} \frac{\omega_{pe}^2}{\hat{\gamma}_e} \left(1 - 2\beta_{ze}\eta \frac{\Delta}{\hat{\gamma}_e m_e c}\right)[1 + \eta Z(\eta)] . \qquad (7.13.7)$$

For a Gaussian distribution function with $\Delta \ll \hat{\gamma}_e m_e V_{ze}$ Eq.(7.13.7) can be substituted into the Compton-regime dispersion relation (7.123) in order to determine the detailed stability properties over a wide range of (complex) values of the dimensionless parameter $\eta = \gamma_z^2(\hat{\gamma}_e m_e c/\Delta) \times [\omega - (k_z + k_0)V_{ze}]/c(k_z + k_0)$. Note that $|\eta| \gg 1$ corresponds to the cold-beam limit [Eq.(7.125)], whereas $|\eta| \ll 1$ corresponds to the warm-beam Compton regime [Eq.(7.127)]. In these limiting cases, $Z(\eta)$ can be approximated by

$$Z(\eta) \simeq -\frac{1}{\eta} - \frac{1}{2\eta^3} - \frac{3}{4\eta^5} - \cdots , \text{ for } |\eta| \gg 1 , \qquad (7.13.8)$$

and

$$Z(\eta) \simeq -2\eta + \frac{4}{3}\eta^3 - \cdots + i\sqrt{\pi}\frac{(k_z + k_0)}{|k_z + k_0|}\exp(-\eta^2), \text{ for } |\eta| \ll 1. \quad (7.13.9)$$

Equations (7.13.8) and (7.13.9) are useful approximations for analytical investigations of the free electron laser instability at large or small values of $|\eta|$. Generally speaking, however, at intermediate values of η, the unexpanded plasma dispersion function $Z(\eta)$ defined in Eq.(7.13.5) should be used to determine the stability behavior numerically.[86]

7.9   Nonlinear Evolution for Multimode Excitation

In circumstances where a single mode (ω_s, k_s) is excited in the free electron laser interaction, a description of the nonlinear evolution of the system requires a coherent nonlinear model[88-91] that follows the development of the wave amplitude and phase as well as the dynamics of the trapped and untrapped electrons interacting with the ponderomotive potential (Sec. 7.10). On the other hand, if the excitation of the free electron laser instability is sufficiently broadband in k_z-space (multimode excitation), then a quasilinear model[92,93] can be developed to describe the nonlinear evolution and saturation of the wave spectrum and the response of the average distribution function $\langle F_e \rangle (p_z, t)$ to the amplifying radiation spectrum. In this section, we summarize the main features of a quasilinear model[92,93] that describes the nonlinear evolution and saturation of the free electron laser instability in the Compton-regime approximation (Sec. 7.8). The analysis assumes a low-density electron beam propagating in the z-direction perpendicular to a constant-amplitude helical wiggler magnetic field [Eq. (7.50)]. The theoretical model is based on the nonlinear Vlasov-Maxwell equations (7.72), (7.75), and (7.76) assuming

$$\frac{\partial}{\partial x} = 0 = \frac{\partial}{\partial y}$$

and negligible longitudinal perturbations $(\delta\phi \simeq 0)$.

The quasilinear model follows the (slow) nonlinear evolution of the average distribution function

$$\langle F_e \rangle (p_z, t) = \frac{1}{L} \int_0^L dz \, F_e(z, p_z, t) \tag{7.141}$$

in response to the amplifying electromagnetic field perturbations. Here, L is the fundamental periodicity length in the z-direction, and the time variation of perturbed quantities in Fourier variables is assumed to be of the form

$$\exp\left\{ -i \int_0^t dt' \, [\omega_r(k_z, t') + i\omega_i(k_z, t')] \right\} . \tag{7.142}$$

For $D^T(k_z + 2k_0, \omega_r + i\omega_i) \neq 0$, the slow nonlinear evolution of the real oscillation frequency $\omega_r(k_z, t)$ and the growth rate $\omega_i(k_z, t)$ are determined adiabatically in terms of the average distribution function $\langle F_e \rangle (p_z, t)$ from the quasilinear dispersion relation[92,93]

$$0 = D^c(k_z, \omega_r + i\omega_i) = (\omega_r + i\omega_i)^2 - c^2 k_z^2 - \omega_{pe}^2 \int \frac{dp_z}{\gamma} \left(1 - \frac{a_w^2}{2\gamma^2}\right) \langle F_e \rangle$$

$$+ \frac{1}{2} a_w^2 m_e c^2 \omega_{pe}^2 \int \frac{dp_z}{\gamma^2} \frac{(k_z + k_0)(\partial/\partial p_z)\langle F_e \rangle(p_z, t)}{\omega_r - (k_z + k_0)v_z + i\omega_i} \tag{7.143}$$

in the Compton-regime approximation. Equation (7.143) is identical in form to Eq.(7.123), with $F_e^0(p_z)$ replaced by $\langle F_e \rangle(p_z, t)$ and $\omega_r(k_z, t) + i\omega_i(k_z, t)$ determined self-consistently in terms of the average distribution function, which changes in response to the instability. In obtaining Eq.(7.143) [or Eq.(7.123)] we have assumed that $D^T(k_z + 2k_0, \omega_r + i\omega_i) \neq 0$ and $D^T(k_z, \omega_r + i\omega_i) \simeq 0$ [see also Eq.(7.98)]. Omitting algebraic details,[92,93] the quasilinear evolution of $\langle F_e \rangle(p_z, t)$ is described by the diffusion equation in momentum space

$$\frac{\partial}{\partial t}\langle F_e \rangle(p_z, t) = \frac{\partial}{\partial p_z} \left[D(p_z, t) \frac{\partial}{\partial p_z} \langle F_e \rangle(p_z, t) \right] . \tag{7.144}$$

Here, the diffusion coefficient $D(p_z, t)$ is defined by

$$D(p_z, t) = \frac{2\pi e^2 a_w^2}{\gamma^2} \sum_{k_z = -\infty}^{\infty} \frac{i|\delta H(k_z, t)|^2}{[\omega_r - (k_z + k_0)v_z + i\omega_i]} , \tag{7.145}$$

where $\omega_i \geq 0$ is assumed, and

$$|\delta H(k_z, t)|^2 = \frac{(k_z + k_0)^2}{8\pi} |\delta A_+(k_z + 2k_0, t) + \delta A_-(k_z, t)|^2 \tag{7.146}$$

is the spectral energy density in the magnetic field perturbations.[92,93] Moreover, $|\delta H(k_z, t)|^2$ evolves according to the wave kinetic equation

$$\frac{\partial}{\partial t}|\delta H(k_z, t)|^2 = 2\omega_i(k_z, t)|\delta H(k_z, t)|^2 , \tag{7.147}$$

and the real oscillation frequency $\omega_r(k_z, t)$ and growth rate $\omega_i(k_z, t)$ are determined adiabatically in time from the dispersion relation (7.143). In Eq.(7.146), the complex amplitudes $\delta A_+(k_z + 2k_0, t) = \delta A_x(k_z + 2k_0, t) + i\delta A_y(k_z + 2k_0, t)$ and $\delta A_-(k_z, t) = \delta A_x(k_z, t) - i\delta A_y(k_z, t)$ are related by the polarization condition in Eq.(7.97). Therefore, for $D^T(k_z, \omega_r + i\omega_i) \simeq 0$, Eq.(7.97) gives $\delta A_+(k_z + 2k_0, t) \simeq 0$, so that Eq.(7.146) can be approximated by $|\delta H(k_z, t)|^2 = (k_z + k_0)^2 |\delta A_-(k_z, t)|^2/8\pi$.

Equations (7.143)–(7.147) provide a closed description of the nonlinear evolution of the system, including the strong interaction between the waves and the particles when $\omega_r - (k_z + k_0)v_z \approx 0$ [see Eq.(7.145)]. For specified initial distribution function $\langle F_e \rangle (p_z, 0)$ and input wave spectrum $|\delta H(k_z, 0)|^2$, the field energy $|\delta H(k_z, t)|^2$ grows [Eq.(7.147)], causing an increase in the diffusion coefficient $D(p_z, t)$ [Eq.(7.145)]. In response to the amplifying field perturbations, the average distribution function $\langle F_e \rangle (p_z, t)$ undergoes significant alterations in momentum space [Eq.(7.144)], typically through plateau formation[93] in the resonant region where $v_z = v_p = \omega_r/(k_z + k_0)$. This is accompanied by a decrease in the average axial momentum $\langle p_z \rangle (t) = \int dp_z \, p_z \langle F_e \rangle (p_z, t)$ and average energy $\langle \gamma \rangle (t) = \int dp_z \, \gamma \langle F_e \rangle (p_z, t)$ of the electron beam. The change in $\langle F_e \rangle (p_z, t)$ leads to a corresponding reduction in the growth rate $\omega_i(k_z, t)$ [Eq.(7.143)] and the eventual stabilization of the free electron laser instability [Eq.(7.147)]. Justification of the quasilinear equations (7.143)–(7.147) for small-amplitude perturbations requires a sufficiently broad spectrum of unstable waves that the inequalities

$$\tau_{\rm ac} \sim |\Delta[\omega_r - (k_z + k_0)v_z]|^{-1} \ll \omega_i^{-1}, \tau_{\rm rel} \qquad (7.148)$$

are satisfied. Here, ω_i is the characteristic growth rate of the instability; $\tau_{\rm rel}$ is the characteristic time for the quasilinear relaxation of $\langle F_e \rangle (p_z, t)$ [Eq.(7.144)]; $\tau_{\rm ac}$ is the autocorrelation time for the waves;[93] and $\Delta[\omega_r - (k_z + k_0)v_z]$ is the characteristic spread in values of $\omega_r - (k_z + k_0)v_z$ over the unstable range of wavenumbers. Estimates of $\tau_{\rm rel}$ show that $\tau_{\rm rel}$ is typically a few times ω_i^{-1}, so that Eq.(7.148) reduces to $\tau_{\rm ac} \ll \omega_i^{-1}$. In addition, we estimate $\Delta[\omega_r - (k_z + k_0)v_z] \sim c(1 - \beta_{ze})\Delta k_z$, where $\beta_{ze}c = V_{ze}$ and Δk_z is the width of the unstable spectrum in k_z-space. Equation (7.148) can then be expressed as

$$\frac{\omega_i}{ck_0} \ll \beta_{ze} \frac{\Delta k_z}{\hat{k}_z^+} , \qquad (7.149)$$

where $\hat{k}_z^+ \simeq k_0 \beta_{ze}/(1 - \beta_{ze})$. For specified distribution function $\langle F_e \rangle$, the growth rate ω_i can be determined from Eq.(7.143) [see, for example, Eq.(7.134)], and the inequality in Eq.(7.149) can be used to quantify the range of validity of the quasilinear model. Although Eq.(7.149) requires a sufficiently broad excitation in k_z-space, note that the inequality $\Delta k_z/\hat{k}_z^+ \ll 1$ is still satisfied in the regimes of practical interest. For example, consider an electron beam with small (but finite) axial momentum spread $\Delta p_z \ll \hat{\gamma}_e m_e V_{ze}$. Then, Eq.(7.140) gives $\Delta k_z/\hat{k}_z^+ = (1+\beta_{ze})\Delta p_z/\hat{\gamma}_e m_e V_{ze} \ll 1$. The quasilinear stabilization process has been examined in detail[93] in the warm-beam Compton regime [Eq.(7.127)], including the process of plateau formation $[\partial\langle F_e\rangle/\partial p_z|_{v_z=v_p} \to 0]$ in the resonant region of momentum space where $v_z = v_p = \omega_r/(k_z + k_0)$, and the concomitant stabilization of the resonant free electron laser instability with $\omega_i \to 0$ [see Eq.(7.134)].

For present purposes, we make use of Eqs.(7.143)–(7.147) to describe the quasilinear stabilization process for the case of a simple waterbag model where $\langle F_e\rangle(p_z, t)$ is assumed to have the rectangular form (Fig. 7.19).[92]

$$\langle F_e\rangle(p_z,t) = \begin{cases} \dfrac{1}{2\Delta(t)} &, |p_z - p(t)| < \Delta(t) , \\ 0 , & |p_z - p(t)| > \Delta(t) . \end{cases} \tag{7.150}$$

Here, $p(t)$ is the average axial momentum, and $\Delta(t)$ is a measure of the momentum spread. In particular, it follows from Eq.(7.150) that

$$\langle p_z\rangle = \int_{-\infty}^{\infty} dp_z\, p_z \langle F_e\rangle(p_z,t) = p(t) ,$$

$$\langle(p_z - \langle p_z\rangle)^2\rangle = \int_{-\infty}^{\infty} dp_z\,(p_z - \langle p_z\rangle)^2 \langle F_e\rangle(p_z,t) = \frac{1}{3}\Delta^2(t) , \tag{7.151}$$

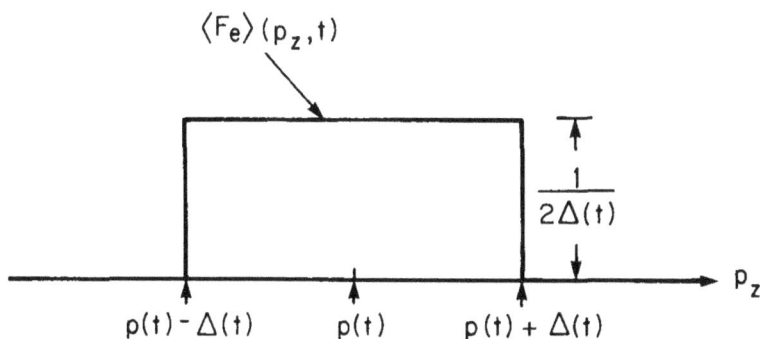

Figure 7.19. Water-bag model assumed for the average distribution function $\langle F_e\rangle(p_z, t)$ in Eq.(7.150). The average momentum $p(t)$ and momentum spread $\Delta(t)$ evolve quasilinearly according to Eqs.(7.155) and (7.156).

It is convenient to define the energy and axial velocity corresponding to $p_z = p(t) \pm \Delta(t)$ by

$$\gamma_{\pm}(t) = \left\{ 1 + a_w^2 + \frac{[p(t) \pm \Delta(t)]^2}{m_e^2 c^2} \right\}^{1/2} , \qquad (7.152)$$

and

$$v_{\pm}(t) = \frac{[p(t) \pm \Delta(t)]}{\gamma_{\pm}(t) m_e} , \qquad (7.153)$$

Substituting Eq.(7.150) into Eq.(7.143) and integrating over p_z then gives the dispersion relation[92]

$$0 = D^c(k_z, \omega_r + i\omega_i)$$

$$= (\omega_r + i\omega_i)^2 - c^2 k_z^2 - \omega_{pe}^2 \int \frac{dp_z}{\gamma} \left(1 - \frac{a_w^2}{2\gamma^2} \right) \langle F_e \rangle + \frac{1}{2} a_w^2 m_e c^2 \omega_{pe}^2 \frac{(k_z + k_0)}{2\Delta}$$

$$\times \left\{ \frac{1}{\gamma_-^2 [\omega_r - (k_z + k_0)v_- + i\omega_i]} - \frac{1}{\gamma_+^2 [\omega_r - (k_z + k_0)v_+ + i\omega_i]} \right\}. \quad (7.154)$$

Similarly, for the choice of $\langle F_e \rangle(p_z, t)$ in Eq.(7.150), the appropriate momentum moments of the quasilinear diffusion equation (7.144) can be taken to determine the nonlinear evolution of $p(t)$ and $\Delta(t)$. Some straightforward algebra gives[92]

$$\frac{d}{dt} p(t) = -2\pi e^2 a_w^2 \sum_{k_z = -\infty}^{\infty} \frac{i |\delta H(k_z, t)|^2}{2\Delta} \qquad (7.155)$$

$$\times \left\{ \frac{1}{\gamma_-^2 [\omega_r - (k_z + k_0)v_- + i\omega_i]} - \frac{1}{\gamma_+^2 [\omega_r - (k_z + k_0)v_+ + i\omega_i]} \right\},$$

and

$$\frac{d}{dt} \Delta(t) = 6\pi e^2 a_w^2 \sum_{k_z = -\infty}^{\infty} \frac{i |\delta H(k_z, t)|^2}{2\Delta} \qquad (7.156)$$

$$\times \left\{ \frac{1}{\gamma_-^2 [\omega_r - (k_z + k_0)v_- + i\omega_i]} + \frac{1}{\gamma_+^2 [\omega_r - (k_z + k_0)v_+ + i\omega_i]} \right\}.$$

For the choice of waterbag distribution function in Eq.(7.150), the wave kinetic equation (7.147) together with Eqs.(7.154)–(7.156) provide a closed description of the nonlinear evolution of the system. In particular, Eqs.(7.155) and (7.156) describe the evolution of $p(t)$ and $\Delta(t)$ in response to the amplifying field perturbations $|\delta H(k_z, t)|^2$, which evolve according to Eq.(7.147). Moreover, the instantaneous growth rate $\omega_i(k_z, t)$ and the real oscillation frequency $\omega_r(k_z, t)$ are determined in terms of $\Delta(t)$ and $p(t)$ from the quasilinear dispersion relation (7.154).

The coupled nonlinear equations (7.147) and (7.154)–(7.156) have been solved numerically[92] over a wide range of system parameters a_w, ω_{pe}/ck_0, $\Delta_0 \equiv \Delta(t = 0)$ and $p_0 \equiv p(t = 0)$. Typical numerical results are illustrated in Fig. 7.20 for the choice of parameters $\omega_{pe}/ck_0 = 0.283$, $a_w = 1.96$, $\gamma_0 \equiv (1 + a_w^2 + p_0^2/m_e^2 c^2)^{1/2} = 10$ and $\Delta_0/p_0 = 0.01$. It is

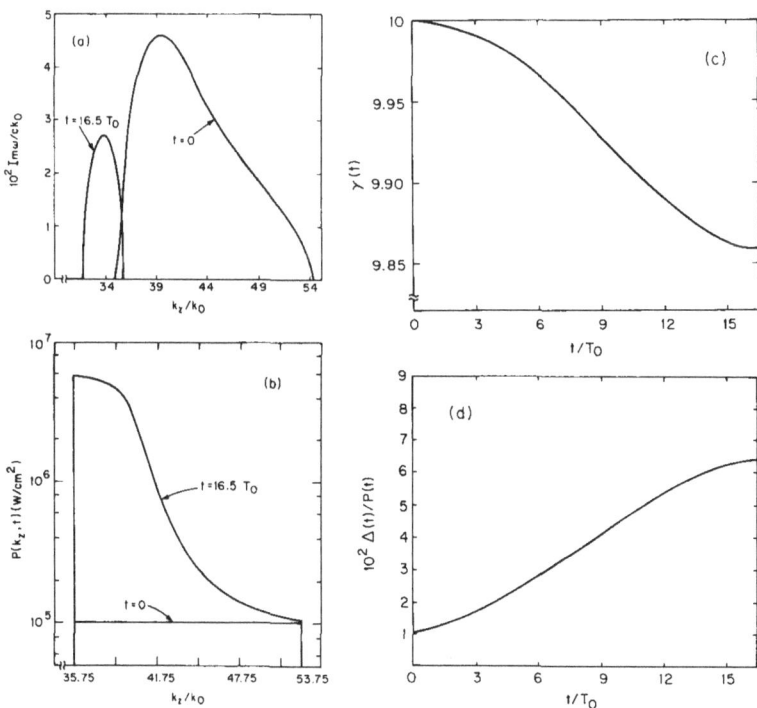

Figure 7.20. Nonlinear evolution of the free electron laser instability obtained numerically from the quasilinear equations (7.147) and (7.154)–(7.156) [R.C. Davidson and Y.-Z. Yin, Phys. Fluids 28, 2524 (1985)]. Input parameters for the system correspond to $a_w = 1.96$, $\omega_{pe}/ck_0 = 0.283$, $\gamma_0 = (1 + a_w^2 + p_0^2/m_e^2 c^2)^{1/2} = 10$ and $\Delta_0/p_0 = 0.01$. Plots show (a) the normalized growth rate $\mathrm{Im}\,\omega/ck_0$ versus k_z/k_0 at $t = 0$ and $t/T_0 = 16.5$, (b) the power spectrum $P(k_z, t)$ versus k_z/k_0 at $t = 0$ and $t/T_0 = 16.5$, (c) the energy $\gamma(t) = [1 + a_w^2 + p^2(t)/m_e^2 c^2]^{1/2}$ versus t/T_0, and (d) the normalized momentum spread $\Delta(t)/p(t)$ versus t/T_0.

useful to introduce the spectral energy density $P(k_z, t)$ associated with the $\delta\mathbf{E} \times \delta\mathbf{B}$ power flow defined by[92,93]

$$P(k_z, t) = \frac{\omega_r(k_z, t)}{8\pi} \left(\frac{m_e c^2}{e}\right)^2 \qquad (7.157)$$

$$\times \left[(k_z + 2k_0)|\delta A_+(k_z + 2k_0, t)|^2 + k_z|\delta A_-(k_z, t)|^2\right] ,$$

where $\delta A^+(k_z+2k_0, t) \simeq 0$ by virtue of the assumption $D^T(k_z, \omega_r + i\omega_i) \simeq 0$ [see Eq.(7.97)]. For the choice of system parameters in Fig. 7.20, it follows from Eq.(7.139) that the upshifted wavenumber at $t = 0$ is given approximately by $\hat{k}_z^+ = \beta_{ze} k_0/(1 - \beta_{ze}) \simeq 41$. Moreover, from Fig. 7.20(a), instability exists at $t = 0$ for k_z/k_0 in the interval $35 < k_z/k_0 < 55$. For the numerical results[92] presented in Fig. 7.20, the input power spectrum $P(k_z, t = 0)$ is taken to be flat and equal to 100 kW/cm^2 in the interval $35.75 < k_z/k_0 < 52.25$, whereas $P(k_z, t = 0) = 0$ outside of this interval [Fig. 7.20(b)]. Within the framework of the quasilinear wave kinetic equation (7.147), the wave spectrum grows only in the region of initial excitation. As the wave spectrum grows according to Eq.(7.147), the average axial momentum $p(t)$ and momentum spread $\Delta(t)$ adjust to the amplifying field perturbations according to Eqs.(7.155) and (7.156). This is illustrated in Figs. 7.20(c) and 7.20(d), where $\gamma(t) \equiv [1 + a_w^2 + p^2(t)/m_e^2 c^2]^{1/2}$ and $\Delta(t)/p(t)$, calculated numerically from Eqs.(7.155) and (7.156), are plotted versus t/T_0 where $T_0 \equiv 2\pi/ck_0$. As would be expected, the axial momentum of the electron beam is reduced, as manifest through a decrease in $\gamma(t)$ and $p(t)$ [Fig. 7.20(c)], and there is a concomitant increase in the axial momentum spread $\Delta(t)$ [Fig. 7.20(d)]. Consistent with the decrease in $p(t)$ and increase in $\Delta(t)$, the growth rate $\omega_i(k_z, t)$ calculated from the quasilinear dispersion relation (7.154) decreases both in bandwidth and magnitude, and the maximum in the growth rate curve shifts to smaller values of k_z [Fig. 7.20(a)]. By the time $t/T_0 = 16.5$, the region of growth [Fig. 7.20(a)] no longer overlaps with the region of initial excitation of $P(k_z, t = 0)$ [Fig. 7.20(b)], and the quasilinear stabilization process is completed. From Fig. 7.20(a) it follows that $16.5\, T_0 \approx 4.7[\omega_i(k_z, t = 0)]_{\max}^{-1}$, which corresponds to a relatively fast stabilization process. For the choice of system parameters in Fig. 7.20, the efficiency η of radiation generation is relatively low. Defining $\eta = \Delta\mathcal{E}_F/[\hat{n}_e(\gamma_0 - 1)m_e c^2]$, where $\Delta\mathcal{E}_F$ is the change in total electromagnetic field energy density, it is found that $\eta \simeq 3\%$ for the choice of parameters in Fig. 7.20.

To summarize, the quasilinear equations (7.143)–(7.147) can be used to investigate the nonlinear evolution of the free electron laser instability over a wide range of system parameters where the Compton-regime approximation is valid. In this regard, validity of the quasilinear model requires multimode excitation with sufficiently large bandwidth in k_z-space that the inequality in Eq.(7.149) is satisfied.

Problem 7.14   Consider the quasilinear kinetic equation (7.144) for $\langle F_e \rangle(p_z, t)$ with diffusion coefficient $D(p_z, t)$ specified by Eq.(7.145). Assume that the average distribution function maintains the waterbag form in Eq.(7.150) as the system evolves nonlinearly.

a. Make use of Eqs. (7.144), (7.145) and (7.150) to show that the average momentum $p(t)$ and momentum spread $\Delta(t)$ defined in Eq.(7.151) evolve according to Eqs.(7.155) and (7.156).

b. Show that the quasilinear dispersion relation (7.143) reduces to Eq.(7.154) for the choice of distribution function in Eq.(7.150).

Problem 7.15

a. For the perturbed electromagnetic field configuration in Eq.(7.65), show that the $\delta \mathbf{E} \times \delta \mathbf{B}$ power flow can be expressed as

$$\frac{c}{4\pi} \left(\delta \mathbf{E} \times \delta \mathbf{B} \right)_z = \frac{1}{4\pi} \sum_{k_z} i k_z \left[\delta A_x^*(k_z, t) \frac{\partial}{\partial t} \delta A_x(k_z, t) \right.$$

$$\left. + \delta A_y^*(k_z, t) \frac{\partial}{\partial t} \delta A_y(k_z, t) \right] , \qquad (7.15.1)$$

where $\delta A_x(k_z, t)$ and $\delta A_y(k_z, t)$ are the Fourier amplitudes of the perturbed vector potential.

b. Assume a (quasilinear) time variation of the form in Eq.(7.142). Show that the spectral energy density for the power flow in Eq.(7.15.1) can be expressed in the form given in Eq.(7.157).

7.10   Sideband Instability

The quasilinear analysis in Sec. 7.9 assumes multimode excitation, in which case there is no phase correlation between the particles and the waves, and the average distribution function $\langle F_e \rangle(p_z, t)$ satisfies the diffusion equation 7.144 in momentum space. In this section, we consider the opposite limit where a single finite-amplitude electromagnetic wave (ω_s, k_s) is excited by the free electron laser interaction, and the particles interact coherently with the wave. Particular emphasis, in the present analysis is placed on the sideband instability,[90,120-126] which is a consequence of the motion of the electrons trapped in the ponderomotive potential.

7.10.1   Nonlinear Wave and Orbit Equations

As in Eq.(7.51), the vector potential for the constant-amplitude helical wiggler field is taken to be

$$\mathbf{A}_w^0(\mathbf{x}) = \frac{m_e c^2}{e} a_w \left[\cos(k_0 z)\hat{\mathbf{e}}_x + sin(k_0 z)\hat{\mathbf{e}}_y\right] , \qquad (7.158)$$

where $a_w = eB_w/m_e c^2 k_0 = $ const. In addition to the static wiggler field, it is assumed that a primary electromagnetic wave with right-circular polarization is excited with vector potential

$$\delta\mathbf{A}(\mathbf{x},t) = -\frac{m_e c^2}{e}\hat{a}_s(z,t)\left\{\cos\left[k_s z - \omega_s t + \delta_s(z,t)\right]\hat{\mathbf{e}}_x\right.$$
$$\left. - \sin\left[k_s z - \omega_s t + \delta_s(z,t)\right]\hat{\mathbf{e}}_y\right\} . \qquad (7.159)$$

Here, ω_s and k_s are the frequency and wavenumber, respectively, and the amplitude $\hat{a}_s(z,t)$ and phase $\delta_s(z,t)$ are treated as slowly varying. In the Compton-regime approximation $(\delta\phi \simeq 0)$, the corresponding electric and magnetic fields are given by $\delta\mathbf{E} = -c^{-1}\partial\delta\mathbf{A}/\partial t$ and $\delta\mathbf{B} = \nabla \times \delta\mathbf{A}$. Although the (dimensionless) amplitude $\hat{a}_s(z,t)$ in Eq.(7.159) is treated as finite in the subsequent analysis, it should be noted that $|\hat{a}_s| \ll 1$ in the regimes of practical interest.

A detailed investigation of the sideband instability simplifies considerably if the analysis is carried out in the ponderomotive frame[120] moving with velocity

$$v_p = \frac{\omega_s}{k_s + k_0} . \qquad (7.160)$$

Therefore, the subsequent analysis is carried out in ponderomotive-frame variables (z', t', γ') defined by the Lorentz transformation

$$z' = \gamma_p(z - v_p t) ,$$
$$t' = \gamma_p(t - v_p z/c^2) ,$$
$$\gamma' = \gamma_p(\gamma - v_p p_z/m_e c^2) , \qquad (7.161)$$

where $\gamma_p = (1 - v_p^2/c^2)^{-1/2}$, $\gamma' m_e c^2 = (m_e^2 c^4 + c^2 p_x'^2 + c^2 p_y'^2 + c^2 p_z'^2)^{1/2}$
is the mechanical energy, and the components of momentum (p_x', p_y', p_z')
are related to the velocity by $\mathbf{p}' = \gamma' m_e \mathbf{v}'$. We introduce the complex
representation of the vector potentials defined by

$$a_w^-(z) = a_{xw}(z) - i a_{yw}(z) \,,$$

$$a_s^-(z, t) = a_{xs}(z, t) - i a_{ys}(z, t) \,, \tag{7.162}$$

where $a_{xw} = a_w \cos(k_0 z)$, $a_{yw} = a_w \sin(k_0 z)$, $a_{xs} = -\hat{a}_s \cos(k_s z - \omega_s t + \delta_s)$,
and $a_{ys} = \hat{a}_s \sin(k_s z - \omega_s t + \delta_s)$. Making use of Eqs.(7.158) and (7.159)
and the inverse transformation

$$z = \gamma_p(z' + v_p t') \quad \text{and} \quad t = \gamma_p(t' + v_p z'/c^2) \,,$$

it is readily shown that

$$a_w^-(z', t') = a_w \exp\left[-i\gamma_p k_0(z' + v_p t')\right] \,,$$

$$a_s^-(z', t') = -\hat{a}_s(z', t') \exp\left[i(k_s' z' - \omega_s' t') + i\delta_s'(z', t')\right] \,, \tag{7.163}$$

in ponderomotive-frame variables. Here, (ω_s', k_s') in the ponderomotive
frame is related to (ω_s, k_s) in the laboratory frame by

$$\omega_s' = \gamma_p(\omega_s - k_s v_p) \,,$$

$$k_s' = \gamma_p\left(k_s - \frac{\omega_s v_p}{c^2}\right) \,. \tag{7.164}$$

We denote the axial position and energy of the k'th electron in the
ponderomotive frame by $z_k'(t')$ and $\gamma_k'(t')$. In addition, it is assumed that
all electrons move on surfaces with zero transverse canonical momentum,
i.e., $P_{zk}' = 0 = P_{yk}'$. This gives for the transverse velocity components of
the k'th electron.

$$v_{xk}' = \frac{c}{\gamma_k'} \left[a_{xw}(z_k', t') + a_{xs}(z_k', t')\right] \,,$$

$$v_{yk}' = \frac{c}{\gamma_k'} \left[a_{yw}(z_k', t') + a_{ys}(z_k', t')\right] \,. \tag{7.165}$$

In the ponderomotive frame, Maxwell's equations for the complex vector potential $a_s^-(z,t) = a_{xs}(z,t) - ia_{ys}(z,t)$ associated with the average electromagnetic fields can be expressed as

$$\left(\frac{1}{c^2} \frac{\partial^2}{\partial t'^2} - \frac{\partial^2}{\partial z'^2} \right) a_s^-(z',t') = -\frac{4\pi e^2}{m_e c^2} \left\langle \sum_k \left(v'_{xk} - iv'_{yk} \right) \delta\left[z' - z'_k(t') \right] \right\rangle,$$

(7.166)

where $\sum_k$ denotes summation over electrons, and $\langle \cdots \rangle$ denotes ensemble average.[122] Making use of Eqs.(7.163) and (7.165), the microscopic current density $J_M^-(z',t') = J_{xM}(z',t') - iJ_{yM}(z',t') = -e\sum_k (v'_{xk} - iv'_{yk}) \times \delta[z' - z'_k(t')]$ occurring in Eq.(7.166) can be expressed as

$$J_M^-(z',t') = -ec \sum_k \left\{ \frac{a_w}{\gamma'_k} \exp\left[i\gamma_p k_0(z' + v_p t') \right] \right. \qquad (7.167)$$

$$\left. - \frac{\hat{a}_s(z',t')}{\gamma'_k} \exp\left[i(k'_s z' - \omega'_s t') + i\delta'_s(z',t') \right] \right\} \delta\left[z' - z'_k(t') \right].$$

Moreover, some straightforward algebra that makes use of Eq.(7.165), and the definitions $p'_{xk} = \gamma'_k m_e v'_{xk}$ and $p'_{yk} = \gamma'_k m_e v'_{yk}$ gives

$$\gamma_k'^2 = 1 + \frac{p'^2_{zk}}{m_e^2 c^2} + a_w^2 + |\hat{a}_s|^2 - 2a_w \hat{a}_s(z'_k, t') \cos\left[\theta'_k(t') + \delta'_s(z'_k, t') \right]. \quad (7.168)$$

Here, $\theta'_k(t')$ is the dimensionless axial coordinate defined by

$$\theta'_k(t') = k'_p z'_k(t'), \qquad (7.169)$$

where

$$k'_p = k'_s + \gamma_p k_0 = \frac{(k_s + k_0)}{\gamma_p}. \qquad (7.170)$$

An important feature of transforming to the ponderomotive frame is evident from Eq.(7.168). Because $\hat{a}_s$, δ'_s and θ'_k are slowly varying, it follows that γ'_k is also slowly varying, and the particle energy does not undergo a rapid temporal modulation in ponderomotive-frame variables.

In the wave equation (7.166), we introduce the complex amplitude

$$a_s(z',t') = \hat{a}_s(z',t') \exp\left[i\delta'_s(z',t') \right]. \qquad (7.171)$$

Consistent with the assumption that $a_s(z', t')$ is slowly varying with z' and t', the second-derivative contributions with respect to z' and t' are neglected in Eq.(7.166), but terms proportional to $\partial a_s/\partial t'$ and $\partial a_s/\partial z'$ are retained (eikonal approximation). Furthermore, we operate on Eq.(7.166) with $\int_0^{L'}(dz'/L')\exp(-ik_s'z' + i\omega_s't')\cdots$, where $L' = 2\pi/k_s'$ is the fundamental periodicity length for the (fast) spatial oscillations in the ponderomotive frame. Treating the variation of a_s, γ_k' and θ_k' as slow, the wave equation (7.166) then gives

$$\left\{\left(\omega_s'^2 - c^2 k_s'^2 - \frac{4\pi e^2}{m_e L'}\left\langle\sum_k\frac{1}{\gamma_k'}\right\rangle\right)a_s + 2i\omega_s'\left(\frac{\partial}{\partial t'} + \frac{c^2 k_s'}{\omega_s'}\frac{\partial}{\partial z'}\right)a_s\right\}$$
$$= -\frac{4\pi e^2 a_w}{m_e L'}\left\langle\sum_k\frac{\exp(-i\theta_k')}{\gamma_k'}\right\rangle. \tag{7.172}$$

Separating Eq.(7.172) into fast and slow contributions gives

$$\omega_s'^2 = c^2 k_s'^2 + \frac{4\pi e^2}{m_e L'}\left\langle\sum_k\frac{1}{\gamma_k'}\right\rangle, \tag{7.173}$$

and

$$2i\omega_s'\left(\frac{\partial}{\partial t'} + \frac{c^2 k_s'}{\omega_s'}\frac{\partial}{\partial z'}\right)a_s = -\frac{4\pi e^2 a_w}{m_e L'}\left\langle\sum_k\frac{\exp(-i\theta_k')}{\gamma_k'}\right\rangle. \tag{7.174}$$

Equation (7.173) determines the real oscillation frequency ω_s' in terms of k_s' and beam dielectric effects (proportional to $\langle\sum_k 1/\gamma_k'\rangle$). On the other hand, Eq.(7.174) describes the (slow) evolution of the complex amplitude $a_s(z', t') = \hat{a}_s(z', t')\exp[i\delta_s'(z', t')]$ induced by the wiggler field a_w. Separating Eq.(7.174) into real and imaginary parts readily gives

$$2\omega_s'\left(\frac{\partial}{\partial t'} + \frac{c^2 k_s'}{\omega_s'}\frac{\partial}{\partial z'}\right)\hat{a}_s = \frac{4\pi e^2 a_w}{m_e L'}\left\langle\sum_k\frac{\sin(\theta_k' + \delta_s')}{\gamma_k'}\right\rangle \tag{7.175}$$

and

$$2\omega_s'\hat{a}_s\left(\frac{\partial}{\partial t'} + \frac{c^2 k_s'}{\omega_s'}\frac{\partial}{\partial z'}\right)\delta_s' = \frac{4\pi e^2 a_w}{m_e L'}\left\langle\sum_k\frac{\cos(\theta_k' + \delta_s')}{\gamma_k'}\right\rangle. \tag{7.176}$$

Equations (7.175) and (7.176) are fully equivalent to the (complex) wave equation (7.174).

The orbit equations for $\theta'_k(t')$ and $\gamma'_k(t')$ in the ponderomotive frame are determined from

$$\frac{d}{dt'}p'_{zk} = -m_e c^2 \frac{\partial}{\partial z'_k}\gamma'_k \,,$$

$$\frac{d}{dt'}\gamma'_k = \frac{\partial}{\partial t'}\gamma'_k \,, \tag{7.177}$$

where $p'_{zk} = \gamma'_k m_e dz'_k/dt'$ and γ'_k is defined in Eq.(7.168). To leading order, we neglect the variation of $\hat{a}_s(z'_k, t')$ and $\delta'_s(z'_k, t')$ with respect to z'_k and t'. Equation (7.168) then gives $\partial\gamma'_k/\partial t' \simeq 0$, and $2\gamma'_k\partial\gamma'_k/\partial z'_k \simeq 2k'_p a_w \hat{a}_s \sin(\theta'_k + \delta'_s)$. The orbit Eq.(7.177) for the evolution of $\theta'_k(t') = k'_p z'_k(t')$ then reduces to the pendulum-like equation

$$\frac{d^2}{dt'^2}\theta'_k + \frac{c^2 a_w \hat{a}_s k'^2_p}{\gamma'^2_k}\sin(\theta'_k + \delta'_s) = 0 \,. \tag{7.178}$$

Here, the (slow) evolution of the wave amplitude $\hat{a}_s(z', t')$ and phase $\delta'_s(z', t')$ is determined self-consistently from Eqs.(7.175) and (7.176). Equations (7.173), (7.175), (7.176), and (7.178) constitute closed equations for the nonlinear evolution of $\hat{a}_s(z', t')$, $\delta'_s(z', t')$ and $\theta'_k(t')$. Several points are noteworthy in this regard.[122]

(a) First, the orbits $\theta'_k(t') = k'_p z'_k(t')(k = 1, \cdots, \bar{N}_e)$ occurring in the statistical averages on the right-hand sides of Eqs.(7.175) and (7.176) are determined self-consistently from Eq.(7.178) in terms of the wiggler and electromagnetic fields.

(b) As the orbits $\theta'_k(t')$ are modified by the electromagnetic fields in Eq.(7.178), the wave amplitude $\hat{a}_s$ and phase δ'_s evolve according to Eqs.(7.175) and (7.176). Indeed, Eqs.(7.175), (7.176), and (7.178) (or versions thereof in the laboratory frame) can form the basis for numerical simulations of the nonlinear evolution of the primary signal and the sideband instability.

(c) Finally, there is some latitude in specifying the precise operational meaning of the statistical averages $\langle \sum_k \cdots \rangle$ occurring in the wave equations (7.175) and (7.176). For present purposes, let us assume that the orbits $\theta'_k(t')$ have been calculated in terms of the initial values $\theta'_k(0)$ and $\gamma'_k(0)$. Then the simplest definition of the statistical average $\langle \sum_k \cdots \rangle$ over some phase function $\psi(\theta'_k(0), \gamma'_k(0))$ is given by

$$\frac{1}{L'} \left\langle \sum_k \psi\left(\theta'_k(0), \gamma'_k(0)\right) \right\rangle = \hat{n}'_e \int_0^{2\pi} \frac{d\theta'_0}{2\pi} \int_1^\infty d\gamma'_0 \, F_e(\theta'_0, \gamma'_0) \psi(\theta'_0, \gamma'_0).$$

(7.179)

Here, $\hat{n}'_e$ is the average density of the beam electrons in the ponderomotive frame, and $F_e(\theta'_0, \gamma'_0)$ is the (probability) distribution of electrons in initial phase θ'_0 and energy γ'_0.

Equations (7.173), (7.175), (7.176), and (7.178) can be used to investigate detailed linear and nonlinear properties of the free electron laser instability for a tenuous electron beam (Compton-regime approximation) in circumstances where a single primary electromagnetic wave (ω_s, k_s) is excited. Because stability properties in the small-signal regime have been discussed in Secs. 7.6–7.8, we focus here on an investigation of the sideband instability[122] for small-amplitude perturbations about an established electromagnetic signal which has evolved to a quasisteady equilibrium $(\partial/\partial t' = 0)$ with finite amplitude $\hat{a}^0_s$ and phase δ^0_s.

For a tenuous electron beam, we approximate Eq.(7.173) by the vacuum dispersion relation $\omega_s^2 = c^2 k_s^2$. Assuming a forward-moving electromagnetic wave, the simultaneous resonance conditions, $\omega_s = +ck_s$ and $\omega_s = (k_s + k_0)v_p$, can be solved to give

$$k_s = \gamma_p^2 \left(1 + \frac{v_p}{c}\right) \frac{v_p}{c} k_0 \,,$$

$$\omega_s = \gamma_p^2 \left(1 + \frac{v_p}{c}\right) k_0 v_p \,,$$

(7.180)

Here, γ_p is defined by $\gamma_p = (1 - v_p^2/c^2)^{-1/2}$, and $v_p = \omega_s/(k_s + k_0)$ is (nearly) synchronous with the average axial velocity V_{ze} of the beam electrons. For future reference, it is useful to introduce the dimensionless gain parameter Γ_e and the bounce frequency $\omega_B(\gamma'_k)$ of the trapped electrons [Eq.(7.178) and Fig. 7.21] defined by

$$\Gamma_e^3 = \frac{a_w^2}{4\hat{\gamma}'^3} \frac{\omega_{pe}^2}{\gamma_p^3 c^2 k_0^2} \frac{(1 + v_p/c)}{v_p/c} \ll 1$$

(7.181)

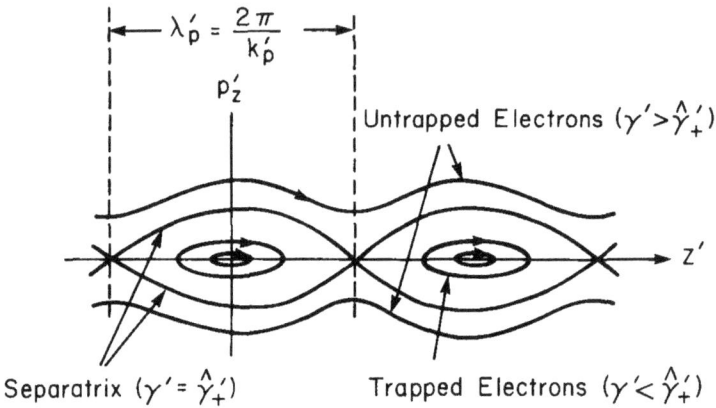

Figure 7.21. Electron trajectories in the ponderomotive-frame phase space (z', p_z'). For constant wave amplitude $\hat{a}_s^0$ and phase δ_s^0, the orbit equation (7.178) is a pendulum equation for $\theta_k' = k_p' z_k'$. Electrons with energy $\gamma' < \hat{\gamma}_+' \equiv [1 + (a_w + \hat{a}_s^0)^2]^{1/2}$ are *trapped* and exhibit periodic motion in the ponderomotive potential. (Here, $a_w > 0$ and $\hat{a}_s^0 > 0$ are assumed.) Electrons with energy $\gamma' \simeq \hat{\gamma}_-' \equiv [1 + (a_w - \hat{a}_s^0)^2]^{1/2}$ are *deeply trapped* near the bottom of the ponderomotive potential (see Problem 7.16).

and

$$\omega_B(\gamma_k') = \left(\frac{c^2 k_p'^2 a_w \hat{a}_s^0}{\gamma_k'^2} \right)^{1/2} . \tag{7.182}$$

In Eqs.(7.181) and (7.182), $k_p' = (k_s + k_0)/\gamma_p = \gamma_p(1 + v_p/c)k_0$, $\omega_{pe}^2 = 4\pi \hat{n}_e e^2/m_e$ is the plasma frequency-squared in the laboratory frame, $\hat{n}_e' = \hat{n}_e/\gamma_p$ is the average electron density in the ponderomotive frame, and the characteristic energy $\hat{\gamma}'$ of an electron trapped in the ponderomotive potential is given approximately by

$$\hat{\gamma}' = \left(1 + a_w^2\right)^{1/2} . \tag{7.183}$$

[See Eq.(7.168) with $p_{zk}' \simeq 0$ and $|\hat{a}_s| \ll a_w$.] Moreover, $\omega_B(\gamma_k')$ is the bounce frequency (in the ponderomotive frame) of an electron with energy

γ'_k trapped near the bottom of the ponderomotive potential. For $\hat{\gamma}'_k \simeq \hat{\gamma}'$, the bounce frequency Ω_B in the laboratory frame is defined by

$$\Omega_B = \frac{\omega_B(\hat{\gamma}')}{\gamma_p} = \left(\frac{c^2 k_p'^2 a_w \hat{a}_s^0}{\hat{\gamma}'^2 \gamma_p^2} \right)^{1/2} = \left(1 + \frac{v_p}{c} \right) \left(\frac{a_w \hat{a}_s^0}{1 + a_w^2} \right)^{1/2} ck_0 \quad (7.184)$$

for deeply trapped electrons.

Problem 7.16   Consider the particle orbit equation (7.178) in ponderomotive frame variables for $\hat{a}_s = \hat{a}_s^0 = \text{const.}$ and $\delta'_s = \delta_s^0 = 0$. Dropping the particle label, Eq.(7.178) can be expressed as

$$\frac{d^2}{dt'^2} \theta' + \frac{c^2 a_w \hat{a}_s^0 k_p'^2}{\gamma'^2} \sin \theta' = 0 , \quad (7.16.1)$$

where $\theta'(t') = k_p' z'(t')$ is the dimensionless axial coordinate, and the energy $\gamma' = \text{const.}$ is defined by

$$\gamma'^2 = 1 + \frac{p_z'^2}{m_e^2 c^2} + a_w^2 + \hat{a}_s^{02} - 2a_w \hat{a}_s^0 \cos \theta' , \quad (7.16.2)$$

where $p_z' = \gamma' m_e dz'/dt'$.

a.   Without loss of generality, assume $\hat{a}_s^0 > 0$ and $a_w > 0$. Show that θ' satisfies

$$\left(\frac{d}{dt'} \theta' \right)^2 = c^2 k_p'^2 \beta'^2 \left[1 - \kappa^2 \sin^2 \left(\frac{\theta'}{2} \right) \right] , \quad (7.16.3)$$

where κ and β' are defined by

$$\kappa^2 = \frac{4 a_w \hat{a}_s^0}{\left(\gamma'^2 - \gamma_+'^2 + 4 a_w \hat{a}_s^0 \right)} , \quad (7.16.4)$$

$$\beta' = \frac{1}{\gamma'} \left(\gamma'^2 - \gamma_+'^2 + 4 a_w \hat{a}_s^0 \right)^{1/2} , \quad (7.16.5)$$

and γ_+' is defined by

$$\gamma_+' = \left[1 + (a_w + \hat{a}_s^0)^2 \right]^{1/2} . \quad (7.16.6)$$

b.   Show that the condition $\kappa^2 < 1$ is equivalent to $\gamma' > \gamma_+'$. In this case the electron motion is "untrapped" but modulated by the ponderomotive potential (Fig. 7.21). Integrate Eq.(7.16.3) for $\kappa^2 < 1$ and show that[120]

$$F \left(\frac{\theta'}{2}, \kappa \right) = \frac{1}{2} c k_p' \beta' t' + F \left(\frac{\theta_0'}{2}, \kappa \right) , \quad (7.16.7)$$

where $\theta' = k_p'z'(t')$ and $\theta_0' = k_p'z'(t' = 0)$. Here, $F(\eta, \kappa)$ is the elliptic integral of the first kind defined by

$$F(\eta, \kappa) = \int_0^\eta \frac{d\eta'}{(1 - \kappa^2 \sin^2 \eta')^{1/2}} . \tag{7.16.8}$$

c. The electrons are "trapped" by the ponderomotive potential (Fig. 7.21) whenever $\kappa^2 > 1$ and $\gamma_-'(\theta') < \gamma' < \gamma_+'$. Here, $\gamma_-'(\theta')$ is defined by

$$\gamma_-'(\theta') = \left[1 + a_w^2 + \hat{a}_s^{02} - 2a_w\hat{a}_s^0 \cos\theta'\right]^{1/2} , \tag{7.16.9}$$

which corresponds to the turning points where $p_z' = 0$ in Eq.(7.16.2). Define $\kappa_T^2 = 1/\kappa^2$ and introduce the new independent variable θ'' defined by

$$\kappa_T \sin\theta'' = \sin\theta' . \tag{7.16.10}$$

Show that the orbit equation (7.16.3) can be expressed as

$$\left(\frac{d}{dt'}\theta''\right)^2 = \omega_B^2(\gamma')\left(1 - \kappa_T^2 \sin^2\theta''\right) \tag{7.16.11}$$

for the trapped electrons. Here, $\omega_B(\gamma')$ is the bounce frequency defined by

$$\omega_B(\gamma') = \frac{ck_p'\kappa\beta'}{2} = \left(\frac{a_w\hat{a}_s^0 c^2 k_p'^2}{\gamma'^2}\right)^{1/2} . \tag{7.16.12}$$

The (periodic) motion described by Eq.(7.16.11) can also be expressed in terms of elliptic integrals.[120]

d. For $\gamma' \simeq \gamma_-' \equiv [1 + (a_w - \hat{a}_s^0)^2]^{1/2}$, the electrons are "deeply" trapped and exhibit simple harmonic oscillations about $\theta' = 0, \pm\pi, \pm2\pi, \cdots$, with frequency $\omega_B(\gamma')$. In this case, show that Eq.(7.16.3) can be approximated by

$$\frac{1}{2}\left(\frac{d}{dt'}\theta'\right)^2 + \frac{1}{2}\omega_B^2\theta'^2 = \frac{1}{2}c^2 k_p'^2 \beta'^2 \tag{7.16.13}$$

for oscillations about $\theta' = 0$.

7.10.2   Macroclump Model of the Sideband Instability

Equations (7.175), (7.176) and (7.178) can be used to investigate the linear and nonlinear evolution of the sideband instability. For present purposes, we assume that all of the electrons are trapped by the ponderomotive potential (Fig. 7.21), and that the trapped electrons can be treated as tightly bunched macroclumps with average density $\hat{n}_e' = \hat{n}_e/\gamma_p$

and axial coordinate $\theta'_k = \theta'_s$ within each period $\lambda'_p = 2\pi/k'_p$ of the ponderomotive potential. For $\gamma'_k \simeq \hat{\gamma}'$, Eqs. (7.175), (7.176) and (7.178) can be approximated by[122]

$$\left(\frac{\partial}{\partial t'} + c\frac{\partial}{\partial z'}\right)\hat{a}_s = \frac{a_w\omega_{pe}^2}{2\omega'_s\hat{\gamma}'\gamma_p}\sin(\theta'_s + \delta'_s) , \qquad (7.185)$$

$$\hat{a}_s\left(\frac{\partial}{\partial t'} + c\frac{\partial}{\partial z'}\right)\delta'_s = \frac{a_w\omega_{pe}^2}{2\omega'_s\hat{\gamma}'\gamma_p}\cos(\theta'_s + \delta'_s) , \qquad (7.186)$$

and

$$\frac{d^2}{dt'^2}\theta'_s + \frac{c^2 k_p'^2 a_w\hat{a}_s}{\hat{\gamma}'^2}\sin(\theta'_s + \delta'_s) = 0 , \qquad (7.187)$$

where use has been made of $\omega'_s \simeq ck'_s$. It is further assumed that the electrons are deeply trapped with $\theta'_s + \delta'_s \simeq 2n\pi(n = 0, \pm1, \pm2, \cdots)$. Without loss of generality, we take $n = 0$ and expand Eqs.(7.185)–(7.187) for small $\theta'_s + \delta'_s$. This readily gives

$$\left(\frac{\partial}{\partial t'} + c\frac{\partial}{\partial z'}\right)\hat{a}_s = \frac{a_w\omega_{pe}^2}{2\omega'_s\hat{\gamma}'\gamma_p}(\theta'_s + \delta'_s) , \qquad (7.188)$$

$$\hat{a}_s\left(\frac{\partial}{\partial t'} + c\frac{\partial}{\partial z'}\right)\delta'_s = \frac{a_w\omega_{pe}^2}{2\omega'_s\hat{\gamma}'\gamma_p} , \qquad (7.189)$$

$$\frac{d^2}{dt'^2}\theta'_s + \frac{c^2 k_p'^2 a_w\hat{a}_s}{\hat{\gamma}'^2}(\theta'_s + \delta'_s) = 0 , \qquad (7.190)$$

correct to sufficient accuracy to develop a linear theory of the sideband instability.

A. Equilibrium Model

We consider perturbations about the quasisteady equilibrium state $(\hat{a}_s^0, \delta_s^0, \theta_s^0)$ described by

$$\theta_s^0 + \delta_s^0 = 0 ,$$

$$\frac{\partial}{\partial t'}\hat{a}_s^0 = 0 = \frac{\partial}{\partial z'}\hat{a}_s^0 , \qquad (7.191)$$

and

$$\frac{\partial}{\partial t'}\delta_s^0 = 0 \, ,$$

$$\hat{a}_s^0 c \frac{\partial}{\partial z'}\delta_s^0 = \frac{a_w \omega_{pe}^2}{2\omega_s' \hat{\gamma}' \gamma_p} \, . \tag{7.192}$$

That is, the equilibrium wave amplitude $\hat{a}_s^0$ is constant (independent of z' and t'), whereas there is a slow variation of the equilibrium phase δ_s^0 with z' described by Eq.(7.192). Making use of Eqs.(7.180)–(7.184), it is convenient to introduce the small parameter $\varepsilon \ll 1$ defined by

$$\varepsilon c k_p' = 2\Gamma_e \left(\frac{\Gamma_e c k_0}{\Omega_B}\right)^2 c k_p' = \frac{a_w \omega_{pe}^2}{2\omega_s' \hat{\gamma}' \gamma_p \hat{a}_s^0} \, . \tag{7.193}$$

Equation (7.192) can be expressed as $\partial \delta_s^0/\partial z' = \varepsilon k_p'$, where $\varepsilon \ll 1$ is required in order that the change in δ_s^0 be small over one wavelength $(\lambda_p' = 2\pi/k_p')$ of the ponderomotive potential. Note that the inequality $\varepsilon \ll 1$ requires that the pump amplitude be above a certain small threshold value $(\Omega_B^2/c^2 k_0^2 \gg 2\Gamma_e^3)$ for the present analysis to be valid.

B. Linear Stability Analysis

We now express

$$\hat{a}_s = \hat{a}_s^0 + \delta\hat{a}_s \, ,$$

$$\delta_s' = \delta_s^0 + \tilde{\delta}_s' \, ,$$

$$\theta_s' = \theta_s^0 + \delta\theta_s' \, , \tag{7.194}$$

where $\delta\hat{a}_s$, $\tilde{\delta}_s'$ and $\delta\theta_s'$ denote small perturbations. Linearizing Eqs.(7.188)–(7.190) about the equilibrium state described by Eqs.(7.191) and (7.192) readily gives[122]

$$\left(\frac{\partial}{\partial t'} + c\frac{\partial}{\partial z'}\right)\delta\hat{a}_s = \varepsilon c k_p' \hat{a}_s^0 (\delta\theta_s' + \tilde{\delta}_s') \, , \tag{7.195}$$

$$\hat{a}_s^0 \left(\frac{\partial}{\partial t'} + c\frac{\partial}{\partial z'}\right)\tilde{\delta}_s' + \varepsilon c k_p' \delta\hat{a}_s = 0 \, , \tag{7.196}$$

$$\frac{d^2}{dt'^2}\delta\theta_s' + \omega_B^2(\hat{\gamma}')(\delta\theta_s' + \tilde{\delta}_s') = 0 \, , \tag{7.197}$$

where $\epsilon c k_p'$ is defined in Eq.(7.193) and $\hat{\omega}_B(\hat{\gamma}') = (c^2 k_p'^2 a_w \hat{a}_s^0 / \hat{\gamma}'^2)^{1/2}$. Assuming that the z'- and t'-dependence in Eqs.(7.195)–(7.197) is proportional to $\exp[-i(\Delta\omega')t' + i(\Delta k_z')z']$, we obtain

$$-i(\Delta\omega' - c\Delta k_z')\delta\hat{a}_s = \epsilon c k_p' \hat{a}_s^0 (\delta\theta_s' + \tilde{\delta}_s') , \tag{7.198}$$

$$-i\hat{a}_s^0 (\Delta\omega' - c\Delta k_z')\tilde{\delta}_s' = -\epsilon c k_p' \delta\hat{a}_s , \tag{7.199}$$

$$\left[(\Delta\omega')^2 - \omega_B^2(\hat{\gamma}')\right] (\delta\theta_s' + \tilde{\delta}_s') = (\Delta\omega')^2 \tilde{\delta}_s' . \tag{7.200}$$

Some straightforward algebraic manipulation of Eqs.(7.198)–(7.200) gives the desired dispersion relation[122]

$$0 = 1 - \frac{\omega_B^2(\hat{\gamma}')}{(\Delta\omega')^2} - \frac{\epsilon^2 c^2 k_p'^2}{(\Delta\omega' - c\Delta k_z')^2} \tag{7.201}$$

in the ponderomotive frame.

The frequency $\Delta\omega'$ and wavenumber $\Delta k_z'$ in the ponderomotive frame are related to $\Delta\omega$ and Δk_z in the laboratory frame by

$$\Delta\omega' = \gamma_p(\Delta\omega - v_p\Delta k_z) ,$$

$$\Delta k_z' = \gamma_p\left(\Delta k_z - \frac{\Delta\omega v_p}{c^2}\right) , \tag{7.202}$$

where $v_p = \omega_s/(k_s + k_0)$ and $\gamma_p = (1 - v_p^2/c^2)^{-1/2}$. Making use of $k_p' = (k_s + k_0)/\gamma_p = \gamma_p(1 + v_p/c)k_0$, it is readily shown that $\Delta\omega' - c\Delta k_z' = (k_p'/k_0)(\Delta\omega - c\Delta k_z)$. Defining $\Delta\Omega \equiv \Delta\omega - v_p\Delta k_z$ and $\Delta K_z \equiv k_0(v_p/c)(\Delta k_z/k_s)$, where $k_s = \gamma_p^2(1 + v_p/c)k_0 v_p/c$, the dispersion relation (7.201) can be expressed in the convenient form

$$0 = 1 - \frac{\Omega_B^2}{(\Delta\Omega)^2} - \frac{4\Omega_B^2(\Gamma_e c k_0/\Omega_B)^6}{(\Delta\Omega - c\Delta K_z)^2} . \tag{7.203}$$

Here, $\Omega_B = \omega_B(\hat{\gamma}')/\gamma_p$ is the bounce frequency of deeply trapped electrons in the laboratory frame. Use has been made of Eq.(7.193) to eliminate ϵ in favor of Γ_e and Ω_B, and the dimensionless gain parameter Γ_e is defined in Eq.(7.181).[122]

Problem 7.17   Consider the linearized equations (7.195)–(7.197) for the perturbed quantities $\delta\hat{a}_s$, $\delta\theta_s'$ and $\tilde{\delta}_s'$.

a. Assume that the z'- and t'-dependence in Eqs.(7.195)–(7.197) is proportional to $\exp[-i(\Delta\omega')t' + i(\Delta k_z')z']$. Derive the dispersion relation (7.201) for the sideband instability. Equation (7.201) relates $\Delta\omega'$ and $\Delta k'_z$ in the ponderomotive frame.

b. Make use of Eqs.(7.201) and (7.202) and $k'_p = (k_s + k_0)/\gamma_p = \gamma_p(1 + v_p/c)k_0$ to obtain the dispersion relation (7.203), which relates $\Delta\omega$ and Δk_z in the laboratory frame.

7.10.3   Analysis of the Sideband Instability

Equation (7.203) has the familiar form of the dispersion relation for the two-stream instability (see Problem 5.13). Here, $\Omega_B^2 \to \hat{\omega}_{p1}^2$ plays the role of the first plasma component, and $\Omega_B^2 4(\Gamma_e ck_0/\Omega_B)^6 \to \hat{\omega}_{p2}^2$ plays the role of the second plasma component which is drifting with velocity c relative to the first component. Equation (7.203) can be solved numerically for the real oscillation frequency $\mathrm{Re}(\Delta\Omega)$ and growth rate $\mathrm{Im}(\Delta\Omega)$ in terms of $c\Delta K_z$, Ω_B and $\Omega_B/\Gamma_e k_0 c$ over a wide range of system parameters. For present purposes, we make use of analytical estimates to determine the instability bandwidth and maximum growth rate from Eq.(7.203).[122]

First, it can be shown from Eq.(7.203) that instability exists [$\mathrm{Im}(\Delta\Omega) = \mathrm{Im}(\delta\omega) > 0$] for ΔK_z in the range

$$-\Delta K_b < \Delta K_z < \Delta K_b, \qquad (7.204)$$

where the bandwidth ΔK_b is given (exactly) by

$$c\Delta K_b = \Omega_B \left\{ 1 + \left[4\left(\frac{\Gamma_e ck_0}{\Omega_B}\right)^6 \right]^{1/3} \right\}^{3/2}. \qquad (7.205)$$

As illustrated schematically in Fig. 7.22, the growth rate $\mathrm{Im}(\Delta\Omega) = \mathrm{Im}(\delta\omega)$ is equal to zero for $\Delta K_z = 0$ and $\Delta K_z = \pm\Delta K_b$, and achieves its maximum value at $\Delta K_z = \pm\Delta K_m$. Equation (7.205) is valid for arbitrary pump strength ranging from the strong-pump regime ($\Omega_B/\Gamma_e k_0 c \gg 1$) to the weak-pump regime ($\Omega_B/\Gamma_e k_0 c \ll 1$). Moreover, it can be shown exactly from Eq.(7.203) that the real oscillation frequency of the unstable branch increases from $\mathrm{Re}(\Delta\Omega) = 0$, for $\Delta K_z = 0$, to

$$\mathrm{Re}(\Delta\Omega) = +\Omega_B \left\{ 1 + \left[4\left(\frac{\Gamma_e ck_0}{\Omega_B}\right)^6 \right]^{1/3} \right\}^{3/2}, \ \text{ for } \Delta K_z = +\Delta K_b.$$

$$(7.206)$$

$$| \text{Im} (\Delta\Omega)$$

$$[\text{Im} (\Delta\Omega)]_{max}$$

$$-\Delta K_b$$
$$-\Delta K_m$$
$$O$$
$$\Delta K_b$$
$$\Delta K_m$$
$$\Delta K_z$$

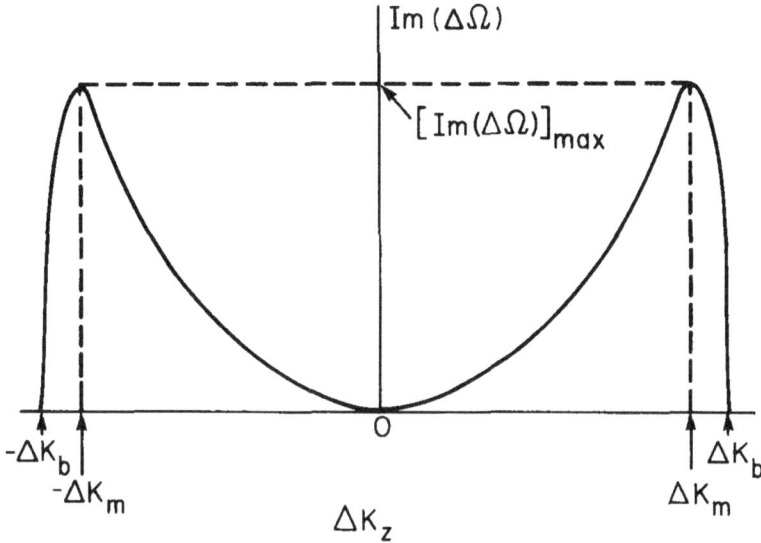

Figure 7.22. Schematic plot of the sideband instability growth rate Im($\Delta\Omega$) versus ΔK_z obtained from the dispersion relation (7.203).

The range of oscillation frequencies described by Eq.(7.206) corresponds to the upper sideband. On the other hand, for ΔK_z in the interval $-\Delta K_b < \Delta K_z < 0$, the lower sideband is unstable, and the polarity of Re($\Delta\Omega$) is reversed relative to Eq.(7.206).[122] Because Im($\Delta\Omega$) is an even function of ΔK_z and Re($\Delta\Omega$) is an odd function of ΔK_z, without loss of generality we limit the subsequent analysis to the interval $0 < \Delta K_z < \Delta K_b$.

Although the bandwidth ΔK_b can be calculated analytically for arbitrary pump strength $\Omega_B/\Gamma_e c k_0$, the growth rate Im($\Delta\Omega$) must generally be determined numerically from Eq.(7.203). However, analytical estimates of the maximum growth rate can be made in both the weak-pump and strong-pump limits. In this regard, it should be kept in mind that $\Gamma_e \ll 1$ is assumed in the present analysis [Eq.(7.181)].

In the weak-pump regime with $\Omega_B/\Gamma_e c k_0 \ll 1$, we also require $(\Omega_B/\Gamma_e c k_0)^2 \gg 2\Gamma_e$ in order to be consistent with the assumption of slowly varying phase δ_s^0, i.e., $\varepsilon \ll 1$ in Eq.(7.193). For $\Omega_B/\Gamma_e c k_0 \ll 1$, it follows from Eq.(7.205) that the instability bandwidth is given approximately by

$$c\Delta K_b = 2\Gamma_e c k_0 \left(\frac{\Gamma_e c k_0}{\Omega_B}\right)^2 \left[1 + \frac{3}{(2)^{5/3}}\left(\frac{\Omega_B}{\Gamma_e c k_0}\right)^2 + \cdots\right] . \quad (7.207)$$

Because $(\Omega_B/\Gamma_e c k_0)^2 \gg 2\Gamma_e$ is required, note from Eq.(7.207) that the instability bandwidth ΔK_b in the weak-pump regime is relatively narrow in units of k_0. It can also be shown from Eq.(7.203) that the maximum growth rate $[\mathrm{Im}(\Delta\Omega)]_{\max} = [\mathrm{Im}(\Delta\omega)]_{\max}$ in the weak-pump regime can be approximated by[122]

$$[\mathrm{Im}(\Delta\Omega)]_{\max} = \frac{(3)^{1/2}}{2}\Gamma_e c k_0 . \qquad (7.208)$$

Moreover, maximum growth occurs for $\Delta K_z = \Delta K_m$, where ΔK_m is defined by

$$c\Delta K_m = 2\Gamma_e c k_0 \left(\frac{\Gamma_e c k_0}{\Omega_B}\right)^2 \left[1 + \frac{3}{16}\left(\frac{\Omega_B}{\Gamma_e c k_0}\right)^4 + \cdots\right] . \qquad (7.209)$$

Comparing Eqs.(7.207) and (7.209), note that ΔK_m is only *slightly* downshifted from ΔK_b. That is, the growth rate $\mathrm{Im}(\Delta\Omega)$ is peaked very close to the upper end of the unstable wavenumber range in Fig. 7.22.

The dispersion relation (7.203) must generally be solved numerically when $\Omega_B/\Gamma_e c k_0 \approx 1$. However, for the special case where

$$\frac{\Omega_B}{\Gamma_e c k_0} = (2)^{1/3} , \qquad (7.210)$$

the dispersion relation (7.203) can be solved exactly. Substituting Eq.(7.210) into Eq.(7.203) gives[122]

$$0 = 1 - \frac{\Omega_B^2}{(\Delta\Omega)^2} - \frac{\Omega_B^2}{(\Delta\Omega - c\Delta K_z)^2} , \qquad (7.211)$$

which is the two-stream dispersion relation for "equidensity" streams with effective plasma frequency Ω_B. It is readily shown from Eq.(7.211) that instability exists for ΔK_z in the range $-\Delta K_b < \Delta K_z < \Delta K_b$ where

$$c\Delta K_b = (2)^{3/2}\Omega_B . \qquad (7.212)$$

Moreover, the growth rate $\mathrm{Im}(\Delta\Omega)$ and real oscillation frequency $\mathrm{Re}(\Delta\Omega)$ of the unstable branch are given by

$$\mathrm{Im}(\Delta\Omega) = \Omega_B \left\{\left[1 + 8\left(\frac{\Delta K_z}{\Delta K_b}\right)^2\right]^{1/2} - 1 - 2\left(\frac{\Delta K_z}{\Delta K_b}\right)^2\right\}^{1/2} \quad (7.213)$$

and

$$\text{Re}(\Delta\Omega) = (2)^{1/2}\Omega_B \left(\frac{\Delta K_z}{\Delta K_b}\right) = \frac{c\Delta K_z}{2} \qquad (7.214)$$

for ΔK_z in the interval $-\Delta K_b < \Delta K_z < \Delta K_b$. The maximum growth rate calculated from Eq.(7.213) is

$$[\text{Im}(\Delta\Omega)]_{\text{max}} = \frac{1}{2}\Omega_B , \qquad (7.215)$$

which occurs for $\Delta K_z = \pm\Delta K_m$, where ΔK_m is defined by

$$\Delta K_m = \left(\frac{3}{8}\right)^{1/2} \Delta K_b . \qquad (7.216)$$

At intermediate pump strengths ($\Omega_B/\Gamma_e ck_0 \approx 1$), it is clear from Eqs.(7.214)–(7.216) that the characteristic oscillation frequency and growth rate of the sideband instability are both of order the bounce frequency Ω_B.

Comparing Eqs.(7.208) and (7.215) for specified $\Gamma_e k_0 c$, it is evident that the maximum growth rate varies only slightly for $\Omega_B/\Gamma_e ck_0$ in the range $2\Gamma_e < \Omega_B/\Gamma_e ck_0 \leq (2)^{1/3}$. On the other hand, in the strong-pump regime with $\Omega_B/\Gamma_e ck_0 \gg 1$, it can be shown from Eq.(7.203) that the maximum growth rate decreases rapidly with $[\text{Im}(\Delta\Omega)]_{\text{max}} = [(3)^{1/2}/(2)^{2/3}]\Gamma_e ck_0 \times (\Gamma_e ck_0/\Omega_B)$. This is illustrated in Fig. 7.23 where the normalized maximum growth rate $[\text{Im}(\Delta\Omega)]_{\text{max}}/[(3)^{1/2}\Gamma_e ck_0/2]$ calculated numerically from the dispersion relation (7.203) is plotted versus the dimensionless pump strength $\Omega_B/\Gamma_e ck_0$.[122]

To summarize, within the framework of a macroclump model in which the (deeply trapped) electrons are treated as tightly bunched, the wave equations (7.175) and (7.176) and the particle orbit equation (7.178) have been used to investigate detailed properties of the sideband instability for small-amplitude perturbations about a primary electromagnetic wave with constant amplitude $\hat{a}_s^0$ [Eq.(7.191)] and slowly varying phase δ_s^0 [Eq.(7.192)]. Both the bandwidth [Eq.(7.205)] and maximum growth rate [Fig. 7.23] are found to exhibit a moderately strong dependence on the dimensionless pump strength $\Omega_B/\Gamma_e ck_0$. The theoretical model based on Eqs. (7.175), (7.176) and (7.178) makes use of the orbit equations for individual particles together with appropriate statistical averages over the microscopic particle currents induced by the wiggler and radiation fields. In this regard, the Vlasov-Maxwell equations (7.72)–(7.76) (with $\delta\phi \simeq 0$)

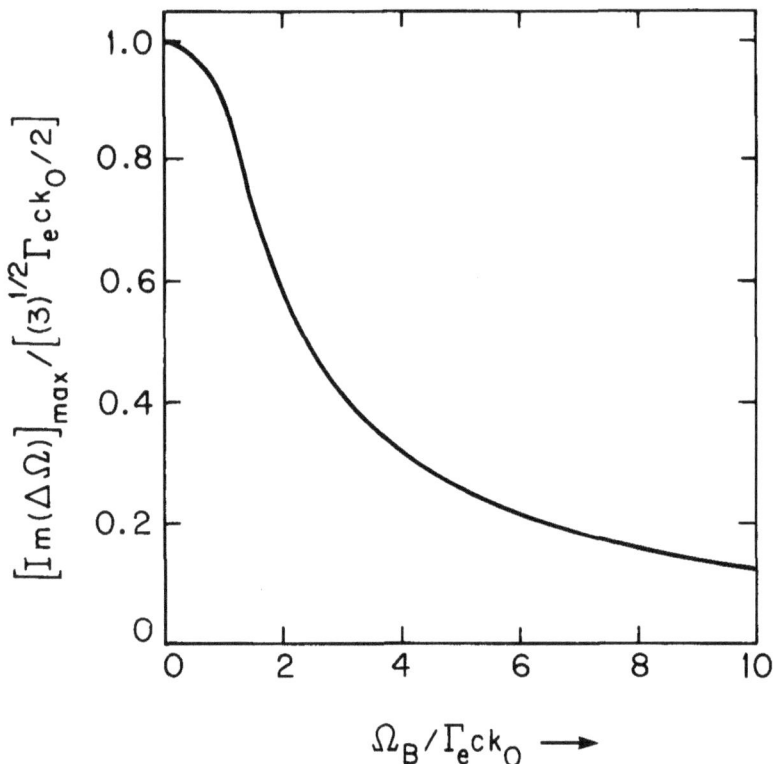

Figure 7.23.   Plot of the normalized maximum growth rate of the sideband instability versus $\Omega_B/\Gamma_e ck_0$ obtained from the dispersion relation (7.203) [R.C. Davidson and J.S. Wurtele, Phys. Fluids **30**, 557 (1987)].

can also be used to develop a self-consistent model[120,121] of the sideband instability, including distribution-function effects for both the trapped electrons and the untrapped electrons.

Problem 7.18   Equation (7.203) has the classic form of the dispersion relation for a two-stream instability (see Problem 5.13).

a.   Show that the bandwidth of the sideband instability calculated from Eq.(7.203) is given exactly by Eq.(7.205).

b.   The special case where $\Omega_B/\Gamma_e ck_0 = (2)^{1/3}$ corresponds to "equidensity" streams. Show that the corresponding growth rate and real frequency of the sideband instability are given by Eqs.(7.213) and (7.214), respectively.

7.10.4   Experimental Results

Experimental investigations of the free electron laser sideband instability
have been carried out at low beam energy (0.8 MeV) including studies
of the effects of tapering and optical guiding,[127] and at moderately high
beam energy (20 MeV) as a function of cavity detuning.[42] In this section,
we summarize briefly the classic experiments by Warren, et al.,[42] who have
made detailed measurements of the sideband instability for a Compton-
regime free electron laser oscillator operating at an output wavelength of
$\lambda_s = 10.35\,\mu$m. The experiment uses an L-band RF linear accelerator
operating at 20 MeV, peak current of 40 A, micropulse width of ~ 35 ps,
and beam energy spread in the range of $\Delta\gamma/\hat{\gamma}_e \sim 2\%$. The electron beam
has radius $r_b = 1$ mm and interacts with a transverse planar wiggler field
with constant wavelength $\lambda_w = 2.73$ cm and constant amplitude $B_w = 3$ kG, which corresponds to a normalized wiggler amplitude $a_w = 0.76$.
The length of the interaction region is $L = 1$ m, and the small-signal gain
at peak output power is about 20% per pass.[42]

Typical experimental results are presented in Fig. 7.24. Figure 7.24(a)
shows the average output power (after passing through a 5% transmit-
ting output coupler) plotted versus small changes in the length (ΔL) of
the optical cavity.[42] The zero of the horizontal scale in Fig. 7.24(a) is
positioned so that $\Delta L = 0$ corresponds to the maximum average output
power of $P = 5$ kW (after 95% attenuation). The corresponding peak
output power inferred from the micropulse width is about 5 MW from the
coupler (or 100 MW in the cavity). Shown in Figs. 7.24(b) and 7.24(c)
are the measured average emission spectra plotted versus the normal-
ized wavelength shift $\Delta\lambda/\lambda_s$, where $\Delta\lambda = 0$ corresponds to emission at
$\lambda_s = 10.35\,\mu$m. The results in Fig. 7.24(b) correspond to an average power
output of 0.8 kW and a moderately large cavity detuning [$\Delta L = -35\,\mu$m
in Fig. 7.24(a)], whereas the results in Fig. 7.24(c) correspond to the maximum average power output of 5 kW [$\Delta L = 0$ in Fig. 7.24(a)]. For the
lower power level in Fig. 7.24(b), there is a strong emission peak centered at $\Delta\lambda = 0$, and the (relatively weak) sideband emission is displaced
from the central peak by about 1%. At the maximum power level in
Fig. 7.24(c), however, the (broadband) sideband excitation is very strong
and irregular, and the total radiated sideband power exceeds the central
emission at $\Delta\lambda = 0$.[42]

While the irregular emission spectrum in Fig. 7.24(c) shows strong
evidence of multimode nonlinear excitation, it is instructive to compare
the measured average displacement of the sideband spectrum [$\Delta\lambda/\lambda_s \simeq 2.4\%$ in Fig. 7.24(c)] with that calculated from the dispersion relation

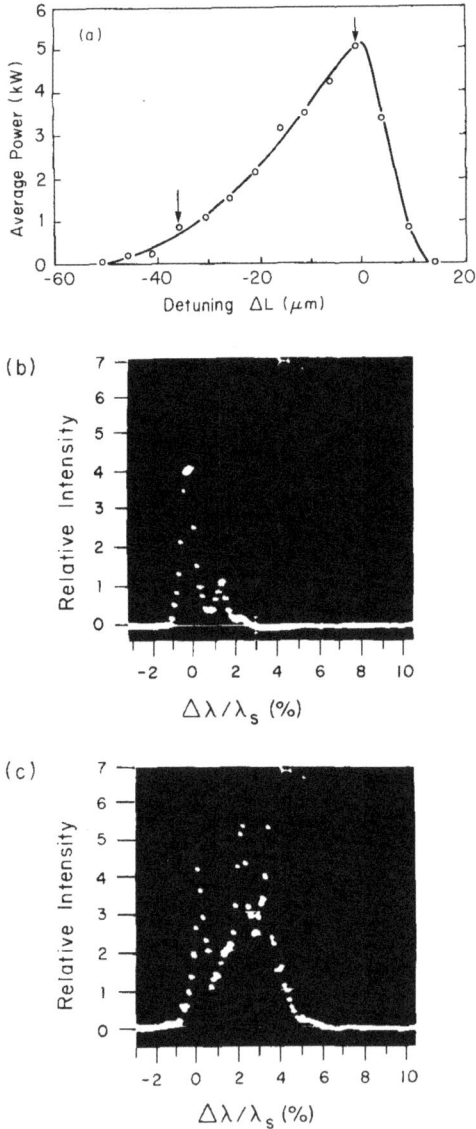

Figure 7.24. Sideband instability measurements for a Compton-regime free electron laser oscillator operating at wavelength $\lambda_s = 10.35\,\mu$m [R.W. Warren, B.E. Newman and J.C. Goldstein, IEEE Journal of Quantum Electronics **QE-21**, 882 (1985)]. Figure 7.24(a) shows a plot of the average output power P versus the change in length ΔL of the optical cavity. The emission spectra are plotted versus the normalized wavelength shift $\Delta\lambda/\lambda_s$ for $P = 0.8\,$kW and $\Delta L = -35\,\mu$m [Fig. 7.24(b)], and for $P = 5\,$kW and $\Delta L = 0$ [Fig. 7.24(c)]. The horizontal scales are chosen so that $\Delta L = 0$ corresponds to maximum average output power, and $\Delta\lambda = 0$ corresponds to emission at $\lambda_s = 10.35\,\mu$m.

(7.203) at maximum growth rate. Taking the peak power in the cavity to be $20 \times 5\,\text{MW} = 100\,\text{MW}$ gives the estimate $\hat{a}_s^0 \simeq 2.4 \times 10^{-4}$ for the normalized amplitude of the radiation field. Making use of the beam current and radius to infer the average beam density $\hat{n}_e = 2.7 \times 10^{11}\text{cm}^{-3}$, the dispersion relation (7.203) can be used to calculate the linear growth characteristics of the sideband instability. At maximum growth rate, it is found that $\text{Re}(\Delta\omega)/\omega_s = \Delta\lambda/\lambda_s \simeq 2.3 \times 10^{-2}$, which is in good qualitative agreement with Fig. 7.24(c). In this regard, keep in mind that Eq.(7.203) was derived for a helical wiggler field, whereas the experiments[42] in Fig. 7.24 employ a planar (linearly polarized) magnetic wiggler.

In conclusion, at sufficiently high power levels (sufficiently large values of $\hat{a}_s$), it is evident from Fig. 7.24 that the sideband instability can play an important role in governing the detailed features of the emission spectrum, leading to irregularities in the spectrum and a substantial increase in the bandwidth of the emitted radiation.

7.11   Harmonic Generation in a Planar Wiggler Free Electron Laser

The analysis in Secs. 7.5–7.10 has featured investigations of the free electron laser instability for a constant-amplitude *helical* wiggler magnetic field [Eq.(7.50)]. This leads to several analytical simplifications. For example, for zero transverse canonical momentum ($P_x = 0 = P_y$) and no radiation field ($\delta\mathbf{A} = 0$), both the axial momentum $p_z = \gamma m_e v_z$ and the energy γ of an individual electron are conserved quantities. In this section, we summarize several properties of the free electron laser instability for the case of a tenuous electron beam (Compton-regime approximation with $\delta\phi \simeq 0$) propagating through the constant-amplitude *planar* wiggler magnetic field (Fig. 7.25)

$$\mathbf{B}_w^0(\mathbf{x}) = -B_w \cos(k_0 z)\hat{\mathbf{e}}_x \ . \tag{7.217}$$

Here, $B_w = \text{const.}$ is the wiggler amplitude, and $\lambda_w = 2\pi/k_0 = \text{const.}$ is the wiggler wavelength. The theoretical model neglects transverse spatial variations ($\partial/\partial x = 0 = \partial/\partial y$), and the beam density and current are assumed to be sufficiently low that the effects of the equilibrium self fields can be neglected. As illustrated in Fig. 7.25, the radiation field is assumed to be plane polarized with electric and magnetic field components $\delta\mathbf{E} = \delta E_y(z,t)\hat{\mathbf{e}}_y$ and $\delta\mathbf{B} = \delta B_x(z,t)\hat{\mathbf{e}}_x$.

Davidson and Wurtele[118,119] have examined detailed properties of the free electron laser instability for a planar magnetic wiggler using a the-

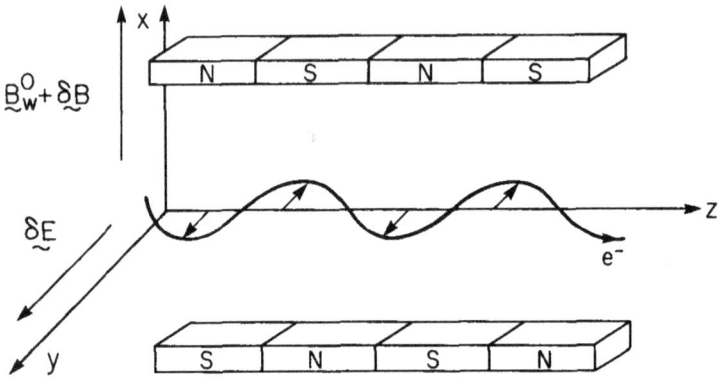

Figure 7.25. Oscillating motion of a relativistic electron in a planar wiggler magnetic field.

oretical model based on the Vlasov-Maxwell equations for the class of self-consistent beam distribution functions

$$f_e(z, \mathbf{p}, t) = \hat{n}_e \delta(p_x) \delta(P_y) F_e(z, p_z, t) . \qquad (7.218)$$

Here, $\mathbf{p} = \gamma m \mathbf{v}$ is the mechanical momentum, and $P_y = p_y - (eB_w/ck_0) \times \sin(k_0 z) - (e/c)\delta A_y(z, t)$ is the canonical momentum in the y-direction, which is exactly conserved. Note from Eq.(7.218) that the transverse motion of the beam electrons in the x- and y-directions is assumed to be "cold," corresponding to zero transverse emittance. In the absence of radiation field ($\delta A_y = 0$), the equilibrium distribution function for the beam electrons, assuming that all electrons are moving in the forward direction with $p_z > 0$, can be expressed as

$$F_e^0(z, p_z) = U(p_z) F_e^+(\gamma) . \qquad (7.219)$$

In Eq.(7.219), $U(p_z)$ is the Heaviside step function defined by $U(p_z) = +1$ for $p_z > 0$ and $U(p_z) = 0$ for $p_z < 0$, and γ is the electron energy defined by

$$\gamma = \left[1 + \frac{p_z^2}{m_e^2 c^2} + a_w^2 \sin^2(k_0 z)\right]^{1/2} \qquad (7.220)$$

for $p_x = 0$, $P_y = 0$ and $\delta A_y = 0$. As before, $a_w = eB_w/m_e c^2 k_0$ is the normalized wiggler amplitude. If we consider the axial motion of an individual electron in the static wiggler field described by Eq.(7.217), then the energy $\gamma(t') = $ const. is conserved, and it is readily shown from Eq.(7.220) that the axial velocity $v_z'(t') = dz'(t')/dt'$ is given by

$$\frac{d}{dt'}z'(t') = c\left\{1 - \frac{[1 + a_w^2\sin^2(k_0z')]}{\gamma^2}\right\}^{1/2}. \tag{7.221}$$

Here, "primed" notation has been used to denote the axial orbit $z'(t')$ in the laboratory frame.

Unlike the case of a helical wiggler field where $dz'/dt' = v_z = $ const. [Eq.(7.56)], it is evident from Eq.(7.221) that the planar wiggler field induces an oscillatory modulation of the axial orbit $z'(t')$. Defining $\kappa^2 = a_w^2/(\gamma^2 - 1)$, the inequality $\kappa^2 < 1$ (or, equivalently, $\gamma^2 > 1 + a_w^2$) assures that none of the beam electrons are "trapped" by the equilibrium wiggler field. It is precisely the oscillatory modulation of the axial orbit by the wiggler field in Eq.(7.221) which leads to harmonic generation[89] in a planar wiggler free electron laser. In particular, it is found that the beam electrons with average axial velocity v_z undergo a strong interaction with the combined wiggler and radiation fields when the harmonic resonance condition[118]

$$\omega \simeq [k_z + k_0(1 + 2\ell)]\,v_z \tag{7.222}$$

is satisfied for $\ell = 0, 1, 2, \cdots$. Here, $\ell = 0$ corresponds to the fundamental resonance $\omega \simeq (k_z + k_0)v_z$.

The linearized Vlasov-Maxwell equations for $\delta F_e(z, p_z, t)$ and $\delta A_y(z, t)$ have been used to investigate detailed properties of the free electron laser instability for a planar magnetic wiggler. For present purposes, it is assumed that the energy spread $\Delta\gamma$ of the beam electrons is small in comparison with the average energy $\hat{\gamma}_e$. It is further assumed that the electron beam is highly relativistic with

$$\hat{\kappa}^2 = \frac{a_w^2}{\hat{\gamma}_e^2 - 1} \ll 1. \tag{7.223}$$

Equation (7.223) corresponds to a relatively weak modulation of the axial orbit [see Eq.(7.221)], and the axial velocity v_z (averaged over a wiggler period) can be approximated by

$$v_z^2 = c^2\left(1 - \frac{1 + a_w^2/2}{\gamma^2}\right). \tag{7.224}$$

Without presenting algebraic details,[118] the linearized Vlasov-Maxwell equations have been solved in the diagonal approximation. For $\hat{\kappa}^2 \ll 1$, the Compton-regime dispersion relation can be expressed as[119]

$$0 = D^c(k_z, \omega)$$

$$= \omega^2 - c^2 k_z^2 - \omega_{pe}^2 \langle S \rangle + \frac{1}{4} \omega a_w^2 \omega_{pe}^2$$

$$\times \sum_{\ell=-\infty}^{\infty} \int_0^{2\pi} \frac{d(k_0 z)}{2\pi} \int_0^\infty \frac{dp_z}{\gamma^2} \frac{K_\ell(b) \partial F_e^+ / \partial \gamma}{\omega - [k_z + k_0(1 + 2\ell)] v_z} . \quad (7.225)$$

Here, $\omega_{pe}^2 = 4\pi \hat{n}_e e^2 / m_e$ is the nonrelativistic plasma frequency-squared, k_z is the axial wavenumber, ω is the complex oscillation frequency with $\text{Im}\,\omega > 0$ corresponding to instability (temporal growth), and v_z is defined in Eq.(7.224). Moreover, the coupling coefficient $K_\ell(b)$ is defined by[118,119]

$$K_\ell(b) = [J_\ell(b) - J_{\ell+1}(b)]^2 , \quad (7.226)$$

where $b = (k_z / 8k_0) a_w^2 / (\gamma^2 - 1)$, and the dielectric factor $\langle S \rangle$ is defined by

$$\langle S \rangle = \int_0^{2\pi} \frac{d(k_0 z)}{2\pi} \int_0^\infty \frac{dp_z}{\gamma} \left[F_e^+(\gamma) - a_w^2 \sin^2(k_0 z) \left(\frac{F_e^+(\gamma)}{\gamma^2} - \frac{1}{\gamma} \frac{\partial}{\partial \gamma} F_e^+(\gamma) \right) \right] . \quad (7.227)$$

Finally, the integrations over z and p_z in Eq.(7.225) can be approximated by

$$\int_0^{2\pi} \frac{d(k_0 z)}{2\pi} \int_0^\infty \frac{dp_z}{\gamma^2} \cdots$$

$$= m_e c \int_0^{2\pi} \frac{d(k_0 z)}{2\pi} \int_1^\infty \frac{d\gamma}{\gamma(\gamma^2 - 1)^{1/2}} \frac{1}{[1 - \kappa^2 \sin^2(k_0 z)]^{1/2}} \cdots$$

$$= m_e c \int_1^\infty \frac{d\gamma}{\gamma(\gamma^2 - 1)^{1/2}} \left(1 + \frac{1}{4} \kappa^2 \right) \cdots , \quad (7.228)$$

for $\kappa^2 = a_w^2 / (\gamma^2 - 1) \ll 1$.

The Compton-regime dispersion relation (7.225) can be used to investigate detailed properties of the free electron laser instability for a wide range of distribution functions $F_e^+(\gamma)$ (subject to $\Delta \gamma \ll \hat{\gamma}_e$) and system parameters a_w and ω_{pe}/ck_0. For a tenuous electron beam, it is clear from Eq.(7.225) that the free electron laser interaction is strongest for frequency and wavenumber (ω, k_z) in the vicinity of $(\hat{\omega}, \hat{k}_z)$ satisfying the simultaneous resonance conditions

$$\hat{\omega} = \left[\hat{k}_z + k_0(1 + 2\ell) \right] V_{ze} ,$$

$$\hat{\omega} = c\hat{k}_z . \tag{7.229}$$

Here, ℓ is the harmonic number, and the average velocity $V_{ze} = \beta_{ze}c$ of the beam electrons is related to a_w and $\hat{\gamma}_e$ by [see Eq.(7.224)]

$$\beta_{ze}^2 = 1 - \frac{1 + a_w^2/2}{\hat{\gamma}_e^2} . \tag{7.230}$$

Solving Eq.(7.229) for the upshifted wavenumber $\hat{k}_z^+$ gives

$$\hat{k}_z^+ = \left(\frac{\beta_{ze}}{1 - \beta_{ze}} \right)(1 + 2\ell)\, k_0 = \frac{\hat{\gamma}_e^2(1 + \beta_{ze})\beta_{ze}}{(1 + a_w^2/2)}(1 + 2\ell)\, k_0 . \tag{7.231}$$

Apart from the replacement $a_w^2 \to a_w^2/2$ and the multiplicative factor, $1 + 2\ell$, the expression for $\hat{k}_z^+$ in Eq.(7.231) is similar to the estimate obtained in Eq.(7.64) for a helical magnetic wiggler. For harmonic numbers $\ell = 1, 2, \cdots$, note from Eq.(7.231) that the additional upshift in wavenumber for a planar magnetic wiggler can be substantial relative to the fundamental ($\ell = 0$) mode. This is offset by the fact that the strength of the free electron laser interaction, as measured by the coupling coefficient $K_\ell(b)$ in Eq.(7.225), is weaker for larger ℓ-values [see Eq.(7.226)]. For specified harmonic number ℓ, we estimate $b \simeq \hat{b}_\ell = (\hat{k}_z^+/8k_0)a_w^2/(\hat{\gamma}_e^2 - 1)$ where $\hat{k}_z^+$ is defined in Eq.(7.231). For $\hat{\gamma}_e \gg 1$ and $\beta_{ze} \simeq 1$, this gives

$$\hat{b}_\ell = \frac{a_w^2}{2(a_w^2 + 2)}(1 + 2\ell) . \tag{7.232}$$

Therefore, for fixed value of a_w, the coupling coefficient $K_\ell(\hat{b}_\ell)$ decreases relatively rapidly with increasing ℓ.

The dispersion relation (7.225) has been solved[119] for distribution functions $F_e^+(\gamma)$ ranging from a cold-beam equilibrium ($\Delta\gamma = 0$) to the warm-beam Compton regime where $|[k_z + k_0(1+2\ell)]\Delta v_z| \gg |\text{Im}\,\omega|$. For present purposes, we consider the (most unstable) case where the electrons are monoenergetic, i.e.,

$$F_e^+(\gamma) = \frac{(\hat{\gamma}_e^2 - 1)^{1/2}}{\hat{\gamma}_e m_e c}\delta(\gamma - \hat{\gamma}_e) , \tag{7.233}$$

where $\hat{\gamma}_e \gg 1$ and $\hat{\kappa}^2 = a_w^2/(\hat{\gamma}_e^2 - 1) \ll 1$. Equation (7.227) gives $\langle S \rangle = \hat{\gamma}_e^{-1}(1 + 3\hat{\kappa}^2/4)$. Substituting Eq.(7.233) into Eq.(7.225) and integrating by parts with respect to γ, the dispersion relation for the ℓ'th harmonic excitation can be expressed as[119]

$$
0 = D^c(k_z, \omega) = \omega^2 - c^2 k_z^2 - \frac{\omega_{pe}^2}{\hat{\gamma}_e} \left(1 + \frac{3}{4}\hat{\kappa}^2 \right)
$$

$$
- \frac{1}{4} a_w^2 \frac{\omega_{pe}^2}{\hat{\gamma}_e^3} \frac{\omega N_1^\ell(\hat{\gamma}_e)}{(\omega - [k_z + k_0(1 + 2\ell)]V_{ze})}
$$

$$
- \frac{1}{4} a_w^2 \frac{\omega_{pe}^2}{\hat{\gamma}_e^5} \frac{\omega[k_z + k_0(1 + 2\ell)]c N_2^\ell(\hat{\gamma}_e)}{(\omega - [k_z + k_0(1 + 2\ell)]V_{ze})^2}. \tag{7.234}
$$

Here, $V_{ze} = \beta_{ze}c$ is defined in Eq.(7.230), and the factors $N_1^\ell(\hat{\gamma}_e)$ and $N_2^\ell(\hat{\gamma}_e)$ are defined by

$$
N_1^\ell(\hat{\gamma}_e) = \left[\gamma^2 \left(\gamma^2 - 1 \right)^{1/2} \frac{\partial}{\partial \gamma} \left(\frac{K_\ell(b)(1 + \kappa^2/4)}{\gamma(\gamma^2 - 1)^{1/2}} \right) \right]_{\gamma = \hat{\gamma}_e}, \tag{7.235}
$$

and

$$
N_2^\ell(\hat{\gamma}_e) = \left[\frac{c}{v_z} \left(1 + \frac{a_w^2}{2} \right) K_\ell(b) \left(1 + \frac{1}{4}\kappa^2 \right) \right]_{\gamma = \hat{\gamma}_e}, \tag{7.236}
$$

where $b = (k_z/8k_0)\kappa^2$, $\kappa^2 = a_w^2/(\gamma^2 - 1)$, and v_z is defined in Eq.(7.224).

Equation (7.234) is a fourth-order algebraic equation for the complex eigenfrequency ω, which can be used to investigate detailed stability properties over a wide range of system parameters ω_{pe}/ck_0, a_w and $\hat{\gamma}_e$. For purposes of obtaining a simple analytical estimate of the characteristic growth rate for the ℓ'th harmonic, we examine Eq.(7.234) for (ω, k_z) closely tuned to $(c\hat{k}_z^+, \hat{k}_z^+)$, where $\hat{k}_z^+$ is the upshifted wavenumber defined in Eq.(7.231). Expressing $\omega = c\hat{k}_z^+ + \delta\omega$ and $k_z = \hat{k}_z^+ + \delta k_z$, then for $|\delta\omega/c\hat{k}_z^+| \ll 1$ the final term on the right-hand side of Eq.(7.234) is balanced approximately by $\omega^2 - c^2 k_z^2$. This gives for $\delta k_z = 0$[119]

$$
(\delta\omega)^3 = \frac{a_w^2}{8} \frac{\omega_{pe}^2}{\hat{\gamma}_e^5} \frac{\hat{k}_z^+ c N_2^\ell(\hat{\gamma}_e)}{\beta_{ze}}, \tag{7.237}
$$

where use has been made of $\hat{k}_z^+ + k_0(1 + 2\ell) = \hat{k}_z^+/\beta_{ze}$. From Eq.(7.237), the characteristic (maximum) growth rate for $\delta k_z = 0$ is

$$[\mathrm{Im}(\delta\omega)]_{\max} = \frac{3^{1/2}}{2}\left[\frac{a_w^2\omega_{pe}^2}{8\hat{\gamma}_e^5}\frac{c\hat{k}_z^+}{\beta_{ze}}N_2^\ell(\hat{\gamma}_e)\right]^{1/3}. \tag{7.238}$$

For $\hat{\gamma}_e \gg 1$ and $\beta_{ze} \simeq 1$, it follows that $\hat{\kappa}^2 \ll 1$, $\hat{k}_z^+ \simeq 2\hat{\gamma}_e^2(1 + a_w^2/2)^{-1} \times$ $(1 + 2\ell)k_0$, and $N_2^\ell(\hat{\gamma}_e)c\hat{k}_z^+ \simeq 2\hat{\gamma}_e^2 K_\ell(\hat{b}_\ell)(1 + 2\ell)ck_0$, where $\hat{b}_\ell$ is defined in Eq.(7.232). Equation (7.238) then becomes

$$[\mathrm{Im}(\delta\omega)]_{\max} = \frac{3^{1/2}}{2}\left[\frac{\omega_{pe}^2}{c^2k_0^2}\frac{a_w^2}{4\hat{\gamma}_e^3}(1 + 2\ell)K_\ell(\hat{b}_\ell)\right]^{1/3}ck_0. \tag{7.239}$$

where $K_\ell(\hat{b}_\ell)$ is defined in Eq.(7.226).

For a cold electron beam, the planar wiggler growth rate in Eq.(7.239) is identical to the helical wiggler growth rate in Eq.(7.124) provided we make the replacement $a_w^2 \rightarrow (a_w^2/2)(1 + 2\ell)K_\ell(\hat{b}_\ell)$ in Eq.(7.124). It is evident from Eq.(7.239) that the ℓ'th harmonic growth rate for a planar magnetic wiggler scales as[119]

$$G_\ell = \left[(1 + 2\ell)K_\ell(\hat{b}_\ell)\right]^{1/3}, \tag{7.240}$$

where $\hat{b}_\ell = [(a_w^2/2)/(2 + a_w^2)](2\ell + 1)$. Although $(1 + 2\ell)K_\ell(\hat{b}_\ell)$ decreases with increasing ℓ, the cube-root dependence in Eq.(7.240) gives a relatively slow decrease in growth rate with increasing ℓ for the case of a cold electron beam. For example, we take $a_w^2 = 2$; then $\hat{b}_0 = 1/4$ and $\hat{b}_1 = 3/4$, which gives $K_0(1/4) = 0.75$ and $K_1(3/4) = 0.081$. From Eq.(7.239), the ratio of the $\ell = 1$ growth rate to the $\ell = 0$ growth rate is equal to $G_1/G_0 = 0.9$. That is, for $a_w^2 = 2$, the $\ell = 1$ growth rate is only slightly smaller than the growth rate for the fundamental ($\ell = 0$) mode.

The situation is very different in the warm-beam Compton regime where $|[k_z + k_0(1 + 2\ell)]\Delta v_z| \gg |\mathrm{Im}\,\omega|$. In this case, thermal effects have a stabilizing influence on the growth rate for higher harmonics, and the ℓ'th harmonic growth rate scales as[119]

$$G_\ell = \frac{K_\ell(\hat{b}_\ell)}{(2\ell + 1)} \tag{7.241}$$

times $[\partial F_e^0/\partial\gamma]_{\gamma=\gamma_r}$. Here the resonant energy γ_r is determined from Eq.(7.224) and $v_z(\gamma_r) = (\mathrm{Re}\,\omega)/[k_z + k_0(1 + 2\ell)]$. Comparing Eqs.(7.240) and (7.241), it is evident that the growth rate in the warm-beam Compton regime decreases much more rapidly with increasing ℓ than in the cold-beam Compton regime.

Problem 7.19   Consider the Compton-regime dispersion relation (7.225) for the case where the (monoenergetic) distribution of electrons is specified by Eq.(7.233). Here, $\hat{\gamma}_e \gg 1$ and $\hat{\kappa}^2 = a_w^2/(\hat{\gamma}_e^2 - 1) \ll 1$ are assumed.

a. Make use of Eq.(7.224) to show that

$$\frac{\partial v_z}{\partial \gamma} = \frac{c^2}{v_z} \frac{1 + a_w^2/2}{\gamma^3} . \tag{7.19.1}$$

b. Substitute Eq.(7.233) into Eq.(7.225) and integrate by parts with respect to γ. Show that the dispersion relation for monoenergetic electrons can be expressed in the form given in Eq.(7.234).

Problem 7.20   Express $\omega = c\hat{k}_z^+ + \delta\omega$ and $k_z = \hat{k}_z^+ + \delta k_z$ in the cold-beam dispersion relation (7.234), where $\hat{k}_z^+$ is the upshifted wavenumber for the ℓ'th harmonic defined in Eq.(7.231).

a. Within the context of the approximations enumerated in the text, show that the dispersion relation (7.234) can be approximated by Eq.(7.237) for $\delta k_z = 0$.

b. Show that the corresponding maximum growth rate is given by Eq.(7.239), where the ℓ'th harmonic coupling coefficient is defined in Eq.(7.226).

7.12   High-Gain Free Electron Laser Amplifier Experiments

It should be emphasized that free electron laser experiments have demonstrated coherent radiation generation in the visible,[41] infrared,[29,42−44] and microwave[45−51] regions of the electromagnetic spectrum. We conclude this chapter with a brief description of the classic experiments by Orzechowski, et al.,[51] in which a high-gain free electron laser amplifier (spatial growth) was operated at 34.6 GHz. Figure 7.26(a) shows the measured microwave power in the fundamental ($\ell = 0$) mode plotted versus the wiggler length z for a planar magnetic wiggler with constant amplitude $B_w = 3.72$ kG and wavelength $\lambda_w = 2\pi/k_0 = 9.8$ cm. The beam energy and current in the experiments[51] are 3.5 MeV and 0.85 kA, respectively, and the input signal used to excite the free electron laser interaction is provided by a 50 kW pulsed magnetron operating at 34.6 GHz. Vertical focusing of the electron beam is provided by the wiggler solenoids, whereas horizontal focusing is achieved with continuous quadrupoles with a field gradient of $30\,\text{G/cm}$. For the experiments shown in Fig. 7.26(a), the amplifier goes into saturation at $z = 1.3\,\text{m}$, where the output power is approximately $180\,\text{MW}$. The exponential gain is approximately $34\,\text{dB/m}$ up to saturation. Beyond $1.3\,\text{m}$, the output power at first decreases and then increases with a period of approximately $1\,\text{m}$. The modulation of the saturated power shown in Fig. 7.26(a) is likely associated with the motion of the electrons trapped in the ponderomotive potential (Sec. 7.10). Indeed, if the spatial bounce period

$$\lambda_B = \frac{2\pi c}{\Omega_B} \tag{7.242}$$

is introduced, where Ω_B is defined in Eq.(7.184), and the measured power level at saturation is used to estimate $\hat{a}_s^0$, then Eq.(7.242) gives $\lambda_B \simeq 0.8\,\text{m}$.[51] Figure 7.26(a) also shows the results of two-dimensional computer simulations (solid curve), which give good agreement with the experiments in both the linear and saturated regimes. Comparing the injected beam power ($3\,\text{GW}$) with the output microwave power ($180\,\text{MW}$) gives a radiation generation efficiency of 6% for the experimental results shown in Fig. 7.26(a), where the wiggler amplitude B_w is constant.

To increase the efficiency of radiation generation, Fig. 7.26(b) shows the results of experiments[51] in which the wiggler amplitude $B_w(z)$ is (spatially) tapered[81] beyond $z = 1.3\,\text{m}$. The radiation field in Fig. 7.26(b) is allowed to saturate at $z = 1.3\,\text{m}$, and then the field amplitude $B_w(z)$ is decreased in each subsequent wiggler segment in such a way as to maximize the output microwave power. Indeed, by $z = 2.4\,\text{m}$, the wiggler amplitude in Fig. 7.26(b) has been reduced to 45% of the maximum value (for $0 < z < 1.3\,\text{m}$). As evident from Fig. 7.26(b), the output microwave power continues to increase beyond $z = 1.3\,\text{m}$, achieving a maximum value of $1\,\text{GW}$ at $z = 2.3\,\text{m}$. The corresponding efficiency of radiation generation in the tapered case is approximately 33%.[51]

The effect of the spatial tapering of $B_w(z)$ is to maintain the axial velocity V_{ze} of the beam electrons in resonance with the ponderomotive velocity $v_p = \omega/(k_z + k_0)$ of the amplifying wave, while the kinetic energy of the beam electrons continues to decrease. As a simple model, we assume that the expression [see Eq.(7.230)]

$$\beta_{ze}^2 = 1 - \frac{1 + a_w^2(z)/2}{\hat{\gamma}_e^2(z)} \tag{7.243}$$

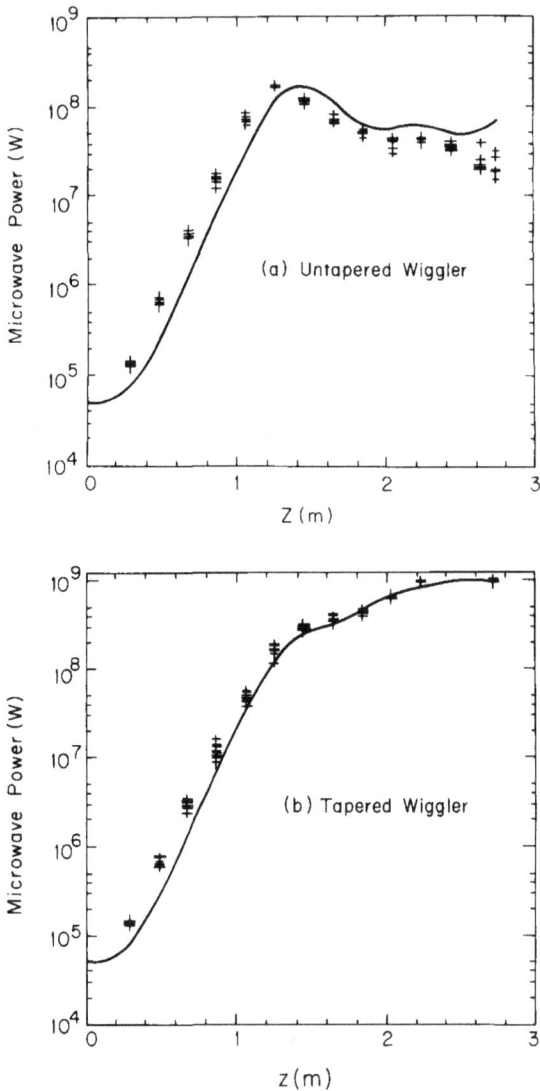

Figure 7.26.   The measured microwave power radiated in the fundamental mode is plotted versus the wiggler length z for a high-gain free electron laser amplifier operating at 34.6 GHz [T.J. Orzechowski, et al., Phys. Rev. Lett. **54**, 889 (1985)]. The beam energy and current are 3.5 MeV and 0.85 kA, respectively, and the wavelength of the (planar) wiggler field is $\lambda_w = 9.8$ cm. In Fig. 7.26(a), the wiggler amplitude $B_w = 3.72$ kG is constant throughout the interaction region. In Fig. 7.26(b), the wiggler amplitude is constant up to $z = 1.3$ m, and then $B_w(z)$ is spatially tapered beyond $z = 1.3$ m in order to maximize the output microwave power. The solid curves correspond to the theoretical predictions by two-dimensional computer simulations.

holds adiabatically for slowly varying $a_w(z)$ and $\hat{\gamma}_e(z)$. Setting $\Delta\beta_{ze} = 0$ then imposes the requirement

$$\frac{1}{2}\frac{a_w}{1 + a_w^2/2}\frac{d}{dz}a_w = \frac{1}{\hat{\gamma}_e}\frac{d}{dz}\hat{\gamma}_e \ . \tag{7.244}$$

As the beam loses energy ($d\hat{\gamma}_e/dz < 0$), if the wiggler amplitude $a_w(z)$ is adjusted to satisfy Eq.(7.244), then the axial velocity $V_{ze} = \beta_{ze}c$ of the electron beam remains synchronous with the velocity v_p of the ponderomotive wave, thereby allowing for continued growth of the radiation field.

Chapter 7 References

1. K.E. Kreisher and R.J. Temkin, "Single-Mode Operation of a High-Power Step-Tunable Gyrotron," Phys. Rev. Lett. **59**, 547 (1987).

2. V.L. Granatstein, "Gyrotron Experimental Studies," in *High-Power Microwave Sources*, eds., V.L. Granatstein and I. Alexeff (Artech House, Boston, Massachusetts, 1987), p. 185, and references therein.

3. J.M. Baird, "Gyrotron Theory," in *High-Power Microwave Sources*, eds., V.L. Granatstein and I. Alexeff (Artech House, Boston, Massachusetts, 1987), p. 103, and references therein.

4. V.L. Bratman, N.S. Ginzburg, G.S. Nusinovich, M.I. Petelin and P.S. Strelkov, "Relativistic Gyrotrons and Cyclotron Autoresonance Masers," Int. J. Electron. **51**, 541 (1981).

5. J. Benford, "Relativistic Magnetrons," in *High-Power Microwave Sources*, eds., V. Granatstein and I. Alexeff (Artech House, Boston, Massachusetts, 1987), p. 351.

6. Y.Y. Lau, "Theory of Crossed-Field Devices and a Comparative Study of Other Radiation Sources," in *High-Power Microwave Sources*, eds., V. Granatstein and I. Alexeff (Artech House, Boston, Massachusetts, 1987), p. 309.

7. T.J. Orzechowski and G. Bekefi, "Microwave Emission from Pulsed, Relativistic Electron Beam Diodes: The Smooth-Bore Magnetron," Phys. Fluids **22**, 978 (1979).

8. A. Palevsky and G. Bekefi, "Microwave Emission from Pulsed, Relativistic Electron Beam Diodes II: The Multiresonator Magnetron," Phys. Fluids **22**, 986 (1979).

9. *Free Electron Laser Handbook*, eds., W. Colson, C. Pellegrini and A. Renieri (North Holland, Amsterdam, 1989).

10. C.W. Roberson and P. Sprangle, "Review of Free Electron Lasers," Phys. Fluids **B1**, 3 (1989), and references therein.

11. T.C. Marshall, *Free Electron Lasers* (Macmillan, New York, 1985).

12. P. Sprangle and T. Coffey, "New Sources of High-Power Coherent Radiation," Physics Today, **37**(3), 44 (1984).

13. R.O. Twiss, "Radiation Transfer and the Possibility of Negative Absorption in Radio Astronomy," Australian Journal of Physics 11, 564 (1958).

14. J. Schneider, "Stimulated Emission of Radiation by Relativistic Electrons in a Magnetic Field," Phys. Rev. 2, 504 (1959).

15. A.V. Gaponov, "Letter to the Editor," Izv. VUZ and Radiofizika 2, 837 (1959).

16. J.L. Hirshfield and J.M. Wachtel, "Electron Cyclotron Maser," Phys. Rev. Lett. **12**, 533 (1964).

17. A.V. Gaponov, M.I. Petelin and V.K. Yulpatov, "The Induced Radiation of Excited Classical Oscillators and Its Use in High-Frequency Electronics," Radiophys. Quant. Electron. **10**, 794 (1967).

18. K.R. Chu and J.L. Hirshfield, "Comparative Study of the Axial and Azimuthal Bunching Mechanisms in Electromagnetic Cyclotron Instabilities," Phys. Fluids 21, 461 (1978).

19. Y.Y. Lau and K.R. Chu, "Electron Cyclotron Maser Instability Driven by a Loss-Cone Distribution," Phys. Rev. Lett. **50**, 243 (1983).

20. P.L. Pritchett, "Electron Cyclotron Maser Radiation from a Relativistic Loss-Cone Distribution," Phys. Fluids **27**, 2393 (1984).

21. P.L. Pritchett, "Electron Cyclotron Maser Instability in Relativistic Plasmas," Phys. Fluids **29**, 2919 (1986).

22. P.H. Yoon and R.C. Davidson, "Closed-Form Analytical Model of the Electron Whistler and Cyclotron Maser Instabilities in Relativistic Plasma with Arbitrary Energy Anisotropy," Phys. Rev. **A35**, 2619 (1987).

23. H. Motz, "Applications of Radiation from Fast Electron Beams," J. Appl. Phys. **22**, 527 (1951).

24. H. Motz, "Undulators and Free Electron Lasers," Contemp. Phys. **20**, 547 (1979).

25. R.M. Phillips, "The Ubitron, a High-Power Traveling-Wave Tube Based on a Periodic Beam Interaction in Unloaded Waveguide," IRE Trans. Electron Devices **ED-7**, 231 (1960).

26. R.M. Phillips, "History of the Ubitron," Nucl. Instrum. Methods **A271**, 1 (1988).

27. T.M.J. Madey, "Stimulated Emission of Bremsstrahlung in a Periodic Magnetic Field," J. Appl. Phys. **42**, 1906 (1971).

28. L.R. Elias, W.M. Fairbank, J.M.J. Madey, H.A. Schwettman and T.I. Smith, "Observation of Stimulated Emission of Radiation by Relativistic Electrons in a Spatially Periodic Transverse Magnetic Field," Phys. Rev. Lett. **36**, 717 (1976).

29. D.A.G. Deacon. L.R. Elias, J.M.J. Madey, G.J. Ramian, H.A. Schwettman and T.I. Smith, "First Operation of a Free Electron Laser," Phys. Rev. Lett. **38**, 892 (1977).

30. W.B. Colson, "Theory of a Free Electron Laser," Phys. Letters **59A**, 187 (1976).

31. T. Kwan, J.M. Dawson and A.T. Lin, "Free Electron Laser," Phys. Fluids **20**, 581 (1977).

32. F.A. Hopf, P. Meystre, M.O. Scully and W.H. Louisell, "Strong-Signal Theory of a Free Electron Laser," Phys. Rev. Lett. **37**, 1342 (1976).

33. A. Hasegawa, "Free Electron Laser," Bell Syst. Tech. J. **57**, 3069 (1978).

34. N.M. Kroll and W.A. McMullin, "Stimulated Emission from Relativistic Electrons Passing Through a Spatially Periodic Transverse Magnetic Field," Phys. Rev. **A17**, 300 (1978).

35. W.H. Louisell, J.F. Lam, D.A. Copeland and W.B. Colson, "Exact Classical Electron Dynamic Approach for a Free Electron Laser Amplifier," Phys. Rev. **A19**, 288 (1979).

36. R.C. Davidson and H.S. Uhm, "Self-Consistent Vlasov Description of the Free Electron Laser Instability in a Relativistic Electron Beam with Uniform Density," Phys. Fluids **23**, 2076 (1980).

37. P. Sprangle and R.A. Smith, "Theory of Free Electron Lasers," Phys. Rev. **A21**, 293 (1980).

38. H. Dreicer, "Kinetic Theory of an Electron-Photon Gas," Phys. Fluids **7**, 735 (1964).

39. V.P. Sukhatme and P.A. Wolff, "Stimulated Compton Scattering as a Radiation Source - Theoretical Limitations," J. Appl. Phys. **44**, 2331 (1973).

40. P. Sprangle, C.M. Tang and I. Bernstein, "Evolution of Spontaneous and Coherent Radiation in the Free Electron Laser Oscillator," Phys. Rev. **A28**, 2300 (1983).

41. M. Billardon, P. Elleume, J.M. Ortega, C. Bazin, M. Bergher, V. Velghe, Y. Petroff, D.A.G. Deacon, K.E. Robinson and J.M.J. Madey, "First Operation of a Storage-Ring Free Electron Laser," Phys. Rev. Lett. **51**, 1652 (1983).

42. R.W. Warren, B.E. Newnam and J.C. Goldstein, "Raman Spectra and the Los Alamos Free Electron Laser," IEEE Journal of Quantum Electronics QE-21, 882 (1985).

43. B.E. Newnam, et al., "Optical Performance of the Los Alamos Free Electron Laser," IEEE Journal of Quantum Electronics QE-21, 867 (1985).

44. T.J. Orzechowski, J.L. Miller, J.T. Weir, Y.-P. Chong, F. Chambers, G.A. Deis, A.C. Paul, D. Prosnitz, E.T. Scharlemann, K. Halbach and J. Edighoffer, "Free Electron Laser Results from the Advanced Test Accelerator," in *Proceedings of the 1988 Linear Accelerator Conference* (Williamsburg, Virginia, 1988).

45. D.B. McDermott, T.C. Marshall, S.P. Schlesinger, R.K. Parker and V. Granatstein, "High-Power Free Electron Laser Based on Stimulated Raman Backscattering," Phys. Rev. Lett. 41, 1368 (1978).

46. A. Grossman, T.C. Marshall and S.P. Schlesinger, "A New Millimeter Free Electron Laser Using a Relativistic Beam with Spiraling Electrons," Phys. Fluids 26, 337 (1983).

47. J. Fajans, G. Bekefi, Y.Z. Yin and B. Lax, "Spectral Measurements from a Tunable, Raman Free Electron Laser," Phys. Rev. Lett. 53, 246 (1984).

48. S.H. Gold, D.L. Hardesty, A.K. Kinkead, L.R. Barnett and V.L. Granatstein, "High-Gain 35 GHz Free Electron Laser Amplifier Experiment," Phys. Rev. Lett. 52, 1218 (1984).

49. J.A. Pasour, R.F. Lucey and C.A. Kapetanakos, "Long-Pulse, High-Power Free Electron Laser with No External Beam Focusing," Phys. Rev. Lett. 53, 1728 (1984).

50. T.J. Orzechowski, B. Anderson, W.M. Fawley, D. Prosnitz, E.T. Scharlemann, S. Yarema, D. Hopkins, A.C. Paul, A.M. Sessler and J.S. Wurtele, "Microwave Radiation from a High-Gain Free Electron Laser Amplifier," Phys. Rev. Lett. 54, 889 (1985).

51. T.J. Orzechowski, B.R. Anderson, J.C. Clark, W.M. Fawley, A.C. Paul, D. Prosnitz, E.T. Scharlemann, S.M. Yarema, D.B. Hopkins, A.M. Sessler and J.S. Wurtele, "High-Efficiency Extraction of Microwave Radiation from a Tapered-Wiggler Free Electron Laser," Phys. Rev. Lett. 57, 2172 (1986).

52. C.S. Wu and L.C. Lee, "A Theory of Terrestrial Kilometric Radiation," Astrophysical Journal 230, 621 (1979).

53. L.C. Lee and C.S. Wu, "Amplification of Radiation Near the Cyclotron Frequency due to Electron Population Inversion," Phys. Fluids 23, 1348 (1980).

54. S.T. Tsai, C.S. Wu, Y.D. Wang and S.W. Kang, "Dielectric Tensor of a Weakly Relativistic, Nonequilibrium, Magnetized Plasma," Phys. Fluids 24, 2186 (1981).

55. C.S. Wu, C.S. Lin, H.K. Wong, S.T. Tsai and R.L. Zhou, "Absorption and Emission of Extraordinary-Mode Electromagnetic Waves Near the Cyclotron Frequency in Nonequilibrium Plasmas," Phys. Fluids **24**, 2191 (1981).

56. R.M. Winglee, "Effects of a Finite Plasma Temperature on Electron Cyclotron Maser Emission," Astrophysical Journal **291**, 160 (1985).

57. H.K. Wong, C.S. Wu and J.D. Gaffey, Jr., "Electron Cyclotron Maser Instability Caused by Hot Electrons," Phys. Fluids **28**, 2751 (1985).

58. A.T. Lin and C.C. Lin, "Competition of Electron Cyclotron Maser and Free Electron Laser Modes with Combined Solenoidal and Longitudinal Wiggler Fields," Phys. Fluids **29**, 1348 (1986).

59. P. Sprangle, C.M. Tang and P. Serafim, "Inverse Resonance Electron Cyclotron Quasioptical Maser in an Open Resonator," Appl. Phys. Lett. **49**, 1154 (1986).

60. A.W. Fliflet, "Linear and Nonlinear Theory of the Doppler-Shifted Cyclotron Resonance Maser Based on TE and TM Waveguide Modes," Int. J. Electron. **61**, 1049 (1986).

61. J.K. Lee, W.B. Bard, S.C. Chiu, R.R. Goforth and R.C. Davidson, "Self-Consistent Nonlinear Evolution of the Cyclotron Autoresonance Maser," Phys. Fluids **31**, 1824 (1988).

62. R.C. Davidson and P.H. Yoon, "Stabilization of the Cyclotron Autoresonance Maser (CARM) Instability," Phys. Rev. **A39**, 2534 (1989).

63. K. Pendergast, B.G. Danly, R. Temkin and J.S. Wurtele, "Self-Consistent Simulation of Cyclotron Autoresonance Maser Amplifiers," IEEE Trans. Plasma Sci. **PS-16**, 122 (1988).

64. H.S. Uhm and R.C. Davidson, "Influence of Intense Equilibrium Self Fields on the Cyclotron Maser Instability in High-Current Gyrotrons," Phys. Fluids **29**, 2713 (1986).

65. P. Sprangle, "Excitation of Electromagnetic Waves from a Rotating Annular Relativistic Electron Beam," J. Appl. Phys. **49**, 2935 (1976).

66. H.S. Uhm, R.C. Davidson and K.R. Chu, "Self-Consistent Theory of Cyclotron Maser Instability for Intense Hollow Electron Beams," Phys. Fluids **21**, 1866 (1978).

67. H.S. Uhm, R.C. Davidson and K.R. Chu, "Cyclotron Maser Instability for General Magnetic Harmonic Number," Phys. Fluids **21**, 1877 (1978).

68. H.S. Uhm and R.C. Davidson, "Cyclotron Maser Instabilityfor Intense Solid Electron Beams," J. Appl. Phys. **50**, 696 (1979).

69. H.S. Uhm and R.C. Davidson, "Influence of Wall Impedance on the Electron Cyclotron Maser Instability," Phys. Fluids **23**, 2538 (1980).

70. H.S. Uhm and R.C. Davidson, "Influence of Energy and Axial Momentum Spreads on the Cyclotron Maser Instability in Intense Hollow Electron Beams," Phys. Fluids 22, 1804 (1979).

71. P. Sprangle and W.M. Manheimer, "Coherent Nonlinear Theory of a Cyclotron Instability," Phys. Fluids 18, 224 (1975).

72. P. Sprangle and A.T. Drobot, "The Linear and Self-Consistent Nonlinear Theory of the Electron Cyclotron Maser Instability," IEEE Trans. Microwave Theory Tech. MTT-25, 528 (1977).

73. H.S. Uhm and R.C. Davidson, "Quasilinear Theory of the Cyclotron Maser Instability," Phys. Fluids 22, 1811 (1979).

74. R.C. Davidson and P.H. Yoon, "Nonlinear Bound on Unstable Field Energy in Relativistic Electron Beams and Plasmas," Phys. Fluids B1, 195 (1989).

75. T. Kwan and J.M. Dawson, "Investigation of the Free Electron Laser with a Guide Magnetic Field," Phys. Fluids 22, 1089 (1979).

76. H.P. Freund and A.T. Drobot, "Relativistic Electron Trajectories in Free Electron Lasers with an Axial Guide Field," Phys. Fluids 25, 736 (1982).

77. I.B. Bernstein and L. Friedland, "Theory of the Free Electron Laser in Combined Helical Pump and Axial Guide Fields," Phys. Rev. A23, 816 (1981).

78. H.P. Freund, "Nonlinear Analysis of Free Electron Laser Amplifiers with Axial Guide Fields," Phys. Rev. A27, 1977 (1983).

79. H.P. Freund and A.K. Ganguly, "Three-Dimensional Theory of the Free Electron Laser in the Collective Regime," Phys. Rev. A28, 3438 (1983).

80. A.K. Ganguly and H.P. Freund, "Nonlinear Analysis of Free Electron Laser Amplifiers in Three Dimensions," Phys. Rev. A32, 2275 (1985).

81. D. Prosnitz, A. Szoke and V.K. Neil, "High-Gain, Free Electron Laser Amplifiers: Design Considerations and Simulation," Phys. Rev. A24, 1436 (1981).

82. H.S. Uhm and R.C. Davidson, "Free Electron Laser Instability for a Relativistic Annular Electron Beam in a Helical Wiggler Field," Phys. Fluids 24, 2348 (1981).

83. H.S. Uhm and R.C. Davidson, "Free Electron Laser Instability for a Relativistic Solid Electron Beam in a Helical Wiggler Field," Phys. Fluids 26, 288 (1983).

84. R.C. Davidson and W.A. McMullin, "Stochastic Particle Instability for Electron Motion in Combined Helical Wiggler, Radiation and Longitudinal Wave Fields," Phys. Rev. A26, 410 (1982).

85. J.A. Davies, R.C. Davidson and G.L. Johnston, "Compton and Raman Free Electron Laser Stability Properties for a Cold Electron Beam Propagating through a Helical Wiggler Magnetic Field," J. Plasma Phys. **33**, 387 (1985).

86. J.A. Davies, R.C. Davidson and G.L. Johnston, "Compton and Raman Free Electron Laser Stability Properties for a Warm Electron Beam Propagating through a Helical Magnetic Wiggler," J. Plasma Phys. **37**, 255 (1987).

87. J.A. Davies, R.C. Davidson and G.L. Johnston, "Pulse Shapes for Absolute and Convective Free Electron Laser Instabilities," J. Plasma Phys. **40**, 1 (1988).

88. P. Sprangle, C.M. Tang and W.M. Manheimer, "Nonlinear Theory of Free Electron Lasers and Efficiency Enhancement," Phys. Rev. **A21**, 302 (1980).

89. W.B. Colson, "The Nonlinear Wave Equation for Higher Harmonics in Free Electron Lasers," IEEE Journal of Quantum Electronics **QE-17**, 1417 (1981).

90. N.M. Kroll, P.L. Morton and M.N. Rosenbluth, "Free Electron Lasers with Variable Parameter Wigglers," IEEE Journal of Quantum Electronics **QE-17**, 1436 (1981).

91. B. Lane and R.C. Davidson, "Nonlinear Traveling-Wave Equilibria for Free Electron Laser Applications," Phys. Rev. **A27**, 2008 (1983).

92. R.C. Davidson and Y.Z. Yin, "Long-Time Quasilinear Evolution of the Free Electron Laser Instability for a Relativistic Electron Beam Propagating Through a Helical Magnetic Wiggler," Phys. Fluids **28**, 2524 (1985).

93. A. Dimos and R.C. Davidson, "Quasilinear Stabilization of the Free Electron Laser Instability for a Relativistic Electron Beam Propagating Through a Transverse Helical Wiggler Magnetic Field," Phys. Fluids **28**, 677 (1985).

94. W.A. McMullin and G. Bekefi, "Coherent Radiation from a Relativistic Electron Beam in a Longitudinal, Periodic Magnetic Field," Appl. Phys. Lett. **39**, 845 (1981).

95. R.C. Davidson and W.A. McMullin, "Higher Harmonic Emission by a Relativistic Electron Beam in a Longitudinal Magnetic Wiggler," Phys. Rev. **A26**, 1997 (1982).

96. W.A. McMullin and G. Bekefi, "Stimulated Emission from Relativistic Electrons Passing Through a Spatially Periodic Longitudinal Magnetic Field," Phys. Rev. **A25**, 1826 (1982).

97. R.C. Davidson and W. McMullin, "Intense Free Electron Laser Harmonic Generation in a Longitudinal Magnetic Wiggler Field," Phys. Fluids **26**, 840 (1983).

98. W.A. McMullin, R.C. Davidson and G.L. Johnston, "Stimulated Emission from a Relativistic Electron Beam in a Variable Parameter Longitudinal Wiggler," Phys. Rev. **A28**, 517 (1983).

99. R.C. Davidson and W.A. McMullin, "Detrapping Stochastic Particle Instability for Electron Motion in Combined Longitudinal Wiggler and Radiation Wave Fields," Phys. Rev. **A29**, 791 (1984).

100. R.C. Davidson, W.A. McMullin and K.T. Tsang, "Cross-Field Free Electron Laser Instability for a Tenuous Electron Beam," Phys. Fluids **27**, 233 (1984).

101. Y.Z. Yin and G. Bekefi, "Generation of Electromagnetic Radiation from a Rotating Electron Ring in a Rippled Magnetic Field," Phys. Fluids **28**, 1186 (1985).

102. G.L. Johnston, F. Hartemann, R.C. Davidson and G. Bekefi, "Resonant Operation of the Cross-Field Free Electron Laser," Phys. Rev. **A38**, 1309 (1988).

103. B.G. Danly, G. Bekefi, R.C. Davidson, R.J. Temkin, T.M. Tran and J.S. Wurtele, "Gyrotron Powered Electromagnetic Wigglers for Free Electron Lasers," IEEE J. Quantum Electronics **QE-23**, 103 (1987).

104. G. Bekefi, J.S. Wurtele and I.H. Deutsch, "Free Electron Laser Radiation Induced by a Periodic Dielectric Medium," Phys. Rev. **A34**, 1228 (1986).

105. D.B. Hopkins, A.M. Sessler and J.S. Wurtele, "The Two-Beam Accelerator," Nucl. Instr. and Methods in Phys. Res. **228**, 15 (1984).

106. J. Fajans, J.S. Wurtele, G. Bekefi, D.S. Knowles and K. Xu, "Nonlinear Power Saturation and Phase (Wave Refractive Index) in the Collective Free Electron Laser," Phys. Rev. Lett. **57**, 579 (1986).

107. J. Fajans and G. Bekefi, "Measurements of Amplification and Phase Shift (Wave Refractive Index) in a Free-Electron Laser," Phys. Fluids **29**, 3461 (1986).

108. F. Hartemann, K. Xu, G. Bekefi, J.S. Wurtele and J. Fajans, "Wave Profile Modification (Optical Guiding) Induced by the Free Electron Laser Interaction," Phys. Rev. Lett. **59**, 1177 (1987).

109. A. Bhattacharjee, S.Y. Cai, S.P. Chang, J.W. Dodd and T.C. Marshall, "Observations of Optical Guiding in a Raman Free Electron Laser," Phys. Rev. Lett. **60**, 1254 (1988).

110. P. Sprangle and C.M. Tang, "Three-Dimensional Nonlinear Theory of the Free Electron Laser," Appl. Phys. Lett. **39**, 677 (1981).

111. E.T. Scharlemann, A.M. Sessler and J.S. Wurtele, "Optical Guiding in a Free Electron Laser," Phys. Rev. Lett. **54**, 1925 (1985).

112. G. Bekefi, R.E. Shefer and W.W. Destler, "Millimeter Wave Emission from a Rotating Electron Ring in a Rippled Magnetic Field," Appl. Phys. Lett. **44**, 280 (1984).

113. W.W. Destler, F.M. Aghamir, D.A. Boyd, G. Bekefi, R.E. Shefer and Y.Z. Yin, "Experimental Study of Millimeter Wave Radiation from a Rotating Electron Beam in a Rippled Magnetic Field," Phys. Fluids **28**, 1962 (1985).

114. F. Hartemann and G. Bekefi, "Microwave Radiation from a Tunable Circular Free Electron Laser," Phys. Fluids **30**, 3283 (1987).

115. J. Ashkenazy and G. Bekefi, "Analysis and Measurement of Permanent Magnet Bifilar Helical Wigglers," IEEE Journal of Quantum Electronics **QE-24**, 812 (1988).

116. G. Bekefi, *Radiation Processes in Plasmas* (Wiley, New York, 1966).

117. J.D. Jackson, *Classical Electrodynamics* (Wiley, New York, 1975).

118. R.C. Davidson and J.S. Wurtele, "Self-Consistent Kinetic Description of the Free Electron Laser Instability in a Planar Magnetic Wiggler," IEEE Trans. Plasma Science **PS-13**, 464 (1985).

119. R.C. Davidson, "Kinetic Description of Harmonic Instabilities in a Planar Wiggler Free Electron Laser," Phys. Fluids **29**, 267 (1986).

120. R.C. Davidson, "Kinetic Description of the Sideband Instability in a Helical Wiggler Free Electron Laser," Phys. Fluids **29**, 2689 (1986).

121. R.C. Davidson, J.S. Wurtele and R.E. Aamodt, "Kinetic Analysis of the Sideband Instability in a Helical Wiggler Free Electron Laser for Electrons Trapped Near the Bottom of the Ponderomotive Potential," Phys. Rev. **A34**, 3063 (1986).

122. R.C. Davidson and J.S. Wurtele, "Single-Particle Analysis of the Free Electron Laser Instability for Primary Electromagnetic Wave with Constant Phase and Slowly Varying Phase," Phys. Fluids **30**, 557 (1987).

123. R.C. Davidson and J.S. Wurtele, "Influence of Untrapped Electrons on the Sideband Instability in a Helical Wiggler Free Electron Laser," Phys. Fluids **30**, 2825 (1987).

124. S. Riyopoulos and C.M. Tang, "Structure of the Sideband Instability in Free Electron Lasers," Phys. Fluids **31**, 1708 (1988).

125. S. Riyopoulos and C.M. Tang, "Chaotic Electron Motion Caused by Sidebands in Free Electron Lasers," Phys. Fluids **31**, 3387 (1988).

126. W.M. Sharp and S.S. Yu, "Two-Dimensional Vlasov Treatment of Free Electron Laser Sidebands," Phys. Fluids **B2**, 581 (1990)

127. J. Masud, T.C. Marshall, S.P. Schlesinger, F.G. Yee, W.M. Fawley, E.T. Scharlemann, S.S. Yu, A.M. Sessler and E.J. Sternbach, "Sideband Control in a Millimeter-Wave Free Electron Laser," Phys. Rev. Lett. **58**, 763 (1987).

The following references, while not cited directly in the text, are relevant to the general subject matter of this chapter.

C. Chen and R.C. Davidson, "Self-Field-Induced Chaoticity in the Electron Orbits in a Helical Wiggler Free Electron Laser with Axial Guide Field," Phys. Fluids **B2**, 171 (1990).

H.P. Freund, R.C. Davidson and G.L. Johnston, "Linear Theory of the Collective Raman Interaction in a Free Electron Laser with a Planar Wiggler and an Axial Guide Field," Phys. Fluids **B2**, 427 (1990).

R.G. Kleva, B. Levush and P. Sprangle, "Radiation Guiding and Efficiency Enhancement in the Cyclotron Autoresonance Maser," Phys. Fluids **B2**, 185 (1990).

T.M. Antonsen, Jr., B. Levush and W.M. Manheimer, "Stable Single Mode Operation of a Quasioptical Gyrotron," Phys. Fluids **B2**, 419 (1990).

D. Shiffler, J.A. Nation and C.B. Wharton, "High-Power Traveling-Wave Tube Amplifier," Appl. Phys. Lett. **54**, 674 (1989).

A. Serbeto, B. Levush and T.M. Antonsen, Jr., "Efficiency Optimization for Free Electron Laser Oscillators," Phys. Fluids **B1**, 435 (1989).

T.M. Antonsen, Jr. and B. Levush, "Mode Competition and Suppression in Free Electron Laser Oscillators," Phys. Fluids **B1**, 1097 (1989).

J.-S. Choi, D.-E. Kim, D.-I. Choi, S.-C. Yang and H.S. Uhm, "Theory of Free-Electron Laser Instability for a Relativistic Electron Beam Propagating in a Dielectric-Loaded Waveguide," Phys. Fluids **B1**, 1316 (1989).

D.A. Kirkpatrick, G. Bekefi, A.C. DiRienzo, H.P. Freund and A.K. Ganguly, "A Millimeter and Submillimeter Wavelength Free-Electron Laser," Phys. Fluids **B1**, 1511 (1989).

K. Xu, G. Bekefi and C. Leibovitch, "Observations of Field Profile Modifications in a Raman Free-Electron Laser Amplifier," Phys. Fluids **B1**, 2066 (1989).

J. Fajans and J.S. Wurtele, "Waveguide Mode Deformation in Free-Electron Lasers," Phys. Fluids **B1**, 2073 (1989).

T.M. Antonsen, Jr. and B. Levush, "Mode Competition in Free Electron Laser Oscillators," Phys. Rev. Lett. **62**, 1488 (1989).

A. Friedman, A. Gover, G. Kurizki, S. Ruschin and A. Yariv, "Spontaneous and Stimulated Emission from Quasifree Electrons," Rev. Mod. Phys. **60**, 471 (1988).

R.A. Kehs, Y. Carmel, V.L. Granatstein and W.W. Destler, "Experimental Demonstation of an Electromagnetically Pumped Free Electron Laser with a Cyclotron Harmonic Idler," Phys. Rev. Lett. **60**, 279 (1988).

S.H. Gold, A.W. Fliflet, W.M. Manheimer, R.B. McCowan, W.M. Black, R.C. Lee, V.L. Granatstein, A.K. Kinkead, D.L. Hardesty and M. Sucy, "High Peak Power K_a-Band Gyrotron Oscillator with Slotted and Unslotted Cavities," IEEE Transactions on Plasma Science **PS-16**, 142 (1988).

H. Weitzner and A. Fruchtman, "Electron Beam Equilibria with Self-Fields for a Free Electron Laser with a Planar Wiggler," Phys. Fluids **31**, 2340 (1988).

S. Riyopoulos and C.M. Tang, "The Structure of the Sideband Spectrum in Free Electron Lasers," Phys. Fluids **31**, 1708 (1988).

L.S. Schuetz, E. Ott and T.M. Antonsen, Jr., "Analysis of a Wide-Band Rotating-Beam Free-Electron Laser," Phys. Fluids **31**, 1720 (1988).

T.M. Antonsen, Jr. and P.E. Latham, "Linear Theory of a Sheet Beam Free Electron Laser," Phys. Fluids **31**, 3379 (1988).

V.K. Tripathi and C.S. Liu, "Sideband Excitation in a Electromagnetic Wiggler Pumped Free Electron Laser," Phys. Fluids **31**, 3799 (1988).

J.H. Booske, et al., "Propagation of Wiggler Focused Relativistic Sheet Electron Beams," J. Appl. Phys. **64**, 6 (1988).

H.P. Freund, "Multimode Nonlinear Analysis of Free Electron Laser Amplifiers in Three Dimensions," Phys. Rev. **A37**, 3371 (1988).

H.P. Freund, J.Q. Dong, C.S. Wu and L.C. Lee, "A Cyclotron-Maser Instability Associated with a Nongyrotropic Distribution," Phys. Fluids **30**, 3106 (1987).

S.A. Prasad, G.J. Morales and B.D. Fried, "Cyclotron Resonance Phenomena in a Nonneutral Plasma," Phys. Fluids **30**, 3093 (1987).

T.M. Tran, K.E. Kreischer and R.J. Temkin, "Self-Consistent Theory of a Harmonic Gyroklystron with a Minimum Q Cavity," Phys. Fluids **29**, 3858 (1986).

B.G. Danly and R.J. Temkin, "Generalized Nonlinear Harmonic Gyrotron Theory," Phys. Fluids **29**, 561 (1986).

T.M. Tran, B.G. Danly, K.E. Kreischer, J.B. Schutkeker and R.J. Temkin, "Optimization of Gyroklystron Efficiency," Phys. Fluids **29**, 1274 (1986).

A. Fruchtman, "Gain Enhancement in a Longitudinal Magnetic Wiggler by Use of a Coherently Gyrophased Electron Beam," Phys. Fluids **29**, 1695 (1986).

A.T. Lin and C.-C. Lin, " Competition of Electron-Cyclotron Maser and Free-Electron Laser Modes with Combined Solenoidal and Longitudinal Wiggler Fields," Phys. Fluids **29**, 1348 (1986).

B.M. Lamb and G.J. Morales, "Ponderomotive Effects in Nonneutral Plasmas," Phys. Fluids **26**, 3488 (1983).

R.K. Parker, R.H. Jackson, S.H. Gold, H.P. Freund, P.C. Efthimion, V.L. Granatstein, M. Herndon and A.K. Kinkead, "Axial Magnetic Field Effects in a Collective-Interaction Free Electron Laser at Millimeter Wavelengths," Phys. Rev. Lett. **48**, 238 (1982).

V.L. Granatstein, P. Sprangle, M. Herndon, R.K. Parker and S.P. Schlesinger, "Microwave Amplification with an Intense Relativistic Electron Beam," J. Appl. Phys. **41**, 3800 (1975).

V.L. Granatstein, M. Herndon, P. Sprangle, Y. Carmel and J. Nation, "Gigawatt Microwave Emission from an Intense Relativistic Electron Beam," Plasma Phys. **17**, 23 (1975).

V.L. Granatstein, M. Herndon, R. Parker and P. Sprangle, "Coherent Synchrotron Radiation from an Intense Relativistic Electron Beam," IEEE Journal of Quantum Electronics **QE-10**, 651 (1974).

CHAPTER 8

EQUILIBRIUM AND STABILITY PROPERTIES OF INTENSE NONNEUTRAL FLOW IN HIGH-POWER DIODES

8.1 Introduction

The technology for generating intense charged particle beams has progressed rapidly during the past twenty years,[1,2] with applications ranging from high-current accelerators,[1-6] to particle-beam fusion,[7,8] to high-power microwave generation,[9-11] to beam propagation through the atmosphere.[12-14] Indeed, using pulsed-power technology at high voltages, intense charged particle beams can be generated with power levels exceeding 10^{13} W for pulse lengths of the order of 100 ns. In this chapter, we investigate several aspects of the equilibrium and stability properties of intense nonneutral electron and ion flow in high power diodes. Because the primary emphasis in this treatise is on the basic physical properties of nonneutral plasmas, including the influence of intense self fields on stability behavior, the reader is referred to the excellent treatment by Miller[1] for a discussion of the technology used to generate intense electron and ion beams, as well as a description of the basic emission processes in high-voltage diodes.

By way of background,[1] electron emission from a cold metal surface (the cathode in a diode, say) can result from several processes, including thermionic emission,[15] photoemission,[16] secondary emission, field emission,[17] and explosive electron emission.[18,19] In thermionic emission and photoemission, the electrons are provided sufficient energy to overcome the surface potential barrier which ordinarily prevents the escape of electrons from the conduction band. In field emission from a cold electrode surface, the potential barrier is sufficiently deformed by the electric field applied across the diode that conduction electrons can tunnel quantum-mechanically through the potential barrier at the surface of the metal. At sufficiently high applied electric field, which is the case considered in Chapter 8, explosive electron emission can be the dominant emission process. In this case, microscopic protrusions (called "whiskers") from the metal surface experience a strong enhancement in the local electric field near the tips of the whiskers, thereby leading to significant field emission from the protrusions. (Typical whisker concentrations range from 1 to 10^4 per square centimeter.[1]) Resistive heating of the protrusions causes their explosive vaporization and leads rapidly to the formation of a thin plasma layer covering the cathode surface. In turn, because the cathode plasma can be viewed effectively as a metal surface with zero work function, the available supply of electrons from the cathode is essentially unlimited. In this case, the amount of electron current flowing from the cathode to the anode (see Fig. 8.1) is limited by the space charge of the electrons in the cathode-anode gap. Under quasi-steady-state conditions $(\partial/\partial t = 0)$, a potential minimum forms just outside the cathode surface. For sufficiently high diode voltage $V (>$ tens of kilovolts, say) the magnitude of the potential minimum is small in comparison with the applied voltage, and the location of the potential minimum is nearly coincident with the cathode surface.[1] Therefore, under steady flow conditions, it is assumed throughout Chapter 8 that the normal electric field vanishes at the cathode surface [i.e., $E_x(x = 0) = 0$ in Fig. 8.1] and that the electrons are emitted from the cathode plasma with zero velocity. This is referred to as *space-charge-limited flow*.

In Secs. 8.2–8.7, we investigate several of the basic equilibrium and stability properties of intense nonneutral electron and ion flow in high-voltage diodes. The analysis in Secs. 8.2–8.4 examines steady-state properties $(\partial/\partial t = 0)$, whereas stability behavior is investigated in Secs. 8.5–8.7. Following a discussion of Child-Langmuir flow[20–24] of electrons from the cathode to the anode in a planar diode (Sec. 8.2), we investigate the effects of an insulating magnetic field $B_0\hat{e}_z$ applied parallel to the cathode and anode surfaces in the diode region (Sec. 8.3). In this case, for sufficiently strong applied magnetic field, the electron flow is parallel to the cathode surface and magnetically insulated[25–28] from contact with the anode (see Fig. 8.5). This diode configuration (the so-called applied-B diode) has important applications in the production[28–34] of intense ion beams (Sec. 8.4) and in the generation of microwaves[35–43] in relativistic and nonrelativistic magnetrons (Secs. 8.5 and 8.6). In magnetically insulated high-voltage diodes used to generate intense ion beams (see Fig. 8.12), a plasma is preformed on the anode surface (e.g., using flash-board techniques[1]). Although the electron flow is magnetically insulated, the (massive) ions accelerate freely from the anode to the cathode, and the resulting ion current can exceed the Child-Langmuir current by a considerable amount (Sec. 8.4). With regard to stability properties, the effects of intense self fields and profile shape on the magnetron instability[44-54] are investigated for relativistic electron flow in a magnetically insulated diode (Sec. 8.5). A brief summary is then presented of the classic magnetron experiments by Palevsky and Bekefi[41] in which microwaves are generated at gigawatt power levels for diode voltages in the operating range of 300–400 kV (Sec. 8.6). Finally, instabilities in magnetically insulated ion diodes are investigated, (Sec. 8.7) with particular emphasis on the ion resonance and transit time instabilities.[55,56] These instabilities result from a strong coupling between the ions and the electrons, which have different equilibrium flow directions in the diode region. Finally, following Desjarlais,[34] a model of the electron and ion flow in a magnetically insulated diode is derived (Sec. 8.8), which assumes uniform electron density and allows for the formation of a virtual cathode.

8.2   Child-Langmuir Flow

8.2.1   Nonrelativistic Planar Diode

As a first example, we consider nonrelativistic electron flow in the planar diode configuration illustrated in Fig. 8.1.[20-22] Here, the cathode is located at $x = 0$, the anode is located at $x = d$, and there is no applied magnetic field ($\mathbf{B}^0 = 0$). The equilibrium electrostatic potential $\phi_0(x)$ satisfies the boundary conditions

$$\phi_0(x = 0) = 0 \quad \text{and} \quad \phi_0(x = d) = V, \tag{8.1}$$

and space-charge-limited flow with

$$E_x(x = 0) = 0 \tag{8.2}$$

is assumed, where $E_x = -\partial\phi_0/\partial x$ is the electric field in the region between the cathode and anode. For nonrelativistic electron flow, $eV/m_e c^2 \ll 1$ is assumed, where $V > 0$ is the voltage. In the absence of applied magnetic field, the average electron flow in Fig. 8.1 is in the positive x-direction, from the cathode to the anode.

To describe the steady-state $(\partial/\partial t = 0)$ electron flow, we make use of a macroscopic cold-fluid model (Sec. 2.3) with average flow velocity $\mathbf{V}_e^0(\mathbf{x}) = V_{xe}^0(x)\hat{\mathbf{e}}_x$. Assuming all quantities have only an x-variation, the electron continuity equation, $(\partial/\partial x)(n_e^0 V_{xe}^0) = 0$, readily gives

$$- en_e^0(x)V_{xe}^0(x) \equiv J_e = \text{const.} \tag{8.3}$$

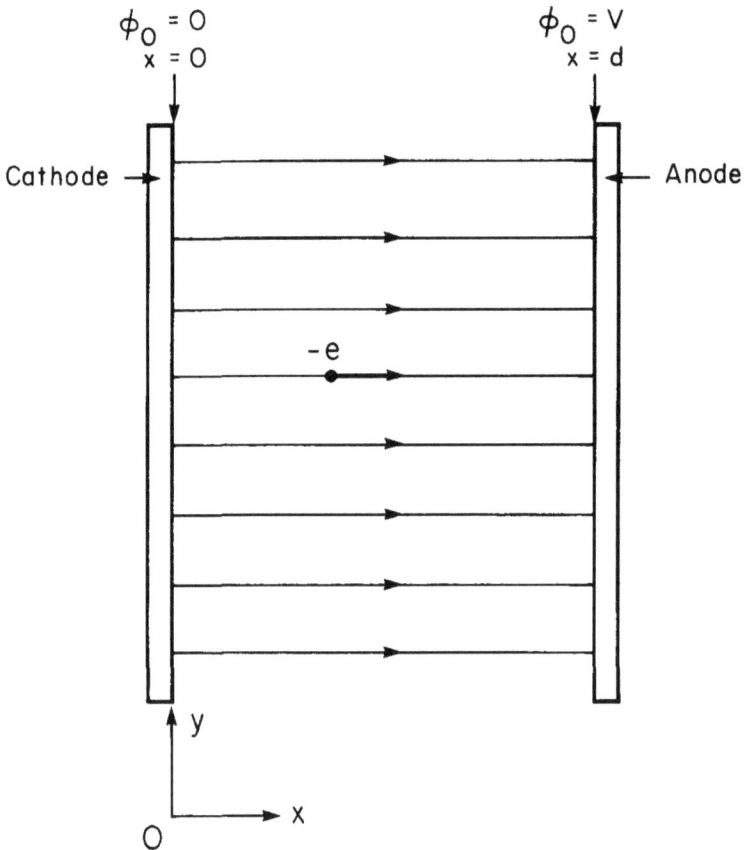

Figure 8.1. Schematic of planar diode configuration. The electrons are emitted from the cathode $(x = 0)$ and accelerate towards the anode $(x = d)$ where the applied voltage is V.

in the region between the cathode and anode $(0 \leq x \leq d)$. Here, $-J_e/e =$ const. is the flux of electrons, and the electron density $n_e^0(x)$ is related self-consistently to the equilibrium electrostatic potential $\phi_0(x)$ by Poisson's equation

$$\frac{\partial^2}{\partial x^2}\phi_0(x) = 4\pi e n_e^0(x) = -\frac{4\pi J_e}{V_{xe}^0(x)}. \tag{8.4}$$

Treating the electrons as cold, equilibrium force balance in the x-direction on an electron fluid element can be expressed as

$$n_e^0 m_e V_{xe}^0 \frac{\partial}{\partial x} V_{xe}^0 = n_e^0 e \frac{\partial \phi_0}{\partial x}, \tag{8.5}$$

which can be integrated to give

$$\frac{m_e}{2} V_{xe}^{02}(x) - e\phi_0(x) = \text{ const.} \tag{8.6}$$

Enforcing $\phi_0(x=0) = 0$ and $V_{xe}^0(x=0) = 0$ at the cathode, the constant in Eq.(8.6) is equal to zero, so that the flow velocity of an electron fluid element can be expressed as

$$V_{xe}^0(x) = \left[\frac{2e\phi_0(x)}{m_e}\right]^{1/2}. \tag{8.7}$$

Poisson's equation (8.4) then reduces to

$$\frac{\partial^2}{\partial x^2}\phi_0(x) = -\frac{4\pi(m_e/2e)^{1/2}J_e}{[\phi_0(x)]^{1/2}}, \tag{8.8}$$

where the constant $J_e < 0$.

We multiply Eq.(8.8) by $\partial \phi_0/\partial x$ and integrate once with respect to x. Enforcing $\phi_0(x=0) = 0$ and $\partial \phi_0/\partial x|_{x=0} = 0$, we obtain

$$\frac{1}{2}\left(\frac{\partial \phi_0}{\partial x}\right)^2 = -8\pi\left(\frac{m_e}{2e}\right)^{1/2}J_e\phi_0^{1/2}. \tag{8.9}$$

Integrating Eq.(8.9) with respect to x gives

$$\phi_0(x) = \left[-9\pi\left(\frac{m_e}{2e}\right)^{1/2}J_e d^2\right]^{2/3}\left(\frac{x}{d}\right)^{4/3} \tag{8.10}$$

in the interval $0 \leq x \leq d$. The boundary condition $\phi_0(x = d) = V$ is enforced at the anode, and Eq.(8.10) can be expressed in the equivalent form

$$\phi_0(x) = V \left(\frac{x}{d}\right)^{4/3}, \quad 0 \leq x \leq d, \tag{8.11}$$

where the voltage V is related to the electron current density J_e by the Child-Langmuir expression[20-22]

$$J_e = J_{\mathrm{CL}} \equiv -\frac{2^{1/2}}{9\pi d^2} \frac{m_e c^3}{e} \left(\frac{eV}{m_e c^2}\right)^{3/2}. \tag{8.12a}$$

For specified diode voltage V and cathode-anode spacing d, Eq.(8.12) determines the steady value of electron current that flows across the diode. The (nonrelativistic) derivation of Eqs.(8.11) and (8.12) necessarily requires $eV/m_e c^2 \ll 1$, which corresponds to applied voltages well below 500 kV. In practical units, Eq.(8.12) can be expressed as

$$J_{\mathrm{CL}} \left(\frac{A}{\mathrm{cm}^2}\right) = -7.6 \times 10^{-2} \frac{[V(\mathrm{kV})]^{3/2}}{[d(\mathrm{cm})]^2}. \tag{8.12b}$$

For example, if $V = 100$ kV and $d = 1$ cm, then the Child-Langmuir current is $J_{\mathrm{CL}} = -76 A/\mathrm{cm}^2$.

Other equilibrium flow properties are readily calculated. For example, substituting Eq.(8.11) into Eq.(8.7) gives for the flow velocity

$$V_{xe}^0(x) = 2^{1/2} c \left(\frac{eV}{m_e c^2}\right)^{1/2} \left(\frac{x}{d}\right)^{2/3}, \quad 0 \leq x \leq d. \tag{8.13}$$

That is, $V_{xe}^0(x)$ increases monotonically from zero at $x = 0$ to $(2eV/m_e)^{1/2}$ at $x = d$. Similarly, the electron density profile $n_e^0(x)$ can be calculated from $n_e^0(x) = -J_e/eV_{xe}^0(x)$, where J_e is defined in Eq.(8.12). Solving for $\omega_{pe}^2(x) = 4\pi n_e^0(x)e^2/m_e$, we obtain

$$\omega_{pe}^2(x) = \frac{4}{9} \left(\frac{eV}{m_e d^2}\right) \left(\frac{d}{x}\right)^{2/3}, \quad 0 \leq x \leq d. \tag{8.14}$$

The electron density in Eq.(8.14) decreases monotonically from an infinite value at the cathode ($x = 0$), to $\omega_{pe}^2(x = d) = (4/9)(eV/m_e d^2)$ at the anode ($x = d$). The divergence of the electron density at $x = 0$ is associated with the fact that the electrons are assumed to be emitted from the cathode with zero velocity. In reality, the initial velocity $V_{xe}^0(x = 0)$ is small but nonzero, and the electron density $n_e^0(x = 0)$ at the cathode is large but finite.

To summarize, presented in Fig. 8.2 are plots of the electrostatic potential $\phi_0(x)$ [Eq.(8.11)], the electron flow velocity $V_{xe}^0(x)$ [Eq.(8.13)], and the plasma frequency-squared $\omega_{pe}^2(x)$ [Eq.(8.14)] in the interval $0 \leq x \leq d$. Here, we have normalized $\phi_0(x), V_{xe}^0(x)$ and $\omega_{pe}^2(x)$, to the values $V, (2eV/m_e)^{1/2}$ and $(4/9)(eV/m_e d^2)$, respectively, at the anode.

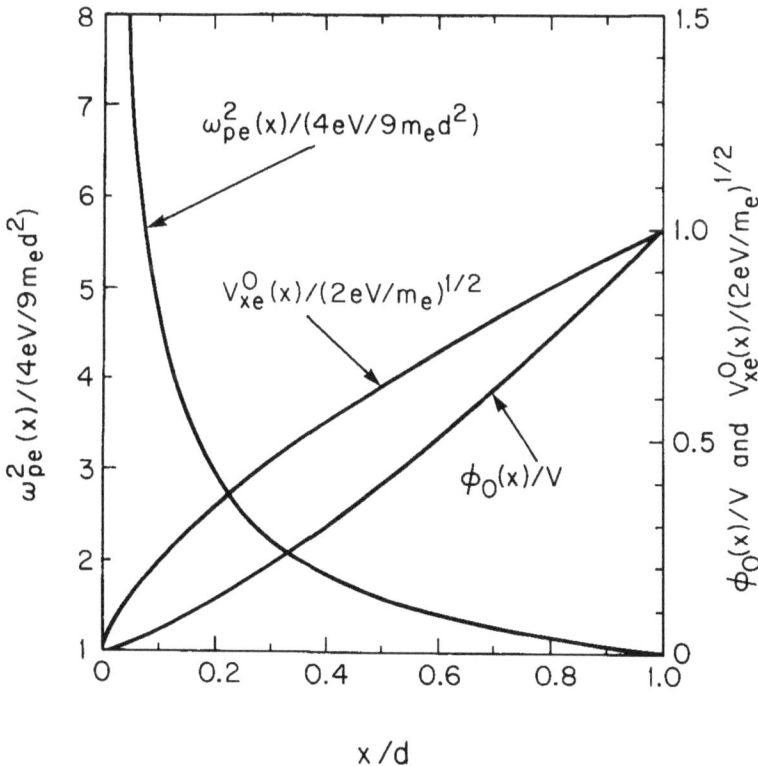

Figure 8.2. Plots versus x/d of the normalized electrostatic potential $\phi_0(x)/V$ [Eq.(8.11)], electron flow velocity $V_{xe}^0(x)/(2eV/m_e)^{1/2}$ [Eq.(8.13)], and electron density $\omega_{pe}^2(x)/(4eV/9m_e d^2)$ [Eq.(8.14)] for non-relativistic Child-Langmuir flow.

Problem 8.1   Equilibrium flow properties in a planar diode can also be characterized in terms of the Lagrangian time variable

$$\tau(x) = \int_0^x \frac{dx}{V_{xe}^0(x)},\tag{8.1.1}$$

which is the time required for an electron fluid element to accelerate from the cathode at $x = 0$ to position x.

a. Make use of Eqs.(8.13) and (8.1.1) to show that

$$\tau(x) = T\left(\frac{x}{d}\right)^{1/3},\tag{8.1.2}$$

where $T \equiv 3d(m_e/2eV)^{1/2}$ is the transit time of a fluid element from the cathode to the anode.

b. In terms of $\tau(x)$, show from Eqs.(8.11), (8.14), and (8.1.2) that $\phi_0(x)$ and $\omega_{pe}^2(x)$ can be expressed as

$$\phi_0(x) = V\left[\frac{\tau(x)}{T}\right]^4,\tag{8.1.3}$$

and

$$\omega_{pe}^2(x) = \frac{2}{\tau^2(x)}.\tag{8.1.4}$$

c. Show that the spatially averaged plasma frequency-squared, $\overline{\omega_{pe}^2(x)} = d^{-1} \times \int_0^d dx\, \omega_{pe}^2(x)$, is given by

$$\overline{\omega_{pe}^2(x)} = \frac{6}{T^2} = \frac{4}{3}\frac{eV}{m_e d^2}.\tag{8.1.5}$$

Problem 8.2   Consider the equilibrium distribution function $f_e^0(x, \mathbf{p})$ specified by

$$f_e^0(x, \mathbf{p}) = \left(\frac{-J_e}{e}\right) U(p_x)\, \delta(H)\, \delta(p_z)\, \delta(p_y)\tag{8.2.1}$$

in the planar diode configuration illustrated in Fig. 8.1. Here, J_e is a negative constant; $U(p_x)$ is the Heaviside step function defined by $U(p_x) = +1$ for $p_x > 0$, and $U(p_x) = 0$ for $p_x < 0$; and $H = (p_x^2 + p_y^2 + p_z^2)/2m_e - e\phi_0(x)$ is the energy of an individual electron.

a. Show directly from Eq.(8.2.1) that the x-directed flux of electrons is given by

$$n_e^0(x) V_{xe}^0(x) = \int d^3p\, \frac{p_x}{m_e}\, f_e^0(x, \mathbf{p}) = -\frac{J_e}{e},\tag{8.2.2}$$

and the electron density can be expressed as

$$n_e^0(x) = \int d^3p\, f_e^0(x, \mathbf{p}) = -\frac{J_e}{e}\left[\frac{m_e}{2e\phi_0(x)}\right]^{1/2}. \tag{8.2.3}$$

Verify that

$$\frac{1}{2m_e}\int d^3p\, \left[p_x - m_e V_{xe}^0(x)\right]^2 f_e^0(x, \mathbf{p}) = 0 \tag{8.2.4}$$

for the choice of distribution function in Eq.(8.2.1).

Comparing Eqs.(8.2.2)–(8.2.4) with Eqs.(8.3), (8.5), and (8.7), we conclude that Eq.(8.2.1) reproduces the "cold" fluid equilibrium assumed in Sec. 8.2.1.

8.2.2   Relativistic Planar Diode

The analysis in Sec. 8.2.1 is readily extended to arbitrary values of $eV/m_e c^2$ and relativistic electron flow.[23] The boundary conditions in Eqs.(8.1) and (8.2) and Poisson's equation (8.4) remain unchanged, but the energy conservation equation (8.6) is modified to become

$$\left[\gamma_e^0(x) - 1\right] m_e c^2 - e\phi_0(x) = \text{const.} \tag{8.15}$$

Here, $\gamma_e^0(x) = [1 - V_{xe}^{02}(x)/c^2]^{-1/2}$ is the relativistic mass factor, and the constant in Eq.(8.15) is equal to zero because $\phi_0(x = 0) = 0$ and $\gamma_e^0(x = 0) = 1$ at the cathode. Making use of $\gamma_e^0(x) = 1 + e\phi_0(x)/m_e c^2$ and $V_{xe}^0(x) = c[1 - 1/\gamma_e^{02}(x)]^{1/2}$, Poisson's equation (8.4) can be expressed in the equivalent form

$$\frac{\partial^2}{\partial x^2}\gamma_e^0(x) = -\frac{4\pi J_e e}{m_e c^3}\frac{\gamma_e^0(x)}{[\gamma_e^{02}(x) - 1]^{1/2}}, \tag{8.16}$$

where $J_e = -e n_e^0(x) V_{xe}^0(x) = \text{const.} < 0$. Equation (8.16) is to be solved subject to the boundary conditions

$$\gamma_e^0(x = 0) = 1, \quad \gamma_e^0(x = d) = 1 + \frac{eV}{m_e c^2},$$

$$\left.\frac{\partial}{\partial x}\gamma_e^0(x)\right|_{x=0} = 0, \tag{8.17}$$

which correspond to Eqs.(8.1) and (8.2). Multiplying Eq.(8.16) by $\partial \gamma_e^0(x)/\partial x$ and integrating once with respect to x, we obtain

$$\frac{1}{2} \left(\frac{\partial \gamma_e^0}{\partial x} \right)^2 = -\frac{4\pi J_e e}{m_e c^3} \left[\gamma_e^{02} - 1 \right]^{1/2}, \qquad (8.18)$$

where use has been made of Eq.(8.17). Taking the square root in Eq.(8.18) and integrating, we obtain the integral relation

$$\int_1^{\gamma_e^0(x)} \frac{d\gamma_e}{(\gamma_e^2 - 1)^{1/4}} = \left(\frac{-8\pi J_e e}{m_e c^3} \right)^{1/2} x, \qquad (8.19)$$

where $\gamma_e^0(x) = 1 + e\phi_0(x)/m_e c^2$. The integral in Eq.(8.19) can be expressed as

$$\int_1^{\gamma_e^0(x)} \frac{d\gamma_e}{(\gamma_e^2 - 1)^{1/4}} = F[\delta(x), 2^{-1/2}] - 2E[\delta(x), 2^{-1/2}]$$

$$+ \frac{2\gamma_e^0(x) \left[\gamma_e^{02}(x) - 1 \right]^{1/4}}{1 + \left[\gamma_e^{02}(x) - 1 \right]^{1/2}}. \qquad (8.20)$$

Here, $\delta(x)$ is defined by

$$\cos \delta(x) = \frac{1 - \left[\gamma_e^{02}(x) - 1 \right]^{1/2}}{1 + \left[\gamma_e^{02}(x) - 1 \right]^{1/2}}, \qquad (8.21)$$

and $F(\xi, \kappa)$ and $E(\xi, \kappa)$ are the elliptic integrals of the first and second kind, respectively, defined by

$$F(\xi, \kappa) = \int_0^{\sin \xi} dx \left[(1 - x^2)(1 - \kappa^2 x^2) \right]^{-1/2}, \qquad (8.22)$$

$$E(\xi, \kappa) = \int_0^{\sin \xi} dx \, (1 - \kappa^2 x^2)^{1/2} (1 - x^2)^{-1/2}. \qquad (8.23)$$

The solution for $\gamma_e^0(x)$ in Eqs.(8.19) and (8.20), which was first obtained by Jory and Trivelpiece,[23] can be used to investigate numerically various properties of the relativistic electron flow between the cathode and anode.

For present purposes, we evaluate Eq.(8.19) at $x = d$, where $\gamma_e^0(x = d) = 1 + eV/m_e c^2$. Solving for the current density J_e, we obtain

$$J_e = -\frac{m_e c^3}{8\pi d^2 e} \left[F(\delta_a, 2^{-1/2}) - 2E\left(\delta_a, 2^{-1/2}\right) + \frac{2\gamma_a \left(\gamma_a^2 - 1\right)^{1/4}}{1 + (\gamma_a^2 - 1)^{1/2}} \right]^2 ,$$

(8.24)

where $\gamma_a \equiv 1 + eV/m_e c^2$ and $\cos\delta_a \equiv [1 - (\gamma_a^2 - 1)^{1/2}]/[1 + (\gamma_a^2 - 1)^{1/2}]$ denote the values of $\gamma_e^0(x)$ and $\delta(x)$ at the anode ($x = d$). In the nonrelativistic limit, Eq.(8.24) reduces to the familiar Child-Langmuir expression

$$J_e = J_{CL} \equiv -\frac{2^{1/2}}{9\pi d^2} \frac{m_e c^3}{e} \left(\frac{eV}{m_e c^2}\right)^{3/2} , \quad \text{for} \quad \frac{eV}{m_e c^2} \ll 1. \quad (8.25)$$

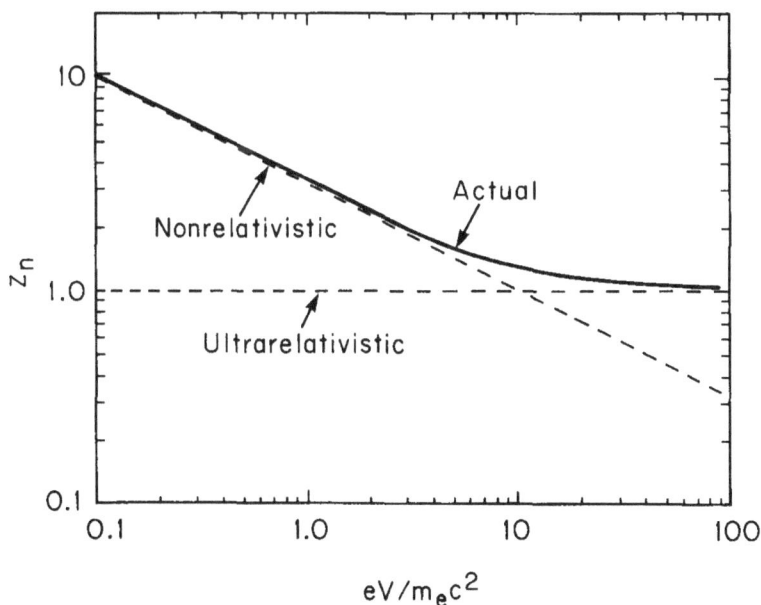

Figure 8.3. Plot of the normalized impedance $Z_n = -cV/2\pi d^2 J_e$ versus $eV/m_e c^2$ obtained numerically from Eqs.(8.24)–(8.27). The dashed curves correspond to the nonrelativistic and ultrarelativistic expressions for J_e in Eqs.(8.25) and (8.26).

On the other hand, in the ultrarelativistic regime, Eq.(8.24) reduces to

$$J_e = -\frac{1}{2\pi d^2}\frac{m_e c^3}{e}\left(\frac{eV}{m_e c^2}\right), \quad \text{for } \frac{eV}{m_e c^2} \gg 1, \quad (8.26)$$

which scales linearly with the applied voltage V.
To display the information in Eq.(8.24), it is convenient to introduce the normalized impedance Z_n defined by

$$Z_n = -\frac{cV}{2\pi d^2 J_e}. \quad (8.27)$$

Note from Eqs.(8.26) and (8.27) that Z_n is normalized such that $Z_n \to 1$ for $eV/m_e c^2 \gg 1$. Shown in Fig. 8.3 is a plot of Z_n versus $eV/m_e c^2$ obtained numerically from Eqs.(8.24) and (8.27). The dashed curves in Fig. 8.3 correspond to the nonrelativistic and ultrarelativistic limits in Eqs.(8.25) and (8.26). Not surprisingly, the impedance Z_n decreases as $eV/m_e c^2$ increases.

Problem 8.3   Make use of the definitions of $\delta(x)$ in Eq.(8.21) and the elliptic integrals in Eqs.(8.22) and (8.23) to verify the integral relation in Eq.(8.20).

Problem 8.4   In both the nonrelativistic and ultrarelativistic limits, show that the elliptic integrals in Eq.(8.24) can be simplified to give the limiting expressions for the electron current density J_e in Eqs.(8.25) and (8.26).
As an alternate approach in the ultrarelativistic regime, show directly that

$$\int_1^{\gamma_a} \frac{d\gamma_e}{[\gamma_e^2 - 1]^{1/4}} \simeq 2\left(\frac{eV}{m_e c^2}\right)^{1/2} \quad (8.4.1)$$

for $eV/m_e c^2 = \gamma_a - 1 \gg 1$. Make use of Eq.(8.4.1) and Eq.(8.19) evaluated at $x = d$ to obtain the ultrarelativistic expression for J_e in Eq.(8.26).

Problem 8.5   The analysis in Sec. 8.2.2 neglects the "pinching" influence of the azimuthal self-magnetic field $B_\theta^s(r)$ produced by the flow of electrons from the cathode to the anode. Assume that the electrons are emitted from a circular cathode surface with radius a_c and that the current density $J_e(< 0)$ is uniform across the (cylindrical) current channel with cross-sectional area πa_c^2.

a.   Show that the self-magnetic field at the outer edge of the electron beam is

$$B_\theta^s(r = a_c) = \frac{2I}{c a_c}, \quad (8.5.1)$$

where $I = | - J_e | \pi a_c^2$ is the electron current.

b. Consider an electron at the outer edge of the electron beam ($r = a_c$) with x-momentum $\gamma m_e v_x$. The self-magnetic field will bend the electron orbit through an effective gyroradius given (approximately) by $r_L \sim \gamma m_e v_x c / e B_\theta^s (r = a_c)$. For $v_x \sim c$, show that the condition for negligible deflection of the electron trajectory by the self-magnetic field ($r_L \gg d$) can be expressed as

$$I \ll I_{cr} \equiv \gamma \frac{mc^3}{e} \frac{a_c}{2d}, \qquad (8.5.2)$$

where d is the cathode-anode spacing.

Here, I_{cr} is the critical current required for the beam to pinch over a distance $d(\sim r_L)$. For a thin gap and large-radius cathode ($a_c/2d \gg 1$), it follows from Eq.(8.5.2) that I_{cr} can exceed $mc^3/e \equiv 17\,\text{kA}$ by a large amount.

8.3   Magnetically Insulated Brillouin Flow

8.3.1   Nonrelativistic Planar Flow

In this section, we consider nonrelativistic electron flow in the planar diode configuration illustrated in Fig. 8.4, where a magnetic field $B_0 \hat{\mathbf{e}}_z$ is applied parallel to the cathode and anode surfaces. It is assumed that the characteristic rise time of the anode voltage to the steady value $\phi_0(x = d) = V$ is long in comparison with the electron gyroperiod $2\pi/\omega_{ce}$. The slowly increasing electric field in the diode will cause the electrons emitted from the cathode to drift slowly towards the anode and establish a quasi-steady state where the thickness of the electron layer is x_b (Fig. 8.5). For sufficiently strong magnetic field B_0, the electron flow will be magnetically insulated from the anode ($x_b < d$).[25-28] Making use of a macroscopic cold-fluid model (Sec. 2.3), the steady-state ($\partial/\partial t = 0$) electron flow velocity

$$\mathbf{V}_e^0(\mathbf{x}) = V_{ye}^0(x)\hat{\mathbf{e}}_y$$

in the crossed electric and magnetic fields, $E_x(x)\hat{\mathbf{e}}_x$ and $B_0\hat{\mathbf{e}}_z$, is determined from the force balance equation

$$0 = -n_e^0(x)\, e \left[E_x(x) + \frac{1}{c} V_{ye}^0(x)\, B_0 \right] . \qquad (8.28)$$

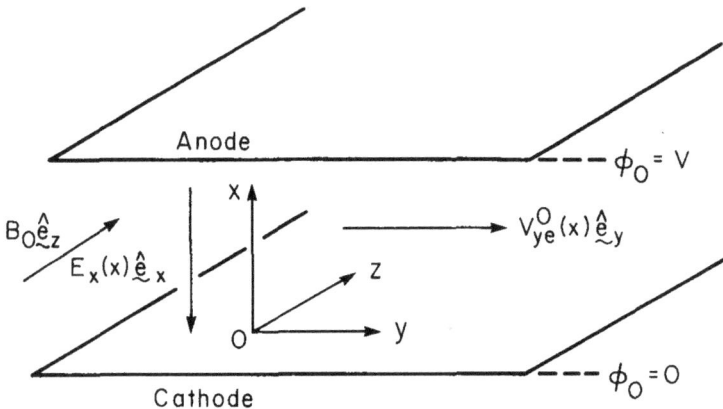

Figure 8.4. Schematic of planar diode configuration with insulating magnetic field $B_0\hat{e}_z$. The electron flow is in the y-direction with average velocity $V_{ye}^0(x) = -cE_x(x)/B_0$.

Here, the electric field $E_x(x) = -\partial\phi_0(x)/\partial x$ is determined self-consistently in terms of the electron density profile $n_e^0(x)$ from Poisson's equation

$$\frac{\partial}{\partial x}E_x(x) = -4\pi e n_e^0(x) \ . \tag{8.29}$$

As in Sec. 8.2, space-charge-limited flow is assumed with $E_x(x = 0) = 0$ [Eq.(8.2)], and $\phi_0(x = 0) = 0$ and $\phi_0(x = d) = V$ [Eq.(8.1)].

The key assumption corresponding to Brillouin flow conditions is that the total energy of an electron fluid element is uniform (independent of x) across the electron layer $(0 \leq x < x_b)$. That is, it is assumed that

$$\frac{1}{2}m_e V_{ye}^{02}(x) - e\phi_0(x) = \text{ const. } , \tag{8.30}$$

where $V_{ye}^0(x) = -cE_x(x)/B_0$. Because $E_x(x = 0) = 0$ and $\phi_0(x = 0) = 0$ at the cathode, it follows that the value of the constant in Eq.(8.30) is equal to zero. Taking the derivative of Eq.(8.30) with respect to x gives

$$m_e V_{ye}^0(x)\frac{\partial}{\partial x}V_{ye}^0(x) + eE_x(x) = 0 \ . \tag{8.31}$$

Expressing $V_{ye}^0(x) = -cE_x(x)/B_0$, Eq.(8.31) reduces to

$$\frac{\partial}{\partial x}V_{ye}^0(x) = \omega_{ce} \ , \tag{8.32}$$

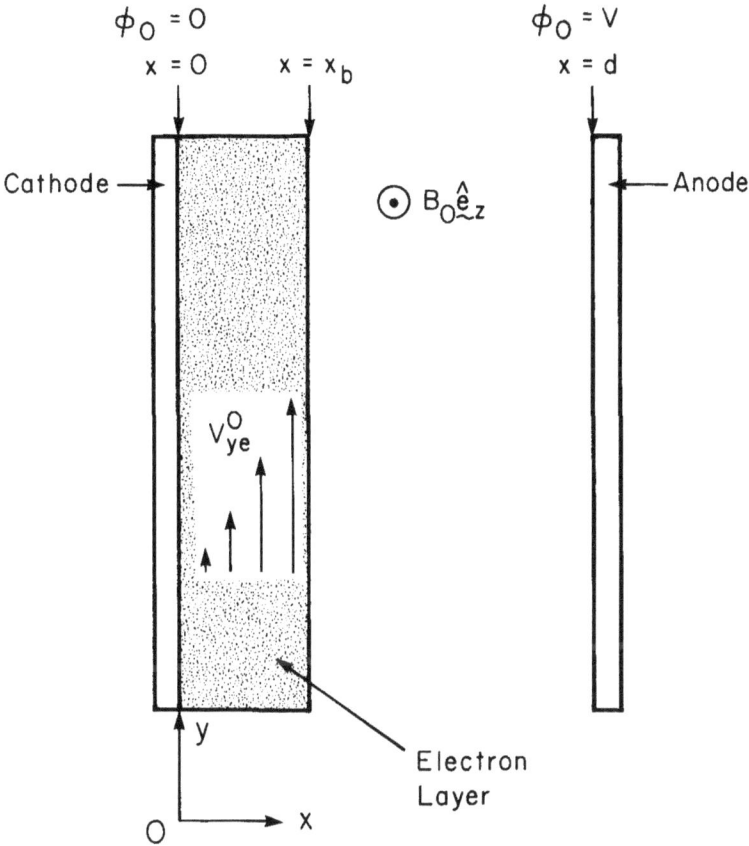

Figure 8.5.   The electron layer in a magnetically insulated planar diode extends from $x = 0$ to $x = x_b < d$.

where $\omega_{ce} = eB_0/m_e c$. Integrating Eq.(8.32) with respect to x, and imposing $V_{ye}^0(x = 0) = 0$, we obtain

$$V_{ye}^0(x) = \omega_{ce}x \, , \qquad (8.33)$$

which corresponds to zero canonical momentum of an electron fluid element in the y-direction.

An important property of the electron flow follows directly from Eq.(8.32). Making use of $\partial V_{ye}^0/\partial x = -(c/B_0)\,\partial E_x/\partial x = 4\pi e n_e^0(x)c/B_0 = \omega_{pe}^2(x)/\omega_{ce}$, Eq.(8.32) yields the condition

$$\omega_{pe}^2(x) = \omega_{ce}^2 \qquad (8.34)$$

within the electron layer ($0 \leq x < x_b$). For the nonrelativistic planar flow considered here, $\omega_{pe}^2(x)$ is uniform and equal to ω_{ce}^2. Equation (8.34) is known as the Brillouin flow condition,[57] and is a direct consequence of the assumption of uniform energy profile in Eq.(8.30).

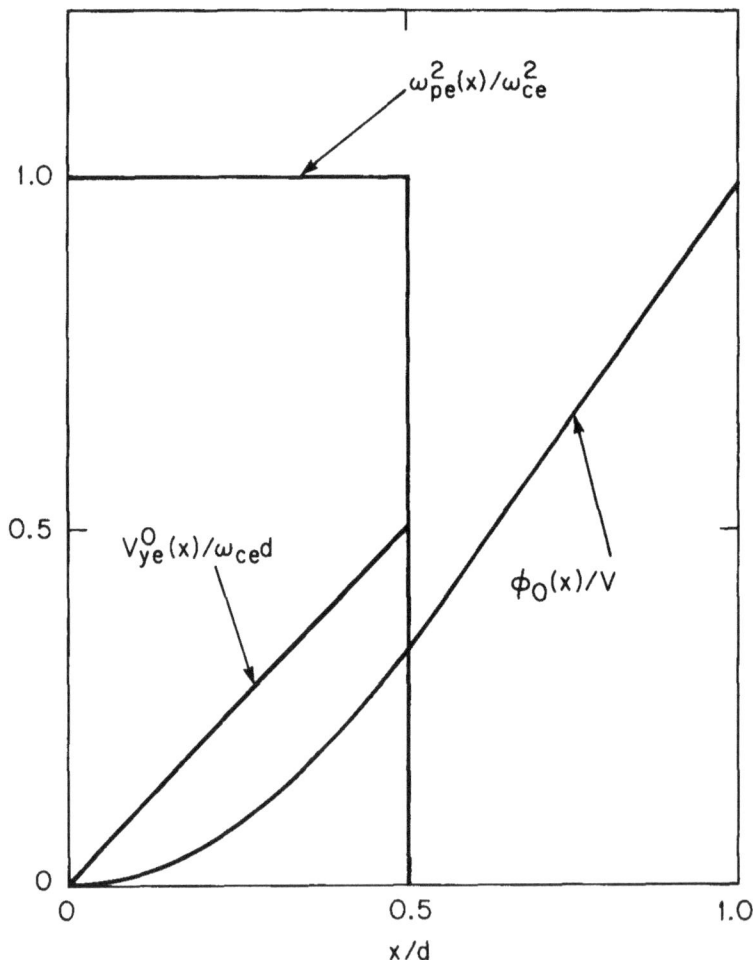

Figure 8.6. Plots versus x/d of the normalized electrostatic potential $\phi_0(x)/V$ [Eq.(8.35)], electron flow velocity $V_{ye}^0(x)/\omega_{ce}d$ [Eq.(8.33)], and electron density $\omega_{pe}^2(x)/\omega_{ce}^2$ [Eq.(8.34)] for the choice of system parameters $V/V_H = 3/4$ and $x_b/d = 1/2$ in a magnetically insulated planar diode.

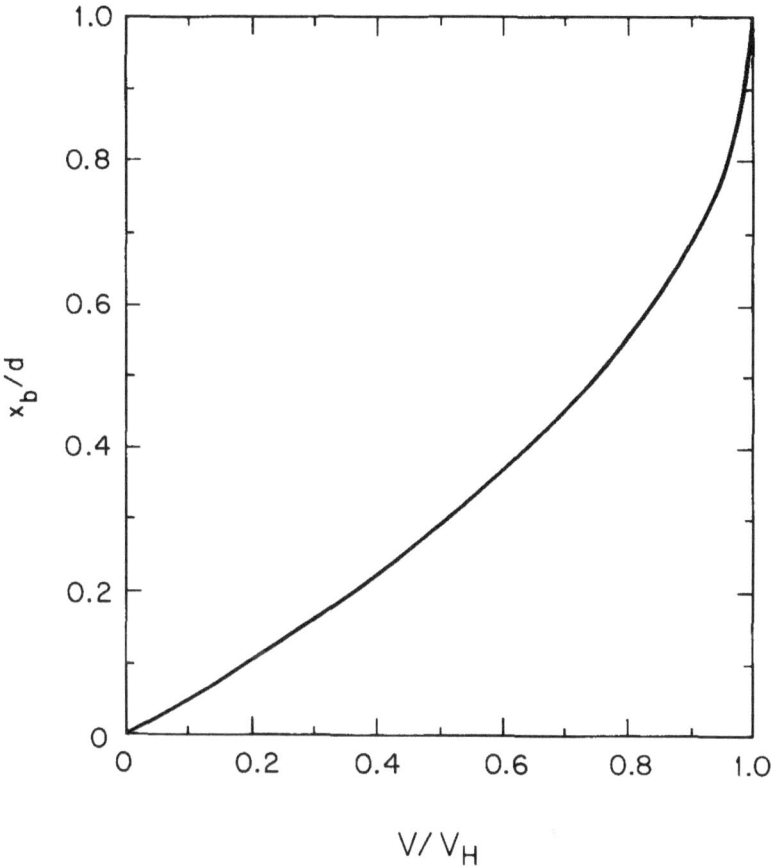

Figure 8.7. Plot of the normalized layer thickness x_b/d versus V/V_H obtained from Eq.(8.37) for a magnetically insulated planar diode.

Equation (8.33) can be expressed as $\partial\phi_0/\partial x = (B_0/c)\omega_{ce}x$ within the electron layer ($0 \leq x < x_b$). Solving for $\phi_0(x)$ then gives

$$\frac{e\phi_0(x)}{m_ec^2} = \begin{cases} \dfrac{\omega_{ce}^2}{c^2}\dfrac{x^2}{2} \, , & 0 \leq x < x_b \, , \\[3mm] \dfrac{\omega_{ce}^2}{c^2}\dfrac{x_b^2}{2} + \dfrac{\omega_{ce}^2}{c^2}x_b(x - x_b) \, , & x_b < x \leq d \, , \end{cases}$$

(8.35)

where we have enforced the continuity of $\phi_0(x)$ and $\partial\phi_0(x)/\partial x$ at $x = x_b$.

Evaluating Eq.(8.35) at the anode where $\phi_0(x = d) = V$, we obtain

$$\frac{eV}{m_e c^2} = \frac{\omega_{ce}^2 d^2}{c^2} \left[\frac{x_b}{d} - \frac{1}{2} \frac{x_b^2}{d^2} \right] . \tag{8.36}$$

For specified values of the normalized voltage $eV/m_e c^2$ and magnetic field strength $\omega_{ce}d/c$, Eq.(8.36) can be viewed as an equation that determines the normalized layer thickness x_b/d. The present nonrelativistic analysis of course requires $eV/m_e c^2 \ll 1$. Solving Eq.(8.36) for x_b/d gives

$$\frac{x_b}{d} = 1 - \left(1 - \frac{2eV/m_e c^2}{\omega_{ce}^2 d^2/c^2} \right)^{1/2} . \tag{8.37}$$

Evidently, magnetically insulated equilibria with $x_b < d$ exist only if

$$\frac{eV}{m_e c^2} < \frac{eV_H}{m_e c^2} \equiv \frac{1}{2} \frac{\omega_{ce}^2 d^2}{c^2} . \tag{8.38}$$

The voltage V_H defined in Eq.(8.38) is known as the Hull cut-off voltage.[43,58] For $V < V_H$, the electron layer partially fills the cathode-anode gap ($x_b < d$). On the other hand, for $V = V_H$ the applied voltage is sufficiently strong that the electrons fill the entire gap ($x_b = d$). For $V > V_H$, steady-state solutions with $V_{xe}^0 = 0$ do not exist.

To summarize, presented in Fig. 8.6 are normalized plots of $V_{ye}^0(x)/\omega_{ce}d$ [Eq.(8.33)], $\omega_{pe}^2(x)/\omega_{ce}^2$ [Eq.(8.34)], and $\phi_0(x)/V$ [Eq.(8.35)] versus x/d for the choice of parameters $V/V_H = 3/4$ and $x_b/d = 1/2$. Figure 8.7 shows a plot of the normalized layer thickness x_b/d versus V/V_H [Eq.(8.37)].

Problem 8.6   Repeat the analysis in Sec. 8.3.1, assuming that the electric field $E_x(x = 0) = E_0 \neq 0$ at the cathode surface. Make use of the boundary conditions $\phi_0(x = 0) = 0$ and $\phi_0(x = d) = V$.

a.   Verify that $\omega_{pe}^2(x) = \omega_{ce}^2$ and $\partial V_{ye}^0(x)/\partial x = \omega_{ce}$ within the electron layer $(0 \leq x < x_b)$.

b.   Show that Eq.(8.36) is modified to become

$$\frac{eV}{m_e c^2} = -\frac{eE_0 d}{m_e c^2} + \frac{\omega_{ce}^2 d^2}{c^2} \left(\frac{x_b}{d} - \frac{1}{2} \frac{x_b^2}{d^2} \right) . \tag{8.6.1}$$

As expected, for specified values of $eV/m_e c^2$ and $\omega_{ce}d/c$, Eq.(8.6.1) predicts that a negative bias field ($E_0 < 0$) decreases the layer thickness x_b/d relative to the value obtained in Eq.(8.37) for $E_0 = 0$.

Problem 8.7   Consider the equilibrium distribution function $f_e^0(x, \mathbf{p})$ specified by

$$f_e^0(x, \mathbf{p}) = \frac{\hat{n}_e}{2\pi m_e} \delta \left(H - m_e \frac{V_0^2}{2} \right) \delta(P_y) \tag{8.7.1}$$

in the magnetically insulated planar diode configuration illustrated in Fig. 8.5. Here, $\hat{n}_e$ and V_0 are constants; $H = (p_x^2 + p_y^2 + p_z^2)/2m_e - e\phi_0(x)$ is the energy of an individual electron; and $P_y = p_y - (eB_0/c)x$ is the canonical momentum in the y-direction. For the choice of distribution function in Eq.(8.7.1), the electrons enter the cathode-anode gap at $x = 0$ with kinetic energy $m_e V_0^2/2$.

a. Show directly from Eq.(8.7.1) that the electron density is given by

$$n_e^0(x) = \int d^3p \, f_e^0(x, \mathbf{p}) = \begin{cases} \hat{n}_e , & 0 \le x < x_b , \\ 0 , & x_b < x \le d , \end{cases} \tag{8.7.2}$$

and the y-directed flux of electrons can be expressed as

$$n_e^0(x) V_{ye}^0(x) = \int d^3p \, \frac{p_y}{m_e} f_e^0(x, \mathbf{p}) = \hat{n}_e \omega_{ce} x \tag{8.7.3}$$

for $0 \le x < x_b$. Here, x_b is related to other system parameters by $\psi(x = x_b) = 0$, where

$$\psi(x) = \left(\frac{m_e}{2} \right) (V_0^2 - \omega_{ce}^2 x^2) + e\phi_0(x) \ge 0 \tag{8.7.4}$$

is the envelope function.

b. Solve Poisson's equation, $\partial^2 \phi_0(x)/\partial x^2 = 4\pi e n_e^0(x)$, in the cathode-anode gap subject to the boundary conditions $\phi_0(x = 0) = 0$ and $\partial\phi_0(x)/\partial x|_{x=0} = 0$. Show that

$$\frac{e\phi_0(x)}{m_e c^2} = \begin{cases} \dfrac{1}{2} \dfrac{\omega_{pe}^2 x^2}{c^2} , & 0 \le x < x_b , \\[2ex] \dfrac{1}{2} \dfrac{\omega_{pe}^2 x_b^2}{c^2} + \dfrac{\omega_{pe}^2}{c^2} x_b(x - x_b) , & x_b < x \le d, \end{cases} \tag{8.7.5}$$

where $\omega_{pe}^2 \equiv 4\pi \hat{n}_e e^2/m_e$. The condition $\psi(x = x_b) = 0$ can then be expressed as

$$V_0^2 = \left(\omega_{ce}^2 - \omega_{pe}^2 \right) x_b^2 , \tag{8.7.6}$$

which requires $\omega_{ce}^2 > \omega_{pe}^2$ for existence of the equilibrium when $V_0^2 \ne 0$. Moreover, enforcing $\phi_0(x = d) = V$ gives

$$\frac{eV}{m_e c^2} = \frac{\omega_{pe}^2 d^2}{c^2} \left[\frac{x_b}{d} - \frac{1}{2} \frac{x_b^2}{d^2} \right] , \tag{8.7.7}$$

where V is the applied voltage. Note that Eqs.(8.7.5) and (8.7.7) are identical to Eqs.(8.35) and (8.36), respectively, with ω_{ce}^2 replaced by ω_{pe}^2.

c. For the choice of distribution function in Eq.(8.7.1), the effective electron temperature perpendicular to the flow direction (the y-direction) is generally nonzero. Show that

$$n_e^0(x)\,T_{e\perp}^0(x) = \frac{1}{2m_e}\int d^3p\,(p_x^2 + p_z^2)\,f_e^0(x,\mathbf{p})$$

$$= \frac{1}{2}\hat{n}_e m_e V_0^2\left(1 - \frac{x^2}{x_b^2}\right) \qquad (8.7.8)$$

for $0 \le x < x_b$. Therefore, $T_{e\perp}^0(x)$ decreases monotonically from the value $m_e V_0^2/2$ at the cathode to zero at the outer edge of the electron layer $(x = x_b)$. That is, $x = x_b$ corresponds to the envelope of turning points for the electron orbits.

To summarize, for the choice of $f_e^0(x,\mathbf{p})$ in Eq.(8.7.1), the electrons enter the cathode-anode gap at $x = 0$ with kinetic energy $m_e V_0^2/2$. For $V_0^2 \neq 0$, the electrons have nonzero temperature $T_{e\perp}^0(x)$ transverse to the flow direction [Eq.(8.7.8)], and the coupled equations (8.7.6) and (8.7.7) must be solved to determine x_b/d self-consistently in terms of $eV/m_ec^2, \omega_{ce}^2 d^2/c^2$ and V_0^2/c^2. For $V_0^2 = 0$, however, the Brillouin flow condition $\omega_{pe}^2 = \omega_{ce}^2$ follows from Eq.(8.7.6), and the voltage relation in Eq.(8.7.7) reduces to the cold-fluid result in Eq.(8.36).

8.3.2   Nonrelativistic Cylindrical Flow

Before extending the preceding analysis to relativistic electron flow in planar geometry (Sec. 8.3.3), we consider the simpler generalization corresponding to nonrelativistic Brillouin flow in a cylindrical diode (Fig. 8.8). Here, the cathode is located at $r = a$ and the anode is located at $r = b$. Space-charge-limited flow is assumed with

$$E_r(r = a) = 0\,, \qquad (8.39)$$

and

$$\phi_0(r = a) = 0 \text{ and } \phi_0(r = b) = V\,. \qquad (8.40)$$

Making use of a macroscopic cold-fluid model (Sec. 2.3) with equilibrium

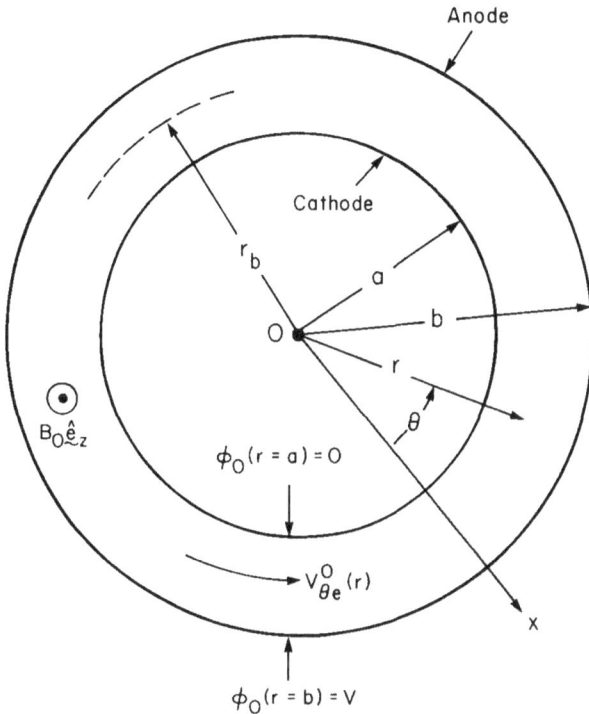

Figure 8.8. Schematic of cylindrical diode configuration with insulating magnetic field $B_0 e_z$. The cathode is located at $r = a$, and the anode is located at $r = b$. The electron layer extends from $r = a$ to $r = r_b < b$.

flow velocity $\mathbf{V}_e^0(\mathbf{x}) = V_{\theta e}^0(r)\hat{\mathbf{e}}_\theta$, the radial force balance on an electron fluid element can be expressed as (for $\partial/\partial t = 0$)

$$-\frac{n_e^0(r)\,m_e V_{\theta e}^{02}(r)}{r} = -n_e^0(r)e\left[E_r(r) + \frac{1}{c}\,V_{\theta e}^0(r)B_0\right] , \qquad (8.41)$$

where the radial electric field $E_r(r)$ is determined self-consistently from Poisson's equation

$$\frac{1}{r}\frac{\partial}{\partial r}rE_r(r) = -4\pi e n_e^0(r) . \qquad (8.42)$$

An important physical effect in cylindrical geometry is the centrifugal force term on the left-hand side of Eq.(8.41).

As in Sec. 8.3.1, it is assumed that the electron layer extends from the cathode $(r = a)$ to some outer radius $r_b < b$. The assumption of uniform energy across the radial extent of the electron layer $(a \le r < r_b)$ can be expressed as

$$\frac{1}{2} m_e V_{\theta e}^{02}(r) - e\phi_0(r) = \text{const.} , \qquad (8.43)$$

where $\phi_0(r)$ is the electrostatic potential. Enforcing $\phi_0(r = a) = 0$ and $V_{\theta e}^0(r = a) = 0$ [see Eq.(8.41)], the value of the constant in Eq.(8.43) is equal to zero. Differentiating Eq.(8.43) with respect to r, and eliminating $E_r(r) = -\partial \phi_0(r)/\partial r$ by means of the equilibrium force balance equation (8.41), we obtain

$$\frac{\partial}{\partial r} V_{\theta e}^0(r) = \omega_{ce} - \frac{V_{\theta e}^0(r)}{r} , \qquad (8.44)$$

where $\omega_{ce} = eB_0/m_e c$. Integrating Eq.(8.44) with respect to r then gives

$$r V_{\theta e}^0(r) - \frac{r^2 \omega_{ce}}{2} = \text{const.} , \qquad (8.45)$$

which corresponds to uniform canonical angular momentum of an electron fluid element across the layer $(a \le r < r_b)$.

Enforcing $V_{\theta e}^0(r = a) = 0$ in Eq.(8.45), we obtain for the azimuthal flow velocity

$$V_{\theta e}^0(r) = \frac{1}{2} \omega_{ce} \left(\frac{r^2 - a^2}{r} \right) . \qquad (8.46)$$

Substituting Eq.(8.46) into Eq.(8.41) and solving for the radial electric field $E_r(r)$, it is readily shown that

$$r E_r(r) = -\frac{B_0 \omega_{ce}}{c} \left(\frac{r^4 - a^4}{4r^2} \right) \qquad (8.47)$$

within the electron layer $(a \le r < r_b)$. Substituting Eq.(8.47) into Poisson's equation (8.42) then gives for $\omega_{pe}^2(r) = 4\pi n_e^0(r) e^2/m_e$

$$\omega_{pe}^2(r) = \omega_{ce}^2 \left[1 - \frac{1}{2} \left(\frac{r^4 - a^4}{r^4} \right) \right] \qquad (8.48)$$

for $a \le r < r_b$. Equation (8.48) is the cylindrical generalization of the planar Brillouin flow condition in Eq.(8.34). For an electron layer with moderate aspect ratio $a/(r_b - a)$, it is clear from Eq.(8.48) that the radial variation of $\omega_{pe}^2(r)$ can be significant. For example, if $r_b/a = 3/2$, then $\omega_{pe}^2(r)$ decreases monotonically from ω_{ce}^2 at $r = a$ to $(97/162)\omega_{ce}^2$ at $r = r_b$.

We now integrate Eq.(8.48) to determine the electrostatic potential $\phi_0(r)$. Enforcing $\phi_0(r = a) = 0$, and the continuity of $\phi_0(r)$ and $\partial\phi_0(r)/\partial r$ at $r = r_b$, we obtain

$$\frac{e\phi_0(r)}{m_e c^2} = \begin{cases} \dfrac{1}{2}\dfrac{\omega_{ce}^2}{c^2}\left(\dfrac{r^2 - a^2}{2r}\right)^2, & a \le r < r_b, \\[3mm] \dfrac{1}{2}\dfrac{\omega_{ce}^2}{c^2}\left[\left(\dfrac{r_b^2 - a^2}{2r_b}\right)^2 + \dfrac{1}{2}\left(\dfrac{r_b^4 - a^4}{r_b^2}\right)\ln\left(\dfrac{r}{r_b}\right)\right], & r_b < r \le b. \end{cases}$$

(8.49)

Evaluating Eq.(8.49) at $r = b$ where $\phi_0(r = b) = V$ gives

$$\frac{eV}{m_e c^2} = \frac{1}{2}\frac{\omega_{ce}^2}{c^2}\left(\frac{r_b^2 - a^2}{2r_b}\right)^2 + \frac{1}{4}\frac{\omega_{ce}^2}{c^2}\left(\frac{r_b^4 - a^4}{r_b^2}\right)\ln\left(\frac{b}{r_b}\right), \qquad (8.50)$$

where the nonrelativistic treatment requires $eV/m_e c^2 \ll 1$.

For specified values of the normalized voltage $eV/m_e c^2$ and magnetic field $\omega_{ce}(b-a)/c$, Eq.(8.50) can be used to determine the normalized layer thickness

$$\Delta_b = \frac{r_b - a}{b - a}. \qquad (8.51)$$

Of course Eq.(8.50) reduces to Eq.(8.36) in the limit of a thin layer and large-aspect-ratio diode with $a \gg r_b - a, b - a$. For a diode with moderate aspect ratio, however, the cylindrical effects incorporated in Eq.(8.50) can be significant. For example, in the limit where the electron layer extends to the anode $(r_b = b)$, the Hull cut-off voltage determined from Eq.(8.50) is given by

$$\frac{eV_H}{m_e c^2} = \frac{1}{2}\frac{\omega_{ce}^2(b - a)^2}{c^2}\left(\frac{b + a}{2b}\right)^2. \qquad (8.52)$$

Equation (8.52) differs from Eq.(8.38) by the factor $[(b + a)/2b]^2 < 1$. If (for example) $b = 2a$, the cut-off voltage in Eq.(8.52) is 9/16 times the planar result in Eq.(8.38). Typical numerical solutions to Eq.(8.50) are presented in Fig. 8.9, where $\Delta_b = (r_b - a)/(b - a)$ is plotted versus $(eV/m_e c^2)/[\omega_{ce}^2(b - a)^2/2c^2]$ for $b/a = 2, b/a = 3/2$, and $a/(b - a) = A_d \to \infty$. The curve labeled $A_d = \infty$ in Fig. 8.9 corresponds to the planar limit in Fig. 8.6. As b/a is increased, we note from Fig. 8.9 that

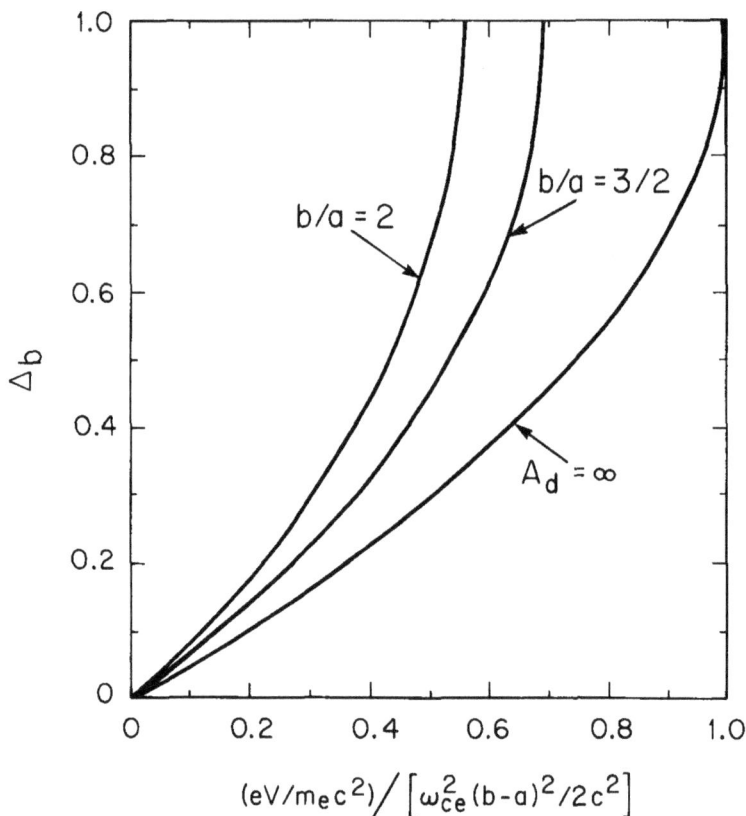

$$(eV/m_e c^2) \Big/ \Big[\omega_{ce}^2 (b-a)^2 / 2c^2\Big]$$

Figure 8.9.   Plots of the normalized layer thickness $\Delta_b = (r_b - a)/(b - a)$ versus $(eV/m_e c^2)/[\omega_{ce}^2(b - a)^2/2c^2]$ obtained from Eq.(8.50) for a magnetically insulated cylindrical diode with $b/a = 2, b/a = 3/2$, and $a/(b - a) = A_d \to \infty$.

the normalized layer thickness Δ_b increases more rapidly with increasing values of the normalized voltage.

Problem 8.8   Repeat the analysis in Sec. 8.3.2, removing the central cathode $(a = 0)$, and assuming that the electron energy is uniform across the radial extent $(0 \leq r < r_b)$ of the nonneutral plasma column [Eq.(8.43)].

a.   Show that the azimuthal flow velocity of an electron fluid element is given by $V_{\theta e}^0(r) = r\omega_{ce}/2$, and that the electron density is uniform with

$$\frac{2\omega_{pe}^2(r)}{\omega_{ce}^2} = \begin{cases} 1, & 0 \leq r < r_b\,, \\ 0, & r_b < r \leq b\,. \end{cases} \tag{8.8.1}$$

Equation (8.8.1) is identical to the Brillouin-flow density limit calculated in Chapter 3 for a nonneutral plasma column with uniform density.

b. Verify that the column radius r_b is related to the voltage V of the conducting cylinder at $r = b$ by

$$\frac{eV}{m_e c^2} = \frac{1}{8} \frac{\omega_{ce}^2 r_b^2}{c^2} \left[1 + 2 \ln \left(\frac{b}{r_b} \right) \right] . \tag{8.8.2}$$

Equation (8.8.2) is identical to Eq.(8.50) with $a = 0$.

8.3.3   Relativistic Planar Flow

In this section, we extend the nonrelativistic analysis in Sec. 8.3.1 to arbitrary values of $eV/m_e c^2$, allowing for relativistic electron flow in the planar diode configuration illustrated in Figs. 8.4 and 8.5.[28] For sufficiently high current, the electron layer produces a diamagnetic depression in the axial magnetic field $B_z(x)$ determined from

$$\frac{\partial}{\partial x} B_z(x) = \frac{1}{c} 4\pi e n_e^0(x) V_{ye}^0(x) = -\frac{4\pi e n_e^0(x) E_x(x)}{B_z(x)} . \tag{8.53}$$

Here, $V_{ye}^0(x) = -cE_x(x)/B_z(x)$ is the equilibrium flow velocity, and $E_x(x)$ is determined self-consistently from

$$\frac{\partial}{\partial x} E_x(x) = -4\pi e n_e^0(x) . \tag{8.54}$$

As in earlier sections, space-charge-limited flow is assumed with $E_x(x = 0) = 0$, and $\phi_0(x = 0) = 0$ and $\phi_0(x = d) = V$. For relativistic Brillouin flow, the condition that the total energy of an electron fluid element is uniform across the electron layer can be expressed as

$$\left[\gamma_e^0(x) - 1 \right] m_e c^2 - e\phi_0(x) = \text{const. ,} \tag{8.55}$$

where $\gamma_e^0(x) = [1 - V_{ye}^{02}(x)/c^2]^{-1/2}$ is the relativistic mass factor. Here, $\phi_0(x = 0) = 0$ and $\gamma_e^0(x = 0) = 1$, so that the value of the constant in Eq.(8.55) is equal to zero.

We eliminate $n_e^0(x)$ in Eq.(8.53) by means of Poisson's equation (8.54). Integrating Eq.(8.53) with respect to x then gives[48]

$$B_z^2(x) - E_x^2(x) = \text{const.} \tag{8.56}$$

within the electron layer ($0 \leq x < x_b$). Equation (8.56) can be expressed in the equivalent form

$$\frac{B_z(x)}{\gamma_e^0(x)} = \text{const.} , \qquad (8.57)$$

where $\gamma_e^0(x) = [1 - E_x^2(x)/B_z^2(x)]^{-1/2}$. Taking the derivative of Eq.(8.55) with respect to x gives $\partial\gamma_e^0/\partial x = (e/m_ec^2)\partial\phi_0/\partial x$, which can also be expressed as

$$\gamma_e^{03}(x)\frac{E_x(x)}{B_z(x)}\frac{\partial}{\partial x}\left[\frac{E_x(x)}{B_z(x)}\right] = -\frac{eE_x(x)}{m_ec^2} , \qquad (8.58)$$

where $E_x(x) = -\partial\phi_0(x)/\partial x$. Using Eqs.(8.53) and (8.54) to eliminate $\partial B_z/\partial x$ and $\partial E_x/\partial x$ in Eq.(8.58), we obtain the relativistic Brillouin flow condition

$$\left[\frac{eB_z(x)}{\gamma_e^0(x)m_ec}\right]^2 = \frac{4\pi n_e^0(x)e^2}{\gamma_e^0(x)m_e} . \qquad (8.59)$$

Because $B_z(x)/\gamma_e^0(x) = \text{const.}$ [Eq.(8.57)], it follows from Eq.(8.59) that $n_e^0(x)/\gamma_e^0(x) = \text{const.}$ within the electron layer ($0 \leq x < x_b$). Therefore, the profiles for $B_z(x), \gamma_e^0(x)$ and $n_e^0(x)$ have identical spatial dependences in the interval $0 \leq x < x_b$.

Taking the derivative of Eq.(8.53) with respect to x, and making use of $n_e^0(x)/B_z(x) = \text{const.}$ and $\partial E_x/\partial x = -4\pi en_e^0$, we obtain

$$\frac{\partial^2}{\partial x^2}B_z(x) - \kappa^2 B_z(x) = 0 , \qquad (8.60)$$

where κ is defined by

$$\kappa \equiv \frac{4\pi en_e^0(x)}{B_x(x)} = \text{const.} \qquad (8.61)$$

The solution to Eq.(8.60) is

$$B_z(x) = \begin{cases} B_0\dfrac{\cosh(\kappa x)}{\cosh(\kappa x_b)} , & 0 \leq x < x_b , \\ B_0 , & x_b < x \leq d , \end{cases} \qquad (8.62)$$

where $B_0 = \text{const.}$ denotes the uniform value of the axial magnetic field in the vacuum region ($x_b < x \leq d$). Similarly, the electron density $n_e^0(x)$, the relativistic mass factor $\gamma_e^0(x)$, the flow velocity $V_{ye}^0(x)$, and the electric field $E_x(x)$ can be expressed as

$$n_e^0(x) = \hat{n}_e \frac{\cosh(\kappa x)}{\cosh(\kappa x_b)} , \tag{8.63}$$

$$\gamma_e^0(x) = \cosh(\kappa x) \tag{8.64}$$

$$V_{ye}^0(x) = c \tanh(\kappa x), \tag{8.65}$$

$$E_x(x) = -B_0 \frac{\sinh(\kappa x)}{\cosh(\kappa x_b)}, \tag{8.66}$$

within the electron layer ($0 \leq x < x_b$). Note from Eq.(8.63) that the electron density $n_e^0(x)$ increases monotonically from the value $\hat{n}_e / \cosh(\kappa x_b)$ at the cathode ($x = 0$) to the value $\hat{n}_e = $ const. at the outer edge of the electron layer ($x = x_b$).

The electrostatic potential $\phi_0(x)$ can be determined from $[\gamma_e^0(x) - 1] \times m_e c^2 - e\phi_0(x) = 0$ within the electron layer ($0 \leq x < x_b$), and from $\partial^2 \phi_0 / \partial x^2 = 0$ in the vacuum region ($x_b < x \leq d$). Making use of Eq.(8.64), and enforcing continuity of $\phi_0(x)$ and $\partial \phi_0(x) / \partial x$ at $x = x_b$, we obtain

$$\frac{e\phi_0(x)}{m_e c^2} = \begin{cases} [\cosh(\kappa x) - 1] , & 0 \leq x < x_b , \\ [\cosh(\kappa x_b) - 1] + \kappa(x - x_b)\sinh(\kappa x_b) , & x_b < x \leq d . \end{cases} \tag{8.67}$$

Evaluating Eq.(8.67) at the anode ($x = d$) where $\phi_0(x = d) = V$ gives

$$1 + \frac{eV}{m_e c^2} = \cosh(\kappa x_b) + \kappa(d - x_b)\sinh(\kappa x_b) , \tag{8.68}$$

which relates the normalized voltage $eV/m_e c^2$ to κx_b and $\kappa(d - x_b)$.

It is useful to relate the vacuum field B_0 in Eq.(8.62) to the initial fill field B_f (assumed uniform in the region $0 \leq x \leq d$) prior to the formation of the electron layer. Assuming magnetic flux conservation with $\int_0^d dx\, B_z = $ const., we obtain

$$B_f d = \frac{B_0}{\kappa} \tanh(\kappa x_b) + B_0(d - x_b) . \tag{8.69}$$

In terms of $B_z(x = 0) = B_0/\cosh(\kappa x_b)$ at the cathode, Eq.(8.69) can be expressed in the equivalent form

$$\frac{eB_f d}{m_e c^2} = \frac{eB_z(x = 0)}{m_e c^2 \kappa} \left[\sinh(\kappa x_b) + \kappa(d - x_b)\cosh(\kappa x_b)\right] . \tag{8.70}$$

The coefficient $eB_z(x = 0)/m_e c^2 \kappa$ in Eq.(8.70) can be further simplified. In terms of the nonrelativistic frequencies $\omega_{ce}(x) = eB_z(x)/m_e c$ and $\omega_{pe}^2(x) = 4\pi n_e^0(x)e^2/m_e$, the Brillouin flow condition in Eq.(8.59) can be expressed at the cathode [where $\gamma_e^0(x = 0) = 1$] as $\omega_{pe}^2(x = 0) = \omega_{ce}^2(x = 0)$. Equation (8.61) then gives $\kappa c = \omega_{pe}^2(x = 0)/\omega_{ce}(x = 0)$, or equivalently,

$$\kappa c = \omega_{pe}(x = 0) = \omega_{ce}(x = 0) . \tag{8.71}$$

Therefore, $eB_z(x = 0)/m_e c^2 \kappa = 1$, and Eq.(8.70) reduces to

$$\frac{eB_f d}{m_e c^2} = \sinh(\kappa x_b) + \kappa(d - x_b)\cosh(\kappa x_b) . \tag{8.72}$$

For specified values of the normalized voltage $eV/m_e c^2$ and fill field $eB_f d/m_e c^2$, Eqs.(8.68) and (8.72) can be solved numerically to determine the self-consistent values of the normalized layer thickness x_b/d and electron density $\kappa^2 d^2 = \omega_{pe}^2(x = 0)d^2/c^2$. Typical numerical results are illustrated in Fig. 8.10 where x_b/d is plotted versus $eV/m_e c^2$ for several values of the magnetic fill field ranging from $eB_f d/m_e c^2 = 0.25$ to 8. As the electron flow becomes increasingly relativistic, we note from Fig. 8.10 that the curves asymptote abruptly to $x_b/d = 1$ as the voltage V approaches the relativistic generalization of the Hull cut-off voltage V_H. To calculate V_H, we set $x_b = d$ in Eqs.(8.68) and (8.72), and make use of $\cosh^2(\kappa x_b) - \sinh^2(\kappa x_b) = 1$. This readily gives

$$\frac{eV_H}{m_e c^2} = \left[1 + \frac{e^2 B_f^2 d^2}{m_e^2 c^4}\right]^{1/2} - 1 . \tag{8.73}$$

In the nonrelativistic regime with $e^2 B_f^2 d^2/m_e^2 c^4 \ll 1$ and $eV_H/m_e c^2 \ll 1$, Eq.(8.73) reduces to the familiar expression in Eq.(8.38). Shown in Fig. 8.11 is a plot of $eV_H/m_e c^2$ versus normalized fill field $eB_f d/m_e c^2$. Magnetically insulated equilibria with $x_b < d$ exist provided the applied voltage satisfies $V < V_H$. For $V \ll V_H$, it is evident from Fig. 8.10 that the layer is thin $(x_b \ll d)$. However, as the voltage is increased to $V = V_H$, the layer thickness increases to $x_b = d$.

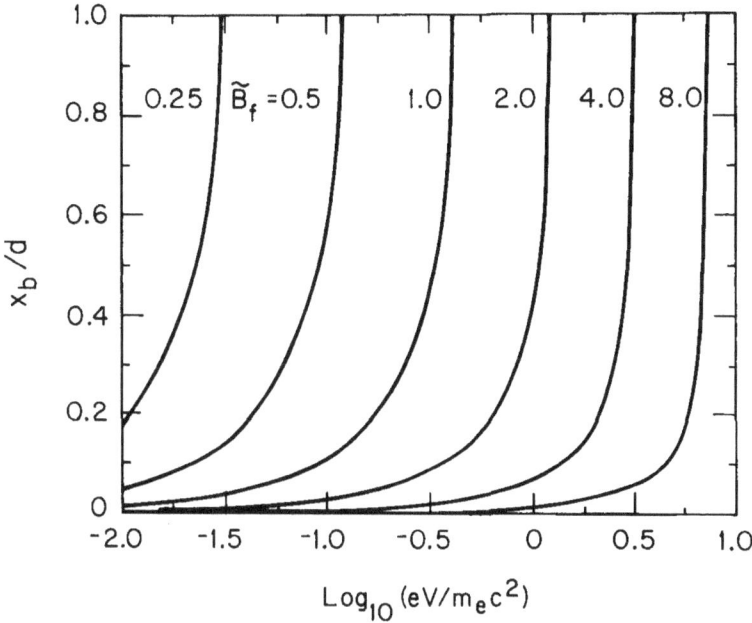

Figure 8.10. Plots of the normalized layer thickness x_b/d versus $eV/m_e c^2$ obtained numerically from Eqs.(8.68) and (8.72) for several values of the normalized fill field ranging from $eB_f d/m_e c^2 = 0.25$ to 8.

For effective interaction between the layer electrons in a relativistic magnetron (Sec. 8.6), the layer thickness x_b should be at least large enough that the (fastest) electrons at $x = x_b$ resonate with the excited wave with phase velocity $v_p = \omega/k_y \equiv \beta_p c$. This requires that the voltage V exceed a value known as the Buneman-Hartree threshold voltage V_{BH}.[43,59] Eliminating $\kappa(d - x_b)$, we combine Eqs.(8.68) and (8.72) to give

$$\frac{eV}{m_e c^2} = \frac{eB_f d}{m_e c^2} \tanh(\kappa x_b) - \left[1 - \frac{1}{\cosh(\kappa x_b)} \right] . \tag{8.74}$$

Setting $V_{ye}^0(x = x_b) = c\tanh(\kappa x_b) = \omega/k_y \equiv \beta_p c$, and $\gamma_e^0(x = x_b) = \cosh(\kappa x_b) = (1 - \beta_p^2)^{-1/2}$, it is readily shown from Eq.(8.74) that the Buneman-Hartree threshold voltage V_{BH} is given relativistically by

$$\frac{eV_{BH}}{m_e c^2} = \frac{eB_f d}{m_e c^2}\beta_p - \left[1 - \left(1 - \beta_p^2\right)^{1/2} \right] . \tag{8.75}$$

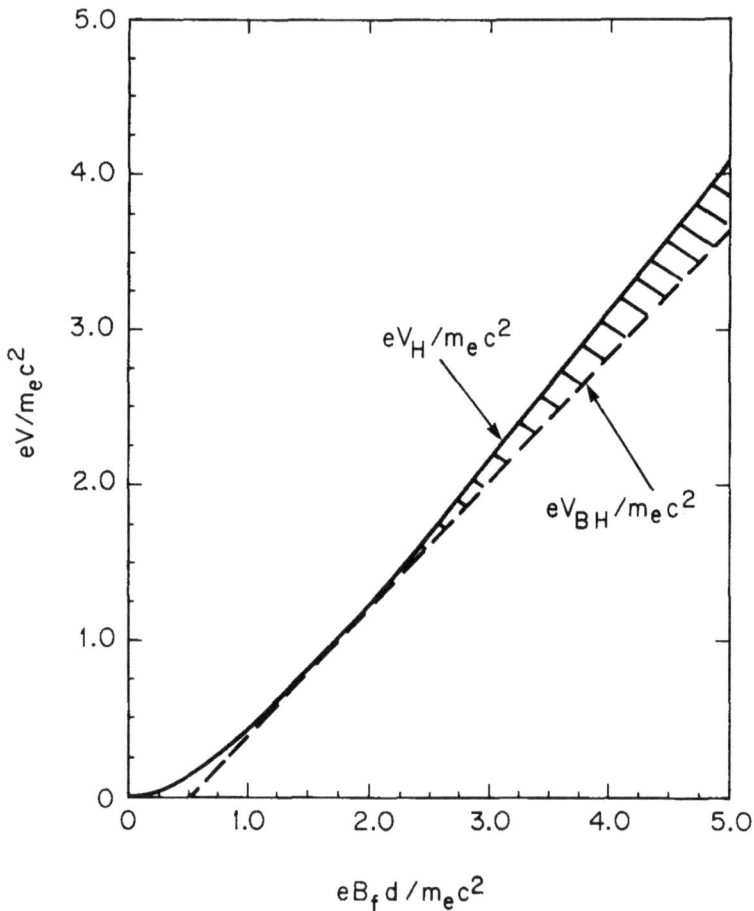

Figure 8.11. Plots versus the normalized fill field $eB_f d/m_e c^2$ of the Hull cut-off voltage $eV_H/m_e c^2$ [Eq.(8.73)], and the Buneman-Hartree threshold voltage $eV_{BH}/m_e c^2$ [Eq.(8.75)] for $\beta_p = \omega/ck_y = 0.8$.

For purposes of illustration, Fig. 8.11 also shows a plot (the dashed curve) of $eV_{BH}/m_e c^2$ versus the normalized fill field $eB_f d/m_e c^2$ obtained from Eq.(8.75) for the choice of phase velocity $\beta_p = \omega/ck_y = 0.8$. The allowed region of magnetron operation ($V_{BH} < V < V_H$) for $\beta_p = 0.8$ then corresponds to the shaded region in Fig. 8.11.

Problem 8.9   Consider the results of Sec. 8.3.3 in the nonrelativistic limit with $\kappa^2 d^2 \ll 1$. Make use of Eqs.(8.59), (8.62)–(8.68), (8.72) and (8.73) to show that the equilibrium profiles, and the expressions for $eV/m_e c^2$ and $eV_H/m_e c^2$, reduce to the corresponding results obtained nonrelativistically in Sec. 8.3.1.

Problem 8.10   For specified values of $eV/m_e c^2$ and $eB_f d/m_e c^2$, Eqs.(8.68) and (8.72) can be used to determine closed analytical expressions for the normalized layer thickness x_b/d and electron density $\kappa^2 d^2 = \omega_{pe}^2(x = 0)d^2/c^2$. Eliminate κ from Eqs.(8.68) and (8.72), and show that x_b/d can be expressed as

$$\frac{x_b}{d} = \cosh^{-1}\left[\frac{(1 + \tilde{V}) - \sqrt{\tilde{V}_H(2 + \tilde{V}_H)}\sqrt{(\tilde{V}_H - \tilde{V})(2 + \tilde{V} + \tilde{V}_H)}}{1 - (\tilde{V}_H - \tilde{V})(2 + \tilde{V} + \tilde{V}_H)}\right]$$

$$\times \left\{\sqrt{(\tilde{V}_H - \tilde{V})(2 + \tilde{V} + \tilde{V}_H)}\right. \tag{8.10.1}$$

$$+ \cosh^{-1}\left[\frac{(1 + \tilde{V}) - \sqrt{\tilde{V}_H(2 + \tilde{V}_H)}\sqrt{(\tilde{V}_H - \tilde{V})(2 + \tilde{V} + \tilde{V}_H)}}{1 - (\tilde{V}_H - \tilde{V})(2 + \tilde{V} + \tilde{V}_H)}\right]\right\}^{-1},$$

where $\tilde{V} \equiv eV/m_e c^2$, $\tilde{V}_H \equiv eV_H/m_e c^2$, and the Hull cut-off voltage V_H is defined in Eq.(8.73). For $V = V_H$, it follows directly from Eq.(8.10.1) that the electron layer fills the cathode-anode gap, with $x_b/d = 1$.

8.4   Magnetically Insulated Ion Diodes

The generation and propagation of intense ion beams is a subject of considerable practical importance, with applications ranging from the heating of laboratory plasmas, to pellet implosion in particle-beam fusion, to plasmoid propagation in the ionosphere. One could envision the generation of intense ion beams by applying high voltage across a planar cathode-anode gap, thus extracting ions from the anode region. The ions then accelerate across the gap and pass through the cathode (a wire mesh, say) to form an ion beam on the other side. The basic problem with this approach, in the absence of an insulating magnetic field, is that the applied voltage also produces an electron flow across the gap from the cathode to the anode (Sec. 8.2). A simple estimate of the corresponding ion current J_i shows that J_i is a factor of $(Z_i m_e/m_i)^{1/2}$ smaller than the electron current J_e. (Here, $+Z_i e$ is the ion charge, and m_i is the ion mass.) Therefore, the corresponding efficiency of ion beam generation is low, because the power delivered to the electrons is a factor of $(m_i/Z_i m_e)^{1/2}$ larger than the power delivered to the ions. Two methods for suppressing the electron flow to the anode include: (a) magnetic insulation, in which an applied magnetic field prevents electron flow from the cathode to the anode (Sec. 8.3),[25-28] but allows ion flow from the anode to the cathode,[28] and (b) the reflex triode, in which a thin-foil anode, largely transparent to the passage of electrons, is located between *two* cathodes, and the electron motion "reflexes" back and forth between the two cathode regions.[1,60,61]

In this section, we consider equilibrium properties of nonrelativistic electron and ion flow in a magnetically insulated ion diode.[28] The planar diode configuration is illustrated in Fig. 8.12. Here, the cathode is located at $x = 0$; the anode is located at $x = d$; the applied magnetic field in the cathode-anode gap is $B_0 \hat{e}_z$; and the boundary conditions on the electrostatic potential are $\phi_0(x = 0) = 0$ and $\phi_0(x = d) = V$, where V is the applied voltage. The electron layer extends from $x = 0$ to $x = x_b$, and the electron flow is in the y-direction with velocity $V_{ye}^0(x) = -cE_x(x)/B_0$. The ions, because of their large mass, are treated as unmagnetized, and flow from the anode to the cathode with average velocity $V_{xi}^0(x) < 0$. For simplicity, $eV/m_e c^2 \ll 1$ is assumed, and the electron (and ion) motions are treated nonrelativistically. The case where the value of $eV/m_e c^2$ is arbitrary and the electron flow is allowed to be relativistic has been analyzed in detail by Antonsen and Ott.[28] As a further point regarding the model, both the emission of electrons from the cathode and the emission of ions from the anode are assumed to be space-charge-limited, so that

$$E_x(x = 0) = 0 = E_x(x = d) \tag{8.76}$$

under steady-state conditions.

In the present analysis, we treat the electrons and ions as macroscopic cold fluids (Sec. 2.3). Indeed, the description of the magnetically insulated electron flow is similar to that in Sec. 8.3.1. On the other hand, the description of the (unmagnetized) ion flow is similar to the Child-Langmuir analysis in Sec. 8.2.1 (with the electrons replaced by ions). Poisson's equation in the cathode-anode gap can be expressed as

$$\frac{\partial^2}{\partial x^2}\phi_0(x) = \begin{cases} 4\pi e\left[n_e^0(x) - Z_i n_i^0(x)\right], & 0 \leq x < x_b, \\ -4\pi e Z_i n_i^0(x), & x_b < x \leq d, \end{cases} \tag{8.77}$$

where $+Z_i e$ is the ion charge, $n_i^0(x)$ is the ion density, and the electron density $n_e^0(x)$ is equal to zero in the region $x_b < x \leq d$ (Fig. 8.12).

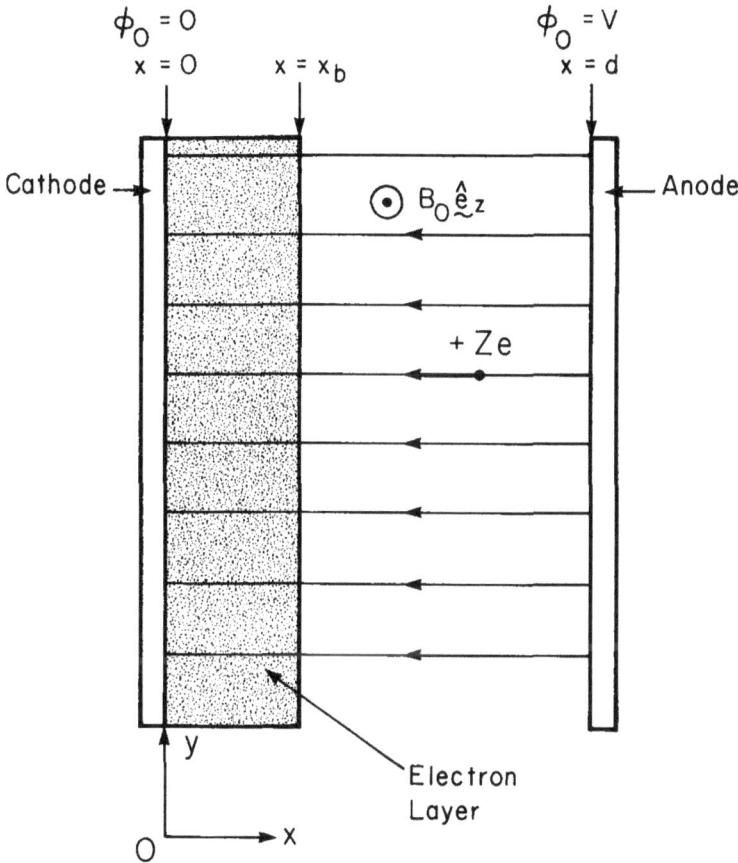

Figure 8.12. Schematic of magnetically insulated ion diode in planar geometry. The (magnetized) electron flow occurs in the layer region $0 \leq x < x_b$ with average velocity $V_{ye}^0(x) = -cE_x(x)/B_0$. The (unmagnetized) ions are emitted from the anode ($x = d$) and accelerate towards the cathode ($x = 0$), which is at lower potential.

Paralleling the analysis in Sec. 8.2.1, the ion continuity equation and conservation of energy can be expressed as

$$Z_i \, en_i^0(x) V_{xi}^0(x) = J_i = \text{const.}, \qquad (8.78)$$

and

$$\frac{m_i}{2} \, V_{xi}^{02}(x) + Z_i e\phi_0(x) = Z_i eV \; . \qquad (8.79)$$

Here, $J_i < 0$ (the ions flow from the anode to the cathode with $V_{xi}^0 < 0$), and use has been made of $\phi_0(x = d) = V$ and $V_{xi}^0(x = d) = 0$. Solving Eq.(8.79) for $V_{xi}^0(x)$ and substituting into Eq.(8.78) gives for the ion charge density

$$Z_i e n_i^0(x) = -\frac{(m_i/2Z_i eV)^{1/2} J_i}{[1 - \phi_0(x)/V]^{1/2}} \qquad (8.80)$$

in the region $0 \leq x \leq d$.

As in Sec. 8.3.1, we assume that the total energy of an electron fluid element is uniform across the electron layer [Eq.(8.30)]. This again leads to the electron flow velocity $V_{ye}^0(x) = \omega_{ce} x$ [Eq.(8.33)]. Making use of $V_{ye}^0(x) = -cE_x(x)/B_0$ then gives

$$E_x(x) = -\frac{B_0}{c}\omega_{ce} x \qquad (8.81)$$

within the electron layer ($0 \leq x < x_b$). Making use of $E_x(x) = -\partial\phi_0(x)/\partial x$, we obtain from Eqs.(8.77) and (8.81) the modified Brillouin flow condition

$$\omega_{ce}^2 = \omega_{pe}^2(x) - \frac{4\pi n_i^0(x)Z_i e^2}{m_e}, \qquad 0 \leq x < x_b, \qquad (8.82)$$

where $\omega_{ce} = eB_0/m_e c$ and $\omega_{pe}^2(x) = 4\pi n_e^0(x)e^2/m_e$. On the other hand, integrating Eq.(8.81) with respect to x and enforcing $\phi_0(x = 0) = 0$ and $\partial\phi_0(x)/\partial x|_{x=0} = 0$ gives

$$\phi_0(x) = \frac{1}{2}\frac{B_0}{c}\omega_{ce} x^2, \qquad 0 \leq x < x_b. \qquad (8.83)$$

Comparing with the case in which the ions are absent [Eq.(8.34)], we note from Eq.(8.82) that $\omega_{pe}^2(x)$ exceeds ω_{ce}^2 by an amount equal to $4\pi n_i^0(x)Z_i e^2/m_e$. However, the electrostatic potential $\phi_0(x)$ within the electron layer ($0 \leq x < x_b$) is identical in both cases [compare Eqs.(8.35) and (8.83)].

We now solve Poisson's equation (8.77) in the region between the electron layer and the anode ($x_b < x \leq d$). Substituting Eq.(8.80) into Eq.(8.77) gives

$$\frac{\partial^2}{\partial x^2}\phi_0(x) = -\frac{K^2 V}{[1 - \phi_0(x)/V]^{1/2}}, \qquad x_b < x \leq d, \qquad (8.84)$$

where the constant K^2 is defined by

$$K^2 = -\frac{4\pi}{V}\left(\frac{m_i}{2Z_ieV}\right)^{1/2} J_i \; . \tag{8.85}$$

Paralleling the analysis in Sec. 8.3.1, Eq.(8.84) is solved for $\phi_0(x)$ subject to the boundary conditions $\phi_0(x = d) = V$ and $\partial\phi_0(x)/\partial x|_{x=d} = 0$ [see Eq.(8.76)]. Some straightforward algebra gives the solution

$$\phi_0(x) = V\left\{1 - \left[\frac{3}{2}K(d-x)\right]^{4/3}\right\} \; , \quad x_b < x \le d \; , \tag{8.86}$$

where K^2 is defined in Eq.(8.85). Making use of the two solutions for $\phi_0(x)$ in Eqs.(8.83) and (8.86), we now enforce the continuity of $\phi_0(x)$ and $\partial\phi_0(x)/\partial x$ at the surface of the electron layer ($x = x_b$). These conditions are readily expressed as[28]

$$1 - \frac{V_H}{V}\frac{x_b^2}{d^2} = \left[\frac{3}{2}Kd\left(1 - \frac{x_b}{d}\right)\right]^{4/3} \; , \tag{8.87}$$

and

$$2\frac{V_H}{V}\frac{x_b}{d} = \frac{4}{3}\left(\frac{3}{2}Kd\right)^{4/3}\left(1 - \frac{x_b}{d}\right)^{1/3} \; , \tag{8.88}$$

Here,

$$V_H \equiv \frac{m_e}{2e}\,\omega_{ce}^2 d^2 \tag{8.89}$$

is the (nonrelativistic) Hull cut-off voltage defined in Eq.(8.38) for the case of magnetically insulated electron flow and zero ion density (Sec. 8.3.1).

For specified value of V/V_H, Eqs.(8.87) and (8.88) can be used to calculate the corresponding values of normalized layer thickness x_b/d and ion current $K^2 d^2$ self-consistently. For example, eliminating Kd and solving Eqs.(8.87) and (8.88) for x_b/d gives

$$\frac{x_b}{d} = \frac{3}{2}\left\{1 - \left(1 - \frac{8}{9}\frac{V}{V_H}\right)^{1/2}\right\} \; . \tag{8.90}$$

It follows from Eq.(8.90) that the electrons are magnetically insulated from the anode ($x_b < d$) provided $V < V_H$. For $V = V_H$, however, the electron layer completely fills the cathode-anode gap ($x_b = d$). It should be noted that the form of Eq.(8.90) differs from Eq.(8.37) because ions

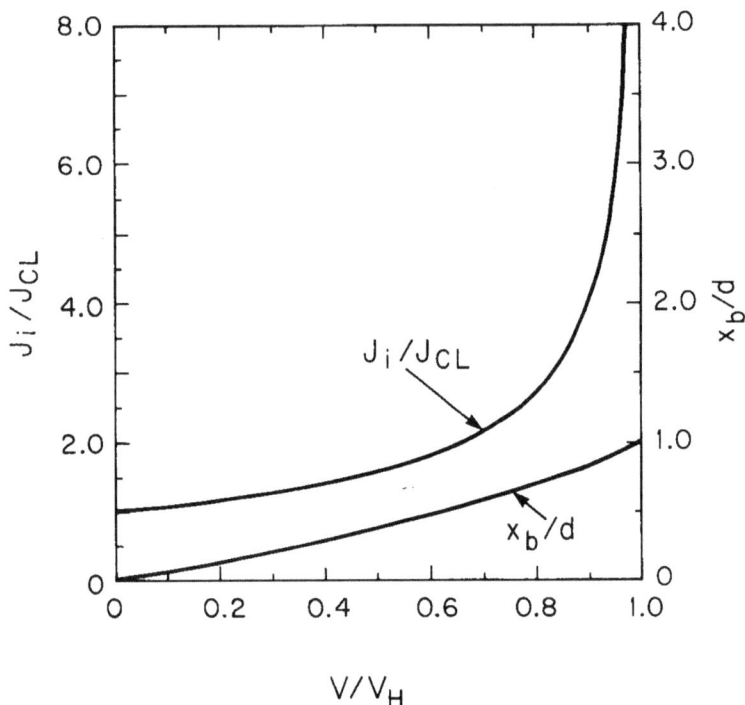

Figure 8.13.   Plots versus V/V_H of the normalized electron layer thickness x_b/d [Eq.(8.90)], and ion current density J_i/J_{CL} [Eq.(8.92)] for a magnetically insulated ion diode.

are present in the gap, and the boundary conditions at the anode include $E_x(x = d) = 0$ [Eq.(8.76)].

It is convenient to measure the ion current density J_i in units of the ion Child-Langmuir current density J_{CL}. Here, J_{CL} is defined by

$$J_{CL} = -\frac{2^{1/2}V}{9\pi d^2} \left(\frac{Z_i eV}{m_i} \right)^{1/2} , \tag{8.91}$$

which is a factor of $(Z_i m_e/m_i)^{1/2}$ smaller than the analogous result for electrons in Eq.(8.12). Combining Eqs.(8.85) and (8.91) then gives $(9/4)K^2 d^2 = J_i/J_{CL}$. Taking the 3/2 power of Eq.(8.88), we obtain for the ion current density[28]

$$J_i = J_{CL} \left(\frac{3}{2} \frac{V_H}{V} \frac{x_b}{d} \right)^{3/2} \left(1 - \frac{x_b}{d} \right)^{-1/2} . \tag{8.92}$$

In Eq.(8.92), J_{CL} is defined in Eq.(8.91), and x_b/d is determined self-consistently in terms of V/V_H from Eq.(8.90).

For $V \ll V_H$, Eq.(8.90) gives $x_b/d \simeq (2/3)(V/V_H) \ll 1$, and Eq.(8.92) reduces to the expected result, $J_i \simeq J_{\mathrm{CL}}$. On the other hand, for increasing values of V/V_H and x_b/d, J_i can be greatly enhanced relative to the Child-Langmuir current J_{CL}, primarily through the geometric factor $(1 - x_b/d)^{-1/2}$ in Eq.(8.92). Physically, the electron sheath reduces the effective gap spacing. As an example, for $V/V_H = 0.9$, we find $x_b/d = 0.83$, and the current enhancement factor is $J_i/J_{\mathrm{CL}} = 3.9$. In Fig. 8.13, the solutions for x_b/d and J_i/J_{CL} obtained from Eqs.(8.90) and (8.92) are plotted versus the normalized voltage V/V_H.

8.5   Magnetron Instability for Magnetically Insulated Electron Flow

There is an extensive literature on the magnetron instability[44-54] for magnetically insulated electron flow in planar and cylindrical geometries, with applications ranging from coherent electromagnetic wave generation to investigations of the stability properties of relativistic electron flow in high-voltage diodes. A discussion of the magnetron instability is particularly appropriate in a treatise on nonneutral plasmas, given the important influence of intense self fields on stability behavior. Indeed, magnetrons used for microwave generation[39-43] typically operate at conditions near Brillouin flow (Sec. 8.3). In this section, we make use of the eigenvalue equation for extraordinary-mode flute perturbations about relativistic electron flow in planar diode geometry (Sec. 8.5.1) to investigate the influence of intense self fields and profile shape on the linear growth properties of the magnetron instability (Sec. 8.5.2).[48,49] Some experimental results on microwave generation in relativistic magnetrons are then summarized (Sec. 8.6).[41]

8.5.1   Extraordinary-Mode Eigenvalue Equation

To illustrate several important features of the magnetron instability in the linear growth regime, we make use of a macroscopic cold-fluid model (Sec. 2.3) to investigate the electromagnetic stability properties of magnetically insulated, relativistic electron flow in the planar diode configuration illustrated in Figs. 8.4 and 8.5. Here, the cathode is located at $x = 0$;

the anode is located at $x = d$; the electron layer extends from $x = 0$ to $x = x_b < d$; and the equilibrium magnetic field in the vacuum region $(x_b < x \leq d)$ is equal to $B_0 \hat{e}_z$. Space-charge-limited flow is assumed with

$$E_x(x = 0) = 0 ,$$

$$\phi_0(x = 0) = 0 \quad \text{and} \quad \phi_0(x = d) = V , \tag{8.93}$$

where V is the applied voltage. Moreover, the equilibrium flow velocity is

$$V_{ye}^0(x) = -c \frac{E_x(x)}{B_z(x)} , \tag{8.94}$$

where the equilibrium electric and magnetic fields, $\mathbf{E}^0(\mathbf{x}) = E_x(x)\hat{e}_x$ and $\mathbf{B}^0(\mathbf{x}) = B_z(x)\hat{e}_z$, are related self-consistently to the electron density profile $n_e^0(x)$ by the steady-state Maxwell equations

$$\frac{\partial}{\partial x} B_z(x) = -\frac{4\pi e n_e^0(x) E_x(x)}{B_z(x)} , \tag{8.95}$$

and

$$\frac{\partial}{\partial x} E_x(x) = -4\pi e n_e^0(x) . \tag{8.96}$$

As discussed in Sec. 8.3.3, it can be shown from Eqs.(8.95) and (8.96) that

$$B_z^2(x) - E_x^2(x) = \text{const.}$$

across the layer profile $(0 \leq x < x_b)$, or equivalently,

$$\frac{B_z(x)}{\gamma_e^0(x)} = \text{const.} \tag{8.97}$$

Here, $\gamma_e^0(x) = [1 - E_x^2(x)/B_z^2(x)]^{-1/2}$ is the relativistic mass factor.

The present stability analysis assumes flute perturbations $(\partial/\partial z = 0)$ about general equilibrium profiles consistent with Eqs.(8.93)–(8.97). In addition, the electromagnetic field perturbations are assumed to have extraordinary-mode polarization with

$$\delta \mathbf{E}(\mathbf{x}, t) = \delta E_x(x, y, t)\hat{e}_x + \delta E_y(x, y, t)\hat{e}_y ,$$

$$\delta \mathbf{B}(\mathbf{x}, t) = \delta B_z(x, y, t)\hat{e}_z . \tag{8.98}$$

The field components $\delta E_y(x, y, t)$, etc., are expressed as

$$\delta E_y(x, y, t) = \sum_{k=-\infty}^{\infty} \delta E_y(x, k) \exp(iky - i\omega t) \,. \qquad (8.99)$$

Here, ω is the complex oscillation frequency with $\mathrm{Im}\,\omega > 0$ corresponding to instability (temporal growth); $k = 2\pi n/L$ is the wavenumber of the perturbation, where n is an integer; and L is the fundamental periodicity length in the y-direction (the equilibrium flow direction). It is convenient to introduce the effective potential amplitude $\Phi_k(x)$ defined by

$$\Phi_k(x) = \frac{i}{k} \delta E_y(x, k) \,. \qquad (8.100)$$

The linearized, relativistic cold-fluid-Maxwell equations (Sec. 2.3) can be used to derive the eigenvalue equation for $\Phi_k(x)$ assuming extraordinary-mode flute perturbations about general equilibrium profiles. Without presenting algebraic details, we obtain the eigenvalue equation (Appendix B)[48,49]

$$\frac{\partial}{\partial x} \left\{ [1 + \chi_\perp(x, k, \omega)] \frac{\partial}{\partial x} \Phi_k(x) \right\} - k^2 \left[1 + \chi_\parallel(x, k, \omega) \right] \Phi_k(x) \qquad (8.101)$$

$$= \frac{k\,\Phi_k(x)}{[\omega - kV_{ye}^0(x)]} \left[1 - \frac{V_{ye}^0(x)}{c} \frac{\omega}{ck} \right] \frac{\partial}{\partial x} \left[\frac{\omega_{pe}^2(x)\omega_{ce}(x)}{\gamma_e^{02}(x)\nu_e^2(x, k, \omega)} \right] ,$$

where $\omega_{ce}(x) = eB_z(x)/m_e c$ and $\omega_{pe}(x) = [4\pi n_e^0(x)e^2/m_e]^{1/2}$ are the non-relativistic electron cyclotron and plasma frequencies, respectively, and

$$\gamma_e^0(x) = \left[1 - \frac{E_x^2(x)}{B_z^2(x)} \right]^{-1/2}$$

is the relativistic mass factor. The susceptibilities occurring in Eq.(8.101) are defined by[48,49]

$$\chi_\perp(x, k, \omega) = -\left[1 - \frac{V_{ye}^0(x)}{c} \frac{\omega}{ck} \right]^2 \frac{\omega_{pe}^2(x)\gamma_e^0(x)}{(1 - \omega^2/c^2 k^2)\,\nu_e^2(x, k, \omega)} \,, \qquad (8.102)$$

$$\chi_\parallel(x, k, \omega) = -\frac{\omega^2}{c^2 k^2} - \frac{\omega_{pe}^2(x)}{\gamma_e^0(x)\nu_e^2(x, k, \omega)} \left[1 - \frac{\omega^2}{c^2 k^2} + \frac{\omega_{pe}^2(x)}{\gamma_e^0(x)c^2 k^2} \right] , (8.103)$$

where

$$\nu_e^2(x,k,\omega) = \gamma_e^{02}(x)\left[\omega - kV_{ye}^0(x)\right]^2 \left[1 + \frac{\omega_{pe}^2(x)/\gamma_e^0(x)c^2k^2}{1 - \omega^2/c^2k^2}\right]$$

$$- \left[\omega_{ce}^2(x)/\gamma_e^{02}(x) - \omega_{pe}^2(x)/\gamma_e^0(x)\right] . \tag{8.104}$$

The eigenvalue equation (8.101) can be used to investigate detailed extraordinary-mode stability properties of the electron flow for a wide range of self-consistent equilibrium profiles $n_e^0(x)$, $E_x(x)$ and $B_z(x)$. In this regard, it should be pointed out that Eq.(8.101) represents the generalization of the electrostatic eigenvalue equation (5.56) to include the full influence of relativistic and electromagnetic effects on stability behavior in planar geometry. The complex eigenfunction $\Phi_k(x) = (i/k)\delta E_y(x,k)$ and the eigenfrequency ω are to be determined from Eq.(8.101) subject to the boundary conditions

$$\Phi_k(x = 0) = 0 = \Phi_k(x = d) , \tag{8.105}$$

which correspond to zero tangential electric field $(\delta E_y = 0)$ at the cathode $(x = 0)$ and the anode $(x = d)$. It is found that the eigenvalue equation (8.101) supports unstable solutions $(\text{Im}\,\omega > 0)$ corresponding to the magnetron instability[44-54] when the Doppler shifted cyclotron resonance condition (Problem 8.11)

$$\left[\text{Re}\,\omega - kV_{ye}^0(x_i)\right]^2 = \omega_{ce}^2(x_i)/\gamma_e^{04}(x_i) \tag{8.106}$$

is satisfied at some internal location $(x = x_i)$ within the electron layer $(0 < x_i < x_b)$. Moreover, the growth rate depends on the strength of the self fields[48] and the shape[49] of the equilibrium profiles.

In Sec. 8.3.3 we discussed several features of relativistic Brillouin flow in a magnetically insulated planar diode, including calculations of the Hull cut-off voltage V_H [Eq.(8.73)] and the Buneman-Hartree threshold voltage [Eq.(8.75)] for magnetron operation. The corresponding equilibrium profiles were determined self-consistently for the particular ansatz that $[\gamma_e^0(x) - 1]m_ec^2 - e\phi_0(x) = 0$ across the electron layer, which led to the equilibrium profiles in Eqs.(8.62)–(8.66) and the relativistic Brillouin flow condition in Eq.(8.59). To illustrate the strong sensitivity of the magnetron instability to the shape of the equilibrium profiles and the intensity of the self fields, in this section we remove the (overly restrictive) assumption of Brillouin flow made in Sec. 8.3.3. In particular, it is assumed that the functional form of $\gamma_e^0(x) = [1 - V_{ye}^{02}(x)/c^2]^{-1/2} = [1 - E_x^2(x)/B_z^2(x)]^{-1/2}$ is specified across the electron layer ($0 \leq x < x_b$), with $\gamma_e^0(x)$ increasing monotonically from $\gamma_e^0(x = 0) = 1$ at the cathode to the value $\gamma_e^0(x = x_b) \equiv \hat{\gamma}_e$ at the outer edge of the electron layer. It then follows from Eqs.(8.95)–(8.97) that the remaining equilibrium profiles are given self-consistently by

$$B_z(x) = \begin{cases} B_0 \dfrac{\gamma_e^0(x)}{\hat{\gamma}_e} \,, & 0 \leq x < x_b \,, \\ B_0 \,, & x_b < x \leq d \,, \end{cases} \tag{8.107}$$

$$n_e^0(x) = \begin{cases} \dfrac{B_0}{4\pi e \hat{\gamma}_e} \left(\dfrac{\partial}{\partial x} \left[\gamma_e^{02}(x) - 1 \right]^{1/2} \right) \,, & 0 \leq x < x_b \,, \\ 0 \,, & x_b < x \leq d \,, \end{cases} \tag{8.108}$$

and

$$E_x(x) = \begin{cases} -\dfrac{B_0}{\hat{\gamma}_e} \left[\gamma_e^{02}(x) - 1 \right]^{1/2} \,, & 0 \leq x < x_b \,, \\ E_x(x_b) \,, & x_b < x \leq d \,. \end{cases} \tag{8.109}$$

Here, $B_0 = B_z(x_b)$ and $-[\hat{\gamma}_e^2 - 1]^{1/2} B_0/\hat{\gamma}_e = E_x(x_b)$ are the fields in the vacuum region ($x_b < x \leq d$), and $\hat{\gamma}_e = \gamma_e^0(x_b) = [1 - E_x^2(x_b)/B_0^2]^{-1/2}$ is the relativistic mass factor at the outer edge of the electron layer ($x = x_b$). Equations (8.107)–(8.109) are valid for general choice of the equilibrium profile $\gamma_e^0(x)$. For the special choice $\gamma_e^0(x) = \cosh(\kappa x)$, where $\kappa \equiv 4\pi e \hat{n}_e/B_0$ and $\hat{n}_e \equiv n_e^0(x_b)$, it can be shown that Eqs.(8.107)–(8.109) recover (exactly) the equilibrium Brillouin flow profiles in Eqs.(8.62), (8.63), and (8.66). For general profile $\gamma_e^0(x)$, we note from Eq.(8.108) that the *average* electron density, $\bar{n}_e = x_b^{-1} \int_0^{x_b} dx \, n_e^0(x)$, can be expressed as

$$\bar{n}_e = \frac{[\hat{\gamma}_e^2 - 1]^{1/2} B_0}{4\pi e x_b \hat{\gamma}_e} = -\frac{E_x(x_b)}{4\pi e x_b} \,, \tag{8.110}$$

which determines $\bar{n}_e$ explicitly in terms of $\hat{\gamma}_e, B_0$ and x_b. As a general remark, it follows from Eq.(8.97) that the relativistic electron cyclotron frequency $\omega_{ce}(x)/\gamma_e^0(x) = eB_z(x)/\gamma_e^0(x)m_e c$, is uniform over the layer cross section. On the other hand, from Eq.(8.108), the relativistic electron plasma frequency-squared, $\omega_{pe}^2(x)/\gamma_e^0(x) = 4\pi n_e^0(x)e^2/\gamma_e^0(x)m_e$, generally varies with x over the layer cross section. The special equilibrium profile $\gamma_e^0(x) = \cosh(\kappa x)$ is the only exception to this, in which case $n_e^0(x)/\gamma_e^0(x) = B_0\kappa/4\pi e\hat{\gamma}_e = \text{const.}$

Problem 8.11   Make use of $\gamma_e^0(x) = [1 - V_{ye}^{02}(x)/c^2]^{-1/2}$ and the definitions of $\chi_\perp(x, k, \omega)$ and $\nu_e^2(x, k, \omega)$ in Eqs.(8.102) and (8.104) to show that

$$1 + \chi_\perp(x, k, \omega) = \frac{\gamma_e^{02}(x)\left[\omega - kV_{ye}^0(x)\right]^2 - \omega_{ce}^2(x)/\gamma_e^{02}(x)}{\nu_e^2(x, k, \omega)}. \tag{8.11.1}$$

Therefore, for sufficiently small growth rate, the Doppler shifted cyclotron resonance condition in Eq.(8.106) corresponds to $1 + \text{Re}\,\chi_\perp(x_i, k, \omega) \simeq 0$, where $x = x_i$ is the location of the resonance within the electron layer.

Problem 8.12   Consider the extraordinary-mode eigenvalue equation (8.101) for the case of nonrelativistic electron flow with $B_z(x) \simeq B_0 = \text{const.}$ and

$$V_{ye}^{02}(x)/c^2 \ll 1. \tag{8.12.1}$$

Also assume slow-wave electrostatic perturbations with

$$\left|\frac{\omega^2}{c^2k^2}\right| \ll 1, \qquad \frac{\omega_{pe}^2(x)}{c^2k^2} \ll 1. \tag{8.12.2}$$

From Maxwell's equations and Eq.(8.12.2), it follows that $\delta\mathbf{B} \simeq 0$ and $\nabla \times \delta\mathbf{E} \simeq 0$.

a. Show from Eqs.(8.12.1) and (8.12.2) that the eigenvalue equation (8.101) can be approximated by

$$\frac{\partial}{\partial x}\left\{\left(1 - \frac{\omega_{pe}^2(x)}{[\omega - kV_{ye}^0(x)]^2 - [\omega_{ce}^2 - \omega_{pe}^2(x)]}\right)\frac{\partial}{\partial x}\Phi_k(x)\right\}$$

$$- k^2\left(1 - \frac{\omega_{pe}^2(x)}{[\omega - kV_{ye}^0(x)]^2 - [\omega_{ce}^2 - \omega_{pe}^2(x)]}\right)\Phi_k(x)$$

$$= \frac{k\,\omega_{ce}\Phi_k(x)}{[\omega - kV_{ye}^0(x)]}\frac{\partial}{\partial x}\left(\frac{\omega_{pe}^2(x)}{[\omega - kV_{ye}^0(x)]^2 - [\omega_{ce}^2 - \omega_{pe}^2(x)]}\right) \tag{8.12.3}$$

in the nonrelativistic electrostatic regime. Here, $\omega_{ce} = eB_0/m_ec$, $\omega_{pe}^2(x) = 4\pi n_e^0(x)e^2/m_e$, and $V_{ye}^0(x) = -cE_x(x)/B_0$.

b. Show that Eq.(8.12.3) is identical to the planar limit of the electrostatic eigenvalue equation (5.56) (derived in cylindrical geometry) provided the replacements $\ell/r \to k$, $\omega_{re}(r)r \to V_{ye}^0(x)$, and $r^{-1}(\partial/\partial r)(r\partial/\partial r) \to \partial^2/\partial x^2$ are made in Eq.(5.56) for a one-component nonneutral plasma consisting only of electrons.

8.5.2   Linear Growth Properties

For purposes of analyzing the extraordinary-mode eigenvalue equation (8.101), we choose $\gamma_e^0(x)$ to have the functional form[49]

$$\gamma_e^0(x) = \lambda \cosh(\kappa_1 x) + (1 - \lambda)\frac{\left[1 - \kappa_2^2\left(x_b^2 - x^2\right)\right]^{1/2}}{\left(1 - \kappa_2^2 x_b^2\right)^{1/2}} \tag{8.111}$$

over the interval $0 \leq x < x_b$. Here, κ_1^{-1} and κ_2^{-1} are constant scale lengths, and λ is a variable parameter in the range $0 \leq \lambda \leq 1$ which can be used to alter the shape of the equilibrium profiles. [The special case $\lambda = 1$ corresponds to $\gamma_e^0(x) = \cosh(\kappa_1 x)$.] Note from Eq.(8.111) that $\gamma_e^0(x)$ increases monotonically from $\gamma_e^0(x = 0) = 1$ at the cathode, to $\gamma_e^0(x = x_b) = \hat{\gamma}_e$ at the outer edge of the electron layer ($x = x_b$), where

$$\hat{\gamma}_e = \lambda \cosh(\kappa_1 x_b) + \frac{(1 - \lambda)}{\left[1 - \kappa_2^2 x_b^2\right]^{1/2}}. \tag{8.112}$$

The equilibrium profiles for $\gamma_e^0(x)/\hat{\gamma}_e$, $B_z(x)/B_0$, and $n_e^0(x)/\gamma_e^0(x)$ calculated from Eqs.(8.107), (8.108), and (8.111) are illustrated in Fig. 8.14 for three values of the shape parameter corresponding to $\lambda = 1, 0.5$ and 0, and fixed values of $\kappa_1 x_b = 1.317, \kappa_2 x_b = 0.866$ and $\hat{\gamma}_e = 2$. Note from Fig. 8.14(b) that the profile for $n_e^0(x)/\gamma_e^0(x)$ exhibits a sensitive dependence on λ, ranging from a uniform profile when $\lambda = 1$ (similar to the case examined in Sec. 8.3.3), to a profile which decreases by a factor of two over the layer cross section when $\lambda = 0$. We will find that the growth rate of the magnetron instability exhibits a sensitive dependence on the shape of the equilibrium profiles (as measured by λ).[49]

A.   Magnetron Instability for $\lambda = 1$

For strong magnetron instability, we first consider the case where $\lambda = 1$. In this regard, it is useful to introduce the self field parameter s_e defined relativistically by

$$s_e = \frac{\omega_{pe}^2(x_b)/\hat{\gamma}_e}{\omega_{ce}^2(x_b)/\hat{\gamma}_e^2} = \frac{4\pi\hat{n}_e\hat{\gamma}_e m_e c^2}{B_0^2}. \tag{8.113}$$

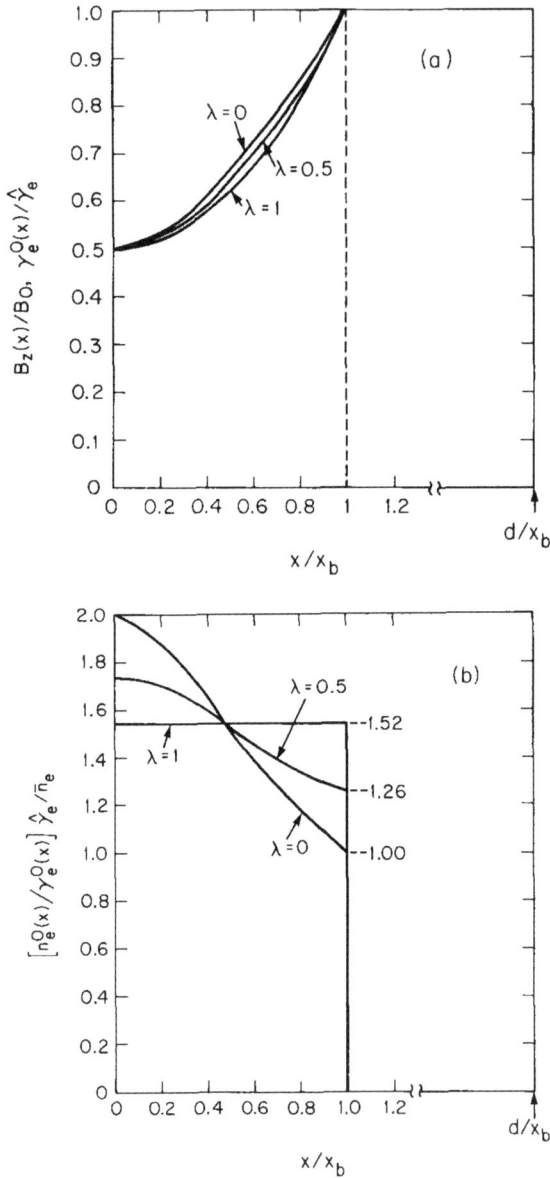

Figure 8.14. Plots versus x/x_b of the normalized profiles for $(a)B_z(x)/B_0$ and $\gamma_e^0(x)/\hat{\gamma}_e$, and $(b)[n_e^0(x)/\gamma_e^0(x)]\hat{\gamma}_e/\bar{n}_e$ calculated from Eqs.(8.107), (8.108), and (8.111) for fixed values of $\kappa_1 x_b = 1.317$, $\kappa_2 x_b = 0.866$ and $\hat{\gamma}_e = 2$, and three values of the profile shape parameter $\lambda = 1, 0.5$ and 0.

Here, $\hat{\gamma}_e = \gamma_e^0(x_b) = \cosh(\kappa_1 x_b), \hat{n}_e = n_e^0(x_b) = B_0\kappa_1/4\pi e$, and $B_0 = B_z(x_b)$ are evaluated at the outer edge of the electron layer ($x = x_b$). Indeed, for $\lambda = 1$ and $\gamma_e^0(x) = \cosh(\kappa_1 x)$, it is readily shown that the profiles for $n_e^0(x)$ and $B_z(x)$ are both proportional to $\cosh(\kappa_1 x)$. Therefore, the local self-field parameter defined by $s_e(x) = [\omega_{pe}^2(x)/\gamma_e^0(x)]/[\omega_{ce}^2(x)/\gamma_e^{0\,2}(x)]$ is uniform over the layer cross section and equal to the value of s_e defined in Eq.(8.113) at $x = x_b$. The condition $s_e = 1$ corresponds exactly to the Brillouin flow condition assumed in Sec. 8.3.3, but is not required in the present equilibrium and stability analysis.

Typical numerical solutions[48] to the extraordinary-mode eigenvalue equation (8.101) are illustrated in Figs. 8.15 and 8.16 for the choice of system parameters corresponding to $\lambda = 1, d = (3/2)x_b$, and $x_b = 2c/\hat{\omega}_{ce}$. Here, $\hat{\omega}_{ce} \equiv \omega_{ce}(x_b)/\hat{\gamma}_e = eB_0/\hat{\gamma}_e m_e c$ is the relativistic cyclotron frequency at the outer edge of the electron layer. Figure 8.15 shows the normalized growth rate $(\text{Im}\,\omega)/\hat{\omega}_{ce}$ and the real oscillation frequency $(\text{Re}\,\omega)/\hat{\omega}_{ce}$ of the magnetron instability plotted versus the normalized wavenumber $kc/\hat{\omega}_{ce}$ for the two cases corresponding to $s_e = 1$ [Fig. 8.15(a)] and $s_e = 0.5$ [Fig. 8.15(b)]. Several points are noteworthy from Fig. 8.15. First, the bandwidth of the magnetron instability is quite broad, with the maximum growth rate occurring for $kc/\hat{\omega}_{ce} \sim 3$ when $s_e = 1$, and for $kc/\hat{\omega}_{ce} \sim 2$ when $s_e = 0.5$. Second, the maximum growth rate decreases by about a factor of four when the self-field parameter s_e is reduced from $s_e = 1$ to $s_e = 0.5$. Finally, the real frequency $\text{Re}\,\omega$ exceeds $\hat{\omega}_{ce}$ and scales approximately linearly with wavenumber k over the entire range of the instability.

The strong dependence of stability properties on the strength of the equilibrium self fields is further illustrated in Fig. 8.16(a), where $(\text{Im}\,\omega)/\hat{\omega}_{ce}$ and $(\text{Re}\,\omega)/\hat{\omega}_{ce}$ are plotted versus the self-field parameter s_e for fixed wavenumber $kc/\hat{\omega}_{ce} = 2$, and system parameters otherwise identical to those in Fig. 8.15. Note from Fig. 8.16(a) that the growth rate of the magnetron instability is negligibly small for $s_e \leq 0.3$, but increases monotonically to $\text{Im}\,\omega = 0.0168\,\hat{\omega}_{ce}$ for $s_e = 1$.

Shown in Fig. 8.16(b) are eigenfunction plots of $\text{Re}\,\Phi_k(x)$ and $\text{Im}\,\Phi_k(x)$ versus x/x_b obtained from Eq.(8.101) for $s_e = 1$ and fixed wavenumber $kc/\hat{\omega}_{ce} = 2$. The corresponding complex oscillation frequency is $\omega = (1.83 + 0.0168i)\,\hat{\omega}_{ce}$. Note from Fig. 8.16(b) that the outer extremum of $\text{Re}\,\Phi_k(x)$ occurs near the boundary of the electron layer ($x = x_b$) where $\partial\omega_{pe}^2(x)/\partial x$ is large, corresponding to a large surface perturbation on the right-hand side of the eigenvalue equation (8.101). The

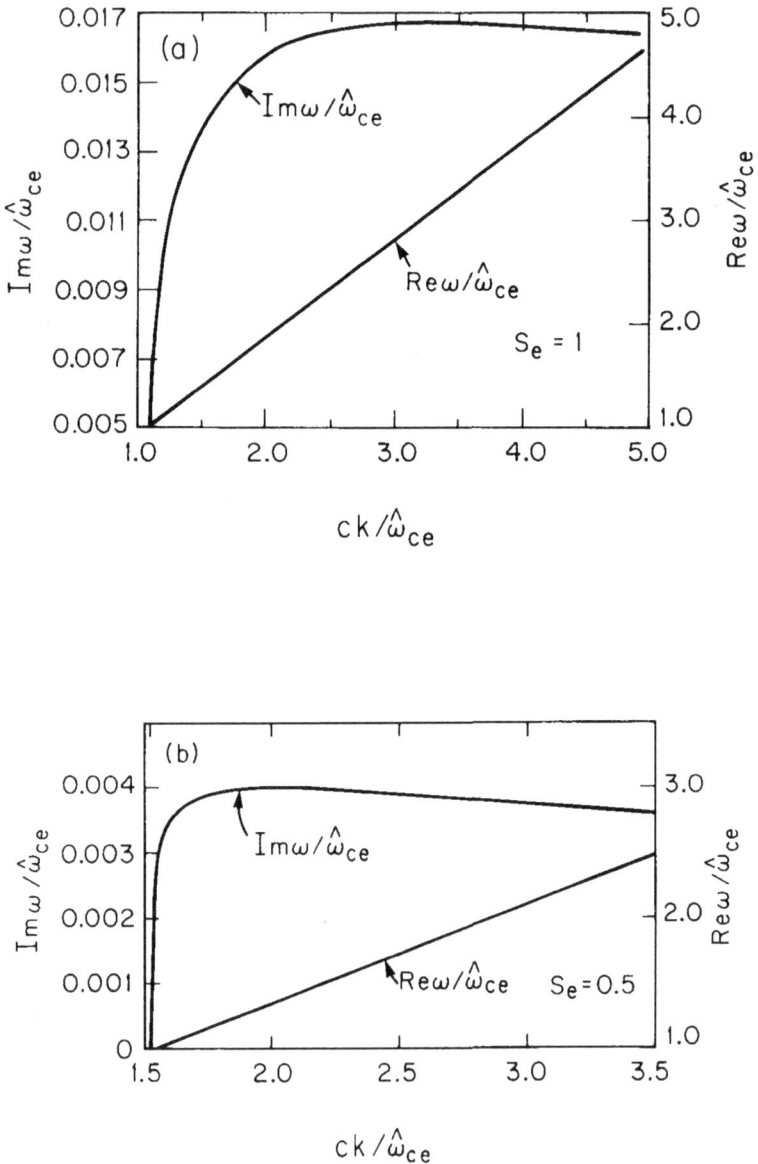

Figure 8.15. Linear growth properties of the magnetron instability. The figures show plots versus $ck/\hat{\omega}_{ce}$ of the normalized growth rate $\mathrm{Im}\,\omega/\hat{\omega}_{ce}$ and the real oscillation frequency $\mathrm{Re}\,\omega/\hat{\omega}_{ce}$ obtained numerically from the eigenvalue equation (8.101). The choice of system parameters corresponds to $\lambda = 1, d = (3/2)x_b$ and $x_b = 2c/\hat{\omega}_{ce}$. The two values of the self-field parameter s_e correspond to $(a)s_e = 1$, and (b) $s_e = 0.5$ [R.C. Davidson, K.T. Tsang and J.A. Swegle, Phys. Fluids **27**, 2332 (1984)].

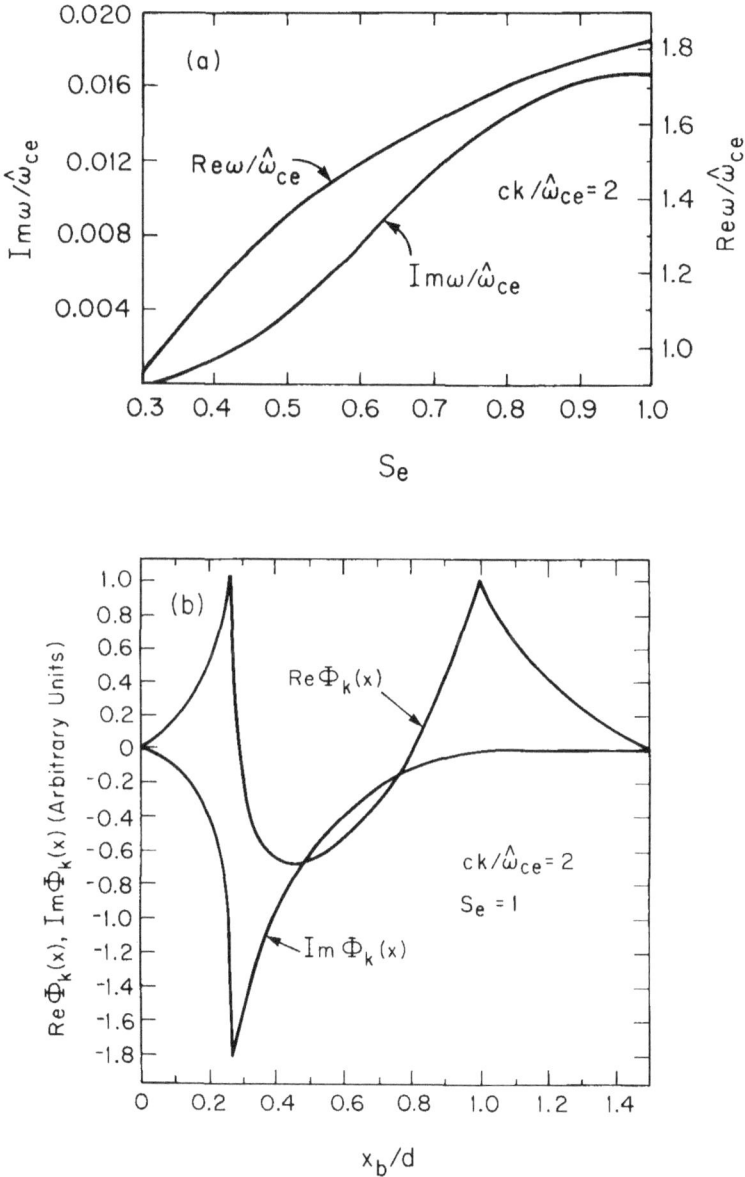

Figure 8.16. Linear growth properties of the magnetron instability. The figures show plots of (a) the normalized growth rate $\text{Im}\,\omega/\hat{\omega}_{ce}$ and the real oscillation frequency $\text{Re}\,\omega/\hat{\omega}_{ce}$ versus s_e, and (b) the eigenfunction components $\text{Re}\,\Phi_k(x)$ and $\text{Im}\,\Phi_k(x)$ versus x/x_b for $s_e = 1$, obtained numerically from the eigenvalue equation (8.101) for fixed wavenumber $ck/\hat{\omega}_{ce} = 2$. The choice of system parameters is otherwise identical to Fig. 8.15 [R.C. Davidson, K.T. Tsang and J.A. Swegle, Phys. Fluids **27**, 2332 (1984)].

strong peaking of $\mathrm{Re}\,\Phi_k(x)$ at $x = x_b$ is expected since it is this large variation of $\omega_{pe}^2(x)$ at the plasma boundary which supports the surface wave which experiences an unstable interaction with the layer electrons. Close examination of Eq.(8.101) and Fig. 8.16(b) shows that the inner extrema of $\mathrm{Re}\,\Phi_k(x)$ and $\mathrm{Im}\,\Phi_k(x)$ occur near the point $(x = x_i)$ where $1 + \mathrm{Re}\,\chi_\perp(x_i, k, \omega) = 0$. After some straightforward algebra that makes use of Eqs.(8.102) and (8.104), it can be shown that (Problem 8.11)

$$1 + \chi_\perp(x, k, \omega) = \frac{\gamma_e^{02}(x)\left[\omega - kV_{ye}^0(x)\right]^2 - \omega_{ce}^2(x)/\gamma_e^{02}(x)}{\nu_e^2(x, k, \omega)} \qquad (8.114)$$

where $\omega_{ce}(x) = eB_z(x)/m_e c$. For small growth rate with $|\mathrm{Im}\,\omega| \ll |\mathrm{Re}\,\omega|$, the solution (x_i) to $1 + \mathrm{Re}\,\chi_\perp(x_i, k, \omega) = 0$ is determined approximately from the Doppler-shifted cyclotron resonance condition

$$\left[\mathrm{Re}\,\omega - kV_{ye}^0(x_e)\right]^2 = \hat{\omega}_{ce}^2/\gamma_e^{02}(x_i), \qquad (8.115)$$

where use has been made of $\omega_{ce}(x)/\gamma_e^0(x) = \hat{\omega}_{ce} = $ const. across the electron layer. It is precisely the occurrence of this cyclotron resonance condition within the electron layer that leads to the magnetron instability.

As a final point, for $\lambda = 1$ and $\gamma_e^0(x) = \cosh(\kappa_1 x)$, it is readily shown that

$$n_e^0(x) = \frac{\hat{n}_e \cosh(\kappa_1 x)}{\cosh(\kappa_1 x_b)} ,$$

where $c\kappa_1 = 4\pi\hat{n}_e ec/B_0 = s_e\hat{\omega}_{ce}$ and $\kappa_1 x_b = s_e(x_b\hat{\omega}_{ce}/c)$. Furthermore, solving Poisson's equation (8.96) for $\phi_0(x)$ and enforcing $\phi_0(x = d) = V$ gives

$$\frac{eV}{m_e c^2} = s_e \frac{\hat{\omega}_{ce}^2}{c^2\kappa_1^2}\left\{[\cosh(\kappa_1 x_b) - 1] + \kappa_1(d - x_b)\sinh(\kappa_1 x_b)\right\}, \qquad (8.116)$$

where $c\kappa_1/\hat{\omega}_{ce} = s_e$ and $\hat{\omega}_{ce} = eB_0/\hat{\gamma}_e m_e c$. For $s_e = 1$ (and therefore $c\kappa_1 = \hat{\omega}_{ce}$), Eq.(8.116) reduces to the familiar voltage relation (8.68), valid for Brillouin flow conditions. For the choice of parameters in Figs. 8.15 and 8.16, however, $x_b = 2c/\hat{\omega}_{ce}$ so that $\kappa_1 x_b = 2s_e$. Therefore, the normalized voltage $eV/m_e c^2$ required to maintain $d = (3/2)x_b$ and $x_b = 2c/\hat{\omega}_{ce}$ in Figs. 8.15 and 8.16 is different for each value of the self-field parameter s_e. For example, for $s_e = 1$ in Figs. 8.15(a) and 8.16(b), the voltage relation in Eq.(8.68) gives $eV/m_e c^2 = 6.39$, corresponding to $V = 3.26$ MV.

B. Influence of Profile Shape on the Magnetron Instability

To illustrate the strong influence of profile shape on the magnetron instability, the eigenvalue equation (8.101) has been solved[49] for the case in which $\gamma_e^0(x)$ is specified by Eq.(8.111) for several values of the shape parameter λ. The corresponding equilibrium profiles for $B_z(x), n_e^0(x)$ and $E_x(x)$ are determined self-consistently from Eqs.(8.107)–(8.109). Typical numerical results are illustrated in Fig. 8.17, where $(\text{Im}\,\omega)/\hat{\omega}_{ce}$ and $(\text{Re}\,\omega)/\hat{\omega}_{ce}$ are plotted versus $kc/\hat{\omega}_{ce}$ for several values of λ. Here, the choice of system parameters corresponds to $d = 3x_b, x_b = c/\hat{\omega}_{ce}, \kappa_1 x_b = 1.319, \kappa_2 x_b = 0.866$, and $\hat{\gamma}_e = \gamma_e^0(x_b) = 2$. As before, $\hat{\omega}_{ce} = eB_0/\hat{\gamma}_e m_e c$ is the relativistic cyclotron frequency at $x = x_b$. Moreover, the average electron density $\bar{n}_e$ defined in Eq.(8.110) is the same for each value of λ in Fig. 8.17, and the value of the *average* self field parameter is $\bar{s}_e = 4\pi\bar{n}_e\hat{\gamma}_e m_e c^2/B_0^2 = 0.866$. For $\lambda = 1$, it is evident from Fig. 8.17(a) that the growth rate of the magnetron instability is quite robust, with general features similar to those in Fig. 8.15(a). On the other hand, as λ is decreased from $\lambda = 1$, and the profile for $n_e^0(x)/\gamma_e^0(x)$ decreases monotonically with increasing x [Fig. 8.14(b)], it is evident from Fig. 8.17(a) that the instability growth rate $\text{Im}\,\omega$ is reduced substantially, at least at longer wavelengths. Indeed, for the range of $kc/\hat{\omega}_{ce}$ shown in the figure $(kc/\hat{\omega}_{ce} \leq 5)$, instability ceases $(\text{Im}\,\omega = 0)$ for $\lambda \leq 0.06$. On the other hand, the plots of $(\text{Re}\,\omega)/\hat{\omega}_{ce}$ versus $ck/\hat{\omega}_{ce}$ for λ in the interval $0.06 \leq \lambda \leq 1$ are virtually identical to the curve plotted in Fig. 8.17(b) for $\lambda = 1$. That is, the real frequency of the magnetron mode is relatively insensitive to the profile shape.

It should be noted that the voltage V required to maintain $d = 3x_b$ and $x_b = c/\hat{\omega}_{ce}$ in Fig. 8.17 is different for each value of λ. For example, for $\lambda = 0$, Poisson's equation (8.96) can be integrated from $x = 0$ to $x = d$ to give $eV/m_e c^2 = 4.33$, which corresponds to $V = 2.2$ MV.

As a general remark, for multimode (broadband) excitation of the magnetron instability in microwave devices or in high-voltage diodes applied to particle-beam fusion, it is reasonable to expect that the nonlinear profile response (quasilinear, say) to the amplifying field perturbations would be in the direction of inducing a monotonically decreasing profile in $\langle n_e/\hat{\gamma}_e \rangle$, thereby reducing the growth rate and stabilizing the instability, at least at long wavelengths. (For comparison, see Sec. 6.5 for a discussion of the quasilinear stabilization of the diocotron instability by profile relaxation.)

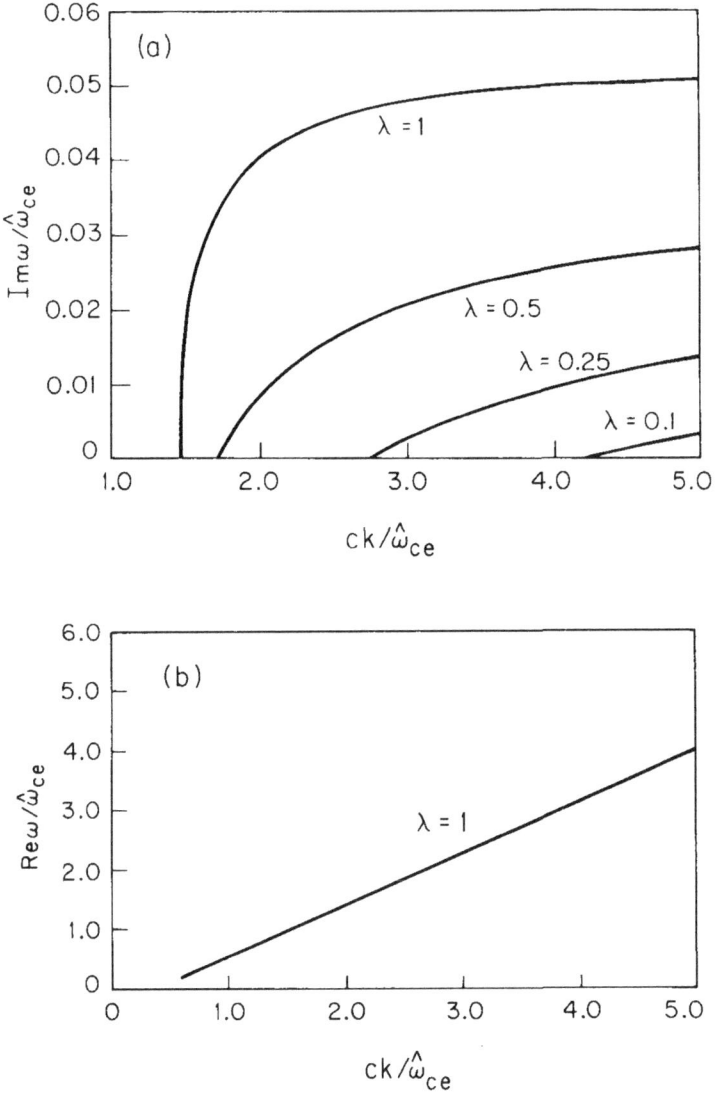

Figure 8.17. Linear growth properties of the magnetron instability. The figures show plots versus $ck/\hat{\omega}_{ce}$ of (a) the normalized growth rate $\text{Im}\,\omega/\hat{\omega}_{ce}$ and (b) the normalized real frequency $\text{Re}\,\omega/\hat{\omega}_{ce}$ obtained numerically from the eigenvalue equation (8.101). The choice of system parameters corresponds to $d = 3x_b$, $x_b = c/\hat{\omega}_{ce}$, $\kappa_1 x_b = 1.319$, $\kappa_2 x_b = 0.866$, $\hat{\gamma}_e = 2$, and $\bar{s}_e = 0.866$, and several values of the profile shape parameter λ. The plots of $\text{Re}\,\omega/\hat{\omega}_{ce}$ versus $ck/\hat{\omega}_{ce}$ for λ in the interval $0.06 \leq \lambda \leq 1$ are virtually identical to the $\lambda = 1$ curve plotted in Fig. 8.17(b) [R.C. Davidson and K.T. Tsang, Phys. Fluids **28**, 1169 (1985)].

Any remnant instability at short wavelengths could be stabilized by kinetic effects,[62-64] which are not included in the cold-fluid model used here.

To conclude Sec. 8.5, it should be pointed out that the extraordinary-mode eigenvalue equation (8.101) can be generalized to cylindrical diode geometry (Fig. 8.8) including the full influence of cylindrical effects,[50-53] such as finite diode and layer aspect ratios, Coriolis and centrifugal acceleration effects, discrete azimuthal mode numbers, etc. As a general remark, detailed properties of the magnetron instability can exhibit a strong dependence on cylindrical effects,[52,53] particularly at moderate values of $(r_b - a)/(b + a)$ and $(b - a)/(b + a)$, where the cathode (anode) is located at $r = a (r = b)$, and $r_b - a$ is the layer thickness.

8.6   Multiresonator Magnetron Experiments

In conventional magnetrons, voltages of a few hundred volts to tens of kilovolts are applied between the anode and a heated, thermionic cathode. Power levels from tens of watts to hundreds of kilowatts can be achieved in the decimeter and centimeter wave length range with conversion efficiencies as high as 80%. In relativistic magnetrons,[39-43] however, pulsed high-voltage diodes (operating in the MV range, say) are used to generate microwaves at gigawatt power levels, although at reduced efficiencies. In this section, we present a brief summary of the classic experiments by Palevsky and Bekefi[41] on the so-called "A6" relativistic magnetron.

Figure 8.18(a) shows a schematic of the cross section of the A6 magnetron.[41] The cold, field-emission, graphite cathode is located at radius $a = 1.58$ cm. The inside radius of the anode block is $b = 2.11$ cm, and six vane-type resonators with outer radius $d = 4.11$ cm are used, with angle $\psi = 20°$ subtended by the resonators on axis. The axial magnetic field $B_f \hat{e}_z$ prior to formation of the circulating electron layer (the so-called "fill" field discussed in Sec. 8.3.3) ranges from 4–10 kG in typical operation. The length of the anode block for the A6 magnetron is $L = 7.2$ cm, and the operating voltage in the experiments is 300–400 kV. The annular interaction space between $r = a$ and $r = b$ together with the periodically-spaced vanes can be viewed as a coaxial microwave resonator. For transverse electric (TE) modes with δE perpendicular to $B_f \hat{e}_z$ and δB parallel to $B_f \hat{e}_z$, the vacuum electric field pattern is illustrated in Fig. 8.18(b) for the $\ell = 3$ and $\ell = 6$ modes (the so-called π and 2π modes, respectively).[41] The circular dashed lines in Fig. 8.18(b) show

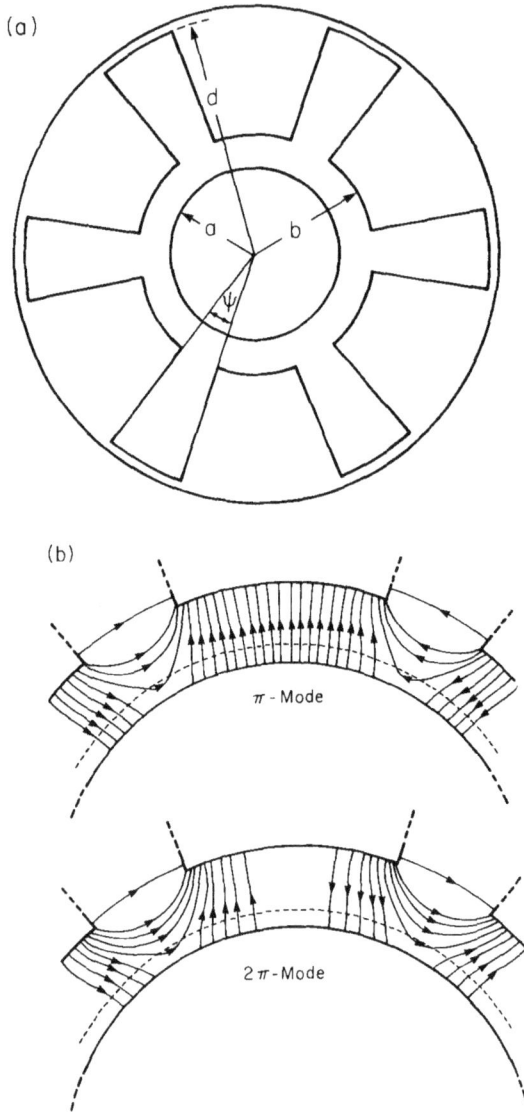

Figure 8.18.   Schematic of the A6 relativistic magnetron [A. Palevsky and G. Bekefi, Phys. Fluids **22**, 986 (1979)]. In Fig. 8.18(a), the emitting cathode is located at $a = 1.58$ cm; the anode is located at $b = 2.11$ cm; six vane-type resonators with outer radius $d = 4.11$ cm subtend an angle of $\psi = 20°$ at the axis; and the fill field $B_f \hat{e}_z$ points into the page. Figure 8.18(b) shows the vacuum electric field pattern for the so-called π and 2π modes of excitation. The outer radial boundary of the electron layer is indicated schematically by the dashed curves.

schematically the outer radial boundary of the electron layer prior to the onset of electromagnetic fluctuations. The A6 magnetron is found to oscillate preferentially in the 2π mode, which is characterized by the fact that the electromagnetic fields in all of the resonators are precisely in phase [Fig. 8.18(b)].

Assuming the applicability of the Brillouin flow model for magnetically insulated electron flow (see Sec. 8.3.3), the cylindrical generalization of the Hull cut-off voltage V_H and the Buneman-Hartree threshold voltage V_{BH} can be expressed relativistically as[41−43]

$$\frac{eV_H}{m_e c^2} = \left[1 + \frac{e^2 B_f^2}{m_e^2 c^4}\left(\frac{b^2 - a^2}{2b}\right)^2\right]^{1/2} - 1 \,, \qquad (8.117)$$

and

$$\frac{eV_{BH}}{m_e c^2} = \frac{eB_f}{m_e c^2}\left(\frac{b^2 - a^2}{2b}\right)\beta_p - \left[1 - \left(1 - \beta_p^2\right)^{1/2}\right] \,, \qquad (8.118)$$

where $\beta_p c$ is the phase velocity of the electromagnetic wave. Note that Eqs.(8.117) and (8.118) are identical in form to the planar expressions derived in Sec. 8.3.3 provided we make the replacement $d = b - a \rightarrow (b^2 - a^2)/2b$ in Eqs.(8.73) and (8.75), respectively. For steady Brillouin flow in a cylindrical diode with specified fill field B_f, the inequality $V < V_H$ is required to assure magnetic insulation of the electron flow from contact with the anode at $r = b$, whereas $V > V_{BH}$ is required for interaction of the outer electrons with the wave field. For specified voltage V, note also that Eq.(8.117) can be inverted to determine the critical magnetic field (B_f^*) defined by $(eB_f^*/m_e c^2)(b^2 - a^2)/2b = [(1 + eV/m_e c^2)^2 - 1]^{1/2}$. That is, for specified voltage V, the inequality $B_f > B_f^*$ is required for the electron flow to be magnetically insulated from contact with the anode.[41]

The region of operation[41] of the A6 magnetron is illustrated by the dashed line in Fig. 8.19(a), with the solid point corresponding to maximum microwave power. Here, the voltages V_H and V_{BH}, calculated from Eqs.(8.117) and (8.118), are plotted versus the fill field B_f, and V_{BH} has been evaluated for β_p corresponding to the $\ell = 6$ mode (2π mode) at frequency $f = \omega/2\pi = 4.55$ GHz. The measured microwave power[41] emitted by the A6 magnetron is plotted versus the fill field B_f in Fig. 8.19(b), where the peak power is approximately 0.45 GW per open port at 7.6 kG. Note from Fig. 8.19(b) that microwave emission occurs for magnetic field in the interval 4.2 kG $< B_f < 10.4$ kG. On the other hand, the intersection

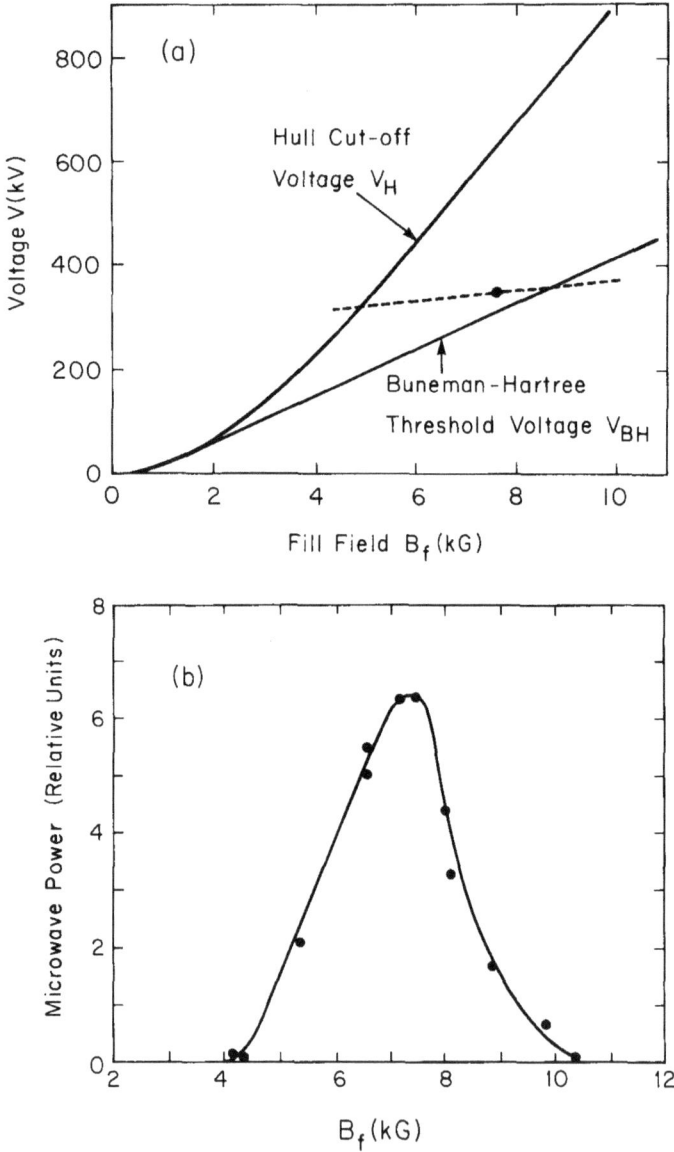

Figure 8.19. Microwave generation by the A6 relativistic magnetron [A. Palevsky and G. Bekefi, Phys. Fluids **22**, 986 (1979)]. The dashed line in Fig. 8.19(a) corresponds to the region of experimental operation, and the solid point corresponds to the maximum microwave power. The measured microwave power is plotted versus the fill field B_f in Fig. 8.19(b), where the peak power is approximately 0.45 GW per open port at $B_f = 7.6$ kG.

points of the dashed line in Fig. 8.19(a) with the two curves corresponding to V_H and V_{BH} would predict that microwave emission occurs in the interval $4.9\,\text{kG} < B_f < 8.8\,\text{kG}$. Since the vane structure was neglected in the derivation of the expressions for V_H and V_{BH} in Eqs.(8.117) and (8.118), and the *a priori* assumption of Brillouin flow may be questionable, the agreement between experiment and the operating range predicted by theory based on the Brillouin flow model is remarkably good.

8.7   Ion Resonance and Transit Time Instabilities in Magnetically Insulated Ion Diodes

In this section, we consider the stability properties of intense nonneutral electron and ion flow in magnetic insulated ion diodes (Sec. 8.4) for the planar diode configuration illustrated in Fig. 8.12. For simplicity, the magnetically insulated electron flow is assumed to be nonrelativistic (diode voltages below 0.5 MV), and the electrons and ions are treated as macroscopic cold fluids with equilibrium density profiles $n_e^0(x)$ and $n_i^0(x)$, and flow velocities $V_{ye}^0(x) = -cE_x(x)/B_0$ and $V_{xi}^0(x) = -(2Z_i eV/m_i)^{1/2} \times [1 - \phi_0(x)/V]^{1/2}$. (Here, $+Z_i e$ is the ion charge.) Particular emphasis in the present analysis is placed on the ion resonance and transit time instabilities investigated by Ott, et al.,[55,56] in which the slow wave excited by the electron layer undergoes an unstable interaction with the ions which transit the anode-cathode gap in a time

$$T = -\int_0^d \frac{dx}{V_{xi}^0(x)}, \tag{8.119}$$

where d is the cathode-anode spacing (Fig. 8.12).

The stability analysis assumes electrostatic flute perturbations ($\delta\mathbf{E} = -\nabla\delta\phi$ and $\partial/\partial z = 0$), and the perturbed electrostatic potential is expressed as

$$\delta\phi(x,y,t) = \sum_{k=-\infty}^{\infty} \Phi_k(x)\exp(iky - i\omega t), \tag{8.120}$$

where $\text{Im}\,\omega > 0$ corresponds to instability (temporal growth), and $k = 2\pi n/L$ is the wavenumber in the y-direction. For nonrelativistic electron flow ($V_{ye}^{0\,2}/c^2 \ll 1$) and electrostatic perturbations satisfying $\omega_{pe}^2/c^2 k^2 \ll 1$ and $|\omega|^2/c^2 k^2 \ll 1$, the extraordinary mode eigenvalue equation (8.101) reduces to (Problem 8.12)

$$
\frac{\partial}{\partial x} \left\{ \left(1 - \frac{\omega_{pe}^2(x)}{\left[\omega - kV_{ye}^0(x)\right]^2 - \left[\omega_{ce}^2 - \omega_{pe}^2(x)\right]} \right) \frac{\partial}{\partial x} \Phi_i(x) \right\}
$$

$$
- k^2 \left(1 - \frac{\omega_{pe}^2(x)}{\left[\omega - kV_{ye}^0(x)\right]^2 - \left[\omega_{ce}^2 - \omega_{pe}^2(x)\right]} \right) \Phi_k(x)
$$

$$
= \frac{k\,\omega_{ce}\Phi_k(x)}{\left[\omega - kV_{ye}^0(x)\right]} \frac{\partial}{\partial x} \left(\frac{\omega_{pe}^2(x)}{\left[\omega - kV_{ye}^0(x)\right]^2 - \left[\omega_{ce}^2 - \omega_{pe}^2(x)\right]} \right)
$$

$$
- 4\pi Z_i e \delta n_i(x, k) \; . \tag{8.121}
$$

Here, $\omega_{pe}^2(x) = 4\pi n_e^0(x)e^2/m_e$ is the local electron plasma frequency-squared, and the perturbed ion charge density $+Z_i\, e\, \delta n_i(x, k)$ is determined self-consistently in terms of $\Phi_k(x)$ from the linearized cold-fluid equations of continuity and momentum transfer (Sec. 2.3). Treating the ions as unmagnetized ($\omega_{ci} \rightarrow 0$) on the time scales of interest, the coupled linearized equations relating the Fourier amplitudes $\delta n_i(x, k), \delta V_{yi}(x, k), \delta V_{xi}(x, k)$ and $\Phi_k(x)$ can be expressed as

$$
- i\omega\,\delta n_i + ik\,n_i^0\,\delta V_{yi} + \frac{\partial}{\partial x}\left(V_{xi}^0\,\delta n_i\right) + \frac{\partial}{\partial x}\left(n_i^0\,\delta V_{xi}\right) = 0 \; , \tag{8.122}
$$

$$
- i\omega\,\delta V_{xi} + \frac{\partial}{\partial x}\left(V_{xi}^0\,\delta V_{xi}\right) = -\frac{Z_i e}{m_i}\frac{\partial}{\partial x}\Phi_k \; , \tag{8.123}
$$

$$
- i\omega\,\delta V_{yi} + V_{xi}^0\frac{\partial}{\partial x}\delta V_{yi} = -ik\frac{Z_i e}{m_i}\Phi_k \; , \tag{8.124}
$$

In Eqs.(8.122)–(8.124), $n_i^0(x)$ and $V_{xi}^0(x)$ are the equilibrium profiles for the ion density and flow velocity, respectively. Only in the high-frequency limit where $|\omega d| \gg |V_{xi}^0|$ can Eqs.(8.122)–(8.124) be used to obtain a closed expression for $\delta n_i(x, k)$ in terms of $\Phi_k(x)$ (Problem 8.13). Therefore, Eqs.(8.121)–(8.124) must generally be solved numerically as coupled equations for the (complex) eigenfrequency ω and the eigenfunction $\Phi_k(x)$. Of course the boundary conditions corresponding to zero tangential electric field $\delta E_y = -ik\Phi_k$ at the cathode and the anode are given by

$$
\Phi_k(x = 0) = 0 = \Phi_k(x = d) \; . \tag{8.125}
$$

A. Stability Properties for $n_i^0(x) = 0$

Before investigating the stability properties when there are ions flowing from the anode to the cathode, we summarize briefly the stability behavior calculated from the electrostatic eigenvalue equation (8.121) in the absence of ions ($n_i^0 = 0 = \delta n_i$). As in the case of relativistic electron flow (Sec. 8.5), it is found that the (fast) cyclotron wave is subject to the magnetron instability for perturbations with sufficiently short wavelength (sufficiently large kx_b). In circumstances where the electron density is uniform, with $n_e^0(x) = \hat{n}_e = $ const. for $0 \le x < x_b$, Buneman, Levy and Linson[44] have carried out an elegant analytical calculation (confirmed by numerical solution) of the asymptotic growth rate and real oscillation frequency of the magnetron instability for large values of $k^2 x_b^2$. Their analysis assumes electrostatic perturbations and nonrelativistic electron flow. Without presenting algebraic details,[44] it is found that

$$\mathrm{Re}\,\omega = \frac{1}{2} s_e \left(k x_b - 1 - \frac{1}{2} s_e \right) \omega_{ce} ,$$

$$\mathrm{Im}\,\omega = \frac{\pi}{2} s_e \omega_{ce} \exp\left(-2/s_e - 1 \right) , \qquad (8.126)$$

for $k^2 x_b^2 \gg 1/s_e^2$. Here, $s_e = \omega_{pe}^2/\omega_{ce}^2 = 4\pi \hat{n}_e e^2/m_e \omega_{ce}^2$ is a measure of the strength of the equilibrium self-electric field. Equation (8.126) has been derived[44] for arbitrary value of s_e. Note that $\mathrm{Re}\,\omega$ scales linearly with the wavenumber k, which is consistent with the results in Sec. 8.5. Moreover, under Brillouin flow conditions ($s_e = 1$), the maximum growth rate of the magnetron instability for short wavelength perturbations corresponds to $\mathrm{Im}\,\omega/\omega_{ce} = \pi/2e^3 \simeq 0.08$.

For $n_i^0 = 0 = \delta n_i$, the eigenvalue equation (8.121) also supports a stable slow-wave solution with $\mathrm{Im}\,\omega = 0$. Typical numerical results[55] are illustrated in Fig. 8.20, where the normalized (real) oscillation frequency ω/ω_{ce} is plotted versus kd for the choice of system parameters $s_e = \omega_{pe}^2/\omega_{ce}^2 = 1$ and $V = (1/4e) m_e \omega_{ce}^2 d^2 = V_H/2$. Here, V_H is the nonrelativistic Hull cut-off voltage defined in Eq.(8.38). Note from Fig. 8.20 that the frequency of the slow-wave mode is negative for $k = 0$, increases with increasing k, and crosses $\omega = 0$ at some critical wavenumber k_c (where $k_c d \simeq 2$ for the system parameters in Fig. 8.20).

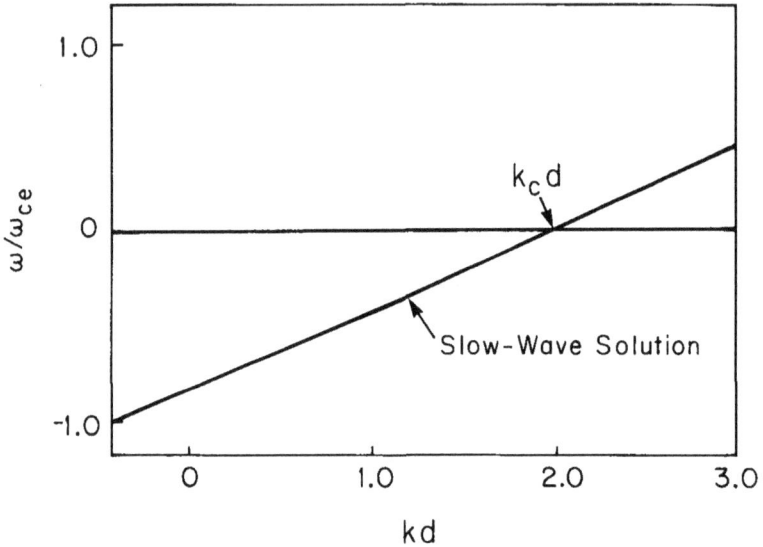

Figure 8.20. Plot of the normalized (real) frequency ω/ω_{ce} versus kd obtained numerically for the slow-wave solution to Eq.(8.121) when $n_i^0 = 0 = \delta n_i$. The choice of system parameters is $\omega_{pe}^2/\omega_{ce}^2 = 1$ and $V = 0.5V_H$ [E. Ott, T.M Antonsen, Jr., C.L. Chang and A.T. Drobot, Phys. Fluids **28**, 1948 (1985)].

B. Ion Resonance and Transit Time Instabilities

While the slow-wave mode in Fig. 8.20 is stable in the absence of ions, the slow-wave solution to Eq.(8.121) exhibits instability when there are ions flowing from the anode to the cathode and the perturbed ion density in Eq.(8.121) is nonzero. Not surprisingly, the instability is strongest in the low frequency region of Fig. 8.20 (in the vicinity of $kd = 2$), where the coupling with the ions is most effective.

Ott, et al.,[55] have solved the eigenvalue problem numerically, using Eqs.(8.122)–(8.124) to determine the perturbed ion charge density $Z_i e \times \delta n_i (x, k)$ occurring in Poisson's equation (8.121). The analysis assumes electrostatic perturbations about the nonrelativistic equilibrium described in Sec. 8.4. To summarize, the equilibrium electron and ion densities are related by the modified Brillouin flow condition in Eq.(8.82); the profiles for the ion density $n_i^0(x)$ and the electrostatic potential $\phi_0(x)$ in the gap region ($0 \leq x \leq d$) are given by Eqs.(8.80), (8.83), and (8.86); and the normalized layer thickness x_b/d and ion current J_i are related self-consistently to the normalized voltage V/V_H by Eqs.(8.90) and (8.92), respectively. Typical numerical results[55] for the slow-wave solution are illustrated in Fig. 8.21 for protons with $Z_i = 1$, and the choice of diode voltage $V = 0.5 V_H$, which corresponds to an electron layer thickness $x_b = 0.38d$ and an ion current $J_i = 1.56 J_{CL}$. Here, the normalized growth rate $\text{Im}\, \omega/\omega_{ce}$ and real oscillation frequency $\text{Re}\, \omega/\omega_{ce}$ are plotted versus kd in Figs. 8.21(a) and 8.21(b), respectively.

Note that there are several modes (labeled $0, 1, 2, \cdots 13$) excited for the range of wavenumbers in Fig. 8.21.[55] Four of the modes (Mode #s 1, 2, 12 and 13) are heavily damped ($\text{Im}\, \omega < 0$). The remaining modes are unstable over selected ranges of kd. For $kd < 1.4$ (Mode #0) and $kd > 2.5$ (Mode #11), the growth rate shows an oscillatory structure such that $(\text{Re}\, \omega)T$ sweeps through 2π on each oscillation of the growth rate as kd is varied. Here, T is the ion transit time from the anode to the cathode defined in Eq.(8.119). (These unstable modes are referred to here as the *ion transit time instability*.) For $1.4 < kd < 2.5$, the ion beam modes (Mode #s 3–10) experience an unstable interaction with the slow-wave branch, giving an instability somewhat analogous to the *ion resonance instability* discussed in Sec. 5.6.3. This interaction is characterized by avoided frequency crossings in the plots of $\text{Re}\, \omega/\omega_{ce}$ versus kd [Fig. 8.21(b)] and by a change in topology in the plots of $\text{Im}\, \omega/\omega_{ce}$ versus kd [Fig. 8.21(a)]. In this regard, the largest growth rate occurs for Mode #7 when $kd \simeq k_c d$, which corresponds to the zero-frequency crossing of the slow-wave solution in the absence of ions (Fig. 8.20). Moreover, the characteristic real frequency and growth rate satisfy

$$|\text{Re}\, \omega|T \sim |\text{Im}\, \omega|T \gg 1, \qquad (8.127)$$

which allows for a simplified estimate of $\delta n_i(x, k)$ (Problem 8.13).

The ion resonance instability is driven by the relative drift (in the y-direction) of the magnetically insulated electrons through the background ions.[65] As indicated in Sec. 5.6.3, it has many features similar to the classical electron-ion two-stream instability. In this regard, one of the most important nonlinear consequences of the fluctuations generated by the ion resonance instability in magnetically insulated ion diodes is likely to be a degradation in the quality of the ion beam. In particular, the fluctuations will induce a transverse velocity spread in the beam ions, typically in a few maximum growth times.

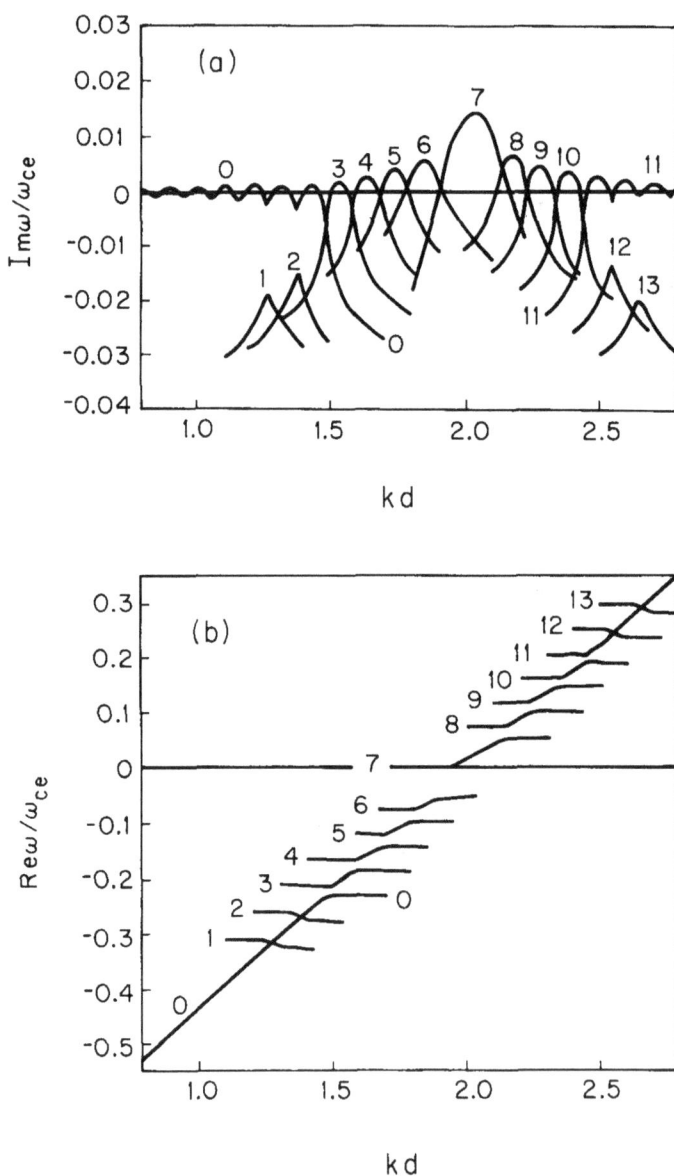

Figure 8.21. Plots versus kd of (a) the normalized growth rate $\text{Im}\,\omega/\omega_{ce}$, and (b) the normalized real frequency $\text{Re}\,\omega/\omega_{ce}$ obtained numerically for the slow-wave solution to Eqs.(8.121)–(8.124) when protons flow from the anode to the cathode. The choice of system parameters is $V = 0.5V_H, x_b = 0.38d$ and $J_i = 1.56J_{\text{CL}}$ [E. Ott, T.M Antonsen, Jr., C.L. Chang and A.T. Drobot, Phys. Fluids **28**, 1948 (1985)].

It should be noted that the fast-wave solution to the eigenvalue equation (8.121) also undergoes an (unstable) modification due to the ions flowing from the anode to the cathode. The corresponding growth rates[55] of the ion-driven modes, however, tend to be small in comparison with those in Fig. 8.21(a) and will not be presented here.

To summarize, the magnetron instability, the ion resonance instability, and the ion transit time instability are likely to be active instability mechanisms affecting the electron and ion flow in magnetically insulated ion diodes. If there is also a neutral plasma in the diode (e.g., near the anode surface), then additional instabilities are possible because of the two additional components of electrons and ions. These include, for example, the lower-hybrid-drift instability,[66] and the diocotron and ion resonance instabilities which involve an unstable coupling of the anode plasma electrons and ions with the magnetically insulated electron layer in contact with the cathode.[67] (The anode plasma is coupled to the layer electrons through the collective fields, and spatial overlap of the profiles is not required for instability to occur.)

Problem 8.13   Consider the linearized equations (8.122)–(8.124) for the ion fluid perturbations $\delta n_i, \delta V_{xi}$ and δV_{yi}. Assume that the frequency of the perturbations is sufficiently high that [see Eq.(8.127)]

$$|\omega d| \gg |V_{xi}^0| \,. \tag{8.13.1}$$

Show that the perturbed ion charge density $Z_i\, e\, \delta n_i\, (x,k)$ can be approximated by

$$Z_i\, e\, \delta n_i\, (x,k) = -\frac{\partial}{\partial x}\left[\frac{\omega_{pi}^2(x)}{\omega^2}\frac{\partial}{\partial x}\Phi_k(x)\right] + k^2\, \frac{\omega_{pi}^2(x)}{\omega^2}\Phi_k(x), \tag{8.13.2}$$

where $\omega_{pi}^2(x) = 4\pi n_i^0(x)\, Z_i^2 e^2/m_i$ is the local ion plasma frequency-squared.

8.8   Applied-B Diode with Virtual Cathode

The magnetically insulated diodes analyzed in Secs. 8.3–8.5 and 8.7 have assumed very simple geometric configurations in which the electron layer is in contact with a planar cathode (see Figs. 8.5 and 8.12). The schematic of an applied-B diode of the type used in particle-beam-fusion applications[32–34] is illustrated in Fig. 8.22. Here, the cylindrical diode is located at the end of a magnetically insulated transmission line, which is connected to a high-voltage pulsed-power accelerator. The anode-cathode spacing $d = b - a$ is typically small in comparison with the anode radius, and an applied axial magnetic field $B_f \hat{e}_z$ provides magnetic insulation of the electron flow, which is assumed to be described by $V_e^0 = (c/B_z)E^0 \times \hat{e}_z$ under steady-state conditions $(\partial/\partial t = 0)$. For present purposes, the diode configuration in Fig. 8.22 is modeled in planar geometry with $(r, \theta) \rightarrow (x, y)$, and the (unmagnetized) ions flow from the anode at $x = d$ in the negative x-direction.

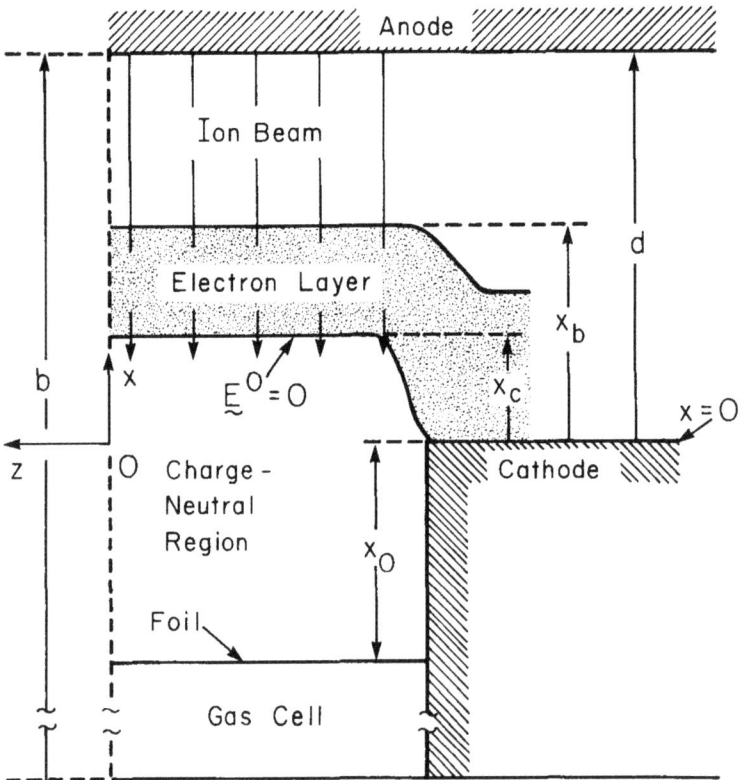

Figure 8.22.   Schematic of the applied-B ion diode configuration studied in Sec. 8.8. The surface of the electron layer labeled $E^0 = 0$ corresponds to the virtual cathode $(x = x_c)$. The gas-cell foil is located a distance $x_0 + x_c$ behind the virtual cathode [M.P. Desjarlais, Phys. Rev. Lett. **59**, 2295 (1987)].

8.8.1   Theoretical Model and Basic Equations

The virtual-cathode surface $(x = x_c)$ is defined by $\mathbf{E}^0 = 0$ in Fig. 8.22, which is approximately coincident with the $A_y^0 = 0$ magnetic flux surface. Here, $B_z(x) = \partial A_y^0(x)/\partial x$ and $E_x(x) = -\partial \phi_0(x)/\partial x$. The electrons are born in the cathode feeds with zero energy, $H = (\gamma - 1)m_e c^2 - e\phi_0 = 0$, and zero y-canonical momentum, $P_y = \gamma m v_y - (e/c)A_y^0 = 0$. Therefore, for $v_y \simeq -cE_x/B_z$, it follows that the $E_x(x = x_c) = 0$ surface in Fig. 8.22 is tied to the $A_y^0(x = x_c) = 0$ magnetic flux surface. Furthermore, computer simulation experiments[68] show that the virtual cathode surface x_c is pushed towards the anode because of the diamagnetic depression in the axial magnetic field $B_z(x)$ produced by the electron current $-n_e^0(x)\,eV_{ye}^0(x)$ in the layer region $x_c \leq x < x_b$. Therefore, to describe the diode configuration in Fig. 8.22, the location x_c of the virtual cathode should be calculated self consistently. Following Desjarlais,[34] because x_c is coincident with a flux surface with $A_y^0(x_c) = 0$, we apply magnetic flux conservation, $\int dx\, B_z(x) = $ const., between the gas-cell foil in Fig. 8.22 and $x = x_c$, and between $x = x_c$ and the anode $(x = d)$. This gives

$$B_c(x_0 + x_c) = B_f x_0 ,\qquad(8.128)$$

and

$$\int_{x_c}^{x_b} dx\, B_z(x) + B_0(d - x_b) = B_f d.\qquad(8.129)$$

Here, B_f is the fill field prior to formation of the electron layer. Moreover, the constants B_c and B_0 correspond to the values of the axial magnetic field behind the virtual cathode $[B_z(x) = B_c, x < x_c]$ and between the outer layer boundary and the anode $[B_z(x) = B_0, x_b < x \leq d]$ after layer formation. In Eqs.(8.128) and (8.129), x_0, d and B_f should be viewed as prescribed quantities, whereas x_c, x_b, and $B_z(x)$ remain to be determined. We now examine the steady-state Maxwell equations

$$\frac{\partial^2}{\partial x^2}\,\phi_0(x) = 4\pi e n_e^0(x) - 4\pi Z_i\, e n_i^0(x),\qquad(8.130)$$

$$\frac{\partial}{\partial x}\,B_z(x) = \frac{4\pi e}{c}\,n_e^0(x)\,V_{ye}^0(x),\qquad(8.131)$$

in the region $x_c \leq x \leq d$. Here, $n_e^0(x)$ is the density profile of the layer electrons, $V_{ye}^0(x) = -cE_x(x)/B_z(x)$ is the electron flow velocity, and $Z_i\, e n_i^0(x)V_{zi}^0(x) = J_i = $ const. is the ion current density $(J_i < 0)$. From energy conservation [Eq.(8.79)], the (nonrelativistic) ion flow velocity is given by $V_{xi}^0(x) = -(2Z_ieV/m_i)^{1/2}[1-\phi_0(x)/V]^{1/2}$, where $V = \phi_0(x = d)$ is the applied voltage, and the ions are treated as a macroscopic cold fluid. Substituting into Eqs.(8.130) and (8.131) gives

$$\frac{\partial^2}{\partial x^2}\phi_0(x) = 4\pi\, en_e^0(x) - \frac{K_i^2 V}{[1 - \phi_0(x)/V]^{1/2}}\,, \qquad (8.132)$$

$$\frac{1}{2}\frac{\partial}{\partial x}B_z^2(x) = 4\pi en_e^0(x)\frac{\partial}{\partial x}\phi_0(x)\,, \qquad (8.133)$$

for $x_c \le x \le d$. Here K_i^2 is related to the ion current and the diode voltage by

$$K_i^2 = -\frac{4\pi}{V}\left(\frac{m_i}{2Z_ieV}\right)^{1/2}J_i\,, \qquad (8.134)$$

where $J_i < 0$ and $K_i^2 > 0$. In the region where the electron density is nonzero $(x_c < x < x_b)$, we make use of Eq.(8.132) to eliminate $n_e^0(x)$ in Eq.(8.133). Integrating once with respect to x, and imposing the boundary conditions $\phi_0(x = x_c) = 0$ and $[\partial\phi_0/\partial x]_{x=x_c} = 0$ at the virtual cathode gives

$$B_z^2(x) = B_c^2 + \left[\frac{\partial}{\partial x}\phi_0(x)\right]^2 + 4K_i^2 V^2\left\{\left[1 - \frac{\phi_0(x)}{V}\right]^{1/2} - 1\right\} \qquad (8.135)$$

within the electron layer. Here, $B_c = B_z(x = x_c)$ is the magnetic field at the virtual cathode, and Eq.(8.135) is valid for general choice of electron density profile $n_e^0(x)$ consistent with Eq.(8.132) and the boundary conditions on $\phi_0(x)$.

As discussed earlier in this book, there is considerable latitude in prescribing equilibrium quantities within the framework of a macroscopic cold-fluid model. In particular, Eqs.(8.132) and (8.133) constitute two equations relating the three unknown quantities $\phi_0(x), n_e^0(x)$ and $B_z(x) = \partial A_y^0(x)/\partial x$. To complete the description and remove the underspecification, one approach would be to assume that the energy of an electron fluid element is uniform across the layer with $[\gamma_e^0(x) - 1]m_ec^2 - e\phi_0(x) = $ const. $= 0$. This is the approach used in Secs. 8.3 and 8.4, and it leads to a modified Brillouin flow condition which includes diamagnetic effects and the effects of the ion charge density in Eq.(8.132). An alternate approach would be to specify the functional form of the electrostatic potential $\phi_0(x)$

within the electron layer [e.g., $\phi_0(x) = (1/2)\kappa_0^2(x - x_c)^2 V$, where κ_0^2 is a constant], and make use of Eq.(8.132) to determine the electron density profile $n_e^0(x)$ self-consistently. A third approach would be to specify the functional form of $n_e^0(x)$ within the layer $(x_c < x < x_b)$, and make use of Eqs.(8.132) and (8.133) [or Eqs.(8.132) and (8.135)] to determine $\phi_0(x)$ and $B_z(x)$ self-consistently. For present purposes, we adopt the latter approach and assume a uniform density profile[34,48] prescribed by

$$n_e^0(x) = \begin{cases} \hat{n}_e, & x_c \leq x < x_b \,, \\ 0, & x_b < x \leq d \,. \end{cases} \tag{8.136}$$

Using such a uniform-density model for the layer electrons, Desjarlais has analyzed[34] Eqs.(8.128), (8.129), (8.132), and (8.133) and shown that the model can be used to describe quantitatively the ion current enhancement [relative to the Child-Langmuir value in Eq.(8.91)] measured at peak power in several ion diode experiments.[69,70] An appealing feature of the choice of flat density profile in Eq.(8.136), in addition to its analytical simplicity, is the fact that the layer electrons are less susceptible to the magnetron instability (at least at long wavelengths) than for the choice of a Brillouin-flow density profile, which is peaked at $x = x_b$ (see Sec. 8.5). Depending on the system parameters, however, the ion resonance and transit time instabilities discussed in Sec. 8.7 are still expected to be present for the diode configuration in Fig. 8.22.

8.8.2 Solution for Uniform-Density Electron Layer

Substituting Eq.(8.136) into Eq.(8.132), Poisson's equation can be expressed as

$$\frac{\partial^2}{\partial x^2} \phi_0(x) = \begin{cases} K_e^2 V - \dfrac{K_i^2 V}{[1 - \phi_0(x)/V]^{1/2}} \,, & x_c \leq x < x_b \,, \\[4mm] -\dfrac{K_i^2 V}{[1 - \phi_0(x)/V]^{1/2}} \,, & x_b < x \leq d \,, \end{cases} \tag{8.137}$$

where the constant K_e^2 is defined by

$$K_e^2 = \frac{4\pi \hat{n}_e e}{V}. \tag{8.138}$$

Similarly, Eq.(8.133) becomes

$$\frac{1}{2}\frac{\partial}{\partial x}B_z^2(x) = \begin{cases} K_e^2 V \dfrac{\partial}{\partial x}\phi_0(x) , & x_c \le x < x_b , \\ 0 , & x_b < x \le d , \end{cases} \tag{8.139}$$

which can be integrated once to give

$$B_z^2(x) = \begin{cases} B_c^2 + 2K_e^2 V\phi_0(x) & x_c \le x < x_b , \\ B_c^2 + 2K_e^2 V\phi_0(x = x_b) \equiv B_0^2 , & x_b < x \le d , \end{cases} \tag{8.140}$$

where use has been made of $\phi_0(x = x_c) = 0$. Here, $B_c = B_z(x = x_c) = $ const. is the (depressed) value of the axial magnetic field behind the virtual cathode ($x < x_c$), and $B_0 = B_z(x = x_b) = $ const. is the (compressed) value of the axial magnetic field between the layer surface and the anode ($x_b < x \le d$).

In the region $x_b < x \le d$, we integrate Poisson's equation (8.137), paralleling the analysis in Sec. 8.4. Assuming space-charge-limited emission of the ions from the anode plasma surface, integration of Eq.(8.137) gives

$$\phi_0(x) = V\left\{1 - \left[\frac{3}{2}K_i(d - x)\right]^{4/3}\right\} , \quad x_b < x \le d , \tag{8.141}$$

where use has been made of the boundary conditions $\phi_0(x = d) = V$ and $[\partial\phi_0/\partial x]_{x=d} = 0$. Within the electron layer, Eq.(8.137) can be integrated once to give

$$\frac{1}{2}\left[\frac{\partial}{\partial x}\left(\frac{\phi_0(x)}{V}\right)\right]^2 = K_e^2\frac{\phi_0(x)}{V} + 2K_i^2\left\{\left[1 - \frac{\phi_0(x)}{V}\right]^{1/2} - 1\right\} ,$$

$$x_c \le x < x_b , \tag{8.142}$$

where $\phi_0(x = x_c) = 0$ and $[\partial\phi_0/\partial x]_{x=x_c} = 0$ at the virtual cathode. It is convenient to introduce the simplified notation

$$\Phi(x) = \frac{\phi_0(x)}{V} ,$$

$$\Phi_b = \frac{\phi_0(x = x_b)}{V} , \tag{8.143}$$

so that $\Phi(x)$ is the electrostatic potential normalized to the diode voltage, and $\Phi_b(< 1)$ is the value of $\Phi(x)$ at the layer boundary $x = x_b$. Evaluating Eq.(8.141) at $x = x_b$, and making use of Eqs.(8.141) and (8.142) to enforce continuity of $\partial\Phi(x)/\partial x$ at $x = x_b$ gives the two conditions

$$K_i\,(d - x_b) = \frac{2}{3}\,(1 - \Phi_b)^{3/4}\,, \tag{8.144}$$

$$K_e^2\Phi_b = 2K_i^2\,, \tag{8.145}$$

which relates K_i^2, K_e^2, Φ_b and $K_i(d - x_b)$. Eliminating $K_i^2 = (1/2)K_e^2\Phi_b$ in Eq.(8.142) and integrating with respect to x, we obtain

$$K_e(x - x_c) = \frac{1}{\sqrt{2}}\int_0^{\Phi(x)} \frac{d\Phi}{[\Phi - \Phi_b + \Phi_b(1 - \Phi)^{1/2}]^{1/2}}\,, \tag{8.146}$$

which determines $\Phi(x)$ in the interval $x_c \leq x < x_b$. Evaluation of Eq.(8.146) at $x = x_b$ gives

$$K_e(x_b - x_c) = \frac{1}{\sqrt{2}}\int_0^{\Phi_b} \frac{d\Phi}{[\Phi - \Phi_b + \Phi_b(1 - \Phi)^{1/2}]^{1/2}}\,, \tag{8.147}$$

which relates the dimensionless parameters $K_e(x_b - x_c)$ and Φ_b. The integrations over Φ in Eqs.(8.146) and (8.147) can be carried out in closed analytical form (Problem 8.14), or the integrals can be evaluated numerically.[34]

Equation (8.140) relates the axial magnetic field $B_z(x)$ to the electrostatic potential $\phi_0(x)$ determined from Eq.(8.142) [or Eq.(8.146)]. Of particular interest in the flux conservation constraint (8.129) is the quantity

$$\int_{x_c}^{x_b} dx\, B_z(x) = K_e V \int_{x_c}^{x_b} dx\, \left[2\Phi(x) + \frac{B_c^2}{K_e^2 V^2}\right]^{1/2}\,, \tag{8.148}$$

where $B_c \equiv B_z(x = x_c)$. Expressing $dx = d\Phi(dx/d\Phi)$, and making use of Eq.(8.142) with $K_i^2 = K_e^2\Phi_b/2$ [Eq.(8.145)], it follows that Eq.(8.148) can be expressed as

$$\int_{x_c}^{x_b} dx\, B_z(x) = \frac{V}{\sqrt{2}}\int_0^{\Phi_b} \frac{d\Phi(2\Phi + \alpha_c^2)^{1/2}}{[\Phi - \Phi_b + \Phi_b(1 - \Phi)^{1/2}]^{1/2}}\,, \tag{8.149}$$

where α_c^2 is the dimensionless parameter defined by

$$\alpha_c^2 = \frac{B_c^2}{K_e^2 V^2} = \frac{B_c^2}{4\pi\hat{n}_e eV}. \tag{8.150}$$

Furthermore, from Eq.(8.140), the uniform field B_0 in the region between the electron layer and the anode ($x_c < x \le d$) can be expressed as

$$B_0 = K_e V (2\Phi_b + \alpha_c^2)^{1/2}. \tag{8.151}$$

Substituting Eqs.(8.149) and (8.151) into the flux conservation constraint (8.129) then gives

$$\frac{B_f d}{V} = K_e (d - x_b)(2\Phi_b + \alpha_c^2)^{1/2} \tag{8.152}$$

$$+ \frac{1}{\sqrt{2}} \int_0^{\Phi_b} \frac{d\Phi (2\Phi + \alpha_c^2)^{1/2}}{[\Phi - \Phi_b + \Phi_b(1 - \Phi)^{1/2}]^{1/2}}.$$

Finally, the flux conservation constraint $B_c(x_0 + x_c) = B_f x_0$ in Eq.(8.128) can be expressed as $x_c = (B_f/B_c - 1)x_0$, or equivalently

$$x_c = \left[\frac{1}{K_e d} \left(\frac{B_f d/V}{\alpha_c} \right) - 1 \right] x_0, \tag{8.153}$$

where $\alpha_c = B_c/K_e V$. Equation (8.153) determines the location x_c of the virtual cathode in terms of x_0 (see Fig. 8.22) and the dimensionless parameters $B_f d/V$ and α_c, which are related by Eq.(8.152).

For future reference, it is convenient to express the ion current J_i in units of the ion Child-Langmuir current $J_{\rm CL} = -(V/9\pi d^2)(2Z_i eV/m_i)^{1/2}$, which would pertain in the absence of an electron layer. Combining Eqs.(8.91), (8.134), and (8.145) readily gives

$$\frac{J_i}{J_{\rm CL}} = \frac{9}{4} K_i^2 d^2 = \frac{9}{8} \Phi_b K_e^2 d^2, \tag{8.154}$$

where $\Phi_b = \phi_0(x = x_b)/V$ and $K_e^2 = 4\pi\hat{n}_e e/V$. Since $\Phi_b \le 1$, it follows from Eq.(8.154) that large values of $K_e^2 d^2$ (compared with unity) are required for the ion current to be enhanced significantly relative to the Child-Langmuir current.

To summarize, the final equations in the present model of the applied-B diode in Fig. 8.22 consist of Eqs.(8.144), (8.145), (8.147), (8.152), and (8.153). These five equations relate the six "unknown" quantities

K_i, Φ_b, K_e, x_c, x_b, and $\alpha_c = B_c/K_e d$, to the four "known" quantities V, B_f, d, and x_0. Therefore, specification of one of the unknown quantities allows a determination of the remaining five. Indeed, Eqs.(8.144), (8.145), (8.147), (8.152), and (8.153) can be used to provide a detailed characterization of diode operation over a wide range of system parameters, at least in circumstances where the uniform-density model (for the layer electrons) is valid.[34]

8.8.3   Ion Current Enhancement at Peak Diode Power

For present purposes, to model an applied-B diode at peak power, we assume that the layer electrons in Fig. 8.22 extend to the anode with $x_b = d$. In this case, $\Phi_b = \phi_0(x = x_b)/V = 1$ [Eq.(8.144)] and Eq.(8.145) reduces to $K_i^2 = (1/2)K_e^2$. Furthermore, for $x_b = d$ and $\Phi_b = 1$, the integral relation in Eq.(8.147) reduces to (Problem 8.14)

$$K_e(d - x_c) = \frac{\pi}{\sqrt{2}} \ . \tag{8.155}$$

The corresponding ion current enhancement in Eq.(8.154) can be expressed as

$$\frac{J_i}{J_{\mathrm{CL}}} = \frac{9}{16}\pi^2 \frac{d^2}{(d - x_c)^2} \ . \tag{8.156}$$

Moreover, for $x_b = d$ and $\Phi_b = 1$, Eq.(8.152) reduces to

$$\frac{B_f d}{V} = \frac{1}{\sqrt{2}} \int_0^1 \frac{d\Phi(2\Phi + \alpha_c^2)^{1/2}}{[\Phi - 1 + (1 - \Phi)^{1/2}]^{1/2}} \ , \tag{8.157}$$

and Eq.(8.153) can be expressed in the equivalent form

$$\frac{d}{d - x_c} = \frac{d}{d + x_0}\left[1 + \frac{\sqrt{2}}{\pi}\frac{(B_f d/V)}{\alpha_c}\frac{x_0}{d}\right] \ . \tag{8.158}$$

Here, the dimensionless parameter $\alpha_c = B_c/K_e V$ is related to $B_f d/V$ by Eq.(8.157), and use has been made of $K_e(d - x_c) = \pi/\sqrt{2}$ in obtaining Eq.(8.158).

Equations (8.156)–(8.158) constitute the final equations describing the ion current enhancement for the case where the layer electrons extend to the anode ($x_b = d$). For specified values of the normalized fill field ($B_f d/V$) and gas-cell location (x_0/d), Eqs.(8.157) and (8.158) can be used to determine the self-consistent values of α_c and the current enhancement factor $d^2/(d - x_c)^2$ required in Eq.(8.156). In the special case where $x_0 = 0$, it follows from Eq.(8.158) that $x_c = 0$, and Eq.(8.156) reduces to $J_i/J_{CL} = 9\pi^2/16 \simeq 5.55$, whatever the values of $B_f d/V$ and α_c consistent with Eq.(8.157).

For specified values of $B_f d/V$, Eq.(8.157) can be solved numerically for the field depression parameter $\alpha_c = B_c/K_e V$ (Fig. 8.23). (Here, $\alpha_c^2 \ll 1$ corresponds to a strong magnetic field depression behind the virtual cathode.) For $B_f d/V \gg 1$, it is readily shown that Eq.(8.157) gives $\alpha_c \simeq (\sqrt{2}/\pi)(B_f d/V) \gg 1$, and Eq.(8.158) reduces to $x_c \simeq 0$. As the value of $B_f d/V$ is reduced, however, the solution for α_c decreases (Fig. 8.23) to smaller values and approaches zero as $B_f d/V$ approaches $(1/2)[3\sqrt{2} - \ln(1 + \sqrt{2})] = 1.6806$. That is, solutions to Eq.(8.157) for α_c exist only for $B_f d/V > 1.681$, or equivalently $V/B_f d < 0.595$. Moreover, the ion current enhancement calculated from Eqs.(8.156) and (8.158) can be large in the regime where α_c is small, i.e., for values of $V/B_f d$ approaching 0.595. This is illustrated in Fig. 8.24, where use has been made of Eqs.(8.156)–(8.158) to obtain self-consistent plots of J_i/J_{CL} versus $V/B_f d$

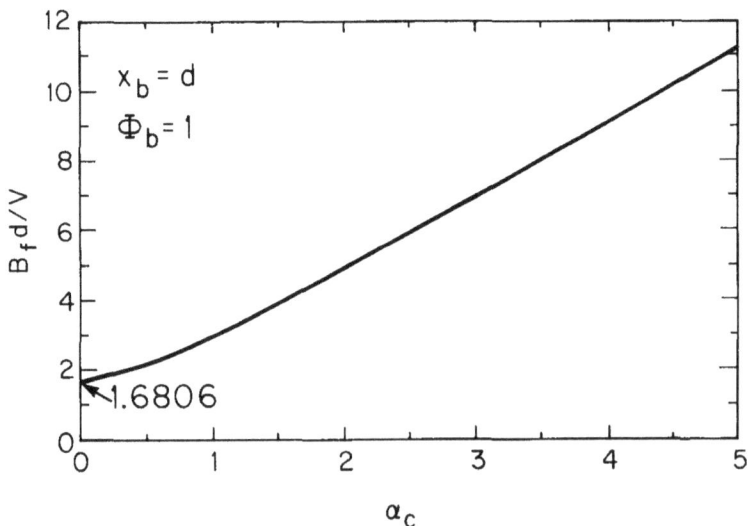

Figure 8.23. Plot of $B_f d/V$ versus $\alpha_c = B_c/K_e d$ obtained numerically from Eq.(8.157).

for several values of x_0/d. Evidently, J_i/J_{CL} increases with increasing values of $V/B_f d$ and x_0/d, relative to the solution $J_i/J_{CL} = 9\pi^2/16 \simeq 5.55$, which pertains as $V/B_f d \to 0$, or $x_0 = 0$ exactly.

As indicated earlier, this model[34] of an applied-B diode predicts values of J_i/J_{CL} which agree with the experimentally measured values of J_i/J_{CL} over a wide range of system parameters and diode experiments (with J_i/J_{CL} as high as 45).[69,70] As a cautionary remark, however, it should be kept in mind that the assumption of a uniform density profile for the layer electrons is somewhat arbitrary. (The density profiles were not measured in the experiments.) Moreover, at very small values of $\alpha_c = B_c/K_e d$

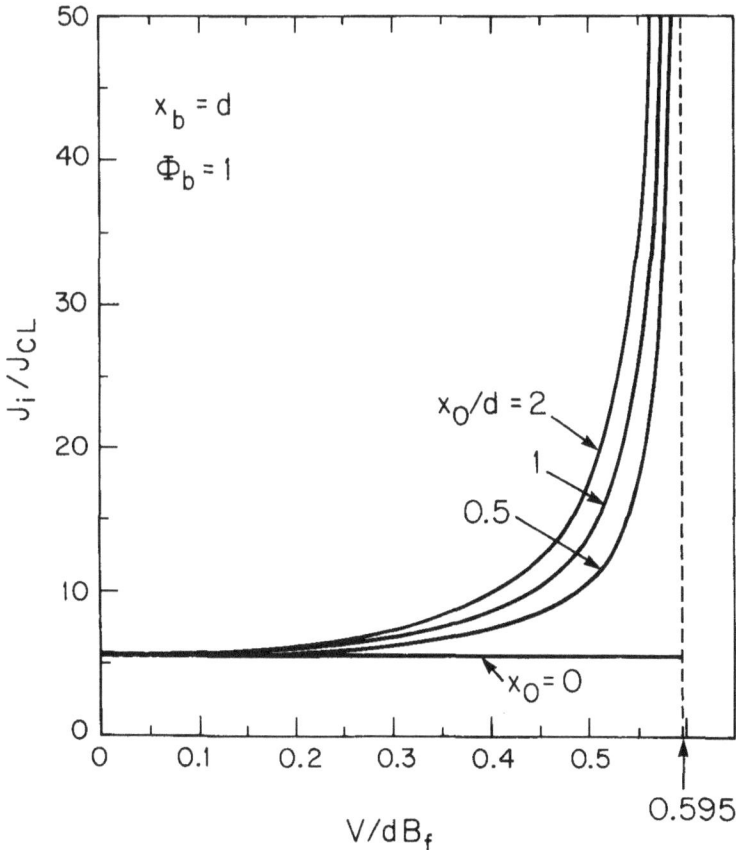

Figure 8.24. Plot of the ion current enhancement J_i/J_{CL} versus V/dB_f obtained numerically from Eqs.(8.156)–(8.158) for several values of the (normalized) gas-cell location x_0/d in Fig. 8.22.

(large values of J_i/J_{CL}), the assumption of laminar electron flow with $V_{ye}^0(x) = -cE_x(x)/B_z(x)$ is likely to break down near the cathode when the depressed magnetic field B_c becomes too weak, and the electrons are no longer strongly magnetized.

Problem 8.14   Consider the integral expression for $\Phi(x)$ in Eq.(8.146), which is valid within the uniform-density electron layer $(x_c \leq x < x_b)$.

a. Change variables to $Y^2 = 1 - \Phi$ and show that the integral in Eq.(8.146) can be evaluated in closed analytical form. Express your answer in the form

$$K_e(x - x_c) = \frac{1}{\sqrt{2}} \left\{ \Phi_b \left[\frac{\pi}{2} - \sin^{-1}\left(\frac{2(1 - \Phi)^{1/2} - \Phi_b}{2 - \Phi_b} \right) \right] \right. \tag{8.14.1}$$

$$\left. + 2[\Phi - \Phi_b + \Phi_b(1 - \Phi)^{1/2}]^{1/2} \right\}.$$

b. Evaluation of Eq.(8.14.1) at $x = x_b$ gives a relation between $K_e(x_b - x_c)$ and $\Phi_b \equiv \Phi(x = x_b) = \phi_0(x = x_b)/V$. For the special case where the electron layer extends to the anode $(x_b = d$ and $\Phi_b = 1)$, show that

$$K_e(d - x_c) = \frac{\pi}{\sqrt{2}}. \tag{8.14.2}$$

Chapter 8 References

1. R.B. Miller, *Intense Charged Particle Beams* (Plenum, New York, 1982).

2. *Advances in Pulsed Power Technology*, eds., A. Guenther and M. Kristiansen (Plenum, New York, 1987).

3. J.D. Lawson, *The Physics of Charged Particle Beams* (Clarendon Press, Oxford, 1988).

4. C.A. Kapetanakos and P. Sprangle, "Ultra-High-Current Electron Induction Accelerators," Physics Today **38** (2), 58 (1985).

5. S. Humphries, Jr., *Principles of Charged Particle Acceleration* (Wiley, New York, 1986).

6. D. Keefe, "Research on High Beam-Current Accelerators," Particle Accelerators **11**, 187 (1987).

7. J.P. VanDevender and D.L. Cook, "Inertial Confinement Fusion with Light Ion Beams," Science **232**, 831 (1986).

8. E.P. Lee and J. Hovingh, "Heavy Ion Induction Linac Drivers for Inertial Confinement Fusion," Fusion Technology **15**, 369 (1989).

9. *High-Power Microwave Sources*, eds., V. Granatstein and I. Alexeff (Artech House, Boston, Massachusetts, 1987), and references therein.

10. T.C. Marshall, *Free Electron Lasers* (Macmillan, New York, 1985).

11. C.W. Roberson and P. Sprangle, "A Review of Free Electron Lasers," Phys. Fluids **B1**, 3 (1989), and references therein.

12. E.P. Lee, "Resistive Hose Instability of a Beam with the Bennett Profile," Phys. Fluids **21**, 1327 (1978).

13. E.J. Lauer, R.J. Briggs, T.J. Fessenden, R.E. Hester and E.P. Lee, "Measurements of Hose Instability of a Relativistic Electron Beam," Phys. Fluids **21**, 1344 (1978).

14. M. Lampe, "Propagation of Charged Particle Beams in the Atmosphere," in *Proceedings of the 1987 Particle Accelerator Conference*, eds. E.R. Lindstrom and L.S. Taylor, IEEE No. 87CH2387-9 (IEEE, New York, 1987), p. 1965.

15. S. Dushman, "Electron Emission from Metals as a Function of Temperature," Phys. Rev. **21**, 623 (1923).

16. A.L. Hughes and L.A. Dubridge, *Photoelectric Phenomena* (McGraw-Hill, New York, 1932).

17. R.H. Good, Jr. and E.W. Müller, in *Handbuch der Physik* (Springer, Berlin, 1956), Volume **21**.

18. H.E. Tomaschke and D. Alpert, "Role of Submicroscopic Projections in Electrical Breakdown," J. Vac. Sci. Technology **4**, 192 (1967).

19. S.P. Bugaev, E.A. Litvinov, G.A. Mesyats and D.J. Proskurovskii, "Explosive Emission of Electrons from a Cathode," Sov. Phys. Usp. **18**, 51 (1975).

20. C.D. Child, "Discharge from Hot CaO," Phys. Rev. **32**, 492 (1911).

21. I. Langmuir, "The Effect of Space Charge and Initial Velocities on the Potential Distribution and Thermionic Current Between Parallel Plane Electrodes," Phys. Rev. **21**, 419 (1923).

22. C.K. Birdsall and W.B. Bridges, *Electron Dynamics of Diode Regions* (Academic Press, New York, 1966).

23. H.R. Jory and A.W. Trivelpiece, "Exact Relativistic Solution for the One-Dimensional Diode," J. Appl. Phys. **40**, 3924 (1969).

24. T.M. Antonsen, Jr., W.H. Miner, E. Ott and A.T. Drobot, "Stability of Space-Charge-Limited Electron Flow," Phys. Fluids **27**, 1257 (1984).

25. O. Buneman, "Self-Consistent Electrodynamics," Proc. Cambridge Phil. Soc. **50**, 77 (1954).

26. R.V. Lovelace and E. Ott, "Theory of Magnetic Insulation," Phys. Fluids **17**, 1263 (1974).

27. E. Ott and R.V. Lovelace, "Magnetic Insulation and Microwave Generation," Appl. Phys. Lett. **27**, 378 (1975).

28. T.M. Antonsen, Jr. and E. Ott, "Theory of Intense Ion Beam Acceleration," Phys. Fluids **19**, 52 (1976).

29. R.N. Sudan and R.V. Lovelace, "Generation of Intense Ion Beams in Pulsed Diodes," Phys. Rev. Lett. **31**, 1174 (1973).

30. P.L. Dreike, C. Eichenberger, S. Humphries, Jr. and R.N. Sudan, "Production of Intense Proton Fluxes in a Magnetically Insulated Diode," J. Appl. Phys. **47**, 85 (1976).

31. S. Humphries, Jr., R.N. Sudan and L. Wiley, "Extraction and Focusing of Intense Ion Beams from a Magnetically Insulated Diode," J. Appl. Phys. **47**, 2382 (1976).

32. D.J. Johnson, P.L. Dreike, S.A. Slutz, R.J. Leeper, E.J.T. Burns, J.R. Freeman, T.A. Mehlhorn and J.P. Quintenz, "Applied-B Ion Dion Studies at 3.5 Terawatts," J. Appl. Phys. **54**, 2230 (1983).

33. D.J. Johnson, R.J. Leeper, W.A. Stygar, R.S. Coats, T.A. Mehlhorn, J.P. Quintenz, S.A. Slutz and M.A. Sweeney, "Time-Resolved Proton Focus of a High-Power Ion Diode," J. Appl. Phys. **58**, 12 (1985).

34. M.P. Desjarlais, "Impedance Characteristics of Applied-B Ion Diodes," Phys. Rev. Lett. **59**, 2295 (1987).

35. *Microwave Magnetrons*, ed., G.B. Collins (McGraw-Hill, New York, 1948).

36. L. Brillouin, "Electronic Theory of the Plane Magnetron," Advan. Electronics **3**, 85 (1951).

37. *Crossed-Field Microwave Devices*, ed., E. Okress (Academic Press, New York, 1961), Volume 1.

38. J. Slater, *Microwave Electronics* (Dover Publications, New York, 1969).

39. G. Bekefi and T.J. Orzechowski, "Giant Microwave Bursts Emitted from a Field-Emission, Relativistic Electron Beam Magnetron," Phys. Rev. Lett. **37**, 379 (1976).

40. T.J. Orzechowski and G. Bekefi, "Microwave Emission from Pulsed, Relativistic Electron Beam Diodes I: The Smooth Bore Magnetron," Phys. Fluids **22**, 978 (1979).

41. A. Palevsky and G. Bekefi, "Microwave Emission from Pulsed, Relativistic Electron Beam Diodes II: The Multiresonator Magnetron," Phys. Fluids 22, 986 (1979).

42. J. Benford, "Relativistic Magnetrons," in *High-Power Microwave Sources*, eds., V. Granatstein and I. Alexeff (Artech House, Boston, Massachusetts, 1987), p. 351.

43. Y.Y. Lau, "Theory of Crossed-Field Devices and a Comparative Study of other Radiation Sources," in *High-Power Microwave Sources*, eds., V. Granatstein and I. Alexeff (Artech House, Boston, Massachusetts, 1987), p. 309.

44. O. Buneman, R.H. Levy and L.M. Linson, "Stability of Crossed Field Electron Beams," J. Appl. Phys. 37, 3203 (1966).

45. J. Swegle and E. Ott, "Instability of the Brillouin-Flow Equilibrium in Magnetically Insulated Diodes," Phys. Rev. Lett. 46, 929 (1981).

46. J. Swegle and E. Ott, "Linear Waves and Instabilities in Magnetically Insulated Gaps," Phys. Fluids 24, 1821 (1981).

47. J. Swegle, "Stability of Relativistic Laminar Flow Equilibria for Electrons Drifting in Crossed Fields," Phys. Fluids 26, 1670 (1983).

48. R.C. Davidson, K.T. Tsang and J.A. Swegle, "Macroscopic Extraordinary-Mode Stability Properties of Relativistic Nonneutral Electron Flow in a Planar Diode with Applied Magnetic Field," Phys. Fluids 27, 2332 (1984).

49. R.C. Davidson and K.T. Tsang, "Influence of Profile Shape on the Extraordinary-Mode Stability Properties of Relativistic Nonneutral Electron Flow in a Planar Diode with Applied Magnetic Field," Phys. Fluids 28, 1169 (1985).

50. D. Chernin and Y.Y. Lau, "Stability of Laminar Electron Layers," Phys. Fluids 27, 2319 (1984).

51. Y.Y. Lau and D. Chernin, "Stabilization of the Negative-Mass Instability in a Rotating Relativistic Electron Beam," Phys. Rev. Lett. 52, 1425 (1984).

52. T.M. Antonsen, Jr., E. Ott, C.L. Chang and A.T. Drobot, "Parametric Scaling of the Stability of Relativistic Laminar Flow Magnetic Insulation," Phys. Fluids 28, 2878 (1985).

53. R.C. Davidson and K.T. Tsang, "Macroscopic Extraordinary Mode Stability Properties of Relativistic Nonneutral Electron Flow in a Cylindrical Diode," Phys. Fluids 29, 3832 (1986).

54. R.C. Davidson and K.T. Tsang, "Analysis of Magnetron Instability for Relativistic Nonneutral Electron Flow in Cylindrical High-Voltage Diodes," Laser and Particle Beams 6, 661 (1988).

55. E. Ott, J.M. Antonsen, Jr., C.L. Chang and A.T. Drobot, "Stability of Magnetically Insulated Ion Diodes," Phys. Fluids **28**, 1948 (1985).

56. C.L. Chang, D.P. Chernin, A.T. Drobot, E. Ott and T.M. Antonsen, Jr., "Electromagnetic Stability of High-Power Ion Diodes," Phys. Fluids **29**, 1258 (1986).

57. L. Brillouin, "A Theorem of Larmor and Its Importance for Electrons in Magnetic Fields," Phys. Rev. **67**, 260 (1945).

58. A.W. Hull, "The Effects of a Uniform Magnetic Field on the Motion of Electrons Between Coaxial Cylinders," Phys. Rev. **18**, 31 (1921).

59. O. Buneman, in *Crossed-Field Microwave Devices*, ed., E. Okress (Academic, New York, 1961), Volume 1, p. 209.

60. S. Humphries, J.J. Lee and R.N. Sudan, "Generation of Intense Pulsed Ion Beams," Appl. Phys. Lett. **25**, 20 (1974).

61. S. Humphries, Jr., J.J. Lee and R.N. Sudan, "Advances in the Efficient Generation of Intense Pulsed Proton Beams," J. Appl. Phys. **46**, 187 (1975).

62. R.C. Davidson, "Kinetic Stability Theorem for Relativistic Nonneutral Electron Flow in a Planar Diode with Applied Magnetic Field," Phys. Fluids **28**, 377 (1985).

63. R.C. Davidson and H.S. Uhm, "Kinetic Extraordinary-Mode Eigenvalue Equation for Relativistic Nonneutral Electron Flow in Planar Geometry," Laser and Particle Beams **7**, 55 (1989).

64. R.C. Davidson and H.S. Uhm, "Kinetic Stability Properties of Relativistic Nonneutral Electron Flow for Low-Frequency Extraordinary-Mode Perturbations," Laser and Particle Beams **7**, 85 (1989).

65. R.H. Levy, J.D. Daugherty and O. Buneman, "Ion Resonance Instability in Grossly Nonneutral Plasmas," Phys. Fluids **12**, 2616 (1969).

66. Y. Maron, E. Sarid, O. Zahavi, L. Perelmutter and M. Sarfaty, "Particle Velocity Distribution and Expansion of a Surface-Flashover Plasma in the Presence of Magnetic Fields," Phys. Rev. **A39**, 5842 (1989).

67. R.C. Davidson, K.T. Tsang and H.S. Uhm, "Collective Instabilities Driven by Anode Plasma Ions and Electrons in a Nonrelativistic Cylindrical Diode with Applied Magnetic Field," Phys. Rev. **A32**, 1044 (1985).

68. S.A. Slutz, D.B. Seidel and R.S. Coats, "Electromagnetic Particle-in-Cell Simulations of Applied-B Proton Diodes," J. Appl. Phys. **59**, 11 (1986).

69. P.A. Miller, "Impedance Scaling of Applied-B Ion Diodes," J. Appl. Phys. **57**, 1473 (1985).

70. P.A. Miller and C.W. Mendel, Jr., "Analytic Model of Applied-B Ion Diode Impedance Behavior," J. Appl. Phys. **61**, 529 (1987).

The following references, while not cited directly in the text, are relevant to the general subject matter of this chapter.

M.P. Desjarlais, "Theory of Applied-B Ion Diodes," Phys. Fluids **B1**, 1709 (1989).

C. Litwin and Y. Maron, "Role of Neutrals in Plasma Expansion in Ion Diodes," Phys. Fluids **B1**, 670 (1989).

P. Martin and G. Donoso, "A New Langmuir-Child Equation Including Temperature Effects," Phys. Fluids **B1**, 247 (1989).

T.M. Antonsen, Jr. and C.L. Chang, "Kinetic Stability Theory of Space-Charge-Limited Flow," Phys. Fluids **B1**, 1728 (1989).

W. Woo, J. Benford, D. Fittinhoff, B. Hartneck, D. Price and H. Sze, "Phase Locking of High-Power Microwave Oscillators," J. Appl. Phys. **65**, 861 (1989).

J. Benford, H. Sze, W. Woo, R.R. Smith and B. Harteneck, "Phase Locking of Relativistic Magnetrons," Phys. Rev. Lett. **62**, 969 (1989).

M.D. Coleman and D.A. Hammer, "Spectroscopic Determination of the Electrostatic Potential Profile in a Plasma-Prefilled Ion Diode," Appl. Phys. Lett. **55**, 1627 (1989).

Y. Maron, M. Coleman, D.A. Hammer and H.S. Peng, "Electric Field and Charge Distribution," Phys. Rev. **A36**, 2818 (1987).

M.P. Desjarlais and R.N. Sudan, "Electron Diffusion and Leakage Currents in Magnetically Insulated Diodes," Phys. Fluids **30**, 1536 (1987).

Y. Maron and C. Litwin, "Local Ion Direction of Motion and Electron Flow in a Magnetically Insulated Diode," Phys. Fluids **30**, 1526 (1987).

R. Kraft and M.W. McGeoch, "Experimental Study of Magnetic Insulation," Phys. Fluids **30**, 1189 (1987).

T. Renk and D.A. Hammer, "Intense Ion Beam Quality and Microwave Emission from the Applied B_θ Diode," J. Appl. Physics **62**, 1655 (1987).

B.B. Godfrey, "Oscillatory Nonlinear Electron Flow in a Pierce Diode," Phys. Fluids **30**, 1553 (1987).

H.W. Chan, R.C. Davidson, K.T. Nguyen and H.S. Uhm, "Relativistic Nonneutral Electron Flow in a Planar Triode," J. Appl. Phys. **61**, 2152 (1987).

J. Benford, D. Price, H. Sze and D. Bromley, "Interaction of a Vircator Microwave Generator with an Enclosing Resonant Cavity," J. Appl. Phys. **61**, 2098 (1987).

Y. Maron, M.D. Coleman, D.A. Hammer and H.S. Peng, "Measurements of Electric Field Distribution in High Power Diodes," Phys. Rev. Lett. **57**, 699 (1986).

M.P. Desjarlais and R.N. Sudan, "A Thermal Distribution Function for Relativistic Magnetically Insulated Electron Flows," Phys. Fluids **29**, 1746 (1986).

K.T. Tsang and R.C. Davidson, "Macroscopic Cold-Fluid Equilibrium Properties of Relativistic Nonneutral Electron Flow in a Cylindrical Diode," Phys. Rev. **A33**, 4284 (1986).

C.L. Chang, E. Ott and T.M. Antonsen, Jr., "Linear Stability of Obliquely Propagating Electromagnetic Waves in Magnetically Insulated Gaps," Phys. Fluids **29**, 3851 (1986).

H.S. Uhm and R.C. Davidson, "Kinetic Equilibrium Properties of Relativistic Nonneutral Electron Flow in a Cylindrical Diode with Applied Magnetic Field," Phys. Rev. **A31**, 2556 (1985).

R.C. Davidson and H.S. Uhm, "Kinetic Stability Properties of Nonrelativistic Nonneutral Electron Flow in a Planar Diode with Applied Magnetic Field," Phys. Rev. **A32**, 3554 (1985).

R.C. Davidson and K.T. Tsang, "Macroscopic Electrostatic Stability Properties of Nonrelativistic Nonneutral Electron Flow in a Cylindrical Diode with Applied Magnetic Field," Phys. Rev. **A30**, 488 (1984).

C.L. Chang, T.M. Antonsen, Jr., E. Ott and A.T. Drobot, "Instabilities in Magnetically Insulated Gaps with Resistive Electrode Plasmas," Phys. Fluids **27**, 2545 (1984).

Y. Maron, "Local Electron Flow to the Anode in a Magnetically Insulated Diode," Phys. Fluids **27**, 285 (1984).

Y. Maron and C. Litwin, "Spectroscopic Doppler-Free Methods for Measuring the Electric and Magnetic Fields in Magnetically Insulated Ion Diodes," J. Appl. Phys. **54**, 2086 (1983).

A. Palevsky, G. Bekefi and A.T. Drobot, "Numerical Simulation of Oscillating Magnetrons," J. Appl. Phys. **52**, 4938 (1981).

M.A. Greenspan, R. Pal, D.A. Hammer, R.N. Sudan and S. Humphries, "An Applied-B_θ Magnetically Insulated Ion Diode," Appl. Phys. Lett. **37**, 248 (1980).

K.D. Bergeron, "One- and Two-Species Equilibria for Magnetic Insulation in Coaxial Geometry," Phys. Fluids **20**, 688 (1977).

CHAPTER 9

PROPAGATION AND STABILITY OF INTENSE CHARGED PARTICLE BEAMS IN A SOLENOIDAL FOCUSING FIELD

9.1 Introduction

As indicated in Chapters 1 and 8, intense charged particle beams have diverse applications ranging from high-current accelerators,[1-5] to particle-beam fusion,[6-9] to high-power microwave generation,[10-12] to beam propagation through the atmosphere.[13,14] In the short space of this treatise on nonneutral plasmas, it is not possible to present a thorough treatment of the propagation and stability of intense electron and ion beams pertaining to this multitude of applications. Therefore, in this chapter, we investigate selected topics on beam propagation and stability which emphasize the important influence of intense self-electric and self-magnetic fields.[15,16] Using well-established theoretical techniques in the study of nonneutral plasmas, the analysis presented here is intended to present a unified treatment of topics as diverse as the space-charge waves excited in a nonneutral electron beam propagating through a one-dimensional drift space (Sec. 9.3), the transverse stability properties of an intense nonneutral heavy ion beam propagating through a solenoidal focusing field (Sec. 9.4), and the resistive hose instability for an intense electron beam propagating through a dense background plasma which provides complete charge neutralization but partial current neutralization of the electron beam (Sec. 9.5). While the details of the physical processes differ in these examples, the theoretical treatments make extensive use of the methodology developed earlier in Chapters 2–8.

To summarize briefly the contents of this chapter, we begin with a description of the steady-state properties $(\partial/\partial t = 0)$ of an intense nonneutral electron beam propagating parallel to a uniform, solenoidal focusing field $B_0 \hat{e}_z$ (Sec. 9.2), including the derivation and analysis of the envelope equation for a cylindrical electron beam,[1,2] the calculation of the space-charge-limiting current for beam propagation through a grounded cylindrical cavity,[17-19] and the analysis of laminar flow equilibria for a cylindrical electron beam subject to the conservation of energy and canonical angular momentum.[20] Detailed stability properties[21-23] are then calculated for longitudinal space-charge perturbations about a nonneutral electron beam propagating in the z-direction through a one-dimensional drift space between grounded planes at $z = 0$ and $z = L$ (Sec. 9.3). The transverse stability properties of an intense nonneutral ion beam[24-26] are determined for low-frequency, azimuthally symmetric perturbations ($\partial/\partial\theta = 0$) and radial mode numbers $n = 0, 1, 2$ (Sec. 9.4). The cases of both a solenoidal focusing field $B_0 \hat{e}_z$[25] and an (averaged) quadrupole focusing field are considered.[26] Then, the stability of an intense electron beam propagating through a dense background plasma[27-43] is analyzed, with particular emphasis on the resistive hose instability (Sec. 9.5), and the resistive sausage and hollowing instabilities (Sec. 9.6). When the background plasma provides sufficient neutralization (f_M) of the beam current, the resistive hose instability occurs for electromagnetic perturbations with azimuthal mode number $\ell = 1$ and radial mode number $n = 0$, whereas the resistive sausage instability and the resistive hollowing instability occur for mode numbers $(\ell, n) = (0, 1)$ and $(\ell, n) = (0, 2)$, respectively. In Chapter 10, the particle orbits, envelope equations, and emittance growth are investigated for an intense nonneutral ion beam propagating through an alternating-gradient focusing field,[44-47] which is a configuration of considerable interest for application to heavy-ion fusion.[8,9]

9.2   Propagation of Intense Nonneutral Electron Beams

In this section, we examine several aspects of the propagation of intense charged particle beams under steady-state conditions ($\partial/\partial t = 0$), with particular emphasis on the propagation of relativistic electron beams parallel to a uniform magnetic field $B_0 \hat{e}_z$. Following an elementary treatment of the envelope equation for a cylindrical electron beam with radius r_b (Sec. 9.2.1), the propagation of an intense nonneutral electron beam through a grounded, cylindrical cavity with radius b and length L is analyzed in the limit of an infinitely strong guide field with $B_0 \to \infty$ (Sec. 9.2.2). The amount of beam current that can propagate through the drift cavity is limited by the electrostatic potential depression $\phi_0(r, z)$ associated with the space charge of the nonneutral electron beam, and the space-charge-limiting current J_ℓ is calculated for the two cases corresponding to an infinitely long drift space with $L \gg r_b$, b (Sec. 9.2.3), and a one-dimensional drift space with $r_b, b \gg L$ (Sec. 9.2.4). Finally, the cold-fluid model based on the moment-Maxwell equations (Chapter 5) is used to investigate the laminar flow equilibria of an intense nonneutral electron beam in which the conservation of energy and the conservation of canonical angular momentum of a fluid element are both imposed as conservation constraints (Sec. 9.2.5). Particular emphasis is placed on calculating the self-consistent equilibrium profiles and space-charge limiting current for an electron beam produced by a shield cathode source (in which case the canonical angular momentum $P_{\theta b}^0 = 0$).

9.2.1   Envelope Equation for Beam Propagation

The development of an equation describing the motion of the outer radial envelope of a charged particle beam is a useful concept in studying the propagation of intense electron and ion beams. Indeed, for several of the kinetic equilibrium examples investigated in Chapter 4, it was convenient to introduce an envelope function $\psi_b(r)$ such that the equilibrium beam radius r_b was the envelope of orbital turning points for which $\psi(r_b) = 0$ [see, for example, Eq.(4.113)]. In this section, we derive a dynamical equation for the envelope of a relativistic electron beam ($j = b$) with characteristic energy $\gamma_b m_e c^2$ propagating axially along a uniform magnetic field $B_0 \hat{e}_z$ through a background plasma.[1,2]

The electron beam is assumed to be infinitely long ($\partial/\partial z = 0$) and cylindrically symmetric ($\partial/\partial\theta = 0$) with density profile $n_b^0(r)$ extending over the interval $0 \leq r < r_b$, and zero density for $r > r_b$. Here, cylindrical polar coordinates (r, θ, z) are introduced as in Fig. 5.1. Moreover, the axial velocity profile is taken to be uniform with $V_{zb}^0(r) = \beta_b c = \text{const.}$ over the beam cross section, and it is assumed that $\nu/\gamma_b \ll 1$, where $\nu = N_b e^2/m_e c^2$ is Budker's parameter[48] and $N_b = 2\pi \int_0^{r_b} dr'\, r' n_b^0(r')$ is the total number of electrons per unit axial length of the electron beam. For present purposes, the background plasma is assumed to provide partial neutralization of the beam charge and axial current, with $\rho_p^0(r) = efn_b^0(r)$ and $J_{zp}^0(r) = ef_M n_b^0(r)\beta_b c$ where f and f_M are constants (independent of r). From Eqs.(5.5) and (5.6), the equilibrium radial electric field $E_r(r)$ and azimuthal self-magnetic field $B_\theta^s(r)$ can be expressed as

$$E_r(r) = -\frac{4\pi e(1-f)}{r} \int_0^r dr' r' n_b^0(r') \,, \qquad (9.1)$$

and

$$B_\theta^s(r) = -\frac{4\pi e(1-f_M)\beta_b}{r} \int_0^r dr' r' n_b^0(r') \,. \qquad (9.2)$$

Note from Eqs.(9.1) and (9.2) that the radial dependence of $E_r(r)$ and $B_\theta^s(r)$ are the same. For simplicity, consistent with $\nu/\gamma_b \ll 1$, it is also assumed that the transverse r-θ motion of the electrons making up the electron beam is nonrelativistic with $v_r^2 + v_\theta^2 \ll \beta_b^2 c^2$ (paraxial approximation),[2] and the axial diamagnetic field $B_z^s(r)\hat{e}_z$ induced by the beam rotation is neglected in comparison with the applied magnetic field $B_0\hat{e}_z$.

Now consider the motion of an individual electron within the azimuthally symmetric beam equilibrium. For an electron with energy γmc^2, the radial force equation for particle motion in the equilibrium field configuration $\mathbf{E}^0(\mathbf{x}) = E_r(r)\hat{e}_r$ and $\mathbf{B}^0(\mathbf{x}) = B_0\hat{e}_z + B_\theta^s(r)\hat{e}_\theta$ can be expressed as

$$\gamma m_e \left[\frac{d^2 r}{dt^2} - r \left(\frac{d\theta}{dt} \right)^2 \right] + m_e \frac{d\gamma}{dt} \frac{dr}{dt} = -e \left[E_r - \frac{1}{c} v_z B_\theta^s + \frac{1}{c} B_0 r \left(\frac{d\theta}{dt} \right) \right].$$

$$(9.3)$$

Here, $-e$ is the electron charge, m_e is the electron rest mass, $v_z = dz/dt$ is the axial velocity, and $r(t)$ and $\theta(t)$ are the particle orbits perpendicular to $B_0\hat{e}_z$. The azimuthal equation of motion for the electron can be integrated once to give

$$r \left[\gamma m_e r \left(\frac{d\theta}{dt} \right) - \frac{1}{2} m_e \omega_{cb} r \right] = P_\theta = \text{const.}, \qquad (9.4)$$

where P_θ is the canonical angular momentum, and $\omega_{cb} = eB_0/m_e c$ is the nonrelativistic electron cyclotron frequency. In Eq.(9.3), for $\nu/\gamma_b \ll 1$ and $v_r^2 + v_\theta^2 \ll \beta_b^2 c^2$, we approximate $m_e c^2 d\gamma/dt = -eE_r dr/dt \simeq 0$ to leading order. Moreover, use is made of Eq.(9.4) to eliminate $rd\theta/dt$ in Eq.(9.3) in terms of r and P_θ . Approximating $\gamma = \gamma_b = (1 - \beta_b^2)^{-1/2}$

and $v_z = \beta_b c$ for a typical beam electron, the orbit equation (9.3) for $r(t)$ then becomes (Problem 9.1)

$$\frac{d^2 r}{dt^2} + \left(\frac{\omega_{cb}}{2\gamma_b}\right)^2 r + \frac{e}{\gamma_b m} (E_r - \beta_b B_\theta^s) = \frac{(P_\theta/\gamma_b m)^2}{r^3} . \qquad (9.5)$$

In the present analysis, we are interested particularly in the motion of an electron near the outer edge of the electron beam $(r = r_b)$. Here, the main radial variation of $E_r(r)$ and $B_\theta^s(r)$ is associated with the $1/r$ factors in Eqs.(9.1) and (9.2). Since the number of beam electrons per unit axial length does not vary as the beam envelope expands or contracts slightly, we denote

$$N_b = 2\pi \int_0^{r_{b0}} dr' \, r' n_b^0(r') \equiv \pi \bar{n}_b r_{b0}^2 = \text{const.} , \qquad (9.6)$$

where $\bar{n}_b$ is the average density, and r_{b0} is the equilibrium beam radius $(r_b = r_{b0}$ for $dr_b/dt = 0)$. Substituting Eqs.(9.1), (9.2), and (9.6) into Eq.(9.5), the equation for the radial envelope $r_b(t)$ of the electron beam becomes

$$\frac{d^2 r_b}{dt^2} + \left\{ \left(\frac{\omega_{cb}}{2\gamma_b}\right)^2 - \frac{\omega_{pb}^2 r_{b0}^2}{2\gamma_b r_b^2} \left[(1-f) - \beta_b^2(1-f_M)\right] \right\} r_b = \frac{(P_\theta/\gamma_b m_e)^2}{r_b^3} , \qquad (9.7)$$

where $\omega_{pb}^2 r_{b0}^2 \equiv 4\pi \bar{n}_b r_{b0}^2 e^2/m_e = (4e^2/m_e)N_b = \text{const.}$

Several important features are evident from Eq.(9.7). First, the axial guide field always acts as a restoring (focusing) force through the term proportional to $(\omega_{cb}/2\gamma_b)^2$ in Eq.(9.7). Second, the self-field contributions proportional to ω_{pb}^2 in Eq.(9.7) are focusing only if

$$\beta_b^2 (1 - f_M) > (1 - f) , \qquad (9.8)$$

which corresponds to the condition that the (confining) force of the self-magnetic field exceeds the (repulsive) electrostatic force. At moderate values of $\beta_b^2(1 - f_M) < 1$, Eq.(9.8) requires some neutralization of the beam space charge with $f > 1 - \beta_b^2(1 - f_M)$. Finally, the term proportional to P_θ^2 in Eq.(9.7) acts as a defocusing force, which can be balanced by the restoring terms on the left-hand side.

Equation (9.7) for the radial envelope $r_b(t)$ of the electron beam can be expressed as

$$\frac{d^2}{dt^2} r_b = -\frac{\partial}{\partial r_b} V(r_b) \,, \tag{9.9}$$

where the effective potential $V(r_b)$ is of the form $V(r_b) = (\alpha/2) r_b^2 + \beta r_{b0}^2 \ln r_b + (1/2) \delta/r_b^2$ (Problem 9.2). Here, α and δ are positive constants, and $\beta > 0$ only when the inequality in Eq.(9.8) is satisfied. Evidently, $V(r_b)$ has a minimum (for which $\ddot{r}_b = 0$ and $r_b = r_{b0}$, say) provided $\alpha + \beta > 0$. For specified value of the average density $\bar{n}_b$ (and therefore ω_{pb}^2), the equilibrium beam radius r_{b0} is determined by setting $\ddot{r}_b = 0$ in Eq.(9.7). This gives

$$r_{b0}^4 = \frac{(2 P_\theta / \gamma_b m_e)^2}{\left(\omega_{rb}^+ - \omega_{rb}^-\right)^2}, \tag{9.10}$$

where $(\omega_{rb}^+ - \omega_{rb}^-)^2$ is defined by

$$\left(\omega_{rb}^+ - \omega_{rb}^-\right)^2 = \left\{ \left(\frac{\omega_{cb}}{\gamma_b}\right)^2 - \frac{2\omega_{pb}^2}{\gamma_b} \left[(1-f) - \beta_b^2(1-f_M)\right] \right\}, \tag{9.11}$$

which should be compared with Eq.(5.33). The condition $\alpha + \beta > 0$ corresponds to the requirement $(\omega_{rb}^+ - \omega_{rb}^-)^2 > 0$, or equivalently,

$$\left(\frac{\omega_{cb}}{\gamma_b}\right)^2 - \frac{2\omega_{pb}^2}{\gamma_b} \left[(1-f) - \beta_b^2(1-f_M)\right] > 0 \,. \tag{9.12}$$

Note that Eq.(9.12) is automatically satisfied whenever $\beta_b^2(1 - f_M) > (1 - f)$.

Equation (9.7) can be used to investigate oscillations of the beam envelope for moderately large perturbations about the equilibrium radius r_{b0}. For present purposes, however, we express $r_b = r_{b0} + \delta r_b(t)$ and examine Eq.(9.7) for small-amplitude oscillations satisfying $|\delta r_b| \ll r_{b0}$. Expanding Eq.(9.7) about the equilibrium radius r_{b0} defined in Eq.(9.10) gives

$$\frac{d^2 \delta r_b}{dt^2} + \left\{ \left(\frac{\omega_{cb}}{2\gamma_b}\right)^2 + \frac{\omega_{pb}^2}{2\gamma_b} \left[(1-f) - \beta_b^2(1-f_M)\right] + \frac{3\left(P_\theta/\gamma_b m_e\right)^2}{r_{b0}^4} \right\} \delta r_b = 0 \,. \tag{9.13}$$

Eliminating $(P_\theta/\gamma_b m_e)^2/r_{b0}^4$ in Eq.(9.13) by means of Eqs.(9.10) and (9.11) readily gives

$$\frac{d^2}{dt^2}\delta r_b + \omega_0^2 \delta r_b = 0 \,, \tag{9.14}$$

where

$$\omega_0^2 = \left(\frac{\omega_{cb}}{\gamma_b}\right)^2 - \frac{\omega_{pb}^2}{\gamma_b}\left[(1-f) - \beta_b^2(1-f_M)\right]\,. \tag{9.15}$$

Here, ω_0 is the natural oscillation frequency of the beam envelope for small-amplitude perturbations. Note that the inequality in Eq.(9.12) assures $\omega_0^2 > 0$, corresponding to a stable oscillation of the beam envelope. Moreover, if the beam envelope is rippled axially with $d/dt \to \beta_b cd/dz$ in Eq.(9.14), then the axial wave number k_{z0} of the oscillation is given by $k_{z0}c = \omega_0/\beta_b$.

As a final point, it is evident from Eq.(9.12) (and similar expressions in earlier chapters) that the equilibrium self fields greatly affect the radial confinement and propagation of intense charged particle beams. In this regard, when the neutralization of the beam space charge by the background plasma is sufficiently large that $f > 1 - \beta_b^2(1 - f_M)$, then the net self-field force is focusing, and the inequality in Eq.(9.12) is always satisfied. In the absence of background plasma, however, Eq.(9.12) can be expressed as (for $f = 0 = f_M$)

$$\omega_{pb}^2 < \frac{1}{2\gamma_b}\omega_{cb}^2 \,, \tag{9.16}$$

where $\omega_{pb}^2 = 4\pi\bar{n}_b e^2/m_e$ and $\omega_{cb} = eB_0/m_e c$. Equation (9.16) places a limit on the magnitude of the beam current $I = |-e\beta_b c|\pi\bar{n}_b r_b^2$ which can propagate in vacuum, i.e.,

$$I < I_\ell \equiv \frac{\beta_b}{8\gamma_b}\frac{m_e c^2}{e}\frac{\omega_{cb}^2 r_b^2}{c^2} \,. \tag{9.17}$$

As an example, for $B_0 = 10$ kG and $r_b = 1$ cm, Eq.(9.17) gives a limiting electron current of $I_\ell = 73.1\,\beta_b/\gamma_b$ kA. Because the right-hand side of Eq.(9.17) scales inversely as the particle mass, the corresponding limiting current for a nonneutral proton beam (again for $B_0 = 10$ kG and $r_b = 1$ cm) would be $I_\ell = 40\beta_b/\gamma_b$ A. Note that an increase in magnetic field from $B_0 = 10$ kG to $B_0 = 100$ kG would only increase the limiting current for protons to $I_\ell = 4\beta_b/\gamma_b$ kA. For this reason, high-current ion beam propagation is often carried out in a plasma channel which provides partial or complete charge neutralization.

Problem 9.1   Make use of the conservation of canonical angular momentum in Eq.(9.4) to eliminate $rd\theta/dt$ in Eq.(9.3). Approximate $d\gamma/dt = 0$, $v_z = \beta_b c$ and $\gamma = \gamma_b = (1 - \beta_b^2)^{-1/2}$, and show that the orbit equation (9.3) for $r(t)$ reduces directly to Eq.(9.5).

Problem 9.2

a. Show that Eq.(9.7) for the radial envelope $r_b(t)$ of the electron beam can be expressed as

$$\frac{d^2 r_b}{dt^2} = -\frac{\partial}{\partial r_b} V(r_b), \qquad (9.2.1)$$

where

$$V(r_b) = \frac{1}{2}\alpha r_b^2 + \beta r_{b0}^2 \ln r_b + \frac{1}{2}\frac{\delta}{r_b^2} \qquad (9.2.2)$$

is the effective potential. Here, the constants α, β and δ are defined by

$$\alpha = \left(\frac{\omega_{cb}}{2\gamma_b}\right)^2, \qquad (9.2.3)$$

$$\beta = -\frac{\omega_{pb}^2}{2\gamma_b}\left[(1 - f) - \beta_b^2(1 - f_M)\right], \qquad (9.2.4)$$

$$\delta = \left(\frac{P_\theta}{\gamma_b m_e}\right)^2, \qquad (9.2.5)$$

where $\omega_{pb}^2 r_{b0}^2 = 4e^2 N_b/m_e$.

b. Sketch $V(r_b)$ as a function of r_b. For specified value of ω_{pb}^2, show that $V(r_b)$ is a minimum for the value of beam radius defined in Eq.(9.10), which corresponds to $r_{b0}^4 = \delta/(\alpha + \beta)$.

c. Calculate the frequency of small-amplitude oscillations of the beam envelope from

$$\omega_0^2 = \left[\frac{\partial^2 V}{\partial r_b^2}\right]_{r_b = r_{b0}} \qquad (9.2.6)$$

Show that $\omega_0^2 = 4\alpha + 2\beta$, which is identical to Eq.(9.15).

d. The analysis leading to the expressions for r_{b0} and ω_0 in Eqs.(9.10) and (9.15) [and therefore the analysis in Parts (b) and (c), above] treated the average beam density (proportional to ω_{pb}^2) as a specified quantity so that βr_{b0}^2 was proportional to r_{b0}^2. A different solution pertains if the value of $\omega_{pb}^2 r_{b0}^2 = 4e^2 N_b/m_e$ is treated as a specified quantity (independent of r_{b0}, which is to be determined). In this case, show that the effective potential $V(r_b)$ can be expressed as

$$V(r_b) = \frac{1}{2}\alpha r_b^2 + \beta' \ln r_b + \frac{1}{2}\frac{\delta}{r_b^2} , \qquad (9.2.7)$$

where α and δ are defined in Eqs.(9.2.3) and (9.2.5), respectively, and β' is defined by

$$\beta' = -\frac{(4e^2 N_b/m_e)}{2\gamma_b}\left[(1-f) - \beta_b^2(1 - f_M)\right] . \qquad (9.2.8)$$

e. Show from Eq.(9.2.7) that $V(r_b)$ is a minimum for $r_b = r_{b0}$, where

$$r_{b0}^2 = \left(\frac{\beta'^2}{4\alpha^2} + \frac{\delta}{\alpha}\right)^{1/2} - \frac{\beta'}{2\alpha} . \qquad (9.2.9)$$

f. Show that the frequency of small-amplitude oscillations of the beam envelope is given by

$$\omega_0^2 = 4\alpha + 2\beta'/r_{b0}^2 , \qquad (9.2.10)$$

which is identical in form to the result in Part (c) with $\beta \to \beta'/r_{b0}^2$.

9.2.2 Space-Charge-Limiting Current

The simple analysis of beam propagation in Sec. 9.2.1 was carried out in the absence of conducting walls, and the limiting beam current for $f = 0 = f_M$ was found to be $I_\ell = (m_e c^3/e)(\beta_b/8\gamma_b)(\omega_{cb}^2 r_b^2/c^2)$ [see Eq.(9.17)]. In this case, the maximum allowed beam density $(\omega_{pb}^2 = \omega_{cb}^2/2\gamma_b)$ was limited by the strength of the applied magnetic field $B_0\hat{e}_z$. In this section, we consider the propagation of an intense nonneutral electron beam (denoted by $j = b$) injected along the axis of a cylindrically symmetric conducting cavity with radius b and axial length L (Fig. 9.1). No background plasma is present, and (for simplicity) the analysis is carried out in the limit of an infinitely strong axial guide field $(B_0 \to \infty)$, which corresponds to zero rotation velocity

$$V_{\theta b}^0(r, z) = 0 \qquad (9.18)$$

for a beam with slow rotational equilibrium (see, for example, Sec. 5.2.3).

$$\phi_0 = 0 \quad \text{Grounded Cylindrical Cavity} \quad \phi_0 = 0$$
$$z = 0 \qquad \qquad \qquad z = L$$
$$\phi_0(r = b, z) = 0$$

$$B_0 \hat{e}_z$$

Electron Beam

$$\phi_0(r = b, z) = 0$$

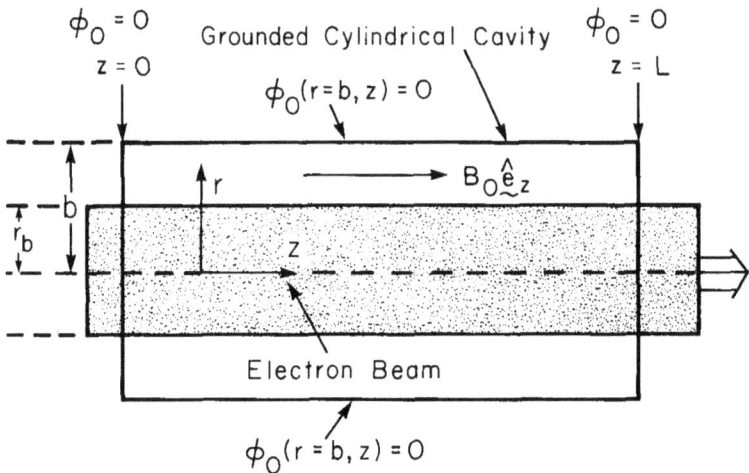

Figure 9.1. A relativistic electron beam propagates in the z-direction through a cylindrical drift cavity with radius b and length L. The walls and ends of the cavity are maintained at zero electrostatic potential ($\phi_0 = 0$).

It is also assumed that the walls of the drift cavity in Fig. 9.1 are maintained at zero electrostatic potential, i.e.,

$$\phi_0(r = b, z) = 0 \,, \quad 0 \le z \le L \,,$$
$$\phi_0(r, z = 0) = 0 = \phi_0(r, z = L) \,, \quad 0 \le r \le b \,. \tag{9.19}$$

The subsequent analysis will show that the amount of beam current that can propagate through the drift cavity is limited by the electrostatic potential depression $\phi_0(r, z)$ associated with the space charge of the non-neutral electron beam.[1,17-19]

For one-dimensional, azimuthally symmetric flow ($B_0 \to \infty, V_{\theta b}^0 = 0 = V_{rb}^0$, and $\partial/\partial\theta = 0$), the steady-state, cold-fluid equations relating the beam density $n_b^0(r, z)$, the axial flow velocity $V_{zb}^0(r, z)$, the electrostatic potential $\phi_0(r, z)$, and the azimuthal self-magnetic field $B_\theta^s(r)$, are given by (Sec. 2.3)

$$\frac{\partial}{\partial z}\left(n_b^0 V_{zb}^0\right) = 0 \,, \tag{9.20}$$

$$\left\{\frac{1}{r}\frac{\partial}{\partial r}r\frac{\partial}{\partial r} + \frac{\partial^2}{\partial z^2}\right\}\phi_0 = 4\pi e n_b^0 \,, \tag{9.21}$$

$$\frac{1}{r}\frac{\partial}{\partial r}r\,B_\theta^s = -\frac{4\pi e}{c}n_b^0 V_{zb}^0\,, \qquad (9.22)$$

within the drift cavity ($0 \le r \le b, 0 \le z \le L$). For present purposes, it is also assumed that the total energy of a beam fluid element is uniform (independent of r and z) throughout the electron beam, i.e.,

$$\left(\gamma_b^0 - 1\right)m_e c^2 - e\phi_0 = \left(\gamma_{b0} - 1\right)m_e c^2 = \text{const.} \qquad (9.23)$$

Here, $\gamma_b^0(r, z) = [1 - V_{zb}^{02}(r, z)/c^2]^{-1/2}$ is the relativistic mass factor associated with the axial flow, and $(\gamma_{b0} - 1)m_e c^2 = \text{const.}$ is the beam kinetic energy as it enters the cavity at $z = 0$ (where $\phi_0 = 0$). From Eq.(9.20), the axial current density

$$-n_b^0(r, z)eV_{zb}^0(r, z) = J_{zb}^0(r) \qquad (9.24)$$

depends only on the radial distance r from the axis of symmetry. Therefore, from Eq.(9.22), the azimuthal self-magnetic field depends only on r and is given by

$$B_\theta^s(r) = \frac{4\pi}{cr}\int_0^r dr'\,r'J_{bz}^0(r') \qquad (9.25)$$

within the drift cavity. We introduce the dimensionless electrostatic potential defined by

$$\Phi(r, z) = \frac{e\phi_0(r, z)}{m_e c^2}\,, \qquad (9.26)$$

and make use of Eqs.(9.23) and (9.24) to express $n_b^0(r, z)$ directly in terms of $J_{zb}^0(r)$ and $\Phi(r, z)$. After some straightforward algebra, Poisson's equation (9.21) can be expressed as

$$\left\{\frac{1}{r}\frac{\partial}{\partial r}r\frac{\partial}{\partial r} + \frac{\partial^2}{\partial z^2}\right\}\Phi(r, z) = -4\pi\frac{e}{m_e c^3}J_{zb}^0(r)\frac{[\gamma_{b0} + \Phi(r, z)]}{\{[\gamma_{b0} + \Phi(r, z)]^2 - 1\}^{1/2}}\,, \qquad (9.27)$$

Equation (9.27) is the final form of Poisson's equation, which is to be solved for $\Phi(r, z)$ within the drift cavity ($0 \le r \le b, 0 \le z \le L$) subject to the boundary conditions in Eq.(9.19). Note that the only variable input parameters in Eq.(9.27) are the current density profile $J_{zb}^0(r)$ and the injection energy $(\gamma_{b0} - 1)m_e c^2$.

Equation (9.27) can be solved numerically for $\Phi(r, z)$ in order to determine the magnitude of the limiting beam current $I_\ell = -2\pi \int_0^b dr'\, r'\, J_{zb}^0(r')$ for specified current density profile $J_{zb}^0(r)$ and specified values of γ_{b0}, b and L. Alternatively, at least in the case where $J_{zb}^0(r)$ is uniform over the radial cross section of the beam, a variational technique using Green's second identity can be developed[17] to place an upper bound on I_ℓ (Problem 9.4). For present purposes, we illustrate the concept of space-charge-limiting current by considering Eq.(9.27) in two limiting regimes where analytical solutions are tractable. These correspond to: (a) a very long drift cavity with $r_b, b \ll L$ and $\partial/\partial z = 0$ (Sec. 9.2.3), and (b) a very-large-radius beam and cavity with $L \ll r_b, b$ and $\partial/\partial r = 0$ (Sec. 9.2.4).

Problem 9.3   Solve the energy conservation equation (9.23) for the axial flow velocity $V_{zb}^0(r)$, and make use of Eq.(9.24) to show that Poisson's equation (9.21) can be expressed in the form given in Eq.(9.27).

Problem 9.4   Consider Green's second identity

$$\int_V d^3x \left(\psi \nabla^2 \Phi - \Phi \nabla^2 \psi \right) = \int_S d^2x \left(\psi \frac{\partial \Phi}{\partial n} - \Phi \frac{\partial \psi}{\partial n} \right). \qquad (9.4.1)$$

Here, $\Phi(r, z) = e\phi_0(r, z)/m_e c^2$ solves the nonlinear Poisson equation (9.27); S represents the bounding surface of the volume V of the cylindrical drift cavity in Fig. 9.1; and $\partial/\partial n$ represents the derivative normal to the surface S. In Eq.(9.4.1), take $\psi(r, z)$ to solve the eigenvalue equation

$$\nabla^2 \psi = \lambda \psi, \qquad (9.4.2)$$

where $\psi(r = b, z) = 0$ for $0 \leq z \leq L$, and $\psi(r, z = 0) = 0 = \psi(r, z = L)$ for $0 \leq r \leq b$. Because $\psi = 0$ and $\Phi = 0$ on the surface S [see Eq.(9.19)], the right-hand side of Eq.(9.4.1) vanishes, and Eq.(9.4.1) reduces to

$$\int_V d^3x\, \psi \left(\nabla^2 \Phi - \lambda \Phi \right) = 0. \qquad (9.4.3)$$

Equation (9.4.3) can be used to place an upper bound on the space-charge-limiting current flowing in the drift cavity in Fig. 9.1 [V.S. Voronin, Yu. T. Zozulya and A.N. Lebedev, Sov. Phys.-Tech. Phys. **17**, 432 (1972)].

a. Show that the eigenfunction $\psi(r, z)$ solving Eq.(9.4.2) is given by

$$\psi(r, z) = A_{n,m} J_0(\kappa_n r) \sin\left(\frac{m\pi z}{L}\right) \qquad (9.4.4)$$

for $0 \leq r \leq b$ and $0 \leq z \leq L$. Here, n and m are integers; $\kappa_n b$ is the n'th zero of $J_0(\kappa_n b) = 0$; and the eigenvalue is $\lambda_{n,m} = -[\kappa_n^2 + (m\pi/L)^2]$.

b. Consider the case where the axial current density occurring in Eq.(9.27) is uniform over the interval $0 \leq r \leq b$, with $J_{zb}^0(r) = -J = \text{const.}$ and $J > 0$. For the lowest nontrivial eigenvalue ($n = 1$ and $m = 1$), show that Eq.(9.4.3) reduces to

$$0 = \int_0^b dr\, r \int_0^L \frac{1}{L} J_0(\kappa_1 r) \sin\left(\frac{\pi z}{L}\right) [F(\Phi) - \lambda_{1,1}\Phi] , \qquad (9.4.5)$$

where $A_{1,1} \neq 0$ has been assumed. Here, $\lambda_{1,1} = -[\kappa_1^2 + (\pi/L)^2]$, and $F(\Phi)$ is defined by

$$F(\Phi) = \frac{4\pi e J}{m_e c^3} \frac{(\gamma_{b0} + \Phi)}{\{(\gamma_{b0} + \Phi)^2 - 1\}^{1/2}} , \qquad (9.4.6)$$

where $(\gamma_{b0} - 1)m_e c^2$ is the injection energy at $z = 0$.

c. In Eq.(9.4.5), the factor $J_0(\kappa_1 r) \sin(\pi z/L)$ does not change sign within the volume V. Therefore, the condition for existence of a solution to Eq.(9.4.5) is that the function

$$G(\Phi) = F(\Phi) - \lambda_{1,1}\Phi \qquad (9.4.7)$$

changes sign within the volume V. Note that $G(\Phi = 0) = F(\Phi = 0) > 0$ on the boundary of the drift cavity. Show that the requirement $G(\Phi) < 0$ can be expressed as

$$J < \frac{1}{4\pi} \frac{m_e c^3}{e} [\kappa_1^2 + (\pi/L)^2] \frac{(-\Phi)\{(\gamma_{b0} + \Phi)^2 - 1\}^{1/2}}{(\gamma_{b0} + \Phi)} . \qquad (9.4.8)$$

Here, $\Phi < 0$ within the cavity because the beam space charge depresses the electrostatic potential relative to its value ($\Phi = 0$) on the boundary.

d. Show that the right-hand side of Eq.(9.4.8) is a maximum for $\Phi = \gamma_{b0}^{1/3} - \gamma_{b0}$. Therefore, show that the magnitude of the beam current $I = \pi b^2 J$ is bounded from above by the limiting value

$$I_\ell = \frac{1}{4} \frac{m_e c^3}{e} \left[\kappa_1^2 b^2 + \left(\frac{\pi b}{L}\right)^2\right] \left(\gamma_{b0}^{2/3} - 1\right)^{3/2} . \qquad (9.4.9)$$

9.2.3   Limiting Current in an Infinitely Long Drift Space

For a very long drift cavity with $L \gg r_b, b$, Poisson's equation (9.27) is solved (approximately) by setting $\partial/\partial z = 0$ and neglecting end effects at $z = 0$ and $z = L$. As a simple example, we consider a thin, annular electron beam with current density profile[1]

$$J_{zb}^0(r) = -\frac{I}{2\pi r}\delta(r - r_b) \ . \tag{9.28}$$

Here, $I = -2\pi \int_0^b dr'\, r' J_{zb}^0(r')$ is the magnitude of the axial electron current. Denoting the normalized electrostatic potential at $r = r_b$ by $\Phi_b \equiv \Phi(r = r_b) = e\phi_0(r = r_b)/m_e c^2$, Poisson's equation (9.27) reduces to

$$\frac{1}{r}\frac{\partial}{\partial r} r \frac{\partial}{\partial r}\Phi(r) = 2\frac{e}{m_e c^3}\frac{I}{r}\frac{(\gamma_{b0} + \Phi_b)}{[(\gamma_{b0} + \Phi_b)^2 - 1]^{1/2}}\delta(r - r_b) \tag{9.29}$$

for $0 \leq r \leq b$. The solution to Eq.(9.29) which satisfies $\Phi(r = b) = 0$, is continuous at $r = r_b$, and remains finite at the origin is given by

$$\Phi(r) = \frac{\Phi_b}{\ln(b/r_b)} \times \begin{cases} \ln(b/r_b), & 0 \leq r \leq r_b, \\ \ln(b/r), & r_b < r \leq b. \end{cases} \tag{9.30}$$

Furthermore, integrating Eq.(9.29) across the annular electron beam at $r = r_b$ gives the boundary condition

$$\lim_{\varepsilon \to 0_+}\left[\frac{\partial \Phi}{\partial r}\right]_{r=r_b(1+\varepsilon)} = 2\frac{e}{m_e c^3}\frac{I}{r_b}\frac{(\gamma_{b0} + \Phi_b)}{[(\gamma_{b0} + \Phi_b)^2 - 1]^{1/2}} \ . \tag{9.31}$$

Substituting Eq.(9.30) into Eq.(9.31) and solving for the beam current I, we obtain (Problem 9.5)

$$I = \frac{(m_e c^3/e)}{2\ln(b/r_b)}G(\Phi_b) \ , \tag{9.32}$$

where $G(\Phi_b)$ is defined by

$$G(\Phi_b) = \frac{-\Phi_b}{(\gamma_{b0} + \Phi_b)}\left[(\gamma_{b0} + \Phi_b)^2 - 1\right]^{1/2} \ . \tag{9.33}$$

Here, it should be kept in mind that $\Phi_b < 0$ because the space charge of the electron beam depresses the electrostatic potential relative to its value at the wall ($\Phi = 0$ at $r = b$).

Denoting $-\Phi_b = |\Phi_b|$, the function $G(\Phi_b)$ is plotted versus $|\Phi_b|$ in Fig. 9.2. Note that $G(\Phi_b) = 0$ for $|\Phi_b| = 0$ and $|\Phi_b| = \gamma_{b0} - 1$. Moreover, $G(\Phi_b)$ passes through a maximum value

$$G_{\max} = \left(\gamma_{b0}^{2/3} - 1\right)^{3/2} \tag{9.34}$$

for $|\Phi_b| = (\gamma_{b0} - \gamma_{b0}^{1/3})$. From Eqs.(9.32) and (9.34), the corresponding space-charge-limiting current I_ℓ is given by[1]

$$I_\ell = \frac{(m_e c^3/e)}{2 \ln(b/r_b)} \left(\gamma_{b0}^{2/3} - 1\right)^{3/2}. \tag{9.35}$$

From Eq.(9.32) and Fig. 9.2, it is evident that two solutions for Φ_b are possible for specified value of beam current $I < I_\ell$. The solution with $0 < |\Phi_b| < \gamma_{b0} - \gamma_{b0}^{1/3}$ corresponds to an electron beam with low density and high axial velocity, whereas the solution with $\gamma_{b0} - \gamma_{b0}^{1/3} < |\Phi_b| < \gamma_{b0} - 1$ corresponds to an electron beam with high density and low axial velocity. It also follows from Eq.(9.32) and Fig. 9.2 that no solutions to Eq.(9.32) exist for beam currents $I > I_\ell$. That is, the steady flow of current in

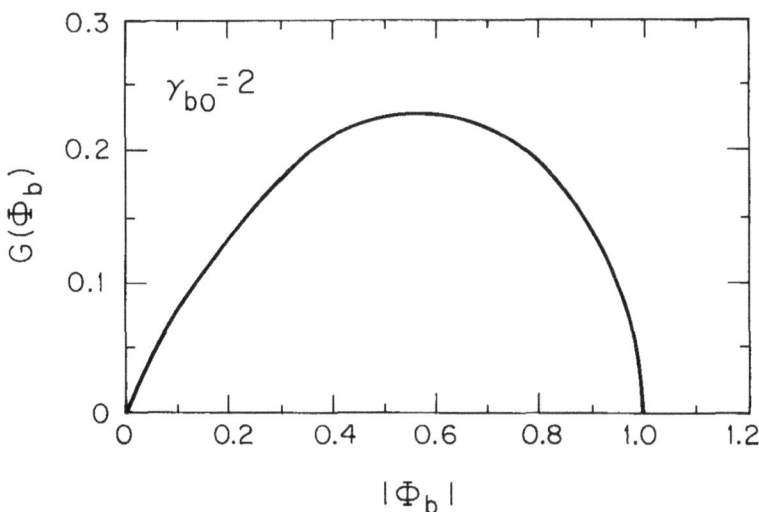

Figure 9.2.   Plot of $G(\Phi_b)$ versus $|\Phi_b| = -\Phi_b$ obtained from Eq.(9.33) for $\gamma_{b0} = 2$. Note from Eq.(9.32) that the magnitude of the beam current (I) is proportional to $G(\Phi_b)$.

excess of the space-charge limit defined in Eq.(9.35) is not possible, at least within the context of the simple cold-fluid model considered here.

Analysis of Poisson's equation (9.27) is also possible for other choices of current density profile $J_{zb}^0(r)$ in the limit of an infinitely long drift space.[1] For example, if the current density is uniform over the interval $0 \leq r < r_b$, then $J_{zb}^0(r)$ can be expressed as

$$J_{zb}^0(r) = \begin{cases} -\dfrac{I}{\pi r_b^2} = \text{const.}, & 0 \leq r < r_b, \\ 0, & r_b < r \leq b, \end{cases} \qquad (9.36)$$

where $I = -2\pi \int_0^b dr' \, r' J_{zb}^0(r')$ is the magnitude of the beam current. For $\partial/\partial z = 0$, Poisson's equation (9.27) then reduces to

$$\frac{1}{r}\frac{\partial}{\partial r} r \frac{\partial}{\partial r} \Phi(r) = \begin{cases} 4\dfrac{e}{m_e c^3}\dfrac{I}{r_b^2} \dfrac{[\gamma_{b0} + \Phi(r)]}{\{[\gamma_{b0} + \Phi(r)]^2 - 1\}^{1/2}}, & 0 \leq r < r_b, \\ 0, & r_b < r \leq b, \end{cases} \qquad (9.37)$$

which is to be solved subject to the boundary condition $\Phi(r = b) = 0$.

Equation (9.37) can be solved numerically for $\Phi(r)$, or alternatively, an iterative approach can be used to determine the space-charge-limiting current (e.g., if r_b/b is small). For example, without presenting algebraic details,[1] it can be shown for $r_b \ll b$ that the limiting beam current I_ℓ is given approximately by Eq.(9.35), the result obtained earlier in Sec. 9.2.3 for a thin, annular electron beam. This is the expected result for $r_b \ll b$ because the variation of $\Phi(r)$ over the interval $0 \leq r < r_b$ is relatively small in comparison with the variation of $\Phi(r)$ over the interval $r_b < r \leq b$.

Problem 9.5   Consider a thin, annular electron beam with current density profile specified by Eq.(9.28).

a. Solve Poisson's equation (9.29) in the two regions $0 \leq r < r_b$ and $r_b < r \leq b$. Show that solution for $\Phi(r) = e\phi_0(r)/m_e c^2$ can be expressed in the form given in Eq.(9.30), where $\Phi_b \equiv \Phi(r = r_b)$ and $\Phi(r = b) = 0$.

b. Integrate Poisson's equation (9.29) across the annular beam at $r = r_b$. Show that the radial electric field at the outer surface of the electron beam is determined from Eq.(9.31).

c. Show that Eqs.(9.30) and (9.31) can be combined to give the expression for the beam current in Eq.(9.32).

9.2.4   Limiting Current in a One-Dimensional Drift Space

We now consider Poisson's equation (9.27) in circumstances where $r_b, b \gg L$ and radial variations can be neglected (Fig. 9.3). Setting $\partial/\partial r = 0$ and $J_{zb}^0(r) = J_b = \text{const.}$, Eq.(9.27) can be expressed as

$$\frac{\partial^2}{\partial z^2}\Phi(z) = -4\pi\frac{e}{m_e c^3} J_b \frac{[\gamma_{b0} + \Phi(z)]}{\{[\gamma_{b0} + \Phi(z)]^2 - 1\}^{1/2}} , \tag{9.38}$$

where $\Phi(z) = e\phi_0(z)/m_e c^2$. Equation (9.38) is to be solved subject to the boundary conditions

$$\Phi(z = 0) = 0 = \Phi(z = L) . \tag{9.39}$$

Here, $(\gamma_{b0} - 1)m_e c^2 = \text{const.}$ is the injected kinetic energy at $z = 0$. Paralleling the analysis in Sec. 8.2.2, Poisson's equation (9.38) can be solved exactly in terms of elliptic integrals.[1] For present purposes, however, it is more instructive to consider the simpler case of nonrelativistic electron flow with

$$(\gamma_{b0} - 1)m_e c^2 \ll m_e c^2 \quad \text{and} \quad |\Phi| \ll 1 .$$

In this case, we introduce the simplified notation $\beta_z(z) \equiv V_{zb}^0(z)/c$, where $\beta_z^2 \ll 1$ and $\beta_{z0}^2 \equiv 2(\gamma_{b0} - 1) \ll 1$. Solving Eq.(9.23) for $\beta_z(z)$ gives

$$\beta_z(z) = \left[\beta_{z0}^2 + 2\Phi(z)\right]^{1/2} , \tag{9.40}$$

where $\beta_z(z) > 0$ is assumed (Fig. 9.3). Furthermore, Poisson's equation (9.38) becomes

$$\frac{\partial^2}{\partial z^2}\Phi(z) = -4\pi\frac{e}{m_e c^3} \frac{J_b}{[\beta_{z0}^2 + 2\Phi(z)]^{1/2}} \tag{9.41}$$

in the nonrelativistic regime.

Equation (9.41) can be solved exactly for $\Phi(z)$ in the interval $0 \leq z \leq L$ subject to the boundary conditions in Eq.(9.39) (Problem 9.6). The beam space charge depresses the electrostatic potential (relative to $\Phi = 0$ at $z = 0$ and $z = L$) to some minimum value $\Phi_m \equiv \Phi(z = L/2)$ at the

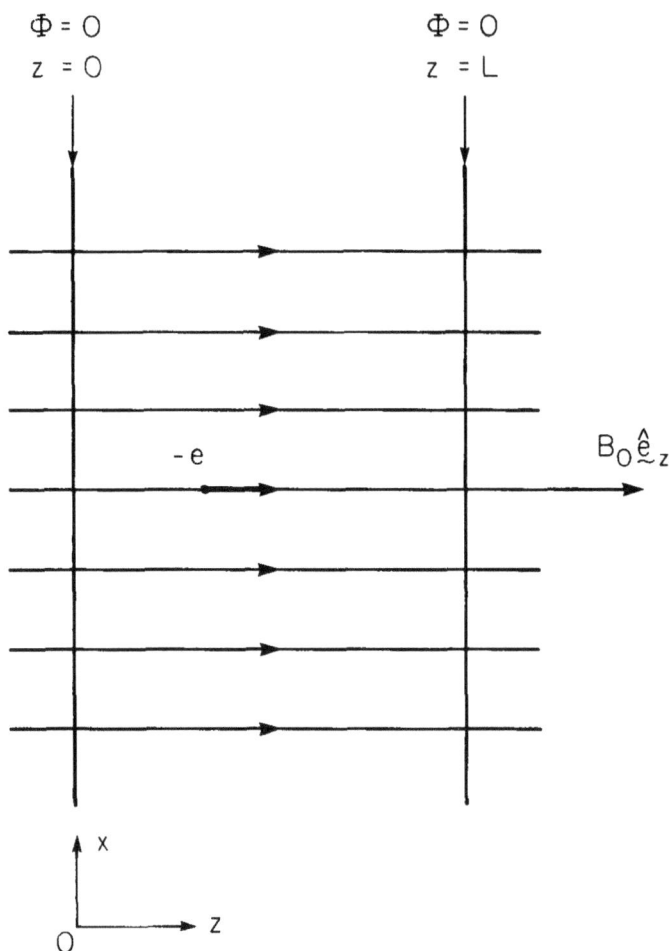

Figure 9.3. A nonrelativistic electron beam propagates in the z-direction through the one-dimensional drift space $0 \leq z \leq L$. All radial variations are neglected $(\partial/\partial r = 0)$, and the planes at $z = 0$ and $Z = L$ are maintained at zero electrostatic potential $(\Phi = 0)$.

midpoint $z = L/2$ where $[\partial\Phi/\partial z]_{z=L/2} = 0$. Integration of Eq.(9.41) gives for $\Phi(z)$

$$\left\{ \left[\beta_{z0}^2 + 2\Phi(z) \right]^{1/2} - \left[\beta_{z0}^2 + 2\Phi_m \right]^{1/2} \right\} \tag{9.42}$$

$$\times \left\{ \left[\beta_{z0}^2 + 2\Phi(z) \right]^{1/2} + 2 \left[\beta_{z0}^2 + 2\Phi_m \right]^{1/2} \right\}^2 = \left(-\frac{18\pi e}{m_e c^3} J_b \right) (z - L/2)^2 \, .$$

Alternatively, in terms of $\beta_z(z) = [\beta_{z0}^2 + 2\Phi(z)]^{1/2}$ and $\beta_{zm} = [\beta_{z0}^2 + 2\Phi_m]^{1/2}$, Eq.(9.42) can be expressed in the equivalent form

$$[\beta_z(z) - \beta_{zm}][\beta_z(z) + 2\beta_{zm}]^2 = \left(-\frac{18\pi e}{m_e c^3} J_b\right)(z - L/2)^2. \tag{9.43}$$

Note from Eqs.(9.39) and (9.43) that the normalized beam velocity $\beta_z(z)$ decreases from the injection value β_{z0} at $z = 0$ to the minimum value $\beta_{zm} = (\beta_{z0}^2 + 2\Phi_m)^{1/2}$ at $z = L/2$, and then $\beta_z(z)$ increases again to the value β_{z0} at $z = L$.

Denoting $J = -J_b > 0$, we evaluate Eq.(9.43) [or Eq.(9.42)] at $z = 0$ or $z = L$ and solve for the current-density magnitude J. This gives

$$J = \frac{2}{9\pi L^2} \frac{m_e c^3}{e} F(\beta_{zm}), \tag{9.44}$$

where $F(\beta_{zm})$ is defined by

$$F(\beta_{zm}) = (\beta_{z0} - \beta_{zm})(\beta_{z0} + \beta_{zm})^2. \tag{9.45}$$

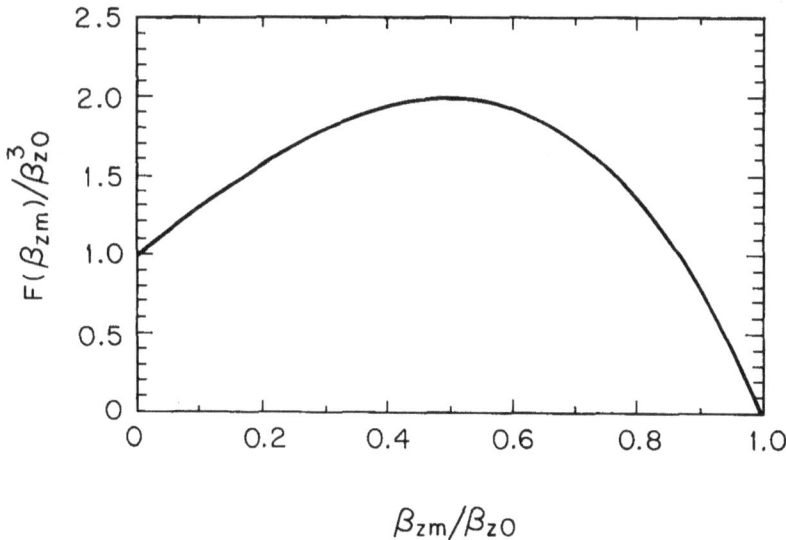

Figure 9.4. Plot of $F(\beta_{zm})/\beta_{z0}^3$ versus β_{zm}/β_{z0} obtained from Eq.(9.45). Note from Eq.(9.44) that the magnitude of the beam current density (J) is proportional to $F(\beta_{zm})$.

The function $F(\beta_{zm})$ is plotted versus β_{zm} in Fig. 9.4. Note that $F(\beta_{zm})$ achieves a maximum value of $2\beta_{z0}^3$ for $\beta_{zm} = \beta_{z0}/2$, and the corresponding space-charge-limiting current density J_ℓ is given by

$$J_\ell = \frac{4}{9\pi L^2} \frac{m_e c^3}{e} \beta_{z0}^3 . \tag{9.46}$$

From Eq.(9.44) and Fig. 9.4, for specified current in the interval $0 < J < J/2$, there is one allowed value of β_{zm} corresponding to high-velocity, low-density flow. On the other hand, for J in the interval $J_\ell/2 < J < J_\ell$, there are two allowed values of β_{zm}. The solution with $0 < \beta_{zm} < \beta_{z0}/2$ corresponds to low-velocity, high-density flow (the so-called "C-overlap" region[21,22]), whereas the solution with $\beta_{zm} > \beta_{z0}/2$ corresponds to high-velocity, low-density flow. Steady-state solutions do not exist for $J > J_\ell$.

The electrostatic stability properties of nonrelativistic electron flow in a one-dimensional drift space will be investigated in Sec. 9.3.

Problem 9.6   Consider Poisson's equation (9.41) which describes nonrelativistic electron flow in the one-dimensional drift space illustrated in Fig. 9.3.

a. Introduce the function $\psi(z) = \beta_{z0}^2 + 2\Phi(z)$ and integrate Eq.(9.41) once from $z = 0$ to $z = L/2$, where $[\partial\psi/\partial z]_{z=L/2} = 0$. Show that

$$\frac{1}{2}\left[\frac{\partial\psi}{\partial z}\right]^2 = -16\pi\frac{e}{m_e c^3}J_b\left(\psi^{1/2} - \psi_m^{1/2}\right) \tag{9.6.1}$$

where $\psi_m = \psi(z = L/2) = \beta_{z0}^2 + 2\Phi_m$.

b. Change variables to $Y^2 = \psi$ and integrate Eq.(9.6.1). Show that the solution can be expressed in the form given in Eq.(9.42).

Problem 9.7   To analyze the one-dimensional flow in Fig. 9.3, it is often convenient to introduce the Lagrangian time variable defined by

$$c\tau(z) = \int_0^z \frac{dz'}{\beta_z(z')} . \tag{9.7.1}$$

Note from Eq.(9.7.1) that $dz/d\tau = \beta_z c$ and $c\beta_z d\beta_z/dz = d\beta_z/d\tau$.

a. Take the derivative of Eq.(9.43) with respect to τ and show that

$$\frac{d\beta_z}{d\tau} = \pm\frac{2}{3}\left(-\frac{18\pi e J_b}{m_e c^3}\right)^{1/2} c\left(\beta_z - \beta_{zm}\right)^{1/2} . \tag{9.7.2}$$

The upper (lower) sign in Eq.(9.7.2) corresponds to $L/2 < z < L (0 < z < L/2)$ where $d\beta_z/d\tau > 0$ $(d\beta_z/d\tau < 0)$.

b. Denote the time for a beam fluid element to transit from $z = 0$ to $z = L$ by t_0, where

$$ct_0 = \int_0^L \frac{dz'}{\beta_z(z')} \, . \tag{9.7.3}$$

By symmetry, the time for a fluid element to transit from $z = 0$ to $z = L/2$ (or from $z = L/2$ to $z = L$) is equal to $t_0/2$. Integrate Eq.(9.7.2) from $\tau = 0$ to $\tau = t_0/2$ and show that

$$(\beta_z - \beta_{zm})^{1/2} = \frac{1}{3} \left(-\frac{18\pi e J_b}{m_e c^3} \right)^{1/2} c \, (t_0/2 - \tau) \, , \tag{9.7.4}$$

where $\beta_z = \beta_{zm}$ at $\tau = t_0/2$.

c. Make use of $\beta_z c = dz/d\tau$ and integrate Eq.(9.7.4). Show that $z(\tau)$ can be expressed as

$$z = \beta_{zm} c\tau - \left(-\frac{2\pi e J_b}{3m_e} \right) \left[(t_0/2 - \tau)^3 - (t_0/2)^3 \right] \tag{9.7.5}$$

for $0 \le \tau \le t_0/2$ $(0 \le z \le L/2)$.

d. Evaluate the expression for β_z in Eq.(9.7.4) at $\tau = 0$ $(z = 0)$ where $\beta_z = \beta_{z0}$. Show that

$$c\beta_{z0} = c\beta_{zm} + \left(-\frac{2\pi e J_b}{m_e} \right) \left(\frac{t_0}{2} \right)^2 \, . \tag{9.7.6}$$

Equation (9.7.6) can be used in Eq.(9.7.5) to eliminate β_{zm} in favor of β_{z0}.

e. Evaluate Eq.(9.7.5) at $z = L/2$ and $\tau = t_0/2$. Show that the transit time t_0 is related to L, β_{z0} and J_b by

$$L = \beta_{z0} c t_0 - \left(-\frac{\pi e J_b}{3m_e} \right) t_0^3 \, . \tag{9.7.7}$$

f. Obtain a closed expression for the transit time t_0 when $-J_b = J_\ell = (4/9\pi L^2)(m_e c^3/e)\beta_{z0}^3$, which corresponds to the space-charge-limiting current in Eq.(9.46). (Answer: $\beta_{z0} c t_0 = 3L/2$.)

9.2.5   Laminar Flow Equilibria in an Infinitely Long Drift Space

In Chapters 4 and 5, we examined a wide range of self-consistent Vlasov equilibria and macroscopic fluid equilibria for intense nonneutral electron beams. These equilibria included: infinitely long electron layers (Sec. 4.2.4); mirror-confined electron layers (Sec. 4.3.2); high-current electron beams in linear geometry (Sec. 4.4); relativistic beam equilibria in a modified betatron configuration (Sec. 4.10); rigid-rotor beam equilibria with strong diamagnetic field (Sec. 5.2.2); rotating beam equilibria with

$\nu/\gamma_b \ll 1$ propagating through a cold background plasma (Sec. 5.2.3); and the classical Bennett pinch equilibrium including the effects of finite beam temperature (Sec. 5.2.4). In this section, we return to the macroscopic cold-fluid model developed in Sec. 5.1, and investigate the equilibrium properties of an intense nonneutral electron beam propagating axially in an infinitely long drift space with $\partial/\partial\theta = 0 = \partial/\partial z$ (Fig. 5.1). Particular emphasis is placed on applying the macroscopic constraint conditions corresponding to the conservation of energy and canonical angular momentum in order to specify the beam equilibria consistent with particular source configurations used to produce the electron beam.[20]

The relativistic electron beam $(j = b)$ propagates parallel to the axial magnetic field $B_z(r)\hat{\mathbf{e}}_z = [B_0 + B_z^s(r)]\hat{\mathbf{e}}_z$, where the self-magnetic field $B_z^s(r)$ is induced by the beam rotation. No plasma background is present $[n_j^0(r) = 0, j = e, i]$. For present purposes, we introduce the simplified notation

$$\beta_\theta(r) = V_{\theta b}^0(r)/c\,, \quad \beta_z(r) = V_{zb}^0(r)/c\,, \tag{9.47}$$

and

$$\gamma(r) = \left[1 - \beta_\theta^2(r) - \beta_z^2(r)\right]^{-1/2}$$
$$= \left[1 + \gamma^2(r)\beta_\theta^2(r) + \gamma^2(r)\beta_z^2(r)\right]^{1/2}\,, \tag{9.48}$$

where r is the radial distance from the axis of symmetry in Fig. 5.1, $\beta_\theta(r)$ and $\beta_z(r)$ are the components of flow velocity normalized to the speed of light c, and $\gamma(r)$ is the relativistic mass factor of a fluid element. Maxwell's equations (5.4) and the radial force balance equation (5.9) can be expressed as

$$\frac{1}{r}\frac{\partial}{\partial r}r\frac{\partial}{\partial r}\phi_0(r) = -\frac{1}{r}\frac{\partial}{\partial r}rE_r(r) = 4\pi en_b^0(r)\,, \tag{9.49}$$

$$\frac{1}{r}\frac{\partial}{\partial r}r\frac{\partial}{\partial r}A_z^s(r) = -\frac{1}{r}\frac{\partial}{\partial r}rB_\theta^s(r) = 4\pi en_b^0(r)\beta_z(r)\,, \tag{9.50}$$

$$\frac{\partial}{\partial r}\frac{1}{r}\frac{\partial}{\partial r}rA_\theta^0(r) = \frac{\partial}{\partial r}B_z(r) = 4\pi en_b^0(r)\beta_\theta(r)\,, \tag{9.51}$$

$$-\frac{\gamma(r)\beta_\theta^2(r)}{r} = -\frac{e}{m_ec^2}\left[E_r(r) + \beta_\theta(r)B_z(r) - \beta_z(r)B_\theta^s(r)\right]\,, \tag{9.52}$$

where $-e$ is the electron charge, and m_e is the electron rest mass. Here, $E_r(r) = -(\partial/\partial r)\phi_0(r)$ is the radial electric field produced by beam space-charge effects, $B_\theta^s(r) = -(\partial/\partial r)A_z^s(r)$ is the azimuthal self-magnetic field, $B_z(r) = B_0 + B_z^s(r) = r^{-1}(\partial/\partial r)[rA_\theta^0(r)]$ is the total axial magnetic field, and $n_b^0(r)$ is the equilibrium density profile of the beam electrons.

As indicated in Chapter 5, there is a certain degree of arbitrariness in a macroscopic, cold-fluid model based on Eqs.(9.49)–(9.52). The radial dependence of any two of the six quantities $E_r(r)$, $B_\theta^s(r)$, $B_z(r)$, $n_b^0(r)$, $\beta_\theta(r)$ and $\beta_z(r)$ can be specified independently, and the remaining four equilibrium profiles determined self-consistently from Eqs.(9.49)–(9.52). Alternatively, paralleling the analysis of Reiser,[20] we can impose the two constraint conditions on Eqs.(9.49)–(9.52) corresponding to the conservation of energy and canonical angular momentum.

Conservation of energy of a fluid element corresponds to the assumption that the kinetic energy plus the electrostatic potential energy is uniform (independent of r) over the cross section of the electron beam, i.e.,

$$[\gamma(r) - 1]\,m_e c^2 - e\phi_0(r) = (\gamma_{b0} - 1)\,m_e c^2 = \text{const.} \tag{9.53}$$

Here, the constant $(\gamma_{b0} - 1)m_e c^2$ is related to the applied diode voltage used to produce the electron beam. Conservation of the canonical angular momentum of a fluid element can be expressed as

$$r\left[\gamma(r)m_e c\beta_\theta(r) - \frac{e}{c}A_\theta^0(r)\right] = P_{\theta b}^0 = \text{const.}, \tag{9.54}$$

where $A_\theta^0(r) = r^{-1}\int_0^r dr'\,r'B_z(r')$ follows from Eq.(9.51).

One source condition of considerable importance corresponds to a shielded cathode in which the cathode is shielded from the axial magnetic field $B_z(r)$ in the downstream region.[20] In this case, the constant $P_{\theta b}^0 = 0$ in Eq.(9.54) because the axial magnetic field at the cathode is equal to zero. For $P_{\theta b}^0 = 0$, Eq.(9.54) reduces to

$$\gamma(r)\beta_\theta(r) = \frac{e}{m_e c^2}A_\theta^0(r)\,, \tag{9.55}$$

and the beam equilibrium can extend from $r = 0$ to some outer radius r_b.

A second important source condition corresponds to an immersed annular cathode which produces a thin, hollow electron beam.[20] In this case, all electrons have the same (constant) value of canonical angular momentum $P_{\theta b}^0 = \text{const.} \neq 0$, and Eq.(9.54) reduces to

$$\gamma(r)\beta_\theta(r) = \frac{e}{m_e c^2} A_\theta^0(r) + \frac{P_{\theta b}^0}{m_e c\, r} \, . \tag{9.56}$$

Note from Eq.(9.56) that motion through $r = 0$ is excluded for $P_{\theta b}^0 \neq 0$, so that the (hollow) electron beam has both an inner radius r_b^- and an outer radius r_b^+ (see also Sec. 4.2.4).

Miller[1] and Reiser[20] have also examined a third type of source condition which corresponds to an immersed cathode in which the axial magnetic field is uniform across the cathode. In this case, $P_{\theta b}^0(r) = \text{const.} \times r^2$ varies across the radial extent of the electron beam, and it is assumed that the radial location r of electrons in the downstream region is related to the emission radius r_c by $r = \alpha r_c$, where the constant α depends on the source geometry.

In the subsequent analysis, we consider only the source conditions corresponding to Eq.(9.55) ($P_{\theta b}^0 = 0$) and Eq.(9.56) ($P_{\theta b}^0 = \text{const.} \neq 0$). In this regard, it is readily verified that the radial force balance equation (9.52) and the two conservation constraints in Eqs.(9.53) and (9.54) necessarily imply that the axial canonical momentum of a fluid element is conserved. That is,

$$\gamma(r)m_e c\beta_z(r) - \frac{e}{c}A_z^s(r) = P_{zb}^0 = \text{const.} \tag{9.57}$$

follows from Eqs.(9.52)–(9.54), where the constant P_{zb}^0 is independent of r (Problem 9.8). Indeed, it can be shown that any one of Eqs.(9.52), (9.53), (9.54) or (9.57) follows from the other three equations. This leads to important simplifications in the equilibrium analysis. For example, in making use of Eqs.(9.49)–(9.54) to investigate equilibrium properties of the electron beam, the radial force balance equation (9.52) can be replaced by the conservation relation (9.57), and the information content in the model remains the same.

It is convenient to introduce the dimensionless potentials $\Phi(r)$, $\chi_\theta(r)$ and $\chi_z(r)$ defined by

$$\Phi(r) = \frac{e\phi_0(r)}{m_e c^2}\,, \quad \chi_\theta(r) = \frac{eA_\theta^0(r)}{m_e c^2}\,, \quad \chi_z(r) = \frac{eA_z^s(r)}{m_e c^2}\, . \tag{9.58}$$

The conservation relations (9.53), (9.54) and (9.57) then become

$$\gamma(r) = \Phi(r) + \gamma_{b0} \, ,$$

$$\gamma(r)\beta_\theta(r) = \chi_\theta(r) + \frac{P_{\theta b}^0}{m_e cr} \, , \tag{9.59}$$

$$\gamma(r)\beta_z(r) = \chi_z(r) + \frac{P_{zb}^0}{m_e c} \, .$$

Making use of Poisson's equation (9.49) to eliminate the density profile $n_b^0(r)$ from Eqs.(9.50) and (9.51) then gives

$$\frac{1}{r}\frac{\partial}{\partial r}r\frac{\partial}{\partial r}\chi_z(r) = \beta_z(r)\frac{1}{r}\frac{\partial}{\partial r}r\frac{\partial}{\partial r}\gamma(r) \, , \tag{9.60}$$

and

$$\frac{\partial}{\partial r}\frac{1}{r}\frac{\partial}{\partial r}r\chi_\theta(r) = \beta_\theta(r)\frac{1}{r}\frac{\partial}{\partial r}r\frac{\partial}{\partial r}\gamma(r) \, , \tag{9.61}$$

where use has been made of $\partial\Phi/\partial r = \partial\gamma/\partial r$ [Eq.(9.59)]. Making use of Eq.(9.59) to eliminate $\chi_z(r)$ and $\chi_\theta(r)$ in Eqs.(9.60) and (9.61), we obtain

$$\frac{1}{r}\frac{\partial}{\partial r}r\frac{\partial}{\partial r}\left[\gamma(r)\beta_z(r)\right] = \beta_z(r)\frac{1}{r}\frac{\partial}{\partial r}r\frac{\partial}{\partial r}\gamma(r) \, , \tag{9.62}$$

and

$$\frac{\partial}{\partial r}\frac{1}{r}\frac{\partial}{\partial r}\left[r\gamma(r)\beta_\theta(r)\right] = \beta_\theta(r)\frac{1}{r}\frac{\partial}{\partial r}r\frac{\partial}{\partial r}\gamma(r) \, , \tag{9.63}$$

where

$$\gamma = (1 - \beta_\theta^2 - \beta_z^2)^{-1/2} = (1 + \gamma^2\beta_\theta^2 + \gamma^2\beta_z^2)^{1/2} \, .$$

Equations (9.62) and (9.63) are coupled differential equations for the flow velocity profiles $\beta_z(r)$ and $\beta_\theta(r)$ which can be used to investigate detailed equilibrium properties in circumstances where the flow is constrained by the conservation of energy [Eq.(9.53)] and the conservation of canonical angular momentum [Eq.(9.54)]. In the remainder of this section, we consider a shielded source ($P_{\theta b}^0 = 0$) so that the beam equilibrium extends from $r = 0$ to some outer radius $r = r_b$. Multiplying Eq.(9.62) by $r\gamma(r)$ and rearranging terms readily gives

$$\frac{\partial}{\partial r}\left[r\gamma^2(r)\frac{\partial}{\partial r}\beta_z(r)\right] = 0 , \tag{9.64}$$

which can be integrated to give $r\gamma^2\partial\beta_z/\partial r = $ const. Requiring that $\beta_z(r)$ be finite at $r = 0$ necessarily implies

$$\beta_z(r) = \beta_{z0} = \text{const.} \tag{9.65}$$

That is, the axial velocity profile is uniform over the entire cross section of the electron beam. From $\gamma = (1 - \beta_\theta^2 - \beta_{z0}^2)^{-1/2}$, we obtain

$$\gamma(r) = \frac{\gamma_{z0}}{[1 - \gamma_{z0}^2\beta_\theta^2(r)]^{1/2}} , \tag{9.66}$$

where $\gamma_{z0} \equiv (1 - \beta_{z0}^2)^{-1/2}$. Note that $\gamma(r = 0) = \gamma_{z0}$ follows from $\beta_\theta(r = 0) = 0$.

Equation (9.63) can be used to determine the azimuthal velocity profile $\beta_\theta(r)$. Carrying out the derivative operations in Eq.(9.63) gives

$$\frac{1}{r}\frac{\partial}{\partial r}r\frac{\partial}{\partial r}\beta_\theta - \frac{\beta_\theta}{r^2} = -\frac{2}{\gamma(r)}\frac{\partial\gamma}{\partial r}\frac{\partial\beta_\theta}{\partial r}$$

$$= -\frac{2\gamma_{z0}^2\beta_\theta(\partial\beta_\theta/\partial r)^2}{1 - \gamma_{z0}^2\beta_\theta^2} , \tag{9.67}$$

where use has been made of Eq.(9.66) relating $\gamma(r)$ and $\beta_\theta(r)$. Note that the radial variable r in Eq.(9.67) can be scaled by a (yet unspecified) constant length scale r_p, and Eq.(9.67) remains invariant under the replacement $r \to r/r_p$. Introducing the variable change

$$X = \frac{r}{r_p}, \quad Y = \gamma_{z0}\beta_\theta, \quad W = X\frac{dY}{dX} , \tag{9.68}$$

some straightforward algebra shows that Eq.(9.67) can be expressed in the equivalent form

$$\frac{d}{dY}W^2 + \frac{4Y}{1 - Y^2}W^2 = 2Y . \tag{9.69}$$

Equations (9.68) and (9.69) are to be solved subject to the boundary conditions that $\beta_\theta(r = 0) = 0$ and that $\partial\beta_\theta/\partial r$ is not singular at $r = 0$. Integrating Eq.(9.69) and imposing $W(Y = 0) = 0$ gives

$$W^2 = \left(X\frac{dY}{dX}\right)^2 = Y^2(1 - Y^2). \tag{9.70}$$

The solution to Eq.(9.70) with $Y(X = 0) = 0$ is given by

$$Y(X) = \frac{2X}{1 + X^2}. \tag{9.71}$$

In terms of the original variables, Eq.(9.71) gives the equilibrium profile

$$\beta_\theta(r) = \frac{2}{\gamma_{z0}} \frac{r/r_p}{(1 + r^2/r_p^2)} \tag{9.72}$$

for the azimuthal flow velocity. Substituting Eq.(9.72) into Eq.(9.66), the profile for $\gamma(r)$ can be expressed as

$$\gamma(r) = \gamma_{z0}\frac{(1 + r^2/r_p^2)}{(1 - r^2/r_p^2)}. \tag{9.73}$$

Moreover, making use of $\partial\gamma/\partial r = \partial\Phi/\partial r = (e/m_e c^2)\partial\phi_0/\partial r$, Poisson's equation (9.49) gives the beam density profile

$$n_b^0(r) = \left(\frac{2\gamma_{z0}m_e c^2}{\pi e^2 r_p^2}\right)\frac{(1 + r^2/r_p^2)}{(1 - r^2/r_p^2)^3} \tag{9.74}$$

over the radial extent of the electron beam $(0 \leq r < r_b)$. Note from Eqs.(9.73) and (9.74) that $r_b < r_p$ is required for nonsingular behavior of $\gamma(r)$ and $n_b^0(r)$.

The constant length scale r_p occurring in Eqs.(9.72)–(9.74) has not yet been specified. If we denote the on-axis $(r = 0)$ value of the beam density by $\hat{n}_b = n_b^0(r = 0)$, then it follows that r_p^2 can be identified with

$$r_p^2 = 8\gamma_{z0}c^2/\omega_{pb}^2, \tag{9.75}$$

where $\omega_{pb}^2 = 4\pi\hat{n}_b e^2/m_e$. Alternatively, note from Eq.(9.73) that $\gamma(r)$ increases as a function of r over the radial cross section of the electron beam. Denoting the value of $\gamma(r)$ at the outer edge of the electron beam by $\gamma_b \equiv \gamma(r = r_b)$, we solve Eq.(9.73) for r_p/r_b. This gives

$$r_p = r_b \left(\frac{\gamma_b + \gamma_{z0}}{\gamma_b - \gamma_{z0}} \right)^{1/2} \tag{9.76}$$

Because $\gamma_b > \gamma_{z0}$, it follows from Eq.(9.76) that $r_p > r_b$.

For nonrelativistic electron flow with $\gamma_b - \gamma_{z0} \ll \gamma_{z0} \simeq 1 - \beta_{z0}^2/2$, it follows from Eq.(9.76) that $r_b \ll r_p$. Therefore, the equilibrium profiles for $\omega_{rb}(r) = \beta_\theta(r)c/r$, $\gamma(r)$ and $n_b^0(r)$ in Eqs.(9.72)–(9.74) are approximately uniform over the entire cross section ($0 \le r < r_b$) of the electron beam. On the other hand, for relativistic electron flow, the equilibrium profiles in Eqs.(9.72)–(9.74) can be highly nonuniform, with $\gamma(r)$ and $n_b^0(r)$ increasing significantly over the interval $0 \le r < r_b$.

Expressed in terms of $\gamma(r)$, the normalized electrostatic potential calculated from the energy conservation relation (9.53) is given by $e\phi_0(r)/m_e c^2 \equiv \Phi(r) = \gamma(r) - \gamma_{b0}$ within the electron beam ($0 \le r < r_b$). [Here, γ_{b0} is the constant occurring in Eq.(9.53).] If we assume that the electron beam propagates through a grounded, cylindrical drift tube with radius $b > r_b$ and $\phi_0(r = b) = 0$, then the solution for $\Phi(r)$ is

$$\Phi(r) = \begin{cases} \Phi_b \left(\dfrac{\gamma(r) - \gamma_{b0}}{\gamma_b - \gamma_{b0}} \right), & 0 \le r < r_b \,, \\[3mm] \Phi_b \dfrac{\ln(b/r)}{\ln(b/r_b)}, & r_b < r \le b \,. \end{cases} \tag{9.77}$$

Here, $\Phi_b \equiv \Phi(r = r_b) = \gamma_b - \gamma_{b0}$ is the normalized electrostatic potential at the surface of the electron beam, and $\gamma_b \equiv \gamma(r = r_b)$. Enforcing the continuity of $\partial\Phi/\partial r$ at $r = r_b$, we obtain from Eqs.(9.76) and (9.77)

$$\Phi_b = -4\gamma_{z0} \ln \left(\frac{b}{r_b} \right) \frac{r_b^2/r_p^2}{(1 - r_b^2/r_p^2)^2}$$

$$= -\gamma_{z0} \ln \left(\frac{b}{r_b} \right) \left(\frac{\gamma_b^2}{\gamma_{z0}^2} - 1 \right). \tag{9.78}$$

When the beam completely fills the drift tube ($r_b = b$), note from Eq.(9.78) that $\Phi_b = 0 = \gamma_b - \gamma_{b0}$.

Making use of the expressions for $\beta_z(r)$, $\beta_\theta(r)$ and $n_b^0(r)$ in Eqs.(9.65), (9.72), and (9.74), the equilibrium profiles for the azimuthal self-magnetic field $B_\theta^s(r)$ and the total axial magnetic field $B_z(r)$ are readily calculated from the Maxwell equations (9.50) and (9.51) (Problem 9.10). For present purposes, we evaluate the magnitude of the total beam current from $I = -2\pi \int_0^{r_b} dr' \, r' J_{zb}^0(r') = 2\pi e \beta_{z0} c \int_0^{r_b} dr' \, r' n_b^0(r')$. Some straightforward algebra that makes use of Eqs.(9.74) and (9.76) gives

$$I = 2\beta_{z0}\gamma_{z0} \frac{m_e c^3}{e} \frac{r_b^2/r_p^2}{(1 - r_b^2/r_p^2)^2}$$

$$= \frac{1}{2} \frac{m_e c^3}{e} (\gamma_{z0}^2 - 1)^{1/2} \left(\frac{\gamma_b^2}{\gamma_{z0}^2} - 1 \right), \qquad (9.79)$$

where $\gamma_{z0} = (1 - \beta_{z0}^2)^{-1/2}$. In Eq.(9.79), we regard the constant γ_b as a specified quantity (related to the diode voltage producing the electron beam). As a function of γ_{z0}, the beam current I in Eq.(9.79) assumes a minimum value (zero) for $\gamma_{z0} = 1$, passes through a single maximum at $\gamma_{z0} = \gamma_m$ where γ_m is defined by

$$\gamma_m^2 = \frac{1}{2}\gamma_b^2 \left[\left(1 + \frac{8}{\gamma_b^2} \right)^{1/2} - 1 \right], \qquad (9.80)$$

and decreases to zero again at $\gamma_{z0} = \gamma_b$. Evaluating Eq.(9.79) for $\gamma_{z0} = \gamma_m$ gives space-charge-limiting current

$$I_\ell = \frac{1}{2\sqrt{2}} \frac{m_e c^3}{e} \left\{ \gamma_b^2 \left[\left(1 + \frac{8}{\gamma_b^2} \right)^{1/2} - 1 \right] - 2 \right\}^{1/2} \frac{[3 - (1 + 8/\gamma_b^2)^{1/2}]}{[(1 + 8/\gamma_b^2)^{1/2} - 1]}. \qquad (9.81)$$

In the ultrarelativistic limit ($\gamma_b \gg 1$), Eq.(9.81) reduces to $I_\ell \simeq (m_e c^3/4e)\gamma_b^2$. This result, obtained for a solid electron beam produced by a shielded source, should be contrasted with Eq.(9.35) for a thin annular electron beam. In particular, note that the limiting current in Eq.(9.35) scales linearly with γ_{b0} in the ultrarelativistic limit.

Problem 9.8   Consider the radial force balance equation (9.52) subject to the two conservation constraints in Eqs.(9.53) and (9.54).

a. Show from the definition of $\gamma(r)$ that

$$\frac{\partial \gamma}{\partial r} = \beta_\theta \frac{\partial}{\partial r}(\gamma \beta_\theta) + \beta_z \frac{\partial}{\partial r}(\gamma \beta_z). \qquad (9.8.1)$$

b. Take the radial derivative of the energy conservation relation (9.53). Make use of $E_r = -\partial\phi_0/\partial r$, $B_z = r^{-1}(\partial/\partial r)[rA_\theta^0]$, and the conservation of canonical angular momentum in Eq.(9.54) to show that

$$-\frac{\gamma m_e c^2 \beta_\theta^2}{r} = -e\left(E_r + \beta_\theta B_z\right) - m_e c^2 \beta_z \frac{\partial}{\partial r}\left(\gamma\beta_z\right) . \tag{9.8.2}$$

c. Compare Eq.(9.8.2) with the radial force balance equation (9.52) and show directly that

$$\frac{\partial}{\partial r}\left(\gamma m_e c\beta_z - \frac{e}{c}A_z^s\right) = 0 , \tag{9.8.3}$$

where $\partial A_z^s/\partial r = -B_\theta^s$. Equation (9.8.3) corresponds to the conservation of axial canonical momentum in Eq.(9.57). Therefore, Eq.(9.57) is a direct consequence of the radial force balance equation (9.52) and the two conservation constraints in Eqs.(9.53) and (9.54).

d. Alternatively, any one of Eqs.(9.52), (9.53), (9.54) or (9.57) follows from the other three equations. For example, show that the radial force balance equation (9.52) follows directly from Eqs.(9.53), (9.54), and (9.57).

Problem 9.9   Consider the differential equation (9.63) which relates the profiles for $\beta_\theta(r)$ and $\gamma(r)$.

a. Show that Eq.(9.63) reduces to Eq.(9.67) when

$$\beta_z(r) = \beta_{z0} = \text{const.}$$

over the cross section of the electron beam.

b. Introduce the variable changes in Eq.(9.68) and show that Eq.(9.69) follows from Eq.(9.67).

c. Solve Eqs.(9.68) and (9.69) subject to the boundary conditions that $\beta_\theta(r = 0) = 0$ and $\partial\beta_\theta/\partial r$ is nonsingular at $r = 0$. Show that $W(Y)$ and $Y(X)$ are given by Eqs.(9.70) and (9.71).

Problem 9.10   Consider the equilibrium profiles for $\beta_\theta(r)$, $\gamma(r)$ and $n_b^0(r)$ specified by Eqs.(9.72)–(9.74). Maxwell's equations (9.50) and (9.51) can be used to determine the self-consistent equilibrium profiles for $B_\theta^s(r)$ and $B_z(r)$.

a. Show that the azimuthal self-magnetic field $B_\theta^s(r)$ is given by

$$B_\theta^s(r) = -\frac{4\gamma_{z0}\beta_{z0}}{r_p}\frac{m_e c^2}{e}\frac{r/r_p}{(1 - r^2/r_p^2)^2} \tag{9.10.1}$$

for $0 \le r < r_b$. Note from Eq.(9.10.1) that $|B_\theta^s(r)|$ increases monotonically from $r = 0$ to $r = r_b < r_p$, where r_b is the outer radius of the electron beam.

b. Show that the total axial magnetic field $B_z(r)$ is given by

$$B_z(r) = B_0 - \frac{4}{r_p}\frac{m_e c^2}{e}\left[\frac{1}{(1 - r_b^2/r_p^2)^2} - \frac{1}{(1 - r^2/r_p^2)^2}\right] \qquad (9.10.2)$$

for $0 \leq r < r_b$. Here, the constant $B_0 = B_z(r = r_b)$ is the value of the (compressed) magnetic field in the outer vacuum region $r_b < r \leq b$.

c. Make use of Eq.(9.10.2) to evaluate

$$A_\theta^0(r) = r^{-1}\int_0^r dr'\, r'B_z(r') \ .$$

Show that the condition

$$\gamma(r)m_e c\beta_\theta(r) - \frac{e}{c}A_\theta^0(r) = 0 \qquad (9.10.3)$$

is satisfied over the entire interval $0 \leq r < r_b$ provided r_b, r_p and the vacuum field B_0 are related by

$$B_0 = \frac{4}{r_p}\frac{m_e c^2}{e}\frac{1}{(1 - r_b^2/r_p^2)^2} \ . \qquad (9.10.4)$$

Comparing Eqs.(9.10.2) and (9.10.4), we note that the on-axis value of the axial magnetic field is $B_z(r = 0) = 4m_e c^2/er_p$.

9.3 Stability of Nonneutral Electron Flow in a One-Dimensional Drift Space

In this section, we investigate stability properties for electrostatic perturbations about the one-dimensional equilibrium flow discussed in Sec. 9.2.4 (see Fig. 9.3). It is found that the low-velocity, high-density flow regime in Fig. 9.4 is unstable to the growth of space-charge waves.[1,21-23]

Following a brief review of the equilibrium flow (Sec. 9.3.1), the linearized fluid-Poisson equations (Sec. 9.3.2) are analyzed by introducing the Lagrangian time variable τ which follows the equilibrium motion of a fluid element (Sec. 9.3.3). The resulting eigenvalue equation is solved analytically (Sec. 9.3.4), and the stability properties are investigated in detail for purely growing modes, including a study of stability behavior for beam propagation near the space-charge-limiting current (Sec. 9.3.5). The analysis in Sec. 9.3 is carried out for the case of nonrelativistic electron flow with

$$\beta_z^2(z) \ll 1 \ .$$

9.3.1 Review of Nonrelativistic Equilibrium Flow

In the limit of an infinitely strong axial guide field ($B_0 \rightarrow \infty$), the governing equations for steady-state electron flow in the one-dimensional drift space $0 \leq z \leq L$ in Fig. 9.3 are given by

$$-en_b^0(z)\beta_z(z)c = J_b = \text{const.} \,, \tag{9.82}$$

$$\beta_z(z)\frac{\partial}{\partial z}\beta_z(z) = \frac{\partial}{\partial z}\Phi(z) \,, \tag{9.83}$$

$$\frac{\partial^2}{\partial z^2}\Phi(z) = \frac{4\pi n_b^0(z)e^2}{m_e c^2} \,. \tag{9.84}$$

Here, $\Phi(z) = e\phi_0(z)/m_e c^2$ is the normalized electrostatic potential, $c\beta_z(z) = V_{zb}^0(z)$ is the axial flow velocity, $n_b^0(z)$ is the electron density, and the constant $J_b < 0$ is the axial current density. In Sec. 9.2.4, Eqs.(9.82)–(9.84) were solved subject to the boundary conditions $\Phi(z = 0) = 0 = \Phi(z = L)$. As illustrated in Fig. 9.5, the electron beam enters the drift space at $z = 0$ with axial velocity $\beta_{z0}c$ and exits at $z = L$ with axial velocity $\beta_{z0}c$. The normalized potential $\Phi(z)$ is determined from Eq.(9.42), and the axial velocity $\beta_z(z)c$ is determined from Eq.(9.43), which can be expressed as

$$[\beta_z(z) - \beta_{zm}]\,[\beta_z(z) + 2\beta_{zm}]^2 = \frac{9}{4}\kappa\,(z - L/2)^2 \,. \tag{9.85}$$

In Eq.(9.85), the (positive) constant κ is defined by

$$\kappa = \frac{8\pi eJ}{m_e c^3} \,, \tag{9.86}$$

where $J \equiv -J_b > 0$.

Note from Eq.(9.85) and Fig. 9.5 that the axial flow velocity slows to the minimum value $\beta_{zm}c$ at $z = L/2$, and there is a corresponding compression of the beam density $n_b^0(z) = J/ec\beta_z(z)$. Moreover, β_{zm} can be related to the injection velocity β_{z0} by evaluating Eq.(9.85) at $z = 0$, which gives $(\beta_{z0} - \beta_{zm})(\beta_{z0} + 2\beta_{zm})^2 = 9\kappa L^2/16$. As discussed in Sec. 9.2.4, for specified value of β_{z0}, the current density assumes the maximum value $J_\ell = (4/9\pi L^2)(m_e c^3/e)\beta_{z0}^3$ when $\beta_{zm} = \beta_{z0}/2$ [Fig. 9.4

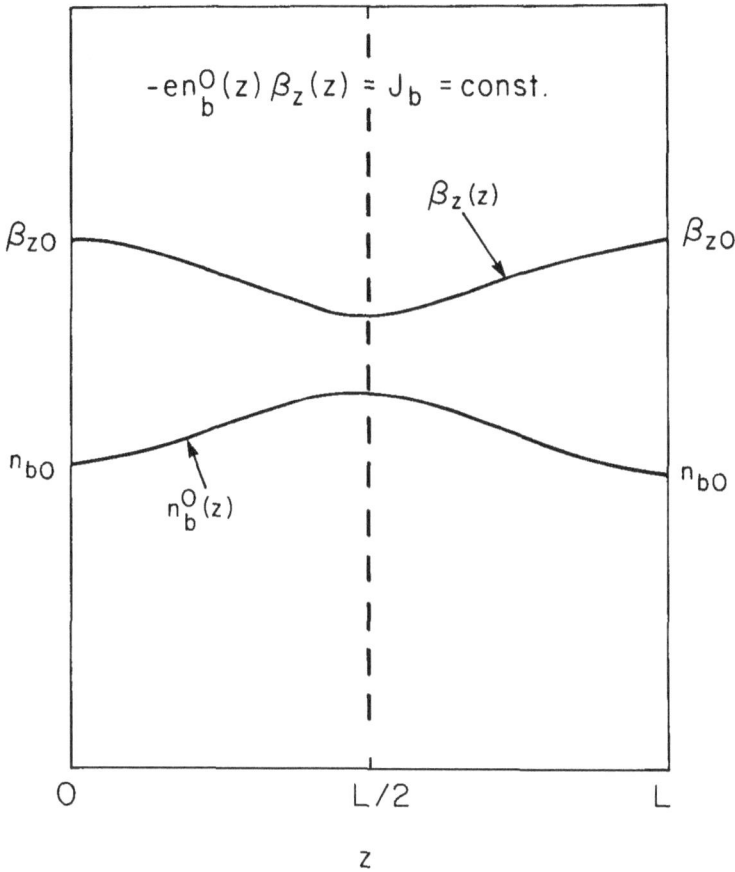

Figure 9.5.   Sketch versus z of $\beta_z(z)$ and $n_b^0(z)$ obtained from Eqs.(9.82) and (9.85) for nonrelativistic electron flow in the one-dimensional drift space in Fig. 9.3. The electron beam slows to a minimum velocity at the midpoint $z = L/2$, where the density compression is a maximum.

and Eq.(9.46)]. The corresponding value of κ at the space-charge-limiting current is

$$\kappa_\ell = \frac{32\beta_{z0}^3}{9L^2} .$$ (9.87)

Note from Eq.(9.87) that κ_ℓ scales as β_{z0}^3 in the present nonrelativistic treatment.

9.3.2   Linearized Cold-Fluid-Poisson Equations

We now consider small-amplitude electrostatic perturbations about the one-dimensional equilibrium described by Eqs.(9.82) and (9.84). Neglecting transverse spatial variations $(\partial/\partial y = 0 = \partial/\partial x)$, the field and fluid quantities

$$\Phi(z,t) = \frac{e\phi(z,t)}{m_e c^2} \ , \quad \beta_z(z,t) = \frac{V_{zb}(z,t)}{c} \ , \quad \text{and} \quad n_b(z,t)$$

are expressed as

$$\Phi(z,t) = \Phi(z) + \delta\Phi(z,t) \ ,$$

$$\beta_z(z,t) = \beta_z(z) + \delta\beta_z(z,t) \ , \qquad (9.88)$$

$$n_b(z,t) = n_b^0(z) + \delta n_b(z,t) \ ,$$

where the equilibrium profiles $\Phi(z)$, $\beta_z(z)$ and $n_b^0(z)$ solve Eqs.(9.82)–(9.84). For small-amplitude perturbations, a normal mode approach is adopted in which perturbed quantities are expressed as

$$\delta\psi(z,t) = \delta\psi(z)\exp(-i\omega t) \ , \qquad (9.89)$$

where $\text{Im}\,\omega > 0$ corresponds to instability (temporal growth).

Within the framework of a macroscopic cold-fluid model, the linearized fluid-Poisson equations relating the perturbation amplitudes $\delta n_b(z)$, $\delta\beta_z(z)$ and $\delta\Phi(z)$ are given by (Sec. 2.3)

$$-i\frac{\omega}{c}\,\delta n_b(z) + \frac{\partial}{\partial z}\left[\beta_z(z)\,\delta n_b(z)\right] + \frac{\partial}{\partial z}\left[n_b^0(z)\,\delta\beta_z(z)\right] = 0 \ , \quad (9.90)$$

$$-i\frac{\omega}{c}\,\delta\beta_z(z) + \frac{\partial}{\partial z}\left[\beta_z(z)\,\delta\beta_z(z)\right] = \frac{\partial}{\partial z}\,\delta\Phi(z) \ , \qquad (9.91)$$

$$\frac{\partial^2}{\partial z^2}\,\delta\Phi(z) = \frac{4\pi e^2}{m_e c^2}\,\delta n_b(z) \ . \qquad (9.92)$$

In Eq.(9.90), we eliminate $n_b^0(z) = J/ec\beta_z(z) = (m_e c^2\kappa/8\pi e^2)/\beta_z(z)$ in favor of $\beta_z(z)$, and make use of Poisson's equation (9.92) to express $\delta n_b(z)$

in terms of $\delta\Phi(z)$. Integrating Eq.(9.90) once with respect to z then gives

$$\left(-i\frac{\omega}{c} + \beta_z(z)\frac{\partial}{\partial z}\right)\frac{\partial}{\partial z}\delta\Phi(z) + \frac{\kappa}{2\beta_z(z)}\delta\beta_z(z) = C , \tag{9.93}$$

where C is an integration constant. Moreover, Eq.(9.91) can be expressed as

$$\left(-i\frac{\omega}{c} + \beta_z(z)\frac{\partial}{\partial z}\right)\delta\beta_z(z) + \frac{\partial\beta_z}{\partial z}\delta\beta_z(z) = \frac{\partial}{\partial z}\delta\Phi(z) . \tag{9.94}$$

Combining Eqs.(9.93) and (9.94) then gives

$$\left(-i\frac{\omega}{c} + \beta_z(z)\frac{\partial}{\partial z}\right)\left(-i\frac{\omega}{c} + \beta_z(z)\frac{\partial}{\partial z}\right)\delta\beta_z(z) \tag{9.95}$$

$$+ \left(-i\frac{\omega}{c} + \beta_z(z)\frac{\partial}{\partial z}\right)\left(\frac{\partial\beta_z}{\partial z}\delta\beta_z(z)\right) + \frac{\kappa}{2\beta_z(z)}\delta\beta_z(z) = C ,$$

which is the desired differential equation for $\delta\beta_z(z)$. Once $\delta\beta_z(z)$ is determined from Eq.(9.95), the perturbations $\delta\Phi(z)$ and $\delta n_b(z)$ can be calculated from Eqs.(9.94) and (9.92), respectively. The dispersion relation for the (complex) eigenfrequency ω is then obtained by imposing the boundary conditions[1,23]

$$\delta\Phi(z = 0) = 0 = \delta\Phi(z = L) ,$$

$$\delta\beta_z(z = 0) = 0 , \tag{9.96}$$

$$\delta n_b(z = 0) = 0 .$$

9.3.3   Lagrangian Representation

To solve the differential equation (9.95) for $\delta\beta_z(z)$, it is convenient to introduce the Lagrangian time variable $\tau(z)$ defined by

$$c\tau(z) = \int_0^z \frac{dz'}{\beta_z(z')} . \tag{9.97}$$

Note that $dz/d\tau = \beta_z c$ and $c\beta_z\partial\beta_z/\partial z = d\beta_z/d\tau$, so that $\tau(z)$ is an effective time variable following the equilibrium motion of a fluid element. Analysis of the equilibrium flow in terms of the Lagrangian time variable

τ is summarized in Problem 9.7. Without presenting algebraic details, the main results are

$$\beta_z(\tau) = \beta_{zm} + \frac{1}{4}\kappa c^2 \left(t_0/2 - \tau\right)^2 ,$$

$$z(\tau) = L/2 + \beta_{zm} c \left(\tau - t_0/2\right) - \frac{1}{12}\kappa c^3 \left(t_0/2 - \tau\right)^3 , \qquad (9.98)$$

which are valid over the intervals $0 \leq \tau \leq t_0$ and $0 \leq z \leq L$. Here, t_0 is the time for the electron beam to transit from $z = 0$ to $z = L$ defined by

$$ct_0 = \int_0^L \frac{dz'}{\beta_z(z')} , \qquad (9.99)$$

and $t_0/2$ is the time for the beam to transit from $z = 0$ to the midpoint $z = L/2$. The minimum velocity $\beta_{zm} c$ in Eq.(9.98) can be eliminated in favor of the injection velocity $\beta_{z0} c$ by evaluating Eq.(9.98) at $\tau = 0$, which gives $\beta_{z0} = \beta_{zm} + \kappa c^2 t_0^2/16$. Equation (9.98) can then be expressed in the equivalent form

$$\beta_z(\tau) = \beta_{z0} + \frac{1}{4}\kappa c^2 \left[\left(t_0/2 - \tau\right)^2 - \left(t_0/2\right)^2\right] , \qquad (9.100)$$

$$z(\tau) = L/2 + \beta_{z0} c \left(\tau - t_0/2\right) - \frac{1}{12}\kappa c^3 \left[\left(t_0/2 - \tau\right)^3 + 3 \left(t_0/2\right)^2 \left(\tau - t_0/2\right)\right] .$$

Evaluating Eq.(9.100) at $z = 0$ and $\tau = 0$ (or at $z = L$ and $\tau = t_0$) readily gives

$$L = \beta_{z0} c t_0 - \frac{1}{24}\kappa c^3 t_0^3 . \qquad (9.101)$$

Equation (9.101) determines the transit time t_0 in terms of L, β_{z0}, and $\kappa = 8\pi e J/m_e c^3$. There are three solutions to Eq.(9.101) for t_0, and the sum of the roots is equal to zero. The smaller, positive root (denote by t_h) corresponds to high-velocity, low-density flow (see Sec. 9.2.4), whereas the larger, positive root (denote by t_{c0}) corresponds to low-velocity, high-density flow (the so-called "C-overlap" region in Fig. 9.4).

From Eq.(9.101) the product of the three roots t_h, t_{c0} and $-(t_h + t_{c0})$ is given by

$$t_h t_{c0}(t_h + t_{c0}) = \frac{24L}{\kappa c^3} . \tag{9.102}$$

From the analysis in Sec. 9.2.4, the parameter κ occurring in Eqs.(9.101) and (9.102) is limited to the range $0 < \kappa < \kappa_\ell$, where $\kappa_\ell = 8\pi e J_\ell / m_e c^3 = (32/9L^2)\beta_{z0}^3$ corresponds to the space-charge limiting current defined in Eq.(9.46). Therefore, from Eqs.(9.102) and (9.103), the roots t_h and t_{c0} satisfy

$$t_h^3 < t_\ell^3 \equiv \frac{12L}{\kappa_\ell c^3} < t_{c0}^3 \tag{9.103}$$

for $\kappa < \kappa_\ell$. For $\kappa = \kappa_\ell$ exactly, the roots t_h and t_{c0} coincide and satisfy $t_h = t_{c0} = t_\ell$.

We now return to Eq.(9.95) for the perturbed flow velocity $c\delta\beta_z(z)$, and introduce the Lagrangian time variable τ defined in Eq.(9.97). Making use of $c\beta_z \partial/\partial z = d/d\tau$, it follows that Eq.(9.95) can be expressed as

$$\left(-i\omega + \frac{d}{d\tau}\right)\left(-i\omega + \frac{d}{d\tau}\right)\delta\beta_z(\tau) \tag{9.104}$$

$$+ \left(-i\omega + \frac{d}{d\tau}\right)\left(\frac{1}{\beta_z(\tau)}\frac{d\beta_z(\tau)}{d\tau}\delta\beta_z(\tau)\right) + \frac{\kappa c^2}{2\beta_z(\tau)}\delta\beta_z(\tau) = c^2 C .$$

Here, $\beta_z(\tau)$ is defined in Eq.(9.100), and Eq.(9.104) is to be solved for $\delta\beta_z(\tau)$ subject to the boundary condition $\delta\beta_z(\tau = 0) = 0$ in Eq.(9.96).

9.3.4   Solution to the Eigenvalue Equation

The analysis of Eq.(9.104) is simplified by introducing the function $Y(\tau)$ defined in terms of $\delta\beta_z(\tau)$ by

$$\delta\beta_z(\tau) = \frac{Y(\tau)}{\beta_z(\tau)}\exp(i\omega\tau) . \tag{9.105}$$

Making use of $(-i\omega + d/d\tau)[F(\tau)\exp(i\omega\tau)] = \exp(i\omega\tau)dF(\tau)/d\tau$, it is readily shown that Eq.(9.104) reduces to

$$\beta_x(\tau)\frac{d^2Y(\tau)}{d\tau^2} - \frac{d\beta_z(\tau)}{d\tau}\frac{dY(\tau)}{d\tau} + \frac{\kappa c^2}{2}Y(\tau) = c^2 C\beta_z^2(\tau)\exp(-i\omega\tau),$$
(9.106)

where $\beta_z(\tau)$ is defined in Eq.(9.100). Once the solution for

$$Y(\tau) = \beta_z(\tau)\exp(-i\omega\tau)\delta\beta_z(\tau)$$

is obtained from Eq.(9.106), the perturbed electrostatic potential $\delta\Phi(\tau)$ and beam density $\delta n_b(\tau)$ can be determined from Eqs.(9.94) and (9.92), respectively. Making use of $c\beta_z\partial/\partial z = d/d\tau$, integration of the linearized force-balance equation (9.94) gives

$$\delta\Phi(\tau) = \int_0^\tau d\tau' \exp(i\omega\tau') \, dY(\tau')/d\tau',$$
(9.107)

where the boundary condition $\delta\Phi(\tau = 0) = 0$ has been imposed [Eq.(9.96)]. Moreover, Poisson's equation (9.92) can be expressed as

$$\delta n_b(\tau) = \frac{m_e}{4\pi e^2}\frac{1}{\beta_z(\tau)}\frac{d}{d\tau}\left(\frac{1}{\beta_z(\tau)}\frac{d}{d\tau}\delta\Phi(\tau)\right),$$
(9.108)

which determines $\delta n_b(\tau)$ in terms of $\delta\Phi(\tau)$, and therefore $Y(\tau)$.

For $\beta_z(\tau)$ specified by Eq.(9.100), some straightforward algebra (Problem 9.11) shows that the general solution to Eq.(9.106) can be expressed as

$$Y(\tau) = \frac{c^2C}{\omega^2}\left(\beta_{z0} + \frac{i}{4}\frac{\kappa c^2 t_0}{\omega}\right)[1 - \exp(-i\omega\tau)]$$

$$+ \alpha_1\tau - \frac{\kappa c^2}{4\beta_{z0}}\left[\frac{c^2C}{\omega^2}\left(\beta_{z0} + \frac{i}{4}\frac{\kappa c^2 t_0}{\omega}\right) + \frac{1}{2}\alpha_1 t_0\right]\tau^2$$

$$+ \frac{c^2C}{\omega^2}\left[\frac{\kappa c^2}{4}\left(t_0 + \frac{2i}{\omega}\right)\tau - \frac{1}{4}\kappa c^2\tau^2\right]\exp(-i\omega\tau).$$
(9.109)

Here, α_1 is a (yet undetermined) constant coefficient, and the boundary condition $Y(\tau = 0) = 0$ has been imposed in Eq.(9.109), which corresponds to $\delta\beta_z(\tau = 0) = 0$ [Eq.(9.96)]. To eliminate the constant α_1 in

Eq.(9.109), and obtain the dispersion relation for the complex eigenfrequency ω, the remaining boundary conditions to be enforced are

$$\delta n_b(\tau = 0) = 0 ,$$

$$\delta\Phi(\tau = t_0) = 0 . \tag{9.110}$$

The solution for $Y(\tau)$ in Eq.(9.109) is substituted into Eqs.(9.107) and (9.108) to determine the perturbed electrostatic potential $\delta\Phi(\tau)$ and beam density $\delta n_b(\tau)$. Enforcing $\delta n_b(\tau = 0) = 0$ gives

$$\alpha_1 = -\frac{i\kappa c^4}{2\omega^3}C , \tag{9.111}$$

which determines α_1 in terms of the constant C. When Eqs.(9.111) and (9.109) are substituted into Eq.(9.107), the boundary condition $\delta\Phi(\tau = t_0) = 0$ gives[1,23]

$$- iL\omega^3 = \frac{1}{2}\kappa c^3 \left\{ i\omega t_0 \left[1 + \exp(i\omega t_0)\right] + 2\left[1 - \exp(i\omega t_0)\right]\right\} \tag{9.112}$$

as the condition for a nontrivial solution with $C \neq 0$ (Problem 9.12).

Equation (9.112) plays the role of the dispersion relation which determines the complex eigenfrequency ω in terms of the length L of the drift region, the beam transit time t_0 determined from Eq.(9.101), and the parameter

$$\kappa = \frac{18\pi e J}{m_e c^3}$$

which is proportional to the beam current density.

Problem 9.11   Consider the differential equation (9.106) for $Y(\tau)$. Here, $c\beta_z(\tau) = c\beta_{z0} + (\kappa c^3/4)[(t_0/2 - \tau)^2 - (t_0/2)^2]$ is the equilibrium flow velocity expressed in terms of the Lagrangian time variable τ. The solution to Eq.(9.106) can be expressed as

$$Y(\tau) = Y_h(\tau) + Y_p(\tau) , \tag{9.11.1}$$

where $Y_h(\tau)$ is the homogeneous solution (for $C = 0$), and $Y_p(\tau)$ is the particular solution (for $C \neq 0$).

a. The homogeneous solution to Eq.(9.106) is of the form

$$Y_h(\tau) = \alpha_0 + \alpha_1 \tau + \alpha_2 \tau^2 \, , \qquad (9.11.2)$$

where α_0, α_1 and α_2 are constant coefficients. Substitute Eqs.(9.100) and (9.11.2) into Eq.(9.106) (for $C = 0$) and show that

$$\alpha_2 = -\frac{\kappa c^2}{4\beta_{z0}} \left(\alpha_0 + \frac{1}{2}\alpha_1 t_0 \right) . \qquad (9.11.3)$$

b. The particular solution to Eq.(9.106) is of the form

$$Y_p(\tau) = \frac{c^2 C}{\omega^2} \left(\alpha_3 + \alpha_4 \tau + \alpha_5 \tau^2 \right) \exp(-i\omega\tau) \, , \qquad (9.11.4)$$

where α_3, α_4 and α_5 are constant coefficients. Substitute Eqs.(9.100) and (9.11.4) into Eq.(9.106) and show that

$$\alpha_3 = - \left(\beta_{z0} + \frac{i}{4}\frac{\kappa c^2 t_0}{\omega} \right) \, ,$$

$$\alpha_4 = \frac{1}{4}\kappa c^2 \left(t_0 + \frac{2i}{\omega} \right) \, , \qquad (9.11.5)$$

$$\alpha_5 = -\frac{1}{4}\kappa c^2 \, ,$$

c. Making use of Eqs.(9.11.2)–(9.11.5), the general solution $Y(\tau) = Y_h(\tau) + Y_p(\tau)$ to Eq.(9.106) can be expressed as

$$Y(\tau) = \alpha_0 + \alpha_1 \tau - \frac{\kappa c^2}{4\beta_{z0}} \left(\alpha_0 + \frac{1}{2}\alpha_1 t_0 \right) \tau^2 \qquad (9.11.6)$$

$$+ \frac{c^2 C}{\omega^2} \left[- \left(\beta_{z0} + \frac{i}{4}\frac{\kappa c^2 t_0}{\omega} \right) + \frac{1}{4}\kappa c^2 \left(t_0 + \frac{2i}{\omega} \right) \tau - \frac{1}{4}\kappa c^2 \tau^2 \right] \exp(-i\omega\tau;) \, .$$

The boundary condition $\delta\beta_z(\tau = 0) = 0$ in Eq.(9.96) corresponds to $Y(\tau = 0) = 0$. Make use of Eq.(9.11.6) to show that this boundary condition implies that the constants α_0 and C are related by

$$\alpha_0 = \frac{c^2 C}{\omega^2} \left(\beta_{z0} + \frac{i}{4}\frac{\kappa c^2 t_0}{\omega} \right) . \qquad (9.11.7)$$

Substituting Eq.(9.11.7) into Eq.(9.11.6) then gives the general solution for $Y(\tau)$ in Eq.(9.109).

Problem 9.12   Consider the general solution for $Y(\tau)$ in Eq.(9.109).

a. Make use of Eqs.(9.107) and (9.108) and $\delta n_b(\tau = 0) = 0$ to show that $\alpha_1 = -(i\kappa c^4/2\omega^3)C$.

b. Make use of Eq.(9.107) and the result in Part (a) to show that $\delta\Phi(\tau = t_0) = 0$ leads to the dispersion relation in Eq.(9.112).

9.3.5   Analysis of Stability Properties

We now examine the dispersion relation (9.112) which determines the complex eigenfrequency ω of the space-charge waves excited in the drift space $0 \leq z \leq L$ for small-amplitude electrostatic perturbations. It is straightforward to show that Eq.(9.112) can be expressed in the equivalent form

$$\exp(i\omega t_0)\,(2 - i\omega t_0) = 2 + i\omega t_0 - \left(\frac{2L}{\kappa c^3 t_0^3}\right)(i\omega t_0)^3. \tag{9.113}$$

Alternatively, recalling from Eq.(9.89) that the time dependence of the perturbations is of the form $\exp(-i\omega t)$, we express $\omega t_0 = P + iQ$ in Eq.(9.113) so that $Q > 0$ corresponds to instability ($\mathrm{Im}\,\omega > 0$). Setting the real and imaginary parts of Eq.(9.113) separately equal to zero then gives

$$Q\frac{2L}{\kappa c^3 t_0^3} = \frac{\exp(-Q)[(2 + Q)\cos P + P\sin P] - (2 - Q)}{(Q^2 - 3P^2)}, \tag{9.114}$$

$$P\frac{2L}{\kappa c^3 t_0^3} = \frac{\exp(-Q)[(2 + Q)\sin P - P\cos P] - P}{(P^2 - 3Q^2)}. \tag{9.115}$$

Equations (9.114) and (9.115) can be used[23] to calculate the real oscillation frequency $P/t_0 = \mathrm{Re}\,\omega$ and growth rate $Q/t_0 = \mathrm{Im}\,\omega$ over a wide range of system parameters as measured by $\kappa/\kappa_\ell = J/J_\ell$, where $J_\ell = (4/9\pi)(m_e c^3/eL^2)\beta_{z0}^3$ is the space-charge-limiting current defined in Eq.(9.46).

Before analyzing Eqs.(9.113)–(9.115), it is instructive to reexamine Eq.(9.101), which relates the transit time t_0 to the system parameters L, β_{z0} and $\kappa = 8\pi e J/m_e c^3$. In this regard, we scale t_0 to the transit time

$t_\ell = (12L/\kappa_\ell c^3)^{1/3} = 3L/2\beta_{z0}c$ at the limiting current when $\kappa = \kappa_\ell = (32/9L^2)\beta_{z0}^3$. Some straightforward algebra shows that Eq.(9.101) can be expressed in the equivalent form

$$\frac{\kappa}{\kappa_\ell}\frac{t_0^3}{t_\ell^3} - 3\frac{t_0}{t_\ell} + 2 = 0 . \qquad (9.116)$$

Equation (9.116) shows directly that the normalized transit time t_0/t_ℓ depends only on the dimensionless parameter $\kappa/\kappa_\ell = J/J_\ell$, and that $t_0 = t_\ell$ whenever $\kappa = \kappa_\ell$. Of course, Eq.(9.116) can be solved exactly for t_0/t_ℓ in terms of κ/κ_ℓ. Alternatively, Fig. 9.6 shows a plot of $\kappa/\kappa_\ell = (t_\ell/t_0)^3(3t_0/t_\ell - 2)$ versus t_0/t_ℓ for positive κ/κ_ℓ. For specified value of κ/κ_ℓ, the smaller root with $2/3 < t_0/t_\ell < 1$ corresponds to the high-velocity, low-density solution ($t_0 = t_h$), whereas the larger root with $t_0/t_\ell > 1$ corresponds to the low-velocity, high-density solution ($t_0 = t_{c0}$), often referred to as the C-overlap region.[21,22]

Detailed stability properties have been determined from Eqs.(9.114) and (9.115) by Paschenko and Rutkevich.[23] For the high-velocity solution corresponding to $t_0 = t_h$ and $2/3 < t_0/t_\ell < 1$ in Fig. 9.6, it is found[23] that the flow is stable ($\text{Im}\,\omega = Q/t_0 < 0$), and there is a rich harmonic

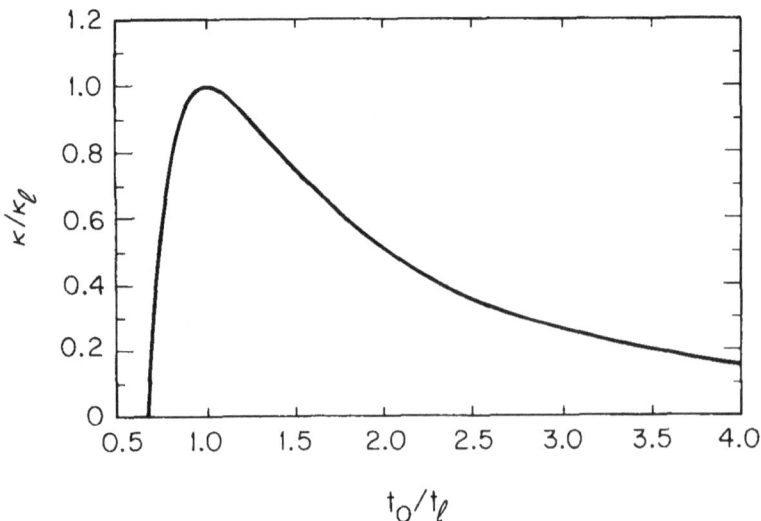

Figure 9.6.   Plot of κ/κ_ℓ versus t_0/t_ℓ obtained from Eq.(9.116) for $t_0/t_\ell \geq 2/3$.

structure with real oscillation frequency $\mathrm{Re}\,\omega = P/t_0 \approx n\pi/t_0$ for $n = 1, 2, 3, \cdots$. On the other hand, the low-velocity solution corresponding to $t_0 = t_{c0}$ and $t_0/t_\ell > 1$ can exhibit strong instability ($\mathrm{Im}\,\omega = Q/t_0 > 0$) for purely growing modes with $\mathrm{Re}\,\omega = 0$ (i.e., $P = 0$).

For $P = 0$ and $Q \neq 0$, it is clear that Eq.(9.115) is satisfied exactly. Setting $P = 0$ in Eq.(9.114) then gives

$$\frac{1}{6}\frac{\kappa_\ell}{\kappa}\frac{t_\ell^3}{t_0^3}Q^3 = \exp(-Q)(2+Q) - (2-Q)\,, \qquad (9.117)$$

which is a transcendental equation for the growth rate $\mathrm{Im}\,\omega = Q/t_0$. Here, we have expressed $2L/\kappa c^3 t_0^3 = (\kappa_\ell/6\kappa)t_\ell^3/t_0^3$ on the left-hand side of Eq.(9.117). Making use of Eq.(9.116) to express $(\kappa_\ell/\kappa)t_\ell^3/t_0^3 = (3t_0/t_\ell - 2)^{-1}$, the dispersion relation (9.117) becomes

$$\frac{Q^3}{18[(t_0/t_\ell - 1) + 1/3]} = \exp(-Q)(2+Q) - (2-Q)\,. \qquad (9.118)$$

Equation (9.118) conveniently parameterizes the stability behavior in terms of t_0/t_ℓ. For $2/3 < t_0/t_\ell < 1$, it can be shown that Eq.(9.118) supports only stable solutions with $Q < 0$. For $t_0/t_\ell > 1$, however, Eq.(9.118) yields instability with $Q = (\mathrm{Im}\,\omega)t_0 > 0$. This is illustrated in Fig. 9.7 where the normalized growth rate $(\mathrm{Im}\,\omega)t_\ell$ obtained numerically from Eq.(9.118) is plotted versus t_0/t_ℓ. Note from the figure that the maximum growth rate $(\mathrm{Im}\,\omega)_{\max} \simeq 1.7t_\ell^{-1}$ occurs for $t_0/t_\ell \simeq 2.5$.

Finally, the dispersion relation (9.118) can be solved analytically in the region near the limiting current where

$$\kappa \approx \kappa_\ell\,, \quad t_0 \approx t_\ell \quad \text{and} \quad |Q| \ll 1\,.$$

Expanding

$$\exp(-Q)(2+Q) - (2-Q) = \frac{Q^3}{6} - \frac{Q^4}{12} + \cdots\,, \qquad (9.119)$$

it is readily shown that Eq.(9.118) can be approximated by

$$Q^3 \frac{(t_0/t_\ell - 1)}{[(t_0/t_\ell - 1) + 1/3]} = \frac{Q^4}{2} + \cdots\,. \qquad (9.120)$$

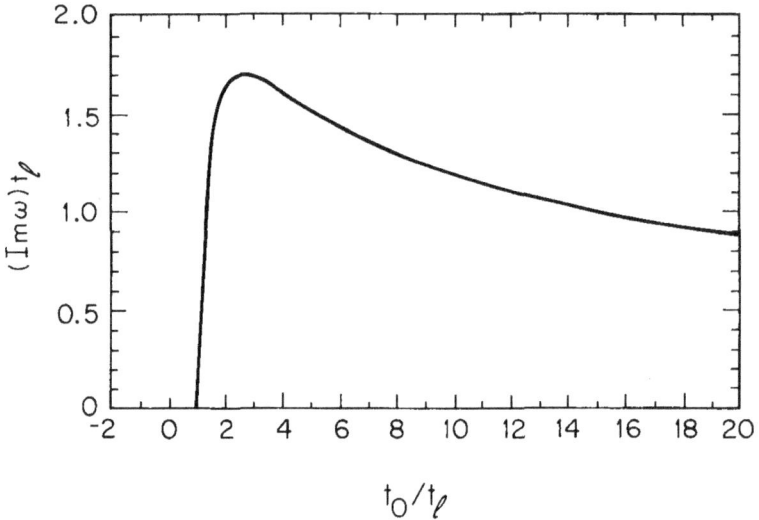

Figure 9.7. Plot versus t_0/t_ℓ of the normalized growth rate $(Im\omega)t_\ell = Qt_\ell/t_0$ obtained numerically from the dispersion relation (9.118) for a purely growing mode with $P = 0$.

For small values of $t_0/t_\ell - 1$, Eq.(9.120) gives for the growth rate

$$(\text{Im}\,\omega)t_\ell = 6\frac{t_\ell}{t_0}\left(\frac{t_0}{t_\ell} - 1\right), \qquad (9.121)$$

where $\text{Im}\,\omega = Q/t_0$. As expected, Eq.(9.121) gives instability ($\text{Im}\,\omega > 0$) when $t_0 > t_\ell$, whereas the waves are damped when $t_0 < t_\ell$.

Problem 9.13   For $\kappa \approx \kappa_\ell$ and $t_0 \approx t_\ell$, Taylor expand the dispersion relation (9.118) for $|Q| \ll 1$. [Equation (9.118) is valid for purely growing modes with $P = 0$ and $\text{Re}\,\omega = 0$.]

a. Show that the approximate dispersion relation which determines $Q = (\text{Im}\,\omega)t_0$ is given by Eq.(9.120).

b. Solve Eq.(9.120), and show that the growth rate $\text{Im}\,\omega$ is given by Eq.(9.121).

Problem 9.14   The electrostatic stability properties of the one-dimensional Child-Langmuir flow described in Sec. 8.2.1 can be analyzed using techniques similar to those developed in Sec. 9.3. The resulting equation for the perturbed flow velocity $c\delta\beta_z(\tau) = [Y(\tau)c/\beta_z(\tau)]\exp(i\omega\tau)$ is identical to Eq.(9.106), with the equilibrium flow velocity $c\beta_z(\tau)$ given by

$$\beta_z(\tau) = \frac{1}{4}c^2\kappa\tau^2, \tag{9.14.1}$$

where $\kappa = 12L/c^3t_0^3 = 8\pi eJ/m_ec^3$. Here, the cathode is located at $z = 0$; the anode (maintained at voltage V) is located at $z = L$; the Lagrangian time variable is $\tau(z) = t_0(z/L)^{1/3}$; and the transit time t_0 is given by

$$t_0 = 3d\left(m_e/2eV\right)^{1/2} = \left(3Lm_ec^2/2\pi eJ\right)^{1/3} \tag{9.14.2}$$

for space-charge-limited flow with $[\partial\Phi/\partial z]_{z=0} = 0$.

a. For $\beta_z(\tau)$ prescribed by Eq.(9.14.1), solve Eq.(9.106) for $Y(\tau)$ using techniques similar to those developed in Sec. 9.3.4.

b. Enforce the boundary conditions $Y(\tau = 0) = 0 = Y(\tau = t_0)$, $\delta\Phi(\tau=0) = 0$, and $\delta n_b(\tau = 0) = 0$, and make use of Eqs.(9.107) and (9.108) to derive the dispersion relation which determines the complex oscillation frequency ω in terms of the system parameters κ, t_0 and L.

c. Make use of the dispersion relation derived in Part (b) of this problem to determine the detailed stability properties of Child-Langmuir flow for one-dimensional spatial variations.

9.4 Stability of Intense Nonneutral Ion and Electron Beams

For an intense nonneutral ion or electron beam propagating through vacuum parallel to a uniform axial guide field $B_0\hat{e}_z$ or through a periodic quadrupole focusing field, one of the most important instabilities that can disrupt beam propagation is the so-called "transverse" or "hollowing" instability.[25,26] For this instability, low-frequency, azimuthally symmetric perturbations ($|\omega|r_b \ll c$ and $\partial/\partial\theta = 0$) with radial mode number $n \geq 2$ can grow and significantly modify the phase-space distribution of beam particles, thereby causing emittance growth. In this section, we make use of the linearized Vlasov-Maxwell equations (Sec. 2.2) to investigate the electromagnetic stability properties of an intense nonneutral ion beam propagating parallel to a uniform magnetic field $B_0\hat{e}_z$. (The results for a nonneutral electron beam are obtained simply by replacing the ions by electrons.) Following a discussion of the theoretical model and assumptions (Sec. 9.4.1), equilibrium properties are summarized for a uniform-density ion beam (Sec. 9.4.2). The linearized Vlasov-Maxwell equations (Sec. 9.4.3) are used to calculate detailed stability properties

(Secs. 9.4.4 and 9.4.5) with particular emphasis on perturbations with mode numbers $(\ell, n) = (0, 2)$, where ℓ and n are the azimuthal and radial mode numbers, respectively. The stability results are then extended to the case where the focusing force associated with the axial guide field is replaced by the (average) focusing of a periodic quadrupole field (Sec. 9.4.6), which is of particular interest for applications to heavy ion fusion.[26] One of the important conclusions of the analysis is that a modest beam rotation can completely stabilize the transverse instability for perturbations with mode numbers $(\ell, n) = (0, 2)$. The present kinetic analysis makes use of several of the theoretical techniques developed in Chapter 4, particularly in Sec. 4.9.

9.4.1   Theoretical Model and Assumptions

The present analysis assumes an intense nonneutral ion beam $(j = b)$ with characteristic energy $\gamma_b m_i c^2$, density $\hat{n}_b$, and radius r_b propagating parallel to a uniform axial guide field $B_0 \hat{e}_z$ (Fig. 9.8). No electrons or background plasma are present, and a perfectly conducting, cylindrical wall is located at radius $r = b$. Under steady-state conditions $(\partial/\partial t = 0)$, all equilibrium properties are assumed to be azimuthally symmetric $(\partial/\partial \theta = 0)$ and uniform in the z-direction $(\partial/\partial z = 0)$. As shown in the figure, the analysis employs cylindrical polar coordinates (r, θ, z), where r is the radial distance from the axis of symmetry. The following simplifying assumptions are made in investigating the equilibrium and stability properties.[25]

(a) We introduce Budker's parameter $\nu = N_b Z_i^2 e^2 / m_i c^2$, where $N_b = 2\pi \int_0^{r_b} dr'\, r' n_b^0(r') = \pi r_b^2 \bar{n}_b$ is the number of ions per unit axial length of the beam, m_i is the ion rest mass, and $+Z_i e$ is the ion charge. The present analysis assumes

$$\frac{\nu}{\gamma_b} = \frac{1}{4\gamma_b} \frac{\bar{\omega}_{pb}^2 r_b^2}{c^2} \ll 1 \,, \tag{9.122}$$

where $\bar{\omega}_{pb}^2 = 4\pi \bar{n}_b Z_i^2 e^2 / m_i$ is the average nonrelativistic plasma frequency-squared.

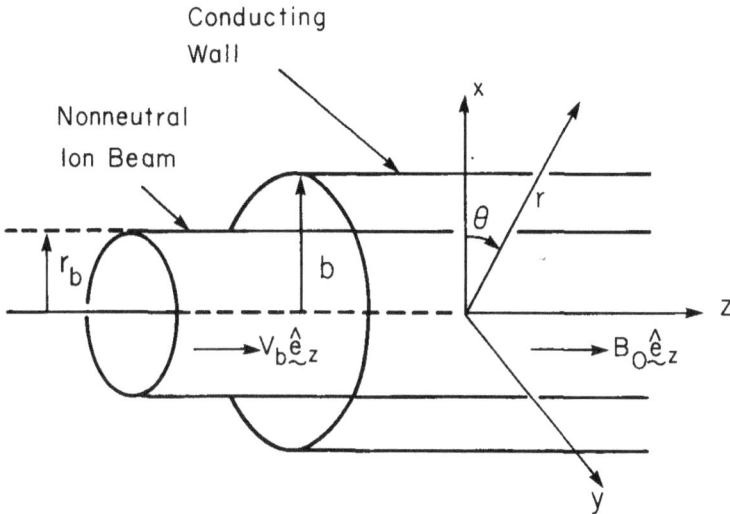

Figure 9.8. An intense nonneutral ion beam with radius r_b propagates with axial velocity $V_b = \beta_b c$ parallel to a solenoidal focusing field $B_0 \hat{e}_z$.

(b) Consistent with Eq.(9.122), it is also assumed that the directed axial energy of the beam ions is large in comparison with the transverse kinetic energy, i.e.,

$$p_z^2 \gg p_r^2 + p_\theta^2 \ . \tag{9.123}$$

Equation (9.123) (consistent with the paraxial approximation[2]) together with Eq.(9.122) permit a closed analytical determination of the particle trajectories in the equilibrium field configuration, at least in circumstances where the beam density is uniform out to radius $r = r_b$.

(c) The ion beam, in addition to propagating axially with average equilibrium velocity $V_{zb}^0(r) = (\int d^3p\, v_z f_b^0)/(\int d^3p\, f_b^0)$, generally rotates with nonzero azimuthal velocity $V_{\theta b}^0(r) = (\int d^3p\, v_\theta f_b^0)/(\int d^3p\, f_b^0)$. Here, $\mathbf{v} = \mathbf{p}/\gamma m_i$ is the particle velocity, $\mathbf{p}$ is the momentum, and $\gamma = (1 + \mathbf{p}^2/m_i^2 c^2)^{1/2}$ is the relativistic mass factor. The axial beam current $Z_i e n_b^0(r) V_{zb}^0(r)$ generates an azimuthal self-magnetic field $B_\theta^s(r)\hat{e}_\theta$. Consistent with Eq.(9.123), however, the axial diamagnetic field $B_z^s(r)\hat{e}_z$ produced by the beam rotation is neglected in comparison with the applied magnetic field $B_0\hat{e}_z$. Therefore, the equilibrium field configuration is of the form

$$\mathbf{E}^0(\mathbf{x}) = E_r(r)\hat{e}_r \,,$$
$$\mathbf{B}^0(\mathbf{x}) = B_0\hat{e}_z + B_\theta^s(r)\hat{e}_\theta \,, \qquad (9.124)$$

where the radial electric field $E_r(r)$ is produced by the space charge of the beam ions.

(d) Consistent with Eq.(9.124), self-consistent Vlasov equilibria for the beam ions are generally of the form (Sec. 4.1)

$$f_b^0(r,\mathbf{p}) = f_b^0(H, P_\theta, P_z) \,. \qquad (9.125)$$

Here,

$$H = (m_i^2 c^4 + c^2\mathbf{p}^2)^{1/2} - m_i c^2 + Z_i e\phi_0(r) \,,$$
$$P_\theta = r\left(p_\theta + m_i r\omega_{cb}/2\right) \,, \qquad (9.126)$$
$$P_z = p_z + \frac{Z_i e}{c} A_z^s(r) \,,$$

are the single-particle constants of the motion corresponding to the energy (H), the canonical angular momentum (P_θ), and the axial canonical momentum (P_z). In Eq.(9.126), $\omega_{cb} = Z_i e B_0/m_i c$ is the nonrelativistic ion cyclotron frequency, and the equilibrium self-field potentials $\phi_0(r)$ and $A_z^s(r)$ are determined from the steady-state Maxwell equations

$$\frac{1}{r}\frac{\partial}{\partial r}r\frac{\partial}{\partial r}\phi_0(r) = -4\pi Z_i e n_b^0(r) = -4\pi Z_i e \int d^3p\, f_b^0(r,\mathbf{p}) \,, \quad (9.127)$$

and

$$\frac{1}{r}\frac{\partial}{\partial r}r\frac{\partial}{\partial r}A_z^s(r) = -\frac{4\pi}{c}Z_i e n_b^0(r)V_{zb}^0(r) = -\frac{4\pi}{c}Z_i e \int d^3p\, v_z f_b^0(r,\mathbf{p}) \,,$$
$$(9.128)$$

where $-\partial\phi_0/\partial r = E_r(r)$, and $-\partial A_z^s/\partial r = B_\theta^s(r)$.

(e) The stability analysis, which is based on the linearized Vlasov-Maxwell equations (Sec. 2.2), assumes low-frequency flute perturbations with

$$|\omega| r_b / c \ll 1 \, ,$$

$$k_z = 0 \, . \tag{9.129}$$

Here, ω is the (complex) oscillation frequency, and $k_z = 0$ corresponds to flute perturbations with $\partial/\partial z = 0$.

(f) For specified equilibrium distribution function $f_b^0(r, \mathbf{p}) = f_b^0(H, P_\theta, P_z)$, the linearized Vlasov equation for the perturbed ion distribution function $\delta f_b(\mathbf{x}_\perp, \mathbf{p}, t)$ is given by

$$\left\{ \frac{\partial}{\partial t} + \mathbf{v}_\perp \cdot \frac{\partial}{\partial \mathbf{x}_\perp} + Z_i e \left(\mathbf{E}^0 + \frac{1}{c} \mathbf{v} \times \mathbf{B}^0 \right) \cdot \frac{\partial}{\partial \mathbf{p}} \right\} \delta f_b$$

$$= -Z_i e \left\{ \left(\delta E_\theta + \frac{1}{c} v_z \, \delta B_r - \frac{1}{c} v_r \, \delta B_z \right) \frac{\partial f_b^0}{\partial p_\theta} \right.$$

$$+ \left(\delta E_r - \frac{1}{c} v_z \, \delta B_\theta + \frac{1}{c} v_\theta \delta B_z \right) \frac{\partial f_b^0}{\partial p_r}$$

$$\left. + \left(\delta E_z + \frac{1}{c} v_r \, \delta B_\theta - \frac{1}{c} v_\theta \delta B_r \right) \frac{\partial f_b^0}{\partial p_z} \right\} \, . \tag{9.130}$$

In Eq.(9.130), $\mathbf{x}_\perp = (r, \theta)$ is the perpendicular spatial coordinate, $\mathbf{v} = \mathbf{p}/\gamma m_i$ is the particle velocity, and $\delta \mathbf{E}(\mathbf{x}_\perp, t)$ and $\delta \mathbf{B}(\mathbf{x}_\perp, t)$ are the perturbed electromagnetic fields. Consistent with the assumptions that $\nu/\gamma_b \ll 1$, $v_r^2 \ll v_z^2$, $v_\theta^2 \ll v_z^2$, $|\omega| r_b / c \ll 1$, and $k_z = 0$, the terms in Eq.(9.130) proportional to $(v_r/c) \delta B_z$, $(v_\theta/c) \delta B_z$, and $(\delta E_z + v_r \delta B_\theta / c - v_\theta \delta B_r / c)$ are treated as negligibly small in comparison with the terms proportional to $(\delta E_\theta + v_z \delta B_r / c)$ and $(\delta E_r - v_z \delta B_\theta / c)$. Therefore, the linearized Vlasov equation (9.130) is approximated by[25]

$$\left\{ \frac{\partial}{\partial t} + \mathbf{v}_\perp \cdot \frac{\partial}{\partial \mathbf{x}_\perp} + Z_i e \left(\mathbf{E}^0 + \frac{1}{c} \mathbf{v} \times \mathbf{B}^0 \right) \cdot \frac{\partial}{\partial \mathbf{p}} \right\} \delta f_b \tag{9.131}$$

$$= -Z_i e \left\{ \left(\delta E_\theta + \frac{1}{c} v_z \delta B_r \right) \frac{\partial f_b^0}{\partial p_\theta} + \left(\delta E_r - \frac{1}{c} v_z \delta B_\theta \right) \frac{\partial f_b^0}{\partial p_r} \right\} \, .$$

The perturbed electromagnetic field components in Eq.(9.131) are to be determined self-consistently from the linearized Maxwell equations.

9.4.2 Kapchinskij-Vladimirskij Beam Equilibrium

Making use of the assumptions outlined in Sec. 9.4.1, the linearized Vlasov-Maxwell equations can be used to investigate detailed stability properties for perturbations about a wide variety of equilibrium distribution functions $f_b^0(r, \mathbf{p}) = f_b^0(H, P_\theta, P_z)$. For purposes of analytical simplification, we specialize to the choice of distribution function

$$f_b^0(r, \mathbf{p}) = \frac{\hat{n}_b}{2\pi\gamma_b m_i} \delta \left[H - \omega_{rb} P_\theta - (\hat{\gamma}_b - 1)m_i c^2 \right] \delta \left(P_z - \gamma_b m_i \beta_b c \right) ,$$

$$(9.132)$$

where H, P_θ and P_z are the single-particle constants of the motion defined in Eq.(9.126), and $\hat{n}_b$, $\gamma_b = (1 - \beta_b^2)^{-1/2}$, ω_{rb} and $\hat{\gamma}_b$ are constants.[25] Equation (9.132) will be recognized as the relativistic generalization of the equilibrium distribution function in Eq.(4.242), assuming zero spread in axial momentum p_z. Note from Eq.(9.132) that f_b^0 has an inverted population in $H - \omega_{rb} P_\theta$, which is peaked at $(\hat{\gamma}_b - 1)m_i c^2$. Moreover, the distribution function in Eq.(9.132) incorporates the influence of the equilibrium self-magnetic field $B_\theta^s(r) = -\partial A_z^s/\partial r$ (through the dependence on P_z), as well as the equilibrium self-electric field $E_r(r) = -\partial \phi_0/\partial r$ (through the dependence on H). As in Sec. 4.9, we will find that the equilibrium profiles calculated from Eq.(9.132) have relatively simple forms. For example, the density profile $n_b^0(r)$ is uniform (and equal to $\hat{n}_b$) out to radius $r = r_b$.

The delta-function dependence in Eq.(9.132) selects $P_z = \gamma_b m_i \beta_b c$, which corresponds to axial momentum $p_z = \gamma_b m_i \beta_b c - Z_i e A_z^s(r)/c$. In addition, we Taylor expand $\gamma m_i c^2 = [m_i^2 c^4 + c^2 p_z^2 + c^2(p_r^2 + p_\theta^2)]^{1/2}$ for $p_r^2 + p_\theta^2 \ll p_z^2$ and $|Z_i e A_z^s/\gamma_b m_i \beta_b c^2| \ll 1$, which is consistent with $\nu/\gamma_b \ll 1$. For $P_z = \gamma_b m_i \beta_b c$, it is readily shown that the combination $H - \omega_{rb} P_\theta$ occurring in Eq.(9.132) can be approximated by (Problem 9.15)

$$H - \omega_{rb} P_\theta = (\gamma_b - 1)m_i c^2 + \frac{1}{2\gamma_b m_i} \left[p_r^2 + (p_\theta - \gamma_b m_i \omega_{rb} r)^2 \right] + \psi_b(r) .$$

$$(9.133)$$

Here, the effective potential $\psi_b(r)$ is defined by

$$\psi_b(r) = \frac{1}{2}\gamma_b m_i \left(-\omega_{rb}\omega_{cb}/\gamma_b - \omega_{rb}^2\right) r^2 + Z_i e \left[\phi_0(r) - \beta_b A_z^s(r)\right] , \quad (9.134)$$

and $(2\gamma_b m_i)^{-1}[p_r^2 + (p_\theta - \gamma_b m_i \omega_{rb} r)^2]$ is the transverse kinetic energy in a frame of reference rotating with angular velocity $\omega_{rb} = \text{const.}$ Consistent with Eqs.(9.122) and (9.123), the equilibrium distribution function in Eq.(9.132) can be expressed in the approximate form

$$f_p^0(r, \mathbf{p}) = \frac{\hat{n}_b}{2\pi\gamma_b m_i} \delta \left[\frac{p_r^2 + (p_\theta - \gamma_b m_i \omega_{rb} r)^2}{2\gamma_b m_i} + \psi_b(r) - (\hat{\gamma}_b - \gamma_b) m_i c^2\right]$$
$$\times \delta(p_z - \gamma_b m_i \beta_b c) , \quad (9.135)$$

where $\psi_b(r)$ is defined in Eq.(9.134).[25] Therefore, for $\nu/\gamma_b \ll 1$ and $p_r^2 + p_\theta^2 \ll p_z^2$, it is evident from Eq.(9.135) that the average axial velocity $V_{zb}^0(r) = (\int d^3 p\, v_z f_b^0)/(\int d^3 p\, f_b^0)$ and the average azimuthal velocity $V_{\theta b}^0(r) = (\int d^3 p\, v_\theta f_b^0)/(\int d^3 p f_b^0)$ can be approximated by

$$V_{zb}^0(r) = \beta_b c = \text{const.} ,$$
$$V_{\theta b}^0(r) = \omega_{rb} r , \quad (9.136)$$

over the radial extent of the ion beam.

Note from Eq.(9.134) that the combination $\phi_0(r) - \beta_b A_z^s(r)$ is required in order to evaluate the effective potential $\psi_b(r)$. Making use of Eq.(9.136), it follows from the equilibrium Maxwell equations (9.127) and (9.128) that $\phi_0(r) - \beta_b A_z^s(r)$ satisfies

$$\frac{1}{r}\frac{\partial}{\partial r} r \frac{\partial}{\partial r} [\phi_0(r) - \beta_b A_z^s(r)] = -4\pi Z_i e \left(1 - \beta_b^2\right) n_b^0(r)$$
$$= -\frac{4\pi Z_i e}{\gamma_b^2} \int d^3 p\, f_b^0(r, \mathbf{p}) , \quad (9.137)$$

where $\gamma_b = (1 - \beta_b^2)^{-1/2}$. For the choice of distribution function in Eq.(9.135), the equilibrium density profile $n_b^0(r) = \int d^3 p f_b^0$ is calculated in a manner completely analogous to Sec. 4.2.3. Expressing $\int d^3 p \cdots = 2\pi \int_0^\infty dp_\perp\, p_\perp \int_{-\infty}^\infty dp_z \cdots$, where $p_\perp^2/2\gamma_b m_i \equiv [p_r^2 + (p_\theta - \gamma_b m_i \omega_{rb} r)^2]/2\gamma_b m_i$

is the transverse kinetic energy in the rotating frame, we obtain

$$
n_b^0(r) = \hat{n}_b \int_0^\infty d\left(\frac{p_\perp^2}{2\gamma_b m_i}\right) \delta\left[\frac{p_\perp^2}{2\gamma_b m_i} + \psi_b(r) - (\hat{\gamma}_b - \gamma_b)m_i c^2\right].
$$

$$(9.138)$$

The integral in Eq.(9.138) is equal to unity for $\psi_b(r) < (\hat{\gamma}_b - \gamma_b)m_i c^2$, and equal to zero for $\psi_b(r) > (\hat{\gamma}_b - \gamma_b)m_i c^2$. Therefore, the density profile $n_b^0(r)$ has the simple rectangular form

$$
n_b^0(r) = \begin{cases} \hat{n}_b, & 0 \le r < r_b, \\ 0, & r_b < r \le b, \end{cases} \tag{9.139}
$$

where the beam radius r_b is determined from

$$
\psi(r_b) = (\hat{\gamma}_b - \gamma_b)m_i c^2, \tag{9.140}
$$

Substituting Eq.(9.139) into Eq.(9.137) and assuming $\phi_0(r = 0) = 0 = A_z^s(r = 0)$ gives

$$
\phi_0(r) - \beta_b A_z^s(r) = -\frac{m_i}{4Z_i e}\omega_{pb}^2 \left(1 - \beta_b^2\right) r^2 \tag{9.141}
$$

for $0 \le r < r_b$. Here, $\omega_{pb}^2 = 4\pi\hat{n}_b Z_i^2 e^2/m_i$ is the nonrelativistic plasma frequency-squared. Therefore, within the ion beam ($0 \le r < r_b$), the effective potential $\psi_b(r)$ defined in Eq.(9.134) can be expressed as

$$
\psi_b(r) = \frac{1}{2}\gamma_b m_i \Omega_\beta^2 r^2, \tag{9.142}
$$

where Ω_β^2 is defined by

$$
\Omega_\beta^2 = -\omega_{rb}\frac{\omega_{cb}}{\gamma_b} - \omega_{rb}^2 - \frac{\omega_{pb}^2}{2\gamma_b}\left(1 - \beta_b^2\right) = \left(\omega_{rb} - \omega_{rb}^+\right)\left(\omega_{rb}^- - \omega_{rb}\right). \tag{9.143}
$$

In Eq.(9.143), ω_{rb}^+ and ω_{rb}^- are defined by

$$
\omega_{rb}^\pm = -\frac{\omega_{cb}}{2\gamma_b}\left\{1 \pm \left(1 - \frac{2\gamma_b\omega_{pb}^2}{\omega_{cb}^2}\left(1 - \beta_b^2\right)\right)^{1/2}\right\}, \tag{9.144}
$$

where $\omega_{cb} = Z_i e B_0/m_i c$ and $1 - \beta_b^2 = \gamma_b^{-2}$. In Eqs.(9.143) and (9.144), the (focusing) contribution proportional to $\omega_{pb}^2\beta_b^2$ arises from the azimuthal self-magnetic field $B_\theta^s(r)$, whereas the (defocusing) contribution proportional to $-\omega_{pb}^2$ arises from the radial self-electric field $E_r(r)$ produced by the beam space charge.

For ions, note from Eq.(9.144) that $\omega_{rb}^+ < \omega_{rb}^- < 0$. Moreover, positive $\Omega_\beta^2 > 0$ requires that ω_{rb} lie in the interval $\omega_{rb}^+ < \omega_{rb} < \omega_{rb}^-$. Substituting Eq.(9.142) into Eq.(9.140) gives $(\hat\gamma_b - \gamma_b)m_i c^2 = \gamma_b m_i \Omega_\beta^2 r_b^2/2$, or equivalently,

$$r_b = \left[\frac{2c^2(\hat\gamma_b - \gamma_b)}{\gamma_b \Omega_\beta^2} \right]^{1/2}, \tag{9.145}$$

which determines the beam radius r_b in terms of $\hat\gamma_b - \gamma_b$ and Ω_β. Making use of Eqs.(9.135), (9.139), and (9.142), the transverse $(r-\theta)$ temperature profile $T_{b\perp}^0(r)$ of the beam ions is readily shown to be

$$T_{b\perp}^0(r) = \left[\int d^3p \frac{[p_r^2 + (p_\theta - \gamma_b m_i \omega_{rb} r)^2]}{2\gamma_b m_i} f_b^0 \right] \left[\int d^3p\, f_b^0 \right]^{-1}$$

$$= \hat T_{b\perp} \left(1 - \frac{r^2}{r_b^2} \right) \tag{9.146}$$

for $0 \le r < r_b$. Here, the constant $\hat T_{b\perp}$ (related to the transverse beam emittance) is defined by

$$\hat T_{b\perp} = (\hat\gamma_b - \gamma_b)m_i c^2 = \frac{1}{2}\gamma_b m_i \Omega_\beta^2 r_b^2. \tag{9.147}$$

Note from Eq.(9.146) that $\hat T_{b\perp}^0(r)$ assumes a maximum value of $\hat T_{b\perp}$ on axis $(r = 0)$ and decreases monotonically to zero at the edge of the ion beam $(r = r_b)$.

Making use of Eqs.(9.142) and (9.147), the equilibrium distribution function in Eq.(9.135) can be expressed as

$$f_b^0(r, \mathbf{p}) = \frac{\hat n_b}{\pi} \delta \left[p_r^2 + (p_\theta - \gamma_b m_i \omega_{rb} r)^2 - \gamma_b^2 m_i^2 \Omega_\beta^2 r_b^2 \left(1 - \frac{r^2}{r_b^2} \right) \right]$$

$$\times \delta\left(p_z - \gamma_b m_i \beta_b c \right), \tag{9.148}$$

where Ω_β^2 is defined in Eq.(9.143). Equation (9.148) is expressed in the familiar form of the Kapchinskij-Vladimirskij distribution function,[49] generalized to include the effects of beam rotation $(\omega_{rb} \ne 0)$, and specialized to the case of a circular ion beam with solenoidal focusing field $B_0 \hat{\mathbf{e}}_z$.

Equation (9.148), appropriately generalized to include the average effects of a quadrupole focusing field (Sec. 9.4.6) or an alternating-gradient focusing field (Chapter 10), is widely used to investigate the stability properties of intense nonneutral ion beams for heavy-ion fusion applications.[45,46] An important feature of the distribution function f_b^0 in Eq.(9.148) is that the radius of the ion beam ($r = r_b$) represents the outer envelope of turning points in the $r - \theta$ orbits for which

$$p_r^2 + (p_\theta - \gamma_b m_i \omega_{rb} r_b)^2 = 0 .$$

As a final point regarding the beam equilibrium described by Eq.(9.148), we calculate the range of beam density (as measured by ω_{pb}^2) which can be confined by the solenoidal focusing field $B_0 \hat{e}_z$. In this regard, it is convenient to introduce the effective ion thermal speed v_{Tb} and thermal

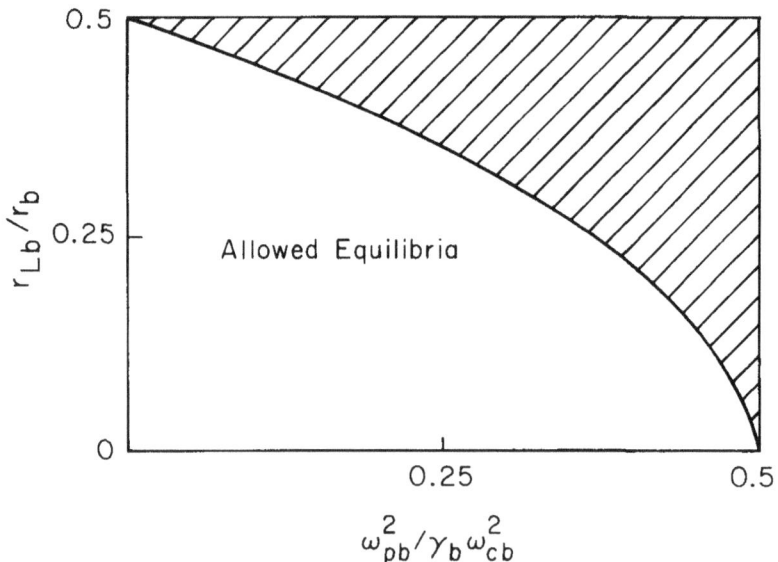

Figure 9.9.   Region of parameter space $(\omega_{pb}^2/\gamma_b \omega_{cb}^2, r_{Lb}/r_b)$ corresponding to allowed ion beam equilibria consistent with Eq.(9.151). Radially confined beam equilibria do not exist in the shaded region.

Larmor radius r_{Lb} defined by $v_{Tb} = (2\hat{T}_{b\perp}/\gamma_b m_i)^{1/2}$ and $r_{Lb} = \gamma_b v_{Tb}/\omega_{cb}$. Equation (9.147) can then be expressed as

$$\frac{r_{Lb}^2}{r_b^2}\frac{\omega_{cb}^2}{\gamma_b^2} = \Omega_\beta^2 = -\omega_{rb}\frac{\omega_{cb}}{\gamma_b} - \omega_{rb}^2 - \frac{\omega_{pb}^2}{2\gamma_b}\left(1 - \beta_b^2\right), \qquad (9.149)$$

which is simply a statement of radial force balance on a beam fluid element. Solving Eq.(9.149) for the rotation frequency ω_{rb} gives the two allowed values

$$\omega_{rb} = -\frac{\omega_{cb}}{2\gamma_b}\left\{1 \pm \left(1 - \frac{4r_{Lb}^2}{r_b^2} - \frac{2\gamma_b\omega_{pb}^2}{\omega_{cb}^2}\left(1 - \beta_b^2\right)\right)^{1/2}\right\}, \qquad (9.150)$$

where $\omega_{rb} < 0$. In the cold-beam limit $(r_{Lb} \ll r_b)$, note that Eq.(9.150) reduces (as expected) to $\omega_{rb} = \omega_{rb}^\pm$ defined in Eq.(9.144). Moreover, it is clear from Eqs.(9.149) and (9.150) that the inequality

$$\frac{2\gamma_b\omega_{pb}^2}{\omega_{cb}^2}\left(1 - \beta_b^2\right) \leq 1 - \frac{4r_{Lb}^2}{r_b^2} \qquad (9.151)$$

must be satisfied for radially confined equilibrium solutions to exist.[25] For specified values of β_b and $r_{Lb}/r_b < 1/2$, the inequality in Eq.(9.151) places a limit on the normalized beam density $\omega_{pb}^2/\omega_{cb}^2$ which can be confined radially by the solenoidal focusing field $B_0\hat{e}_z$ (Fig. 9.9).

Problem 9.15   Consider the combination of variables

$$H - \omega_{rb}P_\theta = \left(m_i^2c^4 + c^2p_z^2 + c^2p_r^2 + c^2p_\theta^2\right)^{1/2} - m_ic^2 + Z_ie\phi_0(r)$$
$$-\omega_{rb}r\left(p_\theta + m_i\omega_{cb}r/2\right), \qquad (9.15.1)$$

which occurs in the equilibrium distribution function f_b^0 in Eq.(9.132). Evaluate Eq.(9.15.1) for $p_z = \gamma_b m_i\beta_b c - Z_ieA_z^s(r)/c$, assuming $p_z^2 \gg p_r^2 + p_\theta^2$ and $|Z_ieA_z^s/\gamma_b m_i\beta_b c^2| \ll 1$. Show that Eq.(9.15.1) can be approximated by Eq.(9.133), where $\gamma_b = (1 - \beta_b^2)^{-1/2}$, and the effective potential $\psi_b(r)$ is defined in Eq.(9.134).

Problem 9.16   Consider the equilibrium distribution function $f_b^0(r, \mathbf{p})$ specified by Eq.(9.148).

a. Show that the ion density profile $n_b^0(r) = \int d^3p\, f_b^0$ has the rectangular form given in Eq.(9.139).

b. Show that the transverse temperature profile $T_{b\perp}^0(r)$ of the beam ions has the parabolic form given in Eq.(9.146).

9.4.3   Linearized Vlasov-Maxwell Equations

We now make use of the linearized Vlasov-Maxwell equations and Assumptions (e) and (f) in Sec. 9.4.1 to investigate detailed stability behavior for electromagnetic perturbations about the equilibrium distribution function $f_b^0(r, \mathbf{p})$ in Eq.(9.148). The analysis is considerably simplified because the ion motion perpendicular to $B_0\hat{\mathbf{e}}_z$ is effectively nonrelativistic when $p_r^2 + p_\theta^2 \ll p_z^2$ and $\nu/\gamma_b \ll 1$, and the perturbations are about the rectangular density profile $n_b^0(r)$ specified by Eq.(9.139). From Eqs.(9.129) and (9.131), the required components of the perturbed electromagnetic field $(\delta E_r, \delta E_\theta, \delta B_r, \delta B_\theta)$ are determined from the scalar electrostatic potential $\delta\phi(r, \theta, t)$ and the z-component of the vector potential $\delta A_z(r, \theta, t)$, where

$$\delta\mathbf{E} = -\nabla\delta\phi - c^{-1}\left(\frac{\partial}{\partial t}\right)\delta\mathbf{A}\,, \quad \delta\mathbf{B} = \nabla \times \delta\mathbf{A}\,.$$

For $\partial/\partial z = 0$, we find

$$\delta E_\theta + \frac{1}{c}v_z\delta B_r = -\frac{1}{r}\frac{\partial}{\partial\theta}\left(\delta\phi - \frac{1}{c}v_z\delta A_z\right)\,,$$

$$\delta E_r - \frac{1}{c}v_z\delta B_\theta = -\frac{\partial}{\partial r}\left(\delta\phi - \frac{1}{c}v_z\delta A_z\right)\,, \qquad (9.152)$$

where

$$\nabla^2\delta\phi = -4\pi Z_i e \int d^3p\, \delta f_b\,,$$

$$\nabla^2\delta A_z = -4\pi Z_i e \int d^3p\, v_z\delta f_b + \frac{1}{c^2}\frac{\partial^2}{\partial t^2}\delta A_z\,, \qquad (9.153)$$

In Eq.(9.153), the perturbed ion distribution function $\delta f_b(\mathbf{x}_\perp, \mathbf{p}, t)$ solves the linearized Vlasov equation (9.131) which can be expressed as

$$\left\{ \frac{\partial}{\partial t} + \mathbf{v}_\perp \cdot \frac{\partial}{\partial \mathbf{x}_\perp} + Z_i e \left(\mathbf{E}^0 + \frac{1}{c} \mathbf{v} \times \mathbf{B}^0 \right) \cdot \frac{\partial}{\partial \mathbf{p}} \right\} \delta f_b$$

$$= Z_i e \left[\frac{\partial}{\partial \mathbf{x}_\perp} \left(\delta \phi - \frac{1}{c} v_z \delta A_z \right) \right] \cdot \frac{\partial f_b^0}{\partial \mathbf{p}_\perp} , \qquad (9.154)$$

where $\mathbf{x}_\perp$ denotes the perpendicular spatial coordinates (r, θ) or (x, y). For low-frequency perturbations with $|\omega| r_b / c \ll 1$ [Eq.(9.129)], the term $c^{-2} \partial^2 \delta A_z / \partial t^2$ in Eq.(9.153) can be neglected in comparison with $\nabla^2 \delta A_z$. Moreover, for $p_r^2 + p_\theta^2 \ll p_z^2$, the delta function $\delta(p_z - \gamma_b m_i \beta_b c)$ in Eq.(9.148) selects $v_z = \beta_b c$ in Eqs.(9.153) and (9.154) when integrations over p_z are carried out. Therefore, for $\partial / \partial z = 0$ and $|\omega| r_b / c \ll 1$, we introduce the effective potential

$$\delta \psi(r, \theta, t) = \delta \phi(r, \theta, t) - \beta_b \delta A_z(r, \theta, t) . \qquad (9.155)$$

Equations (9.153)–(9.155) can be combined to give (Problem 9.17)

$$\left(\frac{1}{r} \frac{\partial}{\partial r} r \frac{\partial}{\partial r} + \frac{1}{r^2} \frac{\partial^2}{\partial \theta^2} \right) \delta \psi = -4\pi Z_i e \left(1 - \beta_b^2 \right) \int d^3 p \, \delta f_b , \qquad (9.156)$$

where $\delta f_b(\mathbf{x}_\perp, \mathbf{p}, t)$ evolves according to

$$\left\{ \frac{\partial}{\partial t} + \mathbf{v}_\perp \cdot \frac{\partial}{\partial \mathbf{x}_\perp} + Z_i e \left(\mathbf{E}^0 + \frac{1}{c} \mathbf{v} \times \mathbf{B}^0 \right) \cdot \frac{\partial}{\partial \mathbf{p}} \right\} \delta f_b = Z_i e \left(\frac{\partial}{\partial \mathbf{x}_\perp} \delta \psi \right) \cdot \frac{\partial f_b^0}{\partial \mathbf{p}_\perp} .$$

$$(9.157)$$

Here, $f_b^0(r, \mathbf{p})$ is the equilibrium distribution function specified in Eq.(9.148), and the equilibrium fields are given by $\mathbf{E}^0(\mathbf{x}) = E_r(r) \hat{\mathbf{e}}_r$ and $\mathbf{B}^0(\mathbf{x}) = B_0 \hat{\mathbf{e}}_z + B_\theta^s(r) \hat{\mathbf{e}}_\theta$, where $E_r = -\partial \phi_0 / \partial r = (m_i / 2 Z_i e) \omega_{pb}^2 r$ and $B_\theta^s = -\partial A_z^s / \partial r = (m_i / 2 Z_i e) \beta_b \omega_{pb}^2 r$ within the ion beam ($0 \leq r < r_b$).

An important feature of Eqs.(9.156) and (9.157) is that they are identical in form (for $\partial / \partial z = 0$) to the linearized Vlasov-Poisson equations (4.182) and (4.183) analyzed in Secs. 4.7 and 4.8 and solved exactly in Sec. 4.9 for the case of surface perturbations (eigenfunction peaked at $r = r_b$) about the uniform-density plasma column described by Eqs.(4.243) and (4.244). The only differences in the present analysis are:

$\delta\phi$ is replaced by $\delta\psi$ in Eqs.(9.156) and (9.157); there is a factor of $1-\beta_b^2 = 1/\gamma_b^2$ multiplying the perturbed ion density $\delta n_b = \int d^3p \, \delta f_b$ in Eq.(9.156), which plays the role of Poisson's equation; and the equilibrium force term $Z_i e(\mathbf{E}^0 + \mathbf{v} \times \mathbf{B}^0/c)$ occurring in the linearized Vlasov equation (9.157) also includes the influence of the azimuthal self-magnetic field $B_\theta^s(r)\hat{\mathbf{e}}_\theta$ produced by the axial beam current.

With straightforward modifications, the theoretical techniques[24] developed in Sec. 4.9 can be applied virtually intact to the analysis of Eqs.(9.156) and (9.157).[25] For example, the orbit equation (4.253) for the transverse ion motion in the equilibrium field configuration is modified to become

$$\left[\frac{d^2}{dt'^2} + i\frac{\omega_{cb}}{\gamma_b}\frac{d}{dt'} - \frac{1}{2}\frac{\omega_{pb}^2}{\gamma_b}\left(1 - \beta_b^2\right) \right] [x'(t') + iy'(t')] = 0 , \qquad (9.158)$$

where $\omega_{cb} = Z_i e B_0/m_i c$, $\omega_{pb}^2 = 4\pi\hat{n}_b Z_i^2 e^2/m_i$, and the contribution proportional to $\omega_{pb}^2\beta_b^2$ is associated with the $v_z\hat{\mathbf{e}}_z \times B_\theta^s(r)\hat{\mathbf{e}}_\theta$ force. Here, the transverse orbits $x'(t')$, $y'(t')$, $p_x'(t') = \gamma_b m_i dx'/dt'$ and $p_y'(t') = \gamma_b m_i dy'/dt'$, which pass through the phase-space point (x, y, p_x, p_y) at time $t' = t$, are required in order to solve the linearized Vlasov equation (9.157) using the method of characteristics. We introduce the perpendicular momentum $p_\perp$ in a frame of reference rotating with angular velocity ω_{rb} defined by

$$p_\perp^2 \equiv p_r^2 + (p_\theta - \gamma_b m_i \omega_{rb} r)^2 = (p_x + \gamma_b m_i \omega_{rb} y)^2 + (p_y - \gamma_b m_i \omega_{rb} x)^2 .$$
$$(9.159)$$

The orbit equation (9.158) is to be solved subject to the "initial" conditions (Sec. 4.9)

$$x'(t' = t) + iy'(t' = t) = x + iy \equiv r\exp(i\theta) ,$$
$$p_x'(t' = t) + \gamma_b m_i \omega_{rb} y'(t' = t) = p_x + \gamma_b m_i \omega_{rb} y \equiv p_\perp \cos\phi ,$$
$$p_y'(t' = t) - \gamma_b m_i \omega_{rb} x'(t' = t) = p_y - \gamma_b m_i \omega_{rb} x \equiv p_\perp \sin\phi , \quad (9.160)$$

where $(p_\perp, \phi)$ are momentum-space variables appropriate to the rotating frame. Integrating Eq.(9.158) and enforcing the boundary conditions in Eq.(9.160), we obtain (Problem 9.18)

$$x'(t') + iy'(t') = (\omega_{rb}^- - \omega_{rb}^+)^{-1} \left\{ i \left(\frac{p_\perp}{\gamma_b m_i} \right) \exp(i\phi) \left[\exp(i\omega_{rb}^+ \tau) - \exp(i\omega_{rb}^- \tau) \right] \right.$$

$$\left. + r \exp(i\theta) \left[(\omega_{rb} - \omega_{rb}^+) \exp(i\omega_{rb}^- \tau) + (\omega_{rb}^- - \omega_{rb}) \exp(i\omega_{rb}^+ \tau) \right] \right\}, \quad (9.161)$$

where $\tau = t' - t$, and $p'_x + ip'_y = \gamma_b m_i (d/dt')(x' + iy')$. Note from Eq.(9.161) that the perpendicular ion motion in the laboratory frame is biharmonic at the frequencies ω_{rb}^+ and ω_{rb}^- defined in Eq.(9.144).

Paralleling the treatment in Sec. 4.9, we analyze Eqs.(9.156) and (9.157) using the method of characteristics to integrate the linearized Vlasov equation (9.157). The perturbed potential $\delta\psi(r, \theta, t)$ is expressed as $\delta\psi(r, \theta, t) = \sum_{\ell=-\infty}^{\infty} \delta\psi^\ell(r) \exp(i\ell\theta - i\omega t)$, where $\text{Im}\,\omega > 0$ corresponds to instability (temporal growth). Some straightforward algebra gives

$$\frac{1}{r}\frac{\partial}{\partial r} r \frac{\partial}{\partial r} \delta\psi^\ell(r) - \frac{\ell^2}{r^2}\delta\psi^\ell(r) \quad (9.162)$$

$$= -4\pi Z_i^2 e^2 \gamma_b m_i \left(1 - \beta_b^2\right) \int d^3p \frac{1}{p_\perp}\frac{\partial f_b^0}{\partial p_\perp} \left[\delta\psi^\ell(r) + (\omega - \ell\omega_{rb}) I_\ell\right],$$

which should be compared with Eq.(4.258). Here, $\int d^3p \cdots = 2\pi \times \int_0^\infty dp_\perp p_\perp \int_{-\infty}^\infty dp_z \cdots$, and the orbit integral $I_\ell(\omega, r, p_\perp)$ is defined by

$$I_\ell = i \int_0^{2\pi} \frac{d\phi}{2\pi} \int_{-\infty}^0 d\tau \delta\psi^\ell[r'(\tau)] \exp\{i\ell[\theta'(\tau) - \theta] - i\omega\tau\}, \quad (9.163)$$

where $r'(\tau)\exp[i\theta'(\tau)] = x'(\tau) + iy'(\tau)$ is determined from Eq.(9.161). Moreover, making use of Eqs.(9.148) and (9.159), the equilibrium distribution function $f_b^0(r, \mathbf{p})$ occurring in the eigenvalue equation (9.162) can be expressed as

$$f_b^0(r, \mathbf{p}) = \frac{\hat{n}_b}{\pi}\delta \left[p_\perp^2 - \gamma_b^2 m_i^2 \Omega_\beta^2 r_b^2 \left(1 - \frac{r^2}{r_b^2}\right)\right] \delta(p_z - \gamma_b m_i \beta_b c), \quad (9.164)$$

where $\Omega_\beta^2 r_b^2 = 2\hat{T}_{b\perp}/\gamma_b m_i$ is related to the transverse beam temperature by Eq.(9.147). For the choice of distribution function in Eq.(9.164), the integration over momentum in Eq.(9.162) can be carried out exactly [compare with Eq.(4.259)]. This gives the eigenvalue equation

$$\frac{1}{r}\frac{\partial}{\partial r}\, r\, \frac{\partial}{\partial r}\delta\psi^\ell(r) - \frac{\ell^2}{r^2}\,\delta\psi^\ell(r)$$

$$= \frac{\omega_{pb}^2 r_b}{\gamma_b^3 v_{Tb}^2}\left\{\delta\psi^\ell(r) + (\omega - \ell\omega_{rb})\,[I_\ell]_{p_\perp=0}\right\}\delta(r - r_b)$$

$$+\frac{\omega_{pb}^2}{\gamma_b^3}U(r_b - r)(\omega - \ell\omega_{rb})\left[\frac{\gamma_b^2 m_i^2}{p_\perp}\frac{\partial I_\ell}{\partial p_\perp}\right]_{p_\perp^2=p_\perp^2(r)}, \quad (9.165)$$

where $p_\perp^2(r) \equiv \gamma_b^2 m_i^2 \Omega_\beta^2 r_b^2(1 - r^2/r_b^2)$, $\omega_{pb}^2 = 4\pi\hat{n}_b Z_i^2 e^2/m_i$, $v_{Tb}^2 = 2\hat{T}_{b\perp}/\gamma_b m_i$, $\gamma_b = (1-\beta_b^2)^{-1/2}$, and $I_\ell(\omega, r, p_\perp)$ is the orbit integral defined in Eq.(9.163). Here, $U(r_b - r)$ is the Heaviside step function defined by $U(r_b - r) = 0$ for $r > r_b$, and $U(r_b - r) = +1$ for $0 \le r < r_b$. Equation (9.165) is the desired eigenvalue equation[25] which describes stability properties for low-frequency flute perturbations ($|\omega|r_b/c \ll 1$ and $k_z = 0$) about the Kapchinskij-Vladimirskij distribution function in Eq.(9.164). Note that the term proportional to $\delta(r - r_b)$ in Eq.(9.165) corresponds to a surface perturbation localized at $r = r_b$, whereas the term proportional to $U(r_b - r)$ corresponds to a body-wave perturbation extending throughout the interior of the ion beam ($0 \le r < r_b$). Analysis of Eq.(9.165) is particularly challenging because the orbit integral I_ℓ defined in Eq.(9.163) involves a time integration over $\delta\psi^\ell[r'(\tau)]\exp[i\ell\theta'(\tau)]$, where the orbits $r'(\tau)$ and $\theta'(\tau)$ sample the entire region $0 \le r'(\tau) < r_b$ and $0 \le \theta'(\tau) \le 2\pi$.

Problem 9.17   For $v_z = \beta_b c$ and low-frequency flute perturbations with $|\omega|r_b/c \ll 1$ and $\partial/\partial z = 0$, show that Eqs.(9.153)–(9.155) can be combined to give Eqs.(9.156) and (9.157), which are similar in form to the linearized Vlasov-Poisson equations analyzed in Sec. 4.9.

Problem 9.18   Solve the orbit equation (9.158) subject to the boundary conditions in (9.160). Show that $x'(t') + iy'(t')$ can be expressed in the form given in Eq.(9.161), where ω_{rb}^+ and ω_{rb}^- are the orbital oscillation frequencies defined in Eq.(9.144).

9.4.4 Solution to the Eigenvalue Equation

The eigenvalue equation (9.165) has been analyzed by Davidson and Uhm.[25] In general, the perturbed potential $\delta\psi^\ell(r)$ that solves Eq.(9.165) is of the form

$$\delta\psi^\ell(r) = Ar^\ell \sum_{j=0}^{n} a_j(\omega) \left(\frac{r}{r_b}\right)^{2j} \tag{9.166}$$

for $0 \leq r < r_b$. Here, ℓ is the azimuthal mode number, n is the radial mode number, A is a constant coefficient, and $a_0(\omega) = 1$ without loss of generality. The complex coefficients $a_j(\omega)$, $j \geq 1$, are determined self-consistently for each choice of mode numbers (ℓ, n). Moreover, for specified mode numbers (ℓ, n), the analysis leads to a dispersion relation of the form

$$D_\ell^n(\omega) = 0, \tag{9.167}$$

which determines the complex eigenfrequency ω.

A. Surface-Wave Perturbations with $n = 0$

Analysis of Eq.(9.165) for general value of ℓ and radial mode number $n = 0$ proceeds in a manner completely analogous to the derivation of the dispersion relation (4.269) in Sec. 4.9. Here, the eigenfunction $\delta\psi^\ell = Ar^\ell$ is peaked at the surface $(r = r_b)$ of the ion beam, and $\partial I_\ell/\partial p_\perp^2 = 0$ for the choice of equilibrium distribution function in Eq.(9.164). The resulting dispersion relation for $n = 0$ is given by

$$D_\ell^0 = 1 + \frac{\omega_{pb}^2 r_b^2}{2\ell\gamma_b^3 v_{Tb}^2} \left[1 - \left(\frac{r_b}{b}\right)^{2\ell}\right] \left\{1 - \left(\frac{\omega_{rb} - \omega_{rb}^+}{\omega_{rb}^- - \omega_{rb}^+}\right)^\ell \right. \tag{9.168}$$

$$\times \sum_{m=0}^{\ell} \frac{\ell!}{m!(\ell-m)!} \frac{\omega - \ell\omega_{rb}}{\omega - \ell\omega_{rb}^- + m(\omega_{rb}^- - \omega_{rb}^+)} \left(\frac{\omega_{rb}^- - \omega_{rb}}{\omega_{rb} - \omega_{rb}^+}\right)^m \left. \right\},$$

which should be compared with Eq.(4.269).[25] In Eq.(9.168) ω_{rb}^+ and ω_{rb}^- are the orbital oscillation frequencies defined in Eq.(9.144), and the angular rotation velocity ω_{rb} of the ion beam is related to other system parameters by Eq.(9.150). As in Sec. 4.9, Eq.(9.168) contains the full influence of the equilibrium self fields and beam thermal effects (as measured by $\hat{T}_{b\perp}$). Moreover, for each value of $\ell \geq 1$, Eq.(9.168) possesses only stable solutions with $\mathrm{Im}\,\omega = 0$. The $\ell + 1$ values of (real) ω determined from Eq.(9.168) of course depend on ω_{pb}, γ_b, ω_{cb}, $\hat{T}_{b\perp}$, etc. Generally speaking, however, the main features of the solutions are similar to those illustrated in Fig. 4.15(b).

B.   Body-Wave Perturbations with $n = 1$

The dispersion relation $D_\ell^1(\omega) = 0$ can also be derived from the eigenvalue equation (9.165) for perturbations with azimuthal mode number ℓ and radial mode number $n = 1$. An important feature of the analysis is that $[\partial I_\ell / \partial p_\perp^2]_{p_\perp^2 = p_\perp^2 (r)} \neq 0$. Moreover, the eigenfunction $\delta\psi^\ell(r) = Ar^\ell[1 + a_1(\omega)(r/r_b)^2]$ can have an internal maximum within the ion beam, corresponding to strong body-wave perturbation in the beam interior. For example, without presenting algebraic details,[25] the dispersion relation for azimuthally symmetric perturbations with $(\ell, n) = (0, 1)$ is given by

$$D_0^1(\omega) = 1 - \frac{\omega_{pb}^2/\gamma_b^3}{\omega^2 - (\omega_{rb}^- - \omega_{rb}^+)^2} = 0 , \qquad (9.169)$$

where $a_1(\omega) = -1$ and the corresponding eigenfunction is

$$\delta\psi^0(r) = \begin{cases} A[1 - (r/r_b)^2], & 0 \leq r < r_b , \\ 0, & r_b < r \leq b . \end{cases} \qquad (9.170)$$

Note from Eq.(9.170) that the eigenfunction vanishes exactly in the vacuum region $(r_b < r \leq b)$, and assumes a maximum value (equal to A) at the beam center $(r = 0)$. From Eq.(9.144), it follows that $(\omega_{rb}^- - \omega_{rb}^+)^2 = \omega_{cb}^2/\gamma_b^2 - 2\omega_{pb}^2/\gamma_b^3$, where $\gamma_b = (1 - \beta_b^2)^{-1/2}$. Therefore, the dispersion relation (9.169) gives

$$\omega^2 = \omega_{cb}^2/\gamma_b^2 - \omega_{pb}^2/\gamma_b^3 \qquad (9.171)$$

for perturbations with $(\ell, n) = (0, 1)$. The condition for existence of radially confined beam equilibria in Eq.(9.151) implies that $\omega_{cb}^2/\gamma_b^2 - \omega_{pb}^2/\gamma_b^3 \geq \omega_{pb}^2/\gamma_b^3 + 4(r_{Lb}^2/r_b^2)(\omega_{pb}^2/\gamma_b^2) > 0$. Therefore, the right-hand side of Eq.(9.171) is positive, which corresponds to stable oscillations with

$\text{Im}\,\omega = 0$. An investigation of stability properties for $n = 1$ and $\ell \geq 1$ also predicts stable oscillations with $\text{Im}\,\omega = 0$ and will not be pursued further here.

C. Body-Wave Perturbations with $n = 2$

For the choice of equilibrium distribution function $f_b^0(r, \mathbf{p})$ in Eq.(9.164), the first evidence of strong instability occurs when the eigenvalue equation (9.165) is solved for azimuthally symmetric perturbations with mode numbers $(\ell, n) = (0, 2)$. In this case, the eigenfunction is of the form

$$\delta\psi^0(r) = A \left[1 + a_1(\omega) \left(\frac{r}{r_b} \right)^2 + a_2(\omega) \left(\frac{r}{r_b} \right)^4 \right] \tag{9.172}$$

for $0 \leq r < r_b$. We substitute Eq.(9.172) into the orbit integral $I_\ell(\omega, r, p_\perp)$ defined in Eq.(9.163) and make use of the radial orbit $[r'(\tau)]^2 = (x' + iy') \times (x' - iy')$ calculated from Eq.(9.161). For azimuthally symmetric perturbations with $\ell = 0$, some algebraic manipulation gives (Problem 9.19)

$$\left[\left(\omega_{rb}^- - \omega_{rb}^+ \right)^2 - \omega^2 \right] \left[\delta\psi^0(r) + \omega I_0(\omega, r, p_\perp) \right]$$

$$= 2A a_1 \left(\Omega_\beta^2 \frac{r^2}{r_b^2} - \frac{p_\perp^2}{\gamma_b^2 m_i^2 r_b^2} \right)$$

$$+ A a_2 \left[\frac{24 p_\perp^2}{\gamma_b^2 m_i^2 r_b^2} \left(\frac{p_\perp^2 / \gamma_b^2 m_i^2 r_b^2 - 4\Omega_\beta^2 r^2 / r_b^2}{\omega^2 - 4(\omega_{rb}^- - \omega_{rb}^+)^2} \right) \right. \tag{9.173}$$

$$\left. - \frac{8 p_\perp^2}{\gamma_b^2 m_i^2 r_b^2} \frac{r^2}{r_b^2} + 4\Omega_\beta^2 \left(\frac{r^2}{r_b^2} \right)^2 \left(1 + \frac{6\Omega_\beta^2}{\omega^2 - 4(\omega_{rb}^- - \omega_{rb}^+)^2} \right) \right].$$

Here, ω_{rb}^+ and ω_{rb}^- are defined in Eq.(9.144), and

$$\Omega_\beta^2 = (\omega_{rb}^+ - \omega_{rb})(\omega_{rb} - \omega_{rb}^-)$$

is defined in Eq.(9.143). In addition, Ω_β^2 can be related to the transverse beam temperature $\hat{T}_{b\perp}$ by Eq.(9.147), which gives $\Omega_\beta^2 r_b^2 = 2\hat{T}_{b\perp}/\gamma_b m_i = v_{Tb}^2$. Note from Eq.(9.173) that $I_0(\omega, r, p_\perp)$ has a nontrivial dependence on r and $p_\perp$. In addition, the terms proportional to Ω_β^2 are related to the finite transverse temperature of the beam particles.

Equation (9.173) is used to evaluate the factors $[I_0]_{p_\perp=0}$ and $[\partial I_0/\partial p_\perp^2]_{p_\perp^2=p_\perp^2}$ (r) occurring on the right-hand side of the eigenvalue equation (9.165). For $\ell = 0$, Eq.(9.165) then becomes

$$
\frac{1}{r}\frac{\partial}{\partial r}r\frac{\partial}{\partial r}\delta\psi^0(r) = -\frac{2\omega_{pb}^2}{\gamma_b^3[\omega^2-(\omega_{rb}^- - \omega_{rb}^+)^2]}\frac{\delta(r-r_b)}{r_b}
$$
$$
\times A\left[a_1 + 2a_2\left(1 + \frac{6\Omega_\beta^2}{\omega^2-4(\omega_{rb}^--\omega_{rb}^+)^2}\right)\right]
$$
$$
+ \frac{4\omega_{pb}^2}{\gamma_b^3[\omega^2-(\omega_{rb}^--\omega_{rb}^+)^2]}\frac{U(r_b-r)}{r_b^2} \tag{9.174}
$$
$$
\times A\left[a_1 + 4a_2\frac{r^2}{r_b^2} - \frac{24a_2\Omega_\beta^2}{\omega^2-4(\omega_{rb}^--\omega_{rb}^+)^2}\left(1-3\frac{r^2}{r_b^2}\right)\right],
$$

where $\omega_{pb}^2 = 4\pi\hat{n}_b Z_i^2 e^2/m_i^2$, $U(r_b-r) = +1$ for $0 \le r < r_b$, and $U(r_b-r) = 0$ for $r > r_b$. Note that the term proportional to $U(r_b-r)$ in Eq.(9.174) corresponds to a body-wave perturbation extending throughout the interior of the ion beam. From Eq.(9.172), it also follows that

$$
\frac{1}{r}\frac{\partial}{\partial r}r\frac{\partial}{\partial r}\delta\psi^0(r) = A\left(4\frac{a_1}{r_b^2} + 16a_2\frac{r^2}{r_b^4}\right) \tag{9.175}
$$

for $0 \le r < r_b$. We equate the coefficients of like powers of r^2 in Eqs.(9.174) and (9.175) within the ion beam ($0 \le r < r_b$). This gives

$$
D_0^2(\omega)a_2 = 0, \tag{9.176}
$$

and

$$
a_1 = \frac{\omega_{pb}^2}{\gamma_b^3[\omega^2-(\omega_{rb}^--\omega_{rb}^+)^2]}\left[a_1 - \frac{24\Omega_\beta^2 a_2}{\omega^2-4(\omega_{rb}^--\omega_{rb}^+)^2}\right], \tag{9.177}
$$

where $D_0^2(\omega)$ is defined by

$$
D_0^2(\omega) = 1 - \frac{\omega_{pb}^2}{\gamma_b^3[\omega^2-(\omega_{rb}^--\omega_{rb}^+)^2]}\left[1 + \frac{18\Omega_\beta^2}{\omega^2-4(\omega_{rb}^--\omega_{rb}^+)^2}\right]. \tag{9.178}
$$

The condition for a nontrivial solution ($a_2 \neq 0$) to Eq.(9.176) gives the desired dispersion relation

$$D_0^2(\omega) = 0 , \qquad (9.179)$$

where $D_0^2(\omega)$ is defined in Eq.(9.178). Substituting Eq.(9.179) into Eq.(9.177) further gives

$$a_2 = -\frac{3}{4}a_1 , \qquad (9.180)$$

which relates the coefficients $a_1(\omega)$ and $a_2(\omega)$ for perturbations with $(\ell, n) = (0, 2)$.

To determine the coefficient a_1, we make use of Eqs.(9.179) and (9.180) to eliminate the factors involving the eigenfrequency ω and the coefficient a_2 from the right-hand side for the eigenvalue equation (9.174). This gives

$$\frac{1}{r}\frac{\partial}{\partial r}r\frac{\partial}{\partial r}\delta\psi^0(r) = A\frac{a_1}{r_b}\delta(r - r_b) + 4A\frac{a_1}{r_b^2}U(r_b - r)\left(1 - 3\frac{r^2}{r_b^2}\right). \quad (9.181)$$

The solution to Eq.(9.181) which is continuous at $r = r_b$ and satisfies $\delta\psi^0(r = b) = 0$ at the conducting wall can be expressed as

$$\delta\psi^0(r) = \begin{cases} A\left(1 + a_1\dfrac{r^2}{r_b^2} - \dfrac{3}{4}a_1\dfrac{r^4}{r_b^4}\right), & 0 \leq r < r_b , \\[3mm] A\left(1 + \dfrac{a_1}{4}\right)\dfrac{\ln(r/b)}{\ln(r_b/b)} , & r_b < r \leq b . \end{cases} \qquad (9.182)$$

Furthermore, multiplying Eq.(9.181) by r, and integrating across the surface of the ion beam from $r_b(1 - \varepsilon)$ to $r_b(1 + \varepsilon)$ with $\varepsilon \to 0_+$ gives

$$r_b\left[\frac{\partial}{\partial r}\delta\psi^0\right]_{r=r_b(1+\varepsilon)} - r_b\left[\frac{\partial}{\partial r}\delta\psi^0\right]_{r=r_b(1-\varepsilon)} = Aa_1 , \qquad (9.183)$$

where $\varepsilon \to 0_+$. Substituting the inner and outer solutions for $\delta\psi^0(r)$ in Eq.(9.182) into Eq.(9.183) and solving for the coefficient $a_1(\omega)$ gives

$$a_1 = -4 . \qquad (9.184)$$

.

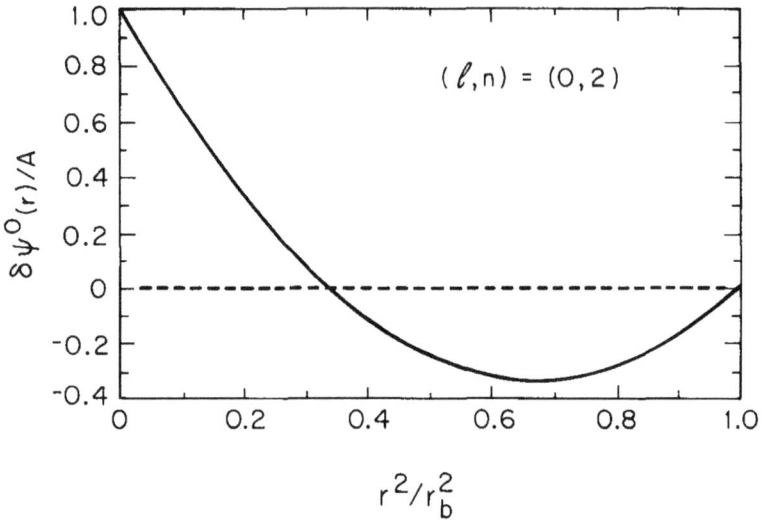

Figure 9.10. Plot of the eigenfunction $\delta\psi^0(r)$ versus r^2/r_b^2 obtained from Eq.(9.185) for transverse perturbations with mode numbers $(\ell, n) = (0, 2)$.

Therefore, the eigenfunction $\delta\psi^0(r)$ in Eq.(9.182) reduces to

$$\delta\psi^0(r) = \begin{cases} A\left(1 - 4\dfrac{r^2}{r_b^2} + 3\dfrac{r^4}{r_b^4}\right), & 0 \le r < r_b\,, \\ 0, & r_b < r \le b\,, \end{cases} \qquad (9.185)$$

which is identically zero in the vacuum region ($r_b < r \le b$). Figure 9.10 shows a plot versus r^2/r_b^2 of the eigenfunction $\delta\psi^0(r)$ defined in Eq.(9.185). Note from the figure that $\delta\psi^0(r)$ decreases monotonically from the value A at $r = 0$ to the value $-A/3$ at $r^2/r_b^2 = 2/3$, and then increases to the value zero at the surface of the ion beam ($r = r_b$). That is, the eigenfunction in Eq.(9.185) corresponds to a strong body-wave perturbation in the beam interior.[25]

We now examine the stability properties predicted by the dispersion relation (9.179).

Problem 9.19   For mode numbers $(\ell, n) = (0, 2)$, the eigenfunction $\delta\psi^0(r)$ which solves Eq.(9.165) has the form given in Eq.(9.172).

a. Substitute Eq.(9.172) into Eq.(9.163) and make use of the radial orbit $r'(\tau)$ calculated from Eq.(9.161) to show that the orbit integral $I_0(\omega, r, p_\perp)$ is given by the expression in Eq.(9.173).

b. Make use of Eq.(9.173) to show that the eigenvalue equation (9.165) reduces to Eq.(9.174) for mode numbers $(\ell, n) = (0, 2)$.

9.4.5 Transverse Stability Properties in a Solenoidal Focusing Field

We now examine the dispersion relation (9.179) which was derived for transverse perturbations with mode numbers $(\ell, n) = (0, 2)$, assuming beam propagation along a uniform solenoidal focusing field $B_0 \hat{e}_z$.[25] Rearranging terms, Eq.(9.179) can be expressed as

$$\left[\omega^2 - (\omega_{rb}^- - \omega_{rb}^+)^2\right]\left[\omega^2 - 4(\omega_{rb}^- - \omega_{rb}^+)^2\right]$$

$$= \frac{\omega_{pb}^2}{\gamma_b^3}\left[\omega^2 - 4(\omega_{rb}^- - \omega_{rb}^+)^2 + 18\Omega_\beta^2\right]. \qquad (9.186)$$

Here, $\gamma_b = (1 - \beta_b^2)^{-1/2}$, $\omega_{pb}^2 = 4\pi\hat{n}_b Z_i^2 e^2/m_i$, and

$$\Omega_\beta^2 = (\omega_{rb}^- - \omega_{rb})(\omega_{rb} - \omega_{rb}^+)$$

is defined in Eq.(9.143). Use is made of Eq.(9.144) to express

$$(\omega_{rb}^- - \omega_{rb}^+)^2 = \frac{\omega_{cb}^2}{\gamma_b^2} - \frac{2\omega_{pb}^2}{\gamma_b^3}, \qquad (9.187)$$

where $\omega_{cb} = Z_i e B_0/m_i c$ is the nonrelativistic ion cyclotron frequency. The dispersion relation (9.186) is a quadratic equation for ω^2. Some straightforward algebra shows that the necessary and sufficient condition for Eq.(9.186) to possess an unstable solution with $\mathrm{Im}\,\omega > 0$ is given by

$$\Omega_\beta^2 > \frac{2}{9}(\omega_{rb}^- - \omega_{rb}^+)^2\left[1 + \frac{\gamma_b^3(\omega_{rb}^- - \omega_{rb}^+)^2}{\omega_{pb}^2}\right]. \qquad (9.188)$$

When the inequality in Eq.(9.188) is satisfied, the unstable solution to Eq.(9.186) is purely growing with $\text{Re}\,\omega = 0$, and the growth rate $\text{Im}\,\omega$ is given by

$$\text{Im}\,\omega = \frac{1}{\sqrt{2}}\left[\left(72\left\{\frac{\omega_{pb}^2}{\gamma_b^3}\Omega_\beta^2 - \frac{2}{9}\left[(\omega_{rb}^- - \omega_{rb}^+)^4 + (\omega_{rb}^- - \omega_{rb}^+)^2\frac{\omega_{pb}^2}{\gamma_b^3}\right]\right\}\right.\right.$$
$$\left.\left. + \left[5(\omega_{rb}^- - \omega_{rb}^+)^2 + \frac{\omega_{pb}^2}{\gamma_b^3}\right]^2\right)^{1/2} - \left(5(\omega_{rb}^- - \omega_{rb}^+)^2 + \frac{\omega_{pb}^2}{\gamma_b^3}\right)\right]^{1/2}. \quad (9.189)$$

Making use of Eq.(9.187) and

$$\Omega_\beta^2 r_b^2 = \frac{2\hat{T}_{b\perp}}{\gamma_b m_i},$$

which follows from Eq.(9.147), the instability criterion in Eq.(9.188) can be expressed in the equivalent form

$$\frac{\omega_{pb}^2}{\gamma_b^3}\frac{2\hat{T}_{b\perp}}{\gamma_b m_i r_b^2} > \frac{2}{9}\left(\frac{\omega_{cb}^2}{\gamma_b^2} - \frac{2\omega_{pb}^2}{\gamma_b^3}\right)\left(\frac{\omega_{cb}^2}{\gamma_b^2} - \frac{\omega_{pb}^2}{\gamma_b^3}\right), \quad (9.190)$$

where r_b is the beam radius, and $\hat{T}_{b\perp}$ is the (maximum) transverse temperature of the beam ions. In addition, from Eq.(9.151), the condition for existence of radially confined beam equilibria requires that the inequality

$$\frac{2\omega_{pb}^2}{\gamma_b^3} \leq \frac{\omega_{cb}^2}{\gamma_b^2} - \frac{8\hat{T}_{b\perp}}{\gamma_b m_i r_b^2} \quad (9.191)$$

be satisfied. For specified values of ω_{cb}, $\hat{T}_{b\perp}$, γ_b and r_b, note that Eq.(9.191) places a limit on the maximum beam density (as measured by ω_{pb}^2) which can be confined by the solenoidal focusing field $B_0\hat{e}_z$.

To characterize the region of parameter space in which both instability exists [Eq.(9.190)] and radially confined beam equilibria exist [Eq.(9.191)], it is convenient to introduce the normalized self-field parameter σ_L (normalized to the transverse beam temperature) defined by

$$\sigma_L = \frac{1}{4}\frac{\omega_{pb}^2 r_b^2}{\gamma_b^3}\frac{\gamma_b m_i}{2\hat{T}_{b\perp}}. \quad (9.192)$$

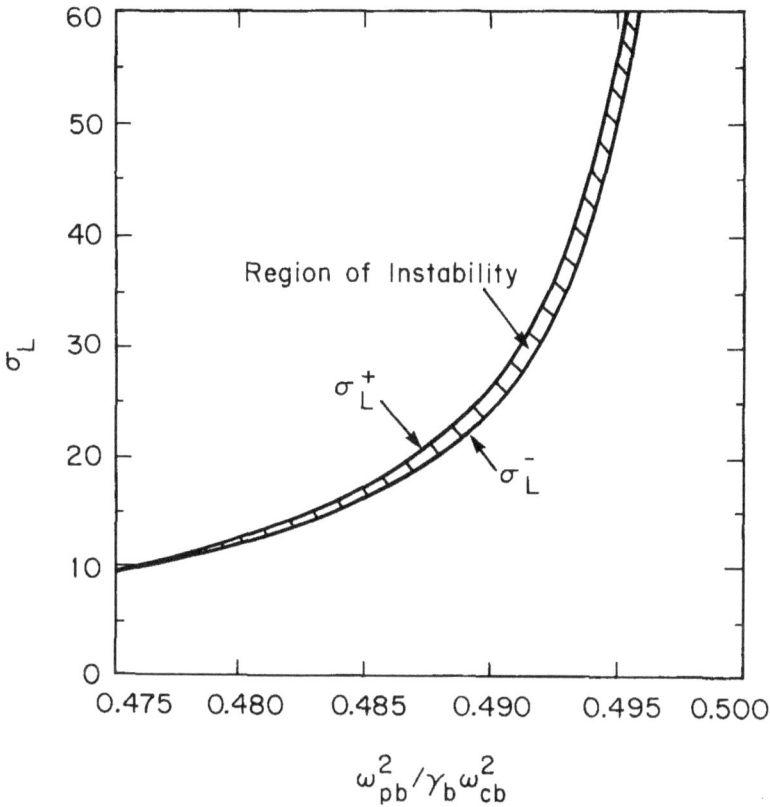

Figure 9.11. The shaded region of parameter space $(\omega_{pb}^2/\gamma_b\omega_{cb}^2, \sigma_L)$ satisfies both of the inequalities in Eqs.(9.194) and (9.195), which assures instability $(\operatorname{Im}\omega > 0)$ and existence of the beam equilibrium.

Here, $\hat{T}_{b\perp}$ is the (maximum) transverse temperature, and $\omega_{pb}^2 r_b^2$ is proportional to the number of particles per unit length of the beam. Note from Eq.(9.192) that σ_L increases as $\hat{T}_{b\perp}$ is reduced or as $\omega_{pb}^2 r_b^2$ is increased. Eliminating $\hat{T}_{b\perp}/r_b^2$ in Eq.(9.191) by means of Eq.(9.192) readily gives

$$\frac{2\omega_{pb}^2}{\gamma_b^3} \leq \frac{\omega_{cb}^2}{\gamma_b^2}\frac{2\sigma_L}{1+2\sigma_L} \ . \tag{9.193}$$

Therefore $2\omega_{pb}^2/\gamma_b^3$ is always bounded from above by ω_{cb}^2/γ_b^2. Equation (9.193) can be expressed in the equivalent form

$$\sigma_L \geq \sigma_L^- \equiv \frac{\omega_{pb}^2/\gamma_b^3}{(\omega_{cb}^2/\gamma_b^2 - 2\omega_{pb}^2/\gamma_b^3)} , \qquad (9.194)$$

which is the condition for radially confined equilibrium solutions to exist. Similarly, making use of Eq.(9.192) to eliminate $\hat{T}_{b\perp}/r_b^2$ in Eq.(9.190) in favor of σ_L readily gives

$$\sigma_L < \sigma_L^+ \equiv \frac{9(\omega_{pb}^2/\gamma_b^3)^2}{8(\omega_{cb}^2/\gamma_b^2 - \omega_{pb}^2/\gamma_b^3)(\omega_{cb}^2/\gamma_b^2 - 2\omega_{pb}^2/\gamma_b^3)} , \qquad (9.195)$$

which is the necessary and sufficient condition for instability ($\text{Im}\,\omega > 0$).

The region of parameter space $(\omega_{pb}^2/\gamma_b\omega_{cb}^2, \sigma_L)$ which satisfies both of the inequalities in Eqs.(9.194) and (9.195) is illustrated by the shaded

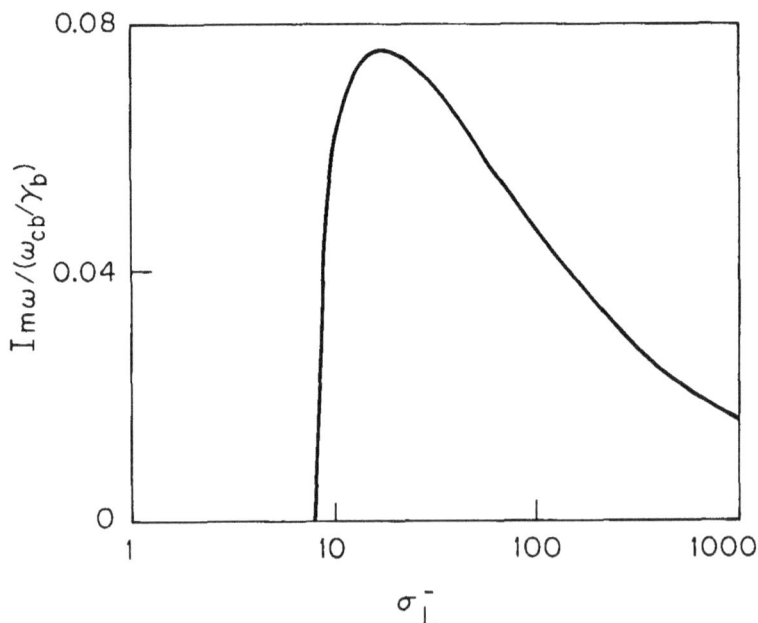

Figure 9.12. Plot of the normalized growth rate $\text{Im}\,\omega/(\omega_{cb}/\gamma_b)$ versus the parameter σ_L^- obtained from Eq.(9.189) at the limiting beam density defined in Eq.(9.197) [H.S. Uhm and R.C. Davidson, Phys. Fluids **23**, 1586 (1980)].

region in Fig. 9.11. Note that the shaded region in the figure is limited
to the intervals

$$\frac{8}{17} < \frac{\omega_{pb}^2}{\gamma_b \omega_{cb}^2} < \frac{1}{2}\ ,\quad \sigma_L > 8\ . \tag{9.196}$$

The maximum growth rate determined from Eq.(9.189) occurs at the
right-most boundary of the shaded region in Fig. 9.11, which corresponds
to beam propagation at the limiting density

$$\left[\frac{\omega_{pb}^2}{\gamma_b^3}\right]_{\max} = \frac{\omega_{cb}^2}{\gamma_b^2}\frac{\sigma_L^-}{1 + 2\sigma_L^-}\ , \tag{9.197}$$

where $\sigma_L = \sigma_L^-$ in Eq.(9.194). Shown in Fig. 9.12 is a plot of the normal-
ized growth rate $(\mathrm{Im}\,\omega)/(\omega_{cb}/\gamma_b)$ versus the parameter σ_L^-. Here, $\mathrm{Im}\,\omega$ is
determined from Eq.(9.189) at the limiting density defined in Eq.(9.197)
(see also Problem 9.21). Note that the growth rate $\mathrm{Im}\,\omega$ increases rapidly
from zero (at $\sigma_L^- = 8$) to the maximum value $0.076\,\omega_{cb}/\gamma_b$ (at $\sigma_L^- = 20$),
and then decreases slowly for $\sigma_L^- > 20$.

Problem 9.20   In the perpendicular phase space $(r, p_\perp)$, the equilibrium
ion motion in Eq.(9.164) occurs on the elliptical surface defined by

$$p_\perp^2 + \gamma_b^2 m_i^2 \Omega_\beta^2 r^2 = \gamma_b^2 m_i^2 \Omega_\beta^2 r_b^2\ , \tag{9.20.1}$$

where $\Omega_\beta^2 r_b^2 = 2\hat{T}_{b\perp}/\gamma_b m_i$, r_b is the beam radius, and $\hat{T}_{b\perp}$ is the (maximum)
transverse temperature of the beam ions.

a. The effective transverse emittance ε_b can be defined as the area in $(r, p_\perp)$
    space of the ellipse in Eq.(9.20.1). Show that

$$\varepsilon_b = \pi(\gamma_b m_i \Omega_\beta r_b)r_b = \pi(2\gamma_b m_i \hat{T}_{b\perp})^{1/2} r_b\ , \tag{9.20.2}$$

which relates ε_b to the transverse temperature $\hat{T}_{b\perp}$. [The definition of beam
emittance, customarily used in accelerator physics, is given in Sec. 10.4.]

b. Introduce the self-field parameter σ_L defined by $\sigma_L = (\omega_{pb}^2 r_b^2/\gamma_b^3) \times$
    $(\gamma_b m_i/8\hat{T}_{b\perp})$. Show that ε_b and σ_L are related by

$$\frac{\varepsilon_b^2 \sigma_L}{(\pi m_i c r_b)^2} = \frac{\nu}{\gamma_b}\ , \tag{9.20.3}$$

where $\nu = N_b e^2/m_i c^2$ is Budker's parameter defined in Eq.(9.122).

Problem 9.21   Consider the dispersion relation (9.186) at the maximum allowed beam density where ω_{pb}^2/γ_b^3 is related to ω_{cb}^2/γ_b^2 and the parameter σ_L^- by [see Eqs.(9.193), (9.194), and (9.197)]

$$\frac{\omega_{pb}^2}{\gamma_b^3} = \frac{\omega_{cb}^2}{\gamma_b^2}\frac{\sigma_L^-}{1+2\sigma_L^-} \ . \tag{9.21.1}$$

a. Make use of the definitions of $(\omega_{rb}^- - \omega_{rb}^+)^2$ and Ω_β^2 to show that

$$(\omega_{rb}^- - \omega_{rb}^+)^2 = \frac{\omega_{cb}^2}{\gamma_b^2}\frac{1}{1+2\sigma_L^-} \ , \tag{9.21.2}$$

$$\Omega_\beta^2 = \frac{1}{4}\frac{\omega_{cb}^2}{\gamma_b^2}\frac{1}{1+2\sigma_L^-} \ , \tag{9.21.3}$$

at the maximum beam density in Eq.(9.21.1).

b. Introduce the dimensionless (complex) frequency

$$\Omega = \frac{\omega}{\omega_{cb}/\gamma_b(1+2\sigma_L^-)^{1/2}}$$

and show that the dispersion relation (9.186) reduces to

$$\Omega^4 - (5 + \sigma_L^-)\Omega^2 - (1/2)(\sigma_L^- - 8) = 0 \ . \tag{9.21.4}$$

c. Show that the necessary and sufficient condition for Eq.(9.21.4) to possess an unstable solution with $\operatorname{Im}\Omega > 0$ is $\sigma_L^- > 8$.

d. When $\sigma_L^- > 8$, show from Eq.(9.21.4) that the unstable solution satisfies $\operatorname{Re}\omega = 0$ and

$$\operatorname{Im}\omega = \frac{\omega_{cb}}{\gamma_b}\left\{\frac{[(5+\sigma_L^-)^2 + 2(\sigma_L^- - 8)]^{1/2} - (5 + \sigma_L^-)}{2(1+2\sigma_L^-)}\right\}^{1/2} \ . \tag{9.21.5}$$

9.4.6   Transverse Stability Properties in a Quadrupole Focusing Field

Within the context of a simple model, the equilibrium and stability analysis in (Secs. 9.4.1–9.4.4) can be extended in a straightforward manner to the case where the solenoidal magnetic field $B_0\hat{e}_z$ is replaced by the average focusing effects produced by periodic quadrupole magnets

(Chapter 10). In this regard, the present analysis assumes that the *average* focusing force associated with the applied quadrupole field is determined from the z-component of the effective vector potential (see Problem 10.10)[25,26]

$$A_z^{\text{ext}}(r) = -\frac{\gamma_b m_i}{2 Z_i e \beta_b} \omega_{qb}^2 r^2 \,, \qquad (9.198)$$

where the oscillation frequency ω_{qb} is related to the quadrupole field gradient. For $B_0 = 0$, the single-particle constants of the motion in Eq.(9.126) are replaced by

$$H = (m_i^2 c^4 + c^2 \mathbf{p}^2)^{1/2} - m_i c^2 + Z_i e \phi_0(r) \,,$$

$$P_\theta = r p_\theta \,,$$

$$P_z = p_z + \frac{Z_i e}{c} A_z^s(r) - \frac{1}{2} \frac{\gamma_b m_i}{\beta_b c} \omega_{qb}^2 r^2 \,. \qquad (9.199)$$

For $\nu/\gamma_b \ll 1$ and $p_z^2 \gg p_r^2 + p_\theta^2$, the analysis of the equilibrium and stability properties for the choice of equilibrium distribution function in Eq.(9.132) proceeds in a manner completely analogous to the treatment in Secs. 9.4.2–9.4.4. Here, for $P_z = \gamma_b m_i \beta_b c$, the linear combination $H - \omega_{rb} P_\theta$ can be approximated by Eq.(9.133), where the effective potential $\psi_b(r)$ is defined by

$$\psi_b(r) = \frac{1}{2} \gamma_b m_i (-\omega_{rb}^2 + \omega_{qb}^2) r^2 + Z_i e[\phi_0(r) - \beta_b A_z^s(r)] \,. \qquad (9.200)$$

As before, the particle density and axial current density are uniform in the beam interior [Eqs.(9.136) and (9.139)], and Eq.(9.200) reduces to

$$\psi_b(r) = \frac{1}{2} \gamma_b m_i \Omega_\beta^2 r^2 \qquad (9.201)$$

for $0 \le r < r_b$. Here, Ω_β^2 is defined by

$$\Omega_\beta^2 = -\omega_{rb}^2 + \omega_{qb}^2 - \frac{\omega_{pb}^2}{2\gamma_b^3} = (\omega_{rb} - \omega_{rb}^+)(\omega_{rb}^- - \omega_{rb}) \,, \qquad (9.202)$$

where

$$\omega_{rb}^\pm = \mp \hat{\omega}_\beta \equiv \mp \left(\omega_{qb}^2 - \frac{\omega_{pb}^2}{2\gamma_b^3} \right)^{1/2} \,. \qquad (9.203)$$

Moreover, Ω_β^2 is related to the transverse temperature of the beam ions by

$$\Omega_\beta^2 r_b^2 = \frac{2\hat{T}_{b\perp}}{\gamma_b m_i} \equiv v_{Tb}^2 \qquad (9.204)$$

Solving Eqs.(9.202) and (9.204) for the angular rotation velocity ω_{rb} gives

$$\omega_{rb} = \mp \left(\omega_{qb}^2 - \frac{\omega_{pb}^2}{2\gamma_b^3} - \frac{2\hat{T}_{b\perp}}{\gamma_b m_i r_b^2} \right)^{1/2} , \qquad (9.205)$$

where

$$\frac{\omega_{pb}^2}{2\gamma_b^3} \le \omega_{qb}^2 - \frac{2\hat{T}_{b\perp}}{\gamma_b m_i r_b^2} \qquad (9.206)$$

is required for existence of a radially confined equilibrium.

To summarize, for the average quadrupole focusing field described by Eq.(9.198), the analysis of the linearized Vlasov-Maxwell equations (9.156) and (9.157), the solution to the eigenvalue equation (9.165), and the resulting dispersion relation (9.179) for transverse perturbations with mode numbers $(\ell, n) = (0, 2)$ are identical to the formulation described in Secs. 9.4.2–9.4.3, provided we make the replacements indicated by Eqs.(9.201)–(9.206). For example, making use of $\omega_{rb}^- - \omega_{rb}^+ = 2\hat{\omega}_\beta = 2(\omega_{qb}^2 - \omega_{pb}^2/2\gamma_b^3)^{1/2}$ and $\Omega_\beta^2 = \hat{\omega}_\beta^2 - \omega_{rb}^2$, the dispersion relation (9.186) can be expressed as

$$(\omega^2 - 4\hat{\omega}_\beta^2)(\omega^2 - 16\,\hat{\omega}_\beta^2) = \frac{\omega_{pb}^2}{\gamma_b^3} \left[\omega^2 - 16\hat{\omega}_\beta^2 + 18(\hat{\omega}_\beta^2 - \omega_{rb}^2) \right] , \quad (9.207)$$

where $\hat{\omega}_\beta^2 - \omega_{rb}^2 = 2\hat{T}_{b\perp}/\gamma_b m_i r_b^2$ follows from Eq.(9.205). The derivation and analysis of Eq.(9.206) is usually carried out for the special case of zero beam rotation ($\omega_{rb} = 0$), which corresponds to beam propagation at the maximum density

$$\left[\frac{\omega_{pb}^2}{2\gamma_b^3} \right]_{\text{max}} = \omega_{qb}^2 - \frac{2\hat{T}_{b\perp}}{\gamma_b m_i r_b^2} \qquad (9.208)$$

consistent with Eqs.(9.205) and (9.206). As in Sec. 9.4.5, for sufficiently high value of $\sigma_L = (\omega_{pb}^2 r_b^2/\gamma_b^3)(\gamma_b m_i/8\hat{T}_{b\perp}) > 8$, the condition in Eq.(9.208) corresponds to the strongest instability for perturbations with mode numbers $(\ell, n) = (0, 2)$.

In the general case, to determine the (stabilizing) influence of beam rotation on stability behavior, we introduce the dimensionless quantities

$$\Omega^2 = \frac{\omega^2}{\omega_{pb}^2/\gamma_b^3}, \quad \Omega_{rb}^2 = \frac{\omega_{rb}^2}{\omega_{pb}^2/\gamma_b^3}, \quad \sigma_L = \frac{1}{4}\frac{\omega_{pb}^2 r_b^2}{\gamma_b^3}\frac{\gamma_b m_i}{2\hat{T}_{b\perp}}, \tag{9.209}$$

in the analysis of the dispersion relation (9.207). In this case, the relation $\hat{\omega}_\beta^2 - \omega_{rb}^2 = 2\hat{T}_{b\perp}/\gamma_b m_i r_b^2$ gives $\hat{\omega}_\beta^2/(\omega_{pb}^2/\gamma_b^3) = \Omega_{rb}^2 + 1/4\sigma_L$, and Eq.(9.207) can be expressed in the dimensionless form

$$\left(\Omega^2 - \frac{1}{\sigma_L} - 4\Omega_{rb}^2\right)\left(\Omega^2 - \frac{4}{\sigma_L} - 16\Omega_{rb}^2\right) = \Omega^2 + \frac{1}{2\sigma_L} - 16\Omega_{rb}^2 .$$

$$\tag{9.210}$$

For zero beam rotation ($\Omega_{rb} = 0$), the necessary and sufficient condition for Eq.(9.210) to possess an unstable solution with $\mathrm{Im}\,\Omega > 0$ is given by $\sigma_L > 8$. In this case, $\mathrm{Re}\,\Omega = 0$ and the growth rate of the unstable mode can be expressed as

$$\mathrm{Im}\,\omega = \frac{\omega_{pb}}{\gamma_b^{3/2}}\left\{\left[\left(\frac{\sigma_L + 5}{2\sigma_L}\right)^2 + \left(\frac{\sigma_L - 8}{2\sigma_L^2}\right)\right]^{1/2} - \left(\frac{\sigma_L + 5}{2\sigma_L}\right)\right\}^{1/2}$$

$$\tag{9.211}$$

for $\Omega_{rb} = 0$ and $\sigma_L > 8$.

Allowing for beam rotation, the necessary and sufficient condition for Eq.(9.210) to possess only stable solutions with $\mathrm{Im}\,\Omega = 0$ is given by

$$4\Omega_{rb}^4 + \Omega_{rb}^2\left(\frac{2 + \sigma_L}{\sigma_L}\right) - \left(\frac{\sigma_L - 8}{32\sigma_L^2}\right) > 0 . \tag{9.212}$$

The inequality in Eq.(9.212) gives the condition

$$\Omega_{rb}^2 > \frac{1}{8}\left\{\left[\left(\frac{2 + \sigma_L}{\sigma_L}\right)^2 + \left(\frac{\sigma_L - 8}{2\sigma_L^2}\right)\right]^{1/2} - \left(\frac{2 + \sigma_L}{\sigma_L}\right)\right\}, \tag{9.213}$$

which assures stability for perturbations with mode numbers $(\ell, n) = (0, 2)$. For normalized self-field parameter $\sigma_L < 8$, the inequality in Eq.(9.213) is automatically satisfied even when $\Omega_{rb} = 0$. For specified value of $\sigma_L > 8$, however, Eq.(9.213) determines the threshold value of beam rotation frequency for which the transverse instability is completely stabilized. Figure 9.13 illustrates the stable and unstable regions of the parameter space $(\sigma_L, \gamma_b^3 \omega_{rb}^2/\omega_{pb}^2)$ consistent with Eq.(9.213). The curve in Fig. 9.13, which corresponds to the right-hand side of the inequality in Eq.(9.213), assumes a maximum value of $1/1152$ for $\sigma_L = 288/17$. Therefore, a *modest* beam rotation is sufficient to completely stabilize the transverse instability for *all* values of the parameter σ_L.[25,26]

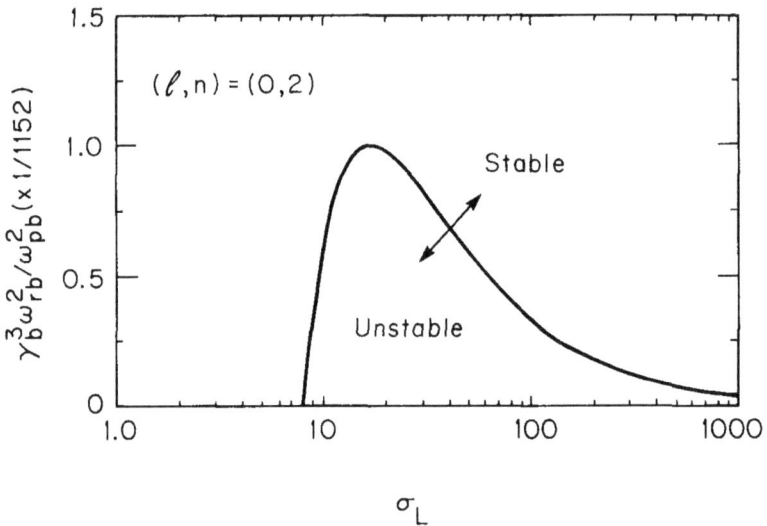

Figure 9.13. The stable and unstable regions of the parameter space $(\sigma_L, \gamma_b^3 \omega_{rb}^2/\omega_{pb}^2)$ consistent with Eq.(9.213) and perturbations with mode numbers $(\ell, n) = (0, 2)$. A *modest* beam rotation with $\gamma_b^3 \omega_{rb}^2/\omega_{pb}^2 > 1/1152$ completely stabilizes the transverse instability for all values of the parameter σ_L.

Problem 9.22 Analyze the dispersion relation (9.210) for the case of a rotating ion beam $(\Omega_{rb} \neq 0)$. Show that the necessary and sufficient condition for Eq.(9.210) to possess only stable solutions with $\text{Im}\,\Omega = 0$ is given by Eq.(9.212), or equivalently, Eq.(9.213). The inequality in Eq.(9.213) is used in Fig. 9.13 to determine the stable and unstable regions in the parameter space $(\sigma_L, \gamma_b^3 \omega_{rb}^2/\omega_{pb}^2)$.

9.5   Resistive Hose Instability for Intense Electron Beam Propagation Through a Background Plasma

For an intense relativistic electron beam propagating through a background plasma, there are two classes of low-frequency resistive instabilities[27-43] that can cause macroscopic distortion of the beam and seriously impair or disrupt beam propagation. These are the resistive hose instability (Sec. 9.5) and the azimuthally symmetric sausage and hollowing instabilities (Sec. 9.6). In this section, we investigate several features of the resistive hose instability using a hybrid model[35,36] in which the beam electrons are treated by the Vlasov equation and the background plasma is treated macroscopically as a dense, collisional plasma. The resistive hose instability, which is driven by the plasma return current, produces a macroscopic snake-like distortion of the electron beam for low-frequency perturbations with azimuthal mode number $\ell = 1$.

The equilibrium and stability analysis presented here employs several of the theoretical techniques developed in Chapter 4 and in Sec. 9.4. Following a description of the theoretical model and assumptions (Sec. 9.5.1), equilibrium properties of the relativistic beam-plasma system are discussed for an intense relativistic electron beam with $\nu/\gamma_b \ll 1$ (Sec. 9.5.2). The linearized Vlasov-Maxwell equations are then used to derive an eigenvalue equation for electromagnetic perturbations with azimuthal mode number $\ell = 1$ (Sec. 9.5.3). Properties of the resistive hose instability are analyzed for the two cases in which the electron beam has a sharp-boundary density profile (Sec. 9.5.4), and a diffuse density profile (Sec. 9.5.5).

9.5.1   Theoretical Model and Assumptions

Paralleling the analysis of Lampe and Uhm,[35] we make the following simplifying assumptions regarding the beam-plasma configuration.

(a) In geometry similar to Fig. 9.8, a relativistic electron beam $(j = b)$ with characteristic energy $\gamma_b m_e c^2$, density $\hat{n}_b$, and radius r_b propagates in the axial direction parallel to a uniform magnetic field $B_0 \hat{e}_z$ through a dense background plasma $(\hat{n}_e \gg \hat{n}_b)$. A perfectly conducting wall is located at radius $r = b$, and the limiting cases of zero guide field $(B_0 = 0)$ and/or no conducting wall $(b \to \infty)$ are not excluded in the analysis. Under quasi-steady-state conditions

($\partial/\partial t = 0$), the beam-plasma system is assumed to be axially uniform ($\partial/\partial z = 0$) and azimuthally symmetric ($\partial/\partial\theta = 0$), with the electron beam centered in the plasma channel and the conducting cylinder. Cylindrical polar coordinates (r, θ, z) are employed in the equilibrium and stability analysis.

(b) The relativistic electron dynamics are treated in the paraxial approximation with

$$p_z^2 \gg p_r^2 + p_\theta^2 . \tag{9.214}$$

That is, the directed axial energy is assumed to be large in comparison with the transverse kinetic energy of the beam electrons.

(c) Introducing Budker's parameter $\nu = N_b e^2/m_e c^2$, where $N_b = 2\pi \int_0^b dr\, r n_b^0(r)$ is the number of beam electrons per unit axial length, it is assumed that

$$\frac{\nu}{\gamma_b} \ll 1 . \tag{9.215}$$

The azimuthal self-magnetic field $B_\theta^s(r)$ produced by the net axial current will play an important role in determining detailed stability behavior. Equation (9.215) assures that the self field readily penetrates into the beam interior (see Sec. 4.4).

(d) The equilibrium return current $J_{zp}^0(r)$ carried by the plasma, if any, is assumed to have the same radial profile as the beam current $J_{zb}^0(r) = -e \int d^3p\, v_z f_b^0(r, \mathbf{p})$, i.e.,

$$J_{zp}^0(r) = -f_M J_{zb}^0(r) , \tag{9.216}$$

where $f_M = \text{const.}$ is the fractional current neutralization by the background plasma.

(e) The electron beam is generally allowed to have an equilibrium rotation velocity ω_{rb}. Consistent with Eqs.(9.214) and (9.215), the axial diamagnetic field $B_z^s(r)$ produced by the beam's rotation is neglected in the present analysis, and the total equilibrium magnetic field is approximated by

$$\mathbf{B}^0(\mathbf{x}) = B_0 \hat{\mathbf{e}}_z + B_\theta^s(r)\hat{\mathbf{e}}_\theta . \tag{9.217}$$

(f) The (collisional) background plasma is described by a real scalar electrical conductivity $\sigma(r)$. It is assumed that the perturbed plasma current $\delta\mathbf{J}_p(\mathbf{x}, t)$ and electric field $\delta\mathbf{E}(\mathbf{x}, t)$ are related by the simple Ohm's law

$$\delta\mathbf{J}_p = \sigma\delta\mathbf{E} \ . \tag{9.218}$$

(g) The density of the background plasma is assumed to be sufficiently large that the beam space charge is completely neutralized, both in equilibrium $[\phi_0(r) \simeq 0]$ and in the perturbed state $[\delta\phi(\mathbf{x}, t) \simeq 0]$. Neutralization of the perturbed beam space charge requires perturbations with sufficiently low frequency ω that

$$|\omega| \ll 4\pi\sigma \tag{9.219}$$

for $0 \le r < r_b$.

(h) The resistive hose instability produces a macroscopic distortion of the electron beam for perturbations with azimuthal mode number $\ell = 1$. The present analysis considers low frequency perturbations with long axial wavelengths satisfying

$$|k_z|r_b \ll 1 \ ,$$
$$|\omega|r_b/c \ll 1 \ . \tag{9.220}$$

(i) Consistent with Eqs.(9.219) and (9.220), the transverse electric (TE) components of the perturbed electromagnetic field, δE_r, δE_θ and δB_z, are treated as negligibly small. Therefore, in the linearized Vlasov equation for the perturbed distribution function $\delta f_b(\mathbf{x}, \mathbf{p}, t)$, we approximate

$$e\left(\delta\mathbf{E} + \frac{\mathbf{v} \times \delta\mathbf{B}}{c}\right) \cdot \frac{\partial}{\partial\mathbf{p}} f_b^0 \tag{9.221}$$

$$= e\left[\frac{v_z\delta B_r}{c}\frac{\partial f_b^0}{\partial p_\theta} - \frac{v_z\delta B_\theta}{c}\frac{\partial f_b^0}{\partial p_r} + \left(\delta E_z + \frac{1}{c}v_r\delta B_\theta - \frac{1}{c}v_\theta\delta B_r\right)\frac{\partial f_b^0}{\partial p_z}\right] ,$$

where $f_b^0(r, \mathbf{p})$ is the equilibrium distribution function of the beam electrons. Here, $\delta B_r(r, \theta, z, t)$, $\delta B_\theta(r, \theta, z, t)$ and $\delta E_z(r, \theta, z, t)$

are related to the z-component of the perturbed vector potential $\delta A_z(r, \theta, z, t)$ by

$$\delta B_r = \frac{1}{r} \frac{\partial}{\partial \theta} \delta A_z , \quad \delta B_\theta = -\frac{\partial}{\partial r} \delta A_z , \quad \delta E_z = -\frac{1}{c} \frac{\partial}{\partial t} \delta A_z ,$$
(9.222)

where $\delta \mathbf{B} = \nabla \times \delta A_z \hat{\mathbf{e}}_z$, and $\delta \phi = 0$ is assumed. Finally, making use of $|\omega| r_b / c \ll 1$, $p_r^2 + p_\theta^2 \ll p_z^2$, and $\nu / \gamma_b \ll 1$, the term proportional to $\partial f_b^0 / \partial p_z$ in Eq.(9.221) can be neglected in comparison with $(e/c) v_z \hat{\mathbf{e}}_z \times \delta \mathbf{B} \cdot \partial f_b^0 / \partial \mathbf{p}$. Therefore, Eq.(9.221) reduces to

$$e \left(\delta \mathbf{E} + \frac{\mathbf{v} \times \delta \mathbf{B}}{c} \right) \cdot \frac{\partial f_b^0}{\partial \mathbf{p}} = \frac{e}{c} v_z \left(\frac{1}{r} \frac{\partial \delta A_z}{\partial \theta} \frac{\partial f_b^0}{\partial p_\theta} + \frac{\partial \delta A_z}{\partial r} \frac{\partial f_b^0}{\partial p_r} \right) .$$
(9.223)

The approximate expression in Eq.(9.223) will be used in the analysis of the linearized Vlasov-Maxwell equations in Sec. 9.5.3.

9.5.2 Equilibrium Properties

Within the context of the assumptions summarized in Sec. 9.5.1, we investigate the equilibrium and stability properties for perturbations about the class of relativistic electron beam equilibria $f_b^0(r, \mathbf{p})$ described by

$$f_b^0(r, \mathbf{p}) = \left(\frac{\hat{n}_b}{2\pi \gamma_b m_e} \right) F_b \left(H - \omega_{rb} P_\theta \right) \delta \left(P_z - \gamma_b m_e \beta_b c \right) , \qquad (9.224)$$

where $\hat{n}_b$, γ_b, ω_{rb} and β_b are constants.[35] Here, the single-particle constants of the motion in the equilibrium field configuration correspond to the energy H, the canonical angular momentum P_θ, and the axial canonical momentum P_z defined by

$$H = \left(m_e^2 c^4 + c^2 p_r^2 + c^2 p_\theta^2 + c^2 p_z^2 \right)^{1/2} - m_e c^2 ,$$
$$P_\theta = r(p_\theta - m_e r \omega_{cb}/2) ,$$
$$P_z = p_z - (e/c) A_z^s(r) . \qquad (9.225)$$

In Eq.(9.225), $-e$ is the electron charge, $\omega_{cb} = eB_0 / m_e c$ is the nonrelativistic electron cyclotron frequency in the applied magnetic field $B_0 \hat{\mathbf{e}}_z$, and $A_z^s(r)$ is the vector potential for the azimuthal self-magnetic field $B_\theta^s(r) = -\partial A_z^s/\partial r$ determined from

$$\frac{1}{r}\frac{\partial}{\partial r}r\frac{\partial}{\partial r}A_z^s(r) = \frac{4\pi e}{c}(1 - f_M)\int d^3p\, v_z f_b^0 \ . \tag{9.226}$$

In Eq.(9.226), $v_z = p_z/\gamma m_e$, and $f_M = $ const. is the fractional current neutralization by the background plasma [Eq.(9.216)]. Note from Eq.(9.225) that $H = (\gamma - 1)m_e c^2$ corresponds to the electron kinetic energy because the beam space charge is completely neutralized by the background plasma [Assumption (g)].

For the class of beam equilibria described by Eq.(9.224), all of the electrons have axial momentum $p_z = \gamma_b m_e \beta_b c + (e/c)A_z^s(r)$, which corresponds to $P_z = \gamma_b m_e \beta_b c = $ const. Similar to the analysis in Sec. 9.3.2, making use of the paraxial assumption [Eq.(9.214)] and $\nu/\gamma_b \ll 1$ [Eq.(9.215)], the combination $H - \omega_{rb}P_\theta$ occurring in Eq.(9.224) can be expanded for small $p_r^2 + p_\theta^2 \ll p_z^2$ and $|eA_z^s/\gamma_b m_e \beta_b c^2| \ll 1$. Evaluating $H - \omega_{rb}P_\theta$ for $P_z = \gamma_b m_e \beta_b c$ readily gives the approximate expression

$$H - \omega_{rb}P_\theta = (\gamma_b - 1)m_e c^2 + \frac{1}{2\gamma_b m_e}\left[p_r^2 + (p_\theta - \gamma_b m_e \omega_{rb}r)^2\right] + \psi_b(r) \ , \tag{9.227}$$

where we have made the identification $\gamma_b = (1 - \beta_b^2)^{-1/2}$ without loss of generality. In Eq.(9.227), the effective potential $\psi_b(r)$ is defined by

$$\psi_b(r) = \frac{1}{2}\gamma_b m_e(\omega_{rb}\omega_{cb}/\gamma_b - \omega_{rb}^2)r^2 + e\beta_b A_z^s(r) \ . \tag{9.228}$$

With the simplified form of $H - \omega_{rb}P_\theta$ in Eq.(9.227), the analysis of equilibrium and stability properties proceeds in a manner similar to that in Secs. 4.2 and 9.4. For example, for $\nu/\gamma_b \ll 1$, the average axial velocity $V_{zb}^0(r) = (\int d^3p\, v_z f_b^0)/(\int d^3p f_b^0)$ and the average azimuthal velocity $V_{\theta b}^0(r) = (\int d^3p\, v_\theta f_b^0)/(\int d^3p\, f_b^0)$ can be approximated by

$$V_{zb}^0(r) = \beta_b c = \text{const.} \ ,$$
$$V_{\theta b}^0(r) = \omega_{rb}r \ . \tag{9.229}$$

Note from Eq.(9.229) that the axial velocity profile is uniform, and the mean azimuthal motion corresponds to a rigid rotation over the radial extent of the electron beam. Introducing the transverse kinetic energy *in the rotating frame* defined by $U = [p_r^2 + (p_\theta - \gamma_b m_e \omega_{rb} r)^2]/2\gamma_b m_e$, the density profile $n_b^0(r) = \int d^3 p f_b^0$ of the beam electrons can be expressed as

$$n_b^0(r) = \hat{n}_b \int_0^\infty dU \, F_b \left[U + (\gamma_b - 1)m_e c^2 + \psi_b(r) \right], \qquad (9.230)$$

which should be compared with Eq.(4.32). Furthermore, making use of $\int d^3 p \, v_z f_b^0 = \beta_b c n_b^0(r)$, the equilibrium Maxwell equation (9.226) can be integrated to give the azimuthal self-magnetic field

$$B_\theta^s(r) = -\frac{4\pi e \beta_b (1 - f_M)}{r} \int_0^r dr' \, r' n_b^0(r') , \qquad (9.231)$$

where $B_\theta^s(r) = -\partial A_z^s / \partial r$, and $n_b^0(r)$ is defined in Eq.(9.230) for general choice of equilibrium distribution function $F_b(H - \omega_{rb} P_\theta)$.

For future reference, it is useful to introduce the effective betatron frequency $\omega_\beta(r)$ for self-field-induced oscillations in the transverse orbits defined by

$$\omega_\beta^2(r) = -\frac{e\beta_b B_\theta^s(r)}{\gamma_b m_e r} = \frac{4\pi e^2 \beta_b^2}{\gamma_b m_e r^2}(1 - f_M) \int_0^r dr' \, r' n_b^0(r') . \qquad (9.232)$$

Making use of Eqs.(9.228) and (9.230), it is readily shown that

$$\begin{aligned}
\frac{\partial}{\partial r} n_b^0(r) &= -\hat{n}_b F_b' \left[(\gamma_b - 1)m_e c^2 + \psi_b(r) \right] \frac{\partial}{\partial r} \psi_b(r) \\
&= -\gamma_b m_e r \hat{n}_b F_b' \left[(\gamma_b - 1)m_e c^2 + \psi_b(r) \right] \qquad (9.233) \\
&\quad \times \left[\omega_{rb}^+(r) - \omega_{rb} \right] \left[\omega_{rb} - \omega_{rb}^-(r) \right] ,
\end{aligned}$$

where

$$\omega_{rb}^\pm(r) = \frac{\omega_{ce}}{2\gamma_b} \left\{ 1 \pm \left[1 + \frac{4\gamma_b^2 \omega_\beta^2(r)}{\omega_{cb}^2} \right]^{1/2} \right\} . \qquad (9.234)$$

The frequencies $\omega_{rb}^\pm(r)$ defined in Eq.(9.234) are the fast and slow rotation frequencies of an electron in an axi-centered circular orbit with radius r. In the special case where $B_0 = 0$, it follows from Eq.(9.234) that $\omega_{rb}^\pm(r) = \pm \omega_\beta(r)$. Also note from Eq.(9.232) that the betatron frequency $\omega_\beta(r)$ generally varies with radius r except in the special case where the density profile $n_b^0(r)$ is uniform out to radius $r = r_b$ [Eq.(9.236)]. In circumstances where $\omega_\beta(r)$ varies with r, there is no single frequency for which the entire beam is resonant with the wave. This can lead to a significant reduction in the growth rate of the resistive hose instability (Sec. 9.5.5) relative to the case where $\omega_\beta = $ const. (Sec. 9.5.4).

Sharp-Boundary Density Profile: As a first equilibrium example, consider the beam distribution function

$$F_b(H - \omega_{rb}P_\theta) = \delta\left[H - \omega_{rb}P_\theta - (\hat{\gamma}_b - 1)m_e c^2\right],\qquad (9.235)$$

which has an inverted population in $H - \omega_{rb}P_\theta$. Here, the constant $\hat{\gamma}_b > \gamma_b = (1 - \beta_b^2)^{-1/2}$. Paralleling the analysis in Sec. 9.4.2, it is readily shown from Eqs.(9.228), (9.230), and (9.231) that the equilibrium density profile consistent with Eq.(9.235) has the rectangular form

$$n_b^0(r) = \begin{cases} \hat{n}_b, & 0 \leq r < r_b\,, \\ 0, & r_b < r \leq b\,. \end{cases}\qquad (9.236)$$

Here, the effective potential is given by $\psi_b(r) = (\gamma_b m_e/2)(\omega_{rb}^+ - \omega_{rb}) \times (\omega_{rb} - \omega_{rb}^-)r^2$, and the beam radius r_b is determined self-consistently from $\psi_b(r_b) = (\hat{\gamma}_b - \gamma_b)m_e c^2$, which gives

$$r_b^2 = \frac{2c^2}{\left(\omega_{rb}^+ - \omega_{rb}\right)\left(\omega_{rb} - \omega_{rb}^-\right)}\left(\frac{\hat{\gamma}_b - \gamma_b}{\gamma_b}\right).\qquad (9.237)$$

It also follows from Eqs.(9.232) and (9.236) that $\omega_\beta^2(r) = \hat{\omega}_\beta^2 = $ const., where $\hat{\omega}_\beta^2$ is defined by

$$\hat{\omega}_\beta^2 = \frac{\beta_b^2}{2\gamma_b}\omega_{pb}^2(1 - f_M)\qquad (9.238)$$

and $\omega_{pb}^2 = 4\pi\hat{n}_b e^2/m_e$. In circumstances where $B_0 = 0$ and $\omega_{rb} = 0$, note from Eq.(9.237) that

$$r_b^2 = \left(\frac{2c^2}{\hat{\omega}_\beta^2}\right)\left(\frac{\hat{\gamma}_b - \gamma_b}{\gamma_b}\right).$$

Therefore, the inequality $\nu/\gamma_b = \omega_{pb}^2 r_b^2/4\gamma_b c^2 \ll 1$ necessarily requires a narrow energy spread with $(\hat{\gamma}_b - \gamma_b)/\gamma_b \ll 1$.

Diffuse Density Profile: As an example of a beam equilibrium with a diffuse density profile, we assume $\omega_{rb} = 0$ and

$$F_b(H) = \frac{1}{k_B T_b} \exp\left\{-\frac{H - (\gamma_b - 1)m_e c^2}{k_B T_b}\right\}, \qquad (9.239)$$

where $T_b = $ const. is the transverse temperature of the beam electrons. Substituting Eq.(9.239) into Eq.(9.230) gives

$$n_b^0 = \hat{n}_b \exp\left\{-\frac{e\beta_b A_z^s(r)}{k_B T}\right\}, \qquad (9.240)$$

where $\psi_b(r) = e\beta_b A_z^s(r)$ for $\omega_{rb} = 0$. Substituting Eq.(9.240) and $B_\theta^s(r) = -\partial A_z^s/\partial r$ into Eq.(9.231) and solving for the density profile $n_b^0(r)$ gives the familiar Bennett pinch equilibrium (see also Sec. 5.2.4)[33,50]

$$n_b^0(r) = \frac{\hat{n}_b}{(1 + r^2/r_b^2)^2}. \qquad (9.241)$$

Here, the effective beam radius r_b is defined by

$$r_b^2 = \frac{8k_B T_b}{m_e \omega_{pb}^2 \beta_b^2 (1 - f_M)}, \qquad (9.242)$$

where $\omega_{pb}^2 = 4\pi \hat{n}_b e^2/m_e$. Moreover, making use of Eqs.(9.232) and (9.241), the betatron frequency can be expressed as

$$\omega_\beta^2(r) = \frac{\hat{\omega}_\beta^2}{(1 + r^2/r_b^2)}, \qquad (9.243)$$

where $\hat{\omega}_\beta^2 = (\omega_{pb}^2 \beta_b^2/2\gamma_b)(1 - f_M)$.

Problem 9.23   As in Sec. 4.2.2, for the class of electron beam equilibria described by Eq.(9.224), assuming $p_z^2 \gg p_r^2 + p_\theta^2$ and $\nu/\gamma_b \ll 1$, a specification of the density profile $n_b^0(r) = \int d^3 p \, f_b^0$ is sufficient information to reconstruct the detailed form of the distribution function $F_b(H - \omega_{rb} P_\theta)$.

a.   Consider the expression for the density profile $n_b^0(r)$ given in Eq.(9.230). Make use of $(\partial/\partial\psi_b - \partial/\partial U)F_b[U + (\gamma_b - 1)m_e c^2 + \psi_b(r)] = 0$ to integrate by parts with respect to U and show that

$$\frac{\partial}{\partial r} n_b^0(r) = -\hat{n}_b F_b\left[(\gamma_b - 1)m_e c^2 + \psi_b(r)\right]\frac{\partial}{\partial r}\psi_b(r). \qquad (9.23.1)$$

b.   Make use of the definition $\psi_b(r) = (\gamma_b m_e/2)(\omega_{rb}\omega_{cb}/\gamma_b - \omega_{rb}^2)r^2 + e\beta_b A_z^s(r)$ in Eq.(9.228) to show that

$$\frac{\partial}{\partial r}\psi_b(r) = \gamma_b m_e r \left[\omega_{rb}^+(r) - \omega_{rb}\right] \left[\omega_{rb} - \omega_{rb}^-(r)\right] , \qquad (9.23.2)$$

where $\omega_{rb}^+(r)$ and $\omega_{rb}^-(r)$ are defined in Eq.(9.234). Equations (9.23.1) and (9.23.2) can be combined to give the result in Eq.(9.233).

c.   A density inversion theorem analogous to that derived in Sec. 4.2.2 can be obtained from Eqs.(9.23.1) and (9.23.2). As a special case, consider a nonrotating electron beam with $\omega_{rb} = 0$. Show that

$$\frac{\partial}{\partial r}\psi_b(r) = e\beta_b \frac{\partial}{\partial r} A_z^s(r) = -e\beta_b B_\theta^s(r) = \frac{4\pi e^2 \beta_b^2(1 - f_M)}{r} \int_0^r dr'\, r' n_b^0(r') .$$

$$(9.23.3)$$

d.   From Eqs.(9.23.1)–(9.23.3), for specified density profile $n_b^0(r)$ (measured, say), it follows that $\psi_b(r)$ is a known function of r, which can be inverted to determine $r = r(\psi_b)$. For $\omega_{rb} = 0$, show from Eq.(9.23.1) that the distribution function can be expressed as

$$\hat{n}_b F_b(H) = -\left[\frac{\partial n_b^0}{\partial \psi_b}\right]_{\psi_b = H - (\gamma_b - 1)m_e c^2} . \qquad (9.23.4)$$

Therefore, knowledge of $n_b^0[r(\psi_b)]$ is sufficient information to reconstruct the distribution function $F_b(H)$.

e.   As an example, assume that $n_b^0(r)$ is specified by the Bennett pinch profile in Eq.(9.241). Make use of the inversion theorem in Eq.(9.23.4) to show that $F_b(H)$ necessarily has the thermal equilibrium form in Eq.(9.239).

9.5.3   Kinetic Eigenvalue Equation for $\ell = 1$ Perturbations

For low-frequency, long-wavelength perturbations satisfying $|\omega| \ll 4\pi\sigma$, $|\omega|r_b/c \ll 1$ and $|k_z|r_b \ll 1$ [Eqs.(9.219) and (9.220)], the transverse electric (TE) components of the perturbed fields, δE_r, δE_θ and δB_z, are negligibly small, and the perturbed field components δE_z, δB_r and δB_θ can be expressed in terms of the axial component of vector potential $\delta A_z(r,\theta,z,t)\hat{\mathbf{e}}_z$ according to Eq.(9.222). Here, the longitudinal electric field $\delta \mathbf{E}_L = -\nabla\delta\phi$ is neglected because it is assumed that the (dense) background plasma completely neutralizes the perturbed beam space charge [Assumption (g)]. The z-component of the Maxwell equation $\nabla \times \delta\mathbf{B} = \nabla \times (\nabla \times \delta A_z \hat{\mathbf{e}}_z)$ can be approximated by

$$\left\{ \frac{1}{r}\frac{\partial}{\partial r}r\frac{\partial}{\partial r} + \frac{1}{r^2}\frac{\partial^2}{\partial\theta^2} - \frac{4\pi\sigma}{c^2}\frac{\partial}{\partial t} \right\} \delta A_z(r,\theta,z,t)$$

$$= \frac{4\pi e}{c} \int d^3p\, v_z \delta f_b(r,\theta,z,\mathbf{p},t) , \qquad (9.244)$$

where terms proportional to $c^{-2}(\partial^2/\partial t^2)\delta A_z$ and $(\partial^2/\partial z^2)\delta A_z$ have been neglected because $|\omega|r_b/c \ll 1$ and $|k_z|r_b \ll 1$ are assumed.[35] In Eq.(9.244), $\delta J_{zb} = -e\int d^3p\, v_z \delta f_b$ is the perturbed axial beam current, and use has been made of $\delta J_{zp} = \sigma \delta E_z$ to express the perturbed axial plasma current in terms of δA_z. Furthermore, making use of Eq.(9.223), the linearized Vlasov equation for the perturbed distribution function $\delta f_b(\mathbf{x},\mathbf{p},t)$ is given by

$$\left\{ \frac{\partial}{\partial t} + \mathbf{v}\cdot\frac{\partial}{\partial\mathbf{x}} - \frac{e}{c}\mathbf{v}\times[B_0\hat{\mathbf{e}}_z + B_\theta^s(r)\hat{\mathbf{e}}_\theta]\cdot\frac{\partial}{\partial\mathbf{p}} \right\} \delta f_b$$

$$= \frac{ep_z}{\gamma mc}\left(\frac{1}{r}\frac{\partial\delta A_z}{\partial\theta}\frac{\partial f_b^0}{\partial p_\theta} + \frac{\partial\delta A_z}{\partial r}\frac{\partial f_b^0}{\partial p_r} \right) , \qquad (9.245)$$

where $p_z = \gamma m v_z$, and $f_b^0(r,\mathbf{p})$ is the equilibrium distribution in Eq.(9.224).
We adopt a normal-mode approach in which the $\ell = 1$ Fourier components of the perturbations δA_z and δf_b are assumed to vary as

$$\delta\psi(r,\theta,z,t) = \delta\psi(r)\,\exp[i(\theta + k_z z - \omega t)] . \qquad (9.246)$$

In the present application, it is convenient to use $\tau = t - z/v_z$ and z as independent variables (rather than t and z) so that Eq.(9.246) becomes (for $v_z = V_b$)

$$\delta\psi(r,\theta,z,t) = \delta\psi(r)\,\exp[i(\theta - \Omega z/V_b - \omega\tau)] , \qquad (9.247)$$

where $V_b = \beta_b c$ is the axial velocity of the electron beam, and $\Omega = \omega - k_z V_b$ is the frequency of the perturbation seen by a beam particle. If each beam segment is assumed to oscillate at a fixed real wavenumber Ω/V_b, then complex ω with $\text{Im}\,\omega > 0$ represents the growth of the wave as one moves backward from the head of the beam. On the other hand, if a perturbation is initialized at $z = z_0$, say, with real perturbation frequency ω, then the complex frequency Ω represents the spatial oscillation and growth of the wave as the beam propagates downstream. To evaluate the perturbed distribution function δf_b, we adopt the latter convention and use the method of characteristics to integrate the linearized Vlasov equation along the unperturbed particle trajectories from $z' = -\infty$ to $z' = z$ assuming $\operatorname{Im}\Omega > 0$.[35]

In terms of the Fourier amplitudes $\delta A_z(r)$ and $\delta f_b(r, \mathbf{p})$ for perturbations with azimuthal mode number $\ell = 1$, the Maxwell equation (9.244) can be expressed as

$$\left\{ \frac{1}{r}\frac{\partial}{\partial r}r\frac{\partial}{\partial r} - \frac{1}{r^2} + \frac{4\pi i\omega\sigma}{c^2} \right\} \delta A_z(r) = \frac{4\pi e}{c} \int d^3p\, v_z \delta f_b(r, \mathbf{p}) . \quad (9.248)$$

An evaluation of $\delta f_b(r, \mathbf{p})$ using the method of characteristics closely parallels the analyses in Secs. 4.9 and 9.4. Making use of Eq.(9.245) and the assumptions summarized in Sec. 9.5.1, it is found that $\delta f_b(r, \mathbf{p})$ can be expressed as (Problem 9.24)

$$\delta f_b(r, \mathbf{p}) = \frac{e}{c}\frac{p_z}{p_\perp}\frac{\partial f_b^0}{\partial p_\perp} \quad (9.249)$$

$$\times \left\{ \delta A_z(r) + \frac{(\Omega - \omega_{rb})}{V_b}i \int_{-\infty}^z dz'\, \delta A_z(r') \exp\left[i(\theta' - \theta) - i\frac{\Omega}{V_b}(z' - z) \right] \right\},$$

where

$$\operatorname{Im}\Omega > 0 \quad \text{and} \quad V_b = \beta_b c .$$

Here, for $p_r^2 + p_\theta^2 \ll p_z^2$ and $\nu/\gamma_b \ll 1$, perturbations are about the equilibrium distribution function

$$f_b^0 = \left(\frac{\hat{n}_b}{2\pi\gamma_b m_e} \right) F_b\left[\frac{p_\perp^2}{2\gamma_b m_e} + (\gamma_b - 1)m_e c^2 + \psi_b(r) \right] \delta(p_z - \gamma_b m_e \beta_b c) , \quad (9.250)$$

where $p_\perp^2 \equiv p_r^2 + (p_\theta - \gamma_b m_e r\omega_{rb})^2$ is the transverse momentum-squared *in the rotating frame*, $v_z = \beta_b c = V_b$ is the axial velocity, $\gamma_b = (1 - \beta_b^2)^{-1/2}$ is the relativistic mass factor, and use has been made of Eqs.(9.224) and (9.227). Moreover, the orbits $\theta'(z')$ and $r'(z')$ occurring in Eq.(9.249) satisfy the "initial" conditions $\theta'(z' = z) = \theta$ and $r'(z' = z) = r$. Substituting Eqs.(9.249) and (9.250) into Eq.(9.248) and carrying out the integration over axial momentum, the eigenvalue equation can be expressed as[35]

$$\left\{ \frac{1}{r}\frac{\partial}{\partial r}r\frac{\partial}{\partial r} - \frac{1}{r^2} + \frac{4\pi i\omega\sigma}{c^2} \right\} \delta A_z(r)$$

$$= m_e\omega_{pb}^2\beta_b^2 \int_0^\infty dU \frac{\partial}{\partial U}F_b \left[U + (\gamma_b - 1)m_ec^2 + \psi_b(r) \right] \quad (9.251)$$

$$\times \left\{ \delta A_z(r) + \frac{(\Omega - \omega_{rb})}{V_b}I_1(\Omega, r, U) \right\} \,,$$

where

$$U \equiv \frac{p_\perp^2}{2\gamma_b m_e} = \frac{p_r^2 + (p_\theta - \gamma_b m_e\omega_{rb}r)^2}{2\gamma_b m_e} \,,$$

$\omega_{pb}^2 = 4\pi\hat{n}_b e^2/m_e$, and the (phase-averaged) orbit integral $I_1(\Omega, r, U)$ is defined by

$$I_1(\Omega, r, U) = i\int_0^{2\pi} \frac{d\phi}{2\pi} \int_{-\infty}^0 d\xi\, \delta A_z(r') \exp\left[i(\theta' - \theta) - i\frac{\Omega}{V_b}\xi \right] \,. \quad (9.252)$$

Here, $\xi = z' - z$, and ϕ is the perpendicular momentum phase in the rotating frame, i.e., $\int d^3p \cdots$ has been expressed as $\int_0^{2\pi} d\phi \int_0^\infty dp_\perp p_\perp \times \int_{-\infty}^\infty dp_z \cdots$ in carrying out the momentum integration on the right-hand side of Eq.(9.248).

The eigenvalue equation (9.251) can be used to determine the eigenfunction $\delta A_z(r)$ and (complex) eigenfrequency Ω of the resistive hose instability for a wide range of beam distribution functions F_b. Note that the orbit integral I_1 involves an integration over the (unknown) eigenfunction $\delta A_z(r')$ along the particle trajectories in the equilibrium field configuration. Therefore, except for certain choices of F_b, analytical solutions to Eq.(9.251) are generally difficult to obtain because the electron orbits span the cross section of the electron beam, and the information regarding the perturbation is communicated from one value of r' to another. For an assumed distribution function F_b, one approach is to solve Eq.(9.251) numerically for $\delta A_z(r)$ and Ω. Alternatively, Lampe and Uhm[35] have developed a variational approach using an "energy-group" model of the beam dynamics.

Problem 9.24   Consider the linearized Vlasov-Maxwell equations (9.244) and (9.245) for the class of self-consistent relativistic electron beam equilibria described by the distribution function [see Eqs.(9.224), (9.227), and (9.228)]

$$f_b^0(r, \mathbf{p}) = \left(\frac{\hat{n}_b}{2\pi \gamma_b m_e} \right) F_b \left[\frac{p_\perp^2}{2\gamma_b m_e} + (\gamma_b - 1)m_e c^2 + \psi_b(r) \right] \delta(p_z - \gamma_b m_e \beta_b c) \ .$$

$$(9.24.1)$$

Here,

$$\psi_b(r) = (\gamma_b m_e / 2)(\omega_{rb}\omega_{cb}/\gamma_b - \omega_{rb}^2)r^2 + e\beta_b A_z^s(r)$$

and $p_\perp^2 = p_r^2 + (p_\theta - \gamma_b m_e \omega_{rb}r)^2$ is the perpendicular momentum-squared in the rotating frame.

a. For the class of beam equilibria in Eq.(9.24.1), assuming $p_z^2 \gg p_r^2 + p_\theta^2$ and $\nu/\gamma_b \ll 1$, the delta function $\delta(p_z - \gamma_b m_e \beta_b c)$ selects the axial velocity $v_z = \beta_b c = V_b$ in Eqs.(9.244) and (9.245). Change variables from (z, t) to (Z, τ) in Eqs.(9.244) and (9.245) according to the transformation

$$\tau = t - z/V_b \ ,$$

$$Z = z \ . \qquad\qquad (9.24.2)$$

Show that $\partial/\partial t$ and $\partial/\partial z$ transform according to

$$\frac{\partial}{\partial t} = \frac{\partial}{\partial \tau} \ ,$$

$$\frac{\partial}{\partial z} = \frac{\partial}{\partial Z} - \frac{1}{V_b}\frac{\partial}{\partial \tau} \ . \qquad\qquad (9.24.3)$$

Therefore, $\partial/\partial t = \partial/\partial \tau$ in the linearized Maxwell equation (9.244), and

$$\frac{\partial}{\partial t} + V_b\frac{\partial}{\partial z} = V_b\frac{\partial}{\partial Z} \qquad\qquad (9.24.4)$$

in the linearized Vlasov equation (9.245).

b. For $f_b^0(r, \mathbf{p})$ of the form in Eq.(9.24.1), show that

$$\frac{1}{r}\frac{\partial \delta A_z}{\partial \theta}\frac{\partial f_b^0}{\partial p_\theta} + \frac{\partial \delta A_z}{\partial r}\frac{\partial f_b^0}{\partial p_r}$$

$$= \frac{1}{p_\perp}\frac{\partial f_b^0}{\partial p_\perp}\left[p_r\frac{\partial \delta A_z}{\partial r} + \frac{(p_\theta - \gamma_b m_e \omega_{rb}r)}{r}\frac{\partial \delta A_z}{\partial \theta} \right] \ . \qquad (9.24.5)$$

c. Make use of Eqs.(9.24.4) and (9.24.5), and employ the method of characteristics (Sec. 2.2) to integrate the linearized Vlasov equation (9.245) from

$z' = -\infty$ to $z' = z$, assuming perturbations that amplify with increasing z'. Neglecting the "initial" perturbation (at $z' = -\infty$), show that

$$\delta f_b(r, \theta, z, \mathbf{p}, \tau) = \frac{e}{c} \frac{p_z}{p_\perp} \frac{\partial f_b^0}{\partial p_\perp} \tag{9.24.6}$$

$$\times \int_{-\infty}^{z} \frac{dz'}{V_b} \left[v_r' \frac{\partial}{\partial r'} + \left(\frac{v_\theta'}{r'} - \omega_{rb} \right) \frac{\partial}{\partial \theta'} \right] \delta A_z[r'(z'), \theta'(z'), z', \tau] \; .$$

Here, use has been made of $p_r = \gamma_b m v_r$, $p_\theta = \gamma_b m v_\theta$, and $\gamma \simeq \gamma_b$. Note that the new time variable τ occurs only parametrically on both sides of Eq.(9.24.6).

d. In Eq.(9.24.6), show that the transverse orbits $x'(z') = r'(z') \cos \theta'(z')$ and $y'(z') = r'(z') \sin \theta'(z')$ satisfy

$$\left[V_b^2 \frac{d^2}{dz'^2} + i \frac{\omega_{cb}}{\gamma_b} V_b \frac{d}{dz'} + \omega_\beta^2[r'(z')] \right] \left[x'(z') + iy'(z') \right] = 0 \; , \tag{9.24.7}$$

where $\omega_\beta^2(r) = -e\beta_b B_\theta^s(r)/\gamma_b m_e r$ is defined in Eq.(9.232). Equation (9.24.7) should be compared with the analogous orbit equations in Secs. 4.9 and 9.4 [see, for example, Eq.(9.158)]. In Eq.(9.24.7), the orbits $r'(z')$ and $\theta'(z')$ satisfy the "initial" conditions $r'(z' = z) = r$ and $\theta'(z' = z) = \theta$.

e. In Eq.(9.24.6), $v_r' = V_b dr'/dz'$ and $v_\theta'/r' = V_b d\theta'/dz'$. For perturbations with azimuthal mode number $\ell = 1$, express

$$\delta A_z(r', \theta', z', \tau) = \delta A_z(r') \exp(i\theta' - i\Omega z'/V_b - i\omega\tau) \; ,$$

$$\delta f_b(r, \theta, z, \mathbf{p}, \tau) = \delta f_b(r, \mathbf{p}) \exp(i\theta - i\Omega z/V_b - i\omega\tau) \; , \tag{9.24.8}$$

where Im $\Omega > 0$ corresponds to spatial growth. Make use of Eqs.(9.24.6) and (9.24.8) to show that

$$\delta f_b(r, \mathbf{p}) = \frac{e}{c} \frac{p_z}{p_\perp} \frac{\partial f_b^0}{\partial p_\perp} \int_{-\infty}^{z} dz' \left[\frac{dr'}{dz'} \frac{\partial}{\partial r'} + \left(\frac{d\theta'}{dz'} - \frac{\omega_{rb}}{V_b} \right) \frac{\partial}{\partial \theta'} \right] \tag{9.24.9}$$

$$\times \delta A_z[r'(z')] \exp \left\{ i[\theta'(z') - \theta] - i(\Omega/V_b)(z' - z) \right\} \; .$$

f. Use the chain rule for differentiation to show that

$$\left[\frac{dr'}{dz'} \frac{\partial}{\partial r'} + \left(\frac{d\theta'}{dz'} - \frac{\omega_{rb}}{V_b} \right) \frac{\partial}{\partial \theta'} \right] \delta A_z(r') \exp \left\{ i(\theta' - \theta) - i(\Omega/V_b)(z' - z) \right\}$$

$$= \left[\frac{d}{dz'} + i \frac{(\Omega - \omega_{rb})}{V_b} \right] \delta A_z(r') \exp \left\{ i(\theta' - \theta) - i(\Omega/V_b)(z' - z) \right\} \; . \tag{9.24.10}$$

in the integrand in Eq.(9.24.9).

g. Make use of Eq.(9.24.10) to integrate by parts with respect to z' in Eq.(9.24.9). Enforce the boundary conditions $\theta'(z' = z) = \theta$, $r'(z' = z) = r$, and $[\delta A_z(r')]_{z' = -\infty} = 0$, and show that Eq.(9.24.9) reduces to

$$\delta f_b(r, \mathbf{p}) = \frac{e}{c} \frac{p_z}{p_\perp} \frac{\partial f_b^0}{\partial p_\perp} \tag{9.24.11}$$

$$\times \left\{ \delta A_z(r) + i \frac{(\Omega - \omega_{rb})}{V_b} \int_{-\infty}^{z} dz' \delta A_z(r') \exp\left[i(\theta' - \theta) - i(\Omega/V_b)(z' - z)\right] \right\}.$$

Substituting Eq.(9.24.11) into Eq.(9.244) then gives the desired eigenvalue equation (9.251).

9.5.4   Resistive Hose Instability for a Sharp-Boundary Density Profile

As an application of Eq.(9.251), we consider the case where $F_b(H - \omega_{rb} P_\theta)$ has the inverted population specified by Eq.(9.235). In this case, the beam density profile is rectangular [Eq.(9.236)] with $n_b^0(r) = \hat{n}_b = \text{const.}$ for $0 \leq r < r_b$. Moreover, the betatron frequency satisfies $\omega_\beta^2(r) = (\beta_b^2 \omega_{pb}^2 / 2\gamma_b)(1 - f_M) = \text{const.}$ over the cross section of the electron beam. For present purposes, we assume that the conductivity is uniform within the electron beam with $\sigma = \sigma_0 = \text{const.}$ for $0 \leq r < r_b$, and define the characteristic resistive decay time τ_d for the perturbed current by

$$\tau_d = \frac{1}{2} \pi \sigma_0 r_b^2 / c^2 . \tag{9.253}$$

Comparing the form of the eigenvalue equation (9.251) with Eq.(4.258) [or Eq.(9.162)] for the choice of distribution function in Eq.(9.235), it is evident from the analysis in Sec. 4.9 [or Sec. 9.4] that in the limit where $4\pi\sigma\omega/c^2 \to 0$ (specifically, $|\omega|\tau_d \ll 1$) the eigenvalue equation (9.251) supports solutions with $\delta A_z(r) = \text{const.} \times r$ within the electron beam ($0 \leq r < r_b$). Such solutions (proportional to r^ℓ with $\ell = 1$) correspond to a rigid displacement of the electron beam. For present purposes, the analysis is restricted to $|\omega|\tau_d < 1$ so that $\delta A_z(r') = [\delta A_z(r)/r]r'$ provides a reasonably good approximation to the eigenfunction within the electron beam. [In actual fact, depending on the size of $|\omega|\tau_d$, the eigenfunction $\delta A_z(r')$ bends over somewhat as r' approaches r_b.] Making use of $r' \exp(i\theta') = [x'(z') + iy'(z')]$, we parallel the analysis in Sec. 4.9 and evaluate the orbit integral I_1 defined in Eq.(9.252). Some straightforward algebra gives (Problem 9.25)

$$I_1(\Omega, r, U) = -\frac{V_b \delta A_z(r)}{\omega_{rb}^+ - \omega_{rb}^-} \left(\frac{\omega_{rb} - \omega_{rb}^-}{\Omega - \omega_{rb}^+} + \frac{\omega_{rb}^+ - \omega_{rb}}{\Omega - \omega_{rb}^-} \right) , \qquad (9.254)$$

where $\omega_{rb}^{\pm} = (\omega_{cb}/2\gamma_b)\{1 \pm (1 + 4\gamma_b^2\hat{\omega}_\beta^2/\omega_{cb}^2)^{1/2}\}$ and $\hat{\omega}_\beta^2 = (\beta_b^2/2\gamma_b)\omega_{pb}^2 \times (1 - f_M) = $ const. Note from Eq.(9.254) that the orbit integral $I_1(\Omega, r, U)$ is independent of $U = p_\perp^2/2\gamma_b m_e$, which greatly simplifies the integration over U on the right-hand side of Eq.(9.251). Substituting Eq.(9.254) into Eq.(9.251) gives

$$\left\{ \frac{1}{r}\frac{\partial}{\partial r}r\frac{\partial}{\partial r} - \frac{1}{r^2} + \frac{4\pi i\omega\sigma}{c^2} \right\} \delta A_z(r)$$

$$= -\frac{m_e\omega_{pb}^2\beta_b^2\delta A_z(r)}{(\Omega - \omega_{rb}^+)(\Omega - \omega_{rb}^-)}(\omega_{rb}^+ - \omega_{rb})(\omega_{rb} - \omega_{rb}^-) \qquad (9.255)$$

$$\times \int_0^\infty dU \frac{\partial}{\partial U}F_b\left[U + (\gamma_b - 1)m_ec^2 + \psi_b(r)\right] .$$

On the right-hand side of Eq.(9.255), we express $\hat{n}_b \int_0^\infty dU\, \partial F_b/\partial U = (\partial/\partial\psi_b)\hat{n}_b \int_0^\infty dU\, F_b$ and make use of $n_b^0(r) = \hat{n}_b \int_0^\infty dU\, F_b$, and $\psi_b(r) = \gamma_b m_e(\omega_{rb}^+ - \omega_{rb})(\omega_{rb} - \omega_{rb}^-)r^2/2$. For the rectangular density profile in Eq.(9.236), $\partial n_b^0/\partial r = -\hat{n}_b\delta(r - r_b)$, and Eq.(9.255) reduces to the eigenvalue equation[35]

$$\left\{ \frac{1}{r}\frac{\partial}{\partial r}r\frac{\partial}{\partial r} - \frac{1}{r^2} + \frac{4\pi i\omega\sigma}{c^2} \right\} \delta A_z(r) = \frac{\delta A_z(r)\omega_{pb}^2\beta_b^2/\gamma_b}{(\Omega - \omega_{rb}^+)(\Omega - \omega_{rb}^-)r}\delta(r - r_b) .$$
$$(9.256)$$

For the case of a rectangular density profile where $\omega_\beta^2(r) = \hat{\omega}_\beta^2 = $ const. over the cross section of the electron beam, note that the perturbed axial beam current on the right-hand side of Eq.(9.256) corresponds to a surface current localized at $r = r_b$.

As indicated earlier, it is assumed that $\sigma = \sigma_0 = $ const. within the electron beam $(0 \le r < r_b)$. To further simplify the analysis, it is also assumed that the plasma conductivity satisfies $\sigma = \sigma_1 \ll \sigma_0/|\omega|\tau_d$ in the region between the electron beam and the conducting wall $(r_b < r \le b)$. (Here, $\tau_d = \pi\sigma_0 r_b^2/2c^2$ is the characteristic resistive decay time within the electron beam.) In this case, the solution to Eq.(9.256) can be approximated by the vacuum solution, $\delta A_z(r) = Br + C/r$, in the region $r_b < r \leq b$. It is convenient to define

$$\kappa^2 = \frac{4\pi i\sigma_0\omega}{c^2} = \frac{8i\omega\tau_d}{r_b^2} , \qquad (9.257)$$

where $|\omega|\tau_d < 1$ is assumed. The solutions to Eq.(9.256), which satisfy the boundary conditions $\delta A_z(r = 0) = 0 = \delta A_z(r = b)$ and are continuous at the surface of the electron beam, can be expressed as

$$\delta A_z^I(r) = AJ_1(\kappa r) , \qquad\qquad 0 \leq r < r_b ,$$

$$\delta A_z^{II}(r) = AJ_1(\kappa r_b)\frac{r_b}{r}\left(\frac{1 - r^2/b^2}{1 - r_b^2/b^2}\right) , \quad r_b < r \leq b , \qquad (9.258)$$

where $A = $ const., and $J_1(x)$ is the Bessel function of the first kind of order unity. We multiply the eigenvalue equation (9.256) by r and integrate across the surface of the electron beam from $r_b(1 - \varepsilon)$ to $r_b(1 + \varepsilon)$ with $\varepsilon \to 0_+$. This gives

$$r_b\left[\frac{\partial}{\partial r}\delta A_z^{II}\right]_{r=r_b} - r_b\left[\frac{\partial}{\partial r}\delta A_z^I\right]_{r=r_b} = \frac{\delta A_z(r = r_b)\omega_{pb}^2\beta_b^2/\gamma_b}{(\Omega - \omega_{rb}^+)(\Omega - \omega_{rb}^-)} . \quad (9.259)$$

Substituting Eq.(9.258) into Eq.(9.259), the condition for a nontrivial solution $(A \neq 0)$ gives the dispersion relation (Problem 9.26)

$$\Omega^2 - \frac{\omega_{cb}}{\gamma_b}\Omega - \hat{\omega}_\beta^2 = -\frac{\omega_{pb}^2}{\gamma_b}\beta_b^2\left[\kappa r_b\frac{J_1'(\kappa r_b)}{J_1(\kappa r_b)} + \frac{1 + r_b^2/b^2}{1 - r_b^2/b^2}\right]^{-1} , \quad (9.260)$$

where $\hat{\omega}_\beta^2 = (\beta_b^2\omega_{pb}^2/2\gamma_b)(1 - f_M)$, and use has been made of $(\Omega - \omega_{rb}^+) \times (\Omega - \omega_{rb}^-) = \Omega^2 - (\omega_{cb}/\gamma_b)\Omega - \hat{\omega}_\beta^2$ [see Eq.(9.234)].[35] Expanding Eq.(9.260) for $|\kappa|^2 r_b^2 \ll 1$ (or, equivalently, $|\omega|\tau_d \ll 1$), the dispersion relation (9.260) can be expressed as (see also Problem 9.27)

$$\omega\tau_d = -i\left(1 - \frac{r_b^2}{b^2}\right)^{-1} - i\frac{\omega_{pb}^2\beta_b^2/2\gamma_b}{(\Omega^2 - \Omega\omega_{cb}/\gamma_b - \omega_\beta^2)} \qquad (9.261)$$

correct to leading order in $\omega\tau_d$.

The formal derivation of the dispersion relation (9.260) [or (9.261)] for the resistive hose instability has been carried out for the case where ω is real and Ω is complex with $\text{Im}\,\Omega > 0$. The dispersion relation can also be derived for the case where Ω is real and ω is complex, and the results are identical in form to Eqs.(9.260) and (9.261). We therefore summarize stability properties for both cases.

Stability Properties for Real Ω and Complex ω: For real Ω, note from Eq.(9.261) that $\mathrm{Re}\,\omega = 0$ and the perturbation is either damped ($\mathrm{Im}\,\omega < 0$) or growing ($\mathrm{Im}\,\omega > 0$) depending on the sign of $\mathrm{Im}\,\omega$. Evidently, the proximity of a conducting wall (increasing r_b/b) has a stabilizing influence through the contribution of the first term on the right-hand side of Eq.(9.261). Expressing $\Omega^2 - \Omega\omega_{cb}/\gamma_b - \hat{\omega}_\beta^2 = -(\omega_{rb}^+ - \Omega)(\Omega - \omega_{rb}^-)$, it also follows that the beam contribution (proportional to $\omega_{pb}^2\beta_b^2$) in Eq.(9.261) is destabilizing only for (real) Ω in the interval $\omega_{rb}^- < \Omega < \omega_{rb}^+$. Moreover, $\mathrm{Im}\,\omega$ becomes infinite [and Eq.(9.261) breaks down] whenever $\Omega \to \omega_{rb}^+$ from below or $\Omega \to \omega_{rb}^-$ from above. For a diffuse beam density profile (Sec. 9.5.5), this singular behavior is absent because the radial spread in the betatron frequency $\omega_\beta(r)$ precludes all of the beam electrons from being in exact resonance with the wave. It should also be noted that the axial magnetic field $B_0\hat{e}_z$ influences stability behavior only through the term $\Omega\omega_{cb}/\gamma_b$ in Eq.(9.261). This has the effect of shifting the range of Ω corresponding to instability relative to the case where $B_0 = 0$, as well as increasing the bandwidth $\omega_{rb}^+ - \omega_{rb}^- = (\omega_{cb}^2/\gamma_b^2 + \hat{\omega}_\beta^2)^{1/2}$ of the instability. Finally, for fixed values of ω_{pb}^2, r_b/b and γ_b, and fixed real Ω in the interval $\omega_{rb}^- < \Omega < \omega_{rb}^+$, we also note from Eq.(9.261) and the definition $\hat{\omega}_\beta^2 = (\beta_b^2\omega_{pb}^2/2\gamma_b)(1 - f_M)$ that increasing the plasma return current (i.e., increasing f_M) has the effect of increasing the growth rate $\mathrm{Im}\,\omega$.

Stability Properties for Real ω and Complex Ω: We now make use of the dispersion relation (9.261) to investigate properties of the resistive hose instability for real ω and complex Ω, with $\mathrm{Im}\,\Omega > 0$ corresponding to instability. Consider first the case $\omega\tau_d \to 0$, which corresponds to very low frequency or very high resistivity of the background plasma. In this regime, Eq.(9.261) reduces to

$$\Omega^2 - \Omega\frac{\omega_{cb}}{\gamma_b} + \frac{\omega_{pb}^2\beta_b^2}{2\gamma_b}\left(f_M - \frac{r_b^2}{b^2}\right) = 0 . \tag{9.262}$$

From Eq.(9.262), it is readily shown that the necessary and sufficient condition for instability ($\mathrm{Im}\,\Omega > 0$) is given by (Fig. 9.14)

$$f_M > \frac{r_b^2}{b^2} + \frac{\omega_{cb}^2}{2\gamma_b\omega_{pb}^2\beta_b^2} . \tag{9.263}$$

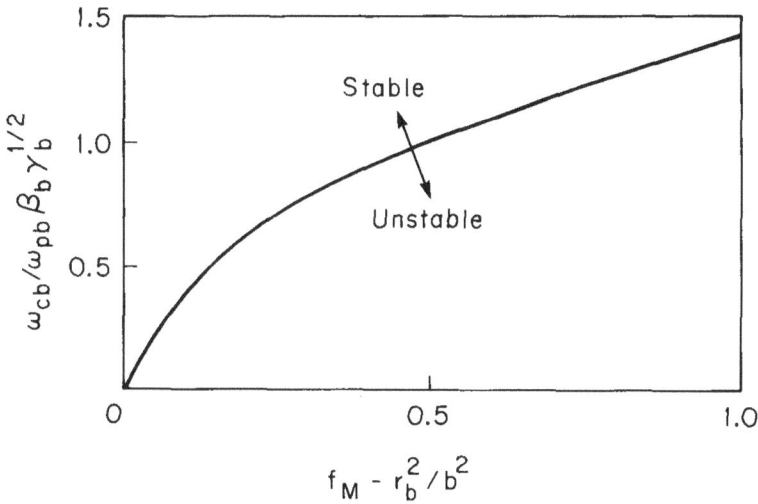

Figure 9.14. The stable and unstable regions of the parameter space $(f_M - r_b^2/b^2,\ \omega_{cb}/\omega_{pb}\beta_b\gamma_b^{1/2})$ consistent with Eq.(9.263) and the dispersion relation (9.262) for the resistive hose instability. Here, $\omega^2\tau_d^2 \ll 1$ is assumed [H.S. Uhm and M. Lampe, Phys. Fluids **23**, 1574 (1980)].

Because $0 < f_M < 1$ is assumed, it follows from Eq.(9.263) that the proximity of a conducting wall and the inclusion of an axial magnetic field $B_0\hat{e}_z$ are both stabilizing effects in the limit $\omega\tau_d \to 0$. It is evident from Eqs.(9.262) and (9.263) that the instability is driven by the plasma return current [the term proportional to f_M in Eq.(9.262)]. An unstable mode with $\omega = 0$ corresponds to the expulsion of the (rigid) electron beam from the conducting plasma channel. Therefore, the instability as viewed in the beam frame is absolute if the unstable range of Ω includes $\omega = 0$. The basic instability mechanism in this regime is due to the repulsion of the beam current by the plasma return current flowing in the conducting channel.

Lampe and Uhm[35] have made use of Eq.(9.261) to investigate detailed stability behavior over a wide range of system parameters f_M, r_b/b, $\omega\tau_d$, and $\omega_{cb}/\omega_{pb}\beta_b$. Typical results are summarized in Fig. 9.15 for the case where the conducting wall is removed to infinity $(r_b/b = 0)$ and there is no axial guide field $(\omega_{cb} = 0)$. Here, the normalized growth rate $(\text{Im}\,\Omega)/(\omega_{pb}\beta_b/\gamma_b^{1/2})$ and the normalized real frequency $(\text{Re}\,\Omega)/(\omega_{pb}\beta_b/\gamma_b^{1/2})$ are plotted versus $\omega\tau_d$ for several values of the current neutralization factor f_M. As f_M is increased to unity, note

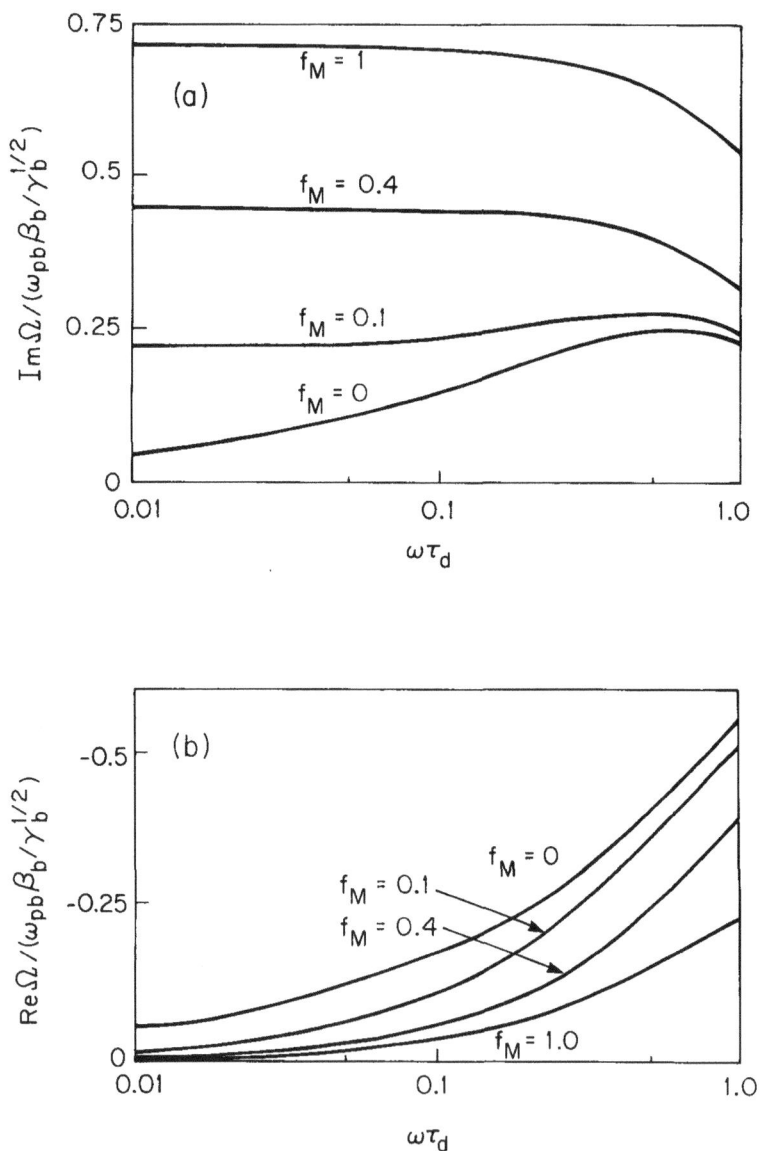

Figure 9.15. Plots versus the magnetic phase lag $\omega\tau_d$ of (a) the normalized growth rate $\operatorname{Im}\Omega/(\omega_{pb}\beta_b/\gamma_b^{1/2})$, and (b) the normalized real oscillation frequency $\operatorname{Re}\Omega/(\omega_{pb}\beta_b/\gamma_b^{1/2})$ for the resistive hose instability obtained from the dispersion relation (9.261) for $\omega_{cb}=0$ and $b\to\infty$ [H.S. Uhm and M. Lampe, Phys. Fluids **23**, 1574 (1980)].

that the growth rate $\text{Im}\,\Omega$ increases, whereas the real frequency $\text{Re}\,\Omega$ decreases. Moreover, for fixed f_M, the maximum growth rate occurs for $\omega\tau_d < 1$. Indeed, for $f_M \gtrsim 0.4$, the instability is driven mainly by the simple repulsion between the· beam current and the plasma return current (described earlier), and the growth rate is nearly constant for $\omega\tau_d \lesssim 0.3$, achieving a maximum value at $\omega\tau_d = 0$. On the other hand, for $f_M \ll 1$, the maximum growth rate occurs for values of $\omega\tau_d$ somewhat less than unity, and the instability is driven mainly by a phase lag between the oscillation of the beam electrons and the oscillation of the magnetic restoring force.

Problem 9.25   Consider the orbit equation (9.24.7) for $x'(z') + iy'(z')$ obtained in Problem 9.24 for the case of a uniform-density electron beam where $\omega_\beta^2(r) = \hat{\omega}_\beta^2 = (\beta_b^2/2\gamma_b)\omega_{pb}^2(1 - f_M)$. Also consider the orbit integral $I_1(\Omega, r, U)$ defined in Eq.(9.252).

a. Show that the solution to Eq.(9.24.7) is given by

$$x'(z') + iy'(z') = \tag{9.25.1}$$

$$(\omega_{rb}^+ - \omega_{rb}^-)^{-1} \left\{ i\left(\frac{p_\perp}{\gamma_b m_e}\right) \exp(i\phi) \left[\exp\left(\frac{i\omega_{rb}^-\xi}{V_b}\right) - \exp\left(\frac{i\omega_{rb}^+\xi}{V_b}\right)\right] \right.$$

$$\left. + r\exp(i\theta)\left[(\omega_{rb} - \omega_{rb}^-)\exp\left(\frac{i\omega_{rb}^+\xi}{V_b}\right) + (\omega_{rb}^+ - \omega_{rb})\exp\left(\frac{i\omega_{rb}^-\xi}{V_b}\right)\right] \right\},$$

where $\xi = z' - z, \omega_{rb}^\pm = (\omega_{cb}/2\gamma_b)\{1 \pm (1 + 4\gamma_b^2\hat{\omega}_\beta^2/\omega_{cb}^2)^{1/2}\}$, and ϕ is the perpendicular momentum phase in the rotating frame. Equation (9.25.1) should be compared with Eq.(9.161), which was derived for a fully nonneutral ion beam.

b. Approximate $\delta A_z(r') = \delta A_z(r)r'/r$ in evaluating the orbit integral defined in Eq.(9.252). Express $r'\exp(i\theta') = x' + iy'$ and carry out the integrations over ϕ and ξ indicated in Eq.(9.252). Show that $I_1(\Omega, r, U)$ reduces to the expression given in Eq.(9.254).

c. Substitute Eq.(9.254) into Eq.(9.251) and derive the form of the eigenvalue equation given in Eq.(9.255), or equivalently, Eq.(9.256).

Problem 9.26   Consider the eigenvalue equation (9.256), which determines the eigenfunction $\delta A_z(r)$ and the complex eigenfrequency Ω for the case of the sharp-boundary density profile in Eq.(9.236).

a. Solve Eq.(9.256) subject to the boundary conditions $\delta A_z(r = 0) = 0 = \delta A_z(r = b)$, and $\delta A_z(r)$ continuous at $r = r_b$. For negligibly small plasma conductivity σ_1 in the outer region $(r_b < r \leq b)$, show that the solution to Eq.(9.256) can be expressed in the form given in Eq.(9.258), where $\kappa^2 = 4\pi i \sigma_0 \omega / c^2 = 8i\omega \tau_d / r_b^2$.

b. Integrate the eigenvalue equation (9.256) across the surface of the electron beam at $r = r_b$. Derive the jump condition in Eq.(9.259), and make use of the solution for $\delta A_z(r)$ in Eq.(9.258) to obtain the dispersion relation (9.260).

c. Expand Eq.(9.260) for $|\kappa|^2 r_b^2 \ll 1$, retaining terms to leading order in $\omega \tau_d$. Show that the dispersion relation can be approximated by Eq.(9.261).

Problem 9.27   For $\omega^2 \tau_d^2 \ll 1$, a variational approach can be developed to obtain an approximate dispersion relation from the eigenvalue equation (9.256) [H.S. Uhm and M. Lampe, Phys. Fluids **23**, 1574 (1980)]. This approach avoids direct use of the Bessel-function solution in Eq.(9.258).

a. Multiply Eq.(9.256) by $r\delta A_z^*(r)$ and integrate from $r = 0$ to $r = b$, enforcing the boundary conditions $\delta A_z(r = 0) = 0 = \delta A_z(r = b)$. Show that

$$\frac{4\pi i \omega}{c^2} \int_0^b dr\, r\sigma(r) \, |\delta A_z(r)|^2 \tag{9.27.1}$$

$$= \frac{\omega_{pb}^2 \beta_b^2 \, |\delta A_z(r_b)|^2}{\gamma_b(\Omega - \omega_{rb}^+)(\Omega - \omega_{rb}^-)} + \int_0^b dr\, r \left\{ \left| \frac{\partial}{\partial r} \delta A_z(r) \right|^2 + \frac{|\delta A_z(r)|^2}{r^2} \right\} .$$

In Eq.(9.27.1), $\delta A_z(r)$ is the exact solution to the eigenvalue equation (9.256).

b. Following Lampe and Uhm,[35] if a trial eigenfunction $\delta A_z^T(r)$ is substituted into the integrals in Eq.(9.27.1), it can be shown that the resulting dispersion relation for the (complex) eigenfrequency is accurate to second order in the error in the trial function $\delta A_z^T(r)$. [Here, the error in the trial function is defined by

$$\delta A_z^E(r) \equiv \delta A_z^A(r) - \delta A_z^T(r) ,$$

where $\delta A_z^A(r)$ is the actual solution to the eigenvalue equation (9.256).] As a trial eigenfunction, take $\omega\sigma(r) \to 0$ in Eq.(9.256) and show that

$$\delta A_z^T(r) = \begin{cases} A\dfrac{r}{r_b} , & 0 \leq r < r_b , \\[2mm] A\dfrac{r_b}{r} \left(\dfrac{1 - r^2/b^2}{1 - r_b^2/b^2} \right) , & r_b < r \leq b . \end{cases} \tag{9.27.2}$$

c. Assume that $\sigma(r) = \sigma_0 = $ const. for $0 \leq r < r_b$, and that $\sigma(r) = \sigma_1$ is negligibly small in the interval $r_b < r \leq b$. Substitute the trial function

$\delta A_z^T(r)$ into Eq.(9.27.1) and show that the approximate dispersion relation is given by

$$i\omega\tau_d = \frac{\omega_{pb}^2\beta_b^2/2\gamma_b}{(\Omega - \omega_{rb}^+)(\Omega - \omega_{rb}^-)} + \left(1 - \frac{r_b^2}{b^2}\right)^{-1}, \qquad (9.27.3)$$

which is identical to Eq.(9.261). Here, $\tau_d = \pi\sigma_0 r_b^2/2c^2$ and $\omega_{rb}^\pm = (\omega_{cb}/2\gamma_b) \times \{1 \pm [1 + 4\gamma_b^2\hat{\omega}_\beta^2/\omega_{cb}^2]^{1/2}\}$.

9.5.5   Resistive Hose Instability for a Diffuse Density Profile

As indicated earlier, for beam distribution functions $F_b(H - \omega_{rb}P_\theta)$ which correspond to a diffuse density profile $n_b^0(r)$, Lampe and Uhm[35] have developed a variational approach using an energy-group model of the beam dynamics to obtain an approximate dispersion relation from the eigenvalue equation (9.251). As a specific example, we consider the thermal equilibrium distribution function $F_b(H)$ in Eq.(9.239). Here, $\omega_{rb} = 0$ is assumed, and the equilibrium profiles for $n_b^0(r)$ and $\omega_\beta(r)$ are given by the familiar Bennett pinch expressions in Eqs.(9.241) and (9.243). Without presenting algebraic details, the corresponding dispersion relation for the resistive hose instability can be expressed as

$$\omega\tau_d = \frac{i}{(1 - f_M)^2}\left[f_M + G\left(\frac{\Omega^2}{\hat{\omega}_\beta^2}\right)\right], \qquad (9.264)$$

where $B_0 = 0$ is assumed, and there is no conducting wall ($b \to \infty$). Here, the magnetic diffusion time τ_d is defined by

$$\tau_d = \frac{6\pi r_b^2}{c^2}\int_0^\infty \frac{dr'}{r_b}\frac{(r'/r_b)^3\sigma(r')}{(1 + r'^2/r_b^2)^2}, \qquad (9.265)$$

and the function $G(\xi)$ is defined by

$$G(\xi) = 6\xi\left\{\frac{1}{2} - \xi + \xi(1 - \xi)\ln\left[\frac{(1 - \xi)}{\xi}\right]\right\}. \qquad (9.266)$$

Moreover, the quantity $\hat{\omega}_\beta = (\beta_b^2\omega_{pb}^2/2\gamma_b)^{1/2}(1 - f_M)^{1/2}$ occurring in

Eq.(9.264) corresponds to the *maximum* betatron frequency (at $r = 0$) for the beam density profile in Eq.(9.241) [see Eq.(9.243)].

Comparing Eq.(9.264) with Eq.(9.261), it is evident that the radial variation of $\omega_\beta(r)$ across the electron beam significantly alters the dielectric response of the beam electrons through the function $G(\Omega^2/\hat{\omega}_\beta^2)$. Indeed, unlike the factor $(\Omega^2 - \hat{\omega}_\beta^2)^{-1}$ occurring in Eq.(9.261) (for $\omega_{cb} = 0$), there is no single frequency Ω for which the entire beam is resonant with the wave,[33] and the dielectric function $G(\Omega^2/\hat{\omega}_\beta^2)$ does not exhibit singular behavior (Fig. 9.16). Note from Eq.(9.266) that $G(\Omega^2/\hat{\omega}_\beta^2)$ has branch points at $\Omega^2 = 0$ and $\Omega^2 = \hat{\omega}_\beta^2$. The appropriate definition of the log-

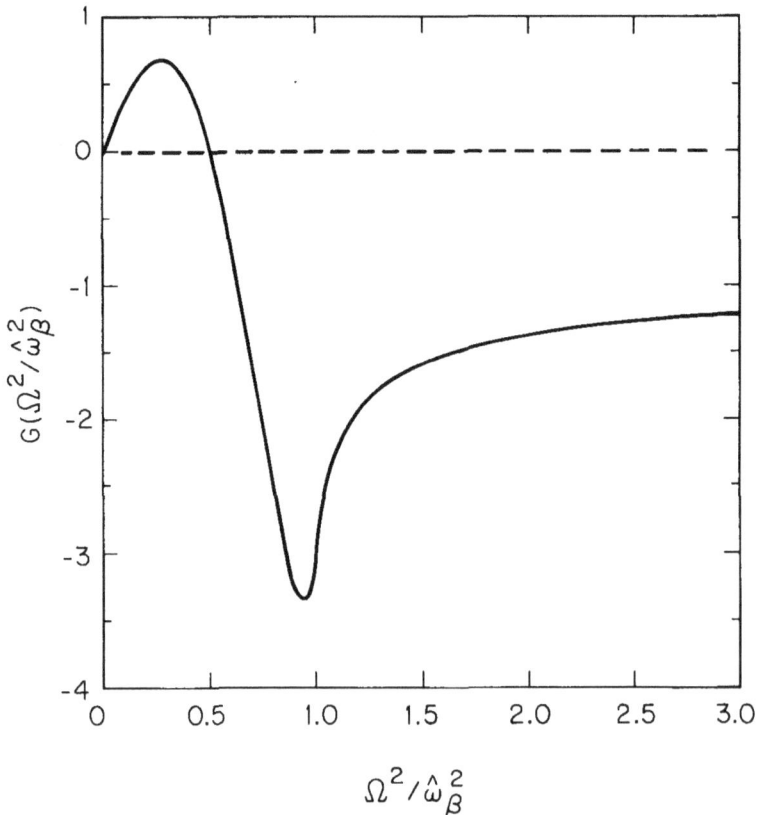

Figure 9.16.   Plot of $G(\Omega^2/\hat{\omega}_\beta^2)$ versus $\Omega^2/\hat{\omega}_\beta^2 > 0$ obtained from Eq. (9.266).

arithmic term[35] along the real-Ω^2 axis [consistent with the assumption $\text{Im}\,\Omega > 0$ used in the derivation of Eq.(9.264)] is given by

$$\ln\left(\frac{\hat{\omega}_\beta^2 - \Omega^2}{\Omega^2}\right) = \ln\left|\frac{\hat{\omega}_\beta^2 - \Omega^2}{\Omega^2}\right| + ig\pi \; , \qquad (9.267)$$

where

$$g = \begin{cases} 0, & \hat{\omega}_\beta^2 < \Omega^2 \; , \\ -1, & -\hat{\omega}_\beta < \Omega < 0 \; , \\ +1, & 0 < \Omega < \hat{\omega}_\beta \; . \end{cases} \qquad (9.268)$$

The function $\text{Re}\,G(\Omega^2/\hat{\omega}_\beta^2)$ is plotted versus real $\Omega^2/\hat{\omega}_\beta^2$ in Fig. 9.16. We examine Eq.(9.264) for real Ω and complex ω. A particularly striking feature of Eq.(9.264) is that instability exists ($\text{Im}\,\omega > 0$) even for $\Omega = 0$ and $\text{Re}\,G = 0$, provided $f_M > 0$. The corresponding growth rate is $\text{Im}\,\omega = f_M(1 - f_M)^{-2}\tau_d^{-1}$. On the other hand, for $f_M = 0$, it follows from Eq.(9.266) and Fig. 9.16 that $\text{Re}\,G > 0$ and therefore $\text{Im}\,\omega > 0$ for Ω^2 in the interval $0 < \Omega^2 < \hat{\omega}_\beta^2/2$. Moreover, maximum growth when $f_M = 0$ occurs for $\Omega^2 \simeq 0.25\,\hat{\omega}_\beta^2$, in which case $(\text{Im}\,\omega)_{\text{max}} \simeq 0.7\,\tau_d^{-1}$ (see Fig. 9.16). For $f_M \neq 0$, it is found[35] that the range of Ω^2 corresponding to instability extends to $0 < \Omega^2 \lesssim 0.65\,\hat{\omega}_\beta^2$ as $f_M \to 1$. Moreover, the growth rate $\text{Im}\,\omega$ predicted by Eq.(9.264) diverges as $f_M \to 1$. As in Sec. 9.5.4, the destabilizing effect of increasing f_M is due to the repulsive interaction between the beam current and the plasma return current. Finally, keep in mind that validity of the dispersion relation (9.264) is restricted to system parameters satisfying $|\omega|\tau_d < 1$.

9.6   Resistive Hollowing and Sausage Instabilities for Intense Electron Beam Propagation Through a Background Plasma

In this section, for an intense electron beam propagating through a background plasma, we investigate low-frequency ($|\omega| \ll 4\pi\sigma$) resistive instabilities for azimuthally symmetric electromagnetic perturbations ($\partial/\partial\theta = 0$). In this case, there are two instabilities that can cause macroscopic distortions of the electron beam and impair or disrupt beam propagation. These are the resistive sausage and resistive hollowing instabilities.[36,37] The resistive sausage instability occurs for perturbations with azimuthal mode number $\ell = 0$ and radial mode number $n = 1$, whereas the resistive hollowing instability occurs for mode numbers $(\ell, n) = (0, 2)$, similar to the transverse instability in an intense nonneutral beam considered in Sec. 9.4. An important feature of both instabilities is that the eigenfunctions have strong internal maxima within the electron beam (unlike the eigenfunction for the resistive hose instability, which is peaked near $r = r_b$). In the nonlinear regime,[37] this feature can lead to a major internal redistribution of beam particles and current.

Many of the theoretical techniques and assumptions used in the present analysis are similar to those in Secs. 9.4 and 9.5. Following a brief summary of the assumptions and theoretical model (Sec. 9.6.1), the kinetic eigenvalue equation (Sec. 9.6.2) is used to derive the dispersion relations for perturbations with mode numbers $(\ell, n) = (0, 1)$ and $(\ell, n) = (0, 2)$. Detailed stability properties are investigated for both the resistive sausage instability (Sec. 9.6.3) and the resistive hollowing instability (Sec. 9.6.4).

9.6.1 Theoretical Model and Assumptions

In the present analysis of the resistive sausage and hollowing instabilities, the assumptions made regarding the beam plasma system are similar to Assumptions (a)–(i) in Sec. 9.5.1.[36]

To summarize briefly, a cylindrically symmetric electron beam $(j = b)$ with characteristic energy $\gamma_b m_e c^2$, density $\hat{n}_b$, and radius r_b propagates parallel to a uniform magnetic field $B_0 \hat{e}_z$ through a dense background plasma $(\hat{n}_e \gg \hat{n}_b)$. A perfectly conducting wall is located at radius $r = b$, and the (collisional) background plasma is described by a real scalar electrical conductivity $\sigma(r)$. The perturbed plasma current $\delta \mathbf{J}_p(\mathbf{x}, t)$ and the perturbed electric field $\delta \mathbf{E}(\mathbf{x}, t)$ are related by the simple Ohm's law $\delta \mathbf{J}_p = \sigma \delta \mathbf{E}$ [Eq.(9.218)], and the density of the background plasma is assumed to be sufficiently large that the beam space charge is completely neutralized in both the equilibrium and the perturbed states, i.e.,

$$\phi_0(r) = 0 = \delta\phi(\mathbf{x}, t) . \tag{9.269}$$

This requires perturbations with sufficiently low frequency ω that $|\omega| \ll 4\pi\sigma$ for $0 \leq r < r_b$ [Eq.(9.219)]. In addition, the equilibrium return current carried by the plasma is assumed to have the same radial profile as the beam current, i.e., $J_{zp}^0(r) = -f_M J_{zb}^0(r)$ where $f_M = $ const. is the fractional current neutralization [Eq.(9.216)].

The electron beam is assumed to satisfy $\nu/\gamma_b \ll 1$, where $\nu = N_b e^2/m_e c^2$ is Budker's parameter [Eq.(9.215)]. Moreover, the dynamics of the beam electrons are treated in the paraxial approximation with $p_z^2 \gg p_r^2 + p_\theta^2$ [Eq.(9.214)]. As in Sec. 9.5, equilibrium and stability properties can be investigated for a wide range of beam equilibria $f_b^0(r, \mathbf{p})$ of the form in Eq.(9.224). For present purposes, however, we specialize to the case where the beam distribution function has the inverted population in $H - \omega_{rb}P_\theta$ specified by Eq.(9.235), and the corresponding self-consistent density profile $n_b^0(r)$ has the simple rectangular form in Eq.(9.236). Therefore, for $\nu/\gamma_b \ll 1$ and $p_z^2 \gg p_r^2 + p_\theta^2$, the equilibrium distribution function of the beam electrons can be expressed as

$$f_b^0(r, \mathbf{p}) = \left(\frac{\hat{n}_b}{2\pi\gamma_b m_e}\right) \delta\left[\frac{p_r^2 + (p_\theta - \gamma_b m_e \omega_{rb} r)^2}{2\gamma_b m_e} - \frac{1}{2}\gamma_b m_e \Omega_\beta^2 r_b^2\left(1 - \frac{r^2}{r_b^2}\right)\right]$$

$$\times \delta(p_z - \gamma_b m_e \beta_b c) , \qquad (9.270)$$

where use has been made of Eqs.(9.224), (9.227), (9.235), and (9.237). Here, Ω_β^2 is defined by

$$\Omega_\beta^2 = \omega_{rb}\omega_{cb}/\gamma_b - \omega_{rb}^2 + \frac{\omega_{pb}^2\beta_b^2}{2\gamma_b}(1 - f_M)$$

$$= (\omega_{rb}^+ - \omega_{rb})(\omega_{rb} - \omega_{rb}^-) , \qquad (9.271)$$

where ω_{rb}^+ and ω_{rb}^- are the orbital oscillation frequencies defined by

$$\omega_{rb}^\pm = \frac{\omega_{cb}}{2\gamma_b}\left\{1 \pm \left(1 + \frac{2\gamma_b\omega_{pb}^2\beta_b^2}{\omega_{cb}^2}(1 - f_M)\right)^{1/2}\right\} . \qquad (9.272)$$

In Eqs.(9.270)–(9.272), the frequency Ω_β is related to the beam radius r_b by

$$\frac{1}{2}\gamma_b m_e \Omega_\beta^2 r_b^2 = (\hat{\gamma}_b - \gamma_b)m_e c^2 \equiv \hat{T}_{b\perp} , \qquad (9.273)$$

where $\hat{T}_{b\perp}$ is the (maximum) transverse temperature of the beam electrons [see also Eq.(9.237)]. Moreover, $\omega_{cb} = eB_0/m_e c$, $\omega_{pb}^2 = 4\pi\hat{n}_b e^2/m_e$, $\gamma_b = (1 - \beta_b^2)^{-1/2}$, and $\omega_{rb} = \text{const.}$ is the angular rotation velocity of the electron beam [Eq.(9.229)]. In Eqs.(9.271) and (9.272), it should also be noted that the contributions proportional to $\omega_{pb}^2\beta_b^2(1 - f_M)$ arise from the focusing force associated with the equilibrium azimuthal self-magnetic field $B_\theta^s(r) = -(m_e\omega_{pb}^2/2e)\beta_b(1 - f_M)r$ for $0 \leq r < r_b$ [Eq.(9.231)]. For the choice of equilibrium distribution function in Eq.(9.270), the beam density $n_b^0(r) = \int d^3p\, f_b^0$ and axial current density $J_{zb}^0(r) = -e \int d^3p\, v_z f_b^0$ are uniform over the interval $0 \leq r < r_b$, and equal to $\hat{n}_b$ and $-\hat{n}_b e\beta_b c$, respectively [Eqs.(9.229) and (9.236)].

In the present stability analysis, carried out for azimuthally symmetric perturbations with $\partial/\partial\theta = 0$, the beam electrons are treated by the linearized Vlasov-Maxwell equations, where the perturbed plasma current is $\delta\mathbf{J}_p = \sigma\delta\mathbf{E}$. As in Sec. 9.5, we consider low-frequency perturbations with long axial wavelengths satisfying $|\omega|r_b/c \ll 1$ and $|k_z|r_b \ll 1$ [Eq.(9.220)]. For $|\omega| \ll 4\pi\sigma$, the longitudinal electric field $\delta\mathbf{E}_L = -\nabla\delta\phi$ is negligibly small because the beam space charge is neutralized by the background plasma. In addition, because $|\omega|r_b/c \ll 1$, the transverse electric field $\delta\mathbf{E}_T = -c^{-1}(\partial/\partial t)\delta\mathbf{A}$ is neglected in comparison with the $v_z\hat{\mathbf{e}}_z \times \delta\mathbf{B}/c$ term in the linearized Vlasov equation for $\delta f_b(\mathbf{x}, \mathbf{p}, t)$. Indeed, for $\nu/\gamma_b \ll 1$ and $p_z^2 \gg p_r^2 + p_\theta^2$, similar arguments to those in Sec. 9.5 lead to Eq.(9.223) with $\partial/\partial\theta = 0$. Therefore, the linearized Vlasov equation is approximated by

$$\left\{ \frac{\partial}{\partial t} + \mathbf{v} \cdot \frac{\partial}{\partial\mathbf{x}} - \frac{e}{c}\mathbf{v} \times [B_0\hat{\mathbf{e}}_z + B_\theta^s(r)\hat{\mathbf{e}}_\theta] \cdot \frac{\partial}{\partial\mathbf{p}} \right\} \delta f_b = e\frac{v_z}{c}\frac{\partial\delta A_z}{\partial r}\frac{\partial f_b^0}{\partial p_r}$$

(9.274)

for perturbations with $\partial/\partial\theta = 0$. Here $f_b^0(r, \mathbf{p})$ is the equilibrium distribution function of the beam electrons defined in Eq.(9.270), and $\delta B_\theta(r, z, t) = -(\partial/\partial r)\delta A_z(r, z, t)$ is the azimuthal component of the perturbed magnetic field. Finally, for $|\omega|r_b/c \ll 1$, $|k_z|r_b \ll 1$, and $\delta J_{zp} = \sigma\delta E_z = -(\sigma/c)(\partial/\partial t)\delta A_z$, the linearized Maxwell equation for $\delta A_z(r, z, t)$ can be expressed as

$$\left\{ \frac{1}{r}\frac{\partial}{\partial r}r\frac{\partial}{\partial r} - \frac{4\pi\sigma}{c^2}\frac{\partial}{\partial t} \right\} \delta A_z = \frac{4\pi e}{c} \int d^3p\, v_z\delta f_b ,$$

(9.275)

where

$$-e \int d^3v\, v_z\delta f_b(r, z, \mathbf{p}, t) = \delta J_{zb}(r, z, t)$$

is the perturbed axial current of the beam electrons. Equations (9.274) and (9.275) are the final forms of the linearized Vlasov-Maxwell equations to be analyzed in Sec. 9.6.2 for azimuthally symmetric perturbations about the equilibrium distribution function in Eq.(9.270).

9.6.2   Kinetic Eigenvalue Equation for Azimuthally Symmetric Perturbations

In the analysis of Eqs.(9.274) and (9.275), we adopt a normal mode approach in which the $\ell = 0$ ($\partial/\partial\theta = 0$) Fourier components of the perturbations δA_z and δf_b are assumed to vary as

$$\delta\psi(r,z,t) = \delta\psi(r)\exp\left[i(k_z z - \omega t)\right] . \qquad (9.276)$$

As in Sec. 9.5 [see Eq.(9.247)], it is convenient to introduce $\tau = t - z/v_z$ and z as independent variables (rather than t and z) so that Eq.(9.276) becomes (for $v_z = V_b$)

$$\delta\psi(r,z,t) = \delta\psi(r)\exp\left[-i(\Omega z/V_b + \omega\tau)\right] . \qquad (9.277)$$

Here, $V_b = \beta_b c$ is the axial velocity of the electron beam, and $\Omega = \omega - k_z V_b$ is the frequency of the perturbation seen by a beam particle. For present purposes, the perturbation frequency ω is treated as a real quantity, so that the complex frequency Ω represents the spatial oscillation and growth ($\text{Im}\,\Omega > 0$) of the wave as the beam propagates downstream. The linearized Vlasov equation (9.274) can be integrated along the unperturbed trajectories from $z' = -\infty$ to $z' = z$ using the method of characteristics. Some straightforward algebra (similar to that in Problem 9.24) gives

$$\delta f_b(r,\mathbf{p}) = \frac{e}{c}\frac{p_z}{p_\perp}\frac{\partial f_b^0}{\partial p_\perp}\left\{\delta A_z(r) + \frac{\Omega}{V_b}i\int_{-\infty}^z dz'\,\delta A_z(r')\,\exp[-i(\Omega/V_b)(z'-z)]\right\},$$

$$(9.278)$$

where $\delta A_z(r)$ and $\delta f_b(r,\mathbf{p})$ are the $\ell = 0$ Fourier amplitudes of the perturbed vector potential and distribution function, respectively. Here, $\text{Im}\,\Omega > 0$ is assumed, and use has been made of the fact that the $\delta(p_z - \gamma_b m_e \beta_b c)$ factor in the equilibrium distribution function $f_b^0(r,\mathbf{p})$ in Eq.(9.270) selects the axial velocity $v_z \simeq \beta_b c = V_b$. Moreover, the orbit $r'(z')$ occurring in the integration over z' in Eq.(9.278) satisfies the "initial" condition $r'(z' = z) = r$, and

$$p_\perp^2 = p_r^2 + (p_\theta - \gamma_b m_e r\omega_{rb})^2$$

is the perpendicular momentum-squared in the rotating frame.

We substitute Eqs.(9.276) and (9.278) into the linearized Maxwell equation (9.275), which gives[36]

$$\left\{ \frac{1}{r} \frac{\partial}{\partial r} r \frac{\partial}{\partial r} + \frac{4\pi i \omega \sigma}{c^2} \right\} \delta A_z(r) \qquad (9.279)$$

$$= 4\pi e^2 \gamma_b m_e \beta_b^2 \int d^3 p \frac{1}{p_\perp} \frac{\partial f_b^0}{\partial p_\perp} \left[\delta A_z(r) + \frac{\Omega}{V_b} I_0(\Omega, r, p_\perp) \right] ,$$

where $f_b^0(r, \mathbf{p})$ is the equilibrium distribution function specified in Eq.(9.270), and the factor $\gamma_b m_e \beta_b^2$ on the right-hand side of Eq.(9.279) results when $(v_z p_z / c^2) \delta(p_z - \gamma_b m_e \beta_b c)$ is integrated over p_z. Here, the (phase-averaged) orbit integral $I_0(\omega, r, p_\perp)$ is defined by

$$I_0(\Omega, r, p_\perp) = i \int_0^{2\pi} \frac{d\phi}{2\pi} \int_{-\infty}^0 d\xi \, \delta A_z(r') \exp[-i(\Omega/V_b)\xi] , \qquad (9.280)$$

where $\xi = z' - z$, and ϕ is the perpendicular momentum phase in the rotating frame.

The eigenvalue equation (9.279), valid for low-frequency azimuthally symmetric perturbations $(\partial/\partial\theta = 0)$ with $|\omega| r_b / c \ll 1$ and $|k_z| r_b \ll 1$, should be compared with the eigenvalue equation (9.162), which was used in Sec. 9.4 to investigate low-frequency transverse instabilities in an intense nonneutral ion beam.[25] If the ions are replaced by electrons in Eq.(9.162), and perturbations with azimuthal mode number $\ell = 0$ are considered, then the following comparisons are particularly noteworthy.

(a) There is no background plasma in Eq.(9.162), whereas the effects of a (resistive) plasma are contained in the term $(4\pi i \omega \sigma / c^2) \delta A_z$ in Eq.(9.279).

(b) The effects of the equilibrium and perturbed space-charge fields are incorporated in Eq.(9.162) (the beam is fully nonneutral), whereas $\phi_0(r) = 0 = \delta\phi(\mathbf{x}, t)$ is assumed in Eq.(9.279) because the background plasma neutralizes the beam space charge ($\hat{n}_e \gg \hat{n}_b$ and $4\pi\sigma \gg |\omega|$).

(c) The factor 1 in the coefficient $1 - \beta_b^2$ on the right-hand side of Eq.(9.162) is associated with the longitudinal potential $\delta\phi$ in the eigenfunction $\delta\psi = \delta\phi - \beta_b\delta A_z$. In Eq.(9.279), $1 - \beta_b^2$ is replaced by $-\beta_b^2$ because $\delta\phi = 0$ in the presence of a dense background plasma. As a consequence, the eigenfunction occurring in Eq.(9.279) is $\delta A_z(r) = -\delta\psi(r)/\beta_b$.

(d) In deriving Eq.(9.162), we have assumed temporal growth $(\text{Im}\,\omega > 0)$ and $k_z = 0$. On the other hand, $\text{Im}\,\omega = 0$ is assumed in Eq.(9.279), and the perturbations grow axially with $\text{Im}\,\Omega = -V_b\,\text{Im}\,k_z > 0$. Here, the perturbations have long axial wavelength with $|k_z|r_b \ll 1$.

(e) Finally, the orbit integral $I_0(\Omega, r, p_\perp)$ occurring in Eq.(9.280) is calculated in a manner completely analogous to Sec. 9.4, with d/dt' replaced by $V_b d/dz'$ and $\omega_{pb}^2(1 - \beta_b^2)$ replaced by $-\omega_{pb}^2\beta_b^2$ in the orbit equation (9.158).

As a consequence of these similarities, apart from the term $(4\pi i\omega\sigma/c^2) \times \delta A_z$ occurring in Eq.(9.279), many of the theoretical techniques and results pertaining to the analysis of Eq.(9.162) also can be employed in the analysis of the eigenvalue equation (9.279) for the resistive sausage and hollowing instabilities.

In the subsequent analysis of Eq.(9.279), it is assumed that the plasma conductivity $\sigma(r)$ is specified by the step-function profile

$$\sigma(r) = \begin{cases} \sigma_0 = \text{const.}, & 0 \leq r < r_b\,, \\ \sigma_1 = \text{const.}, & r_b < r \leq b\,, \end{cases} \qquad (9.281)$$

where

$$\sigma_1 \ll \frac{\sigma_0 r_b^2}{b^2}\,.$$

The simple form of the conductivity profile in Eq.(9.281) greatly facilitates the analysis of Eq.(9.279), and provides a reasonable approximation for an intense electron beam propagating through (and ionizing) a neutral gas or weakly preionized plasma.[36] As in Sec. 9.5, it is convenient to introduce the characteristic resistive decay time τ_d defined by

$$\tau_d = \frac{1}{2c^2}\pi\sigma_0 r_b^2 \qquad (9.282)$$

so that the factor $4\pi i\omega\sigma(r)/c^2$ occurring in the eigenvalue equation (9.279) can be expressed as

$$\frac{4\pi i\omega\sigma}{c^2} = \frac{8i\omega\tau_d}{r_b^2} \tag{9.283}$$

in the beam interior ($0 \le r < r_b$). Note that the assumption $|\omega| \ll 4\pi\sigma$ [Eq.(9.219)] can be expressed as

$$\frac{|\omega|^2 r_b^2}{c^2} \ll 8|\omega|\tau_d , \tag{9.284}$$

where $|\omega|r_b/c \ll 1$ is also assumed for the low-frequency perturbations considered here [Eq.(9.220)].

For $|\omega|\tau_d \ll 1$, the eigenvalue equation (9.279) can be solved iteratively, and the lowest-order solutions for the eigenfunction are similar to those obtained in Sec. 9.4. In particular, for azimuthal mode number $\ell = 0$ and radial mode number n, the solution to Eq.(9.279) can be expressed as [see Eq.(9.166)]

$$\delta A_z(r) = A \sum_{j=0}^{n} a_j(\omega) \left(\frac{r}{r_b}\right)^{2j} \tag{9.285}$$

for $0 \le r < r_b$ and $\omega\tau_d \to 0$. Here, A is a constant coefficient, and $a_0(\omega) = 1$ without loss of generality. Treating the term proportional to $i(\omega\tau_d/r_b^2)\delta A_z$ as a small correction term in Eq.(9.279), the resistive sausage instability can occur for perturbations with radial mode number $n = 1$ (Sec. 9.6.3), whereas the resistive hollowing instability can occur for perturbations with radial mode number $n = 2$ (Sec. 9.6.4).

9.6.3   Resistive Sausage Instability ($n = 1$)

For perturbations with mode numbers $(\ell, n) = (0, 1)$ and $\omega\tau_d \to 0$, the eigenfunction solution to Eq.(9.279) is of the form

$$\delta A_z(r) = \begin{cases} A[1 - (r/r_b)^2] , & 0 \le r < r_b , \\ 0 , & r_b < r \le b , \end{cases} \tag{9.286}$$

which should be compared with Eq.(9.170). Note from Eq.(9.286) that $a_1(\omega) = -1$, and $\delta A_z(r)$ vanishes in the region $r_b < r \le b$. Moreover, $\delta A_z(r)$ assumes a maximum value (equal to A) at $r = 0$, which corresponds to a strong internal perturbation within the electron beam. We now iterate about the solution in Eq.(9.286), retaining the term

$(4\pi i\sigma\omega/c^2)\delta A_z$ in Eq.(9.279) as a small correction. Evaluating the orbit integral $I_0(\Omega, r, p_\perp)$ in Eq.(9.280), and paralleling the analysis in Sec. 9.4.4, we obtain the eigenvalue equation[36]

$$\frac{1}{r}\frac{\partial}{\partial r}\left\{r\left[1 + \frac{\omega_{pb}^2\beta_b^2/\gamma_b}{\Omega^2 - (\omega_{rb}^+ - \omega_{rb}^-)^2}\right]\frac{\partial}{\partial r}\delta A_z(r)\right\} + \frac{8i\omega\tau_d}{r_b^2}\delta A_z(r) = (9.287)$$

for $0 \le r < r_b$, and

$$\frac{1}{r}\frac{\partial}{\partial r}\left(r\frac{\partial}{\partial r}\delta A_z(r)\right) + \frac{\sigma_1}{\sigma_0}\frac{8i\omega\tau_d}{r_b^2}\delta A_z(r) = 0 \qquad (9.288)$$

for $r_b < r \le b$. Here, $\sigma_0 r_b^2 \gg \sigma_1 b^2$ is assumed, τ_d is defined in Eq.(9.282), $\omega_{pb}^2 = 4\pi\hat{n}_b e^2/m_e$ is the nonrelativistic plasma frequency-squared, and $(\omega_{rb}^+ - \omega_{rb}^-)^2$ is defined by [Eq.(9.272)]

$$(\omega_{rb}^+ - \omega_{rb}^-)^2 = \frac{\omega_{cb}^2}{\gamma_b^2} + \frac{2\omega_{pb}^2\beta_b^2}{\gamma_b}(1 - f_M) . \qquad (9.289)$$

The solutions to Eqs.(9.287) and (9.288) for $\delta A_z(r)$ can be expressed in terms of Bessel functions. Alternatively, following Lampe and Uhm,[36] a variational technique can be used to derive an approximate dispersion relation, valid for $|\omega|\tau_d < 1$. Multiplying Eqs.(9.287) and (9.288) by $r\delta A_z^*(r)$, and integrating from $r = 0$ to $r = b$, we obtain

$$-\left[1 + \frac{\omega_{pb}^2\beta_b^2/\gamma_b}{\Omega^2 - (\omega_{rb}^+ - \omega_{rb}^-)^2}\right]\int_0^{r_b} dr\, r\left|\frac{\partial}{\partial r}\delta A_z\right|^2 - \int_{r_b}^b dr\, r\left|\frac{\partial}{\partial r}\delta A_z\right|^2$$

$$+ \frac{8i\omega\tau_d}{r_b^2}\left[\int_0^{r_b} dr\, r\,|\delta A_z|^2 + \frac{\sigma_1}{\sigma_0}\int_{r_b}^b dr\, r\,|\delta A_z|^2\right] = 0 , \qquad (9.290)$$

where the boundary conditions $\delta A_z(r = 0) = 0$ and $\delta A_z(r = b) = 0$ have been enforced. In general, the eigenfunction $\delta A_z(r)$ occurring in Eq.(9.290) is the exact solution to Eqs.(9.287) and (9.288). However, Lampe and Uhm[36] have shown that if a trial eigenfunction $\delta A_z^T(r)$ is substituted into the integrals in Eq.(9.290) then the resulting dispersion relation for the (complex) eigenfrequency Ω is accurate to second order in the error in the trial function $\delta A_z^T(r)$. [Here, the error in the trial function is defined by $\delta A_z^E(r) \equiv \delta A_z^A(r) - \delta A_z^T(r)$, where $\delta A_z^A(r)$ is the actual solution to the eigenvalue equation.] For the choice of trial function, we take $\delta A_z^T(r) = A[1 - (r/r_b)^2]$ for $0 \le r < r_b$, and $\delta A_z^T(r) = 0$ for

$r_b < r \leq b$, which corresponds to the solution in Eq.(9.286) obtained for $\omega \tau_d \to 0$. Evaluating the integrals in Eq.(9.290) with $\delta A_z(r) = \delta A_z^T(r)$ then gives the approximate dispersion relation (Problem 9.28)

$$1 + \frac{\omega_{pb}^2 \beta_b^2 / \gamma_b}{\Omega^2 - (\omega_{rb}^+ - \omega_{rb}^-)^2} = \frac{4}{3} i \omega \tau_d . \tag{9.291}$$

Strictly speaking, Eq.(9.291) should be applied only when $|\omega| \tau_d < 1$. As expected, in the absence of the term $4i\omega\tau_d/3$, Eq.(9.291) is identical in form to the dispersion relation (9.169) obtained in Sec. 9.4 provided we make the replacements $\omega^2 \to \Omega^2$ and $\omega_{pb}^2 / \gamma_b^2 = \omega_{pb}^2 (1 - \beta_b^2) \to -\omega_{pb}^2 \beta_b^2$ in Eq.(9.169). [In the present analysis, keep in mind that the beam space charge is completely neutralized by the background plasma.]

Making use of Eq.(9.289), we solve Eq.(9.291) for Ω^2, which gives

$$\Omega^2 = \frac{\omega_{cb}^2}{\gamma_b^2} + \frac{\omega_{pb}^2 \beta_b^2 / \gamma_b}{1 + (4\omega\tau_d/3)^2} \left[(1 - 2f_M) + 2(1 - f_M) \left(\frac{4\omega\tau_d}{3} \right)^2 - i \left(\frac{4\omega\tau_d}{3} \right) \right] .$$

$$\tag{9.292}$$

Here, real ω has been assumed, and Eq.(9.292) determines the complex eigenfrequency Ω, where $(\mathrm{Im}\,\Omega)/V_b = -\mathrm{Im}\,k_z$ corresponds to the spatial growth rate of the perturbation in the z-direction. It is evident from Eq.(9.292) that the axial guide field $B_0 \hat{e}_z$ has a stabilizing influence on the resistive sausage instability (through the term $\omega_{cb}^2 / \gamma_b^2$).

As a simple application of Eq.(9.292), consider a self-pinched electron beam with $\omega_{cb} = 0$. In this case, it follows that the sausage instability is driven by the plasma return current (f_M) and the magnetic phase lag $(\omega \tau_d)$. For $\omega^2 \tau_d^2 \ll 1$, Eq.(9.292) has a purely growing solution $(\mathrm{Re}\,\Omega = 0$ and $\mathrm{Im}\,\Omega > 0)$ provided the fractional current neutralization by the background plasma satisfies $f_M > 0.5$. In this case, the growth rate is given by

$$\mathrm{Im}\,\Omega = \left(\frac{2}{\gamma_b} \right)^{1/2} \omega_{pb} \beta_b (f_M - 0.5)^{1/2} \tag{9.293}$$

for $\omega^2 \tau_d^2 \ll 1$ and $f_M \geq 0.5$. Measured in units of $\omega_{pb}\beta_b/\gamma_b^{1/2}$, it is clear from Eq.(9.293) that the growth rate of the resistive sausage instability can be substantial, and comparable to the growth rate of the resistive hose instability when $f_M < 0.5$ [compare with Eq.(9.262) for $\omega_{cb} = 0$].

Figure 9.17 illustrates the influence of the magnetic phase lag $(\omega \tau_d)$ on the resistive sausage instability. Here, for $\omega_{cb} = 0$, the normalized growth

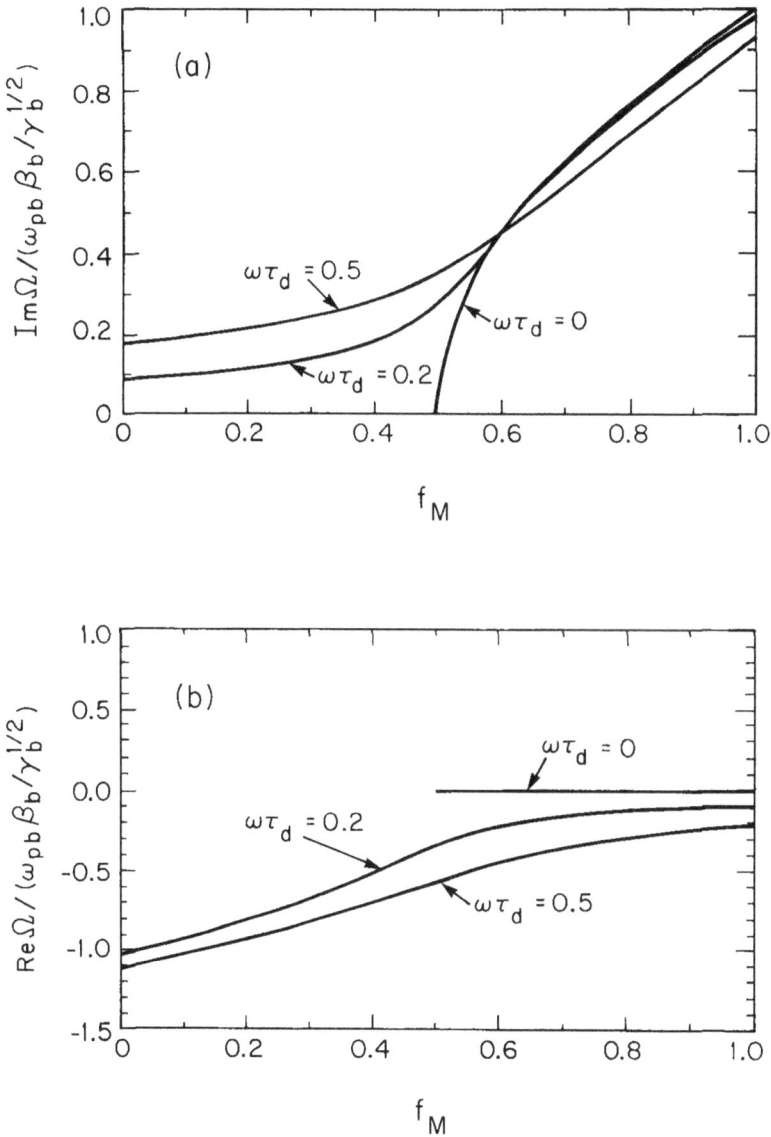

Figure 9.17. Plots versus the fractional current neutralization f_M of (a) the normalized growth rate $\operatorname{Im}\Omega/(\omega_{pb}\beta_b/\gamma_b^{1/2})$, and (b) the normalized real oscillation frequency $\operatorname{Re}\Omega/(\omega_{pb}\beta_b/\gamma_b^{1/2})$ for the resistive sausage instability obtained from the dispersion relation (9.292) for $\omega_{cb}=0$ and several values of $\omega\tau_d$.

rate $\operatorname{Im}\Omega/(\omega_{pb}\beta_b/\gamma_b^{1/2})$ and real frequency $\operatorname{Re}\Omega/(\omega_{pb}\beta_b/\gamma_b^{1/2})$ calculated from Eq.(9.292) are plotted versus the fractional current neutralization f_M for several values of $\omega\tau_d$. For $\omega\tau_d = 0$, note from the figure that $\operatorname{Re}\Omega = 0$ for $f_M \geq 0.5$. As $\omega\tau_d$ is increased, however, $\operatorname{Re}\Omega \neq 0$ in the region of f_M corresponding to instability. Moreover, the threshold value of f_M for the onset of instability shifts to larger values as $\omega\tau_d$ is increased, and the characteristic growth rate in the unstable region decreases relative to the case where $\omega\tau_d = 0$.

Problem 9.28   Consider the integral relation in Eq.(9.290), which is an exact consequence of Eqs.(9.287) and (9.288) for perturbations with mode numbers $(\ell, n) = (0, 1)$.

a. As a trial eigenfunction, assume that $\delta A_z^T(r)$ is specified by Eq.(9.286) (valid for $\sigma\omega \to 0$). Substitute $\delta A_z^T(r)$ into Eq.(9.290), and show that the resulting (approximate) dispersion relation is given in Eq.(9.291).

b. Solve Eq.(9.291) for Ω^2 and show that the solution can be expressed in the form given in Eq.(9.292).

c. For $\omega_{cb} = 0$, $\omega^2\tau_d^2 \ll 1$, and $f_M \geq 0.5$, make use of Eq.(9.292) to show that $\operatorname{Re}\Omega = 0$ and the growth rate $\operatorname{Im}\Omega$ is given by Eq.(9.293).

9.6.4 Resistive Hollowing Instability ($n = 2$)

For perturbations with mode numbers $(\ell, n) = (0, 2)$ and $\omega\tau_d \to 0$, the eigenfunction solution to Eq.(9.279) is of the form

$$\delta A_z(r) = \begin{cases} A\left(1 - 4\dfrac{r^2}{r_b^2} + 3\dfrac{r^4}{r_b^4}\right), & 0 \leq r < r_b, \\ 0, & r_b < r \leq b, \end{cases} \tag{9.294}$$

which should be compared with Eq.(9.185). Note that the eigenfunction $\delta A_z(r)$ in Eq.(9.294) has two internal extrema (at $r = 0$ and $r^2/r_b^2 = 2/3$) within the electron beam, and is equal to zero in the region $r_b < r \leq b$. As in Sec. 9.6.3, we iterate about the solution in Eq.(9.294), retaining the term $(4\pi i\sigma\omega/c^2)\delta A_z$ in Eq.(9.279) as a small correction. Paralleling the analysis in Sec. 9.4.4, we obtain the eigenvalue equation[36]

$$\frac{1}{r}\frac{\partial}{\partial r}\left\{ r\left[1+\frac{\omega_{pb}^2\beta_b^2/\gamma_b}{\Omega^2-(\omega_{rb}^+-\omega_{rb}^-)^2}\left(1+\frac{18\Omega_\beta^2}{\Omega^2-4(\omega_{rb}^+-\omega_{rb}^-)^2}\right)\right]\frac{\partial}{\partial r}\delta A_z(r)\right\}$$

$$+\frac{8i\omega\tau_d}{r_b^2}\delta A_z(r)=0 \qquad (9.295)$$

for $0\le r < r_b$. Here, $\Omega_\beta^2=(\omega_{rb}^+-\omega_{rb})(\omega_{rb}-\omega_{rb}^-)$ is defined in Eq.(9.271), and $\delta A_z(r)$ solves Eq.(9.288) in the interval $r_b < r \le b$. As in Sec. 9.6.3, an approximate dispersion relation can be obtained from Eq.(9.295) by using Eq.(9.294) as the trial eigenfunction $\delta A_z^T(r)$. Multiplying Eq.(9.295) by $r\delta A_z^{T*}(r)$, and integrating from $r=0$ to $r=r_b$, gives the dispersion relation (Problem 9.29)

$$1+\frac{\omega_{pb}^2\beta_b^2/\gamma_b}{\Omega^2-(\omega_{rb}^+-\omega_{rb}^-)^2}\left[1+\frac{18\Omega_\beta^2}{\Omega^2-4(\omega_{rb}^+-\omega_{rb}^-)^2}\right]=\frac{4}{15}i\omega\tau_d , \qquad (9.296)$$

which is valid for $\omega\tau_d < 1$. As expected, in the absence of the term $4i\omega\tau_d/15$, Eq.(9.296) is identical in form to the dispersion relation (9.179) obtained in Sec. 9.4 provided we make the replacements $\omega^2 \to \Omega^2$ and $\omega_{pb}^2/\gamma_b^2=\omega_{pb}^2(1-\beta_b^2) \to -\omega_{pb}^2\beta_b^2$ in Eq.(9.178).

Equation (9.296) can be used to investigate detailed properties of the resistive hollowing instability over a wide range of system parameters, $\omega\tau_d$, ω_{cb}, f_M, ω_{rb}, etc. For present purposes, we examine Eq.(9.296) for $\omega^2\tau_d^2 \ll 1$, $\omega_{cb}=0$, $\omega_{rb}=0$, and

$$\Omega_\beta^2=\frac{\omega_{pb}^2\beta_b^2}{2\gamma_b}(1-f_M) .$$

In this case, Eq.(9.296) reduces to

$$\Omega^4-[10(1-f_M)-1]\Omega^2\frac{\omega_{pb}^2\beta_b^2}{\gamma_b}+(1-f_M)[16(1-f_M)+1]\frac{\omega_{pb}^4\beta_b^4}{\gamma_b^2}=0 , \qquad (9.297)$$

which is applicable to a nonrotating, self-pinched electron beam with negligibly small magnetic decay time. Equation (9.297) is a quadratic equa-

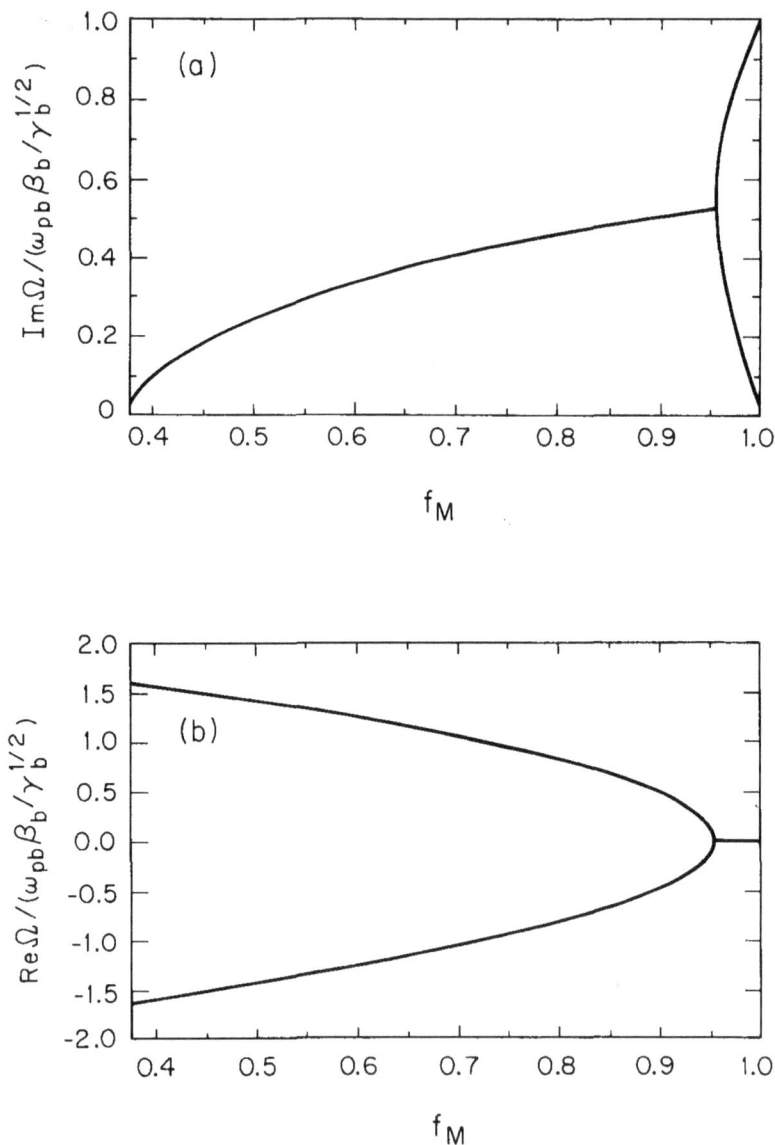

Figure 9.18.   Plots versus the fractional current neutralization f_M of (a) the normalized growth rate $\mathrm{Im}\,\Omega/(\omega_{pb}\beta_b/\gamma_b^{1/2})$ and (b) the normalized real oscillation frequency $\mathrm{Re}\,\Omega/(\omega_{pb}\beta_b/\gamma_b^{1/2})$ for the resistive hollowing instability obtained from the dispersion relation (9.297), valid for $\omega^2\tau_d^2 \ll 1$, $\omega_{cb} = 0$ and $\omega_{rb} = 0$.

tion for Ω^2, and the necessary and sufficient condition for instability can be expressed as

$$f_M > (4 - \sqrt{3})/6 = 0.38 \, , \qquad (9.298)$$

which is a lower threshold value than that obtained for the resistive sausage instability [Eq.(9.293)]. The unstable solution to Eq.(9.297) is illustrated in Fig. 9.18. Here, the normalized growth rate $\mathrm{Im}\,\Omega/(\omega_{pb}\beta_b/\gamma_b^{1/2})$ and real frequency $\mathrm{Re}\,\Omega/(\omega_{pb}\beta_b/\gamma_b^{1/2})$ are plotted versus the fractional current neutralization f_M for $0.38 \leq f_M \leq 1$. Note that there are generally two unstable branches over the unstable range of f_M. Moreover, the maximum growth rate $(\mathrm{Im}\,\Omega)_{\max} = \omega_{pb}\beta_b/\gamma_b^{1/2}$ occurs for $f_M = 1$, when the beam current is completely neutralized by the plasma return current. Similar to the resistive hose and sausage instabilities, it is found that the growth rate of the resistive hollowing instability can be substantial. Moreover, unstable perturbations with $(\ell, n) = (0, 2)$ can reduce the beam density on axis in the nonlinear regime,[37] thereby leading to a hollow profile for the beam density and current.

Problem 9.29   Consider the eigenvalue equation (9.295) for perturbations with mode numbers $(\ell, n) = (0, 2)$.

a. For $\sigma_1 \ll \sigma_0$, assume that $\delta A_z(r) = 0$ in the interval $r_b < r \leq b$. Multiply Eq.(9.295) by $r\delta A_z^*(r)$, and integrate from $r = 0$ to $r = r_b$, enforcing the boundary conditions $\delta A_z(r = 0) = 0 = \delta A_z(r = r_b)$.

b. In the integral relation derived in Part (a), assume that the trial eigenfunction $\delta A_z^T(r)$ is specified by Eq.(9.294) (valid for $\sigma\omega \to 0$). Carry out the required integrations and show that the resulting (approximate) dispersion relation is given by Eq.(9.296).

c. For $\omega_{cb} = 0$, $\omega_{rb} = 0$, $\Omega_\beta^2 = (\omega_{pb}^2\beta_b^2/2\gamma_b)(1 - f_M)$, and $\omega^2\tau_d^2 \ll 1$, show that the dispersion relation (9.296) reduces to Eq.(9.297).

d. Show from Eq.(9.297) that the necessary and sufficient condition for instability (existence of a solution with $\mathrm{Im}\,\Omega > 0$) is given by

$$f_M > (4 - \sqrt{3})/6 = 0.38 \, . \qquad (9.29.1)$$

Chapter 9 References

1. R.B. Miller, *Intense Charged Particle Beams* (Plenum, New York, 1982).

2. J.D. Lawson, *The Physics of Charged Particle Beams* (Clarendon Press, Oxford, 1988).

3. C.A. Kapetanakos and P. Sprangle, "Ultra-High-Current Electron Induction Accelerators," Physics Today **38** (2), 58 (1985).

4. S. Humphries, Jr., *Principles of Charged Particle Acceleration* (Wiley, New York, 1986).

5. D. Keefe, "Research on High Beam-Current Accelerators," Particle Accelerators **11**, 187 (1987).

6. S. Humphries, Jr., "Intense Pulsed Ion Beams for Fusion Applications," Nuclear Fusion **20**, 1549 (1980).

7. J.P. VanDevender and D.L. Cook, "Inertial Confinement Fusion with Light Ion Beams," Science **232**, 831 (1986).

8. *Heavy Ion Inertial Fusion*, eds., M. Reiser, T. Godlove and R. Bangerter, IAP Conference Proceedings No. 152 (American Institute of Physics, New York, 1986), and references therein.

9. E.P. Lee and J. Hovingh, "Heavy Ion Induction Linac Drivers for Inertial Confinement Fusion," Fusion Technology **15**, 369 (1989).

10. High-Power Microwave Sources, eds., V. Granatstein and I. Alexeff (Artech House, Boston, Massachusetts, 1987).

11. T.C. Marshall, *Free Electron Lasers* (Macmillan, New York, 1985).

12. C.W. Roberson and P. Sprangle, "A Review of Free Electron Lasers," Phys. Fluids **B1**, 3 (1989).

13. M. Lampe, "Propagation of Charged Particle Beams in the Atmosphere," in *Proceedings of the 1987 Particle Accelerator Conference*, eds. E.R. Lindstrom and L.S. Taylor, IEEE No. 87CH2387-9 (IEEE, New York, 1987), p. 1965.

14. J.E. Brandenburg and E.P. Lee, "A Model of Hose Instabilities in Rotating Electron Beams," Phys. Fluids **31**, 3403 (1988).

15. R.C. Davidson, *Theory of Nonneutral Plasmas* (Addison-Wesley, Reading, Massachusetts, 1989).

16. R.C. Davidson, "Relativistic Electron Beam-Plasma Interaction with Intense Self Fields," in *Handbook of Plasma Physics—Basic Plasma Physics*, eds., M.N. Rosenbluth and R.Z. Sagdeev (North Holland, Amsterdam, 1984), Volume 2, pp. 729–819.

17. V.S. Voronin, Y.T. Zozulya and A.N. Lebedev, "Self-Consistent Stationary State of a Relativistic Electron Beam in a Drift Space," Sov. Phys.-Tech. Phys. **17**, 432 (1972).

18. B. Breizman and D.D. Ryutov, "Powerful Relativistic Electron Beams in a Plasma in a Vacuum (Theory)," Nuclear Fusion **14**, 873 (1974).

19. R.B. Miller and D.C. Straw, "Propagation of an Unneutralized Intense Relativistic Electron Beam in a Magnetic Field," J. Appl. Phys. **48**, 1061 (1977).

20. M. Reiser, "Laminar-Flow Equilibria and Limiting Currents in Magnetically Focused Relativistic Beams," Phys. Fluids **20**, 477 (1977).

21. F.B. Llewellyn, *Electron Inertial Effects* (Cambridge University Press, London, 1941).

22. C.K. Birdsall and W.B. Bridges, *Electron Dynamics of Diode Regions* (Academic Press, New York, 1966).

23. A.V. Pashchenko and B.N. Rutkevich, "Stability of an Electron Beam in a Diode," Sov. J. Plasma Phys. **3**, 437 (1977).

24. R.C. Davidson and H.S. Uhm, "Influence of Finite Ion Larmor Radius Effects on the Ion Resonance Instability in a Nonneutral Plasma Column," Phys. Fluids **21**, 265 (1978).

25. H.S. Uhm and R.C. Davidson, "Low-Frequency Flute Perturbations in Intense Nonneutral Electron and Ion Beams," Phys. Fluids **23**, 1586 (1980).

26. H.S. Uhm and R.C. Davidson, "Stability Properties of Intense Nonneutral Ion Beams for Heavy-Ion Fusion," Particle Accelerators **11**, 65 (1980).

27. M.N. Rosenbluth, "Long-Wavelength Beam Instability," Phys. Fluids **3**, 932 (1960).

28. S. Weinberg, "The Hose Instability Dispersion Relation," J. Math. Phys. **5**, 1371 (1964).

29. S. Weinberg, "General Theory of Resistive Beam Instabilities," J. Math. Phys. **8**, 614 (1967).

30. K.G. Moses, R.W. Bauer and S.D. Winter, "Behavior of Relativistic Beams Undergoing Hose Motion in a Plasma Channel," Phys. Fluids **16**, 436 (1973).

31. E.P. Lee and L.D. Pearlstein, "Hollow Equilibrium and Stability of a Relativistic Electron Beam Propagating in a Preionized Channel," Phys. Fluids **16**, 904 (1973).

32. E.P. Lee, "Kinetic Theory of a Relativistic Beam," Phys. Fluids **19**, 60 (1976).

33. E.P. Lee, "Resistive Hose Instability of a Beam with the Bennett Profile," Phys. Fluids **21**, 1327 (1978).

34. E.J. Lauer, R.J. Briggs, T.J. Fessenden, R.E. Hester and E.P. Lee, "Measurements of Hose Instability of a Relativistic Electron Beam," Phys. Fluids **21**, 1344 (1978).

35. H.S. Uhm and L. Lampe, "Theory of the Resistive Hose Instability in Relativistic Electron Beams," Phys. Fluids **23**, 1574 (1980).

36. H.S. Uhm and M. Lampe, "Stability Properties for Azimuthally Symmetric Perturbations in an Intense Electron Beam," Phys. Fluids **24**, 1553 (1981).

37. G. Joyce and M. Lampe, "Numerical Simulation of the Axisymmetric Hollowing Instability of a Propagating Beam," Phys. Fluids **26**, 3377 (1983).

38. H.S. Uhm and M. Lampe, "Return-Current-Driven Instabilities of Propagating Electron Beams," Phys. Fluids **25**, 1444 (1982).

39. M. Lampe and G. Joyce, "Stabilizing Effect of Gas Conductivity Evolution on the Resistive Sausage Mode of a Propagating Beam," Phys. Fluids **26**, 3371 (1983).

40. M. Lampe, W. Sharp, R.F. Hubbard, E.P. Lee and R.J. Briggs, "Plasma Current and Conductivity Effects on the Hose Instability," Phys. Fluids **27**, 2821 (1984).

41. G. Joyce and M. Lampe, "SIMM 1: A Linearized Particle Code," J. Computational Physics **63**, 398 (1986).

42. R.F. Fernsler, R.F. Hubbard, B. Hui, G. Joyce, M. Lampe and Y.Y. Lau, "Current Enhancement for Hose-Unstable Electron Beams," Phys. Fluids **29**, 3056 (1986).

43. S.P. Slinker, R.F. Hubbard and M. Lampe, "Variable Collision Frequency Effects on Hose and Sausage Instabilities in Relativistic Electron Beams," J. Appl. Phys. **62**, 1171 (1987).

44. M. Reiser, "Periodic Focusing of Intense Beams," Particle Accelerators **8**, 167 (1978).

45. I. Hofman, L.J. Laslett, L. Smith and I. Haber, "Stability of the Kapchinskij-Vladimirskij Distribution in Long Periodic Transport Systems," Particle Accelerators **13**, 145 (1983).

46. J. Struckmeier, J. Klabunde and M. Reiser, "On the Stability and Emittance Growth of Different Particle Phase-Space Distributions in a Long Magnetic Quadrupole Channel," Particle Accelerators **15**, 47 (1984).

47. M. Reiser, C.R. Chang, D. Kehne, K. Low, T. Shea, H. Rudd and I. Haber, "Emittance Growth and Image Formation in a Nonuniform Space-Charge-Dominated Electron Beam," Phys. Rev. Lett. **61**, 2933 (1988).

48. G.J. Budker, "Relativistic Stabilized Electron Beam," in *Proceedings of the Symposium on High-Energy Accelerators and Pion Physics* (CERN Scientific Information Service, Geneva, 1956), Volume 1, p. 68.

49. I.M. Kapchinskij and V.V. Vladimirskij, "Limitations of Proton Beam Current in a Strong-Focusing Linear Accelerator Associated with the Beam Space Charge," in *Proceedings of the International Conference on High Energy Accelerators and Instrumentation* (CERN Scientific Information Service, Geneva, 1959), p. 274.

50. W. Bennett, "Magnetically Self-Focussing Streams," Phys. Rev. **45**, 890 (1934).

The following references, while not cited directly in the text are relevant to the general subject matter of this chapter.

D. Chernin and A. Mondelli, "Effect of Energy Spread on the Single Bunch Dipole Beam Breakup Instability in a High Energy RF Linac," Particle Accelerators **24**, 177 (1989).

F.J. Wessel, R. Hong, J. Song, A. Fisher, N. Rostoker, A. Ron, R. Li, and R.Y. Fan, "Plasmoid Propagation in a Transverse Magnetic Field and in a Magnetized Plasma," Phys. Fluids **31**, 3778 (1988).

B.B. Godfrey, "Electron Beam Hollowing Instability Threshold," Phys. Fluids **30**, 570 (1987).

B.B. Godfrey, "Electron Beam Hollowing Instability Simulations," Phys. Fluids **30**, 575 (1987).

A. Mankofsky, R.N. Sudan and J. Denavit, "Hybrid Simulation of Ion Beams in Background Plasma," J. Comp. Phys. **70**, 89 (1987).

K.T. Nguyen, R.F. Schneider, J.R. Smith and H.S. Uhm, "Transverse Instability of an Electron Beam in Beam-Induced Ion Channel," Appl. Phys. Lett. **50**, 239 (1987).

H.L. Buchanan, "Electron Beam Propagation in the Ion-Focused Regime," Phys. Fluids **30**, 221 (1987).

K.O. Busby, J.B. Greenly, Y. Nakagawa, P.D. Pedrow and D.A. Hammer, "Propagation of Intense Ion Beams Across a Plasma-Filled Magnetic Cusp," J. Appl. Phys. **60**, 4095 (1986).

E.F. Chrien, E.J. Valeo, R.M. Kulsrud and C.R. Oberman, "Propagation of Ion Beams Through a Tenuous Magnetized Plasma," Phys. Fluids **29**, 1675 (1986).

R.L. Gluckstern, R.K. Cooper and P.J. Channell, "Cumulative Beam Breakup in RF Linacs," Particle Accelerators **16**, 125 (1985).

A. Mankofsky and R.N. Sudan, "Numerical Simulation of Ion Beam Propagation in Z-Pinch Plasma Channels," Nucl. Fusion **24**, 827 (1984).

H.S. Uhm and R.C. Davidson, "Kinetic Description of Coupled Transverse Oscillations in an Intense Relativistic Electron Beam-Plasma System," Phys. Fluids **23**, 813 (1980).

R.C. Davidson and H.S. Uhm, "Coupled Dipole Oscillations in an Intense Relativistic Electron Beam," J. Appl. Phys. **51**, 885 (1980).

V.K. Neil, L.S. Hall and R.K. Cooper, "Theoretical Studies of the Beam Breakup Instability," Particle Accelerators **9**, 213 (1979).

R.J. Briggs, "Space-Charge Waves on a Relativistic, Unneutralized Electron Beam and Collective Ion Acceleration," Phys. Fluids **19**, 1257 (1976).

D.G. Koshkarev and P.R. Zenkevich, "Resonance of Coupled Transverse Oscillations in Two Circular Beams," Particle Accelerators **3**, 1 (1972).

L.I. Rudakov, V.P. Smirnov and A.M. Spektor, "Behavior of a Large-Current Electron Beam in a Dense Gas," JETP Letters **15**, 382 (1972).

Ya. B. Fainberg, V.D. Shapiro and V.I. Shevchenko, "Nonlinear Waves in Uncompensated Electron Beams," Sov. Phys. JETP **34**, 103 (1972).

A.A. Rukhadze and V.G. Rukhlin, "Injection of a Relativistic Electron Beam into a Plasma," Sov. Phys. JETP **34**, 93 (1972).

B.N. Breizman and D.D. Ryutov, "Quasilinear Relaxation of an Ultrarelativistic Electron Beam in a Plasma," Sov. Phys. JETP **33**, 220 (1971).

R. Lee and R.N. Sudan, "Return Current Induced by a Relativistic Beam Propagating in a Magnetized Plasma," Phys. Fluids **14**, 1213 (1971).

D.A. McArthur and J.W. Poukey, "Return Current Induced by a Relativistic Electron Beam Propagating into Neutral Gas," Phys. Rev. Lett. **27**, 1765 (1971).

L.I. Rudakov, "Collective Slowing Down of an Intense Beam of Relativistic Electrons in Dense Plasma Target," Sov. Phys. JETP **32**, 1134 (1971).

Ya. B. Fainberg, V.D. Shapiro and V.I. Shevchenko, "Nonlinear Theory of Interaction Between a Monochromatic Beam of Relativistic Electrons and a Plasma," Sov. Phys. JETP **3**, 528 (1970).

W.K.H. Panofsky and M. Bander, "Asymptotic Theory of Beam Breakup in Linear Accelerators," Rev. Sci. Instr. **39**, 206 (1968).

CHAPTER 10

PROPAGATION AND STABILITY OF INTENSE CHARGED PARTICLE BEAMS IN AN ALTERNATING-GRADIENT FOCUSING FIELD

10.1 Introduction

Periodic (alternating-gradient) focusing fields, stimulated by the classic work of Courant, Livingston and Snyder,[1,2] and Christofilos,[3] have been used for the propagation and acceleration of charged particle beams[4,5] since the early 1950s, with applications ranging from cyclic accelerators, such as the synchrotron,[6] to periodic quadrupole lattices used in heavy ion fusion.[7-9] In this chapter, we summarize several basic properties of intense beam propagation in periodic focusing systems (Fig. 10.1), including the important influence of equilibrium self fields. Similar to the case of beam propagation parallel to a uniform solenoidal field $B_0 \hat{\mathbf{e}}_z$, it is expected that equilibrium self-field effects and the detailed form of the distribution function $f_b^0(\mathbf{x}, \mathbf{p})$ can have a large influence on the beam dynamics,[10-16] stability behavior,[17-23] and emittance growth.[22-30] It is not the purpose of the present chapter to describe state-of-the-art stability results for intense beam propagation in periodic focusing systems. Rather the analysis here is intended to introduce and develop basic concepts in this area, and to present a consolidated treatment of the single-particle orbits, the beam envelope equations, and the kinetic stability of the Kapchinskij-Vladimirskij distribution function[10] for a uniform-density beam including self-field effects.

Within the context of a simple physical model discussed in Secs. 10.2 and 10.3, we consider a uniform-density ion beam with elliptical cross section and characteristic energy $\gamma_b m_i c^2$ propagating in the z-direction through a periodic focusing field with axial velocity $\beta_b c$. One possible

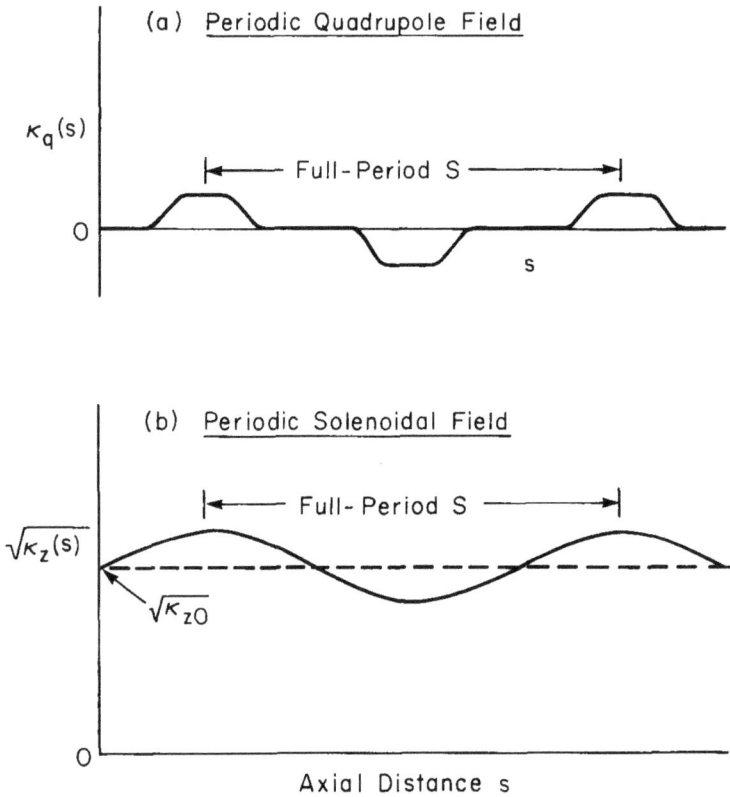

Figure 10.1. Illustrative plots versus axial distance s of (a) the coupling coefficient $\kappa_q(s) = (Z_i e / \gamma_b m_i \beta_b c^2) B_q'(s)$ for an alternating-gradient quadrupole field [Eq.(10.24)], and (b) the coupling coefficient $\sqrt{\kappa_z(s)} = (Z_i e / 2 \gamma_b m_i \beta_b c^2) B_z(s)$ for an alternating-gradient solenoidal field [Eq.(10.25)]. Here, the x- and y-components of the quadrupole magnetic field are proportional to $\kappa_q(s)y$ and $\kappa_q(s)x$, respectively, where x and y are measured from the beam center [Eq.(10.1)]. In addition, $\kappa_q(s + S) = \kappa_q(s)$ and $\int_{s_0}^{s_0+S} ds \, \kappa_q(s) = 0$ in Fig. 10.1(a), where S is the periodicity length of the lattice. Therefore, in the transverse orbit equations for $x(s)$ and $y(s)$, the periodic quadrupole force is alternately focusing and defocusing at different axial positions [Eqs.(10.26) and (10.27)]. In general, the solenoidal coupling coefficient $\sqrt{\kappa_z(s)}$ illustrated in Fig. 10.1(b) is allowed to oscillate about a nonzero value $\sqrt{\kappa_{z0}} = Z_i e \, B_0 / 2 \gamma_b m_i \beta_b c^2$, where $B_0 = \int_{s_0}^{s_0+S} ds \, B_z(s)/S$ is the average magnetic field. The case where $B_0 = 0$ is not excluded in the present analysis.

periodic focusing system is illustrated in Fig. 10.1(a), where the transverse quadrupole field components, $B_x^q = B_q'(s)y$ and $B_y^q = B_q'(s)x$, can be expressed as [see Eqs.(10.4) and (10.24)]

$$B_x^q(s) = \frac{\gamma_b m_i \beta_b c^2}{Z_i e} \kappa_q(s)y ,$$

$$B_y^q(s) = \frac{\gamma_b m_i \beta_b c^2}{Z_i e} \kappa_q(s)x . \tag{10.1}$$

In Eq.(10.1), x and y denote the transverse spatial coordinates measured from the beam center, and the quadrupole coupling coefficient $\kappa_q(s) = \kappa_q(s + S)$ has periodicity length S in the z-direction and oscillates about zero average value.[2] (Here, s is the axial coordinate measured along the direction of beam propagation.) Another periodic focusing system is illustrated in Fig. 10.1(b), where the axial magnetic field $B_z(s) = 2(\gamma_b m_i \beta_b c^2/Z_i e)\sqrt{\kappa_z(s)}$ oscillates with axial periodicity length S about some average value $B_0 = \int_{s_0}^{s_0+S} ds\, B_z(s)/S = $ const. Although Fig. 10.1(b) is drawn for the case where the average axial magnetic B_0 is nonzero, it should be emphasized that the analysis of the orbit equations presented later in Chapter 10 applies equally well to the case where $B_0 = 0$. Finally, superimposed on the applied periodic field configuration are the self-electric and self-magnetic fields defined in Eq.(10.12) for an elliptical beam with uniform charge density and current density (Sec. 10.2).

In subsequent sections, we examine the orbit equations for an individual test ion moving in the total (applied plus self) field configuration. Following a simplification of the transverse equations of motion in the paraxial approximation (Sec. 10.3), the orbit equations and the beam envelope equations are analyzed for the case of beam propagation through a periodic quadrupole lattice with $B_z = 0$ and nonzero B_x^q and B_y^q (Sec. 10.4), and for beam propagation through a periodic solenoidal field with $B_x^q = 0 = B_y^q$ and nonzero B_z (Sec. 10.5). Approximate estimates of the average focusing force and phase advance are then made for the case of beam propagation through a periodic quadrupole field in circumstances where the transverse modulation of the beam envelope is treated as a weak effect (Sec. 10.6). By introducing the appropriate generating function,[31] a Hamiltonian formalism is developed (Sec. 10.7) which describes the transverse particle orbits in a periodic quadrupole field in terms of pseudoharmonic oscillations. Self-field effects are also included in the analysis (assuming a uniform-density beam), and it is shown

that the Kapchinskij-Vladimirskij distribution function $f_b^0(X, Y, X', Y')$
self-consistently generates such a beam equilibrium. Finally, the kinetic
eigenvalue equation is derived for (small-amplitude) transverse perturbations about the Kapchinskij-Vladimirskij beam equilibrium (Sec. 10.8),
and the emittance growth and saturation observed in numerical simulations of transverse stability properties are discussed (Sec. 10.9).

As a final point, it should be emphasized that the analysis of the orbit
equations and beam propagation properties presented in Chapter 10 for
a periodic quadrupole magnetic field $B_x^q \hat{e}_x + B_y^q \hat{e}_y$ can be extended in a
straightforward manner to the case of a periodic quadrupole electric field
$E_x^q \hat{e}_x + E_y^q \hat{e}_y$ (see Problem 10.2).

10.2 Alternating-Gradient Field Configuration

In the present analysis, we consider an intense ion beam with characteristic energy $\gamma_b m_i c^2$ propagating in the z-direction through a periodic focusing lattice with average axial velocity $\beta_b c$. (The analysis for an electron
beam proceeds in a similar manner.) The ion dynamics are investigated
in the paraxial approximation with[4]

$$p_z^2 \gg p_x^2 + p_y^2 \ ,$$

$$\frac{\nu}{\beta_b^2 \gamma_b^3} = \frac{Z_i^2 e^2}{\beta_b^2 \gamma_b^3 m_i c^2} N_b \ll 1. \tag{10.2}$$

Here, $+Z_i e$ and m_i are the ion charge and rest mass, respectively, ν is Budker's parameter, and $N_b = \int\int dx\, dy\, n_b^0(\mathbf{x})$ is the number of ions per unit
axial length. In the x-y plane, the cross section of the ion beam is generally elliptical with dimension $2a$ in the x-direction and dimension $2b$ in the
y-direction. The applied focusing field consists of a transverse quadrupole
field $B_x^q \hat{e}_x + B_y^q \hat{e}_y$ and/or a solenoidal focusing field $B_z \hat{e}_z + B_x^{sol} \hat{e}_x + B_y^{sol} \hat{e}_y$
which are periodic in the z-direction with fundamental periodicity length
S. Here, for a simple quadrupole configuration, B_x^q and B_y^q oscillate about
zero average value [Fig. 10.1(a)], whereas the solenoidal field B_z generally
oscillates about a nonzero average value B_0 [Fig. 10.1(b)]. However, the
case where $B_0 = 0$ is not excluded in the present analysis. The analysis
assumes a thin beam with

$$a, b \ll S \ . \tag{10.3}$$

Taylor expanding the applied magnetic field components about the beam center $(x, y) = (0, 0)$ and making use of $\nabla \cdot \mathbf{B}_0^{\text{ext}} = 0$ and $\nabla \times \mathbf{B}_0^{\text{ext}} = 0$ gives

$$B_x^q = B_q'(s)y \,, \qquad B_y^q = B_q'(s)x \,,$$
$$B_x^{\text{sol}} = -\tfrac{1}{2}B_z'(s)x, \quad B_y^{\text{sol}} = -\tfrac{1}{2}B_z'(s)y, \tag{10.4}$$

to leading order. In Eq.(10.4), $B_q'(s)$ and $B_z'(s)$ are defined by $B_q'(s) \equiv (\partial B_x^q/\partial y)_0 = (\partial B_y^q/\partial x)_0$, and $B_z'(s) = (\partial B_z/\partial s)_0$, where s is the axial distance. Therefore, over the cross section of the ion beam, the applied magnetic field is approximated by

$$\mathbf{B}^0(\mathbf{x}) = B_q'(s)\,[y\hat{\mathbf{e}}_x + x\hat{\mathbf{e}}_y] + B_z(s)\hat{\mathbf{e}}_z - \frac{1}{2}B_z'(s)\,[x\hat{\mathbf{e}}_x + y\hat{\mathbf{e}}_y]\,, \tag{10.5}$$

where $B_q'(s)$, $B_z'(s)$ and $B_z(s)$ have periodicity length S in the z-direction.

Under steady-state conditions $(\partial/\partial t = 0)$, the space charge and axial current of the ion beam produce an equilibrium self-electric field $\mathbf{E}^0(\mathbf{x}) = -\nabla\phi_0(\mathbf{x})$ and a self-magnetic field $\mathbf{B}^s(\mathbf{x}) = \nabla \times \mathbf{A}^s(\mathbf{x})$. In terms of the equilibrium beam density $n_b^0(\mathbf{x})$, Poisson's equation for $\phi_0(\mathbf{x})$ is given by

$$\nabla^2\phi_0(\mathbf{x}) = -4\pi Z_i e n_b^0(\mathbf{x}) \,. \tag{10.6}$$

For $a, b \ll S$, we approximate $\nabla^2 = \partial^2/\partial x^2 + \partial^2/\partial y^2$ to leading order in Eq.(10.6), and the transverse electric field components are

$$E_x = -\partial\phi_0/\partial x \,, \quad E_y = -\partial\phi_0/\partial y \,. \tag{10.7}$$

In addition, for an ion beam with negligible axial velocity spread about $v_z = p_z/\gamma m_i = \beta_b c$, it follows from the $\nabla \times \mathbf{B}^s$ Maxwell equation that

$$\nabla^2 A_z^s(\mathbf{x}) = -4\pi Z_i e \beta_b n_b^0(\mathbf{x}) \,. \tag{10.8}$$

Comparing Eq.(10.6) with Eq.(10.8) gives $A_z^s(\mathbf{x}) = \beta_b\phi_0(\mathbf{x})$. Therefore, the transverse components of the self-magnetic field, $B_x^s = \partial A_z^s/\partial y$ and $B_y^s = -\partial A_z^s/\partial x$, can be expressed as

$$B_x^s = \beta_b\partial\phi_0/\partial y \,, \quad B_y^s = -\beta_b\partial\phi_0/\partial x \,. \tag{10.9}$$

It is assumed that the beam density $n_b^0(\mathbf{x})$ is constant on elliptical surfaces with $x^2/a^2 + y^2/b^2 =$const. For $a, b \ll S$, the self-field potential $\phi_0(\mathbf{x})$ is determined to leading order from

$$\left(\frac{\partial^2}{\partial x^2} + \frac{\partial^2}{\partial y^2} \right) \phi_0(\mathbf{x}) = -4\pi Z_i e n_b^0(\xi). \tag{10.10}$$

where $\xi = x^2/a^2 + y^2/b^2$. For the special case of a uniform density ion beam with $n_b^0 = \hat{n}_b =$const. for $x^2/a^2 + y^2/b^2 < 1$, the solution to Eq.(10.10) is given by (Problem 10.1)

$$\phi_0(\mathbf{x}) = -2\pi Z_i e \hat{n}_b \left[\frac{b}{a+b} x^2 + \frac{a}{a+b} y^2 \right] \tag{10.11}$$

for $x^2/a^2 + y^2/b^2 < 1$. Here, it is assumed that the transverse beam dimensions (a and b) are small in comparison with the radius of the conducting wall.[4] For a circular ion beam with $a = b$, note that Eq.(10.11) reduces to $\phi_0 = -\pi Z_i e \hat{n}_b r^2$, which is the expected result. Making use of Eqs.(10.7), (10.9), and (10.11), the transverse components of the self-electric and self-magnetic fields can be expressed as

$$
\begin{aligned}
E_x &= \frac{4\pi Z_i e \hat{n}_b b}{a+b} x, & E_y &= \frac{4\pi Z_i e \hat{n}_b a}{a+b} y, \\
B_x^s &= -\frac{4\pi Z_i e \hat{n}_b \beta_b a}{a+b} y, & B_y^s &= \frac{4\pi Z_i e \hat{n}_b \beta_b b}{a+b} x,
\end{aligned}
\tag{10.12}
$$

within the uniform-density beam.

The components of the transverse force on an individual ion produced by the equilibrium self fields in Eq.(10.12) are linearly proportional to the transverse displacement, x and y, from the beam axis. This is a consequence of the assumption of uniform beam density, which leads to considerable simplification in the orbit analysis (Secs. 10.3–10.7). If the density profile is more complicated (e.g., parabolic, or Gaussian), the self-field forces are nonlinear (Problem 10.1), and the corresponding orbit equations generally require numerical solution.

Problem 10.1   Consider an infinitely long ion beam with elliptical cross section and uniform charge density $+Z_i e \hat{n}_b =$ const. in the interior region $x^2/a^2 + y^2/b^2 < 1$. Poisson's equation (10.10) for the self-field potential $\phi_0(x, y)$

can be expressed as

$$\left\{ \frac{\partial^2}{\partial x^2} + \frac{\partial^2}{\partial y^2} \right\} \phi_0(x, y) = \begin{cases} -4\pi Z_i e\hat{n}_b, & \dfrac{x^2}{a^2} + \dfrac{y^2}{b^2} < 1, \\ \\ 0, & \dfrac{x^2}{a^2} + \dfrac{y^2}{b^2} > 1. \end{cases} \tag{10.1.1}$$

Assume that the transverse dimensions of the ion beam are small in comparison with the radius of the conducting wall $(a, b \ll r_w)$.

a. The complete solution to Eq.(10.1.1) has been given by Landau and Lifshitz [*The Classical Theory of Fields* (Pergamon, New York, 1971)] and by Chrien, et al. [Phys. Fluids **29**, 1675 (1986)]. Show that the solution to Eq.(10.1.1) can be expressed as

$$\phi_0(x, y) = -\pi Z_i e\hat{n}_b ab \left[\int_0^\xi \frac{ds}{[(a^2 + s)(b^2 + s)]^{1/2}} \tag{10.1.2} \right.$$

$$\left. + \int_\xi^\infty \frac{ds}{[(a^2 + s)(b^2 + s)]^{1/2}} \left(\frac{x^2}{a^2 + s} + \frac{y^2}{b^2 + s} \right) \right].$$

Here, the quantity ξ is defined by

$$\frac{x^2}{a^2 + \xi} + \frac{y^2}{b^2 + \xi} = 1 \tag{10.1.3}$$

outside the beam $(x^2/a^2 + y^2/b^2 > 1)$, and by $\xi = 0$ inside the beam $(x^2/a^2 + y^2/b^2 < 1)$.

b. Show that the solution in Eq.(10.1.2) reduces to

$$\phi_0(x, y) = -2\pi Z_i e\hat{n}_b \left[\frac{b}{a + b} x^2 + \frac{a}{a + b} y^2 \right] \tag{10.1.4}$$

for $x^2/a + y^2/b < 1$. Note from Eq.(10.1.4) that the self electric field components, $E_x = -\partial\phi_0/\partial x$ and $E_y = -\partial\phi_0/\partial y$, are linearly proportional to the transverse displacements x and y, respectively, within the ion beam [Eq.(10.12)].

c. Consider a circular ion beam with Gaussian density profile $n_b^0(r) = \hat{n}_b \times \exp(-r^2/a^2)$, where $r^2 = x^2 + y^2$ and $a = $ const. Integrate Poisson's equation (10.10) and show that the transverse self-electric field components are given by

$$E_x = 2\pi\hat{n}_b Z_i ef(r/a)x ,$$

$$E_y = 2\pi\hat{n}_b Z_i ef(r/a)y , \tag{10.1.5}$$

where

$$f(r/a) = \frac{a^2}{r^2} \left[1 - \exp(-r^2/a^2) \right]. \tag{10.1.6}$$

Note that only for $r^2/a^2 \ll 1$ [where $f(r/a) \simeq 1$] are the electric field components in Eq.(10.1.5) linearly proportional to x and y.

10.3   Equations of Motion in the Paraxial Approximation

We consider the motion of an individual test ion in the equilibrium field configuration described by Eqs.(10.5), (10.7), and (10.9). The orbit equations in Cartesian coordinates can be expressed as

$$\frac{d}{dt}p_x = Z_i e \left(E_x - \frac{1}{c}v_z B_y^s + \frac{1}{c}v_y B_z - \frac{1}{c}v_z \left(B_y^q + B_y^{sol} \right) \right), \qquad (10.13)$$

$$\frac{d}{dt}p_y = Z_i e \left(E_y + \frac{1}{c}v_z B_x^s - \frac{1}{c}v_x B_z + \frac{1}{c}v_z \left(B_x^q + B_x^{sol} \right) \right), \qquad (10.14)$$

$$\frac{d}{dt}p_z = Z_i e \left(E_z + \frac{1}{c}v_x \left(B_y^s + B_y^q + B_y^{sol} \right) - \frac{1}{c}v_y \left(B_x^s + B_x^q + B_x^{sol} \right) \right).$$

$$(10.15)$$

Here, the particle momentum $\mathbf{p}(t)$ and velocity $\mathbf{v}(t) = d\mathbf{x}(t)/dt$ are related by $\mathbf{p}(t) = \gamma(t)m_i\mathbf{v}(t)$, where $\gamma(t) = [1 + \mathbf{p}^2(t)/m_i^2 c^2]^{1/2}$ is the relativistic mass factor. Equations (10.13)–(10.15) possess the exact single-particle constant of the motion corresponding to the total energy $H = (\gamma - 1)m_i c^2 + Z_i e\phi_0 = $ const. Consistent with $p_z^2 \gg p_x^2 + p_y^2$ and $\nu/\beta_b^2\gamma_b^3 \ll 1$, we approximate $dp_x/dt = m_i v_x d\gamma/dt + \gamma m_i d^2x/dt^2 \simeq \gamma m_i d^2x/dt^2$ in Eq.(10.13), and similarly for the y-motion in Eq.(10.14). Equations (10.13) and (10.14) then reduce to

$$\frac{d^2x}{dt^2} = \frac{Z_i e}{\gamma m_i} \left[-\left(1 - \frac{1}{c}v_z\beta_b \right) \frac{\partial\phi_0}{\partial x} + \frac{1}{c}B_z\frac{dy}{dt} - \frac{1}{c}v_z \left(B_q'x - \frac{1}{2}B_z'y \right) \right],$$

$$(10.16)$$

$$\frac{d^2y}{dt^2} = \frac{Z_i e}{\gamma m_i} \left[-\left(1 - \frac{1}{c}v_z\beta_b \right) \frac{\partial\phi_0}{\partial y} - \frac{1}{c}B_z\frac{dx}{dt} + \frac{1}{c}v_z \left(B_q'y - \frac{1}{2}B_z'x \right) \right],$$

$$(10.17)$$

where use has been made of Eqs.(10.5), (10.7), and (10.9), and $d\gamma/dt \simeq 0$. For an ion beam with $p_z^2 \gg p_x^2 + p_y^2$ and negligible spread in axial velocity about $v_z = p_z/\gamma m = \beta_b c$, it is valid to approximate $v_z = \beta_b c$

and $\gamma = \gamma_b = (1 - \beta_b^2)^{-1/2}$ in the coefficients in Eqs.(10.16) and (10.17). Therefore, the transverse orbit equations (10.16) and (10.17) can be expressed as

$$\frac{d^2x}{dt^2} = -\frac{Z_i e}{\gamma_b^3 m_i}\frac{\partial\phi_0}{\partial x} + \frac{Z_i e B_z}{\gamma_b m_i c}\frac{dy}{dt} - \frac{Z_i e \beta_b}{\gamma_b m_i}\left(B_q'x - \frac{1}{2}B_z'y\right), \quad (10.18)$$

$$\frac{d^2y}{dt^2} = -\frac{Z_i e}{\gamma_b^3 m_i}\frac{\partial\phi_0}{\partial y} - \frac{Z_i e B_z}{\gamma_b m_i c}\frac{dx}{dt} + \frac{Z_i e \beta_b}{\gamma_b m_i}\left(B_q'y - \frac{1}{2}B_z'x\right). \quad (10.19)$$

It is customary in accelerator physics to express[4]

$$\frac{d}{dt} = \beta_b c\frac{d}{ds} \quad (10.20)$$

in the analysis of Eqs.(10.18) and (10.19). Here, the coordinate $s = s_0 + \beta_b ct$ follows the axial motion of a beam ion. Specializing to the case of a uniform-density ion beam where $E_x = -\partial\phi_0/\partial x$ and $E_y = -\partial\phi_0/\partial y$ are given by Eq.(10.12), the orbit equations (10.18) and (10.19) reduce to

$$\frac{d^2}{ds^2}x(s) = \left[\frac{4Z_i e I_b}{\gamma_b^3\beta_b^3 m_i c^3}\frac{1}{a(a+b)} - \frac{Z_i e B_q'(s)}{\gamma_b m_i \beta_b c^2}\right]x(s)$$

$$+ \frac{Z_i e B_z(s)}{\gamma_b m_i \beta_b c^2}\frac{d}{ds}y(s) + \frac{Z_i e B_z'(s)}{2\gamma_b m_i \beta_b c^2}y(s), \quad (10.21)$$

$$\frac{d^2}{ds^2}y(s) = \left[\frac{4Z_i e I_b}{\gamma_b^3\beta_b^3 m_i c^3}\frac{1}{b(a+b)} + \frac{Z_i e B_q'(s)}{\gamma_b m_i \beta_b c^2}\right]y(s)$$

$$- \frac{Z_i e B_z(s)}{\gamma_b m_i \beta_b c^2}\frac{d}{ds}x(s) - \frac{Z_i e B_z'(s)}{2\gamma_b m_i \beta_b c^2}x(s). \quad (10.22)$$

Here, $I_b = Z_i e \beta_b c \hat{n}_b \pi ab$ is the axial current (assumed constant) carried by the ion beam, which is related to the self-field perveance K defined by[13]

$$K = \frac{I_b}{(m_i c^2/Z_i e)}\frac{2}{\beta_b^3\gamma_b^3} = \frac{2\nu}{\beta_b^2\gamma_b^3}. \quad (10.23)$$

In Eq.(10.23), $\nu = \omega_{pb}^2 ab/4c^2$ is Budker's parameter, where $\omega_{pb}^2 = 4\pi\hat{n}_b Z_i^2 \times e^2/m_i$ is the nonrelativistic plasma frequency-squared. It is convenient to introduce the abbreviated notation

$$\kappa_q(s) = \frac{Z_i e B_q'(s)}{\gamma_b m_i \beta_b c^2}, \quad (10.24)$$

and

$$\sqrt{\kappa_z(s)} = \frac{Z_i e B_z(s)}{2\gamma_b m_i \beta_b c^2} \; . \tag{10.25}$$

The orbit equations (10.21) and (10.22) can then be expressed in the compact form

$$\frac{d^2}{ds^2}x(s) + \left\{ \kappa_q(s) - \frac{2K}{a(s)[a(s)+b(s)]} \right\} x(s)$$

$$- 2\sqrt{\kappa_z(s)}\frac{d}{ds}y(s) - \left(\frac{d}{ds}\sqrt{\kappa_z(s)} \right) y(s) = 0 \; , \tag{10.26}$$

$$\frac{d^2}{ds^2}y(s) + \left\{ -\kappa_q(s) - \frac{2K}{b(s)[a(s)+b(s)]} \right\} y(s)$$

$$+ 2\sqrt{\kappa_z(s)}\frac{d}{ds}x(s) + \left(\frac{d}{ds}\sqrt{\kappa_z(s)} \right) x(s) = 0 \; , \tag{10.27}$$

where the envelope functions $a(s)$ and $b(s)$ generally vary with s for a periodic focusing system. Equations similar in form to Eqs.(10.26) and (10.27) pertain when a periodic quadrupole electric field $E_x^q \hat{e}_x + E_y^q \hat{e}_y$ is applied (Problem 10.2).

For a periodic quadrupole lattice, the coupling coefficient $\kappa_q(s)$ satisfies[2]

$$\kappa_q(s + S) = \kappa_q(s) \; , \tag{10.28}$$

and the sign of $\kappa_q(s)$ alternates over one period S with average value $\int_{s_0}^{s_0+S} ds\,\kappa_q(s) = 0$. Therefore, the quadrupole-field contributions in Eqs.(10.26) and (10.27) alternately produce focusing and defocusing of the x- and y-orbits, and the x- and y-forces have opposite polarity for each value of s. In contrast, the self-field contributions in Eqs.(10.26) and (10.27), which are proportional to $-Kx(s)$ and $-Ky(s)$, are always defocusing. Moreover, the self-field forces are linearly proportional to $x(s)$ and $y(s)$ because of the assumption of uniform beam density (Sec. 10.2). Finally, in Eqs.(10.26) and (10.27), the solenoidal coefficient $\sqrt{\kappa_z(s)} = (Z_i e/2\gamma_b m_i \beta_b c^2) B_z(s)$ can oscillate about a nonzero average value [the case illustrated in Fig. 10.1(b)] or about zero average value. In both cases, it is readily shown that the resulting transverse force on the ion orbit is focusing (Sec. 10.5.1).

Problem 10.2　　In the orbit equations (10.13)–(10.15), the effects of the quadrupole magnetic field $B_x^q \hat{e}_x + B_y^q \hat{e}_y$ can be replaced by a periodic quadrupole electric field $E_x^q \hat{e}_x + E_y^q \hat{e}_y$, where

$$E_x^q = E_q'(s)x , \quad E_y^q = -E_q'(s)y . \tag{10.2.1}$$

Here $E_q'(s) \equiv (\partial E_x^q/\partial x)_0 = -(\partial E_y^q/\partial y)_0$ for a thin beam with $a, b \ll S$. Set $B_x^q = 0 = B_y^q$ in Eqs.(10.13)–(10.15), and assume that the quadrupole electric field components in Eq.(10.2.1) are applied. Show that the resulting orbit equations for $x(s)$ and $y(s)$ in the paraxial approximation are again given by Eqs.(10.26) and (10.27), with $\kappa_q(s)$ replaced by

$$\kappa_q(s) = -\frac{Z_i e E_q'(s)}{\gamma_b m_i \beta_b^2 c^2}. \tag{10.2.2}$$

Therefore, the analysis of the orbit equations and beam propagation properties in Chapter 10 applies equally well to the case in which a periodic quadrupole electric field is applied.

10.4　Orbit and Envelope Equations for a Periodic Quadrupole Field

10.4.1　Particle Orbit Equations

We first analyze the transverse orbit equations (10.26) and (10.27) for beam propagation through a periodic quadrupole lattice with $B_z(s) = 0$ and $\kappa_z(s) = 0$. In this case, the x- and y-motions are decoupled, and Eqs.(10.26) and (10.27) reduce to

$$\frac{d^2}{ds^2}x(s) + \kappa_x(s)x(s) = 0 , \tag{10.29}$$

and

$$\frac{d^2}{ds^2}y(s) + \kappa_y(s)y(s) = 0 , \tag{10.30}$$

where $\kappa_x(s)$ and $\kappa_y(s)$ are defined by

$$\kappa_x(s) = \kappa_q(s) - \frac{2K}{a(s)[a(s) + b(s)]} ,$$

$$\kappa_y(s) = -\kappa_q(s) - \frac{2K}{b(s)[a(s) + b(s)]} . \tag{10.31}$$

Following Courant and Snyder,[2,4] it is useful to express the x- and y-orbits as

$$x(s) = A_x w_x(s) \cos\left[\psi_x(s) + \Phi_{x0}\right],$$

$$y(s) = A_y w_y(s) \sin\left[\psi_y(s) + \Phi_{y0}\right], \qquad (10.32)$$

where A_x, A_y, Φ_{x0} and Φ_{y0} are constants. Here, $\psi_x(s)$ and $\psi_y(s)$ are called the betatron phase functions, and $w_x(s)$ and $w_y(s)$ are called the envelope functions. For present purposes, because Eq.(10.30) is similar in form to Eq.(10.29), it is sufficient to analyze Eq.(10.29) for the x-motion. From Eq.(10.32) it follows that

$$\frac{d}{ds}x(s) = A_x w_x' \cos(\psi_x + \Phi_{x0}) - A_x w_x \psi_x' \sin(\psi_x + \Phi_{x0}) , \quad (10.33)$$

$$\frac{d^2}{ds^2}x(s) = A_x \left[w_x'' - w_x(\psi_x')^2\right] \cos(\psi_x + \Phi_{x0})$$

$$-A_x \left[w_x' \psi_x' + (w_x \psi_x')'\right] \sin(\psi_x + \Phi_{x0}), \qquad (10.34)$$

where the "prime" notation denotes derivative with respect to s, i.e., $w_x' = dw_x(s)/ds$, etc. Substituting Eqs.(10.32) and (10.34) into Eq.(10.29) and setting the coefficient of $\sin(\psi_x + \Phi_{x0})$ equal to zero gives $w_x' \psi_x' + (w_x \psi_x')' = 0$, or equivalently

$$\psi_x(s) = \int_{s_0}^{s} \frac{ds}{w_x^2(s)}. \qquad (10.35)$$

Setting the coefficient of $\cos(\psi_x + \Phi_{x0})$ equal to zero and making use of $d\psi_x/ds = 1/w_x^2$ then gives[4]

$$\frac{d^2}{ds^2}w_x(s) + \kappa_x(s)w_x(s) = \frac{1}{w_x^3(s)}. \qquad (10.36)$$

Apart from the constant factor A_x occurring in Eq.(10.32), note that Eq.(10.36) is a closed equation for the amplitude $w_x(s)$ of the x-orbit. Following an identical procedure, it is readily shown that the equation of motion for $w_y(s)$ is given by[4]

$$\frac{d^2}{ds^2}w_y(s) + \kappa_y(s)w_y(s) = \frac{1}{w_y^3(s)}. \qquad (10.37)$$

where $\psi_y(s) = \int_{s_0}^{s} ds/w_y^2(s)$.

The constant factors A_x and A_y in Eq.(10.32) can be viewed as constants of the motion for the x- and y-orbits. For example, making use of Eqs.(10.32) and (10.33), it can be shown that (Problem 10.3)

$$A_x^2 = \frac{x^2}{w_x^2} + \left(w_x \frac{dx}{ds} - \frac{dw_x}{ds} x \right)^2 . \qquad (10.38)$$

It is customary in accelerator physics to refer to $(x, dx/ds)$ [rather than (x, p_x)] as "phase space." As illustrated in Fig. 10.2, Eq.(10.38) represents an ellipse in the phase space $(x, dx/ds)$ with area πA_x^2 (Problem 10.3). The size, orientation and eccentricity of the ellipse are determined by A_x and the values of w_x and dw_x/ds (and hence the "initial" conditions at

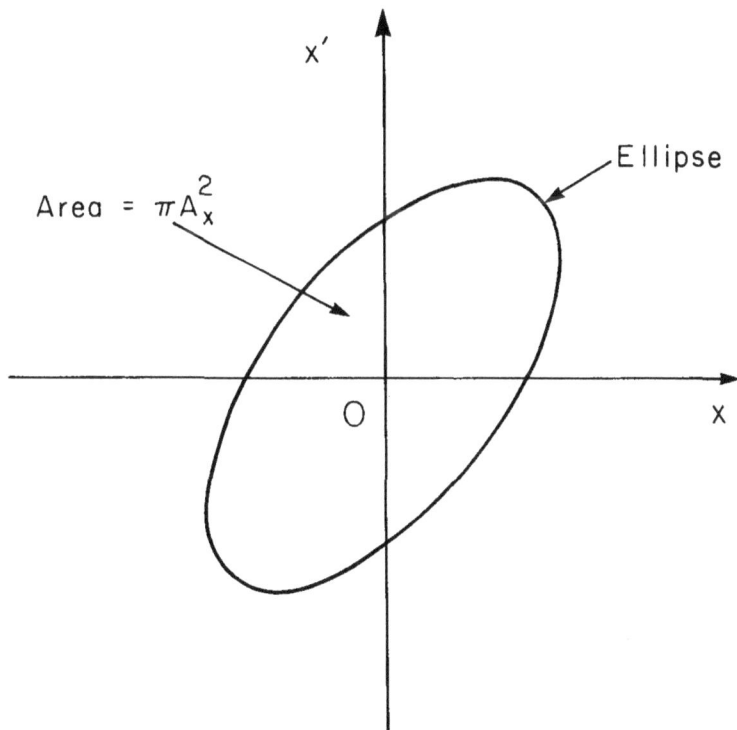

Figure 10.2. Equation (10.38) is the equation for an ellipse with area πA_x^2 in the phase space (x, x'), where $x' = dx/ds$ denotes the normalized momentum. The size, orientation and eccentricity of the ellipse are determined by A_x, and the values of w_x and dw_x/ds which vary as s varies.

$s = s_0$, say). As s varies, so do w_x and dw_x/ds, and the shape and orientation of the ellipse change accordingly.[4] For specified value of A_x, however, the area of the ellipse remains constant and equal to πA_x^2.

The fact that the quantity A_x^2 defined in Eq.(10.38) is a constant of the motion constitutes a very powerful constraint condition on the evolution of the orbit $x(s)$. Similarly, the y-motion is constrained by the analogous condition $A_y^2 = $ const. It should be emphasized, however, that the existence of these constants of the motion is a consequence of the fact that the forcing terms in Eqs.(10.29) and (10.30) are linearly proportional to $x(s)$ and $y(s)$, respectively. This is not the case if the ion density profile $n_b^0(x,y)$ is spatially nonuniform in the beam interior.

10.4.2   Beam Envelope Equations

The outer envelope of the ion beam is defined by $x^2/a^2(s) + y^2/b^2(s) = 1$. The maximum excursion of an ion in the x-direction corresponds to $x = a(s)$, which occurs when $\psi_x + \Phi_{x0} = 0, \psi_y + \Phi_{y0} = 0$ (i.e., $y = 0$), and A_x assumes a maximum value $A_x = A_x^{\max}$. The area of the corresponding ellipse in the phase space $(x, dx/ds)$ is $\pi(A_x^{\max})^2 = \pi\varepsilon_x$, where ε_x denotes the unnormalized "edge" emittance in the x-direction. [In this notation, the "normalized" emittance in the phase space $(x, p_x/m_i c)$, which is often used in accelerator design, is equal to $\beta_b\gamma_b\varepsilon_x$.] Therefore, for $\psi_x + \Phi_{x0} = 0$ and $A_x = A_x^{\max} = \sqrt{\varepsilon_x}$, it follows from Eq.(10.32) and $x = a(s)$ that $w_x(s) = a(s)/\sqrt{\varepsilon_x}$. Equation (10.36) then gives the differential equation

$$\frac{d^2}{ds^2}a(s) + \kappa_x(s)\,a(s) - \frac{\varepsilon_x^2}{a^3(s)} = 0, \tag{10.39}$$

which determines the beam half-width $a(s)$ in the x-plane. Substituting Eq.(10.31) into Eq.(10.39) gives

$$\frac{d^2}{ds^2}a(s) + \kappa_q(s)\,a(s) - \frac{2K}{a(s) + b(s)} - \frac{\varepsilon_x^2}{a^3(s)} = 0. \tag{10.40}$$

A similar analysis of Eqs.(10.32) and (10.37) for $\psi_y + \Phi_{y0} = \pi/2, \psi_x + \Phi_{x0} = \pi/2, A_y^{\max} = \sqrt{\varepsilon_y}$, and $y = b(s) = \sqrt{\varepsilon_y}w_y(s)$ gives

$$\frac{d^2}{ds^2}b(s) - \kappa_q(s)\,b(s) - \frac{2K}{a(s) + b(s)} - \frac{\varepsilon_y^2}{b^3(s)} = 0\,, \tag{10.41}$$

which determines the beam half-width $b(s)$ in the y-plane. Here, use has been made of the definition of $\kappa_y(s)$ in Eq.(10.38). We refer to Eqs.(10.40) and (10.41) as the "envelope" equations for $a(s)$ and $b(s)$. The orbit equations (10.36) and (10.37) for $w_x(s)$ and $w_y(s)$, and the envelope equations (10.40) and (10.41) for $a(s)$ and $b(s)$ can be solved numerically over a wide range of system parameters for specified periodic quadrupole field.[13,14] By taking appropriate averages over the periodicity length S, use is made of Eqs.(10.40) and (10.41) in Sec. 10.6 to obtain analytical estimates of the phase advance and the beam radius. For purposes of comparison, it is instructive to analyze first the orbit equations (10.26) and (10.27) for the case of a solenoidal focusing field with $\kappa_z(s) \neq 0$ and $\kappa_q(s) = 0$.

Problem 10.3   Consider the expressions for $x(s)$ and $dx(s)/ds$ in Eqs.(10.32) and (10.33), where A_x and Φ_{x0} are constants, $w_x(s)$ evolves according to Eq.(10.36), and the phase function $\psi_x(s)$ is defined in Eq.(10.35).

a. Show that the amplitude factor A_x, which is a constant of the motion, is given exactly by Eq.(10.38).

b. Equation (10.38) can be expressed as

$$A_x^2 = \left(\frac{1}{w_x^2} + w_x'^2\right) x^2 - 2w_x w_x' x \left(\frac{dx}{ds}\right) + w_x^2 \left(\frac{dx}{ds}\right)^2, \tag{10.3.1}$$

where $w_x' = dw_x(s)/ds$. Equation (10.3.1) is the equation for an ellipse in the phase space $(x, dx/ds)$. When $w_x' \neq 0$, the major axis of the ellipse is rotated relative to the x-axis, as illustrated in Fig. 10.2. Make use of Eq.(10.3.1) to show that the area of the ellipse in the phase space $(x, dx/ds)$ is

$$\text{Area} = \pi A_x^2. \tag{10.3.2}$$

10.5   Orbit and Envelope Equations for a Solenoidal Focusing Field

10.5.1   Particle Orbit Equations

In this section, the transverse orbit equations (10.26) and (10.27) are analyzed for an ion beam with circular cross section propagating in the z-direction through the periodic solenoidal focusing field $B_z(s)\hat{e}_z - (1/2)B_z'(s)[x\hat{e}_x + y\hat{e}_y]$, with zero quadrupole field components, i.e.,

$\kappa_q(s) = 0$. [The case in which $B_z(s) = B_0 = $ const. is not excluded.] For $\kappa_q(s) = 0$, Eqs.(10.26) and (10.27) reduce to

$$\frac{d^2}{ds^2}x(s) - \kappa_s(s)x(s) - 2\sqrt{\kappa_z(s)}\frac{d}{ds}y(s) - \left(\frac{d}{ds}\sqrt{\kappa_z(s)}\right)y(s) = 0,$$

(10.42)

$$\frac{d^2}{ds^2}y(s) - \kappa_s(s)y(s) + 2\sqrt{\kappa_z(s)}\frac{d}{ds}x(s) + \left(\frac{d}{ds}\sqrt{\kappa_z(s)}\right)x(s) = 0.$$

(10.43)

Here, the self-field coupling coefficient $\kappa_s(s)$ is defined by

$$\kappa_s(s) = \frac{K}{r_b^2(s)}$$

(10.44)

for a circular ion beam with

$$a(s) = b(s) \equiv r_b(s) .$$

(10.45)

In Eqs.(10.42) and (10.43), the x- and y-motions are coupled through the terms proportional to $2\sqrt{\kappa_z(s)} = Z_i e B_z(s)/\gamma_b m_i \beta_b c^2$. It is convenient to express

$$x(s) = Aw(s)\cos\left[\psi(s) - \int_{s_0}^{s} ds\sqrt{\kappa_z(s)} + \Phi_0\right],$$

$$y(s) = Aw(s)\sin\left[\psi(s) - \int_{s_0}^{s} ds\sqrt{\kappa_z(s)} + \Phi_0\right],$$

(10.46)

where A and Φ_0 are constants, and $w(s)$ and $\psi(s)$ are amplitude and phase functions. Substituting Eq.(10.46) into Eqs.(10.42) and (10.43) gives (Problem 10.4)

$$\frac{d^2}{ds^2}w(s) + [\kappa_z(s) - \kappa_s(s)]w(s) - \left(\frac{d\psi}{ds}\right)^2 w(s) = 0,$$

(10.47)

$$2\frac{dw(s)}{ds}\frac{d\psi(s)}{ds} + w(s)\frac{d^2\psi(s)}{ds^2} = 0.$$

(10.48)

Integrating Eq.(10.48) then gives

$$\psi(s) = \int_{s_0}^{s} \frac{ds}{w^2(s)}. \tag{10.49}$$

Finally, substituting Eq.(10.49) into Eq.(10.47), we obtain the orbit equation

$$\frac{d^2}{ds^2}w(s) + [\kappa_z(s) - \kappa_s(s)]\,w(s) = \frac{1}{w^3(s)}, \tag{10.50}$$

which determines the evolution of the amplitude function $w(s)$.

Equations (10.49) and (10.50), derived for a solenoidal focusing field, should be compared with Eqs.(10.35) and (10.36), derived for a periodic quadrupole focusing field. While the self-field contributions are similar in the two cases, the magnetic focusing contributions are different. In Eq.(10.36), the quadrupole coupling coefficient $\kappa_q(s)$ alternates sign over one period S of the lattice, with zero average value $\int_{s_0}^{s_0+S} ds\,\kappa_q(s) = 0$ [Fig. 10.1(a)]. On the other hand, the solenoidal coefficient $\kappa_z(s) = Z_i^2 e^2 B_z^2(s)/4\gamma_b^2 m_i^2 \beta_b^2 c^4$ in Eq.(10.50) oscillates about a *positive* average value $\bar{\kappa}_z = \int_{s_0}^{s_0+S} ds\,\kappa_z(s)/S$, thereby providing a focusing force at every value of s.

We denote the average axial magnetic field by

$$B_0 = \int_{s_0}^{s_0+S} \frac{ds}{S}\, B_z(s). \tag{10.51}$$

It should be emphasized that Eq.(10.50) constitutes a valid description of the evolution of the amplitude function $w(s)$ in circumstances where the average axial field is zero ($B_0 = 0$), or when $B_z(s)$ oscillates about some constant nonzero value of B_0 [the case illustrated in Fig. 10.1(b)].

Problem 10.4   Consider the orbit equations (10.42) and (10.43), which describe the transverse particle motion in the combined self fields of the beam and the periodic focusing field $B_z(s)\hat{e}_z - (1/2)\,B_z'(s)[x\hat{e}_x + y\hat{e}_y]$.

a.   The amplitude function $w(s)$ and the phase function $\psi(s)$ are related to the transverse orbits $x(s)$ and $y(s)$ by Eq.(10.46). Make use of Eqs.(10.42), (10.43), and (10.46) to show that $w(s)$ and $\psi(s)$ evolve according to Eqs.(10.47) and (10.48).

b.   Show that $\psi(s)$ and $w(s)$ satisfy Eqs.(10.49) and (10.50).

10.5.2   Beam Envelope Equation

An equation for the outer envelope $r_b(s) \equiv a(s) = b(s)$ of the (circular) ion beam is readily derived from Eq.(10.50). Paralleling the analysis leading to Eq.(10.39), the envelope equation is obtained by substituting $w(s) = r_b(s)/\sqrt{\varepsilon}$ into Eq.(10.50), which gives[13]

$$\frac{d^2}{ds^2}r_b(s) + \kappa_z(s)r_b(s) - \frac{K}{r_b(s)} - \frac{\varepsilon^2}{r_b^3(s)} = 0, \qquad (10.52)$$

where use has been made of Eq.(10.44). Here, ε is the unnormalized emittance, and $\kappa_z(s)$ is defined in Eq.(10.25). Because $\int_{s_0}^{s} ds\sqrt{\kappa_z(s)} = (1/2)\int_{s_0}^{s} ds[Z_i eB_z(s)/\gamma_b m_i \beta_b c^2]$ occurs in the phase factors in Eq.(10.46), it should be noted that ε is the unnormalized emittance in the *Larmor* frame (rotating at *one-half* of the cyclotron frequency).[4]

As expected, Eq.(10.52) is identical in form to the envelope equation (9.7) derived in Sec. 9.2 for a cylindrical electron beam, provided we replace electrons by ions in Eq.(9.7), and assume that the beam propagates through vacuum $(f = 0 = f_M)$ parallel to a uniform solenoidal field $B_z(s) = B_0 = \text{const}$. In this case, $2\sqrt{\kappa_z(s)} = eB_0/\gamma_b m_i \beta_b c^2 = \omega_{cb}/\gamma_b \beta_b c$, and the self-field perveance can be expressed as $K = (\omega_{pb}^2 r_b^2/2\gamma_b \beta_b^2 c^2) \times (1 - \beta_b^2)$, where $\omega_{pb}^2 = 4\pi \hat{n}_b Z_i^2 e^2/m_i, \omega_{cb} = Z_i eB_0/m_i c$, and $1 - \beta_b^2 = \gamma_b^{-2}$. Making the identification $c\beta_b d/ds = d/dt$ and $\varepsilon^2 = (P_\theta/\gamma_b m_i)^2$, it follows that Eq.(10.52) reduces exactly to Eq.(9.7). Furthermore, examination of the particle orbit equation (10.50) for $w(s)$ shows that stable (oscillatory) solutions exist only for $\kappa_z > \kappa_s = K/r_b^2$, which gives the familiar restriction on beam density

$$\frac{2\omega_{pb}^2}{\gamma_b \omega_{cb}^2} < 1, \qquad (10.53)$$

which is required for the existence of a radially confined beam equilibrium. [Compare with Eq.(9.12) for $f = 0 = f_M$.]

10.5.3   Beam Radius and Current for a Matched Beam

For future reference, it is instructive to examine the envelope equation (10.52) for a uniform solenoidal field $B_z(s) = B_0 = \text{const}$., and a "matched" beam with $d^2 r_b/ds^2 = 0$ and $r_b(s) = \text{const}$.[13] In this case,

$\kappa_z(s) = \kappa_{z0} = \text{const.}$, where $\kappa_{z0} = (\omega_{cb}/2\gamma_b\beta_b c)^2$ and $\omega_{cb} = Z_i e B_0/m_i c$. For a matched beam, Eq.(10.52) gives

$$K = \kappa_{z0}r_b^2\left(1 - \frac{\varepsilon^2}{\kappa_{z0}r_b^4}\right), \tag{10.54}$$

which relates the self-field perveance K [defined in terms of the beam current I_b by Eq.(10.23)] to the beam radius r_b, the emittance ε, and the focusing coefficient κ_{z0}. For negligibly small beam current $(K \ll \kappa_{z0}r_b^2)$, Eq.(10.52) gives the beam radius

$$r_{b0} = \left(\frac{\varepsilon}{\sqrt{\kappa_{z0}}}\right)^{1/2} \tag{10.55}$$

in the absence of self-field effects. For nonzero K, the solution to Eq.(10.54) for the beam radius r_b is

$$r_b = \left\{\left[\frac{\varepsilon^2}{\kappa_{z0}} + \left(\frac{K}{2\kappa_{z0}}\right)^2\right]^{1/2} + \frac{K}{2\kappa_{z0}}\right\}^{1/2}. \tag{10.56}$$

For $K \ll 2\varepsilon^2$, Eq.(10.56) reduces to $r_b = (\varepsilon/\sqrt{\kappa_{z0}})^{1/2} = r_{b0}$, as expected. As K is increased, however, the beam radius expands due to the (defocusing) self-field effects. [Compare with Part (e) of Problem 9.2.]

For specified values of the beam radius r_b, emittance ε, and focusing coefficient κ_{z0}, Eq.(10.54) can also be used to determine the beam current I_b. Substituting Eq.(10.23) into Eq.(10.54) gives

$$I_b = \frac{1}{8}\beta_b\gamma_b\left(\frac{m_i c^3}{Z_i e}\right)\left(\frac{\omega_{cb}r_b}{c}\right)^2\left[1 - \left(\frac{\varepsilon}{\alpha}\right)^2\right], \tag{10.57}$$

where $\omega_{cb} = Z_i e B_0/m_i c$. Here, we have introduced the "acceptance" α of the channel defined by [13] $\alpha = r_b^2\sqrt{\kappa_{z0}}$. Note that $(\varepsilon/\alpha)^2 = \varepsilon^2/\kappa_{z0}r_b^4 = (r_{b0}/r_b)^4$, so that propagation at high current requires $(\varepsilon/\alpha)^2 \ll 1$, or equivalently $(r_{b0}/r_b)^4 \ll 1$. For ions with mass number A and charge state Z_i, Eq.(10.57) can be expressed as

$$I_b(A) = 0.4\beta_b\gamma_b\frac{Z_i}{A}\left[B_0(\text{kG})r_b(\text{cm})\right]^2\left[1 - \left(\frac{\varepsilon}{\alpha}\right)^2\right], \tag{10.58}$$

where the beam current is in amperes. For example, if $Z_i = 3$ and $A = 200$ (corresponding to Hg^{+++} ions), $B_0 = 40$ kG, $r_b = 0.5$ cm, and

$(\varepsilon/\alpha)^2 \ll 1$, then Eq.(10.58) gives $I_b = 2.4\,\beta_b\gamma_b$ amperes. Evidently, use of a purely solenoidal magnetic field $B_0\hat{\mathbf{e}}_z$ to transport a nonneutral ion beam is limited to relatively low currents.

10.5.4   Phase Advance σ

It is also useful, particularly for periodic focusing systems (Sec. 10.6), to introduce the phase advance σ over one axial period S defined by [see Eqs.(10.46) and (10.49)][2,4,13,16]

$$\sigma = \int_{s_0}^{s_0+S} \frac{ds}{w^2(s)}. \tag{10.59}$$

The quantity σ is also called the "tune." Here, for the beam envelope, $w(s)$ and $r_b(s)$ are related by $w(s) = r_b(s)/\sqrt{\varepsilon}$. For a matched cylindrical beam with $r_b(s) = \text{const.}$, Eq.(10.59) reduces to

$$\sigma = \frac{\varepsilon S}{r_b^2}. \tag{10.60}$$

Eliminating $r_b^2 = \varepsilon S/\sigma$, Eq.(10.54) can be expressed in the equivalent form

$$\kappa_{z0} = \left(\frac{\sigma}{S}\right)^2 + \frac{\sigma}{S}\left(\frac{K}{\varepsilon}\right) \tag{10.61}$$

for a uniform solenoidal field $B_z(s) = B_0 = \text{const.}$ For a uniform solenoidal field note that the phase advance per unit length (σ/S) occurs as a natural parameter in Eq.(10.61). In the zero-current limit $(K/2\varepsilon\sqrt{\kappa_{z0}} \to 0)$, Eq.(10.61) gives $\sigma/S = \sqrt{\kappa_{z0}} \equiv \sigma_0/S$. For general value of the self-field perveance K, the solution to Eq.(10.61) is

$$\frac{\sigma}{S} = \left[\left(\frac{\sigma_0}{S}\right)^2 + \left(\frac{K}{2\varepsilon}\right)^2\right]^{1/2} - \frac{K}{2\varepsilon}. \tag{10.62}$$

Therefore, as K is increased, self-field effects "depress" the phase advance σ relative to the free-space value σ_0.[16] Indeed, for $K/2\varepsilon \gg \sigma_0/S$, Eq.(10.62) can be approximated by

$$\frac{\sigma}{S} = \frac{\varepsilon}{K}\left(\frac{\sigma_0}{S}\right)^2. \tag{10.63}$$

That is, in the high-current regime, σ/S increases with increasing ε, and decreases with increasing beam current.

The phase advance σ can also be related to the ripple wavelength of the beam envelope.[16] We consider the envelope equation (10.52) for small-amplitude perturbations $\delta r_b(s)$ about the equilibrium radius determined from Eq.(10.54). Linearization of Eq.(10.52) gives (Problem 10.5)

$$\frac{d^2}{ds^2}\delta r_b(s) + \left(\kappa_{z0} + \frac{K}{r_b^2} + \frac{3\varepsilon^2}{r_b^4}\right)\delta r_b(s) = 0 \qquad (10.64)$$

for a uniform solenoidal field with $\kappa_z(s) = \kappa_{z0} = \text{const.}$ Eliminating ε^2/r_b^2 and K/ε by means of Eqs.(10.54) and (10.61), some straightforward algebra shows that Eq.(10.64) can be expressed in the equivalent form

$$\frac{d^2}{ds^2}\delta r_b(s) + 2\left[\left(\frac{\sigma_0}{S}\right)^2 + \left(\frac{\sigma}{S}\right)^2\right]\delta r_b(s) = 0, \qquad (10.65)$$

where $(\sigma_0/S)^2 = \kappa_{z0}$. For sinusoidal perturbations of the form $\delta r_b(s) = \delta\hat{r}_b\exp(iks)$, Eq.(10.65) gives

$$k^2 S^2 = 2\sigma_0^2 + 2\sigma^2 , \qquad (10.66)$$

which relates the ripple wavelength $\lambda = 2\pi/k$ of the beam boundary directly to σ_0 and σ. In the low-current regime, $\sigma \simeq \sigma_0$ and Eq.(10.66) reduces to $kS \simeq 2\sigma_0$. On the other hand, for sufficiently high beam current, $\sigma \ll \sigma_0$ and Eq.(10.66) reduces to $kS \simeq \sqrt{2}\sigma_0$.

Finally, for a uniform solenoidal field, the quantities σ_0 and σ are readily expressed in notation familiar in plasma physics. Making use of Eqs.(10.60) and (10.61) to express $(\sigma/S)^2 = \kappa_{z0} - K/r_b^2$, we obtain

$$\frac{\sigma_0}{S} = \frac{1}{2}\frac{\omega_{cb}}{\gamma_b\beta_b c} ,$$

$$\frac{\sigma}{S} = \frac{1}{2}\left[\left(\frac{\omega_{cb}}{\gamma_b\beta_b c}\right)^2 - \frac{2\omega_{pb}^2}{\gamma_b^3\beta_b^2 c^2}\right]^{1/2} , \qquad (10.67)$$

where $\omega_{cb} = Z_i eB_0/m_i c$ and $\omega_{pb}^2 = 4\pi\hat{n}_b Z_i^2 e^2/m_i$. Here, $\sqrt{\kappa_{z0}} = \omega_{cb}/2\gamma_b\beta_b c$, and use has been made of Eq.(10.23) to express K/r_b^2 in terms of the plasma frequency ω_{pb}. Note from Eq.(10.67) that the depression of the phase advance σ by space-charge effects is large ($\sigma \ll \sigma_0$) whenever the beam density approaches the limiting value $2\omega_{pb}^2/\gamma_b\omega_{cb}^2 \rightarrow 1$ [see Eq.(10.53)].

Problem 10.5   Linearize the envelope equation (10.52) for small-amplitude perturbations $\delta r_b(s)$ of the envelope about the equilibrium beam radius $r_b(s) =$ const. determined from Eq.(10.54) for $\kappa_z(s) = \kappa_{z0} =$ const.

a.   Show that the envelope perturbation $\delta r_b(s)$ satisfies Eq.(10.64).

b.   Eliminate ε^2/r_b^2 and K/ε in Eq.(10.64) by means of Eqs.(10.54) and (10.61). Show that Eq.(10.64) reduces to Eq.(10.65), where $(\sigma_0/S)^2 = \kappa_{z0}$, and the phase advance σ is determined from Eq.(10.61).

c.   Make use of Eq.(10.23) to show that the phase advance σ can be expressed in the form given in Eq.(10.67), which uses notation familiar in plasma physics.

10.6   Average Focusing Force and Phase Advance for a Periodic Quadrupole Field

We now return to the case of a periodic quadrupole focusing field discussed in Sec. 10.4. Of particular interest are estimates of the phase advance over one period S of the lattice,[13] and estimates of the average quadrupole focusing force.[2] These can be calculated numerically for an individual test ion by integration of the orbit equations (10.36) and (10.37) for $w_x(s)$ and $w_y(s)$, or for the beam envelope by integration of the envelope equations (10.40) and (10.41) for $a(s)$ and $b(s)$. For present purposes, use is made of the envelope equations (10.40) and (10.41) to obtain simple analytical estimates, approximately valid when the phase advances σ_x and σ_y are less than 90°.

10.6.1   Envelope Equations for Small-Amplitude Modulation

For the envelope ions, $w_x(s) = a(s)/\sqrt{\varepsilon_x}$ in Eq.(10.35), and the phase advance per period is defined by

$$\sigma_x = \varepsilon_x \int_{s_0}^{s_0+S} \frac{ds}{a^2(s)}, \quad \sigma_y = \varepsilon_y \int_{s_0}^{s_0+S} \frac{ds}{b^2(s)}, \tag{10.68}$$

in the x- and y-planes. Here, the emittances ε_x and ε_y are treated as constants. In the analysis of Eqs.(10.40) and (10.41), we express

$$a(s) = \bar{a}(s) + \delta a(s) ,$$
$$b(s) = \bar{b}(s) + \delta b(s) , \qquad (10.69)$$

where the average envelope dimensions $\bar{a}(s)$ and $\bar{b}(s)$ are assumed to vary slowly over one period S of the focusing lattice in Fig. 10.1(a), and the modulations $\delta a(s)$ and $\delta b(s)$ of the beam envelope are treated as small-amplitude perturbations with $|\delta a| \ll \bar{a}$ and $|\delta b| \ll \bar{b}$. It is also assumed that the envelope modulations are periodic, with the same periodicity length S as the focusing lattice. That is, $\delta a(s + S) = \delta a(s), \delta b(s + S) = \delta b(s)$, and

$$\langle \delta a \rangle = 0 = \langle \delta b \rangle , \qquad (10.70)$$

where $\langle \cdots \rangle$ denotes the spatial average

$$\langle \cdots \rangle = \frac{1}{S} \int_{s_0}^{s_0+S} ds \cdots \qquad (10.71)$$

over one period. Expanding the envelope equation (10.40) correct to first order in $\delta a(s)$ and $\delta b(s)$ gives

$$\frac{d^2}{ds^2}\bar{a} + \frac{d^2}{ds^2}\delta a + \kappa_q(s)\bar{a} + \kappa_q(s)\delta a$$
$$- \frac{2K}{\bar{a} + \bar{b}}\left(1 - \frac{\delta a + \delta b}{\bar{a} + \bar{b}}\right) - \frac{\varepsilon_x^2}{\bar{a}^3}\left(1 - \frac{3\delta a}{\bar{a}}\right) = 0. \qquad (10.72)$$

We take the average of Eq.(10.72) over one period S, treating $\bar{a}(s)$ and $\bar{b}(s)$ as constant on this length scale, and making use of $\langle \kappa_q(s) \rangle = 0$ and $\langle d^2\delta a/ds^2 \rangle = 0$. This readily gives[13]

$$\frac{d^2}{ds^2}\bar{a}(s) + \langle \kappa_q(s)\delta a(s) \rangle - \frac{2K}{\bar{a}(s) + \bar{b}(s)} - \frac{\varepsilon_x^2}{\bar{a}^3(s)} = 0, \qquad (10.73)$$

which describes the slow evolution of the average envelope $\bar{a}(s)$. An equation for $\bar{b}(s)$ can be derived from Eq.(10.41), which is identical in form to Eq.(10.73) provided we make the replacements $\bar{a}(s) \rightarrow \bar{b}(s)$ and $\langle \kappa_q(s)\delta a(s) \rangle \rightarrow -\langle \kappa_q(s)\delta b(s) \rangle$ in Eq.(10.73). Note from Eq.(10.73) that the average focusing force $\bar{F}_x$ on the beam envelope produced by the quadrupole field is given by

$$\bar{F}_x = -\langle \kappa_q(s)\delta a(s)\rangle , \qquad (10.74)$$

which involves the average of the product of the quadrupole coupling coefficient $\kappa_q(s)$ and the envelope modulation $\delta a(s)$. Moreover, for the average quadrupole force $\bar{F}_x$ to be restoring (i.e., "focusing"), $\langle \kappa_q(s)\delta a(s)\rangle > 0$ is required. It will be shown that this is indeed the case (Sec. 10.6.3) following an estimate of the average phase advance σ per period S of the quadrupole lattice (Sec. 10.6.2).

10.6.2   Phase Advance σ

To make quantitative estimates of the average quadrupole force and the phase advance σ_x and σ_y per period S, it is convenient to introduce the Courant-Snyder amplitude functions[2] $\beta_x(s) = a^2(s)/\varepsilon_x$ and $\beta_y(s) = b^2(s)/\varepsilon_y$ in the analysis of the envelope equations (10.40) and (10.41). In addition, for simplicity, it is assumed that the emittances and phase shifts are the same in the two transverse planes, i.e.,

$$\varepsilon_x = \varepsilon_y \equiv \varepsilon ,$$

$$\sigma_x = \sigma_y \equiv \sigma . \qquad (10.75)$$

For a matched beam, Eq.(10.75) is consistent with the assumption of an ion beam with circular cross section $(\bar{a} = \bar{b})$. Substituting $a(s) = \sqrt{\varepsilon\beta_x(s)}$ and $b(s) = \sqrt{\varepsilon\beta_y(s)}$ into Eq.(10.40) and carrying out the required derivative operations readily gives

$$\frac{1}{2}\beta_x\beta_x'' - \frac{1}{4}\left(\beta_x'\right)^2 + \kappa_q(s)\beta_x^2 - \frac{2K\beta_x}{\varepsilon\left[1 + (\beta_y/\beta_x)^{1/2}\right]} = 1, \qquad (10.76)$$

where "prime" denotes d/ds. Similarly, Eq.(10.41) gives an equation for $\beta_y(s)$ which is identical in form to Eq.(10.76) provided we make the replacements $\beta_x \to \beta_y, \kappa_q(s) \to -\kappa_q(s)$ and $\beta_y/\beta_x \to \beta_x/\beta_y$ in Eq.(10.76). Equation (10.76) provides a nonlinear normalization constraint on $\beta_x(s)$. Assuming that $\beta_x(s)$ and $\beta_y(s)$ have the same periodicity length as the focusing lattice, i.e., $\beta_x(s + S) = \beta_x(s)$, it then follows that $\langle \beta_x\beta_x''\rangle = -\langle(\beta_x')^2\rangle$ where spatial averages are defined according to

Eq.(10.71). Averaging Eq.(10.76) over one period S then gives the normalization condition

$$\left\langle -\frac{3}{4}\left(\beta_x'\right)^2 + \kappa_q(s)\beta_x^2 \right\rangle - \frac{2K}{\varepsilon}\left\langle \frac{\beta_x}{1+(\beta_y/\beta_x)^{1/2}} \right\rangle = 1. \qquad (10.77)$$

Similarly, averaging the equation for $\beta_y(s)$ [analogous to Eq.(10.76) for $\beta_x(s)$] gives

$$\left\langle -\frac{3}{4}\left(\beta_y'\right)^2 - \kappa_q(s)\beta_y^2 \right\rangle - \frac{2K}{\varepsilon}\left\langle \frac{\beta_y}{1+(\beta_x/\beta_y)^{1/2}} \right\rangle = 1. \qquad (10.78)$$

Equation (10.77) [or Eq.(10.78)] provides a powerful constraint condition which can be used to estimate the phase advance defined in Eq.(10.68). In this regard, it is convenient to introduce the functions $f_x(s)$ and $f_y(s)$ defined by

$$\beta_x(s) = \bar{\beta}[1 + f_x(s)] ,$$
$$\beta_y(s) = \bar{\beta}[1 + f_y(s)] , \qquad (10.79)$$

for a circular ion beam with $\bar{a} = \bar{b}$ and $\varepsilon_x = \varepsilon_y = \varepsilon$. Here, $\bar{\beta}$ is the average amplitude, and the oscillatory modulations $f_x(s)$ and $f_y(s)$ have periodicity length S, with zero average value $\langle f_x \rangle = 0 = \langle f_y \rangle$. Substituting $a^2 = \varepsilon\beta_x$ and $b^2 = \varepsilon\beta_y$ into Eq.(10.68), it follows that

$$\sigma = \frac{1}{\bar{\beta}}\int_{s_0}^{s_0+S} \frac{ds}{[1+f_x(s)]} = \frac{1}{\bar{\beta}}\int_{s_0}^{s_0+S} \frac{ds}{[1+f_y(s)]}, \qquad (10.80)$$

where $\sigma = \sigma_x = \sigma_y$ for an ion beam with circular cross section. For small-amplitude modulations of the beam envelope with $|f_x|, |f_y| \ll 1$, note from Eq.(10.80) and $[1+f_x]^{-1} = 1 - f_x + f_x^2 \cdots$ that $\sigma \simeq S/\bar{\beta}$ correct to $O(f_x)$. [Here, use has been made of $\langle f_x \rangle = 0$.] Therefore, a determination of $\bar{\beta}$ provides a lowest-order estimate of the phase advance σ.

Substituting $\beta_x = \bar{\beta}(1 + f_x)$ and $\beta_y = \bar{\beta}(1 + f_y)$ into Eq.(10.77) gives

$$\left\langle -\frac{3}{4}\left(f_x'\right)^2 + \kappa_q(s)\left(2f_x + f_x^2\right) \right\rangle \bar{\beta}^2 - \frac{2K}{\varepsilon}\left\langle \frac{(1+f_x)^{3/2}}{(1+f_x)^{1/2}+(1+f_y)^{1/2}} \right\rangle \bar{\beta} = 1 , \qquad (10.81)$$

where use has been made of $\langle\kappa_q(s)\rangle = 0$. A similar equation involving averages of $(f_y')^2$, etc., can be obtained from Eq.(10.78). Insofar as the averages in Eq.(10.81) are independent of $\bar{\beta}$ (or depend only weakly on $\bar{\beta}$), the normalization condition in Eq.(10.81) constitutes a quadratic equation for the average amplitude

$$\bar{\beta} = \frac{S}{\sigma} \langle (1 + f_x)^{-1} \rangle. \tag{10.82}$$

Eliminating $\bar{\beta}$ in favor of the phase advance σ by means of Eq.(10.82), it follows that Eq.(10.81) can be expressed in the equivalent form

$$\frac{\sigma^2}{S^2} + \frac{2K}{\varepsilon} \langle (1 + f_x)^{-1} \rangle \left\langle \frac{(1 + f_x)^{3/2}}{(1 + f_x)^{1/2} + (1 + f_y)^{1/2}} \right\rangle \frac{\sigma}{S}$$

$$= \langle (1 + f_x)^{-1} \rangle^2 \left\langle -\frac{3}{4} (f_x')^2 + \kappa_q(s)(2f_x + f_x^2) \right\rangle. \tag{10.83}$$

Equation (10.83) is a fully nonlinear result which can be used to determine the phase advance σ over a wide range of system parameters. To evaluate the average coefficients, Eq.(10.83) must of course be solved in conjunction with the dynamical equations for $f_x(s)$ and $f_y(s)$. For future reference, assuming small-amplitude modulations with $|f_x|$, $|f_y| \ll 1$, we expand $(1 + f_x)^{-1} = 1 - f_x + \cdots$, $(1 + f_x)^{3/2} = 1 + (3/2)f_x + \cdots$, etc. in Eq.(10.83) and make use of $\langle f_x \rangle = 0$. To leading order, Eq.(10.83) can be approximated by

$$\frac{\sigma^2}{S^2} + \frac{K}{\varepsilon} \frac{\sigma}{S} = \left\langle -\frac{3}{4} (f_x')^2 + 2\kappa_q(s)f_x \right\rangle, \tag{10.84}$$

where $\langle \kappa_q f_x^2 \rangle$ has been neglected in comparison with $2 \langle \kappa_q f_x \rangle$ for $|f_x| \ll 1$. Equation (10.84) will be used later in Sec. 10.6.2 to estimate the phase advance σ for a periodic quadrupole lattice. Equation (10.84) should be compared with the analogous result in Eq.(10.61), derived for a uniform solenoidal focusing field $B_0 \hat{e}_z$.[13] Note that the right-hand side of Eq.(10.84) plays the same role as the solenoidal focusing parameter $\kappa_{z0} = (\omega_{cb}/2\gamma_b \beta_b c)^2$ occurring in Eq.(10.61).

To determine the averages in Eq.(10.83) [or in Eq.(10.84)], we now return to the dynamical equation (10.76) for $\beta_x(s) = \bar{\beta}[1 + f_x(s)]$. Taking the derivative of Eq.(10.76) with respect to s gives

$$\beta_x''' + 4\kappa_q(s)\beta_x' + 2\kappa_q'(s)\beta_x - \frac{4K}{\varepsilon \beta_x} \left[\frac{\beta_x^{3/2}}{\beta_x^{1/2} + \beta_y^{1/2}} \right]' = 0, \tag{10.85}$$

or equivalently,

$$f_x''' + 4\kappa_q(s)f_x' + 2\kappa_q'(s)(1 + f_x)$$
$$- \frac{4K}{\varepsilon\bar{\beta}(1 + f_x)} \left[\frac{(1 + f_x)^{3/2}}{(1 + f_x)^{1/2} + (1 + f_y)^{1/2}} \right]' = 0 . \qquad (10.86)$$

No small-amplitude approximation has yet been made in Eqs.(10.85) or (10.86). Moreover, the equations for $\beta_y(s)$ and $f_y(s)$ are similar in form, provided we make the replacements $\beta_x \rightarrow \beta_y, f_x \rightarrow f_y$ and $\kappa_q(s) \rightarrow -\kappa_q(s)$ in Eqs.(10.85) and (10.86). In the limit of zero beam current ($K \rightarrow 0$), a striking feature of Eq.(10.85) is that Eq.(10.85) is a *linear* differential equation for the amplitude function $\beta_x(s)$ (Problem 10.6).[2] [Here, $\kappa_q(s)$ is a (prescribed) periodic function of s with periodicity length S.]

Equations (10.83) and (10.86) can be solved numerically to determine the phase advance σ in circumstances where the modulation of the beam envelope is relatively strong. For present purposes, however, we examine Eq.(10.86) for small-amplitude modulations with $|f_x|, |f_y| \ll 1$. Use will then be made of Eq.(10.84) to obtain a leading-order estimate of σ. Expanding the self-field term in Eq.(10.86) for small f_x and f_y gives

$$f_x''' + 4\kappa_q(s)f_x' + 2\kappa_q'(s)(1 + f_x) - \frac{K}{\varepsilon\bar{\beta}} \left(\frac{5}{2}f_x' - \frac{1}{2}f_y' \right) = 0, \qquad (10.87)$$

where terms quadratic and higher order have been neglected. In analyzing Eq.(10.87), we treat $\kappa_q(s)$ and $K/\varepsilon\bar{\beta}$ as "small" quantities which are comparable in magnitude with the amplitude modulation f_x. Therefore, to leading order, it follows from Eq.(10.87) that

$$f_x''' = -2\kappa_q'(s) . \qquad (10.88)$$

Here, $\langle \kappa_q(s) \rangle = 0, \langle d\kappa_q/ds \rangle = 0$, and $\langle d^n f_x/ds^n \rangle = 0$ for $n = 0, 1, 2 \cdots$. For prescribed quadrupole lattice function $\kappa_q(s)$, Eq.(10.88) can be integrated directly to determine $f_x(s)$. Making use of $f_x'' = -2\kappa_q(s)$, it follows that

$$\langle \kappa_q(s)f_x \rangle = -\frac{1}{2}\langle f_x'' f_x \rangle = \frac{1}{2}\langle (f_x')^2 \rangle, \qquad (10.89)$$

where the periodic properties of f_x have been utilized. Substituting Eq.(10.89) into Eq.(10.84) then gives

$$\frac{\sigma^2}{S^2} + \frac{K}{\epsilon}\frac{\sigma}{S} = \frac{1}{4}\langle(f_x')^2\rangle. \tag{10.90}$$

Furthermore, from $f_x' = -2\int^s ds\,\kappa_q(s)$, it follows that Eq.(10.90) can be expressed as

$$\frac{\sigma^2}{S^2} + \frac{K}{\epsilon}\frac{\sigma}{S} = \frac{\sigma_0^2}{S^2}, \tag{10.91}$$

where σ_0 is defined by

$$\frac{\sigma_0^2}{S^2} = \left\langle\left(\int^s ds\,\kappa_q(s)\right)^2\right\rangle. \tag{10.92}$$

In Eqs.(10.91) and (10.92), the quantity σ_0 is the phase advance per period induced by the (periodic) quadrupole field which would pertain in the absence of self-field effects ($K \to 0$). Note that Eq.(10.91) is similar in form to Eq.(10.61), which was derived in Sec. 10.5.4 for a uniform solenoidal field $B_0\hat{e}_z$, provided we make the replacement $\langle(\int^s ds\,\kappa_q(s))^2\rangle \to \kappa_{z0}$ in Eq.(10.91). Similar to Eq.(10.61), the self-field contribution in Eq.(10.91) depresses the value of σ relative to σ_0.

An important feature of the analysis leading to Eq.(10.91) is the fact that the nonlinear constraint condition (10.77) [and its derivative versions in Eqs.(10.81), (10.83), and (10.84)] naturally includes self-field effects, whereas the determination of f_x to lowest order from Eq.(10.88) depends only on the quadrupole coupling coefficient $\kappa_q(s)$. Although validity of Eq.(10.91) is restricted to moderately small values of the phase advance σ (less than 90°, say), this suggests that a numerical analysis of the more precise equations (10.83) and (10.86) is likely to converge rapidly.

Problem 10.6   Consider Eq.(10.86) for a low-current ion beam with $K \to 0$. Dropping the subscript x on $f_x(s)$, Eq.(10.86) can be expressed as

$$f''' + 4\kappa_q(s)f' + 2\kappa_q'(s)(1+f) = 0, \tag{10.6.1}$$

which is a linear differential equation for $f(s)$. Following Courant and Snyder [Annals of Physics **3**, 1 (1958)], Eq.(10.6.1) can be solved iteratively, treating the quadrupole coupling coefficient $\kappa_q(s)$ as a known, "small" quantity.[2]

a. Express $\kappa_q(s)$ and $f(s)$ as

$$\kappa_q(s) = \varepsilon K_q(s) \,, \tag{10.6.2}$$

$$f(s) = \varepsilon f_1(s) + \varepsilon^2 f_2(s) + \cdots , \tag{10.6.3}$$

where $\varepsilon \ll 1$ is a small parameter (not to be confused with the emittance), and $K_q(s)$, $f_1(s)$, $f_2(s)$, $\cdots$ are functions of order unity. Substitute Eqs.(10.6.2) and (10.6.3) into Eq.(10.6.1) and equate the coefficients of successive powers of ε equal to zero. Show that

$$f_1''' = -2K_q' \,, \tag{10.6.4}$$

and derive the recursion relation

$$f_n''' = -4K_q f_{n-1}' - 2K_q' f_{n-1} \tag{10.6.5}$$

for $n \geq 2$. For specified $K_q(s)$, Eq.(10.6.4) can be integrated to determine $f_1(s)$, and then $f_2(s)$, $f_3(s)$, $\cdots$ are calculated from Eq.(10.6.5).

b. Make use of $f_1'' = -2K_q$ to eliminate K_q and K_q' in Eq.(10.6.5) for $n = 2$. Integrate Eq.(10.6.5) once with respect to s and show that

$$f_2'' = \frac{1}{2}(f_1')^2 + f_1'' f_1 . \tag{10.6.6}$$

c. For $K \to 0$, expand the right-hand side of Eq.(10.83) correct to order ε^3 and show that

$$\frac{\sigma^2}{S^2} = \left\langle \varepsilon^2 \left(-\frac{3}{4} \left(f_1' \right)^2 + 2K_q f_1 \right) \right. $$
$$\left. + \varepsilon^3 \left(-\frac{3}{2} f_1' f_2' + 2K_q f_2 + K_q f_1^2 \right) + \cdots \right\rangle , \tag{10.6.7}$$

where the spatial average $\langle \cdots \rangle$ over one period S is defined in Eq.(10.71).

d. Make use of Eq.(10.6.4) to express $K_q = (-1/2)f_1''$ in Eq.(10.6.7). Because $f_n(s)$ has periodicity length S, it follows upon integrating by parts that $\langle f_1'' f_1 \rangle = -\langle (f_1')^2 \rangle$, $\langle f_1' f_2' \rangle = -\langle f_1 f_2'' \rangle$, $\langle f_1'' f_2 \rangle = \langle f_1 f_2'' \rangle$, etc. Show that Eq.(10.6.7) can be expressed in the equivalent form

$$\frac{\sigma^2}{S^2} = \left\langle \frac{1}{4}\varepsilon^2 \left(f_1' \right)^2 + \frac{1}{2}\varepsilon^3 \left(f_1 f_2'' - f_1^2 f_1'' \right) + \cdots \right\rangle . \tag{10.6.8}$$

Eliminating f_2'' in Eq.(10.6.8) by means of Eq.(10.6.6) then gives for the phase advance σ

$$\frac{\sigma^2}{S^2} = \frac{1}{4}\varepsilon^2 \langle (f_1')^2 \rangle + \frac{1}{4}\varepsilon^3 \langle f_1 (f_1')^2 \rangle + \cdots , \tag{10.6.9}$$

where $f_1(s)$ is determined in terms of the quadrupole coupling coefficient $K_q(s)$ from $f_1'' = -2K_q(s)$.

Note that Eq.(10.6.9) determines σ^2/S^2 correct to order ε^3, which is one higher order than Eq.(10.90). Moreover, the iterative procedure used in Problem 10.6 can be extended to even higher order.

10.6.3   Average Quadrupole Focusing Force

The procedure leading to Eq.(10.91) can also be used to estimate the average quadrupole focusing force $\bar{F}_x = -\langle \kappa_q(s)\,\delta a(s) \rangle$ in the envelope equation (10.73) for $\bar{a}(s)$. From $(\bar{a} + \delta a)^2 = \varepsilon \beta_x = \varepsilon \bar{\beta}(1 + f_x)$, we make the identifications $\bar{a}^2 = \bar{\beta}\varepsilon$ and $\delta a = (1/2)f_x \bar{a}$ for small-amplitude modulations with $|\delta a| \ll \bar{a}$ and $|f_x| \ll 1$. The average quadrupole focusing force $\bar{F}_x = -\langle \kappa_q(s)\,\delta a(s) \rangle$ is then given by

$$\bar{F}_x = -\frac{1}{2}\,\langle \kappa_q(s) f_x(s) \rangle\,\bar{a}(s), \qquad (10.93)$$

where the variation in $\bar{a}(s)$ over one period S of the lattice is assumed to be slow. Making use of $f_x'' = -2\kappa_q(s)$ [see Eq.(10.88)] gives $\bar{F}_x = -(1/4)\langle (f_x')^2 \rangle \bar{a}$, or equivalently,

$$\bar{F}_x = -\left\langle \left(\int^s ds\, \kappa_q(s) \right)^2 \right\rangle \bar{a} = -\frac{\sigma_0^2}{S^2}\,\bar{a}. \qquad (10.94)$$

Note that the average quadrupole force in Eq.(10.94) is indeed restoring (i.e., "focusing"). Substituting Eq.(10.94) into Eq.(10.73) for a circular ion beam with $\bar{a} = \bar{b}$ and $\varepsilon_x = \varepsilon_y = \varepsilon$, we obtain[13]

$$\frac{d^2}{ds^2}\bar{a}(s) + \frac{\sigma_0^2}{S^2}\bar{a}(s) - \frac{K}{\bar{a}(s)} - \frac{\varepsilon^2}{\bar{a}^3(s)} = 0, \qquad (10.95)$$

which can be used to determine the (slow) evolution of the average beam envelope $\bar{a}(s)$ on a length scale long in comparison with the lattice period S. For completeness, it should be noted that an identical equation for $\bar{b}(s)[= \bar{a}(s)]$ can be derived from the analogous equations for $f_y(s), b(s) = \bar{b}(s) + \delta b(s)$, etc.

10.6.4   Beam Radius and Current for a Matched Beam

Equation (10.95) is identical in form to the envelope equation (10.52) analyzed in Sec. 10.5 for the case of a uniform solenoidal focusing field, provided κ_{z0} is replaced by the average quadrupole focusing coefficient $\sigma_0^2/S^2 = $ const. defined in Eq.(10.92). Therefore, most of the results derived from Eq.(10.52) can be applied virtually intact to Eq.(10.95). For example, for a matched beam with $d^2\bar{a}/ds^2 = 0$, the equilibrium beam radius $\bar{a}$ determined from Eq.(10.95) satisfies $(\sigma_0^2/S^2)\bar{a} = K/\bar{a} + \varepsilon^2/\bar{a}^3$. Solving for $\bar{a}$ gives the beam radius

$$\bar{a} = \bar{a}_0 \left\{ \left[1 + \left(\frac{K\bar{a}_0^2}{2\varepsilon^2} \right)^2 \right]^{1/2} + \frac{K\bar{a}_0^2}{2\varepsilon^2} \right\}^{1/2} , \qquad (10.96)$$

which should be compared with Eq.(10.56). Here, $\bar{a}_0 \equiv (\varepsilon S/\sigma_0)^{1/2}$ is the equilibrium beam radius in the limit of zero beam current $(K \to 0)$. Similarly, if the envelope equation (10.95) is analyzed for small-amplitude perturbations $\delta\bar{a}$ about the equilibrium radius $\bar{a}$ in Eq.(10.96), then the wavenumber k of the perturbation is given self-consistently by

$$k^2 = 2\frac{\sigma_0^2}{S^2} + 2\frac{\sigma^2}{S^2}, \qquad (10.97)$$

where

$$\frac{\sigma}{S} = \left[\left(\frac{\sigma_0}{S} \right)^2 + \left(\frac{K}{2\varepsilon} \right)^2 \right]^{1/2} - \frac{K}{2\varepsilon} \qquad (10.98)$$

is the phase advance calculated from the quadratic equation (10.91) with $(\sigma_0/S)^2 = \langle (\int^s ds\, \kappa_q(s))^2 \rangle$. As expected, Eqs.(10.97) and (10.98) are analogous to Eqs.(10.62) and (10.66), derived for a solenoidal focusing field. Validity of Eq.(10.97) for a quadrupole focusing field requires small values of $k^2 S^2 = 2(\sigma_0^2 + \sigma^2)$ in comparison with $(2\pi)^2$.

Finally, making use of Eq.(10.23), the ion beam current I_b consistent with the equilibrium force condition $K/\bar{a} = (\sigma_0/S)^2\bar{a} - \varepsilon^2/\bar{a}^3$ can be expressed as

$$I_b = \frac{1}{2}\beta_b^3\gamma_b^3\frac{m_i c^3}{Z_i e} \left(\frac{\sigma_0\bar{a}}{S} \right)^2 \left[1 - \left(\frac{\varepsilon}{\alpha} \right)^2 \right]. \qquad (10.99)$$

Here, $\alpha = \bar{a}^2 \sigma_0 / S$ is the acceptance of the quadrupole channel,[13] and $(\varepsilon/\alpha)^2 = (\bar{a}_0/\bar{a})^4$ where $\bar{a}_0 = (\varepsilon S/\sigma_0)^{1/2}$ is the beam radius in the limit of negligibly small self fields $(K \to 0)$. In Eq.(10.99), $m_i c^3 / Z_i e = 3.29 \times 10^7 A/Z_i$ amperes, where A is the ion mass number and Z_i is the charge state. Evidently, $(\varepsilon/\alpha)^2 \ll 1$ is required for propagation at moderately high ion current. Even for $\sigma_0 \lesssim \pi/2$ (say), the factor $(\bar{a}/S)^2$ in Eq.(10.99) is small $(\lesssim 10^{-6})$ in practical applications,[7-9] and the ion beam current in Eq.(10.99) is limited to the range of amperes.

10.7 Hamiltonian Formulation and Relationship to the Kapchinskij-Vladimirskij Distribution Function

In this section, we make use of a Hamiltonian formalism[22] to examine the transverse orbit equations (10.26) and (10.27) for $x(s)$ and $y(s)$ for the case of a uniform-density ion beam with elliptical cross section propagating through a periodic quadrupole field with $B_z(s) = 0$. No *a priori* restriction is made to weak quadrupole focusing or to small phase advance σ per period S of the quadrupole lattice. Setting $\kappa_z(s) = 0$ in Eqs.(10.26) and (10.27) gives

$$\frac{d^2}{ds^2} x(s) + \left\{ \kappa_q(s) - \frac{2K}{a(s)[a(s) + b(s)]} \right\} x(s) = 0, \qquad (10.100)$$

$$\frac{d^2}{ds^2} y(s) + \left\{ -\kappa_q(s) - \frac{2K}{b(s)[a(s) + b(s)]} \right\} y(s) = 0, \qquad (10.101)$$

where $\kappa_q(s)$ is the quadrupole coupling coefficient defined in Eq.(10.24) [see also Fig. 10.1(a)], and K is the self-field perveance defined in Eq.(10.23). In this regard, it is important to keep in mind that the derivation of the self-field contributions in Eqs.(10.100) and (10.101) [the terms proportional to $Kx(s)$ and $Ky(s)$, respectively] assumed an ion beam with *uniform* density and elliptical cross section (Sec. 10.3). Therefore, for the s-dependent focusing manifest in Eqs.(10.100) and (10.101), the important question arises as to whether or not there is any choice of equilibrium distribution function $f_b^0(x, y, x', y')$ with $\partial/\partial s = 0$ which generates self-consistently an ion beam equilibrium with uniform density satisfying $n_b^0(x, y) = \hat{n}_b$ for $x^2/a^2 + y^2/b^2 < 1$, an assumption implicit in

the analysis in Secs. 10.3–10.6. The answer is "yes," and the subsequent
analysis will show that the distribution function f_b^0 corresponds to the
Kapchinskij-Vladimirskij equilibrium specified by Eq.(10.113).[10]

As a notational point, throughout the remainder of Chapter 10,
the "primed" notation (x', y') refers to the phase-space variables
$(dx/ds, dy/ds)$, which correspond to the (normalized) components of the
transverse momentum.

10.7.1 Hamiltonian Formulation

It is instructive to address this question by considering the Hamiltonian
$H_\perp$ for the transverse motion described by Eqs.(10.100) and (10.101).
Here, in units scaled by a constant factor $\gamma_b m_i \beta_b^2 c^2$, the Hamiltonian $H_\perp$
is defined by

$$H_\perp(x, y, x', y', s) = \frac{1}{2}\left[x'^2 + \kappa_x(s)x^2\right] + \frac{1}{2}\left[y'^2 + \kappa_y(s)y^2\right] , \quad (10.102)$$

where (x, y, x', y') are the phase-space variables and $\kappa_x(s)$ and $\kappa_y(s)$ are
the coupling coefficients defined by

$$\kappa_x(s) = \kappa_q(s) - \frac{2K}{a(s)[a(s) + b(s)]},$$

$$\kappa_y(s) = -\kappa_q(s) - \frac{2K}{b(s)[a(s) + b(s)]}. \quad (10.103)$$

The equations of motion obtained from Eq.(10.102) are $d^2x/ds^2 =$
$-\partial H_\perp/\partial x = -\kappa_x(s)x$, and $d^2y/ds^2 = -\partial H_\perp/\partial y = -\kappa_y(s)y$, which are
identical to Eqs.(10.100) and (10.101), as expected. Because $\kappa_x(s)$ and
$\kappa_y(s)$ depend on s, and $\kappa_q(s)$ alternates sign, note from Eq. (10.102) that
the Hamiltonian $H_\perp$ is neither a constant of the motion nor a sum of
positive-definite terms.

The analysis of the orbit equations in Sec. 10.4 showed that $A_x^2 =$
$x^2/w_x^2 + (w_x x' - x\, dw_x/ds)^2$ and $A_y^2 = y^2/w_y^2 + (w_y y' - y\, dw_y/ds)^2$ are
independently constants of the motion [see Eq.(10.38)], where the amplitude functions $w_x(s)$ and $w_y(s)$ solve the nonlinear differential equations (10.36) and (10.37), respectively. This suggests that the natural
phase-space variables in the present problem are likely to be $x/w_x(s)$,

$w_x(s)x' - x\,dw_x(s)/ds$, etc. With this in mind, following Hofmann, et al., we introduce the generating function $F_2(x, y, X', Y')$ defined by[22,31]

$$F_2(x, y, X', Y') = \frac{x}{w_x(s)} \left[X' + \frac{1}{2}x\frac{dw_x(s)}{ds} \right] + \frac{y}{w_y(s)} \left[Y' + \frac{1}{2}y\frac{dw_y(s)}{ds} \right],$$
(10.104)

where $w_x(s)$ and $w_y(s)$ solve Eqs.(10.36) and (10.37), respectively. In terms of the new phase-space variables (X, Y, X', Y'), if follows that

$$X = \frac{\partial F_2}{\partial X'} = \frac{x}{w_x},$$

$$Y = \frac{\partial F_2}{\partial Y'} = \frac{y}{w_y},$$

$$x' = \frac{\partial F_2}{\partial x} = \frac{1}{w_x}\left(X' + x\frac{dw_x}{ds} \right),$$

$$y' = \frac{\partial F_2}{\partial y} = \frac{1}{w_y}\left(Y' + y\frac{dw_y}{ds} \right).$$
(10.105)

Solving for X' and Y' gives

$$X' = w_x x' - x\frac{dw_x}{ds},$$

$$Y' = w_y y' - y\frac{dw_y}{ds}.$$
(10.106)

Moreover, the new Hamiltonian function is given by $H'_\perp = H_\perp + \partial F_2/\partial s$, where $H_\perp$ is defined by Eq.(10.102) and $\partial F_2/\partial s$ is calculated from Eq.(10.104). Making use of Eqs.(10.105) and (10.106) to express x, y, x' and y' in terms of X, Y, X' and Y', and Eqs.(10.36) and (10.37) to eliminate $d^2 w_x/ds^2$ and $d^2 w_y/ds^2$ in the evaluation of $\partial F_2/\partial s$, some straightforward algebra shows that the new Hamiltonian function $H'_\perp$ can be expressed in the simple form (Problem 10.7)[22]

$$H'_\perp(X, Y, X', Y', s) = \frac{1}{2w_x^2(s)}\left(X'^2 + X^2 \right) + \frac{1}{2w_y^2(s)}\left(Y'^2 + Y^2 \right).$$
(10.107)

Although $H'_\perp$ is not a constant of the motion (nor is it expected to be), an important feature of Eq.(10.107) is that the new Hamiltonian is a sum of positive-definite quadratic terms.

In terms of the new variables, the equations of motion for $X(s)$, $X'(s)$, $Y(s)$ and $Y'(s)$ calculated from the Hamiltonian $H'_\perp$ in Eq.(10.107) are given by $dX'/ds = -\partial H'_\perp/\partial X$ and $dX/ds = \partial H'_\perp/\partial X'$, with similar equations for the Y-motion. This readily gives

$$\frac{d}{ds}X'(s) = -\frac{1}{w_x^2(s)}X(s) , \qquad \frac{d}{ds}X(s) = \frac{1}{w_x^2(s)}X'(s) , \qquad (10.108)$$

and

$$\frac{d}{ds}Y'(s) = -\frac{1}{w_y^2(s)}Y(s) , \qquad \frac{d}{ds}Y(s) = \frac{1}{w_y^2(s)}Y'(s) , \qquad (10.109)$$

where the amplitude functions $w_x(s)$ and $w_y(s)$ solve the nonlinear differential equations (10.36) and (10.37). In solving Eqs.(10.108) and (10.109), analogous to the method of characteristics discussed in Sec. 2.2, we denote the effective length variable in the axial direction by s' [i.e., s is replaced by s' in the orbit equations (10.108) and (10.109)]. The solutions to Eq.(10.108) for $X(s')$ and $X'(s')$ can be expressed as the pseudoharmonic oscillations[22]

$$X(s') = X \cos[\psi'_x(s',s)] + X' \sin[\psi'_x(s',s)] ,$$
$$X'(s') = X' \cos[\psi'_x(s',s)] - X \sin[\psi'_x(s',s)] , \qquad (10.110)$$

where

$$\psi'_x(s',s) = \int_s^{s'} \frac{ds}{w_x^2(s)}. \qquad (10.111)$$

Here, $X(s')$ and $X'(s')$ are the particle orbits which pass through the phase-space point (X,X') at $s' = s$, i.e., $X'(s' = s) \equiv X'$ and $X(s' = s) \equiv X$. The solutions to Eq.(10.109) for $Y(s')$ and $Y'(s')$ are identical in form to Eq.(10.110) provided we make the replacements $(X,X') \rightarrow (Y,Y')$ and $\psi'_x \rightarrow \psi'_y(s',s) = \int_s^{s'} ds\, w_y^{-2}(s)$ in Eqs.(10.110) and (10.111). One very appealing feature of the new phase space variables (X,Y,X',Y') is that there are two natural constants of the motion, i.e.,

$$[X'(s')]^2 + [X(s')]^2 = X'^2 + X^2 = \text{const.},$$
$$[Y'(s')]^2 + [Y(s')]^2 = Y'^2 + Y^2 = \text{const.}, \qquad (10.112)$$

which are analogous to the constants of the motion derived in Sec. 10.4 [see Eq.(10.38)]. In particular, the first constant of the motion in Eq.(10.112) follows directly by inspection of Eq.(10.110), and similarly for the constant of the motion in the Y-direction.

Problem 10.7   Consider the new Hamiltonian function $H'_\perp = H_\perp + \partial F_2/\partial s$, where $H_\perp$ is defined in Eq.(10.102), and the generating function F_2 is defined in Eq.(10.104).

a. Make use of Eqs.(10.36) and (10.37) for the evolution of the amplitude functions $w_x(s)$ and $w_y(s)$, together with the definitions in Eqs.(10.104)–(10.106) to show that

$$\frac{\partial F_2}{\partial s} = -\frac{x}{w_x^2}\frac{dw_x}{ds}\left(X' + \frac{1}{2}x\frac{dw_x}{ds}\right) + \frac{x^2}{2w_x^4} - \frac{1}{2}\kappa_x(s)x^2$$

$$-\frac{y}{w_y^2}\frac{dw_y}{ds}\left(Y' + \frac{1}{2}y\frac{dw_y}{ds}\right) + \frac{y^2}{2w_y^4} - \frac{1}{2}\kappa_y(s)y^2 , \quad (10.7.1)$$

where $X = x/w_x$ and $Y = y/w_y$.

b. Make use of Eqs.(10.102), (10.104), (10.105), and (10.7.1) to show that the new Hamiltonian $H'_\perp = H_\perp + \partial F_2/\partial s$ reduces exactly to the result given in Eq.(10.107).

10.7.2 Kapchinskij-Vladimirskij Distribution Function

Consistent with the assumption that the self-field forces in the orbit equations (10.100) and (10.101) are linearly proportional to $Kx(s)$ and $Ky(s)$, respectively, the analysis leading to Eqs.(10.108)–(10.112) implicitly assumes a uniform beam density with $n_b^0(x,y) = \hat{n}_b = $ const. within the elliptical cross section $x^2/a^2 + y^2/b^2 < 1$. Such a self-consistent Vlasov equilibrium is readily constructed from the constants of the motion defined in Eq.(10.112). Making use of the transformation relating the phase-space variables (X, Y, X', Y') to (x, y, x', y') in Eqs.(10.103) and (10.104), it is readily shown that the volume elements transform according to $dx'dy' = (w_x w_y)^{-1}dX'dY', dxdy = (w_x w_y)dXdY$, and $dxdydx'dy' = dXdYdX'dY'$. Analogous to the uniform-density Vlasov equilibria constructed in Secs. 4.2, 4.7, and 9.4, we choose the equilibrium distribution function $(\partial/\partial s = 0)$ to be[4,22]

$$f_b^0(X, Y, X', Y') = \frac{\hat{n}_b}{\pi(\varepsilon_x \varepsilon_y)^{1/2}} \delta \left[\frac{1}{\varepsilon_x}(X'^2 + X^2) + \frac{1}{\varepsilon_y}(Y'^2 + Y^2) - 1 \right],$$

(10.113)

which corresponds to a hyperellipsoidal shell in the phase space (X, Y, X', Y'). Here, the constants ε_x and ε_y are the transverse beam emittances introduced in Sec. 10.4, and $\hat{n}_b = N_b/\pi ab =$ const., where N_b is the number of ions per unit axial length of the beam. Note that the argument of the delta-function in Eq.(10.113) is a linear combination of the exact constants of the motion defined in Eq.(10.112). Equation (10.113) is referred to as the Kapchinskij-Vladimirskij distribution function.[10] The equilibrium density profile $n_b^0(X, Y)$ is calculated from

$$n_b^0(X, Y) = \int\int dX' dY' \, f_b^0(X, Y, X', Y'),$$

(10.114)

where $w_x = a/\sqrt{\varepsilon_x}$ and $w_y = b/\sqrt{\varepsilon_y}$ for a matched elliptical beam [see the discussion following Eq.(10.38) in Sec. 10.4]. Introducing the transverse kinetic energy variable $U_\perp^2 = X'^2/\varepsilon_x + Y'^2/\varepsilon_y$, and transforming $dX' dY' = (\varepsilon_x \varepsilon_y)^{1/2} d\phi \, U_\perp dU_\perp$, we substitute Eq.(10.113) into Eq.(10.114). This yields

$$n_b^0(X, Y) = \hat{n}_b \int_0^\infty dU_\perp^2 \, \delta \left[U_\perp^2 - \left(1 - \frac{1}{\varepsilon_x} X^2 - \frac{1}{\varepsilon_y} Y^2 \right) \right],$$

(10.115)

which can be integrated to give

$$n_b^0(X, Y) = \begin{cases} \hat{n}_b = \text{const.,} & X^2/\varepsilon_x + Y^2/\varepsilon_y < 1, \\ 0, & X^2/\varepsilon_x + Y^2/\varepsilon_y > 1. \end{cases}$$

(10.116)

Making the identifications $X = x/w_x = (x/a)\sqrt{\varepsilon_x}$ and $Y = y/w_y = (y/b)\sqrt{\varepsilon_y}$, the density profile in the configuration space (x, y) can be expressed as

$$n_b^0(x, y) = \begin{cases} \hat{n}_b = \text{const.,} & x^2/a^2 + y^2/b^2 < 1, \\ 0, & x^2/a^2 + y^2/b^2 > 1, \end{cases}$$

(10.117)

which is the expected result.

Several important conclusions pertain to the preceeding analysis:

(a) Exact analytical solutions for the orbits $X'(s')$, $X(s)$, etc., are not generally tractable for particle motion in an alternating-gradient quadrupole field. For example, a determination of

$$\psi_x'(s',s) = \int_s^{s'} \frac{ds}{w_x^2(s)}$$

in Eq.(10.110) requires integration of the nonlinear differential equation (10.36) for $w_x(s)$. Nonetheless, the quantities $X'^2 + X^2$ and $Y'^2 + Y^2$ defined in Eq.(10.112) are *exact* constants of the motion for the case of a uniform-density ion beam with elliptical cross section.

(b) The Kapchinskij-Vladimirskij distribution function defined in Eq.(10.113) self-consistently generates the uniform density profile in Eq.(10.117). Indeed, Eq.(10.113) is the *only* equilibrium distribution function which recovers a uniform-density beam equilibrium with elliptical cross section. [A density inversion theorem can be demonstrated analogous to that proved in Sec. 4.2.2 for a uniform solenoidal field $B_0\hat{e}_z$.]

(c) As indicated earlier, the analysis leading to Eqs.(10.108)–(10.112), and therefore the very existence of the constants of the motion defined in Eq.(10.112), assume a uniform-density ion beam. At this writing, there is no known generalization of the constants of the motion in Eq.(10.112) to circumstances where the beam density profile is nonuniform, and the self-field forces are no longer linearly proportional to $Kx(s)$ and $Ky(s)$, as in Eqs.(10.100) and (10.101), respectively. Stated another way, for alternating-gradient focusing systems, including beam self-field effects, there is no known exact kinetic equilibrium other than the Kapchinskij-Vladimirskij distribution function in Eq.(10.113), which corresponds to a uniform-density ion beam. This is in contrast to the case of a uniform solenoidal field $B_0\hat{e}_z$, where a wide variety of self-consistent Vlasov equilibria can be constructed (see Chapter 4).

(d) The one exception to the conclusion in (c) above occurs in the limit of low beam density and current ($K \rightarrow 0$), where self-field effects are neglected in the original equations of motion, Eqs.(10.16) and (10.17). In this case, Eqs.(10.100) and (10.101) are replaced by

$$\frac{d^2}{ds^2}x(s) + \kappa_q(s)x(s) = 0,$$

$$\frac{d^2}{ds^2}y(s) - \kappa_q(s)y(s) = 0, \qquad (10.118)$$

and the equations for the envelope functions $w_x(s)$ and $w_y(s)$ become

$$\frac{d^2}{ds^2}w_x(s) + \kappa_q(s)w_x(s) = \frac{1}{w_x^3(s)},$$

$$\frac{d^2}{ds^2}w_y(s) - \kappa_q(s)w_y(s) = \frac{1}{w_y^3(s)}. \qquad (10.119)$$

Note that self-field effects are neglected in Eqs.(10.118) and (10.119). Within the context of Eqs.(10.118) and (10.119), the analysis leading to Eqs.(10.108)–(10.111) is the same as before, and there are two independent constants of the motion given by Eq.(10.112). Therefore, in this regime, there is considerable latitude in constructing (low-density) self-consistent Vlasov equilibria from the constants of the motion $X'^2 + X^2$ and $Y'^2 + Y^2$. As an example, the choice of $f_b^0(X, Y, X', Y')$ which generates a Gaussian density profile is considered in Problem 10.8.

Problem 10.8   Consider the equilibrium distribution function specified by

$$f_b^0(X, Y, X', Y') = \frac{\hat{n}_b}{\pi(\epsilon_x \epsilon_y)^{1/2}} \exp\left[-\frac{1}{\epsilon_x}\left(X'^2 + X^2 \right) - \frac{1}{\epsilon_y}\left(Y'^2 + Y^2 \right) \right]$$
$$(10.8.1)$$

for a tenuous ion beam with negligibly small self fields. Make use of Eqs.(10.114) and (10.8.1) to show that the corresponding density profile is given by

$$n_b^0(x, y) = \hat{n}_b \exp\left(-\frac{x^2}{a^2} - \frac{y^2}{b^2} \right) \qquad (10.8.2)$$

in the configuration-space variables (x, y). Note from Eq.(10.8.2) the beam density is constant on elliptical surfaces with $x^2/a^2 + y^2/b^2 = $ const.

10.7.3   Definition of Statistical Averages and RMS Emittance

Returning to the equilibrium distribution function f_b^0 in Eq.(10.113) for a uniform-density ion beam including self-field effects, the average of the function $\chi(X, Y, X', Y')$ over a subspace Γ of the phase space (X, Y, X', Y') is defined by

$$\langle \chi \rangle_\Gamma = \frac{\int d\Gamma\, \chi f_b^0}{\int d\Gamma\, f_b^0}. \tag{10.120}$$

For example, $\langle \chi \rangle_{X', Y'} = \left(\iint dX'\, dY'\, \chi f_b^0 \right) / \left(\iint dX'\, dY'\, f_b^0 \right)$, and it readily follows from Eq.(10.113) that

$$\left\langle \frac{1}{\varepsilon_x} X'^2 \right\rangle_{X', Y'} = \left\langle \frac{1}{\varepsilon_y} Y'^2 \right\rangle_{X', Y'} = \frac{1}{2} \left\langle \frac{1}{\varepsilon_x} X'^2 + \frac{1}{\varepsilon_y} Y'^2 \right\rangle_{X', Y'}$$

$$= \frac{1}{2} \left(1 - \frac{x^2}{a^2} - \frac{y^2}{b^2} \right) \tag{10.121}$$

for $x^2/a^2 + y^2/b^2 < 1$. Therefore, for $\varepsilon_x = \varepsilon_y = \varepsilon$ (say), it follows that

$$\left\langle X'^2 + Y'^2 \right\rangle_{X', Y'} = \varepsilon \left(1 - \frac{x^2}{a^2} - \frac{y^2}{b^2} \right) \tag{10.122}$$

for $x^2/a^2 + y^2/b^2 < 1$. That is, the emittance ε is directly proportional to the maximum transverse "temperature" $\hat{T}_{b\perp}$ of the beam ions at $x = 0 = y$ (see also Problem 9.20 and the discussion in Sec. 9.4.2).

Following Lapostolle,[24] it is often convenient to introduce the rms emittance $\bar{\varepsilon}_x$ defined by

$$\bar{\varepsilon}_x = 4 \left[\langle X^2 \rangle_{\Gamma_T} \langle X'^2 \rangle_{\Gamma_T} - \langle XX' \rangle_{\Gamma_T}^2 \right]^{1/2}, \tag{10.123}$$

with a similar definition for $\bar{\varepsilon}_y$. Here, $\langle \cdots \rangle_{\Gamma_T}$ denotes the average over the entire available phase space $\Gamma_T = (X, Y, X', Y')$. The form of $\bar{\varepsilon}_x$ in Eq.(10.123) gives a result which is independent of the orientation of the axes in phase space.[4] For the particular choice of distribution function f_b^0 in Eq.(10.113), it is readily shown from Eqs.(10.120) and (10.123) that $\langle XX' \rangle_{\Gamma_T} = 0$, $\langle X^2 \rangle_{\Gamma_T} = \varepsilon_x/2$, $\langle X'^2 \rangle_{\Gamma_T} = \varepsilon_x/2$, and therefore $\bar{\varepsilon}_x = \varepsilon_x$. This is the expected result, because the major and minor axes of the beam cross section were assumed *ab initio* to coincide with x- and y-axes.

Finally, consistent with the discussion following Eq.(10.38), it is evident from Eq.(10.113) that $\pi \varepsilon_x$ is the area of the available (X, X') phase space when $Y = 0$, $Y' = 0$ and $X^2/\varepsilon_x + X'^2/\varepsilon_x = 1$. Similarly, $\pi\varepsilon_y$ is the area of the available (Y, Y') phase space when $X = 0$, $X' = 0$ and $Y^2/\varepsilon_y + Y'^2/\varepsilon_y = 1$.

10.8 Transverse Stability of the Kapchinskij-Vladimirskij Distribution Function

10.8.1 Linearized Vlasov-Poisson Equations

In this section, we examine the linearized Vlasov-Maxwell equations for low-frequency transverse perturbations about the Kapchinskij-Vladimirskij beam equilibrium defined by

$$f_b^0(X, Y, X', Y') = \frac{\hat{n}_b}{\pi}\,\delta(X'^2 + Y'^2 + X^2 + Y^2 - \varepsilon). \tag{10.124}$$

Here, $\varepsilon_x = \varepsilon_y = \varepsilon$ has been assumed [see Eq.(10.113)], and Eq.(10.124) corresponds to uniform beam density within the elliptical cross section $X^2 + Y^2 < \varepsilon$, or equivalently $x^2/a^2 + y^2/b^2 < 1$. Similar to the analysis in Sec. 9.4, the perturbed potential occurring in the linearized Vlasov equation is $\delta\phi - \beta_b\delta A_z = (1 - \beta_b^2)\delta\phi$, where the electrostatic potential $\delta\phi(X, Y, s)$ satisfies the linearized Poisson equation. Consistent with the units used earlier in Chapter 10 [see Eqs.(10.18) and (10.19)], we introduce the normalized potential

$$\delta\Phi(X, Y, s) = \frac{Z_i e\,\delta\phi(X, Y, s)}{\gamma_b^3 m_i \beta_b^2 c^2}. \tag{10.125}$$

In the phase space (X, Y, X', Y'), the perturbed distribution function $\delta f_b(X, Y, X', Y', s)$ evolves according to the linearized Vlasov equation

$$\left\{ \frac{\partial}{\partial s} + \frac{X'}{w_x^2}\frac{\partial}{\partial X} + \frac{Y'}{w_y^2}\frac{\partial}{\partial Y} - \frac{X}{w_x^2}\frac{\partial}{\partial X'} - \frac{Y}{w_y^2}\frac{\partial}{\partial Y'} \right\} \delta f_b$$

$$= \left(\frac{\partial\delta\Phi}{\partial X}\frac{\partial}{\partial X'} + \frac{\partial\delta\Phi}{\partial Y}\frac{\partial}{\partial Y'} \right) f_b^0, \tag{10.126}$$

where the equilibrium distribution function f_b^0 is defined in Eq.(10.124). Note that the characteristics of the differential operator on the left-hand side of Eq.(10.126) correspond to the particle trajectories in the equilibrium field configuration [see Eqs.(10.108) and (10.109)]. Because f_b^0

depends on X' and Y' only through the combination $X'^2 + Y'^2 \equiv V_\perp^2$, the linearized Vlasov equation (10.126) can be expressed in the equivalent form[22]

$$\left\{ \frac{\partial}{\partial s} + \frac{X'}{w_x^2}\frac{\partial}{\partial X} + \frac{Y'}{w_y^2}\frac{\partial}{\partial Y} - \frac{X}{w_x^2}\frac{\partial}{\partial X'} - \frac{Y}{w_y^2}\frac{\partial}{\partial Y'} \right\} \delta f_b$$

$$= 2\left(X'\frac{\partial}{\partial X}\delta\Phi + Y'\frac{\partial}{\partial Y}\delta\Phi \right) \frac{\partial f_b^0}{\partial V_\perp^2}. \qquad (10.127)$$

Moreover, because $X'^2 + X^2 + Y'^2 + Y^2 = V_\perp^2 + X^2 + Y^2$ is a constant of the motion, note that the factor $\partial f_b^0/\partial V_\perp^2$ occurring on the right-hand side of Eq.(10.127) is also constant (independent of s') following a particle trajectory in the equilibrium field configuration.

Poisson's equation for the perturbed electrostatic potential $\delta\phi$ is given by

$$\left(\frac{\partial^2}{\partial x^2} + \frac{\partial^2}{\partial y^2} \right) \delta\phi = -4\pi Z_i\, e\, \delta n_b, \qquad (10.128)$$

where $\delta n_b = \iint dX' dY' \delta f_b$ is the perturbed ion density. Scaling the x and y variables in Eq.(10.128) according to $x = w_x X = aX/\sqrt{\varepsilon}$ and $y = w_y Y = bY/\sqrt{\varepsilon}$, and introducing the normalized potential $\delta\Phi(X,Y,s)$ defined in Eq.(10.125), some straightforward algebra shows that the linearized Poisson equation (10.128) can be expressed in the equivalent form

$$\left(\frac{1}{a^2}\frac{\partial^2}{\partial X^2} + \frac{1}{b^2}\frac{\partial^2}{\partial Y^2} \right) \delta\Phi = -\frac{2K}{\hat{n}_b abe} \iint dX' dY' \delta f_b. \qquad (10.129)$$

Here, $K = (2/\gamma_b^3\beta_b^3)(Z_i e/m_i c^3)I_b$ is the self-field perveance defined in Eq.(10.23), and $I_b = \pi\hat{n}_b ab Z_i e\beta_b c$ is the beam current.

Paralleling the analysis in Secs. 4.7, 4.9, and 9.4, the linearized Vlasov equation (10.127) can be integrated formally using the method of characteristics. That is, Eq.(10.127) is integrated along the unperturbed particle trajectories $X(s')$, $X'(s')$, etc., determined from Eqs.(10.108) and (10.109). For growing perturbations, the "initial" value (at $s' = s_0$, say) of δf_b is neglected, and the solution to Eq.(10.127) can be expressed as

$$\delta f_b(X, Y, X', Y', s) \qquad (10.130)$$

$$= 2\frac{\partial f_b^0}{\partial V_\perp^2} \int_{s_0}^{s} ds' \left[X'(s')\frac{\partial}{\partial X(s')} + Y'(s')\frac{\partial}{\partial Y(s')} \right] \delta\Phi\left[X(s'), Y(s'), s' \right].$$

Here, the particle trajectories $X(s')$, $Y(s')$, $X'(s')$ and $Y'(s')$ solve the orbit equations (10.108) and (10.109), and are chosen to pass through the phase-space point (X, Y, X', Y') at $s' = s$ [see Eq.(10.110)]. Note that $(\partial/\partial V_\perp^2)f_b^0(V_\perp^2 + X^2 + Y^2)$ has been taken outside the integration over s' in Eq.(10.130) because $\partial f_b^0/\partial V_\perp^2$ is constant (independent of s') following a particle trajectory in the equilibrium field configuration. In terms of the ancillary phase variables $\psi_x'(s', s) = \int_s^{s'} ds/w_x^2(s)$ and $\psi_y'(s', s) = \int_s^{s'} ds/w_y^2(s)$ introduced in Eqs.(10.110) and (10.111), the solution for δf_b in Eq.(10.130) can be expressed in the equivalent form (Problem 10.9)

$$\delta f_b(X, Y, X', Y', s) \tag{10.131}$$

$$= 2\frac{\partial f_b^0}{\partial V_\perp^2} \int_{s_0}^s ds' \left[\frac{\partial}{\partial \psi_x'(s', s)} + \frac{\partial}{\partial \psi_y'(s', s)} \right] \delta\Phi\left[X(s'), Y(s'), s'\right].$$

In this regard, the amplitude functions $w_x(s)$ and $w_y(s)$ occurring in the definitions of ψ_x' and ψ_y' solve the nonlinear differential equations (10.36) and (10.37), where the coupling coefficients $\kappa_x(s)$ and $\kappa_y(s)$ are defined in Eq.(10.103) for an alternating gradient quadrupole field and a uniform-density ion beam.

The expression for the perturbed distribution function $\delta f_b(X, Y, X', Y', s)$ is now substituted into the linearized Poisson equation (10.129) for $\delta\Phi(X, Y, s)$. Introducing the new integration variables $(V_\perp, \phi)$ defined by $X' = V_\perp \cos\phi$ and $Y' = V_\perp \sin\phi$, we express $\int\int dX' \, dY' \cdots = \int_0^{2\pi} d\phi \int_0^\infty dV_\perp \, V_\perp \cdots$ on the right-hand side of Eq.(10.129). Moreover, it is convenient to introduce the phase-averaged orbit integral $I(X, Y, V_\perp^2, s)$ defined by

$$I(X, Y, V_\perp^2, s) = \int_0^{2\pi} \frac{d\phi}{2\pi} \int_{s_0}^s ds' \left(\frac{\partial}{\partial \psi_x'} + \frac{\partial}{\partial \psi_y'} \right) \delta\Phi\left[X(s'), Y(s'), s'\right].$$
$$\tag{10.132}$$

Substituting Eqs.(10.131) and (10.132) into Eq.(10.129) then gives[22]

$$\left(\frac{1}{a^2} \frac{\partial^2}{\partial X^2} + \frac{1}{b^2} \frac{\partial^2}{\partial Y^2} \right) \delta\Phi(X, Y, s)$$

$$= -\frac{4\pi K}{\hat{n}_b abe} \int_0^\infty dV_\perp^2 \frac{\partial f_b^0}{\partial V_\perp^2} I(X, Y, V_\perp^2, s). \tag{10.133}$$

Equation (10.133) is the desired equation for $\delta\Phi(X, Y, s)$. Note that $\delta\Phi[X(s'), Y(s'), s']$ occurs in the integrand of the orbit integral I defined in Eq.(10.132). Therefore, as in Secs. 4.9 and 9.4, Eq.(10.133) represents a complicated integral equation in which the solution for $\delta\Phi(X, Y, s)$ depends on the orbit history through an s'-integration over the very function $\delta\Phi[X(s'), Y(s'), s']$ which is to be determined. The situation is further complicated by the fact that the phase functions $\psi'_x(s', s)$ and $\psi'_y(s', s)$ occurring in the orbit solutions for $X(s')$, $X'(s')$, etc., in Eq.(10.110) cannot be determined in closed analytical form from the differential equations (10.36) and (10.37) for $w_x(s)$ and $w_y(s)$. This is in contrast to the case of a uniform solenoidal field $B_0\hat{e}_z$ considered in Secs. 4.9 and 9.4, where the particle orbits in the equilibrium field configuration can be evaluated in closed form. Nonetheless, Eq.(10.133) is a very significant result.[22] In particular, Eq.(10.133) is formally similar to the eigenvalue equations (4.257) and (9.162) derived for a uniform density ion beam in a solenoidal focusing field $B_0\hat{e}_z$. Therefore, the simple form of Eq.(10.133) illustrates the considerable value of transforming to the new phase-space variables (X, Y, X', Y'). Moreover, numerical techniques developed to solve Eqs.(4.257) or (9.162) can be applied to Eq.(10.133) by appropriate modification of the orbit equations to correspond to Eqs.(10.108) and (10.109), supplemented by Eqs.(10.36) and (10.37) for $w_x(s)$ and $w_y(s)$. Finally, because of the similarity between Eq.(10.133) and Eqs.(4.257) and (9.162), a sufficiently simple model of the particle orbits in a quadrupole focusing field would allow us to apply (with straightforward modifications) many of the techniques and stability results developed for the case of a solenoidal focusing field.[19] The stability behavior obtained from such a simplified model (Problem 10.10) is investigated in Sec. 9.4.6.

10.8.2 Transverse Stability Properties

For the choice of distribution function in Eq.(10.124), the integration over $V_\perp^2$ in Eq.(10.133) can be carried out exactly. Substituting $f_b^0 = (\hat{n}_b/\pi)\delta[V_\perp^2 - (\varepsilon - X^2 - Y^2)]$ into Eq.(10.133) and integrating by parts, we obtain

$$\left(\frac{1}{a^2} \frac{\partial^2}{\partial X^2} + \frac{1}{b^2} \frac{\partial^2}{\partial Y^2} \right) \delta\Phi$$

$$= \frac{4K}{ab\varepsilon} U(\varepsilon - X^2 - Y^2) \left(\frac{\partial}{\partial V_\perp^2} I(X, Y, V_\perp^2, s) \right)_{V_\perp^2 = V_{\perp 0}^2(X,Y)}$$

$$+ \frac{4K}{ab\varepsilon} \delta(\varepsilon - X^2 - Y^2) \left[I(X, Y, V_\perp^2, s) \right]_{V_\perp^2 = 0}. \qquad (10.134)$$

Here, $V_{\perp 0}^2(X, Y) \equiv \varepsilon - X^2 - Y^2$, and $U(x)$ is the Heaviside step function defined by $U(x) = +1$ for $x > 0$, and $U(x) = 0$ for $x < 0$. Making the identification $\varepsilon - X^2 - Y^2 = \varepsilon(1 - x^2/a^2 - y^2/b^2)$, it is evident that the first term on the right-hand side of Eq.(10.134) corresponds to a body-charge perturbation extending throughout the interior of the ion beam $(x^2/a^2 + y^2/b^2 < 1)$, whereas the second term on the right-hand side of Eq.(10.134) corresponds to a surface-charge perturbation localized at $x^2/a^2 + y^2/b^2 = 1$. The form of Eq.(10.134) should be compared with the eigenvalue equations (4.259) and (9.165) derived for a uniform-density ion beam (with circular cross section) in a solenoidal focusing field $B_0 \hat{e}_z$. Moreover, from Eq.(10.23) note that the factor $4K/ab$ occurring in Eq.(10.134) can be expressed in the familiar notation $4K/ab = 2\omega_{pb}^2/\gamma_b^3 \beta_b^2 c^2$, where $\omega_{pb}^2 = 4\pi \hat{n}_b Z_i^2 e^2/m_i$ is the nonrelativistic plasma frequency-squared.

Elegant analytical and numerical studies of the eigenvalue equation (10.134) have been carried out by Hofmann, Laslett, Smith and Haber[22] for the periodic quadrupole lattice illustrated in Fig. 10.3. Here, $L = S/2$ is the half-period of the lattice, $\hat{\kappa}_q = $ const. is the amplitude of the quadrupole field, which is taken to be an alternating step function, and η denotes the fraction of the lattice occupied by lens elements (η is also called the occupancy factor). Potential perturbations are assumed to be of the form[22]

$$\delta\Phi(X, Y, s) = \sum_{m=0}^{n} A_m(s) X^{n-m} Y^m + \sum_{m=0}^{n-2} A_m^{(1)}(s) X^{n-m-2} Y^m + \cdots$$

$$(10.135)$$

in the beam interior, where n denotes the order of the mode. For a given value of n, a mode is said to be "even" if the index m in Eq.(10.135) is restricted to even integer values, and "odd" if m is restricted to odd

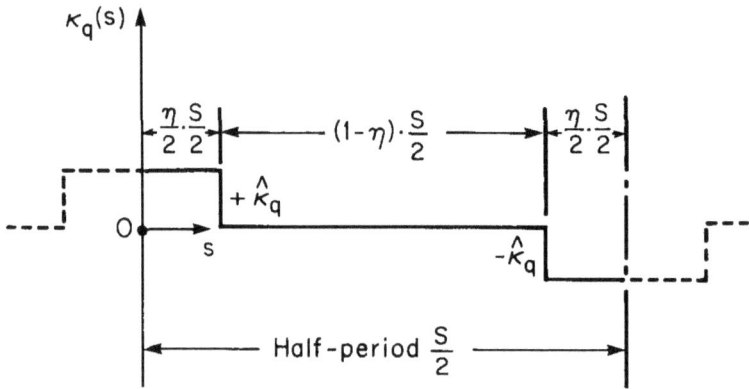

Figure 10.3. Alternating step-function model of a periodic quadrupole lattice with occupancy factor η for the lens elements. The figure shows a plot of the quadrupole coupling coefficient $\kappa_q(s)$ versus s for one half-period $(S/2)$ of the lattice. Such a configuration is often called a FODO transport lattice (acronym for focusing-off-defocusing-off).

integer values. In the stability analysis,[22] Eq.(10.134) provides a coupled set of algebraic equations that relate the functions $A_m(s)$ in Eq.(10.135) to integrals $I_m(k, \ell, s)$ of the form

$$I_m(k, \ell, s) = \int_{s_0}^{s} ds'\, A_m(s') \sin\left[k\psi_x' + \ell\psi_y'\right],\qquad (10.136)$$

where $\psi_x' = \int_s^{s'} ds/w_x^2(s)$ and $\psi_y' = \int_s^{s'} ds/w_y^2(s)$. Integration of Poisson's equation (10.134) across the surface of the ion beam gives a second set of equations, which, when taken together with Eq.(10.136) can be solved at each value of s, $a(s)$ and $b(s)$. In this way, the coefficients $A_m(s)$ can be determined in terms of the integrals defined in Eq.(10.136).[22] Furthermore, it can be shown from Eq.(10.136) that

$$C(k, \ell, s)\frac{d}{ds}\left[C(k, \ell, s)\frac{d}{ds}I_m(k, \ell, s)\right] + I_m(k, \ell, s) = -C(k, \ell, s)A_m(s),$$

$$\qquad (10.137)$$

where

$$C(k, \ell, s) = \left[\frac{k}{w_x^2(s)} + \frac{\ell}{w_y^2(s)}\right]^{-1},\qquad (10.138)$$

and the amplitude functions $w_x(s)$ and $w_y(s)$ solve the nonlinear orbit equations (10.36) and (10.37). With $A_m(s)$ determined in terms of $I_m(k, \ell, s)$, Eq.(10.137) constitutes a coupled set of second-order differential equations for $I_m(k, \ell, s)$. Therefore, numerical integration of Eq.(10.137) through one period S will provide the matrix elements that advance I_m and dI_m/ds through one period of the lattice. The eigenvalues λ of the matrix are then determined, and the occurrence of any eigenvalue with magnitude greater than unity corresponds to instability for the particular mode under consideration. Moreover, the factor $|\lambda| - 1$ is a measure of the instability growth per period of the lattice.[22]

It is convenient to characterize the detailed stability properties of the Kapchinskij-Vladimirskij distribution function in terms of the occupancy factor η in Fig. 10.3, the phase advances σ and σ_0 per lattice period (with and without space charge effects, respectively) defined in Sec. 10.6, and the normalized self-field perveance $K' = K/\varepsilon\hat{\kappa}_q^{1/2}$. Here, K is defined in

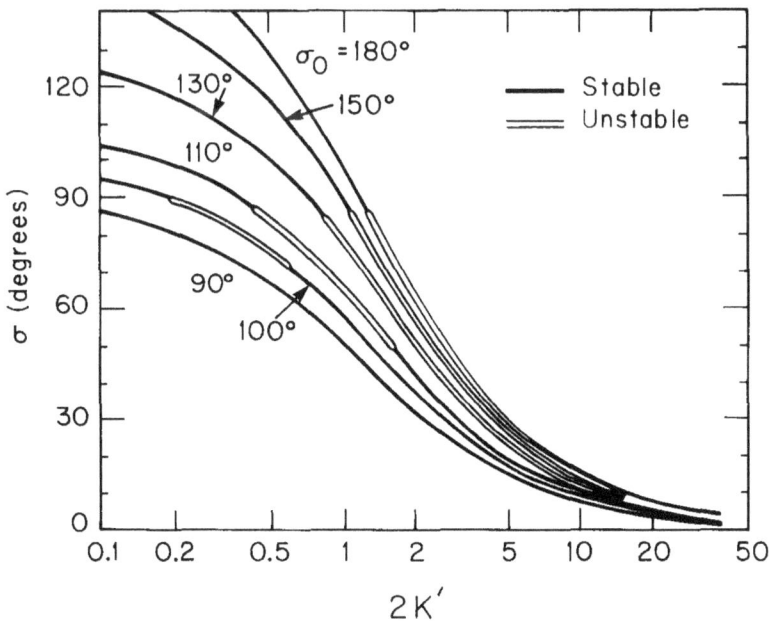

Figure 10.4. Bands of stability and instability in the parameter space $(2K', \sigma)$ obtained numerically from Eqs.(10.134) and (10.137) for the second-order (even) envelope mode. Here, the periodic quadrupole channel has occupancy factor $\eta = 1/2$, and the phase advance σ_0 ranges from $90°$ to $180°$. The envelope instability is suppressed whenever $\sigma_0 \leq 90°$ [I. Hofmann, L.J. Laslett, L. Smith and I. Haber, Particle Accelerators **13**, 145 (1983)].

Eq.(10.23) and $\hat{\kappa}_q$ is the amplitude of the quadrupole coupling coefficient in Fig. 10.3. Note that K' is a dimensionless measure of beam intensity. To summarize briefly, Hofmann et al.[22] conclude that the restriction $\sigma_0 \leq 90°$ should be imposed to avoid the dangerous second-order (even) envelope instability. This is illustrated in Fig. 10.4 where bands of stability and instability calculated numerically from Eqs.(10.134) and (10.137) are illustrated in the parameter space $(2K', \sigma)$ for $\eta = 1/2$ and σ_0 ranging from $90°$ to $180°$. Note from the figure that the second-order envelope instability is absent for $\sigma_0 = 90°$, but covers broad bands of K' for $\sigma_0 \geq 100°$. It is further concluded,[22] in order to avoid regions of pronounced instability associated with the third-order mode ($\delta\Phi \sim XY^2$, say), the restriction $\sigma_0 \leq 60°$ should be imposed.

The behavior of the third-order mode for a periodic quadrupole channel with $\eta = 1/10$ and $\sigma_0 = 90°$ is illustrated in Fig. 10.5. Here, the

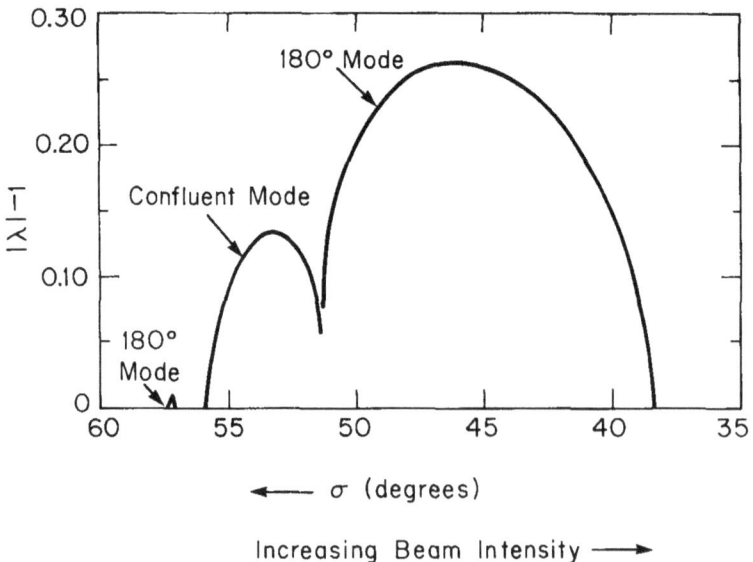

Figure 10.5.   Plot of the instability growth $|\lambda| - 1$ versus σ obtained numerically from Eqs.(10.134) and (10.137) for the third-order mode. Here, the periodic quadrupole channel has occupancy factor $\eta = 1/10$, and the phase advance is $\sigma_0 = 90°$. The strong instability centered near $\sigma = 45°$ is a structure resonance. The third-order instability is suppressed whenever $\sigma_0 \leq 60°$ [I. Hofmann, L.J. Laslett, L. Smith and I. Haber, Particle Accelerators **13**, 145 (1983)].

strong instability centered near $\sigma = 45°$ (the so-called 180°-mode) corresponds to a structure resonance, whereas the instability centered near $\sigma = 53°$ is associated with a confluence of eigenvalues. By using a lattice with $\sigma_0 \leq 60°$, it is found[22] that instability is suppressed for both the 180-degree mode and the confluent third-order modes. Finally, if $\sigma_0 \leq 60°$ and the beam intensity is low enough that $\sigma \geq 24°$, then it is found[22] that the fourth-order and higher-order instabilities associated with the Kapchinskij-Vladimirskij distribution can be avoided, at least in the context of the linearized Vlasov-Poisson equations. However, this latter restriction to $\sigma \geq 24°$ may be too conservative, because numerical simulation studies[22] for $\sigma_0 \leq 60°$ indicate that the higher-order modes saturate at low levels. Indeed, the rms emittance does not appear to be affected by the higher-order modes for values of σ as low as 6°.

10.8.3   Generalization of the Eigenvalue Equation in the Low-Density Regime

In the short space of this book, it is not our intention to analyze Poisson's equation (10.133) in detail. In concluding Sec. 10.8, however, some important observations are in order. It was noted in Sec. 10.7 that the Kapchinskij-Vladimirskij distribution function in Eq.(10.124) [or Eq.(10.113)] is the only known equilibrium solution to the steady-state Vlasov equation $(\partial/\partial s = 0)$ which includes both self-field effects and the effects of the periodic focusing field. Moreover, the corresponding density profile $n_b^0(X, Y)$ is uniform [Eq.(10.116)], and the equilibrium distribution function is somewhat pathological in that f_b^0 has an inverted population in the phase space (X, Y, X', Y') which is peaked at $X'^2 + Y'^2 + X^2 + Y^2 = \varepsilon$. Therefore, because Eq.(10.124) is the only rigorously known equilibrium solution (for $\varepsilon_x = \varepsilon_y = \varepsilon$), an analysis of Eq.(10.133) is necessarily restricted to the choice of distribution function f_b^0 in Eq.(10.124), which leads to the simplified form of Poisson's equation in Eq.(10.134) for a uniform-density beam. The exception to this restrictive application of the linearized Vlasov-Poisson equations is the same as that discussed in Paragraph (d) following Eq.(10.117) in Sec. 10.7. In particular, whenever the beam density and current are sufficiently low that the equilibrium self-field effects can be neglected in comparison with the applied quadrupole forces, the orbit equations in the equilibrium field configuration can be approximated by Eqs.(10.118) and (10.119). The formal analysis leading to Eqs.(10.108)–(10.111) is the same as before, and there are two independent constants of the motion given by Eq.(10.112).

As noted earlier, there is considerable latitude in constructing (low-density) self-consistent Vlasov equilibria $f_b^0(X, Y, X', Y')$ from the constants of the motion $X'^2 + X^2$ and $Y'^2 + Y^2$. For present purposes, we consider the class of isotropic equilibrium distribution functions $f_b^0(X'^2 + Y'^2 + X^2 + Y^2)$. The solution to the linearized Vlasov equation (10.127) is given again by Eq.(10.130), and Poisson's equation (10.128) for the normalized potential $\delta\Phi$ becomes

$$\left(\frac{\partial^2}{\partial x^2} + \frac{\partial^2}{\partial y^2}\right) \delta\Phi = -\frac{8\pi^2 Z_i^2 e^2}{\gamma_b^3 m_i \beta_b^2 c^2} \int_0^\infty dV_\perp^2 \frac{\partial f_b^0}{\partial V_\perp^2} I(X, Y, V_\perp^2, s). \quad (10.139)$$

Here, $V_\perp^2 = X'^2 + Y'^2$, and the orbit integral I is defined in Eq.(10.132). In the tenuous-beam limit, the particle trajectories $X(s')$, $X'(s')$, etc., occurring in Eq.(10.132) are given again by Eqs.(10.110) and (10.111). However, the amplitude functions $w_x(s)$ and $w_y(s)$ solve Eq.(10.119), valid in the limit of weak self fields.

Therefore, for a low-density ion beam propagating through a periodic quadrupole field, Eq.(10.139) can be used to investigate detailed stability behavior for a wide variety of equilibrium distribution functions $f_b^0(X'^2 + Y'^2 + X^2 + Y^2)$. By analogy with the stability theorem developed in Sec. 4.5, it is anticipated that monotonically decreasing equilibria with

$$\frac{\partial f_b^0}{\partial V_\perp^2} \le 0$$

have more favorable stability properties than the Kapchinskij-Vladimirskij equilibrium in Eq.(10.124), which has an inverted population in $X'^2 + Y'^2 + X^2 + Y^2$.

Problem 10.9 Consider the linearized Vlasov equation (10.127) for $\delta f_b(X, Y, X', Y', s)$.

a. Make use of the orbit equations (10.108) and (10.109) to show that Eq.(10.127) is equivalent to

$$\frac{d}{ds'}\delta f_b\left[X(s'), Y(s'), X'(s'), Y'(s'), s'\right] \quad (10.9.1)$$

$$= 2\frac{\partial f_b^0}{\partial V_\perp^2}\left[X'(s')\frac{\partial}{\partial X(s')} + Y'(s')\frac{\partial}{\partial Y(s')}\right]\delta\Phi\left[X(s'), Y(s'), s'\right]$$

evaluated at $s' = s$. Here, $X(s' = s) = X$, $Y(s' = s) = Y$, $X'(s' = s) = X'$, $Y'(s' = s) = Y'$, and $\partial f_b^0/\partial V_\perp^2$ is independent of s' by virtue of Eq.(10.112).

b. Integrate Eq.(10.9.1) with respect to s' and obtain the formal solution for $\delta f_b(X, Y, X', Y', s)$ given in Eq.(10.130), where the "initial" value (at $s' = s_0$, say) is neglected.

c. Consider the pseudoharmonic oscillator solutions for $X(s')$ and $X'(s')$ given in Eq.(10.110), and the analogous solutions for $Y(s')$ and $Y'(s')$. Show that

$$\left[X'(s') \frac{\partial}{\partial X(s')} + Y'(s') \frac{\partial}{\partial Y(s')} \right] F[X(s'), Y(s'), s']$$

$$= \left[\frac{\partial}{\partial \psi_x'(s', s)} + \frac{\partial}{\partial \psi_y'(s', s)} \right] F[X(s'), Y(s'), s] \qquad (10.9.2)$$

for general function $F[X(s'), Y(s'), s']$. Here $\psi_x(s', s) = \int_s^{s'} ds/w_x^2(s)$ and $\psi_y(s', s) = \int_s^{s'} ds/w_y^2(s)$. Equation (10.131) then follows directly from Eqs.(10.130) and (10.9.2).

Problem 10.10 Consider a matched ion beam with uniform density and circular cross section $(a = b = r_b)$ propagating through a periodic quadrupole lattice [Fig. 10.1(a)]. For moderately small phase advance (σ less than $90°$, say), the analysis in Sec. 10.6 suggests that the periodic quadrupole force on the envelope ions can be replaced by an average *focusing* force proportional to $-(\sigma_0^2/S^2)r_b$ [see Eq.(10.95)]. A similar conclusion pertains for the motion of an ion in the beam interior. As a simplified *model*, assume that the transverse particle orbits $x(s)$ and $y(s)$ satisfy

$$\frac{d^2}{ds^2} x(s) + \frac{\omega_{qb}^2}{\beta_b^2 c^2} x(s) - \frac{K}{r_b^2} x(s) = 0, \qquad (10.10.1)$$

$$\frac{d^2}{ds^2} y(s) + \frac{\omega_{qb}^2}{\beta_b^2 c^2} y(s) - \frac{K}{r_b^2} y(s) = 0, \qquad (10.10.2)$$

where $\omega_{qb}^2/\beta_b^2 \equiv \sigma_0^2/S^2 = $ const. represents the average quadrupole focusing, and $K = (\omega_{pb}^2/2\beta_b^2 c^2 \gamma_b^3)r_b^2$ is the self-field perveance.

a. Show that the condition for stable (oscillatory) solutions to Eqs.(10.10.1) and (10.10.2) is given by

$$\omega_{qb}^2 > \frac{\omega_{pb}^2}{2\gamma_b^3}. \qquad (10.10.3)$$

b. Because the coefficients of $x(s)$ and $y(s)$ are constants in Eqs.(10.10.1) and (10.10.2), there is no advantage to transforming to the new phase space coordinates (X, Y, X', Y') introduced in Sec. 10.7. Show that the solutions to Eq.(10.10.1) for $x(s')$ and $dx(s')/ds'$ which pass through the phase-space point (x, x') at $s' = s$ are given by

$$x(s') = x\cos\left[\frac{\hat{\omega}_\beta(s'-s)}{\beta_b c}\right] + x'\left(\frac{\beta_b c}{\hat{\omega}_\beta}\right)\sin\left[\frac{\hat{\omega}_\beta(s'-s)}{\beta_b c}\right], \quad (10.10.4)$$

$$\frac{dx(s')}{ds'} = x'\cos\left[\frac{\hat{\omega}_\beta(s'-s)}{\beta_b c}\right] - x\left(\frac{\hat{\omega}_\beta}{\beta_b c}\right)\sin\left[\frac{\hat{\omega}_\beta(s'-s)}{\beta_b c}\right], \quad (10.10.5)$$

where $\hat{\omega}_\beta$ is defined by

$$\hat{\omega}_\beta = \left(\omega_{qb}^2 - \frac{\omega_{pb}^2}{2\gamma_b^3}\right)^{1/2}. \quad (10.10.6)$$

Similar solutions for $y(s')$ and $dy(s')/ds'$ can be obtained from Eq.(10.10.2).

c. The (average) quadrupole force terms proportional to $(\omega_{qb}^2/\beta_b^2 c^2)x(s)$ and $(\omega_{qb}^2/\beta_b^2 c^2)y(s)$ in Eqs.(10.10.1) and (10.10.2) can be derived from an equivalent axial vector potential

$$A_z^{\text{ext}}(r) = -\frac{\gamma_b m_i}{2Z_i e \beta_b}\,\omega_{qb}^2(x^2 + y^2). \quad (10.10.7)$$

Calculate $B_x = \partial A_z^{\text{ext}}/\partial y$ and $B_y = -\partial A_z^{\text{ext}}/\partial x$ from Eq.(10.10.7). Parallel the analysis in Sec. 10.3 and show that the resulting force components on a beam ion (with $v_z \simeq \beta_b c$) in the x- and y-directions are given by

$$F_x = -\frac{\omega_{qb}^2}{\beta_b^2 c^2}\,x, \quad (10.10.8)$$

$$F_y = -\frac{\omega_{qb}^2}{\beta_b^2 c^2}\,y, \quad (10.10.9)$$

where F_x and F_y are scaled by $\beta_b^2 c^2$. Equations (10.10.8) and (10.10.9) are identical to the quadrupole force components in Eqs.(10.10.1) and (10.10.2).

Because of the simple harmonic form of the orbit solutions in Eqs.(10.10.4) and (10.10.5) and the fact that the restoring forces in Eqs.(10.10.1) and (10.10.2) are linearly proportional to $x(s)$ and $y(s)$, respectively, many of the theoretical techniques and transverse stability properties developed in Secs. 4.9 and 9.4 (for a uniform-density beam propagating in a solenoidal focusing field) can be applied to the present problem with straightforward modifications. The stability behavior obtained from such a simplified model of ion beam propagation in a periodic quadrupole field is investigated in Sec. 9.4.6 for perturbations with mode numbers $(\ell, n) = (0, 2)$.[19]

10.9   Emittance Growth

Emittance growth is a topic of fundamental importance to advanced accelerator applications requiring the transport of high-brightness beams. These applications range from heavy-ion fusion, to free electron lasers, to relativistic klystrons for electron-positron colliders, to mention a few examples. As a consequence of experimental and theoretical investigations,[20-30] including computer simulation studies,[22,23,30] there have been several advances in understanding and reducing the emittance growth for intense beam propagation in alternating-gradient focusing systems. In this section, we summarize selected aspects of this important problem.

10.9.1   Instability-Induced Emittance Growth

It is important to distinguish two potential causes of emittance growth. One is associated with collective instabilities,[22,23] and the other is associated with a spatially nonuniform charge distribution.[23-30] Emittance growth associated with collective instabilities can occur if the properties of the injected beam distribution and the focusing channel (as measured by σ, K, σ_0, etc.) are chosen in such a way that the system is linearly unstable (as discussed in Sec. 10.8). The nonlinear reaction of the particles to the amplifying field perturbations leads to a redistribution in the phase space (X, Y, X', Y') and a concomitant growth and saturation of the rms emittance $\bar{\varepsilon}$ [Eq.(10.123)]. Such an example of "instability-induced" emittance growth is illustrated in Figs. 10.6 and 10.7, which show the results of computer simulation studies by Struckmeier, Klabunde and Reiser.[23] Here, a beam with Kapchinskij-Vladimirskij beam distribution function is injected into a periodic quadrupole channel with phase advance $\sigma_0 = 90°$ (without self-field effects) and depressed tune $\sigma = 41°$ (including self-field effects), which is near the maximum growth rate of the third-order mode (described in Sec. 10.8.2) for the particular choice of lattice parameters[23] in the simulation studies. Figure 10.6 shows the nonlinear evolution in phase space of an initial Kapchinskij-Vladimirskij distribution function for a circular ion beam $(a = b)$ propagating through 100 channel periods. After about 50 periods, the distribution in phase space becomes relatively quiescent (Fig. 10.6), and the rms emittance $\bar{\varepsilon}$ saturates at a level which is about 2.2 times the initial emittance $\bar{\varepsilon}_i$ (Fig. 10.7). Such a large instability-induced emittance growth can of course be avoided for

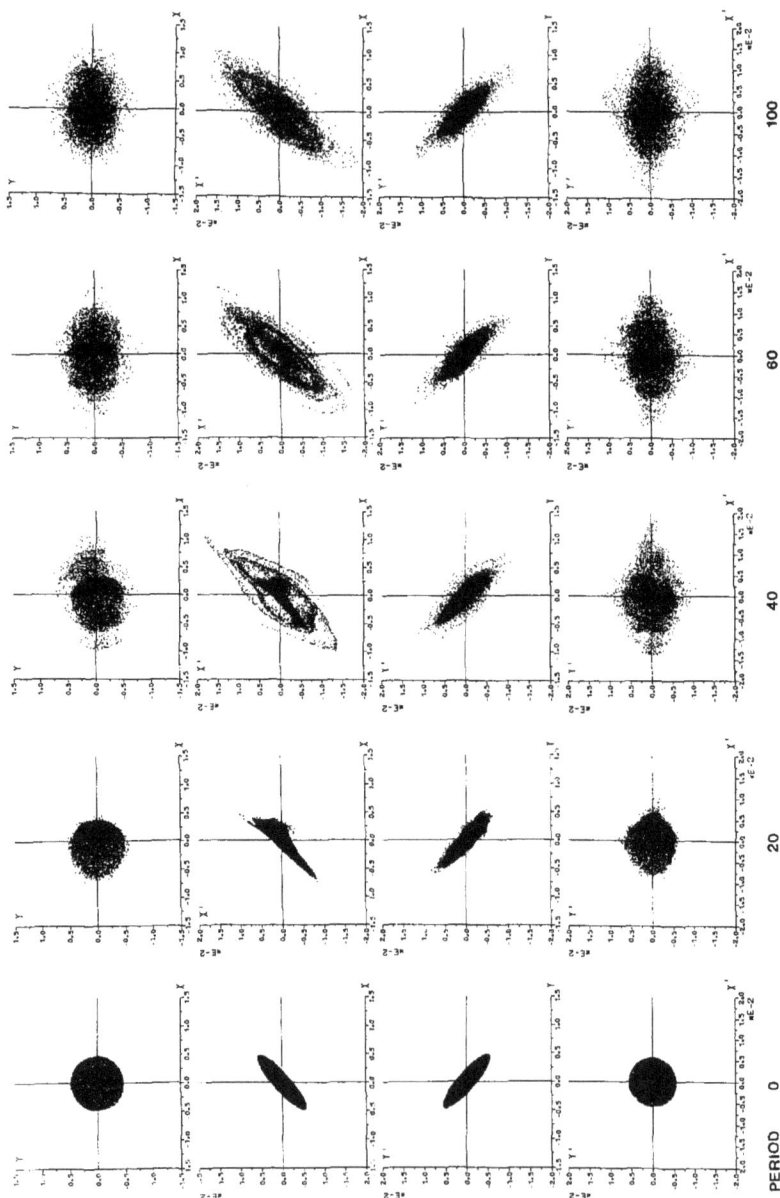

Figure 10.6. Computer simulation results showing the evolution in phase space of an initial Kapchinskij-Vladimirskij distribution function which is unstable to third-order transverse perturbations (such as $\delta\Phi \simeq XY^2$). The propagation of the circular beam $(a = b)$ is followed through one hundred lattice periods $(100\ S)$ of a periodic quadruple channel with $\sigma_0 = 90°$ and $\sigma = 41°$. The distribution in phase space becomes relatively quiescent after about 50 lattice periods [J. Struckmeier, J. Klabunde and M. Reiser, Particle Accelerators **15**, 47 (1984)].

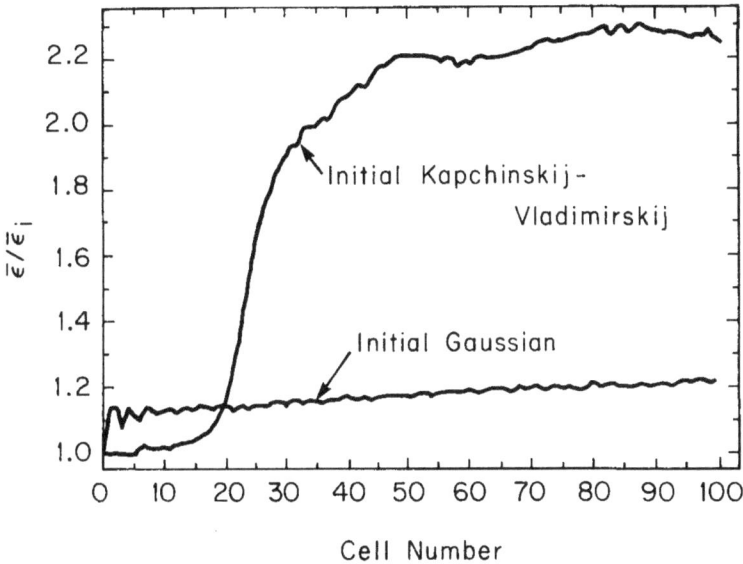

Cell Number

Figure 10.7. Plot of the normalized rms emittance $\bar{\varepsilon}/\bar{\varepsilon}_i$ versus cell number obtained from computer simulations of beam propagation through a periodic quadrupole channel with $\sigma_0 = 90°$ and $\sigma = 41°$. The curve labeled *initial Kapchinskij-Vladimirskij* corresponds to the simulation in Fig. 10.6, and the emittance growth induced by the third-order instability is substantial. The curve labeled *initial Gaussian* corresponds to an injected distribution function which is initially a (truncated) Gaussian in phase space [J. Struckmeier, J. Klabunde and M. Reiser, Particle Accelerators **15**, 47 (1984)].

the Kapchinskij-Vladimirskij distribution function by choosing the channel and beam parameters appropriately (e.g., $\sigma_0 \leq 60°$, as discussed in Sec. 10.8.2) in order to preclude the onset of instability. Even for $\sigma_0 = 90°$ and $\sigma = 41°$, if the injected distribution function is a truncated Gaussian in phase space rather than a Kapchinskij-Vladimirskij distribution, then the emittance growth is reduced substantially (Fig. 10.7). However, there is a rapid initial increase in the emittance (by about 14%) due to the homogenization of the nonuniform charge distribution (see Sec. 10.9.2).[23]

10.9.2 Emittance Growth Caused by Charge Nonuniformity

The second major cause of emittance growth is associated with spatial nonuniformities in the charge distribution of the propagating beam.[23-30]

As discussed in Sec. 10.8, at moderate beam intensity, the only known steady-state solution to the Vlasov equation for beam propagation through a periodic focusing lattice is the Kapchinskij-Vladimirskij distribution function, which has a uniform charge density in the beam interior. Therefore, an intense beam with a nonuniform density profile (e.g., a Gaussian density profile, or multiple beamlets) inherently is *not* in an "equilibrium" state with $\partial/\partial t = 0$. Analytical investigations and computer simulation studies,[23,30] as well as experiments on the merging of multiple electron beams[30] and multiple ion beams[32] have demonstrated that such nonuniform charge distributions can be a major source of emittance growth for intense beam propagation in alternating-gradient focusing systems. For beams with the same mean-square radius $\overline{r^2}$, analytical estimates show that an intense beam with a nonuniform density profile has a higher electric field energy than an equivalent beam with uniform density. Moreover, computer simulation studies and laboratory experiments[23,30] indicate that the self-consistent response of an initially nonuniform density profile is such that the charge density becomes approximately uniform through a collective homogenization process which occurs on a characteristic (short) length scale

$$L_h = \frac{1}{4} \frac{2\pi\gamma_b^{1/2}}{\omega_{pb}} \beta_b c, \qquad (10.140)$$

where ω_{pb} is the nonrelativistic plasma frequency. The difference in field energy (ΔU_E) between beam configurations with nonuniform charge density and uniform charge density is converted into an increase in the transverse kinetic energy and potential energy of the beam particles, which is manifest as an increase in the rms emittance $\bar{\varepsilon}$. Such a rapid increase in $\bar{\varepsilon}$ (by about 14%) is evident from the lower curve in Fig. 10.7 during the initial stages of evolution of an injected Gaussian distribution function. About one lattice period S is required for homogenization of the charge density for the choice of system parameters corresponding to the lower curve in Fig. 10.7.

A differential equation for emittance growth induced by charge nonuniformity has been derived by Lapostolle[24] and by Wangler, et al.[25] Moreover, Struckmeier, Klabunde and Reiser[23] have obtained a simple analytical estimate of the ratio of the final rms emittance $(\bar{\varepsilon}_f)$ after charge homogenization to the initial rms emittance $(\bar{\varepsilon}_i)$ using energy conservation arguments in which the decrease in field energy is balanced by an

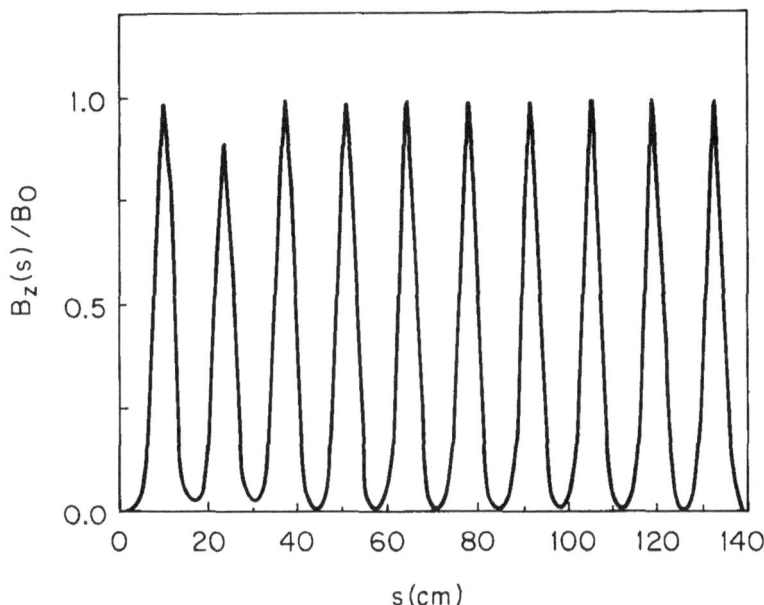

Figure 10.8. Plot versus axial distance s of the periodic solenoidal focusing field $B_z(s)/B_0$ used in experiments to investigate beam merging and emittance growth due to charge homogenization [M. Reiser, et al., Phys. Rev. Lett. **61**, 2933 (1988)].

increase in the emittance. Without presenting algebraic details,[23] it is found that

$$\frac{\bar{\varepsilon}_f}{\bar{\varepsilon}_i} = \left[1 + \frac{1}{2} g \left(\frac{\sigma_0^2}{\sigma^2} - 1 \right) \right]^{1/2} \qquad (10.141)$$

for a cylindrical beam with $a = b$. Here, g is a geometric factor which measures the deviation from charge uniformity of the injected beam ($g = 0$ for a uniform-density beam); σ_0 is the phase advance in the absence of self-field effects; and σ is the phase advance including self-field effects. The ratio σ_0/σ in Eq.(10.141) is to be estimated for the injected beam in the smooth-beam approximation.

In the range of validity of Eq.(10.141), Reiser, et al.,[30] have carried out an elegant set of merging-beam experiments and computer simulation studies in which five beamlets are injected into a periodic solenoidal focusing channel with phase advance $\sigma_0 = 70°$ (Fig. 10.8) and observed to merge into a single beam with nearly uniform density (Fig. 10.9). In the experiments,[30] a 5 keV electron beam from a one-inch-diameter

Experiment   Simulation

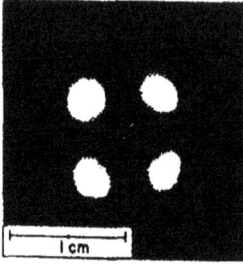   s = 3.4 cm 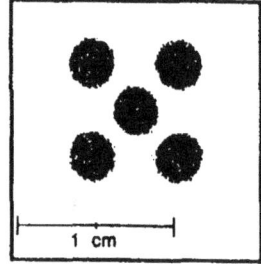

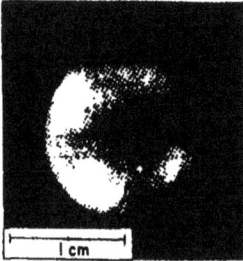   s = 17.0 cm 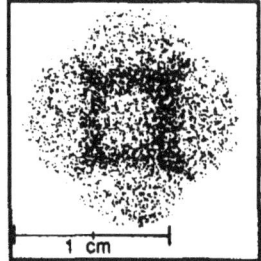

   s = 101.0 cm 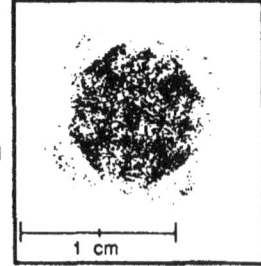

Figure 10.9. Phosphor-screen images of the beam cross section at various axial locations along the transport channel. Also shown are computer simulation plots of the beam cross section at the same axial locations as the experimental data. By $s = 17$ cm, the five beamlets have merged into a single cylindrical beam with nearly uniform charge density. Faint images of the initial five-beamlet configuration reappear between $s = 85$ cm and $s = 115$ cm, indicating that the beam retains some phase coherence and "memory" of its initial conditions [M. Reiser, et al., Phys. Rev. Lett. **61**, 2933 (1988)].

thermionic cathode is transported along a periodic channel of 38 solenoidal lenses with period $S = 13.6$ cm. Five beamlets are masked out of the beam at the anode, leading to a beam current of 44 mA and a self-field perveance of $K = (2/\beta_b^3\gamma_b^3)I_b/(m_e c^3/e) = 1.88 \times 10^{-3}$ [Eq.(10.23)]. The beam emittance is measured at the end of the transport channel by a split-pinhole diagnostic with a Faraday cup,[33] and a moveable phosphor screen is able to intercept the beam image at various axial positions along the solenoidal focusing channel (Fig. 10.9).

Several features of the experimental results and simulation studies illustrated in Fig. 10.9 are noteworthy. Consistent with Eq.(10.140), by $s = 17$ cm the five beamlets have merged into a single cylindrical beam with nearly uniform charge density. An unexpected result in Fig. 10.9 is the appearance (between $s = 85$ cm and $s = 115$ cm from the aperture plate) of images of the initial five-beamlet configuration. Although the computer-simulation images in Fig. 10.9 are fainter than the phosphor-screen images in the experiment, it is evident that the beam retains some phase coherence and "memory" of its initial conditions.[34]

Chapter 10 References

1. E.D. Courant, M.S. Livingstone and H. Snyder, "The Strong Focusing Synchrotron—a New High Energy Accelerator," Phys. Rev. **88**, 1190 (1952).

2. E.D. Courant and H.S. Snyder, "Theory of the Alternating Gradient Synchrotron," Annals of Physics **3**, 1 (1958).

3. N.C. Christofilos, unpublished (1950).

4. J.D. Lawson, *The Physics of Charged Particle Beams* (Clarendon Press, Oxford, 1988).

5. S. Humphries, Jr., *Principles of Charged Particle Acceleration* (Wiley, New York, 1986).

6. J.J. Livingood, *Principles of Cyclic Particle Accelerators* (Van Nostrund, New York, 1961).

7. *Heavy Ion Inertial Fusion*, eds., M. Reiser, T. Godlove and R. Bangerter, IAP Conference Proceedings No. 152 (American Institute of Physics, New York, 1986), and references therein.

8. D. Keefe, "Research on High Beam-Current Accelerators," Particle Accelerators **11**, 187 (1987).

9. E.P. Lee and J. Hovingh, "Heavy Ion Induction Linac Drivers for Inertial Confinement Fusion," Fusion Technology 15, 369 (1989).

10. I.M. Kapchinskij and V.V. Vladimirskij, "Limitations of Proton Beam Current in a Strong-Focusing Linear Accelerator Associated with the Beam Space Charge," in *Proceedings of the International Conference on High Energy Accelerators and Instrumentation* (CERN Scientific Information Service, Geneva, 1959), p. 274.

11. G.R. Lambertson, L.J. Laslett and L. Smith, "Transport of Intense Ion Beams," IEEE Transactions on Nuclear Science NS-24, 993 (1977).

12. I. Haber and A.W. Maschke, "Steady-State Transport of High Current Beams in a Focused Channel," Phys. Rev. Lett. 42, 1479 (1979).

13. M. Reiser, "Periodic Focusing of Intense Beams," Particle Accelerators 8, 167 (1978).

14. J. Struckmeier and M. Reiser, "Theoretical Studies of Envelope Oscillations and Instabilities of Mismatched Intense Charged-Particle Beams in Periodic Focusing Channels," Particle Accelerators 14, 227 (1984).

15. M. Reiser, "Current Limits in Linear Accelerators," J. Appl. Phys. 52, 555 (1981).

16. M.G. Tiefenback, "Space-Charge Limits on the Transport of Ion Beams in a Long Alternating-Gradient System," Ph.D. Thesis (University of California, Berkeley, 1986).

17. R.L. Gluckstern, R. Chasman and K. Crandall, "Stability of Phase-Space Distributions in Two-Dimensional Beams," in *Proceedings of the 1970 Proton Linear Accelerator Conference* (Batavia, Illinois, 1970), p. 823.

18. R.L. Gluckstern, "Oscillation Modes in Two-Dimensional Beams," in *Proceedings of the 1970 Proton Linear Accelerator Conference* (Batavia, Illinois, 1970), p. 811.

19. H.S. Uhm and R.C. Davidson, "Stability Properties of Intense Nonneutral Ion Beams for Heavy-Ion Fusion," Particle Accelerators 11, 65 (1980).

20. M.G. Tiefenback and D. Keefe, "Measurements of Stability Limits for a Space-Charge-Dominated Ion Beam in a Long Alternating-Gradient Transport Channel," IEEE Transactions on Nuclear Science NS-32, 2483 (1985).

21. D. Keefe, "Summary of High Beam-Current Transport," in *High-Current, High-Brightness, and High-Duty Ion Injectors*, eds. G.H. Gillespie, Y.-Y. Kuo, D. Keefe and T.P. Wangler, AIP Conference Proceedings No. 139 (American Institute of Physics, New York, 1986), p. 165.

22. I. Hofmann, L.J. Laslett, L. Smith and I. Haber, "Stability of the Kapchinskij-Vladimirskij Distribution in Long Periodic Transport Systems," Particle Accelerators **13**, 145 (1983).

23. J. Struckmeier, J. Klabunde and M. Reiser, "On the Stability and Emittance Growth of Different Particle Phase-Space Distributions in a Long Magnetic Quadrupole Channel," Particle Accelerators **15**, 47 (1984).

24. P.M. Lapostolle, "Possible Emittance Increase Through Filamentation Due to Space Charge in Continuous Beams," IEEE Transactions on Nuclear Science **NS-18**, 1101 (1971).

25. T.P. Wangler, K.R. Crandall, R.S. Mills and M. Reiser, "Relation Between Field Energy and RMS Emittance in Intense Particle Beams," IEEE Transactions on Nuclear Science **NS-32**, 2196 (1985).

26. F.W. Guy, P.M. Lapostolle and T.P. Wangler, "The Influence of Density Distribution on the Stability Behavior of Beams," in *Proceedings of the 1987 Particle Accelerator Conference*, eds. E.R. Lindstrom and L.S. Taylor, IEEE No. 87CH2387-9 (IEEE, New York, 1987), p. 1149.

27. T.P. Wangler, R.S. Mills and K.R. Crandall, "Emittance Growth in Intense Beams," in *Proceedings of the 1987 Particle Accelerator Conference*, eds. E.R. Lindstrom and L.S. Taylor, IEEE No. 87CH2387-9 (IEEE, New York, 1987), p. 1006.

28. O.A. Anderson, "Internal Dynamics and Emittance Growth in Space-Charge-Dominated Beams," Particle Accelerators **21**, 197 (1987).

29. I. Hofmann and J. Struckmeier, "Generalized Three Dimensional Equations for the Emittance and Field Energy of High-Current Beams in Periodic Focusing Structures," Particle Accelerators **21**, 69 (1987).

30. M. Reiser, C.R. Chang, D. Kehne, K. Low, T. Shea, H. Rudd and I. Haber, "Emittance Growth and Image Formation in a Nonuniform Space-Charge-Dominated Electron Beam," Phys. Rev. Lett. **61**, 2933 (1988).

31. H. Goldstein, *Classical Mechanics* (Addison-Wesley, New York, 1981).

32. C.M. Celata, A. Faltens, D.L. Judd, L. Smith and M.G. Tiefenback, "Transverse Combining of Nonrelativistic Beams in a Multiple-Beam Induction Linac," in *Proceedings of the 1987 Particle Accelerator Conference*, eds. E.R. Lindstrom and L.S. Taylor, IEEE No. 87CH2387-9 (IEEE, New York, 1987), p. 1167.

33. M.J. Rhee and R.F. Schneider, "The Root-Mean-Square Emittance of an Axisymmetric Beam with a Maxwellian Velocity Distribution," Particle Accelerators **20**, 133 (1986).

34. J.D. Lawson, P.M. Lapostolle and R.L. Gluckstern, "Emittance, Entropy and Information," Particle Accelerators **5**, 61 (1973).

The following references, while not cited directly in the text, are relevant to the general subject matter of this chapter.

D. Chernin, "Evolution of RMS Beam Envelopes in Transport Systems with Linear X-Y Coupling," Particle Accelerators 24, 29 (1988).

T.P. Wangler, K.R. Crandall and R.S. Mills, "Field Energy and RMS Emittance in Intense Particle Beams," in *High-Current, High-Brightness, and High-Duty Factor Ion Injectors*, eds., G.H. Gillespie, Y.-Y. Kuo, D. Keefe and T.P. Wangler, AIP Conference Proceedings No. 139 (American Institute of Physics, New York, 1986), p. 133.

M. Reiser, "High-Current Beam Transport and Charge-Neutralization Effects," in *High-Current, High-Brightness, and High-Duty Factor Ion Injectors*, eds., G.H. Gillespie, Y.-Y. Kuo, D. Keefe and T.P. Wangler, AIP Conference Proceedings No. 139 (American Institute of Physics, New York, 1986), p. 45.

D. Keefe, "Highlights of the Heavy Ion Fusion Symposium," in *Proceedings of the 1986 Linear Accelerator Conference*, Stanford University Report No. SLAC-303 (1986), p. 171.

J. Struckmeier, "Self-Consistent Non-Kapchinskij-Vladimirskij Distributions for Periodic Focusing Systems," IEEE Transactions on Nuclear Science **NS-32**, 2516 (1985).

E.P. Lee, T.J. Fessenden and L.J. Laslett, "Transportable Charge in a Periodic Alternating Gradient System," IEEE Transactions on Nuclear Science **NS-32**, 2489 (1985).

C.M. Celata, I. Haber, L.J. Laslett, L. Smith and M.G. Tiefenback, "The Effect of Induced Charge at Boundaries on Transverse Dynamics of a Space-Charge Dominated Beam," IEEE Transactions on Nuclear Science **NS-32**, 2480 (1985).

E.P. Lee, S.S. Yu and W.A. Barletta, "Phase-Space Distortion of a Heavy-Ion Beam Propagating Through a Vacuum Reactor Vessel," Nuclear Fusion 21, 961 (1981).

SOLUTION TO THE EIGENVALUE EQUATION FOR LONGITUDINAL STABILITY OF THE MODIFIED BETATRON

In this appendix, use is made of the eigenvalue equation (4.351) to derive the dispersion relation for longitudinal perturbations about an intense relativistic electron beam circulating in a modified betatron (or conventional betatron in the absence of the toroidal field). Making use of the assumption of large aspect ratio ($a, a_c \ll r_0$), the eigenvalue equation (4.351) is solved in the straight-beam approximation. Since the perturbed density on the right-hand side of Eq.(4.351) is nonzero inside the electron beam, the eigenfunction $\Phi^\ell(\rho, \phi)$ can be determined in terms of the appropriate Green's function for the left-hand side of Eq.(4.351). Assuming that the Green's function $G(\rho, \rho', \phi, \phi')$ satisfies

$$\left(\frac{1}{\rho}\frac{\partial}{\partial\rho}\rho\frac{\partial}{\partial\rho} + \frac{1}{\rho^2}\frac{\partial^2}{\partial\phi^2} - q^2\right)G(\rho, \rho', \phi, \phi') = \frac{1}{\rho}\delta(\rho - \rho')\delta(\phi - \phi') ,\quad \text{(A-1)}$$

the eigenfunction $\Phi^\ell(\rho, \phi)$ can be expressed as

$$\Phi^\ell(\rho, \phi) = \int_0^{a_c} d\rho'\rho' \int_0^{2\pi} d\phi' G(\rho, \rho', \phi, \phi')C(\rho', \phi') ,\qquad \text{(A-2)}$$

where the source term $C(\rho', \phi')$ is defined by

$$C(\rho, \phi) = -\frac{S\hat{\Phi}_0^\ell}{a}\cos\phi\,\delta(\rho - a) + \frac{1}{a^2}\left[N_0\hat{\Phi}_0^\ell - N_1\hat{\Phi}_0^{\ell\prime}\right]U(a - \rho) ,\quad \text{(A-3)}$$

and $S(k, \omega)$, $N_0(k, \omega)$ and $N_1(k, \omega)$ are defined in Eqs.(4.352) and (4.354). Making use of the identity

$$\delta(\phi - \phi') = \frac{1}{2\pi}\sum_{m=-\infty}^{\infty}\exp\left[im(\phi - \phi')\right] ,\qquad \text{(A-4)}$$

the Green's function $G(\rho, \rho', \phi, \phi')$ can be expressed as

$$G(\rho, \rho', \phi, \phi') = \frac{1}{2\pi} \sum_{m=-\infty}^{\infty} g_m(\rho, \rho') \exp\left[im(\phi - \phi')\right] , \qquad (A\text{-}5)$$

where the radial Green's function $g_m(\rho, \rho')$ is the solution to the differential equation

$$\left\{\frac{1}{\rho}\frac{\partial}{\partial\rho}\rho\frac{\partial}{\partial\rho} - \left(\frac{m^2}{\rho^2} + q^2\right)\right\} g_m(\rho, \rho') = \frac{1}{\rho}\delta(\rho - \rho') . \qquad (A\text{-}6)$$

After some straightforward algebraic manipulation, it is found that the appropriate radial Green's function that solves Eq.(A-6) is given by

$$g_m(\rho, \rho') = \begin{cases} g_m^+(\rho, \rho') = \dfrac{K_m(qa_c)I_m(q\rho')}{I_m(qa_c)}\left[I_m(q\rho) - \dfrac{I_m(qa_c)}{K_m(qa_c)}K_m(q\rho)\right] , \\[2mm] \qquad\qquad\qquad \text{for } \rho' < \rho \le a_c , \\[3mm] g_m^-(\rho, \rho') = \dfrac{K_m(qa_c)I_m(q\rho')}{I_m(qa_c)}\left[1 - \dfrac{I_m(qa_c)K_m(q\rho')}{I_m(q\rho')K_m(qa_c)}\right]I_m(q\rho) , \\[2mm] \qquad\qquad\qquad \text{for } 0 \le \rho < \rho' , \end{cases} \qquad (A\text{-}7)$$

where use has been made of $g_m^+(a_c, \rho') = 0$, $g_m^+(\rho', \rho') = g_m^-(\rho', \rho')$ and $(\partial g_m^+/\partial\rho)_{\rho'} - (\partial g_m^-/\partial\rho)_{\rho'} = 1/\rho'$. In Eq.(A-7), $I_m(x)$ and $K_m(x)$ are the modified Bessel functions of the first and second kinds, respectively, of order m. Making use of Eq.(A-6), we obtain $g_m(\rho, \rho') = g_{-m}(\rho, \rho')$. In the subsequent discussion, the stability analysis is restricted to relatively low azimuthal mode numbers satisfying

$$ka_c = \frac{\ell a_c}{r_0} \ll 1 , \qquad (A\text{-}8)$$

which is consistent with Eq.(4.326). Within the context of Eq.(A-8), making use of the small-argument expansions of $I_m(x)$ and $K_m(x)$, the radial Green's function is approximated to lowest order by

$$g_m^+(\rho, \rho') = \begin{cases} \ln\left(\dfrac{\rho}{a_c}\right) , & m = 0 , \\[3mm] \dfrac{1}{2m}\left[\left(\dfrac{\rho\rho'}{a_c^2}\right)^m - \left(\dfrac{\rho'}{\rho}\right)^m\right] , & m \ge 1 , \end{cases} \qquad (A\text{-}9)$$

and

$$g_m^-(\rho, \rho') = \begin{cases} \ln\left(\dfrac{\rho'}{a_c}\right), & m = 0, \\ \dfrac{1}{2m}\left[\left(\dfrac{\rho\rho'}{a_c^2}\right)^m - \left(\dfrac{\rho}{\rho'}\right)^m\right], & m \geq 1. \end{cases} \qquad \text{(A-10)}$$

Substituting Eqs.(A-5) and (A-7) into Eq.(A-2), the eigenfunction is given by

$$\Phi^\ell(\rho, \phi) = \sum_{m=-\infty}^{\infty} \Phi_m^\ell(\rho)\exp(im\phi), \qquad \text{(A-11)}$$

where

$$\Phi_m^\ell(\rho) = \frac{1}{2\pi}\int_0^{2\pi} d\phi'\exp(-im\phi') \qquad \text{(A-12)}$$

$$\times \left[\int_0^\rho d\rho'\rho' g_m^+(\rho, \rho')C(\rho', \phi') + \int_\rho^{a_c} d\rho'\rho' g_m^-(\rho, \rho')C(\rho', \phi')\right].$$

The source term in Eq.(A-3) can also be expressed as

$$C(\rho, \phi) = \sum_{m=-\infty}^{\infty} C_m(\rho)\exp(im\phi), \qquad \text{(A-13)}$$

where the coefficients $C_m(\rho)$ are defined by

$$C_0(\rho) = \frac{1}{a^2}[N_0\hat{\Phi}_0^\ell - N_1\hat{\Phi}_0^{\ell\prime}]U(a - \rho), \qquad \text{(A-14)}$$

$$C_{\pm 1}(\rho) = -\frac{S\hat{\Phi}_0^\ell}{2a}\delta(\rho - a), \qquad \text{(A-15)}$$

$$C_m(\rho) = 0, \text{ for } m = \pm 2, \pm 3, \cdots. \qquad \text{(A-16)}$$

We now substitute Eq.(A-13) into Eq.(A-12) and integrate over ϕ' using the identity

$$\int_0^{2\pi} d\phi'\exp(-in\phi') = \begin{cases} 2\pi, & n = 0, \\ 0, & n \neq 0. \end{cases} \qquad \text{(A-17)}$$

Then, Eq.(A-12) can be simplified to give

$$\Phi_m^\ell(\rho) = \int_0^\rho d\rho'\rho' g_m^+(\rho, \rho')C_m(\rho') + \int_0^{a_c} d\rho'\rho' g_m^-(\rho, \rho')C_m(\rho'). \qquad \text{(A-18)}$$

It is evident that $\Phi_1^\ell(\rho) = \Phi_{-1}^\ell(\rho)$ in Eq.(A-18). Therefore, the eigenfunction in Eq.(A-11) can be expressed as

$$\Phi^\ell(\rho, \phi) = \Phi_0^\ell(\rho) + 2\Phi_1^\ell(\rho)\cos\phi. \tag{A-19}$$

It is evident from Eqs.(A-12)–(A-18) that the eigenfunction $\Phi^\ell(\rho, \phi)$ in Eq.(A-19) is determined in terms of $\hat{\Phi}_0^\ell \equiv \Phi^\ell(r_0, 0)$ and $\hat{\Phi}_0^{\ell\prime} \equiv [r\partial\Phi^\ell/\partial r]_{(r_0,0)}$. To derive the dispersion relation from Eq.(A-19), we evaluate $\Phi^\ell(r, z)$ and $r(\partial/\partial r)\Phi^\ell(r, z)$ at $(r, z) = (r_0, 0)$. Some straightforward algebraic manipulation gives

$$\Phi^\ell(\rho, \phi) = \frac{1}{2}S\hat{\Phi}_0^\ell\left(1 - \frac{a^2}{a_c^2}\right)\left(\frac{r - r_0}{a}\right)$$

$$- \frac{1}{4}(N_0\hat{\Phi}_0^\ell - N_1\hat{\Phi}_0^{\ell\prime})\left[2\ln\left(\frac{a_c}{a}\right) + 1 - \frac{\rho^2}{a^2}\right] \tag{A-20}$$

for $0 \le \rho < a$. Upon evaluating $\Phi^\ell(r, z)$ and $r(\partial/\partial r)\Phi^\ell(r, z)$ at $(r, z) = (r_0, 0)$, we obtain two homogeneous equations relating the two amplitudes $\hat{\Phi}_0^\ell$ and $\hat{\Phi}_0^{\ell\prime}$. The condition for a nontrivial solution is that the determinant of the coefficients of $\hat{\Phi}_0^\ell$ and $\hat{\Phi}_0^{\ell\prime}$ be equal to zero. Setting the determinant equal to zero gives the dispersion relation

$$1 + \frac{1}{4}\left[1 + 2\ln\left(\frac{a_c}{a}\right)\right]\left[N_0 - \frac{SN_1r_0}{2a}\left(1 - \frac{a^2}{a_c^2}\right)\right] = 0, \tag{A-21}$$

where the dielectric coefficients $S(k, \omega)$, $N_0(k, \omega)$ and $N_1(k, \omega)$ are defined in Eqs.(4.352)–(4.354). The term proportional to S in Eq.(A-21) originates from the surface perturbation in Eq.(4.351). It is evident from Eq.(A-21) that the contribution from the surface term vanishes as the conducting wall radius ($\rho = a_c$) approaches the outer radius of the electron beam ($\rho = a$). Indeed, surface-driven instabilities obtained from Eq.(A-21) are easily stabilized when the conducting wall radius is in sufficiently close proximity to the surface of the electron beam.

DERIVATION OF THE EXTRAORDINARY-MODE EIGENVALUE EQUATION FOR RELATIVISTIC ELECTRON FLOW IN PLANAR GEOMETRY

In this appendix, the extraordinary-mode eigenvalue equation (8.101) is derived for flute perturbations about a relativistic nonneutral electron layer in the planar geometry illustrated in Fig. 8.5. The assumptions and notation are the same as in Eqs.(8.93)–(8.100).

The linearized cold-fluid equations for the perturbed electron density $\delta n_e(\mathbf{x}, t)$ and the flow velocity $\delta \mathbf{V}_e(\mathbf{x}, t)$ are given by (Sec. 2.3)

$$\left(\frac{\partial}{\partial t} + V_{ye}^0(x) \frac{\partial}{\partial y} \right) \delta n_e(\mathbf{x}, t) + \frac{\partial}{\partial \mathbf{x}} \cdot [n_e^0(x) \delta \mathbf{V}_e(\mathbf{x}, t)] = 0 , \quad \text{(B-1)}$$

$$\left(\frac{\partial}{\partial t} + V_{ye}^0(x) \frac{\partial}{\partial y} \right) [\gamma_e^0(x) m_e \delta \mathbf{V}_e(\mathbf{x}, t) + \delta \gamma_e(\mathbf{x}, t) m_e V_{ye}^0(x) \hat{\mathbf{e}}_y]$$

$$+ \delta V_{xe}(\mathbf{x}, t) \frac{\partial}{\partial x} [\gamma_e^0(x) m_e V_{ye}^0(x) \hat{\mathbf{e}}_y] \quad \text{(B-2)}$$

$$= -e \left(\delta \mathbf{E}(\mathbf{x}, t) + \frac{V_{ye}^0(x) \hat{\mathbf{e}}_y \times \delta \mathbf{B}(\mathbf{x}, t)}{c} + \frac{\delta \mathbf{V}_e(\mathbf{x}, t) \times B_z(x) \hat{\mathbf{e}}_z}{c} \right) .$$

Here, $V_{ye}^0(x) = -cE_x(x)/B_z(x), \gamma_e^0(x) = [1 - E_x^2(x)/B_z^2(x)]^{-1/2}$, and $\delta \gamma_e(\mathbf{x}, t) = \gamma_e^{03}(x)[V_{ye}^0(x)/c^2]\delta V_{ye}(\mathbf{x}, t)$ in the linearized approximation. Flute perturbations with $\partial/\partial z = 0$ are assumed, and perturbed quantities $\delta \psi(x, y, t)$ are expressed as

$$\delta \psi(x, y, t) = \sum_{k=-\infty}^{\infty} \delta \psi(x, k) \exp(iky - i\omega t) , \quad \text{(B-3)}$$

where $\text{Im}\, \omega > 0$ corresponds to instability (temporal growth). Making use of Eq.(B-3), the linearized equations (B-1) and (B-2) give the coupled equations for $\delta n_e(x,k), \delta V_{xe}(x,k), \delta V_{ye}(x,k), \delta E_x(x,k), \delta E_y(x,k)$ and $\delta B_z(x,k)$,

$$-i[\omega - kV_{ye}^0(x)]\delta n_e = -ikn_e^0(x)\delta V_{ye} - \frac{\partial}{\partial x}[n_e^0(x)\delta V_{xe}] , \quad \text{(B-4)}$$

$$-i[\omega - kV_{ye}^0(x)]\delta V_{xe} + [\omega_{ce}(x)/\gamma_e^0(x)]\delta V_{ye}$$

$$= -\frac{e}{\gamma_e^0(x)m_e}\left(\delta E_x + \frac{1}{c}V_{ye}^0(x)\delta B_z\right) , \quad \text{(B-5)}$$

$$\left(-\frac{\omega_{ce}(x)}{\gamma_e^0(x)} + \frac{1}{\gamma_e^0(x)}\frac{\partial}{\partial x}[\gamma_e^0(x)V_{ye}^0(x)]\right)\delta V_{xe}$$

$$-i[\omega - kV_{ye}^0(x)]\gamma_e^{02}(x)\delta V_{ye} = -\frac{e}{\gamma_e^0(x)m_e}\delta E_y . \quad \text{(B-6)}$$

Here, $\omega_{ce}(x) = eB_z(x)/m_e c$ is the nonrelativistic cyclotron frequency, and

$$\frac{1}{\gamma_e^0}\frac{\partial}{\partial x}(\gamma_e^0 V_{ye}^0) = \gamma_e^{02}\frac{\partial V_{ye}^0}{\partial x} = \frac{\omega_{pe}^2(x)}{\omega_{ce}(x)} , \quad \text{(B-7)}$$

where $\omega_{pe}^2(x) = 4\pi n_e^0(x)e^2/m_e$ and use has been made of the equilibrium Poisson equation (8.96).

The perturbed field components $\delta E_x, \delta E_y$ and δB_z are related self-consistently to $\delta n_e, \delta V_{xe}$ and δV_{ye} by the linearized Maxwell equations. In this regard, it is convenient to introduce the effective potential $\Phi_k(x)$ defined by

$$\Phi_k(x) = \frac{i}{k}\delta E_y(x,k) . \quad \text{(B-8)}$$

After some straightforward algebra, the linearized Maxwell equations give

$$\delta E_x(x,k) = -\frac{1}{(1 - \omega^2/c^2k^2)}\left[\frac{\partial}{\partial x}\Phi_k(x) + \frac{i\omega}{c^2k^2}4\pi en_e^0(x)\delta V_{xe}(x,k)\right] , \quad \text{(B-9)}$$

$$\delta B_z(x,k) = \frac{1}{(1 - \omega^2/c^2k^2)}\left[\frac{\omega}{ck}\frac{\partial}{\partial x}\Phi_k(x) + \frac{i}{ck}4\pi en_e^0(x)\delta V_{xe}(x,k)\right] . \quad \text{(B-10)}$$

In addition, Poisson's equation becomes

$$\frac{\partial}{\partial x}\delta E_x(x,k) - k^2\Phi_k(x) = -4\pi e\delta n_e(x,k) . \quad \text{(B-11)}$$

Note from Eqs.(B-8)–(B-10) that the perturbations are electrostatic with $\delta E_x \simeq -\partial \Phi_k / \partial x, \delta E_y = -ik\Phi_k$ and $\delta B_z \simeq 0$, only in the limit where $|\omega / k| \ll c \to \infty$.

Substituting Eqs.(B-4) and (B-9) into Poisson's equation (B-11) gives

$$\left\{ \frac{\partial^2}{\partial x^2} - k^2 \left(1 - \frac{\omega^2}{c^2 k^2} \right) \right\} \Phi_k(x) = -\frac{4\pi e i}{[\omega - kV_{ye}^0(x)]} \tag{B-12}$$

$$\times \left[\left(1 - \frac{V_{ye}^0(x)}{c} \frac{\omega}{ck} \right) \frac{\partial}{\partial x} \left[n_e^0(x) \delta V_{xe} \right] + ik \left(1 - \frac{\omega^2}{c^2 k^2} \right) \left[n_e^0(x) \delta V_{ye} \right] \right].$$

Furthermore, substituting Eqs.(B-8)–(B-10) into Eqs.(B-5) and (B-6), we can express the perturbed fluid velocities $\delta V_{xe}(x, k)$ and $\delta V_{ye}(x, k)$ directly in terms of the effective potential $\Phi_k(x)$. After considerable algebraic manipulation and rearrangement of terms, Poisson's equation (B-12) can be expressed in the form of the eigenvalue equation given in Eq.(8.101).

SUBJECT INDEX